Abiogenesis

Laurel O. Sillerud

Abiogenesis

The Physical Basis for Living Systems

 Springer

Laurel O. Sillerud
Department of Neurology
University of New Mexico
Albuquerque, NM, USA

ISBN 978-3-031-56689-9   ISBN 978-3-031-56687-5   (eBook)
https://doi.org/10.1007/978-3-031-56687-5

This Springer imprint is published by the registered company Springer Nature Switzerland AG
The registered company address is: Gewerbestrasse 11, 6330 Cham, Switzerland

Paper in this product is recyclable.

Introduction

The Self-Assembly of Living Systems

The time has come, the walrus said, to speak of many things, of shoes, of ships, of sealing wax, of cabbages and kings and why the sea is boiling hot and whether pigs have wings. (Lewis Carrol, *Alice's Adventures in Wonderland*)

There is an obvious, but often overlooked, maxim that the evolution of life first required an *ordered physical basis* on which natural selection could operate, much as quantum mechanical operators function on the basis vectors that span the Hilbert space of states. Stewart Kauffman (1995) argued persuasively that this *ordered basis* arose spontaneously, science suggesting

...that the order is not all accidental, that vast veins of spontaneous order lie at hand. Laws ... spontaneously generate much of the order of the natural world. It is only <u>then</u> that selection comes into play, further molding and refining. We have all known that simple physical systems exhibit spontaneous order: an oil droplet in water forms a sphere; snow-flakes exhibit their evanescent sixfold symmetry. What is new is that the range of spontaneous order is enormously greater than we have supposed. Profound order is being discovered in large, complex, and apparently random systems. I believe that this emergent order underlies not only the origin of life itself, but much of the order seen in organisms today. (S. Kauffman, 1995, in *At Home in the Universe*)

Advances in astronomy, physics, and chemistry in the last few centuries have shown how this *ordered basis* arose. It is abundantly clear that the Universe and the Life within it were generated in processes of energy-directed, probabilistic self-assembly, or *Abiogenesis*.

Who among us has not gazed at the brilliant stars blazing in a clear night sky and wondered how the Universe and Life arose? What grand organizing principles led to the self-assembly of matter that generated its myriad present living forms? What is Kauffman's physical basis upon which Life originated? The simplest answer, one firmly supported by observation, is that the universe and living systems self-assembled in a probabilistic manner organized by flows of energy. Our task here then is to develop a summary of the Grand Organizing Principles that led to the development of this *ordered physical basis*, which served as the template for the evolution of living systems. This *basis* began with the origin of matter in the Big Bang; the decay of this primordial material into atoms and molecules, and their coalescence into stars, galaxies, and planets, set the stage upon which these subsequent chemical actors evolved nascent biology. As we progress in our understanding of the development and evolution of living systems, we inevitably move up the length scale of how matter is organized, from the smallest elementary particles, such as quarks and leptons, to the condensation of these primordial objects into nuclei, atoms, molecules, and biomolecular nanomachines.

Faust contemplating the macrocosm (Universe) in a drawing from 1789 by Johann Heinrich Lips, for Johann Wolfgang von Göthe's *Faust*.

Faust contemplating the macrocosm (Universe) in a drawing from 1789 by Johann Heinrich Lips, for Johann Wolfgang von Göthe's *Faust*

A synthesis of this nature has been sought for centuries past. One of the first literary explorations of our scientific search for knowledge arose in Das Faustbuch from 1587, inspiration for Göthe's later Faust, where, in the *Urfaust*, we find our hero sitting alone in his medieval study late at night reflecting and complaining to himself that he has spent a life vigorously studying Medicine, Philosophy, Jurisprudence, "*. . .und leider auch Die Theologie*", in short, all the available contemporary sources of knowledge and speculation. But, this path has left him no wiser than before, even though all he sought was "*. . .Das Ich erkenne, was die Welt im Innersten zusammenhält, . . .*", i.e., what holds the innermost parts of the World together. A twenty-first-century scholar would, on the other hand, be most intrigued now by what we *don't know* concerning the structure of matter and the forces which hold it together, for we can now answer Faust's questions with certainty, painting, in broad strokes, the canvas of knowledge with the details of the origin of the Universe, the synthesis of the chemical elements summarized in the Periodic Table, the nature of forces and particles which make up living systems, and the grand organizing principles of Energy and Entropy that guide the self-assembly of the Universe and the Life within it. To have come so far in this understanding in only 500 years, most of it within the last 100, is a breathtaking feat of logic and experimentation previously unmatched in the human experience.

At every step on this path many theories, which appeared logical enough when conceived, were found wanting when their predictions were compared to observations; other theories, which appeared fanciful at first, were found to correctly describe the innermost workings of the World. In this give and take between theory and observation in the twentieth century fundamental limits on human knowledge have been discovered. The process of science has given us not only knowledge in itself but also an understanding of what is both possible and impossible to know. Science has therefore rendered moot many old philosophical puzzles concerning how and what we can know about the world around us.

These limits on knowledge occur most acutely in the microworld of atoms and subatomic particles where our ability to make measurements involves those same objects. Our descriptions of the properties of atoms cannot be separated from the limits imposed by the physical world. We make measuring instruments from those very things that we seek to measure; there are no better methods. In particular, up until the twentieth century, knowledge was derived from the scattering of light from objects. Light particles (photons) have a finite size given by their wavelength. Observing objects with a characteristic size smaller than the wavelength of light was only possible within certain limits, and this fact gave rise to the laws of quantum mechanics. In a similar fashion, the fact that there is no method for transmitting information faster than the speed of light gave rise to the laws of Relativity and to the equivalence of mass and energy. Both of these revolutions in human knowledge resulted from the realization that it is not possible to make instantaneous measurements with arbitrary precision, and therefore, emphasized the intimate relationship between knowledge and measurement. In the case of Relativity, the naïve concept of simultaneity of events between local and distant observers was shown to depend on the ability to measure such events. In quantum mechanics, the ability to describe the state of an atom was shown to depend on measurement, and the very act of measurement influenced the atomic state one was attempting to measure. These examples, and there are myriads more, illustrate the central role that measurement has assumed in the last 100 years in science. Not only must theories provide testable predictions which must match observations, it is now well-understood that measurements are subject to the same finite principles as the system to be measured. We have discovered the limits of our tools and those tools are the only ones available. Furthermore, quantum mechanics has shown that we can only understand the world in probabilistic terms.

Reference

S. Kauffman, *At Home in the Universe: The Search for the Laws of the Self-Organization and Complexity* (Oxford University Press, Oxford, 1995).

Contents

Chapter 1
Energy-Directed, Probabilistic, Self-Assembly

"Der liebe Gott spielt nicht mit Würfeln."

Albert Einstein

"It looks as if chance must once more be reckoned with."
W. Somerset Maugham in "The Summing Up"

1.1 The Probabilistic Underpinnings of Living Systems

It is perhaps surprising that the deterministic-appearing macroscopic world that surrounds us is governed purely by chance at the nuclear, atomic, and molecular level. Classical determinism required that for each cause there was a unique effect, with Newtonian mechanics in the seventeenth century providing the solid link between them. In the middle of the nineteenth century, James Clerk Maxwell (in 1859), Ludwig Boltzmann (in 1872), and Josiah Willard Gibbs (in 1889) realized that molecules in gases could be understood in probabilistic terms, and developed the basis for a highly successful statistical theory of thermodynamics. At the beginning of the twentieth century, a probabilistic understanding of atomic structure naturally emerged from "quantum mechanics." Statistical mechanics and quantum theory have withstood more than a century of rigorous critical examination and now form part of the basis for a triumphant, predictive, yet probabilistic, modern scientific worldview.

The physical basis of determinism required a "Complete description of Nature." Simone Pierre Laplace (in 1812) used Isaac Newton's laws of motion to suggest that this "Complete description of Nature" consisted of the positions and momenta of every particle in the Universe, i.e., the set of vectors $\{x_i, p_i\}$ in six-dimensional phase space, where the index i runs over all the matter in the Universe. It was assumed, of course, that x and p commute so that each could be measured with arbitrary precision. Determinism was based on the assumption that once the initial conditions were specified, systems would evolve in strictly prescribed sequences according to Newton's laws. One only needed to specify the initial conditions and the Universe would evolve, like a clockwork, in a perfectly deterministic fashion. If a given system did not so evolve, it was thought that the fault resided in the imprecision

L. O. Sillerud, *Abiogenesis*, https://doi.org/10.1007/978-3-031-56687-5_1

with which the phase space coordinates were known. Measurements more precise would surely remedy the situation, for, it was firmly believed that measurements could be conducted to arbitrary precision.

This classical description of nature was overthrown in the early twentieth century through the discovery of quantum mechanics by Neils Bohr, Werner Heisenberg, and Erwin Schrödinger. Instead of the distribution of points in phase space, now the complex wave function contained all the knowledge of a system, and only in a probabilistic sense. Furthermore, due to the finite wavelength of probing radiation with respect to the sizes of quantum systems, the process of measurement was now understood to strongly perturb the system under study. Heisenberg described this process quantitatively through the uncertainty principle, which showed that arbitrary precision was not possible in a measurement. The minimum volume in phase space was on the order of h^3, the cube of Planck's constant. For the first time in the history of science, it was now understood that there were fundamental, physical limits to human knowledge imposed by the fact that the Universe contains a known but finite set of quantum fields and particles.

How does this apply to the origin of Life on Earth? Another of the developments of the twentieth century was the discovery that the laws of Physics and Chemistry also apply to living systems. Cells carry out chemical reactions. They swell and shrink in response to physical osmotic forces. Energy stored in substrates can be transferred to the products of biochemical reactions. The laws of probability could be as applied equally to biochemical reactions as to the molecules of a gas. The probabilistic underpinning of the properties of biological systems does not ordinarily manifest itself in the examination of macroscopic measurements because these measurements are concerned with a small set of average quantities, such as temperature, salt concentration, and dielectric constant. These averages are taken over such large numbers of particles that the mean values of macroscopic quantities are extremely well-defined. For example, a single mammalian cell, modeled as a cube with an edge measuring 10µm contains more than 30 billion ($\sim3.4 \times 10^{10}$) water molecules. We shall see that the uncertainty in the properties of these molecules is of the order of only one part in 10^5. Therefore, the mean value, say the chemical potential of water, can be defined with a relative error of only one part in 100,000.

We will find that, even though understood only in a statistical sense, the law of large numbers (convergence to the mean) supports the concept that the Universe self-assembled under the influence of the two grand organizing principles of matter: entropy (*probability*) directed by energy. This assembly can now be traced through the evolution of matter beginning with the Big Bang, progressing into nucleosynthesis leading to the periodic table, and the generation of primordial biomolecules. While it is certainly true that some pieces of this giant puzzle still await our comprehensive understanding, it is clear that the main features of the physical basis for the self-assembly of Life are well-enough known to attempt a synthesis based on our current state of science. We must begin with the study of elementary probability theory in order to understand the physical and chemical foundations of life. As we progress in our understanding of the development and evolution of living systems, we inevitably move up the length scale of matter organization, from the smallest elementary particles, such as quarks and leptons, to the condensation of these primordial objects into nuclei, atoms, molecules, and biomolecular nanomachines.

Nuclei, atoms, biomolecules, and cells are *Systems*, that exist in a variety of *States* characterized by physical and chemical *State Variables*, such as energy, entropy, spin, ionic strength, and temperature. The properties of Systems are determined by the probability of the occupancy of the accessible States and this is governed by the influence of the State variables on this probability. We then seek to understand the relationships between States and State variables: what is the distribution of States over the State variables? Once we know this distribution, we can calculate the sought-after average properties of the Systems. The self-assembly of matter into organized structures is therefore governed by probability distributions. These distributions are obtained from an examination of the relative probabilities of the States accessible to the Systems. In this case, a distribution is equivalent to an arrangement of objects, and this is the subject of probability theory. It is, however, important to keep in mind that State Variables can directly alter the probability of States.

1.2   The Definition of Probability

The a priori probability $P(n)$ of a State n was defined by Blaise Pascal and Pierre de Fermat, in 1654, as

$$P(n) = N(n)/N_\mathrm{T},$$

where $N(n)$ is the number of times the System is found in this particular State, n, out of a total of N_T distinguishable possible States. In this definition of *a priori* probability, it was assumed that each of the N_T states was equally accessible, and that we could know, without ambiguity, the total number of states because they could be found by counting permutations or through combinatorics. These seventeenth century thinkers were primarily concerned with ideal situations, such as games with cards or dice, where the total number of states could be known in advance (a priori). It was later realized, as experimental science became more sophisticated, that many situations existed in nature in which one could not know a priori the total number of accessible states of a system. For example, what is the probability that an asteroid would hit the earth based on measurements of both the orbits of the earth and the asteroid? Each of these measurements would contain errors and the resulting probability of impact would depend on the sizes of these errors.

Suppose then, that one can measure the probabilities of each State by preparing N_T identical systems and determining $N(n)$ by measurements on each one. The same result given above for $P(n)$ would hold here, except that the probabilities of the States would have been determined after the fact, or a posteriori, and this empirical set of $P(n)$ would contain both random and measurement errors. It is this approach that is taken today in most of experimental science, even though it is often a great challenge to prepare N_T identical systems. For systems, such as water molecules in a liver cell, it would seem hopeless to prepare a sufficient quantity of identical cells so that a meaningful set of $P(n)$ could be measured. Nevertheless, for many systems of biophysical importance, it is possible to use simple counting methods to find the a priori probabilities of the accessible states. Furthermore, since these counting methods are so simple, they must have general validity. We will see their power in the next chapter.

Let us begin, then, with a consideration of a priori probability. We can generalize the concept of a State in this context to include its synonyms commonly used when discussing probability: a State can be a distinguishable arrangement, outcome, or result of some operation on a System. For example, consider a *System* consisting of a pair of standard, six-faced, cubic dice. Let a *State* of the dice be labeled by $n = n_1 + n_2$, where n_1 and n_2 are the numbers (*State variables*) appearing on the uppermost face of Die #1 and Die #2, respectively, after a completed roll of the pair. The probability of the macrostate "Snake eye's" ($n = 2$) is given by $N(2)/N_T$, where $N(2)$ is the number of States such that $n_1 + n_2 = 2$ and N_T is the total number of possible simultaneous microstates of the two dice. There exist six faces of each die, so that the number of microstates, D_i, of the ith die must be six. Since the microstates of one die are uncorrelated with those of the other die, $N_T = D_1 \times D_2 = 6 \times 6 = 36$; the total number of distinguishable macrostate of the pair of dice is the product of the number of microstates of each die. Another way of illustrating this is by assigning a box to each die and to write in each box the number of ways of picking each face (Fig. 1.1).

The result of this symbol is the product of the numbers in each box. This is a common method for diagrammatically illustrating permutations. Thus, there are 6 possible ways of picking the first face and 6 ways for the second face. There is only a single distinguishable macrostate of this System characterized by the State variable $n = 2$, and this is given by $n_1 + n_2 = 1 + 1 = 2$. Then $P(2) = P(n_1 + n_2 = 2) = N(2)/N_T = 1/36$. In this case, the assumption is made that the dice are not loaded (biased) so that the a priori probability of the microstate of each Die is 1/6. We can consider the System as consisting of the two dice. The *microstates* of the System are characterized by the set of numbers $\{n_1, n_2\}$ appearing on the faces of the dice, while the *macrostate* is labeled by n. Then n_1, n_2, and n are the State variables. We can illustrate this concretely by constructing a state table (Table 1.1), listing all of the possible States accessible to our System of two dice. The total number, N_T, of values of the State variables n_1 and n_2 is seen by counting to be 36, as we found using logic above. This process may become quite difficult as N_T becomes very large.

$$N_T \;\; = \;\; \boxed{D_1 \;\;\vert\;\; D_2} \;\; = \;\; \boxed{6 \;\;\vert\;\; 6} \;\; = 36$$

Fig. 1.1 The number of arrangements of two six-sided dice; the total number of arrangements is the product of the numbers is each box

Table 1.1 The distinguishable States of a System of two 6-faced dice

i	n_1	n_2	N	i	n_1	n_2	n
1	1	1	2	19	1	4	5
2	2	1	3	20	2	4	6
3	3	1	4	21	3	4	7
4	4	1	5	22	4	4	8
5	5	1	6	23	5	4	9
6	6	1	7	24	6	4	10
7	1	2	3	25	1	5	11
8	2	2	4	26	2	5	7
9	3	2	5	27	3	5	8
10	4	2	6	28	4	5	9
11	5	2	7	29	5	5	10
12	6	2	8	30	6	5	11
13	1	3	4	31	1	6	7
14	2	3	5	32	2	6	8
15	3	3	6	33	3	6	9
16	4	3	7	34	4	6	10
17	5	3	8	35	5	6	11
18	6	3	9	36	6	6	12

Table 1.2 The probabilities that a System of two dice exists in each of its macrostates

b	$N(n)$	$P(n)$
2	1	1/36
3	2	2/36
4	3	3/36
5	4	4/36
6	5	5/36
7	6	6/36
8	5	5/36
9	4	4/36
10	3	3/36
11	2	2/36
12	1	1/36

Now, what are the probabilities that the System exists in each of its distinguishable macrostates? From the entries in Table 1.1 we can find $N(n)$, the number of times the particular State, n, appears out of a total of N_T possible States, by counting the number of times a particular value of n appears. The probability is found by dividing that by $N_T = 36$. Table 1.2 summarizes the results of this enumeration of States and probabilities. In Table 1.2, the State variable n labels the macrostate of the System, $N(n)$ gives the number of ways of finding the System in macrostate n, and $P(n)$ is the probability of finding the System in macrostate n.

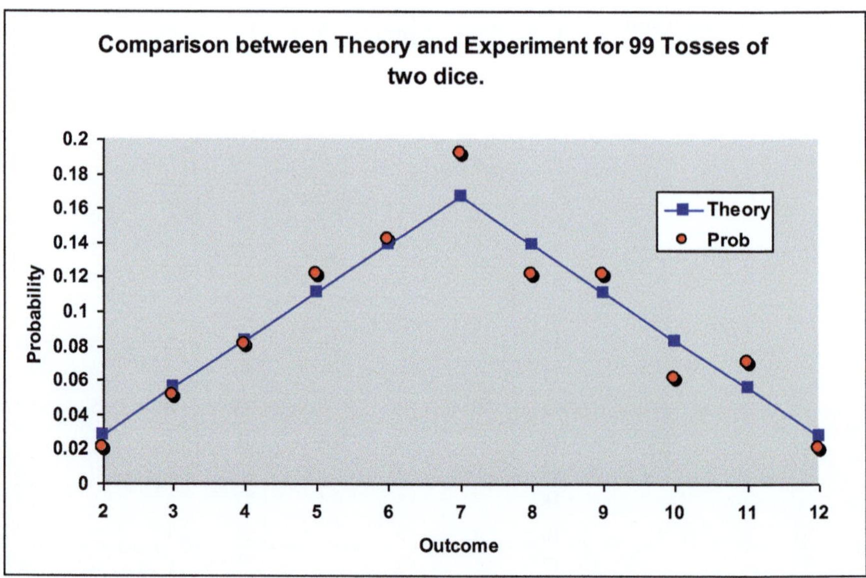

Fig. 1.2 Comparison between the calculated, a priori probability distribution $P(n)$ (Theory) for the outcome, n, for the rolls of a pair of six-sided dice, and the measured, a posteriori distribution for 99 rolls (Prob)

1.3   Probability Distributions

How does the probability of a macrostate depend upon its State variables? A concise graphical method for rapidly assimilating this is to plot the probability $P(n)$ versus n (Fig. 1.2). Here, we see an example of an elementary, discrete probability distribution generated, first, from an a priori calculation based on state counting, and second, from a measurement of the outcomes from 99 rolls of a pair of six-sided dice. One can observe that even though we used only 99 rolls of the dice, the two distributions agree well with each other, except for the fact that the a posteriori distribution contains random deviations due to the finite sampling. One can explore the distributions resulting from different numbers of dice rolls in the Problems at the end of this chapter.

1.4   Properties of Probabilities

Suppose that there exists a System, **S**, and that we can determine by means of measurement the macrostate, r, in which this System exists. Let the measurements be repeated $\alpha \gg 1$ times, to produce α independent, uncorrelated, mutually exclusive outcomes, $N(r)$. This process builds up a set of $N(r)$ results of the State measurements, where $r \in \{1, 2, 3, \ldots, \alpha\}$ using Mathematica's (Appendix 1) notation for lists, *to be read as: r is contained in the set of integers ranging from 1 to α*. That is, the set of outcomes can be listed as

$$N(1), N(2), N(3), \ldots, N(\alpha),$$

where $N(r)$ is the number of times the System, **S**, was found in State r. It is clear that the total number of results, N_T, of the measurements is given by:

$$N_T = \sum_{i=1}^{\alpha} N(i) = N(1) + N(2) + \cdots + N(\alpha).$$

Dividing both sides of this equation by N_T, which is just a number, independent of α, gives

$$\frac{N_T}{N_T} = 1 = \sum_{i=1}^{\alpha} \frac{N(i)}{N_T} = \frac{N(1)}{N_T} + \frac{N(2)}{N_T} + \cdots + \frac{N(\alpha)}{N_T}.$$

However, it is known that (vide supra) $P(i) = N(i)/N_T$, so we have that

$$1 = P(1) + P(2) + \cdots + P(\alpha),$$

or that

$$1 = \sum_{i=1}^{\alpha} P(i).$$

We have reached the logical conclusion that the sum of all the individual probabilities must add up to unity. That is, for any measurement there must exist at least one result. Probabilities are therefore normalized; their sum adds to 1. A corollary is that $\forall i, 0 \leq P(i) \leq 1$, *to be read as: for any i, the probability of the ith state must lie between 0 and 1*. We can observe this behavior in Table 1.2. Adding the probabilities in column 3 gives

$$\sum_{i=2}^{12} P(i) = \frac{1+2+3+4+5+6+5+4+3+2+1}{36} = \frac{36}{36} = 1.$$

Although we have so far only mentioned discrete probability distributions, this normalization condition also applies to continuous probability distributions, $P(x)$, where it is expressed as:

$$\int_{-\infty}^{\infty} P(x)dx = 1.$$

That is, in any experiment, the probability of achieving a result is certain.

1.4.1 Exclusive Probabilities

Often in a series of measurements on Systems, one is interested in whether the System is in either of two or more States. Since the exclusive *or* implies that the system cannot be in both states r and s simultaneously, this combination of probabilities is called an *exclusive probability*, because residence in state s excludes residence in state r. How then do we combine probabilities in order to produce the probability of the result?

To calculate the probability of finding the System in either state r or state s from a measurement, if $P(r)$ and $P(s)$ is known, we need to consider that the System exists in $N(r)$ states giving the measurement result r, so that the System is in state r, and $N(s)$ states giving s for the System to be in state s. Therefore, there are $N(r) + N(s)$ results for the System to be in either state r or s. Since there are still N_T total states of the System, the probability of finding the System in either state r or s is given by:

$$P(r, s) = [N(r) + N(s)]/N_T$$

where we have defined $P(r \text{ or } s) \equiv P(r, s)$. By distributing the denominator, we obtain

$$P(r, s) = \frac{N(r)}{N_T} + \frac{N(s)}{N_T}$$

but we know that $P(r) = N(r)/N_T$, so that

$$P(r, s) = P(r) + P(s).$$

The probability of finding the system in either state r or state s is the *sum* of the probabilities of these states. The keyword *or* here translates into the summation of probabilities. If the system is in state r, then it cannot be in state s and vice versa. The states r and s are mutually exclusive. For example, what is the probability that on a single roll of a pair of six-sided dice we find the outcome to be either a 6 or a 7? From Table 1.2 we find that $P(6,7) = P(6) + P(7) = 6/36 + 5/36 = 11/36$. An interesting corollary is to ask for the probability that the result is neither a 6 nor a 7.

The probability normalization condition,

$$\sum_{i=1}^{\alpha} P(i) = 1,$$

implies that

$$P(1) + P(2) + \ldots + P(\alpha) = 1,$$

which can be read as

$$P(1) \text{ or } P(2) \text{ or} \ldots \text{or } P(\alpha) = 1,$$

that is, the probability of finding the System in either state 1, or 2, or ... or α, is unity. Note that the probability of finding the System in at least one state is 1. The probability of *an* outcome is certainty. For this dice problem, we know that $P(1) = 0$ and that there are 11 distinguishable outcomes ($\alpha \in \{2,12\}$), so this sum reads

$$P(2) + P(3) + P(4) + P(5) + P(6) + P(7) + P(8) + P(9) + P(10) + P(11) + P(12) = 1$$

or, by bringing all terms, except $P(6) + P(7)$ to the right, we have that

$$P(6) + P(7) = 1 - [P(2) + P(3) + P(4) + P(5) + P(8) + P(9) + P(10) + P(11) + P(12)]$$

and, by subtracting this from 1, we obtain

$$1 - [P(6) + P(7)] = [P(2) + P(3) + P(4) + P(5) + P(8) + P(9) + P(10) + P(11) + P(12)]$$

which, with the insertion of the values from Table 1.2 gives

$$1 - [P(6) + P(7)] = [1/36 + 2/36 + 3/36 + 4/36 + 5/36 + 4/36 + 3/36 + 2/36 + 1/36]$$

and, by computing the sum on the right, we find that the probability of the dice displaying neither a 6 nor a 7 is

$$1 - [P(6) + P(7)] = 25/36.$$

An easier way to determine this is to use the fact that we know the values of $P(n)$ from Table 1.2:

$$1 - P(6,7) = 1 - [P(6) + P(7)] = 25/36.$$

1.4.2 *Inclusive Probabilities*

There is another interesting situation that arises during measurements of the states in which a System might find itself; that is the probability of finding the System in both states r and s. This is referred to as a *joint* or *inclusive* probability because it includes several states at the same time. Let the states of interest of the System be designated by r *and* s, such that $r \in \{1, 2, 3, \ldots, \alpha\}$ and $s \in \{1, 2, 3, \ldots, \beta\}$, and denote the probability of finding the System in both states r *and* s by $P(r \cdot s) = N(r \cdot s)/N_T$. The total number of accessible States of the System might be thought to be given by:

$$N_T = \sum_{i=1}^{\alpha} N(i) + \sum_{j=1}^{\beta} N(j),$$

but it is clear that this expression overcounts States because some of the α and β states are identical. We also note that, since we assume that states r and s are uncorrelated, as well as statistically independent, the probability that the system is in state r, $P(r) = N(r)/N_T$, does not depend upon state s, and vice versa for state r.

Now, consider the $N(r)$ states in which the System is in state r; of these a fraction $P(s)$ will also be in state s, so that

$$N(r \cdot s) = N(r)\, P(s) = N(r)N(s)/N_T.$$

Then

$$P(r \cdot s) = N(r \cdot s)/N_T = [N(r)/N_T]\, P(s) = P(r)P(s);$$

therefore,

$$P(r \cdot s) = P(r)P(s)$$

Hence, if r and s are uncorrelated states, the probability of finding the System in state r and s is given by the product of the probabilities of each of the individual states. Here, the keyword *and* translates into the mathematical operation of multiplication.

We can use this equation to find the probability that Die #1 shows a 4 and Die #2 shows a 5. This is given by $P(n_1 \cdot n_2) = P(4 \cdot 5) = P(4)P(5) = (1/6)(1/6) = 1/36$, where, in this case

$$P(n_i) = 1/6, \forall (n, i) \in \{1, 2\}$$

because we have specified the outcome for each Die separately and the probability of a given face appearing in a roll is simply the number of ways a given face can appear (one) divided by the total number of faces (six). Exclusive and inclusive probabilities can be combined as follows to compute the a priori probability that, in a given roll of a pair of six-sided dice, the result is $n_1 + n_2 = n = 9$. This outcome can be achieved in four ways:

$$P(9) = P(4 \cdot 5) + P(5 \cdot 4) + P(3 \cdot 6) + P(6 \cdot 3).$$

We can understand this because the probability of rolling a 9 is given by the probability that Die #1 is a 4 *and* Die #2 is a 5, *or* Die #1 is a 5 *and* Die #2 is a 4, *or* Die #1 is a 3 *and* Die #2 is a 6, *or* Die #1 is a 6 *and* Die #2 is a 3. The result is $P(9) = 1/36 + 1/36 + 1/36 + 1/36 = 4/36$, which agrees with the entry for $n = 9$ in Table 1.2.

1.4.3   Mean Values of a Probability Distribution

One of the most important properties of a System governed by a probability distribution is the mean, or average value, of the State of the System. This is given by the mean value of the State variables, since we have made the explicit assumption that the State of a System is determined by its State variables. Suppose that a State variable, u, is found from a measurement to attain any of j values, $u(j)$, such that $j \in \{1,2,3,\ldots, \alpha\}$ and $u(j) \in \{u(1), u(2), \ldots, u(\alpha)\}$ with corresponding a posteriori probabilities $P(j)$, $P(j) \in \{P(1), P(2),\ldots, P(\alpha)\}$. Then for N_T similar Systems, when $N_T \rightarrow \infty$, i.e., when N_T is very large, the number of times you will find the System in state $u(r)$ is the occupation number $N(r)$, which is given by $N(r) = P(r) N_T$. The number of systems in state r is the total number of systems times the probability of the state r. The average value of a State variable, u, which we denote by $\bar{u}$ (or, equivalently, $<u>$) is given by the sum of the products of the occupation numbers, $N(j)$, and the State variable $u(j)$ as (from here on we drop the subscript T on N_T for brevity):

$$\bar{u} = \frac{1}{N}[N(1)u(1) + N(2)u(2) + \ldots + N(\alpha)u(\alpha)]$$

or, we can write this more compactly as

$$\bar{u} = \frac{1}{N}\sum_{j=1}^{\alpha} N(j)u(j) = \sum_{j=1}^{\alpha} \frac{N(j)u(j)}{N},$$

where we have taken the constant N inside the summation. Notice that this sum contains the term $N(j)/N$, and we will remember from earlier sections that $P(j) = N(j)/N$ is the a posteriori probability that we will find the System in State j by means of measurement. Therefore, we can write the *mean value equation* as:

Table 1.3 A priori probability distribution for a System consisting of two six-sided dice

N	P(n)	nP(n)	N(n)
2	0.028	0.056	1
3	0.056	0.167	2
4	0.083	0.333	3
5	0.111	0.556	4
6	0.139	0.833	5
7	0.167	1.167	6
8	0.139	1.111	5
9	0.111	1.000	4
10	0.083	0.833	3
11	0.056	0.611	2
12	0.028	0.333	1
Sum =	**1.00**	**7.00**	**36.00 = N**

$$\bar{u} = \sum_{j=1}^{\alpha} u(j)P(j) = \,<u>.$$

This is the probability-weighted sum of the State variable u. We can check this by referring to our System of two dice (Table 1.3) where we see that the mean value of the State variable n is 7, and from the plot of the a priori probability distribution (Fig. 1.2), we also see that this is the most probable State of the System.

Three important properties of the mean value of a probability distribution are described below. Let f and g be two State variables of a system. Then we have that:

1. The average value of the sum of two State variables f and g is the sum of the average values:

$$<\mathbf{f}+\mathbf{g}> \, = \, <\mathbf{f}> + <\mathbf{g}>,$$

   where $<...>$ means the average of the quantity within the brackets.
2. A constant, **c**, times the average of a sum distributes among the components of the sum:

$$\mathbf{c}(<\mathbf{f}+\mathbf{g}>) = \mathbf{c}<\mathbf{f}> + \mathbf{c}<\mathbf{g}>,$$

   and therefore,

$$<\mathbf{cf}> \, = \mathbf{c}<\mathbf{f}>.$$

3. What is the mean value of the product of two State variables, $<\mathbf{fg}>$?

Let us consider two State variables, $u(j)$ and $v(k)$, $j \in \{1, 2, ..., \alpha\}$, $k \in \{1, 2, ..., \beta\}$, which are the results of separate, uncorrelated measurements: $u \in \{u(1), u(2),...,u(\alpha)\}$, $v \in \{v(1), v(2), ..., v(\beta)\}$. We will also assume that the probability of finding the System in States $u(j)$ and $v(k)$ is given by $P(u(j)) = u(j)/u_T$ and $P(v(k)) = v(k)/v_T$, where u_T and v_T are the total number of j and k States accessible to the System. Then for uncorrelated results of a measurement, the joint probability of finding the System in States u *and* v is given by:

$$P(u \cdot v) = P(u(j))\, P(v(k))$$

Let $f(u)$ and $g(v)$ be functions of the State variables u and v. Then the average value of the product of f and g is:

$$<f(u)g(v)> \; = \; \sum_{j=1}^{\alpha} \sum_{k=1}^{\beta} P(u \cdot v) f(u(j)) g(v(k))$$

Now, we invoke the fact that the measurements are uncorrelated to replace $P(u \cdot v)$ with $P(u(j))\, P(v(k))$

$$<fg> \; = \; \sum_{j=1}^{\alpha} \sum_{k=1}^{\beta} P(u(j)) P(v(k)) f(u(j)) g(v(k))$$

or, by regrouping terms

$$<fg> \; = \; \sum_{j=1}^{\alpha} \sum_{k=1}^{\beta} [P(u(j)) f(u(j))] [P(v(k)) g(v(k))]$$

The first sum is only a function of $u(j)$, while the second is only a function of $v(k)$, so we can rewrite this as:

$$<fg> \; = \; \sum_{j=1}^{\alpha} P(u(j)) f(u(j)) \sum_{k=1}^{\beta} P(v(k)) g(v(k))$$

which is the product of the separate means. Therefore, we have proved that the average value of a product of uncorrelated State variables is given by the product of the average values:

$$<\mathbf{fg}> \; = \; <\mathbf{f}> <\mathbf{g}>.$$

Once again, using our example of a pair of six-sided dice, one can ask for the average value of the product of the means of two separate probability distributions $<\mathbf{fg}> = <\mathbf{f}><\mathbf{g}> = 7 * 7 = 49.$

1.4.4 The Maximum Value of a Probability Distribution

We have seen from our simple example of a System composed of a pair of dice that the probability that the System is found to be in a given State is a function of the State variables, and that this probability distribution is not constant as a function of the State variables, but that it displays a maximum value at some particular value of the State variable (Fig. 1.2). This maximum is the most probable State of the System. Since all nontrivial probability distributions found in nature display a maximum, we would like to find this maximum. This is not difficult in the case of a continuous probability distribution where we can differentiate the distribution and set the derivative equal to zero. For discrete distributions, such as with our pair of dice, the maximum would need to be found empirically. In the case of a discrete distribution, in the limit of large numbers of outcomes, one can treat the distribution as essentially continuous.

1.4.5 The Dispersion of a Probability Distribution About the Mean

It will perhaps be no surprise that probability distributions for biological States also display a maximum. In addition to a maximum, probability distributions are characterized by a width, which is also referred to as the dispersion (or standard deviation) of the probability distribution. The width of a distribution gives a quantitative measure of the precision with which one can determine the State of a System. A narrower probability distribution implies that the mean values of the State variables are more precisely defined. It is for this reason that we seek a positive definite parameter that will provide us with a quantitative measure of the width of the probability distribution for State variables.

Let us parameterize the probability distribution $P(u)$ for the State variable u by the mean, $<u>$ and the deviation, Δu, from the mean (Fig. 1.3) where the deviation from the mean is given by the difference,

$$\Delta u = u - <u>.$$

Note that the mean of $\Delta u = 0$, which we can see by computing

$$<\Delta u> = <(u - <u>)> = <u> - <u> = 0.$$

Therefore, Δu does not satisfy our requirement for a positive definite measure of the width of the probability distribution. As perhaps a better measure, let us try to use the average value of the square of the deviation of the State variable from its mean, $<(\Delta u)^2>$, which must be ≥ 0, i.e., positive definite. What is the average value of the mean of the square of the deviation, $<(\Delta u)^2>$? It is defined as the *variance* σ^2:

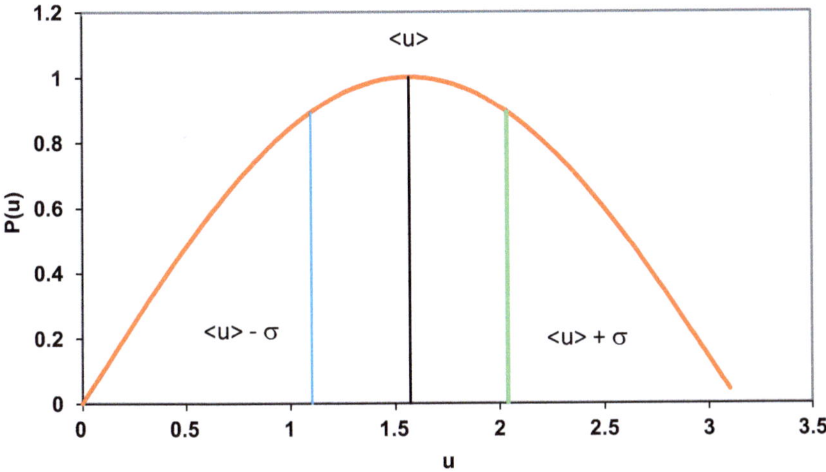

Fig. 1.3 A hypothetical, not-normalized probability distribution $P(u)$ showing the mean, $<u> = 1.57$, and the dispersion, $\sigma = 0.47$

$$\sigma^2 = <(\Delta u)^2> = \sum_{i=1}^{\alpha} P(i)(\Delta u(i))^2 = \sum_{i=1}^{\alpha} P(i)(u(i) - <u>)^2$$

And the square root, **σ**, is often referred to as the *standard deviation* of u. Note that $<(\Delta u)^2>$ is not equal to $(<\Delta u>)^2$. It is very important to remember that any measurement contains two parts: the mean and the variance (uncertainty or fluctuations). The specification of just the mean does not allow one to judge the precision of the measurement, and for comparison of a theory or model with data, the experimental precision must be known.

1.5   The Binomial Probability Distribution Function

When a biophysical system is described by a probability distribution function, we can use the mathematical tools just developed to compute several useful properties of the distribution. Our next task is to discover the nature of those probability distributions that describe nature so that we can apply these tools. Let us imagine a system, S, consisting of N objects, each of which can occupy one of a discrete set of r states, $r \in \{2, 3, \ldots, M\}$. The trivial cases where r is 0 or 1 are excluded for obvious reasons. The smallest, interesting value of r is 2, and this is referred to as the binomial (two-number, or two-state) distribution.

An example of one such important two-state system is that of a collection of N spin one-half nuclei in a magnetic field; a given spin can only be aligned either parallel (up) or antiparallel (down) to the applied magnetic field due to the quantum nature of spin. The magnetic moment of such a system is proportional to the difference between the mean number of spin up and spin down nuclei, and this can be found once the probability distribution is known. When $r \geq 3$, this probability distribution is called the multinomial (many-number) distribution, and we shall deal with this case later in our study of entropy.

1.5.1   The Random Walk

Of great importance in Biophysics is the System known as the Random Walk, which gives rise to the binomial distribution. We will see later that the biophysical process of the self-diffusion of water, or solute biomolecules in cells, is accurately described as a random walk in three-dimensions. In order to develop the binomial distribution, we require that the N objects are not coupled to one another, i.e., their states are uncorrelated and random, and that the probability of each of the r (in this case, 2) states is a constant.

A common way of describing the random walk is to consider a drunk leaving a bar so inebriated that she cannot remember which direction she has just walked so that each step taken is random and uncorrelated with the previous step. Let us suppose that there are pedestrian barricades in front of the bar, much like found in Picadilly Circus in London, that prevent the drunk from walking in any direction but along a single, linear sidewalk. This gives rise to the one-dimensional random walk. Each step taken has the identical length, L, with a probability of p that it is taken to the right, and a probability of $q = 1 - p$ that it is taken to the left. In general, p does not have to be equal to q, but $p + q = 1$ by the properties of probability that we have just encountered above. We can choose the direction of the sidewalk in front of the bar as lying along the x-axis of our coordinate system, with the door situated at $x = 0$, the origin. How far from the front door of the bar will the patron be after N steps? The position of the patron after N steps will be denoted by $x = \text{mL}$, where m is an integer of positive, negative or zero magnitude, lying between $-N \leq m \leq N$. In this case our System is the patron, the State is the position after N steps, and the State variable is m, the resulting position along the sidewalk. We seek the distribution $P_N(m)$ that describes the probability of finding the patron at the position $x = \text{mL}$ after N steps of equal length, L. Let n_1 correspond to the number of steps to the right and n_2 the number of steps to the left. The total number of steps then must be equal to N; $N = n_1 + n_2$. The actual movement to the right can be denoted as $m = n_1 - n_2$ as measured in units of L ($m = x/L$). If, in some sequence of N steps, the patron has taken n_1 steps to the right, then the movement away (displacement) from the front door of the bar can be found as $m = n_1 - n_2 = n_1 - (N - n_1) = 2n_1 - N$. We have made the explicit assumption that the steps to the right and left are uncorrelated, so that $p = 1 - q$, and we can find the probability of a given sequence of right and left steps by multiplying the individual probabilities:

$$p\,p\,p\bullet\,\bullet\,\bullet\,p\,q\,q\,q\,\bullet\,\bullet\,\bullet\,q = p^{n1}q^{n2}$$

Here, we have used the definition of joint probabilities (i.e., the combination rule *and*), and this is to be read as the probability of n_1 steps to the right *and* n_2 steps to the left. Given n_1 steps to the right, and n_2 steps to the left, the probability of a particular combination is described by this product.

There are many different ways of producing the State characterized by N steps, since both n_1 and n_2 can take on any values subject only to the constraint that $N = n_1 + n_2$. We are then faced with the task of finding the number of unique ways in which N steps can be distributed among $N = n_1 + n_2$ total steps. We can solve this problem *via* induction. Let us pick the first way; there are N possibilities for this choice. Once this choice is made, there remain only $N - 1$ choices for the second way, and $N - 2$ choices for the third way, ..., down to only a single choice for the last way. Since we are dealing here with joint probabilities, we know that the total number of choices is the product of the individual choices:

$$N(N{-}1)(N{-}2)(N{-}3)\dots 1 = N!$$

or the number of permutations of N objects, which is N-factorial ($N!$). This discussion considers each choice as *distinguishable*, but the steps to the right, n_1 are all the same, as are all of the n_2 steps to the left. Therefore, all of the $n_1!$ permutations of the steps to the right lead to the same result, and the same is true for the $n_2!$ permutations of the steps to the left. The result is that we have overcounted the total number of distinct permutations of right and left steps by the number of indistinguishable permutations. To compensate for this, we need to divide $N!$ by the number of indistinguishable right and left permutations; the result is that the total number of distinct ways in which N steps can be arranged if n_1 correspond to the number of steps to the right, and n_2 the number of steps to the left is given by $N!/(n_1!\,n_2!)$. The probability $P_N(n_1)$ of taking n_1 steps to the right out of a total of N steps is then given by multiplying the probability of this sequence by the number of possible sequences:

$$P_N(n_1) = \frac{N!}{n_1!n_2!}p^{n_1}q^{n_2}$$

This probability distribution is called the "binomial distribution" because the equation above describes the binomial coefficients of the expansion of $(p + q)^N$.

$$(p + q)^N = \sum_{n=0}^{N} \frac{N!}{n!(N - n)!}p^n q^{N-n}$$

There is another way of looking at the binomial distribution, which is perhaps more transparent. In our random walk with N total steps and n_1 to the right, what is the probability $P(n_1 \cdot n_2)$ of taking n_1 steps to the right *and* n_2 steps to the left? We

Table 1.4 The possible arrangements of $N = 4$ steps

Step Number				n_1	n_2	$P(n_1)$	$C_N(n_1)$	m
1	2	3	4					
R	R	R	R	4	0	1/16	1	4
R	R	R	L	3	1			
R	R	L	R	3	1			
R	L	R	R	3	1			
L	R	R	R	3	1	4/16	4	2
R	R	L	L	2	2			
R	L	L	R	2	2			
L	L	R	R	2	2			
L	R	R	L	2	2			
R	L	R	L	2	2			
L	R	L	R	2	2	6/16	6	0
R	L	L	L	3	1			
L	R	L	L	3	1			
L	L	R	L	3	1			
L	L	L	R	3	1	4/16	4	-2
L	L	L	L	0	4	1/16	1	-4

use the joint probability result from above, because the right and left steps are uncorrelated, to find that

$$P(n_1 \cdot n_2) = p\,p\,p \bullet\ \bullet\ \bullet p\,q\,q\,q \bullet\ \bullet\ \bullet q = p^{n1} q^{n2}.$$

But there are many ways in which we can arrange n_1 steps to the right out of a total of N steps. Let's build a table (Table 1.4) of the possible arrangements of steps for $N = 4$. The total number of arrangements of 4 steps is $N_T = 2^4 = 16$. This can be seen if we note that there are 2 ways (R or L) of picking the first step, *and* 2 ways (R or L) of picking the second step, *and* 2 ways (R or L) of picking the third step, *and* 2 ways (R or L) of picking the fourth step, or N_T is given by:

2	2	2	2

Table 1.4 shows all of these 16 states of our system in detail.

Here, we see, for example, that there are 6 ways that one can take two steps to the right; we can go R, R, and then L, L: or we can go L, R, L, R, etc. However, if we are just interested in the probability of moving two steps to the right, all of these are equivalent and indistinguishable. That is, it does not matter how we traverse the various microstates of the System to reach the final, macrostate. But also note that since there are the most (six) ways of moving 2 steps to the R, the probability of this move is larger than all of the other moves because there are simply fewer ways to reach the other macrostates.

We can incorporate these ideas into our mathematical reasoning by introducing the binomial coefficients, $C_N(n_1)$, which are defined as the number of distinct configurations of N steps where any n_1 are to the right. The C_Ns are also called the number of distinct combinations of N objects taken n_1 at a time. Then the probability, $P_N(n_1)$, that n_1 steps are to the right out of N steps is given by the probability that either the first, *or* the second, *or* the third, ..., *or* the last of the $C_N(n_1)$ alternative possibilities occurs. This [either...*or*] implies that we sum the probabilities over the $C_N(n_1)$ possible configurations with n_1 steps to the right.

This is found by multiplying $P(n_1 \cdot n_2)$ by $C_N(n_1)$. Then

$$
\begin{aligned}
P_N(n_1) &= C_N(n_1)\, P(n_1 \cdot n_2) \\
&= C_N(n_1)\, p^{n1} q^{n2} \\
&= C_N(n_1)\, p^{n1} q^{N-n1}
\end{aligned}
$$

since $N = n_1 + n_2$. It remains to find $C_N(n_1)$ for any $\{N, n_1\}$. Consider Table 1.4 in which we list all the configurations of $N = 4$ steps. The number $C_N(n_1)$ is then the number of entries in which R appears N times. How many entries are there? Look at n_1 steps to the right, $n_1 \in \{R_1, R_2, \ldots, Rn_1\}$. In how many ways can we list these? The possibilities are shown in Table 1.5 for $N = 4$, $n_1 = 2$. There are N ways to pick the first way, and $N - 1$ ways to pick the second, and so on, to $N - n_1 + 1$ ways to pick the last one. That is, we can place R_1 into any of N boxes, but once we have picked the box into which to put R_1, there are only $N - 1$ boxes into which we can place R_2, and only $N - 2$ boxes into which we can place R_3, and so on. The possible number of distinct entries is $J_N(n_1)$, which is defined as the number of permutations of $R_1, R_2, \ldots, R_{n1}$ in Table 1.5. We find the value of $J_N(n_1)$ by multiplying together the numbers of possible locations of the $R_1, R_2, \ldots, R_{n1}$:

$$
J_N(n_1) = N(N-1)(N-2)(N-3)\ldots(N-n_1+1),
$$

or, by using factorials [$m! = m(m-1)(m-2)\ldots(1)$] and from the fact that

Table 1.5 The possible arrangements for obtaining 2 steps to the right out of a total of $N = 4$ steps

1	2	3	4	S
R_1	R_2	L	L	A
R_1	L	R_2	L	B
R_1	L	L	R_2	C
R_2	R_1	L	L	A
L	R_1	R_2	L	D
L	R_1	L	R_2	E
R_2	L	R_1	L	B
L	R_2	R_1	L	D
L	L	R_1	R_2	F
R_2	L	L	R_1	C
L	R_2	L	R_1	E
L	L	R_2	R_1	F

$$\frac{(N-n_1)(N-n_1-1)(N-n_1-2)\ldots(1)}{(N-n_1)(N-n_1-1)(N-n_1-2)\ldots(1)}=1$$

which we can rewrite as $\frac{(N-n_1)!}{(N-n_1)!}=1$, we then have

$$J_N(n_1)=\frac{N(N-1)(N-2)\ldots(N-n_1+1)(N-n_1)\ldots(1)}{(N-n_1)\ldots(1)}$$

which reduces to

$$J_N(n_1)=\frac{N!}{(N-n_1)!}.$$

Now, the R_{n1} are not all distinct, so we have included too many. Those entries in Table 1.5 that only differ by a permutation of n_1 are equivalent and physically indistinguishable. The number of permutations of n_1 is $n_1!$, so we have counted $n_1!$ too many entries, and we must correct for this by dividing the $J_N(n_1)$ by $n_1!$. Therefore, the binomial coefficients are given by:

$$C_N(n_1)=J_N(n_1)/n_1!,$$

which gives the binomial coefficients, $C_N(n_1)$, as:

$$C_N(n_1)=\frac{N!}{n_1!(N-n_1)!}.$$

Then the probability $P_N(n_1)$ becomes

$$P_N(n_1)=\frac{N!}{n_1!(N-n_1)!}p^{n1}q^{N-n1},$$

or since $N=n_1+n_2$,

$$P_N(n_1)=\frac{N!}{n_1!n_2!}p^{n1}q^{n2}.$$

This is once again the *binomial distribution*. If we have equal left and right probabilities, then $p=q=\frac{1}{2}$, and

$$P_N(n_1)=\frac{N!}{n_1!n_2!}\left(\frac{1}{2}\right)^N.$$

A convenient mnemonic useful for remembering the binomial coefficients is Pascal's Triangle, which is shown in Table 1.6 up to $N=6$. The entries in the n_1 column are

Table 1.6 Binomial coefficients and Pascal's Triangle

N	$n_1 \rightarrow$							2^N	$P_N(n_1)$
0			1					1	1
1			1 1					2	½ ½
2			1 2 1					4	¼ ½ ¼
3			1 3 3 1					8	1/8 3/8 3/8 1/8
4			1 4 6 4 1					16	1/16 4/16 6/16 4/16 1/16
5			1 5 10 10 5 1					32	1/32 5/32 10/32 10/32 5/32 1/32
6	1	6	15	20	15	6	1	64	1/64 6/64 15/64 20/64 15/64 6/64 1/64

formed by adding together the two entries directly above them. For example, $1 + 2 = 3$, in row 3 for $N = 3$. The entries in the n_1 column are the binomial coefficients, $C_N(n_1)$. For example, for the 6th row, $C_6(6) = 1$; $C_6(5) = 6$; $C_6(4) = 15$; $C_6(3) = 20$; etc.

In the column labeled 2^N, we have the sum of the binomial coefficients, $C_N(n_1)$, from which we can see that

$$\sum_{n_1=0}^{N} C_N(n_1) = 2^N$$

which says that the sum of the binomial coefficients for a given value of N is simply 2^N. The probability of a set of n_1 steps to the right, out of a total of N steps, is

$$P_N(n_1) = C_N(n_1) p^{n_1} q^{N-n_1}$$

And we note that the probability must be normalized,

$$\sum_{n_1=0}^{N} P_N(n_1) = \sum_{n_1=0}^{N} \frac{N!}{n_1!(N-n_1)!} = p^{n_1} q^{N-n_1} = (p+q)^N$$

This last equality comes from the binomial theorem, and, since $p + q = 1$, $(p + q)^N = 1^N = 1$, we have shown that this probability is normalized:

$$\sum_{n_1=0}^{N} P_N(n_1) = 1$$

This can be verified, row by row, in Table 1.6.

What does the *binomial distribution* look like, if we plot it for, say $N = 10$? Figure 1.4 shows that this distribution is strictly defined only for integer values of N, m. However, by using the gamma function

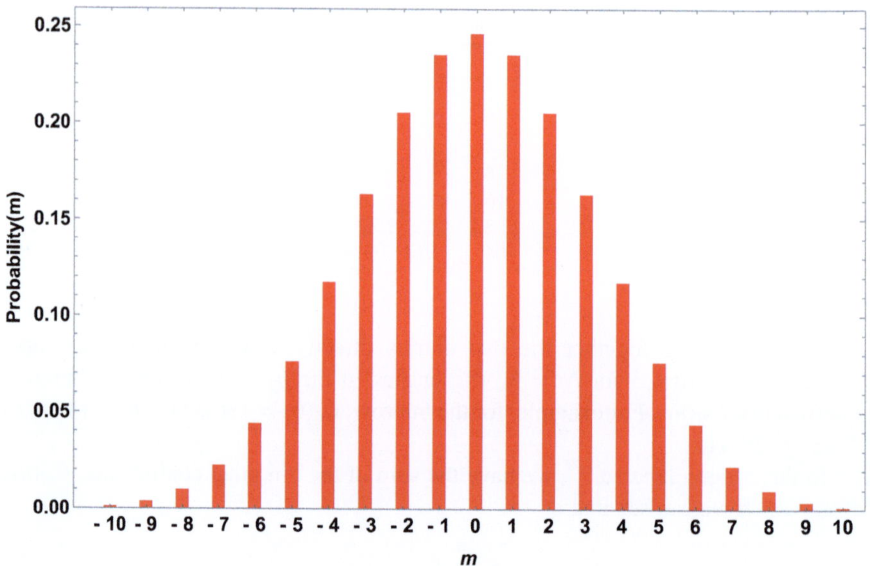

Fig. 1.4 A plot of the binomial distribution of the net displacement, m, from the origin for a random walk with $N = 10$ steps

$$\Gamma(n) = (n-1)!$$

or
$$n\Gamma(n) = n!$$

one can obtain n-factorial as a continuous function of n.

Let us return to the random walk problem. We will find our bar patron at a position $x = mL$ after N steps, where m is the net displacement to the right after N steps. Let us assume that $p = q = \frac{1}{2}$, that is, that the probabilities of a right or a left step are equal. Remember that $m = n_1 - n_2$ and that $N = n_1 + n_2$. It is useful to reexpress these equations in the new variables, m and N: $N + m = 2n_1$ and $N - m = 2n_2$. Then the probability that our patron has moved m steps to the right

$$P_{\mathrm{N}}(n_1) = \frac{N!}{n_1! n_2!} \left(\frac{1}{2}\right)^N$$

is given by $P_{\mathrm{N}}(m)$

$$P_{\mathrm{N}}(m) = \frac{N!}{\left(\frac{N+m}{2}\right)! \left(\frac{N-m}{2}\right)!} \left(\frac{1}{2}\right)^N$$

and we have plotted $P_{10}(m)$ versus m, for $N = 10$ in Fig. 1.4. It is also useful to include a table of the data values (Table 1.7; for $N = 20$).

Table 1.7   Random walk in one-dimension, $N = 20$ steps, $p = q = 1/2$

N	m	$C_N(m)$	$P_N(m)$	$(N + m)/2$	$(N - m)/2$	$mP_N(m)$	$P_N(m)m^2$
20	-20	1	9.54E$-$07	0	20	-1.91E-05	3.81E-04
20	-18	20	1.91E-05	1	19	-3.43E-04	6.18E-03
20	-16	190	1.81E-04	2	18	-2.90E-03	4.64E-02
20	-14	1140	1.09E-03	3	17	-1.52E-02	2.13E-01
20	-12	4845	4.62E-03	4	16	-5.54E-02	6.65E-01
20	-10	15,504	1.48E-02	5	15	-1.48E-01	1.48E+00
20	-8	38,760	3.70E-02	6	14	-2.96E-01	2.37E+00
20	-6	77,520	7.39E-02	7	13	-4.44E-01	2.66E+00
20	-4	125,970	1.20E-01	8	12	-4.81E-01	1.92E+00
20	-2	167,960	1.60E-01	9	11	-3.20E-01	6.41E-01
20	0	184,756	1.76E-01	10	10	0.00E+00	0.00E+00
20	2	167,960	1.60E-01	11	9	3.20E-01	6.41E-01
20	4	125,970	1.20E-01	12	8	4.81E-01	1.92E+00
20	6	77,520	7.39E-02	13	7	4.44E-01	2.66E+00
20	8	38,760	3.70E-02	14	6	2.96E-01	2.37E+00
20	10	15,504	1.48E-02	15	5	1.48E-01	1.48E+00
20	12	4845	4.62E-03	16	4	5.54E-02	6.65E-01
20	14	1140	1.09E-03	17	3	1.52E-02	2.13E-01
20	16	190	1.81E-04	18	2	2.90E-03	4.64E-02
20	18	20	1.91E-05	19	1	3.43E-04	6.18E-03
20	20	1	9.54E-07	20	0	1.91E-05	3.81E-04
$2^{20} =$	1,048,576				**Mean (m)=**	0.00	
$20! =$	2.43×10^{18}				Dispersion	$\sigma^2 =$	**20**

What is the average displacement of our patron from the front door of the bar? The last column in Table 1.7 shows that:

$$<m> = \sum_{m=-20}^{20} mP_{20}(m) = 0;$$

the patron's average distance from the front door is zero steps! We could have guessed this because the mean number of right steps away from the door is the probability of a right step times the total number of steps,

$$<n_1> = Np,$$

and

$$<n_2> = Nq,$$

and for $p = q = \frac{1}{2}$, $<n_1> = <n_2>$, so that $<m> = n_1 - n_2 = 0$. She is too drunk to get home, so the best strategy is for the bartender to call a cab for her.

For the case where p and q are not equal (perhaps the sidewalk runs uphill to the right), we can still find the average displacement $<x> = <m>L$ by computing $<m>$ ($m = n_1 - n_2$), which is given by $<m> = <n_1 - n_2> = <n_1> - <n_2>$, because the mean of a sum is the sum of the means. We have found the individual means, $<n_1>$ and $<n_2>$, and substituting them into the express for $<m>$ gives $<m> = Np - Nq$, or

$$<m> = N(p-q).$$

The average displacement is given by the number of steps times the difference between the left and right-step probabilities.

1.5.2   *The Dispersion of the Random Walk Distribution*

Even though our bar patron never gets far from the front door, a glance at the plot of the binomial distribution (Fig. 1.4) shows that she wanders around quite a bit. On average, how far does our patron wander from the front door? How much space does she sample? What is the width of our random walk distribution? That is found from the dispersion for, e.g., $N = 20$ steps

$$<(\Delta m)^2> = \sum_{m=-20}^{20} P_{20}(m)[m - <m>]^2,$$

but note that $<m>$ is zero in the case that $p = q = 1/2$, so that

$$<(\Delta m)^2> = \sum_{m=-20}^{20} P_{20}(m)m^2 = 20,$$

and the width is the square root of 20, or 4.47. This can be verified by inspection of a plot of the distribution (Fig. 1.4) where the width is defined as the full width at half maximum.

What is a general expression for the dispersion of m? A useful general theorem involving the dispersion is found by expanding the definition of the dispersion for the number of steps to the right:

$$\begin{aligned}
<(\Delta n_1)^2> = <(n_1 - <n_1>)^2> &= <n_1^2 - 2n_1<n_1> + <n_1>^2> \\
&= <n_1^2> - 2<n_1><n_1> + <n_1>^2 \\
&= <n_1^2> - 2<n_1>^2 + <n_1>^2 \\
&= <n_1^2> - <n_1>^2.
\end{aligned}$$

This result is read to mean that the dispersion is equal to the average value of $n_1{}^2$ minus the square of the average value of n_1. We already know that $<n_1> = Np$, so what remains is to find $<n_1{}^2>$. I will state without proof that

$$<n_1{}^2> = p[N + pN(N-1)]$$

(but for those interested, $np^n = p\frac{\partial}{\partial p}(p^n)$ is very useful.)

$$= Np[1 + pN - p]$$

or, since $1 - p = q$,

$$<n_1{}^2> = (Np)^2 + Npq = <n_1>^2 + Npq,$$

which can be rearranged to read:

$$<n_1{}^2> - <n_1>^2 = Npq,$$

which is the definition of $<(\Delta n_1)^2>$ as given above, therefore, the dispersion in n_1 is given by σ^2 as

$$\sigma^2 = <(\Delta n_1)^2> = Npq.$$

We are often more interested in the width of the distribution, or the square root of σ^2, which is called the root-mean-square (rms) deviation (or the standard deviation in statistics) and is a linear measure of the width of the distribution,

$$\sigma(n_1) = \sqrt{Npq}.$$

Now, let us go back and use this result to compute the dispersion of the mean displacement from the door of the bar, $<(\Delta m)^2>$. Remember that $m = n_1 - n_2 = 2n_1 - N$. Hence, one obtains

$$\Delta m = m - <m> = \quad (2n_1 - N) - (2<n_1> - N) = 2(n_1 - <n_1>)$$
$$\Delta m = 2\Delta n_1$$

and, by squaring this we have that

$$(\Delta m)^2 = 4(\Delta n_1)^2$$

and by taking averages, we have the important result that

$$\sigma^2 = \; <(\Delta m)^2> \; = 4 <(\Delta n_1)^2> \; = 4Npq$$

and for the *special case* where $p = q = \frac{1}{2}$

$$\sigma^2 = \; <(\Delta m)^2> \; = N,$$

and we have the very important result that

$$\sigma(m) = \sqrt{N}.$$

We can check this for our binomial distribution example, where Table 1.7 shows that $\sigma^2 = <(\Delta m)^2> = 20 = N$, and the rms deviation, $\sigma = \mathrm{Sqrt}(20) = 4.47$.

1.5.3   *The Random Walk in Science*

The random walk serves as an excellent, often exact, model for a number of biochemical processes. A few examples include nuclear magnetic resonance (NMR), diffusion, signal averaging, and protein dimensions.

1.5.4   *Signal-to-Noise Ratio in Experiments*

In signal averaging for Fourier transform NMR and infrared spectroscopy, the rms deviation from the mean is such an important quantity that it is given a special name: *Noise*. One can generalize the binomial distribution results to generate a theory of measurement. Since we derived the binomial distribution based on an analysis of the random walk, it must apply to any random process, such as the noise or error present in every measurement. Let σ be the noise and n_1 the value of the measured quantity of interest. The relative width of the binomial distribution can be written as the ratio of the noise bandwidth to the mean as

$$\frac{\sigma}{<n_1>} = \frac{\sqrt{Npq}}{Np} = \sqrt{\frac{p}{q}}\frac{1}{\sqrt{N}},$$

and when $p = q = \frac{1}{2}$,

$$\frac{\sigma}{<n_1>} = \frac{1}{\sqrt{N}},$$

we observe that the relative noise decreases as the inverse square root of the number of observations, N, as long as the noise is completely random and uncorrelated.

This observation has many applications, perhaps the most direct is when measurements are repeated many times, as in signal averaging. We can understand the growth of the signal-to-noise ratio in, e.g., an NMR experiment, by noting that the time-dependent signal, $S(t)$, is proportional to the magnetization of the sample, which we can denote by M, $S(t) = kM$, where k is an arbitrary constant. When we repeat the experiment a number of times the noise, η, however, is proportional to the fluctuations, $\sigma(M)$, in the magnetization, $\eta = k'\sigma(M)$, where k' is another unspecified constant. The signal-to-noise ratio, which determines the actual quality of the NMR data, then goes like $S(t)/\eta = kM/k'\sigma(M)$. Now the sample magnetization $M = n<m>$, where n is the number of nuclei in the sample and $<m>$ is the average magnetic moment per nucleon. The net magnetization of simple, spin ½ nuclei, such as the proton, is accurately described by a biased random walk with only two states, spin up and spin down, with $p > q$. Therefore, $<m>$ and its fluctuations $\sigma(m)$ are well modeled by the binomial distribution. The signal-to-noise ratio is then $S(t)/\eta = kn<m>/k'N\sigma(m)$, or $K<m>/\sigma(m)$, where $K = nk/k'$ is a constant that we set equal to one without any loss in generality. From the discussion above, we know that

$$\frac{<m>}{\sigma(m)} = \sqrt{\frac{Nq}{p}}$$

and, if the number of samples is $N = ct$, where c is the sampling rate and t is the total sampling time, we then have that

$$\frac{S(t)}{\eta} \propto \frac{<m>}{\sigma(m)} = \sqrt{\frac{qc}{p}}\sqrt{t} = \text{const}\sqrt{t}$$

the signal-to-noise ratio is proportional to the square root of the total observation time. If one achieves a signal-to-noise ratio of, say, 4 in 1 h, it would take 4 h to double the S/N to 8, imposing a practical limit to signal averaging. The signals and the total cost of an experiment increase with the number of samples, a quantity proportional to the time, while the data quality (signal-to-noise ratio) only increases according to the square root of the time, so that one ultimately reaches a point of diminishing returns. Nevertheless, signal averaging has fundamentally and profoundly expanded the applications of NMR to biochemistry because of this vast increase in sensitivity.

1.5.5   The Structure of Random-Coil Polymers

Another application of the random walk is to compute the average structure of a random-coil polymer, such as a protein. One fixes one end of an N-residue molecule at the center of a coordinate system and then lets the other end experience a random walk as in Fig. 1.5.

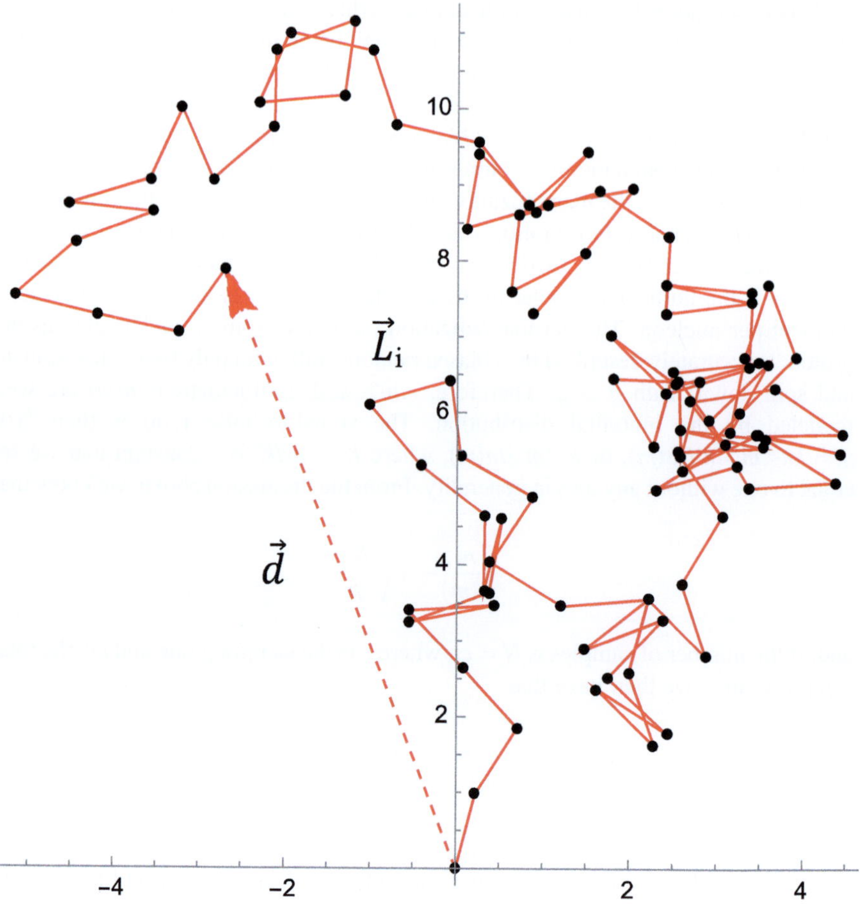

Fig. 1.5 The end-to-end distance vector $\vec{d}$ for an amino acid chain illustrated as the sum of the N individual vectors $\vec{L_i}$ representing peptide bonds. The bonds are assumed to all have the same length, $L = |\vec{L_i}|$

The average end-to-end distance, $|\vec{d}|$, is then the length of the vector sum of the N individual bond vectors $\vec{L_\mathrm{i}}$

$$\vec{d} = \sum_{i=1}^{N} \vec{L_\mathrm{i}}$$

The magnitude of the vector $\vec{d}$ is $d = \sqrt{\vec{d} \cdot \vec{d}}$, where

$$d^2 = \sum_{i=1}^{N} \vec{L_i} \cdot \sum_{j=1}^{N} \vec{L_j} = \sum_{i=1}^{N} \sum_{j=1}^{N} \vec{L_i} \cdot \vec{L_j}$$

and in the double sum on the right, the only terms that survive are for $i = j$ because we are modeling the chain as a 2D random walk in which the individual bond vectors (steps) are uncorrelated with each other. Therefore, for N residues, we have that

$$d^2 = \sum_{i=1}^{N} \vec{L_i} \cdot \vec{L_i} = NL^2$$

or

$$\mathbf{d = \sqrt{N}L.}$$

The average end-to-end distance is given by the square root of N, the number of residues times the length of a residue. In terms of the random walk, the mean displacement from the origin is the step length times the square root of the number of steps.

One can also use randomness to compute the entropy of unfolding of a protein. The average end-to-end distance, d, gives a statistical estimate of the volume occupied by N amino acid residues. This problem can be solved as a random walk; the solution is the width of the distribution of net displacement from the origin after N steps of length L, which grows as the Sqrt[N]. The unfolded state can be modeled by placing the N-terminus of the polypeptide chain at the origin of three-space, and then asking how much volume the chain could sample after unfolding. This would be the volume of a sphere with a radius equal to that of the total length of the polypeptide chain, given by NL, where L is the length of a peptide bond. The value of d is $d = $ Sqrt[N] L, and the volume is $(4\pi/3)d^3$, while the volume of the unfolded protein would be like $(4\pi/3)(NL/2)^3$. The volume change is $(1/8)N^{3/2}$, and the entropy change is then $(3k/2)$ (Ln[N] $-$ Ln[8]). Compare this value with that from a model based on the expansion of an ideal gas given in Chap. 2.

1.6 The Binomial Distribution for Large N

The limit, as N becomes large, of the binomial distribution was derived by Gauss in the nineteenth century, and is often named the Gaussian distribution. It is one of the most important probability distributions that we will encounter in our studies of Biophysics. As N becomes large, the binomial probability distribution,

$$P_{\mathrm{N}}(n_1) = \frac{N!}{n_1!(N - n_1)!} p^{n1} q^{N - n1}$$

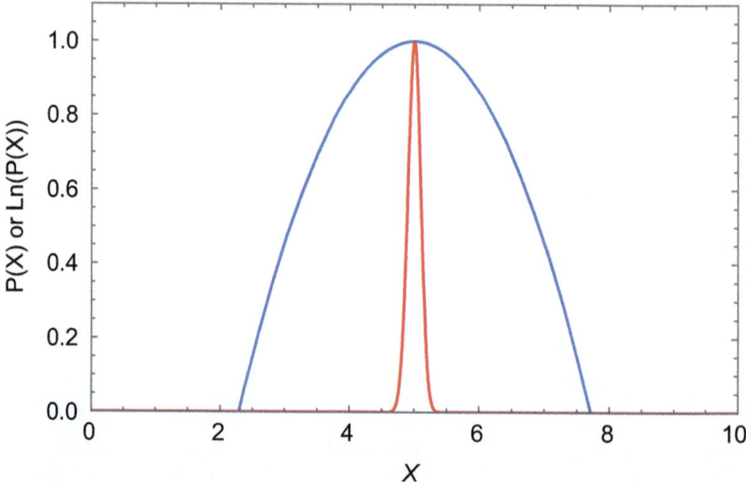

Fig. 1.6 Behavior of a probability distribution (red) near its maximum and the logarithm (blue) of this distribution

becomes tedious to compute due to the large factorials involved; it also tends to display a very sharp maximum because the width of the distribution increases only as fast as the square root of N, whereas the mean number of steps to either the right or left increases linearly with N (e.g., $<n_1> = pN$). This peaking of the distribution (Fig. 1.6) also means that the probability falls rapidly away from the mean value of n. We can use these facts to derive an approximate expression for $P_N(n_1)$ in the limit of large N. In doing so, we will also introduce two concepts that will be extensively used in our subsequent discussions of Entropy, Energy, and quantum mechanics; these are (1) a "complete orthonormal set" of functions and (2) the calculation of the maximum of a probability distribution.

In the vicinity of $<n_1>$, which is also large for large N, the fractional change in $P_N(n_1)$ for a unit change in n_1 is small,

$$|P_N(n_1 + 1) - P_N(n_1)| \ll P_N(n_1)$$

and $P_N(n_1)$ can be considered as a continuous, rather than a discrete, function of the variable n_1, although we know that only integral values of n_1 are physically possible. The location $n_1 = n$ of the maximum of P is then given by the value of P where its derivative is zero:

$$\frac{dP(n = n_1)}{dn_1} = 0,$$

or equivalently, we can replace P by the logarithm of P without losing generality, because if P is a maximum at $n_1 = n$, then $\ln[P]$ also reaches a maximum there, so that we also have that

$$\frac{d \ln P(n = n_1)}{dn_1} = 0$$

where the derivative is evaluated for $n_1 = n$. The reason for using $\ln P(n)$, rather than $P(n)$, is because the logarithm is a much more slowly varying function of n than $P(n)$, and any series expansion will therefore converge much faster (see Fig. 1.6). Now we let $n_1 = n + \varepsilon$, where $\varepsilon \ll n$ and we will examine the behavior of P near its maximum, n_1. Expanding $\ln P$ in a Taylor's series, we find

$$\ln P_N(n_1) = \ln P_N(n) + D_1\varepsilon + \frac{1}{2}D_2\varepsilon^2 + \frac{1}{6}D_3\varepsilon^3 + \dots$$

where $D_k \equiv \frac{d^k \ln P}{dn_1{}^k}$ is the kth derivative of $\ln P$ evaluated at $n_1 = n$.

Why use a Taylor's series here? All of the derivatives of a function comprise a "set of complete orthonormal" functions; complete in the mathematical sense that one can expand any function in terms of this set without recourse to any other functions, and orthonormal, because each order of derivative is orthogonal to every other order of derivative and is normalized so that its integral over all space is unity. For example, if the set $\{f_i(x)\}$, $i \in \{1, 2, 3, \dots, \infty\}$ is an orthonormal set of functions, then the $f_i(x)$ obey the following equations:

$$\int_{-\infty}^{\infty} f_i(x)f_j(x)dx = \delta_{ij}$$

where the Kronecker delta, $\delta_{ij} = 0$, if $i \neq j$, and $\delta_{ij} = 1$ if $i = j$. The set $\{f_i(x)\}$ is also complete if any function can be written as a linear combination of the $f_i(x)$, without using other functions. We will use many other "complete orthonormal sets" of functions when we use Fourier transforms in NMR, and in quantum mechanics, where the solutions of Schrödinger's equation form such a set. One easily verified complete orthonormal set is the set of unit vectors $\{x_i\}$ $i \in \{1, n\}$ in n-dimensions, where one uses the inner product $<|>$ instead of the integral to show that they are orthogonal;

$$< \vec{x_i} \mid \vec{x_j} > = \delta_{ij} = \begin{cases} 1, & i = j \\ 0, & i \neq j \end{cases}$$

their normalization is self-evident, as is the fact that any vector in n-dimensions can be written in terms of this set.

Now, in our Taylor's series for $\ln P_N(n_1)$ elementary calculus tells us that since we are expanding about a maximum, the first derivative, $D_1(n_1) = 0$, and that the second derivative, $D_2(n_1) < 0$. In the region close to the maximum in P, ε is small, so we can neglect higher terms in ε to obtain an approximate form for $\ln P(n_1)$ as

$$\ln P(n_1) = \ln P(n) + \frac{1}{2} D_2 \varepsilon^2$$

and, by exponentiating both sides, we find that

$$P(n_1) = P(n) e^{\frac{1}{2} D_2 \varepsilon^2}.$$

This would have the functional form of a Gaussian if indeed $D_2(n_1) < 0$. We need to compute D_1 and D_2. Consider the expansion above for $\ln P$. We also have an expression for P in terms of factorials, which we can use to find $\ln P$.

$$P_N(n_1) = \frac{N!}{n_1!(N-n_1)!} p^{n_1} q^{N-n_1}$$

then

$$\ln P_N(n_1) = \ln N! - \ln n_1! - \ln (N-n_1)! + n_1 \ln p + (N-n_1) \ln q.$$

But if n is any large integer, $n \gg 1$, then $\ln(n!)$ is an almost continuous function of n, and, for $\Delta n = 1$

$$\frac{d \ln n!}{dn} \approx \frac{\ln (n+1)! - \ln n!}{1} = \ln \frac{(n+1)!}{n!} = \ln (n+1)$$

where the second term arises from the definition of the derivative, $\frac{df}{dx} \approx \frac{f(x+\Delta x) - f(x)}{\Delta x}$. Thus, for $n \gg 1$, $\ln(n+1) \sim \ln(n)$, and

$$\frac{d \ln n!}{dn} \approx \ln n.$$

Returning to our Taylor's series expansion, we can use the above equation to evaluate the first derivative term, first noting that $d \ln N!/dn = 0$ since N is a constant,

$$D_1 = \frac{d \ln P}{d n_1} = - \ln n_1 + \ln (N-n_1) + \ln p - \ln q,$$

but this term must be zero since we are expanding about a maximum, so

$$- \ln n_1 + \ln (N-n_1) + \ln p - \ln q = 0$$

at $n_1 = n$. Therefore, we can find the value of n where P is a maximum by solving

$$\ln\left[\frac{(N-n)p}{nq}\right] = 0$$

or, by exponentiating both sides,

$$\left[\frac{(N-n)p}{nq}\right] = 1.$$

This gives $(N-n)\,p = n\,q$, and with $p + q = 1$, we have that

$$n = Np,$$

which states what we already know, that the average number of steps to the right, n_1, is found by multiplying the number of steps N by the probability of a step to the right, p.

Now that we know D_1, we can find an expression for the second derivative, D_2, by differentiating D_1 with respect to n_1

$$D_2 = \frac{d^2 \ln P}{dn_1{}^2} = \frac{dD_1}{dn_1} = -\frac{1}{n_1} - \frac{1}{N-n_1}$$

and, by evaluating this at the maximum where $n_1 = n$, one gets that

$$D_2 = -\frac{1}{Np} - \frac{1}{N-Np} = -\frac{1}{N}\left[\frac{1}{p} + \frac{1}{q}\right]$$

or, since $p + q = 1$,

$$D_2 = -\frac{1}{Npq}$$

and, we see that D_2 is indeed negative, as required for P to exhibit a maximum, since N, p, and q are all >0.

Now, we return to our expression for the Taylor's expansion of $P(n_1)$

$$P(n_1) = P(n)e^{\frac{1}{2}D_2\varepsilon^2},$$

And we insert the value of D_2 found above to give

$$P(n_1) = P(n)e^{-\frac{\varepsilon^2}{2Npq}}.$$

Note that this is now looks like a Gaussian distribution function, named after *Carl Friedrich Gauss* (1777–1855; Fig. 1.7). We already know from above that

Fig. 1.7 A pre-Euro 10 Mark note showing the homage to Gauss, including his famous Gaussian probability equation

$$\sigma^2 = \;<(\Delta n_1)^2> \;= Npq$$

is the dispersion, whose square root is the width of the distribution.

What is $P(n)$? We can find it by using the fact that any physical probability distribution must be normalized, so the integral of $P(n_1)$ must be 1.

$$\int_{-\infty}^{\infty} P(n_1)dn_1 = 1$$

with $n_1 = n + \varepsilon$. Then

$$\int_{-\infty}^{\infty} P(n+\varepsilon)d\varepsilon = 1 = \int_{-\infty}^{\infty} P(n\,)e^{\frac{D_2\varepsilon^2}{2}}dn = P(n)\int_{-\infty}^{\infty} e^{-\frac{\varepsilon^2}{2Npq}}d\varepsilon$$

therefore, the integral over epsilon is the inverse of the probability at the maximum of the distribution

$$\int_{-\infty}^{\infty} e^{-\frac{\varepsilon^2}{2Npq}}d\varepsilon = \frac{1}{P(n)}.$$

Note that for a Gaussian integral, such as this, over the range from $-\infty$ to ∞, the integrand is symmetric about zero. We can write this integral in the form found in tables as

$$\int_0^\infty e^{-a^2 x^2}\, dx = \frac{\sqrt{\pi}}{2a},$$

and we can then extend the integration from 0 to $-\infty$. This just multiplies the right-hand side by a factor of 2. Now, here

$$a^2 = 1/(2Npq),$$

and

$$\varepsilon = n_1 - n,$$

with

$$n = Np$$

so that this gives

$$P(n) = \frac{1}{\sqrt{2\pi Npq}}$$

and we have our result that

$$P(n) = \frac{1}{\sqrt{2\pi Npq}}\, e^{-\frac{(n-Np)^2}{2Npq}}.$$

This probability function is the Gaussian distribution. We can rewrite it in a more familiar form using $n = Np$ and $\sigma^2 = Npq$ as

$$P(n_1) = \frac{1}{\sigma\sqrt{2\pi}}\, e^{-\frac{(n_1-n)^2}{2\sigma^2}}$$

where n is the mean value. The half width (variance) of the Gaussian is σ^2, and the dispersion (standard deviation) is σ, but this is not Γ the full width at half maximum that is commonly measured. The relationship between the two measures of the dispersion can be seen if we let $P(x) = P_0\, e^{-\frac{x^2}{2\sigma^2}}$, where P_0 is a constant. Assume that h is the value of x at which $P(h) = P_0/2$, then $P(h) = P_0\, e^{-\frac{h^2}{2\sigma^2}} = P_0/2$, or $e^{-\frac{h^2}{2\sigma^2}} = 1/2$. Take the logarithm of both sides, to give $\frac{h^2}{2\sigma^2} = -\ln(0.5) = \ln 2$. Solving for h we have that

$$h = \sqrt{2\,\ln 2}\,\sigma$$

$$h = 1.177\,\sigma,$$

but the full width at half maximum is $2h$, so we find that

$$\Gamma = 2h = 2\sqrt{2\,\ln 2}\,\sigma = 2.355\,\sigma.$$

Although this function represents only one of Gauss's many accomplishments in mathematics, it was famous enough to be used on the zehn Deutsch Mark note prior to the unification of European currency into the Euro (Fig. 1.7).

How well does the Gaussian probability distribution describe a random process? Many examples could be thought of, such as the distribution of 200 pennies after a toss. It would be tedious to toss 200 pennies enough times to satisfy oneself that enough data had been obtained to enable an adequate test of the Gaussian, but many electronic devices produce random events that can be examined for their statistical behavior. One such device is a CCD (charge coupled device) camera in which photons strike a silicon substrate and liberate electrons in a pixel array. The fluctuations in the number of electrons per pixel are governed by a random walk of thermal phonons in the silicon chip. Electrons are then counted in each pixel, and the result is a two-dimensional array of the number of electrons generated in each pixel. This is a random walk because there is equal probability of a pixel having a charge above or below the mean charge, and so it is a well-known two-state system. A 60-second dark-frame exposure of a Santa Barbara Instruments Group STL11000XCM camera, operating at $-15\,°C$, produced the noise histogram shown in Fig. 1.8. A total of $N = 361,000$ electrons were counted here to give the data points (Fig. 1.8). This is

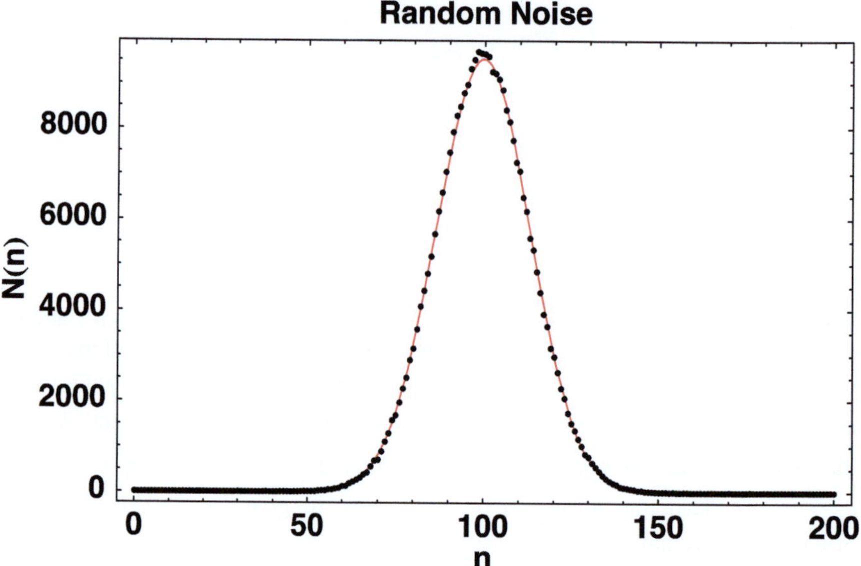

Fig. 1.8 Comparison between the measured thermal noise distribution from a CCD camera (dots) and that from a Gaussian function (red line) computed with the same empirical parameters as derived from the experimental data. The computed mean was 99.6, compared to the a priori mean of 100; the computed width of 14 compared favorably with the a priori width of Sqrt[200] = 14.1

the equivalent of tossing 200 pennies ~180 times. The mean number of thermal electrons per pixel accumulated within the CCD chip was found to be $n = 99.6$, the width $\sigma = 14$; while the expected, a priori mean was 100, with a width of Sqrt [200] = 14.1. This good agreement is hardly surprising since our derivation of the binomial distribution was simply based on counting. We are seeing that randomness is beginning to resemble determinism, when the number of objects, or particles, or trials becomes large. It is even more interesting that the properties of randomness are amenable to measurement for anyone who has a cell phone camera, a computer, and a modicum of imaging software.

1.7   The Exponential Probability Distribution

While the binomial and the Gaussian probability distributions arise from simple assumptions about the nature of events, this is also true for another probability distribution, the Exponential Distribution. The nature of this distribution is important to understand because it describes how the density of states drives an important result. This distribution is the basis for the Boltzmann distribution, perhaps the most important probability distribution to understand in order to grasp how Life profits from it. The exponential distribution simply assumes that the density of states is inversely proportional to a scale parameter, which we will call L_0 here. Later we will identify it with kT, the average thermal background energy. It is against this scale parameter that the mean of the distribution is compared; the scale parameter indicates how rapidly the probability decreases. In the case of Life on Earth, this scale is set by the temperature of mammals, i.e., ~310 K.

So, let's look at this distribution and get a feel for its nature. I will derive it based on a model for the density of states. Suppose that we relax the requirement in the random walk that the steps need to be the same length, and rather allow them to vary in a random fashion. We seek to find the probability distribution $P(L)$ for finding a step of length L between L and dL, i.e., $P(L + dL)$. This will equally serve if we subtract L to examine the interval from 0 to dL, then $P(dL)$ will be proportional to dL and we can write

$$P(dL) \; \alpha \; dL$$

so that $P(dL) = f(L)dL$, where $f(L)$ is the density of probability states for this distribution function, the probability per unit length. Since $f(L)$ is a probability density, it must be normalized so that

$$\int_0^\infty f(L)dL = 1.$$

The mean step length is L_o as given by

$$L_0 = <L> = \int_0^\infty L f(L) dL$$

and the variance is

$$\sigma^2(L) = \int_0^\infty (L - <L>)^2 f(L) dL$$

A continuous probability distribution like this is describes a random sequence of steps along the x-axis, with each step independent of the previous or subsequent step. The total number of steps, N, could correspond to the displacement from the origin or the distance, d, diffused by a water molecule after a time, t. Then

$$d = N \, L_0 t / \Delta t,$$

where $L_0/\Delta t$ is the velocity. A step of length L would be equally likely to be found in any element, dx, of the x-axis. The mean number of steps contained in a unit length of the x-axis is then $1/L_0$ and the mean step length is also the mean distance between steps so that the step density (the density of states in the state space of steps, the probability space) is simply the reciprocal of the average step length. For this reason, the probability that an element dx of the x-axis contains a step of length dL is dx/L_0 and this quantity does not depend on where dx occurs along the x-axis.

We have previously used the conservation of probability, $p + q = 1$, in a binary fashion to find $p = P(n)$ the chances of a return to the origin of the random walk after N steps, and we know that the opposite, the chances, q, of not returning to the origin are $q = 1 - p$. In the present case, $f(L)$ must have the limits, $f(0) = 1$ and $f(\infty) = 0$; the probability that a zero-length segment $dx = 0$ contains a line segment of length L is 0, while the probability that the entire x-axis contains a line segment of length L is 1, certainty. In a similar vein we now ask for the probability that a step of length L will be found in the interval of length L_0 along the x-axis. Call this probability $P(x_0)$ where $= x_0/L_0$ and $dx_0 = dx/L_0$ divides the x-axis into units of the mean step length, L_0.

The probability $P(x_0 + dx_0)$ that no steps are in the interval is equal (I refer the reader to the rules for *and* as a joint probability) to the probability that no steps are of length L_0 times the probability that no steps also have length dx, which is $1 - dx_0$ or, in symbols as a Taylor's series expansion, keeping only the first-order derivatives

$$P(x_0 + dx_0) = P(x_0) + dP(x_0)/dx$$
$$= P(x_0)(1 - dx_0) = P(x_0)(1 - dx/L_0)$$

which can be rearranged to read

$$\frac{dP(x_0)}{dx} = -\frac{1}{L_0} P(x_0)$$

and can be integrated as

$$\int \frac{dP(x_o)}{P(x_o)} = -\frac{1}{L_o} \int \frac{dP(x_o)}{P(x_o)} = -\frac{1}{L_o} \int dx$$

which has the solution

$$P(x_o) = e^{-\frac{x_o}{L_o}}$$

and the probability that no step of length L is present within any distance x anywhere on the x-axis diminishes as x increases and decays as $\exp(-x/L_o)$, where the scale factor L_o determines the spatial decay rate. Note that for a zero-length interval along x that $P(0) = 1$, one can rest assured that no step of finite length can be found in a zero-length interval. The probability that no step of length L_o occurs in an x interval decays exponentially as the x interval grows. For this reason it is called the exponential probability distribution. Here, we have introduced an important concept for continuous probability distributions, the density of states, or how many outcomes exist per unit outcome interval.

As a final touch, we need to ask if the probability is normalized; is the sum (the integral in this case) of all the probabilities unity? Let's compute the integral and find out.

$$\int_0^\infty P(x)dx = \int_0^\infty e^{-\frac{x}{L_o}} dx = L_o.$$

The answer is no, because we need to normalize our probability distribution by dividing by this factor. The complete exponential probability density function is therefore

$$P(x) = \frac{1}{L_o} e^{-\frac{x}{L_o}}$$

where the scale parameter is L_o, the mean step length. In the next chapter, we will see how we can use this model to equate the mean step length, using a similar density of states, to the average energy, kT, available from the thermal background environment, and to thereby derive the important Boltzmann probability distribution.

What remains is to calculate the mean and variance of the exponential distribution. Using the definitions we derived earlier, the mean value of L is

$$\int_0^\infty L\,P(L)\,dL = \int_0^\infty x\,e^{-\frac{x}{L_o}} dx = L_o$$

and the variance is

$$\sigma^2(L) = \int_0^\infty (x - L_o)^2\, e^{-\frac{x}{L_o}} dx = L_o^2$$

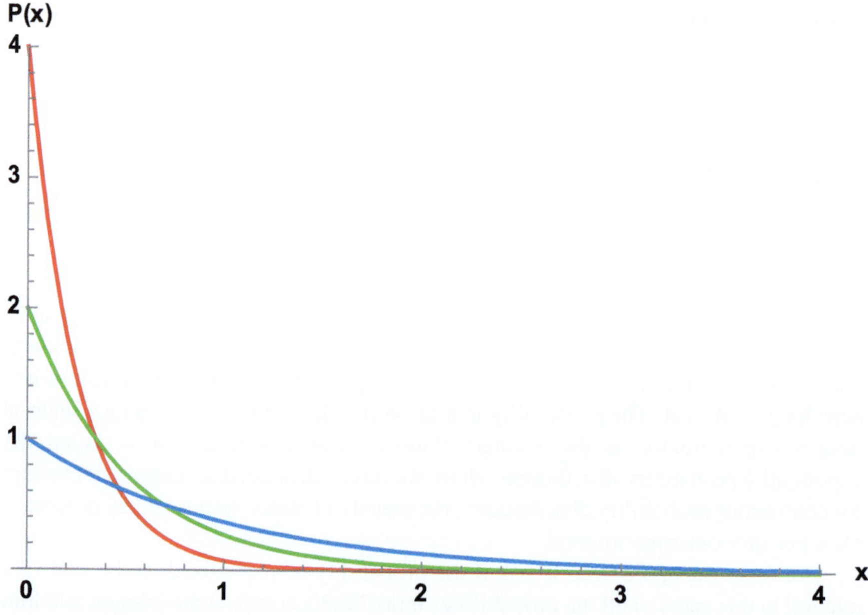

Fig. 1.9 Behavior of the exponential probability distribution for $L_o = \{0.25, 0.5, 1;$ Red, Green, Blue, respectively$\}$

therefore, the standard deviation (width) of the exponential distribution is the same as L_o, the mean step length, $\sigma = L_o$. A plot of the exponential distribution for $L_o = \{0.25, 0.5, 1\}$ is shown in Fig. 1.9.

1.8 The Poisson Distribution

The next probability distribution that we need to be familiar with is the "Poisson distribution." This will be derived, much like the Gaussian, as a limit of the "binomial distribution." The Poisson distribution is often used to determine features of time series, such as the arrival times of nuclear decays, or the waiting times in queues. We can also use it to model double strand DNA breaks caused by radiation, or the arrival times of substrates to enzymes in a cell. Its main usage is for events where the probability of an event in a certain temporal or spatial interval is very small. In the case of the random walk, one can think of our poor bar patron attempting to find her way home when the sidewalk is severely tilted uphill to the left so that the probability q of a left-hand step is much smaller than that of p, a right-hand step; $q \ll p \sim 1$. This bias anticipates how living systems use energy to manipulate probability. We assume that because there is an energetic cost associated with walking uphill the probability of an uphill (left) step is reduced with respect to a right step, downhill. In any event, the mean number of left steps is still $\mu = Nq$, but in a long sequence of N steps, the appearance of a left step is a rare event. We earlier wrote the binomial distribution as

$$P_N(n_1) = \frac{N!}{n_1! n_2!} p^{n_1} q^{n_2}$$

where the number of right steps was n_1 and that of left steps was n_2, but we can rewrite this in terms of just the total number of steps, N, and the number of left-hand steps, n $(= n_2)$, as

$$P_N(n) = \frac{N!}{(N-n)! n!} q^n (1-q)^{N-n}.$$

Now, since $q \ll 1$, the term $(1-q) \sim 1$ and $N - n \sim N$ so that these two terms can be approximated using natural logarithms, as we did earlier (vide supra):

$$\ln\left[(1-q)^{N-n}\right] = (N-n)\ \ln\ (1-q)$$

and for $q \ll 1$,

$$\ln\ (1-q) \sim -q,$$

so that

$$\ln\left[(1-q)^{N-n}\right] \sim -q(N-n) \sim -Nq$$

or, after exponentiating both sides, we have that

$$(1-q)^{N-n} \sim e^{-Nq}.$$

Now, the second term, $N!/(N-n)!$, can be approximated using Sterling's first approximation as

$$\mathrm{Ln}\left[N!/(N-n)!\right] \sim N \ln\ (N) - N - (N-n) \ln\ (N-n) + (N-n).$$

Here note that $\ln(N-n)$ can be rewritten as

$$\ln\ [N(N-n)/N] = \ln\ [N(1-n/N)] = \ln\ (N) + \ln\ (1-n/N)$$

or

$$\ln\ (N-n) \sim \ln\ (N) - n/N$$

and

$$(N-n)[\ln\ (N) - n/N] = N \ln\ (N) - \ln\ (N) - n + n^2/N$$

The term n^2/N can be neglected since $n \ll N$, and this gives

$$\mathrm{Ln}\left[N!/(N-n)!\right] \sim n\,\ln(N)$$

or

$$N!/(N-n)! \sim N^n.$$

Putting these results together gives the Poisson distribution

$$P_N(n) = \frac{(Nq)^n}{n!}e^{-Nq}.$$

The mean number of events (left steps) is $\mu = Nq$ so that we can write this as

$$\boldsymbol{P_N(n) = \frac{\mu^n}{n!}e^{-\mu}.}$$

This is the product of two dimensionless numbers, the number of steps, which is an integer, and the probability of a left step, which is a rational number. Any other combination of parameters can be tolerated as long as the units cancel; often the Poisson distribution is used to model time series, such as the radioactive decay of nuclei, or the distribution of arrival times in queuing theory, where Nq is replaced by λt consisting of a decay rate, λ, and an arrival time, t. Examples of the Poisson distribution with various values of $\mu = Nq = \{1, 5, 10\}$ are shown in Fig. 1.10, where we have plotted it as if it were a continuous function of n, whereas it is really only defined for integer values of n.

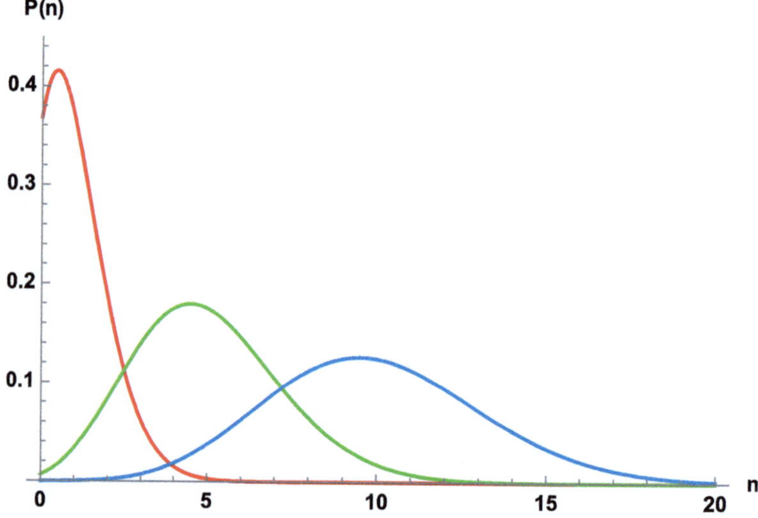

Fig. 1.10 Behavior of the Poisson probability distribution for $\mu = \{1, 5, 10;$ Red, Green, Blue, respectively$\}$

The final properties of the Poisson distribution that we require for subsequent work are its mean, variance, and normalization. The Poisson distribution is normalized:

$$\sum_{n=0}^{\infty} P(n) = \sum_{n=0}^{\infty} \frac{\mu^n}{n!} e^{-\mu} = 1.$$

We can find the mean using the definition of the mean (Eq. W) as

$$\sum_{n=0}^{\infty} nP(n) = \sum_{n=0}^{\infty} n\frac{\mu^n}{n!} e^{-\mu} = \mu$$

while the variance is

$$\sigma^2 = \sum_{n=0}^{\infty} (n-\mu)^2 P(n) = \sum_{n=0}^{\infty} (n-\mu)^2 \frac{\mu^n}{n!} e^{-\mu} = \mu$$

so that the width of the Poisson distribution is the square root of the mean, $\sigma = \sqrt{\mu}$. Like the binomial distribution, the Poisson distribution is strictly defined only for integral values of n due to the appearance of n! in the denominator; a more accurate representation is shown as a bar graph (Fig. 1.11).

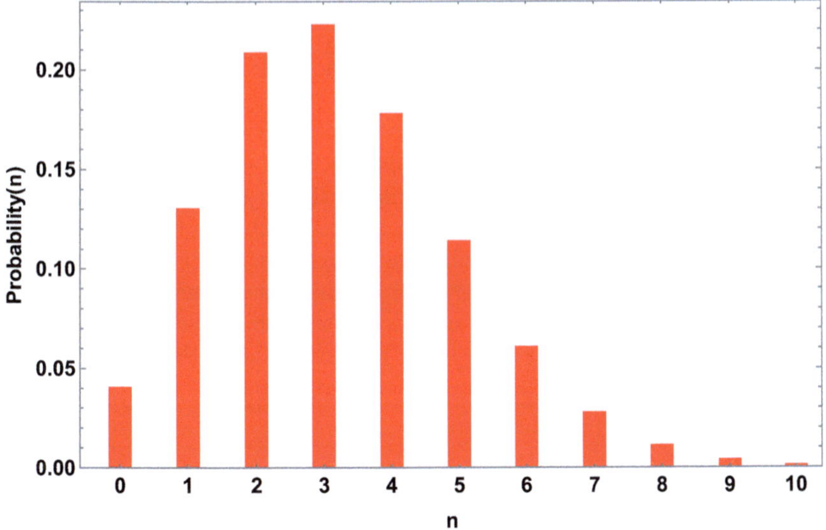

Fig. 1.11 A bar graph representation of the Poisson distribution for $\mu = 3.2$

1.9   The Fermi–Dirac Distribution

When we encounter quantum mechanics in Chap. 3, we will find that one of the fundamental assumptions of our treatment of probability is invalid. The Exponential (Boltzmann) distribution is founded on the assumption that particles are solely distributed according to the energies of the states. We will see from our investigation of the Poincare group of relativistic coordinate transformations that the only objects that can exist in the Universe are characterized by two fundamental quantities, their *masses* and their intrinsic angular momentum quantum number, *spin*. The spins can only take on integral or half-integral values: 0, ½, 1, 3/2, 2, ... Since elementary particles of a given class, such as electrons, protons, and photons, are identical, e.g., any given electron is indistinguishable from any other electron, and the same is true for protons, photons, and all the rest, one might expect that the wave functions for multiparticle states would be invariant under the interchange of any two particles. While this is true for *Bosons*, whose spins are integral, 0, 1, 2, 3, . . ., it is not true for the other class of particles, the *Fermions*, whose spins are half-integral, ½, 3/2, 5/2, ... The Pauli Principle (the Spin-Statistics theorem) places restrictions on the occupancy of quantum states for Fermions; no two Fermions can occupy the same quantum state, their wave functions must be antisymmetric under particle interchange. Boson states must be symmetric under particle interchange. Bosons are therefore distributed solely according to their energies.

Fermions fill the available states singly, until all the states are filled up to the maximum energy. The Fermi–Dirac probability distribution is given by

$$P_{\mathrm{FD}}(E) = \frac{1}{e^{(E-f)/kT} + 1}$$

where f is the energy of the highest filled state, known as the Fermi energy. For low temperatures (~10 K), the distribution is almost a step function; few of the particles have enough energy to exceed kT, while as the temperature increases, more particles can move up in energy and occupy higher energy states, depleting the population in the lower states (Fig. 1.12). A little reflection is all that is necessary to understand that Fermi–Dirac statistics are responsible for the structures of nuclei, atoms, and molecules, the Periodic Table and ultimately, *life itself*. For if electrons in atoms, protons, and neutrons in nuclei obeyed Bose–Einstein statistics, all the particles would hover near the ground state and nuclei and atoms would essentially be all featureless and distinguished only by their masses and not their nuclear physics or their chemistry.

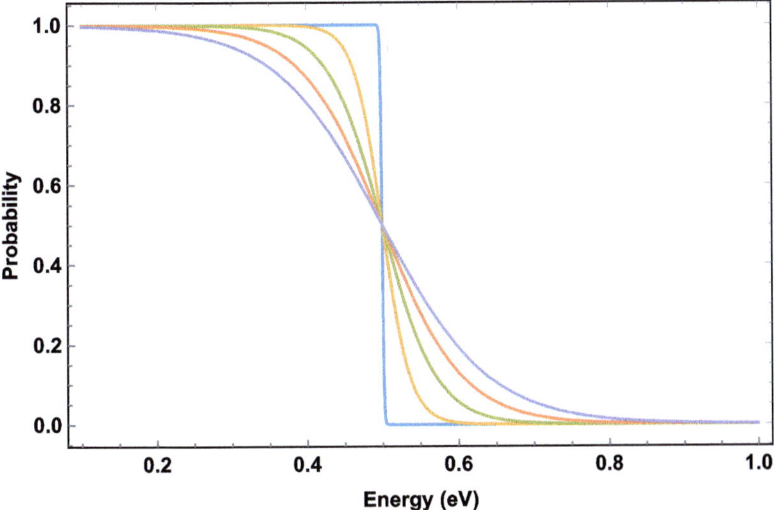

Fig. 1.12   A plot of the Fermi–Dirac probability distribution with the Fermi energy $f = 0.5$ eV. The various curves are for $T = 10$ K (blue) to 1000 K (purple) in steps of 200 K

Fig. 1.13   A plot of the Bose–Einstein probability distribution with the chemical potential $f = 0.5$ eV. The various curves are for $T = 10$ K (blue) to 1000 K (purple) in steps of 200 K

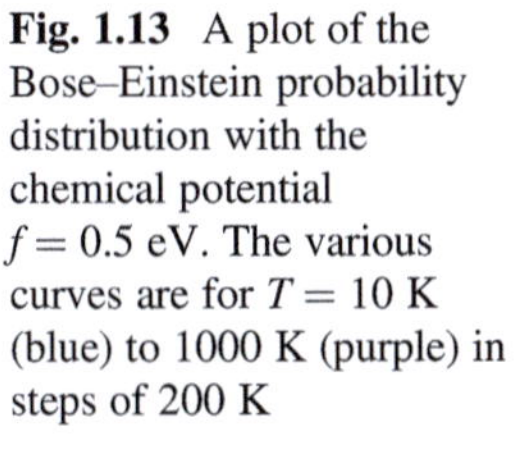

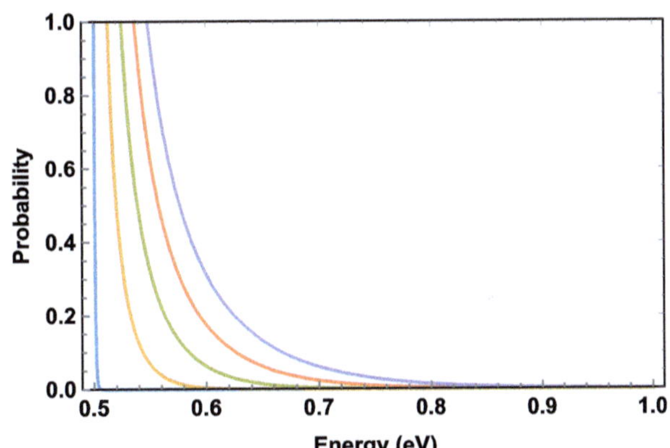

1.10   The Bose–Einstein Distribution

The integral spin Bosons obey the exponential probability distribution given by

$$P_{\mathrm{BE}}(E) = \frac{1}{e^{(E-f)/kT} - 1}$$

where in this case f (as is also true for the Fermi–Dirac distribution) is the chemical potential of the Bosons. This probability distribution (Fig. 1.13) shows the expected behavior with the Bosons piling up at the lowest energies, and only filling the higher states as the temperature, and hence the energy from kT, increases.

1.11   The Maxwell–Boltzmann Distribution

Massive particles, such as atoms and molecules, are also indistinguishable, but their large masses lead to a classical treatment where the occupation of states is governed by the relationship between their kinetic energies and the scale factor kT. We will investigate the Maxwell–Boltzmann probability distribution in Chap. 2. Meanwhile, in concluding this brief introduction to probability distributions, we need to emphasize a modern paradox that is hardly appreciated by the layman; the world appears deterministic, when viewed at the macroscopic scale, and yet the microscopic world is completely random. Indeed, we will find that in quantum mechanics, which governs the atomic and subatomic realms, only probabilities are computable and when computing the interactions between quantum particles, one must use *Unitary* operators, ones that explicitly conserve probability.

Problems

These problems use Probability and Physical Biochemistry to introduce you to the usage of *Mathematica*®. This computer algebra system is an essential tool for exploring the concepts in this text, so it makes sense for you to become familiar with this software. Appendix 1 gives a very brief introduction to the use of this software. The syntax for unknown functions can be found using the *Mathematica* Help system. The solutions for selected problems are given in italics. (However, devoted students will want to solve these separately and then check their answers.)

1. *Card shuffling*: In all of human history, the probability that the exact order of the cards in a properly shuffled deck has been repeated is essentially nil.

   (a) Solve for how many combinations a standard 52 card deck can be arranged to 5 significant figures.
   (b) Assuming the world population is at 7.5 billion, and that each person was able to entirely shuffle a deck once per minute, how many years would it take for the first repeat to statistically appear?
   (c) But now imagine that each one of these 7.5 billion people got their deck of cards mixed up. Picking up 64 cards at random, what are the chances of having 5 aces of spades?
   (d) With all of this variation in just a simple deck of cards, imagine the probability of a protein randomly forming from amino acids. Think of a primordial soup, for instance, filled with all 20 amino acids in an ideal solution that promotes the formation of peptides. What is the probability that the smallest subunit of RNA polymerase II, RPB9 at 125 amino acids, would randomly form?

2. *Water in a liver cell.* Assume a human liver cell is a cube 10^{-5} m on a side.

(a) If 80% of this cell's mass consists of H_2O, how many water molecules reside in this cell?

(b) If you could measure the momentum of each water molecule, what would you expect the relative width of this distribution to be?

2(a) *The volume of a cell is $(10^{-5}$ m$)^3 = 10^{-15}$ m.*

The density of water is 10^3 kg/m^3, so the cell contains 10^{-12} kg water. At 80% the mass of water this is 8×10^{-13} kg. With the molecular mass of water $= 1.8 \times 10^{-2}$ kg/mole, this amount of water is $8/1.8 = 4.44 \times 10^{-11}$ moles, or $(4.44 \times 10^{-11}$ moles $\times 6.023 \times 10^{23}$ mole$^{-1}) = 2.677 \times 10^{13}$ water molecules/cell.

This is a large number when thinking about the statistical properties of intracellular water!

2(b) *If you could measure the momentum of each water molecule, what would you expect the relative width of this distribution to be?*

$$\sigma/N = \mathrm{Sqrt}(N)/N = 5.17397 \times 10^6/2.677 \times 10^{13}\ 3 = 1.93275 \times 10^{-7}$$

3. *Gaussian probability distribution.* Nanoparticles have found numerous applications in our everyday lives. Important biomedical applications require nanoparticles that are specific sizes. The size range of synthesized nanoparticles is characterized by a probability distribution, similar to many random processes in nature. One of the common probability distributions is the Gaussian, named after the German mathematician, Karl Friedrich Gauss. If $P(x)$ is the probability of x, then the Gaussian probability distribution is given by:

$$P(x) = \frac{1}{\sigma\sqrt{2\pi}} e^{-\frac{(x-\mu)^2}{2\sigma^2}}$$

where σ is the width (standard deviation) of the distribution and μ is the mean. Assume that you have synthesized nanoparticles in your lab with an average diameter of $<x> = \mu = 42$ nm and a dispersion of 8 nm.

Using *Mathematica*:

(a) Write the Gaussian distribution using *Mathematica* syntax for a user-defined function p (see Appendix) and `Plot[p(x, μ, σ),{x,0,100}]` for $\mu = 42$ nm and a dispersion of $\sigma = 8$ nm.

(b) Use `Manipulate` to explore what happens as σ varies from 2 to 32.

(c) How about μ? Plot the normal distributions on the same graph with μ values of -1, 0, and 1.

(d) Compute and plot the derivative $p'(x) = dp/dx$. `Plot[p'(x), {x,0,100}]`

(e) What is the most probable value of x in a?

(f) The frequency of the first four chemical elements H, He, Li, and Be of the periodic table can be represented very roughly by a Gaussian distribution. Refer to the following website:

http://www.angelo.edu/faculty/kboudrea/periodic/physical_abundances.htm

and using the `ListPlot` function of *Mathematica*, create the Gaussian-like graph of the abundance in the Universe of the first four elements on the periodic table: H, He, Li, and Be. Make sure to include axes labels and connect the points in the graph so that they show as a continuous line instead of isolated points. Use atomic number as the x-axis.

3(a, b, c) *Write the Gaussian distribution using* Mathematica *syntax for a user-defined function (see Appendix) and* `Plot[p(x, μ, σ), {x,0,100}]` *for* $\mu = 42$ nm *and a dispersion of* $\sigma = 8$ nm.

```
Manipulate[Plot[gauss[x,μ,σ]],
          {x,0,100},
          PlotStyle->Hue[0],
          PlotRange->All],
      {μ,11,84},{σ,2,32}]
```

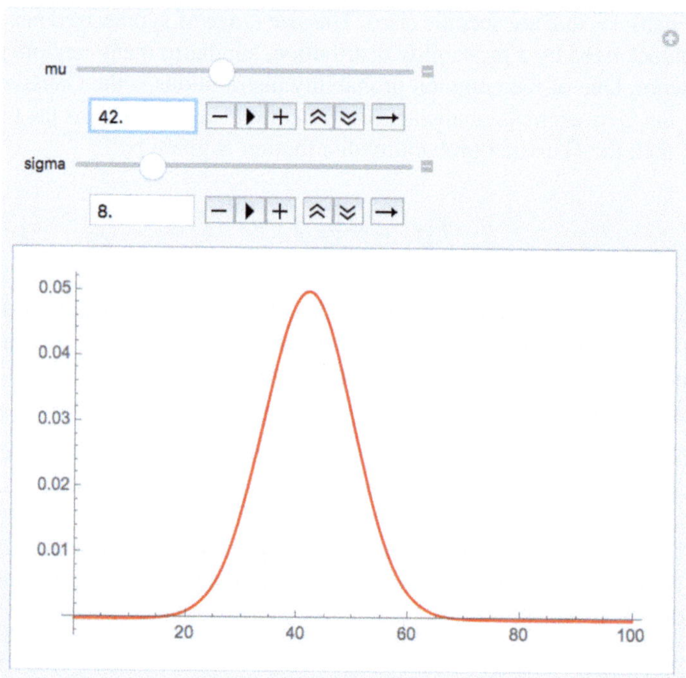

3(d)   *Compute and plot the derivative p′(x) = dp/dx.*

$$\text{deriv}[x\_, \mu\_, \sigma\_] := -\frac{e^{-\frac{(x-\mu)^2}{2\sigma^2}}\,(x-\mu)}{\sqrt{2\pi}\,\sigma^3}$$

```
Plot[deriv[x,42,8],{x,0,100},
    PlotStyle->Hue[0],
    PlotRange->All ,
    AxesLabel -> {"x", "P'(x)"}]
```

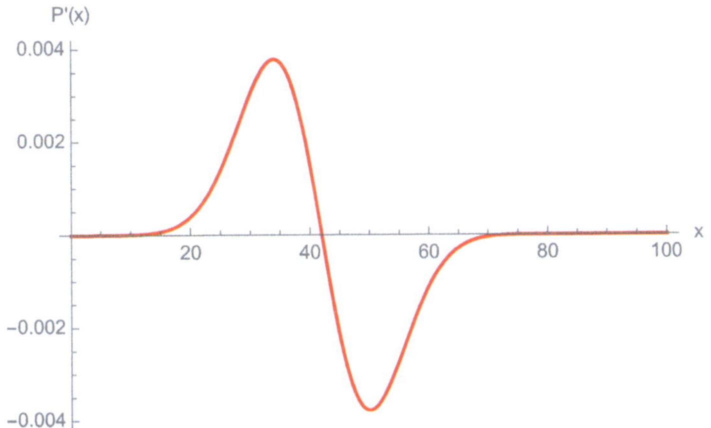

3(e)   *What is the most probable value of x in a? This is given by the peak, which
is at the mean, or from the place where the derivative is zero, is this case,
at x = 42 nm.*

4. *Probabilities are normalized.* The sum of all probabilities must equal 1, therefore
probability distributions must be normalized. Show that $\int_{-\infty}^{\infty} P(x)dx = 1.$ for the
nanoparticles in problem 3.

5. *Average values.* The average value of a variable x governed by a probability
distribution P(x) is given by:

$$\overline{x} = \frac{\int_{-\infty}^{\infty} xP(x)dx}{\int_{-\infty}^{\infty} P(x)dx},$$ what is the average value of x for the Gaussian in problem 3?

6. *Nuclear spins in a magnetic field.* The binomial distribution arising from the
random walk in one-dimension is an excellent model for the behavior of nuclear
spins in a large magnetic field. For a set of $N = 400$ spins of $I = ½$ each, with the
probability of an up spin $= p = ½$:

(a) Find n_1 the average number of up spins.
(b) Make a plot of $\{n_1, P(n_1)\}$.
(c) Find the standard deviation of this distribution.
(d) Repeat this for $p' = 0.500005$. This spin probability difference is about that for two protons in a magnetic field of 10 Tesla.

6(a) *Find n_1 the average number of up spins.*

   Use $<n_1> = Np = 0.5 * 400 = 200$ up spins.

6(b & d) *The plots are shown below, as is the difference between the plots, for $p = 0.5$ and for $p' = 0.500005$. Note that the two plots are almost indistinguishable, but the difference plot shows nonzero values at the 5 ppm level.*

6(c) *Find the standard deviation of this distribution.*

   $$\sigma = Sqrt(Npq) = 3.5355 \ \ldots, for\ p = 0.5, and$$

6(d) $<n> = 25.00025$, and

$\sigma = Sqrt(Npq) = 3.5355 \ \ldots, p' = 0.500005.$ *These differ by only* 1.77×10^{-10}.

7. *Properties of the binomial coefficients.* In the random walk for N total steps, with n_1 Right and n_2 Left steps, and $N = n_1 + n_2$:

(a) Show that $C_N(n_1) = C_N(n_2)$.
(b) Show that $C_N(n + 1)/C_N(n) = [N - n]/[n + 1]$.

7 (a) *Show that $C_N(n_1) = C_N(n_2)$.*

   $C_N(n) = N\ !\ /[n\ !\ (N - n)!]$,
   let $N = n_1 + n_2$,
   then $n_2 = N - n_1$ and $C_N(n_2) = N!/[n_2!(N - n_2)!] = N!/[n_2!n_1!] = N!/[n_1!n_2!]$.

7 (b) *Show that $C_N(n + 1)/C_N(n) = [N - n]/[n + 1]$.*

   $$C_N(n + 1)/C_N(n) = \{N!/[(n + 1)!(N - n - 1)!]\}/\{N!/[n!(N - n)!]\}$$
   $$= n!(N - n)!/[(n + 1)!(N - n - 1)!]$$

   Now, expand the numerator and denominator to give:

   $n!(N - n)!/[(n + 1)!(N - n - 1)!]$
   $$= \frac{(n)(n-1)(n-2)\ldots(1)\ [N - n](N-n-1)(N-n-2)\ldots(1)}{[n + 1](n)(n-1)(n-2)\ldots(1)\qquad (N-n-1)(N-n-2)\ldots(1)}$$

   all the terms cancel except those in square brackets, giving

$$C_N(n+1)/C_N(n) = n!(N-n)!/[(n+1)!(N-n-1)!] = [N-n]/[n+1],$$

Therefore, $C_N(n+1)/C_N(n) = [N-n]/[n+1]$.

8. *Stirling's first approximation.* Use Stirling's first approximation for $x!$, given as:

$$x! = \text{Sqrt}(2\pi)x^{x+1/2}e^{-x}$$

(a) To find an approximate form for $C_N(n)$ for large N, which does not involve any factorials.
(b) How large must N be for the relative error between the exact and approximate expressions for $C_N(n)$ to be less than 1%?

8(a) *We are given that Stirling's First approximation for $x!$ is $x! = \text{Sqrt}(2\pi)$ $x^{x+1/2}e^{-x}$, so plug this into the definition of the binomial coefficients:*

$$C_N(n) = N!/[n!(N-n)!] = \frac{\cancel{\text{Sqrt}(2\pi)}N^{N+1/2}e^{-N}}{\text{Sqrt}(2\pi)n^{n+1/2}e^{-n}\cancel{\text{Sqrt}(2\pi)}(N-n)^{N-n+1/2}e^{-N+n}}.$$

Which can be written, after collecting factors and exponentials, as

$$C_N(n) = \frac{1}{\text{Sqrt}(2\pi)}\frac{N^{N+1/2}}{n^{n+1/2}(N-n)^{N-n+1/2}}\frac{e^{-N}}{e^{-n}e^{-N+n}}.$$

The last term is $e^{-n}e^{-N+n}e^{-N} = e^{-n-N+n-N} = e^{0} = 1$, *so*

$$C_N(n) = \frac{1}{\text{Sqrt}(2\pi)}\frac{N^{N+1/2}}{n^{n+1/2}(N-n)^{N-n+1/2}}.$$

This is much easier to deal with for large N than taking N!

8(b) *How large must N be for the relative error between the exact and approximate expressions for $C_N(n)$ to less than 1%? Use* Mathematica *to compute the difference as a function of N.*

9. *Nuclear spins.* For our NMR spin problem (#6, above), let the probability of spin up be $p = 0.7$ and spin down be $q = 0.3$.

(a) Find average value of the magnetization $<m>$, $\sigma(m)$, and the ratio $\sigma(m)/$ $<m>$ for $N = 10, 50, 100, 10^3, 10^4, 10^5, 10^{10}, 10^{15}, 10^{20}$, and 10^{25} spins.
(b) If we can determine fluctuations by NMR to one part in 10^{10}, then how many spins must we have in our sample in order to NOT measure any fluctuations?

9(a)   *Let $p = 0.7$ and $q = 0.3$. Find $<m>$, $\sigma(m)$, and the ratio $\sigma(m)/<m>$ for $N = 10, 50, 100, 10^3, 10^4, 10^5, 10^{10}, 10^{15}, 10^{20}$, and 10^{25} spins. Use units of μ_o, so μ_o never appears explicitly.*

$p =$	0.7		
$q =$	0.3		
N	$<m>/\mu_o$	$\sigma(m)/\mu_o$	$\sigma(m)/<m>$
10	4	2.90E+00	7.25E−01
50	20	6.48E+00	3.24E−01
100	40	9.17E+00	2.29E−01
1000	400	2.90E+01	7.25E−02
10,000	4000	9.17E+01	2.29E−02
1.00E+05	40,000	2.90E+02	7.25E−03
1.00E+10	4E+09	9.17E+04	2.29E−05
1.00E+15	4E+14	2.90E+07	7.25E−08
1.00E+20	4E+19	9.17E+09	2.29E−10
5.25E+20	2.1E+20	2.10E+10	1.00E−10
1.00E+25	4E+24	2.90E+12	7.25E−13

9(b)   *Let $\sigma(m)/<m> = 10^{-10}$, then we will find the lower limit for the number of spins N_{min} with fluctuations in the magnetization less than or equal to this value.*

   Now, $\sigma(m)/<m> = 2\ Sqrt(Npq)\ \mu_o/[N(p - q)\mu_o] = 10^{-10}$. Let $K = 10^{-10}$, then solve for N to give

$$N = 4pq/\left(K^2(p{-}q)^2\right) = (4^*\ 0.7^*\ 0.3)/\left[10^{-20}(0.7{-}0.3)^2\right] = 5.25 \times 10^{20}\ spins.$$

   Therefore, $N_{min} \geq 5.25 \times 10^{20}$ spins.
   For water protons, this corresponds to 4.36×10^{-4} moles, or 7.84 mg ($= 7.84\ \mu L$) because there are two protons per molecule.

10. *Random walk.* For N total steps, with n_1 Right and n_2 Left steps, and $N = n_1 + n_2$:

(a) The following is an equation to generate a random walk in *Mathematica*:

```
RandomWalk[n_]:=NestList[(# + ( − 1)^R andom[Integer])&, 0, n]
```

   Using the ListPlot function, graph the equation with 200 steps (your plot will be different every time *Mathematica* computes this). This is written as a user-defined function containing a pure function, as denoted by the & with # as its formal parameter. (See the *Mathematica* Help system for further details.)

(b) In your specific plot, does the random walker end with a final position to the left or right of the bar? How many steps away is he from his starting point?

10(a)
```
RandomWalk[n_] = NestList[(# + (-1)^Random[Integer])
      &, 0, n]
```

```
ListPlot[RandomWalk[n1 = 200],
      PlotLabel -> {"A Random Walk of ", n1, "Steps."},
      AxesLabel -> {"Step Number", "Displacement(Steps)"},
      Joined -> True,
      PlotStyle -> Hue[0]]
```

11. *Protein synthesis.* The rate at which amino acids can be added to the growing polypeptide chain in the ribosome is 20 residues per second. In his laboratory, a scientist monitors such a synthesis and acquires the following data:

Time (s)	Number of amino acids added
0	15
1	35
1.5	42
3	60
4.5	91

(a) Generate two different lists, one for the time (x) function, and one for the number of amino acids in the protein (y). Use the `Transpose` function of *Mathematica* to combine them into one list.

(b) Graph the list using the `ListPlot` function. Use the help function of *Mathematica* to investigate the different styles of plots you can do, and do not forget the axis labels!

(c) Fit the data (create an equation from the data points) and plot it on the same graph as the one provided in part (c).

12. *Dice problem.* The following *Mathematica* code will generate a State Table for the arrangements of two six-sided dice:

```
list = Table[Table[{i, i - (k - 1)6, k, i - (k - 1) 6 + k}, {k, 1, 6}],
{i,1,36}];
statetable = Table[Table[list[[i]][[k]], {k, 1, 6}], {i, 6 k - 5, 6
k}];
```

where `statetable` is a list with elements $\{i, d1, d2, s\} = \{$state number, value of Die #1, value of Die #2, sum of the values on $D1$ and $D2\}$, such as:

```
statetable={{1,1,1,2},{2,2,1,3},{3,3,1,4}, ... , {36,6,6,12}}
```

(although the above code generates this list with the states in a different order).

The various outcomes are the sums, s, that are the fourth elements of this list, which can be extracted from `statetable` by using

```
outcomes = Flatten[Table[Table[statetable[[i]][[k]][[4]], {i,
1, 6}], {k, 1, 6}]];
```

(a) Use the function `BinCounts` to count the number of times a given outcome occurs.

(b) Compute and plot the a priori probability distribution of each outcome *versus* the value of the outcome.

(c) Use `RandomInteger` to generate a Table of a priori probability distribution random values of the sums of two six-sided dice for N rolls of the dice, where N takes on the values $N \in \{10, 25, 50, 100, 250, 500, 1000\}$.

(d) Compute and plot the a posteriori probability distribution for the outcomes in (c) and compare them with the results in (b).

(e) Compute the means and standard deviations for the distributions in (b) and (d).

(f) Compute the root-mean-squared deviation between the a priori and a priori probability distributions as a function of N and plot this as a function of N.

(g) In the text, and in Problem 8 above, we said that the root-mean-squared deviation decreases as a function of N. Use the `Fit` function to fit the a posteriori data to this function of N and plot this function along with the root-mean-squared deviation found in (f) as a function of N.

(h) Many games, such as Dungeons&Dragons, that use dice do not limit the number of faces on each die of a pair to six. How would you modify the *Mathematica* code that generates the `statetable` to handle pairs of dice with $m \times n$ faces where m and n are arbitrary integers greater than 4?

13. *Peptides.* Covalent and non-covalent bonds within a peptide help determine a peptide's structure. Cysteine residues can covalently bond to each other forming disulfide bonds:

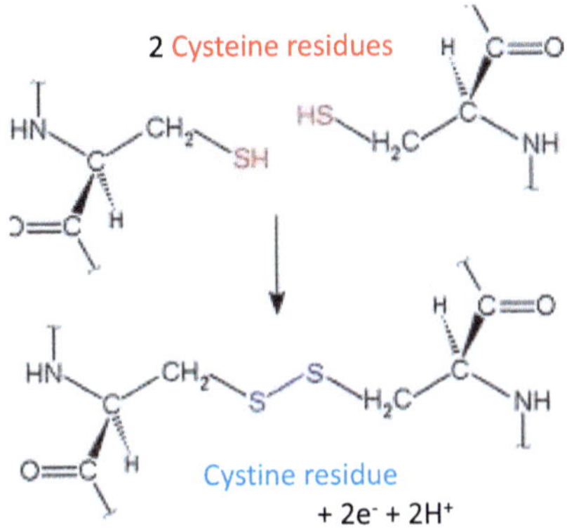

2 Cysteine residues

Cystine residue
$+ 2e^- + 2H^+$

Assume that each of the 20 amino acids has an equal probability of appearing in a peptide chain.

(a) What is the probability of observing 1 cysteine in a 20-residue peptide?
(b) What is the average number of cysteines in a 407 amino acid residue peptide?
(c) What is the probability of observing 6 cysteines in a 20-residue peptide?

14. *Particle probabilities.* Consider N total water molecules in equilibrium within a container of volume V_0. Divide the container into two subvolumes, V_L on the left and V_R on the right, such that $V_0 = V_L + V_R$. Let n_R (n_L) be the number of water molecules located in the right (left) subvolume V_R (V_L) of this container. Express your answers to the following questions in terms of $\{n_i, V_i\}$, where $i \in \{L, R\}$.

(a) Find p, the probability that a given water molecule is located within the subvolume V_R.
(b) Use p to calculate $<n_R>$, the mean number of water molecules located in the subvolume V_R.
(c) What is the probability, q, that a given molecule is *not* located within this subvolume V_R?
(d) What is $m = n_R - n_L$, i.e., the average excess number of water molecules in the right volume, V_R?
(e) Find the dispersion, σ^2, in the number of molecules within V. Express your answer as a ratio of $\sigma^2/<n>^2$ in terms of N, V, and V_0.

(f) What happens to the ratio of $\sigma^2/<n_R>^2$ as $V_R \rightarrow V_0$? How does this make physical sense?

(g) Use the binomial distribution to find an expression for the probability of finding n_R molecules in V_R *and* n_L molecules in V_L.

15. *The digits of π.* The digits of π are random.

   (a) What is the a priori probability of obtaining a 5 from the first 100 digits of π?
   (b) If you calculated the first 100 digits of π, and found that 8 occurred 13 times, what then is the a posteriori probability of obtaining an 8?
   (c) What is the a posteriori probability of NOT obtaining an 8?
   (d) From a sample of 287 digits, what is the expected number of sevens?
   (e) *Mathematica* can generate π to many accurate digits. Use the *Mathematica* function $N[\pi, n]$ to generate the digits of π to n significant figures and test your findings against what you found for (a)–(d).

16. *Nuclear transcription factors.* The random walk is a model for the binding of nuclear transcription factors to the origin of transcription on DNA. Let transcription factor X recognize the dinucleotide **TG** and say that it binds to the G in **TG** in the following sequence but X has to diffuse along the strand until it finds an **A** to start transcription:

CGCGTGGTCCAGCCTCCGT

   (a) Find an expression for the probability that X finds **A** after 10 steps up and down the strand.
   (b) On average, how many steps will it take until X reaches the **A**?
   (c) What if there is another transcription factor Y that initially binds to the T in **TC** upstream and it has to diffuse downstream until it too finds the **A**.
   (d) On average, how many steps must it take until it reaches the **A**?
   (e) If we start with 100 molecules each of X, Y, and the DNA, what would be the equilibrium ratio of X/Y at position **A**?

17. *Parkinson's disease.* Your lab has discovered a new protein called PBC, which is the product of a gene *PBC* involved in the development of Parkinson's disease. Within the *PBC* gene is a large segment of DNA that contains an intron (a segment of DNA that is transcribed, but normally removed before translation) that is retained in the translated PBC. Three stages of Parkinson's disease are possible, and the stage of disease is dependent on the base sequence of the intron. A table of disease onset (stage) *versus* retained intron content is shown below. Assume the appearance of each base of the set C,G,A,T is random, with a probability of 1/4.

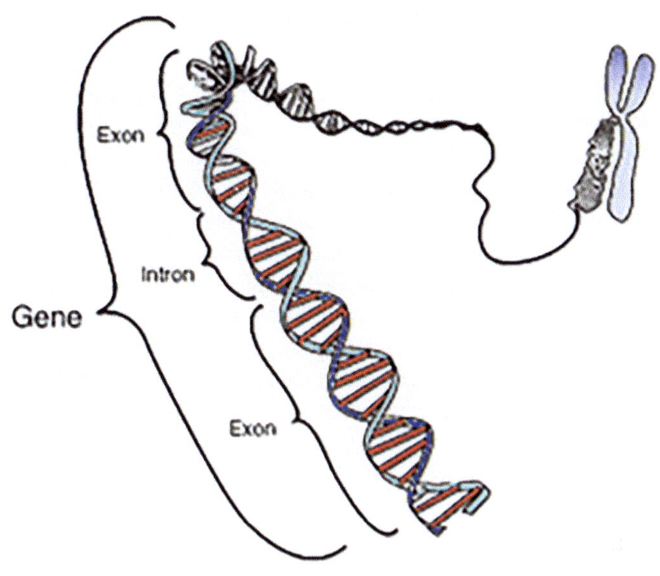

Disease onset	Intron version/bases
Early (before age 40)	GCC CCA TA
Common (after age 60)	GCC GCA
Late (after age 80)	GCC

(a) What is the probability of observing each of the sequences (early, common, and late onset) anywhere in the genome?

(b) If 52 individuals were found to possess the retained intron, but do not yet know its base sequence, what is the probability of *early onset* Parkinson's disease in this sample?

(c) What percentage of the 52 individuals will *NOT* get *early onset* disease?

18. *An alpha-helical peptide.* Let a peptide contain N amino acid residues within a folded structure, S. Let n be the number of residues located within an alpha-helical stretch, $\mathbf{S}_\alpha$, of the structure.

(a) What is the probability, $\mathbf{p}$, that a given residue is located within an alpha-helical stretch, $\mathbf{S}_\alpha$?

(b) What is $<\mathbf{n}>$, the mean number of amino acids located in the alpha-helix?

(c) Find the dispersion, σ^2, in number of amino acid residues within the alpha-helix, $\mathbf{S}_\alpha$.

19. *Exercise in probability distributions.* There are two kinds of probabilities: *a priori* and *a posteriori*. They differ in the way that they are determined. *A priori* probabilities can be determined in advance, usually by counting or by combinatorics, while *a posteriori* probabilities must be determined empirically, often

through measurement or experimentation. This exercise is designed to illustrate the practical difference between them.

(a) Begin by picking two dice from a collection of dice with 4–10 sides each. Note the number of sides on each die and compute the total number of outcomes as their product, N_T. Use the *Mathematica* code from problem 10 to prepare a State Table, such as the one shown in Table 1.1 in the text (generated for two, six-sided dice) where i is the state number, which ranges from 1 to N_T, n_1 and n_2 are the possible numbers on Die #1 and Die #2, and $n = n_1 + n_2$ is the outcome for each state, i.

(b) Construct an *a priori* Probability Table of $P(n) = N(n)/N_T$ (such as shown in Table 1.2 in the text) by counting the number of times, $N(n)$, a particular outcome, n, occurs in the State Table. Use the function `BinCounts` to count the number of times a given outcome occurs.

(c) Now, roll the dice $N_T' = 100$ times and record the outcome, $n' = n_1' + n_2'$, for each roll in a `list`. Use `BinCounts` to find $N'(n')$, the number of times an outcome n' appears in `list`, as the a posteriori outcomes.

(d) Compute the a posteriori probabilities $P'(n') = N'(n')/N_T'$ as decimal fractions. (Note: Here $P'(n')$ is not the derivative, but just another function.)

(e) Use these measured $P'(n')$ data and your calculated $P(n)$ Table from (b) to make a plot, like that shown in Fig. 1.1 in the text of the expected (a priori) probabilities, $P(n)$, and the observed (a posteriori) probabilities $P'(n')$, versus outcome, n. In order to guide the eye, you may connect the dots with a line for the expected (a priori) probabilities, but only show the data points for the observed (a posteriori) probabilities.

(f) Compute and report the mean and standard deviation for both the expected (a priori) probabilities, and the observed (a posteriori) probabilities, and compare them.

(g) Compute the root-mean-squared deviation between $P(n)$ and $P'(n')$. Is it compatible with $\frac{\sigma(m)}{m} = \frac{1}{\sqrt{N}}$ found from the binomial distribution?

20. *A genuine random distribution?* Use your digital camera or cell phone to take a photograph of a uniformly illuminated gray surface, such as a piece of posterboard, or the clear blue sky. Download this image to your computer and `Import` it into *Mathematica* as `Data`. Use the `Histogram` function to create and plot a histogram of the pixel intensities.

(a) Are the intensities distributed according to a Gaussian?
(b) Find the mean and width of this distribution.
(c) What is the rms difference between a Gaussian and the measured distribution?

21. *Rat eradication.* Prior to introduction of rats and mice to the island of South Georgia by the landing of Captain James Cooke in 1775, this remote South Atlantic island was home to millions of birds, including the endemic South

Georgia pipit. However, the rodents ate most of the bird eggs and decimated the native avian species.

(a) If a survey finds 1 rodent per thousand square meters, what is the probability of finding 3 rodents in a nesting area of 50 square meters?

(b) If you wish to eradicate the rodents with warfarin pellets dropped from helicopters, but the probability of killing a rodent with a single pellet is only 0.0001, how many pellets per square meter would you need?

22. *Mutations in hemoglobin.* The rate of DNA mutations in the general population is 75 per 3 billion base pairs. The gene for hemoglobin is 3000 base pairs long. What is the probability of a mutation in the hemoglobin gene?

23. *The average end-to-end distance for a polypeptide.* The average end-to-end distance for a polypeptide of N residues can be modeled using *Mathematica* with the following code (with $L = 1$, for simplicity):

```
step[position_] =
  With[{t = 2 Pi RandomReal[]}, position + {Cos[t], Sin[t]}]
  walk[n_, origin_: {0, 0}] = NestList[step, origin, n]
  distance[aWalk_] = EuclideanDistance @@ aWalk[[{1, -1}]]
  draw[aWalk_] =
  Graphics[{Red, Line[aWalk], Dashed, Arrowheads[Large],
  Arrow[aWalk[[{1, -1}]]], Black, PointSize[Medium], Point
  [aWalk]},
  ImageSize -> Small, Axes -> True, AspectRatio -> 1]
```

Now, run this as:

```
draw[walk[100, {0, 0}]]
```

to see the distribution in 2D of the vectors.

    a. Make a `Table` of $N = 1000$ distances:

```
dist = Table[distance[walk[i, {0, 0}]], {i, 1, 1000}];
```

    b. Use `ListPlot` to display this table.

    c. Fit the distance *versus* step number to Sqrt[N]. Is the statement in the text that $d = $ Sqrt[N] correct?

24. *The thieving mathematician.* A man leaves a bar so intoxicated that he is unaware of which direction he is going. He drops his wallet on the sidewalk outside the bar and wanders around randomly up and down the level sidewalk taking 0.5 steps per second. A starving mathematician in dire need of money is waiting at the bus stop by the bar entrance and observes the first man dropping

his wallet. The mathematician wants to pick up the wallet, but does not want to be caught doing so, thereby ruining his Stirling reputation. However, the mathematician knows about one-dimensional random walks, and that the probability of the drunk returning to his wallet will decrease with the number of steps the drunk has taken. The mathematician decides to wait until the probability of the drunk returning to his wallet is less than 1 chance in 10. He also has a table of factorials of only even integers up to 156 on Excel in his laptop. The bus schedule posted at the bus stop says that the next bus will arrive in 3 min. Will the mathematician be able to grab the wallet in time to hop on the bus and get away without being caught by the first man?

Hint: You will most likely have to solve this problem using numerical values of the probability of *not* returning to the origin of a random walk.

25. *The length of the PBC protein.* **N** amino acid residues are in equilibrium within a folded structure, **S**, of PBC (From #18). Let **n** be the number of residues located within an alpha-helical stretch, $\mathbf{S_\alpha}$, of the structure. The probability, **p**, that a given residue is located within an alpha-helical stretch, $\mathbf{S_\alpha}$, is then given by $\mathbf{p = S_\alpha/S}$.

Hint: You do not need to use any sums here and I am not looking for numerical values.

(a) What is $\mathbf{<n>}$, the mean number of amino acids located in the alpha-helix? Express your answers in terms of **N**, $\mathbf{S_\alpha}$, and **S**.
(b) Find the dispersion, $\mathbf{\sigma^2}$, in number of amino acid residues within the alpha-helix, $\mathbf{S_\alpha}$. Express your answer as a ratio of $\mathbf{\sigma^2/<n>^2}$, in terms of **N**, $\mathbf{S_\alpha}$, and **S**. Simplify your answer as far as possible.
(c) Upon further research, you find that PBC, once internalized into cells, has a random coil configuration at a pH of 6.0 inside endosomes. The root-mean-square (rms) distance between the ends of the random coil is $<r_N^2> = L\sqrt{3N}$, where L is the length of an amino acid and N is the number of amino acids. You found that the rms distance of the random coil was 75.4 Angstroms. How many amino acids are in the PBC protein. Assume an average amino acid length of 1.5 Angstroms.

26. *The dispersion of the random walk*

Show that $<n_1^2> = p[N + pN(N - 1)]$
Hint: $np^n = p\frac{\partial}{\partial p}(p^n)$ is very useful.

27. Normalization of the Poisson distribution
    Show that the Poisson distribution is normalized, i.e., the sum or integral of the distribution is unity.

Bibliography

P.R. Bevington, D.K. Robinson, *Data Reduction and Error Analysis for the Physical Sciences*, 2nd edn. (McGraw-Hill, New York, 1992)

T.W. Gray, J. Glynn, *The Beginnner's Guide to Mathematica. Version 4* (Cambridge University Press, New York, 2000)

H. Pishro-Nik, *Introduction to Probability, Statistics, and Random Processes* (Kappa Research LLC, Blue Bell, PA, 2014)

P. Tam, *A Physicist's Guide to Mathematica* (Academic Press, San Diego, 1997)

R.L. Zimmerman, F.I. Olness, *Mathematica for Physics*, 2nd edn. (Addison Wesley, 2002)

Chapter 2
Can Energy Direct Probability? Energy, Entropy, and the Boltzmann Distribution

"Thermodynamic miracles. . .Events with odds against so astronomical they're effectively impossible. . .I long to observe such a thing."

Dr. Manhattan, from
Watchmen
By A. Moore and D. Gibbons

The main thrust of the first chapter has been a detailed examination of how that devil *probability* can be captured in Faust's Pentagram of Organizing Principles. We have seen that the macrostate of a System can be well-defined, even though the microstates are defined only on a purely probabilistic basis. However strange it may seem, probability theory is a fundamental organizing principle of the self-assembly of living systems; it not only describes how well we can know the State of a System in terms of averages over State variables, but it governs the distributions of molecules among their accessible states, and it even predicts the time evolution of biophysical processes. Other organizing principles of life are the State functions and their behavior under transformations of time and space.

The two most important State functions are Energy and Entropy, but entropy as we shall see in this chapter is really a sheep in wolf's clothing in that it is simply probability once again. Hamilton's view of the physical world assigns a State function, the Hamiltonian, H, to the sum of the kinetic T and potential energy V of a System, $H(x, y, z, t) = T + V$. Like all State functions, the Hamiltonian is invariant under a number of symmetry operations. According to Noether's theorem (Chap. 3), any symmetry (or invariance) of the Hamiltonian gives rise to a conserved quantity. For example, in the case of the Hamiltonian, it is well established that the results of a measurement of the total energy of an isolated System do not depend upon when the measurements are taken. We can express this symbolically by writing $\frac{\partial H(x,y,z,t)}{\partial t} = 0$. Integration of this produces the extraordinarily profound property of the Hamiltonian State function, that

L. O. Sillerud, *Abiogenesis*, https://doi.org/10.1007/978-3-031-56687-5_2

Fig. 2.1 In any spontaneous biological process, the energy of a System tends to a minimum. The change in energy is $\Delta E = E_f - E_i < 0$. The blue curve shows that enzymes lower the transition state energy barrier from E_B to $E_B{}'$

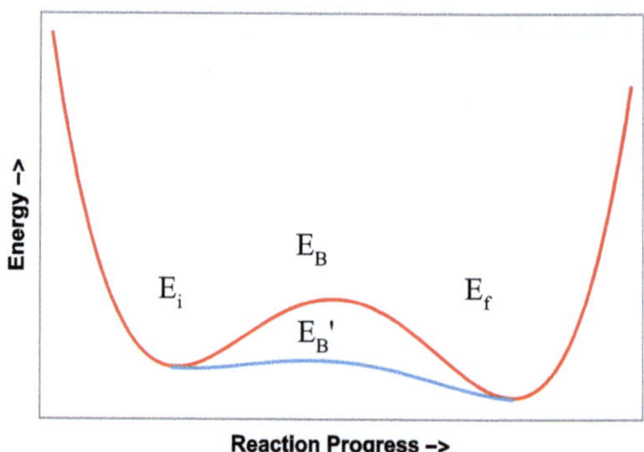

$$H(x, y, z, t) = \text{constant},$$

which is of course the conservation of energy; restated, we have the first law of thermodynamics,

$$\Delta E = Q - W,$$

where ΔE is the change in the internal energy of a System, Q is the heat exchanged with a reservoir, and W is the work done by the System. In any spontaneous process, a System evolves to the lowest accessible State of energy. If this State is time independent, then we say that the System is in equilibrium. This is shown graphically in Fig. 2.1.

The time evolution of the State, S, of the System will be from $S_i(E_i) \rightarrow S_f(E_f)$ because $E_i > E_f$. At equilibrium the System will tend to remain in State $S(E_f)$ because this is the lowest energy State accessible to the System (Fig. 2.1).

The change in energy does not, however, drive all biophysical processes. There are many situations in which there is no net change in energy, $\Delta E = 0$, and yet the System evolves toward equilibrium just the same. What if there were simply more isoenergetic States accessible to the System in its final configuration than in its initial configuration? We have already seen that the probability of a given State is the number of arrangements of the State variables accessible to a given State divided by the total number of arrangements of the State variables. It is both logical and correct to assume that Systems will evolve from States of lower probability to ones of higher probability.

2.1 Entropy

Therefore, a State function, Entropy, related to the probability of a State is needed to describe this alternative. Let us consider a total of N particles distributed among n States. We seek the probability of a given arrangement. We know that N particles can be arranged in $N!$ permutations, but permuting the N_n particles in the nth state does not produce a new arrangement, so $N!$ overestimates the number of unique arrangements by $N_n!$. The number of arrangements W of N total particles, with N_n particles in each of n states is therefore

$$W = \frac{N!}{N_1!N_2!N_3!\ldots N_n!}$$

Then the probability of finding N indistinguishable particles distributed among n States is the number of arrangements times the probabilities of each arrangement using the *and* probability combination rule, which gives the multinomial distribution:

$$P_N(n) = \frac{N!}{N_1!N_2!N_3!\ldots N_n!}\, p_1^{N1} p_2^{N2} \ldots p_n^{Nn}$$

where p_n^{Nn} is the probability of finding n particles in the N_nth state. We assume that all states are equally likely so that all of the p_i are equal $p_i = p$, $\forall i$, and only contribute an overall constant factor of p^N:

$$P_N(n) = \frac{N!}{N_1!N_2!N_3!\ldots N_n!}\, p^N,$$

so that in the following we concentrate our attention on W, the number of arrangements, and without loss of generality can set $p = 1$.

Spontaneous processes occur not only because the energy of the final State might be lower than that of the initial State but also because the final State is more probable than the initial State. At equilibrium, an isolated System is found with equal probability in each one of its accessible States. This is a rigorous definition of the equilibrium State of a System, because if an isolated System is not found with equal probability in each of its accessible States, it will tend to change with time until it does occupy each of its accessible States with equal probability. Thus, the State of the System was time dependent to begin with, and not in equilibrium.

We will define the Entropy, S, of a System as

$$S = k \ln W,$$

where W is the number of accessible States and k is Boltzmann's constant (1.38×10^{-23} Joule/Kelvin).

2.1.1   The State of Maximum Entropy

A spontaneous process will then be the most probable, the state where the entropy is a maximum. We can determine this state of maximum entropy by taking the derivative of W and setting it to zero at $W = W_o$ (Fig. 2.2). Then $S_{max} = S(W_o)$ if $\frac{\partial S}{\partial W}\big|_{W=W_o} = 0$. Consider a System of N particles arranged among n States, with N_i particles per State I,

$$N_i \in \{N_1, N_2, N_3, \ldots, N_n\}.$$

As we saw at the end of the last chapter, the logarithm of the probability is a more slowly varying function than that of the probability itself, so let's begin by taking the logarithm of W. The Entropy is defined in terms of the logarithm of the number of accessible states,

$$S = k \ln W = k \ln \frac{N!}{N_1! N_2! N_3! \ldots N_n!}$$

and the maximum of the logarithm of W will also be the maximum of the entropy. The log of W is

$$\ln W = \ln N! - \sum_{i=1}^{n} \ln N_i!$$

and we will find the maximum of this quantity subject to the constraint that the total number of particles is a constant,

$$N = \sum_{i=1}^{n} N_i.$$

Here we introduce <u>one form</u> of **Stirling's approximation** for $N!$ when N is a large number,

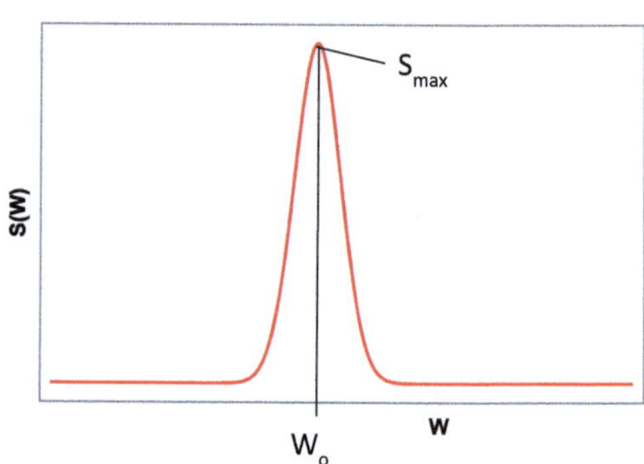

Fig. 2.2 The most probable state of a system is W_o where the entropy, S, is a maximum, S_{max}

$$\ln N! = N \ln N - N,$$

to give:

$$\ln W = N \ln N - N - \sum_{i=1}^{n} N_i \ln N_i + \sum_{i=1}^{n} N_i$$

We can use the particle number constraint to cancel $-N + \sum_{i=1}^{n} N_i = 0$, leaving

$$\ln W = N \ln N - \sum_{i=1}^{n} N_i \ln N_i.$$

We next need to compute $d(\ln W)$, first noting that $d(N \ln N) = 0$, because $N \ln N$ is a constant, so that

$$d(\ln W) = - d\left[\sum_{i=1}^{n} N_i \ln N_i \right].$$

Bring the differential operator inside the sum and use the product rule from calculus

$$d(\ln W) = - d\left[\sum_{i=1}^{n} N_i \ln N_i \right] = \sum_{i=1}^{n} d(N_i \ln N_i) = \sum_{i=1}^{n} \ln N_i dN_i + \sum_{i=1}^{n} N_i d(\ln N_i).$$

In this expression, we can use the fact that $d(\ln x) = \frac{1}{x} dx$ to find that the last term is zero.

$$\sum_{i=1}^{n} N_i d(\ln N_i) = \sum_{i=1}^{n} \frac{N_i}{N_i} dN_i = \sum_{i=1}^{n} dN.$$

We can take the differential out of the sum and find that

$$\sum_{i=1}^{n} dN_i = d \sum_{i=1}^{n} N_i = dN = 0$$

because the last sum is once again our constraint that the total number of particles is conserved so that the sum is just a constant, and has a differential of zero. Therefore, we have that

$$d(\ln W) = -\sum_{i=1}^{n} \ln N_i dN_i.$$

In order to continue, we need to know how to handle the right-hand side of this equation. For this, we turn to the expansion of our constraint equation

$$N = N_1 + N_2 + \ldots + N_n$$

The differential of this equation is $dN = dN_1 + dN_2 + \ldots + dN_n = 0$, as a result of the fact that N is a constant, once again. We can solve this equation for dN_n and then use this result in

$$d(\ln W) = -\sum_{i=1}^{n} \ln N_i dN_i.$$

$$dN_n = -(dN_1 + dN_2 + \ldots + dN_{n-1})$$

and then we can introduce this solution for dN_n, and write out the terms in

$$d(\ln W) = -\sum_{i=1}^{n} \ln N_i dN_i$$

as

$$d(\ln W) = -\sum_{i=1}^{n-1} \ln N_i dN_i + \ln N_n(dN_1 + dN_2 + \ldots + dN_{n-1}).$$

This reduces the dimension of the sum by 1, resulting in $(n-1)$ terms. Now we write out the terms from this last equation:

$$d(\ln W) = -(\ln N_1\, dN_1 + \ln N_2\, dN_2 + \ldots + \ln N_{n-1}\, dN_{n-1})$$
$$+ \ln N_n dN_1 + \ln N_n dN_2 + \ldots + \ln N_n dN_{n-1}$$

and then group terms involving dN_i

$$d(\ln W) = (\ln N_n dN_1 - \ln N_1 dN_1)$$
$$+(\ln N_n dN_2 - \ln N_2\, dN_2) + \ldots$$
$$+(\ln N_n dN_{n-1} - \ln N_{n-1}\, dN_{n-1}).$$

We can write each term in parentheses as

$$(\ln N_n - \ln N_1)dN_1 = \ln \frac{N_n}{N_1}dN_1$$

or, for $i \in \{1, 2, 3, \ldots, n-1\}$, we have that

$$(\ln N_n - \ln N_i)dN_i = \ln \frac{N_n}{N_i}dN_i$$

so that

$$d(\ln W) = \ln \frac{N_n}{N_1}dN_1 + \ln \frac{N_n}{N_2}dN_2 + \ldots + \ln \frac{N_n}{N_{n-1}}dN_{n-1}$$

Collecting terms gives

$$d(\ln W) = -\sum_{i=1}^{n-1} \ln \frac{N_n}{N_i}dN_i.$$

Now, our extremum equation states that $d(\ln W) = 0$, so this sum becomes

$$d(\ln W) = -\sum_{i=1}^{n-1} \ln \frac{N_n}{N_i}dN_i = 0$$

and this can only happen in a sum if all of the coefficients of dN_i are identically zero, term by term. Then,

$$\forall i \in \{1, 2, 3, \ldots, n-1\}, \ \ln \frac{N_n}{N_i} = 0,$$

or

$$e^{\ln \frac{N_n}{N_i}} = \frac{N_n}{N_i} = 1$$

which implies that $N_n = N_i$, $\forall i \in \{1, 2, 3, \ldots, n-1\}$, and all of the N_i are equal to each other:

$$N_1 = N_2 = N_3 = \ldots = N_{n-1} = N_n$$

So that in accord with our constraint equation,

$$N = \sum_{i=1}^{n} N_i$$

the sum

$$N = N_1 + N_2 + N_3 + \ldots + N_{n-1} + N_n$$

Then consists of n identical terms of N_n each

$$N = N_n + N_n + N_n + \ldots + N_n + N_n$$

Or

$$N_n = \frac{N}{n}$$

At equilibrium, the number of particles $\mathbf{N_n}$ per State n is just the total number of particles divided by the total number of States. This is the condition of *maximum Entropy* in which *all accessible States are equally occupied.* It corresponds to the most probable arrangement of N particles among n States (Fig. 2.2) and the value of W_o at S_{max} is given by:

$$W_o = \frac{N!}{\left[\left(\frac{N}{n} \right)! \right]^n}$$

We can again use Stirling's approximation to show that the number of arrangements at the maximum entropy is

$$W_o = n^N$$

and that the maximum Entropy S_{max}

$$S_{max} = k \ln W_o,$$

or, inserting the above for W_o, we have that the maximum is given by this important result:

$$S_{max} = N k \ln n.$$

We begin this derivation by taking the logarithm of both sides of the equation for S_{max}:

$$\text{Ln } W_o = \ln N! - n \, \ln(N/n)!$$

and using Sterling's approximation, we have that

$$\text{Ln } W_o = N \text{ Ln } N - N - n\left[\frac{N}{n}\text{Ln}\left(\frac{N}{n}\right) - \frac{N}{n}\right]$$

or

$$\text{Ln } W_o = N \text{ Ln } N - N - N \text{ Ln}\left(\frac{N}{n}\right) + N$$

$$\text{Ln } W_o = N \text{ Ln } n$$

which gives upon exponentiation, our result:

$$\mathbf{W_o = n^N}.$$

2.1.2 The Isothermal Expansion of an Ideal Gas Is Due to Entropy

A simple example of the state of maximum entropy is a quantity of N gas molecules, such as oxygen, in a container (Fig. 2.3) with a total volume of V, which is partitioned into an equal number n of smaller volumes, V_i, $i \in \{1, 2, 3, \ldots, n\}$. The most probable state of the gas is the state with equal numbers of gas molecules in each cell, V_i. The probability of finding a molecule in a cell of volume V_i is V_i/V, while the average number of molecules per cell is $N\,V_i/V$.

Let us compute the entropy change if we have N gas molecules on the left-hand side of an insulated container of volume V_1, separated from the empty right-hand

Fig. 2.3 A box of volume V enclosing N gas molecules. We divide the box into two parts of volumes V_1 and V_2 by means of a removable partition (Red). Further divide each volume into small cells of number n_1 and n_2, respectively

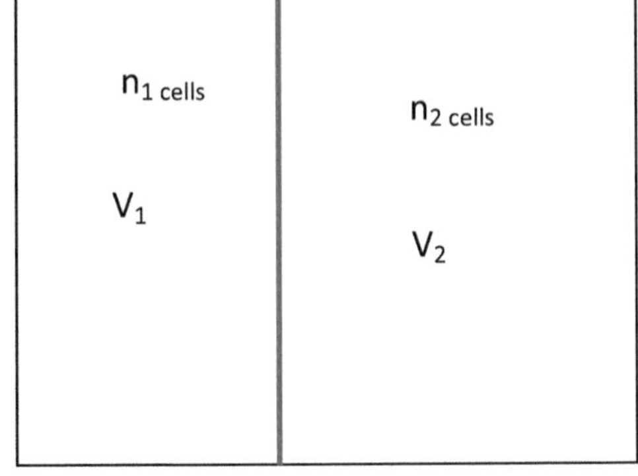

side of volume V_2, by a movable partition, such that the total volume is $V = V_1 + V_2$ (Fig. 2.3). What is the entropy change after removing the partition? From a probabilistic point of view, we need to find the number of accessible States of the System prior to and after the removal of the partition. Divide the left volume V_1 into n_1 cells of equal volume, v. Then $V_1 = n_1 v$ is the total volume of the left side of the box. Similarly, we will divide the right side into n_2 cells of equal volume, v, and $V_2 = n_2 v$. How many ways can we put 1 molecule into n_1 boxes? We can put it into either box 1 *or* box 2 *or* ... *or* box n_1, which gives n_1 ways. Then how many ways can we put the second molecule into n_1 boxes? Since the number of molecules per box is not constrained, there are again n_1 ways. The number of ways we can put two molecules into n_1 boxes is the product, $n_1 * n_1$, because we are putting molecule 1 *and* molecule 2 into n_1 boxes, so we multiply the number of individual ways. For N molecules we have $W_1 = (n_1)^N$ ways. Now for the right-hand volume, V_2, there are $W_2 = (n_2)^N$ ways. After we remove the partition, there are a total of $n = n_1 + n_2$ cells of volume v, in a total volume of V, so there are $W = n^N$ ways of arranging the molecules in the large volume, V. The entropy change, ΔS, is therefore

$$\Delta S = k \left[\ln (n)^N - \ln (n_1)^N \right] = k \ln \left(\frac{n}{n_1} \right)^N .$$

Suppose we are working with a mole of oxygen molecules, then $N = N_o$, which is Avogadro's Number (6.023×10^{23} molecules/mole), and we know that $N_o k = R$, the gas constant. The entropy change can be rewritten as

$$\Delta S = N_o\, k \ln \left(\frac{n}{n_1} \right) = R \ln \left(\frac{n}{n_1} \right),$$

but $n\, v = V$ and $n_1 V = V_1$, so

$$\Delta S = R \ln \left(\frac{Vv}{V_1 v} \right) = R \ln \left(\frac{V}{V_1} \right).$$

The entropy change is given by the logarithm of the volume ratio and is positive, indicating that the final state is more probable than the initial state. After we removed the partition, more states were accessible to the gas, so the system evolved to occupy them.

2.1.3 The Entropy of Mixing

Biomolecules do not exist in the gaseous phase, but rather conduct all of their functions in an aqueous milieu. We can develop the concept of entropy in solutions in a manner similar to that just seen for a gas in a box. We saw how the increase in entropy (probability) drove the gas in the previous example from the left side of the box to fill the entire box volume once the partition was removed. A system of more biological relevance is a box with two compartments of volumes V_1 and V_2, as before, but this time each volume is filled with water, and, in addition, only the left volume, V_1, contains a solute, e.g., glucose, at the beginning of the experiment. The moveable partition could be replaced with a dialysis membrane which is permeable to both water and glucose. As time increases after the preparation of the system, the glucose will diffuse through the membrane until its concentration is the same in both volumes. This approach to equilibrium is formally identical to the expansion of an ideal gas, but its analysis requires the introduction to the concept of the Entropy of Mixing.

Let there be a solution made up of n solutes with N_i molecules of each solute, $i \in \{1, 2, 3, \ldots, n\}$, so that there are N_1 molecules of solute 1, N_2 molecules of solute 2, and so on. The total number of solute molecules is N_T, where

$$N_T = \sum_{i=1}^{n} N_i.$$

Assume that each molecule is approximately the same size (i.e., volume), so each molecule can occupy a cell in the solution of the same small volume. Let there also be N_T cells, one cell per solute molecule. The Entropy of Mixing is then defined to be

$$\Delta S_m = S_{sol} - S_o,$$

where S_{sol} is the entropy of the solution and S_o is the entropy of the pure components. The entropy of mixing can be written as

$$\Delta S_m = k \ln \frac{W_{sol}}{W_o},$$

where W_o is the number of distinguishable ways of putting N_T molecules into N_T cells. However, all of the W_o are the same, since there is only one way of arranging the pure components; therefore, $W_o = 1$, and we have that $\Delta S_m = k \ln W_{sol}$. It remains to find the entropy of the solution, $k \ln W_{sol}$, which is the logarithm of the number of ways of arranging N_T molecules into N_i groups:

$$W_{sol} = \frac{N_T!}{N_1! N_2! \ldots N_n!} = \frac{N_T!}{\prod\limits_{i=1}^{n} N_i!}$$

The entropy of the solution is proportional to the logarithm of W_{sol}, and since $N_i \gg 1$, we can again use Stirling's approximation to find an expression for ΔS_m.

$$\Delta S_m = k \ln W_{\text{sol}} = k(N_T \ln N_T - N_T) - k \sum_{i=1}^{n} N_i \ln N_i + k \sum_{i=1}^{n} N_i$$

Now, we recognize that we have both $k \sum_{i=1}^{n} N_i$ and $-kN_T$ on the right side of this equation, which cancel, giving

$$\Delta S_m = kN_T \ln N_T - k \sum_{i=1}^{n} N_i \ln N_i.$$

Here, we can replace the first factor of kN_T on the right with its defining sum, to find that

$$k \sum_{i=1}^{n} N_i$$

and we can now factor out this sum

$$\Delta S_m = k \sum_{i=1}^{n} N_i \ln N_T - k \sum_{i=1}^{n} N_i \ln N_i$$

$$\Delta S_m = k \sum_{i=1}^{n} N_i (\ln N_T - \ln N_i) = -k \sum_{i=1}^{n} N_i (\ln N_i - \ln N_T)$$

to give this expression for the entropy of mixing

$$\Delta S_m = -k \sum_{i=1}^{n} N_i \ln \frac{N_i}{N_T}$$

We notice that the quotient in the logarithm is X_i the mole fraction of the ith compound,

$$\text{where } X_i = \frac{N_i}{N_T}.$$

The entropy of mixing per particle upon dissolution is

$$\Delta S_m = -k \sum_{i=1}^{n} N_i \ln X_i.$$

Let us multiply the right side of this equation by $\frac{N_o}{N_o} = 1$, where No is Avogadro's number, to give

$$\Delta S_m = -N_o k \sum_{i=1}^{n} \frac{N_i}{N_o} \ln X_i,$$

or, since the number of moles of the *i*th compound is defined to be $n_i = \frac{N_i}{N_o}$, and the gas constant is $R = N_o \, k$, we have that the molar entropy of mixing is given by

$$\Delta S_m = -R \sum_{i=1}^{n} n_i \ln X_i.$$

This entropy is positive, $\Delta S_m \geq 0$, because $X_i \leq 1$, and $\ln(X_i \leq 1) < 0$. The mixture is more probable than that of the pure components, so it has a higher entropy, and the equilibrium State of the System is one in which there is an equal number of glucose molecules in each cell on both sides of the membrane, even though the energy of the System does not change in the process.

2.1.4   A Puzzle: The Entropic Instability of Proteins

Before we run off with the idea that entropy is the only State function of a biological System that matters, let us examine the stability of proteins in cells at 310 K. It is an empirical fact that the proteins in living systems are folded into stable configurations and retain function at the temperature of a human body. It is also known that proteins can be denatured, or unfolded, at elevated temperatures. Frying an egg is a good example, because the egg white contains a high concentration of ovalbumin, which changes from a soluble form to an insoluble form as the temperature is raised as evidenced by its change in optical properties from a transparent to an opaque material. We can develop a simple, toy model of the entropy change of protein unfolding by placing the polypeptide chain on a three-dimensional lattice. This model is based on the isothermal expansion of an ideal gas.

Let there be N residues in the protein, with a volume of V_o per residue. In the native, folded state, the volume of the protein is then just NV_o. In the unfolded state, however, the volume accessible to the protein is approximately a cube of side N units with cells of volume V_o. An approximate value for the entropy change, ΔS, on protein denaturation is

$$\Delta S = k \ln W_d - k \ln W_n$$
$$= k \ln (W_d / W_n),$$

where $W_{d(n)}$ is the number of arrangements of the protein in the denatured (native) state,

$$W_d \sim N^3 V_{\mathrm{o}}, \text{ and } W_n \sim N V_{\mathrm{o}}.$$

The entropy change upon protein unfolding is then approximately

$$\Delta S \sim k \ln \left[\frac{N^3 V_o}{N V_o} \right] = k \ln N^2 = 2k \ln N$$

The right-hand side of this equation is always >0 for $N > 1$ residue, independent of temperature.

Earlier, we used the properties of the random walk to find a similar estimate of the entropy change upon the unfolding of a protein as $\sim (3k/2) \ln N$. Both of these estimates indicate that the probability of the unfolded state is always greater than that of the folded state. Why? Because there are many more ways of producing the unfolded state. So why do not proteins spontaneously unfold? Obviously, they do not or life itself would be impossible. The answer, as we shall see, is that there are other organizing principles which can and do compensate for this entropy increase and stabilize proteins at biological temperatures.

2.2   The Boltzmann Distribution

We have just seen that the change in Entropy is alone sufficient to drive some isoenergetic processes of a System to the State of equilibrium. In our examination of the properties of the random walk, we worked with cases in which p and q were either equal or very similar to each other. What if the sidewalk outside the bar were to be tilted so that one end was considerably higher than the other. There would then be an energy difference between right and left steps due to changes in elevation in the Earth's gravitational field and our bar patron would find it rather easier to wander downhill than up. A similar situation occurs for a set of nonzero nuclear spins when placed into a large magnetic field, or for the particle density in the Earth's atmosphere, which decreases with altitude above the Earth.

All of these processes involve, in addition to the State function, Entropy, another State function, Energy, which a particle possesses due to its interaction with a Field. Even though the total energy of a system is conserved, this alone does not reveal how that total energy might be distributed among the States accessible to the particles of a System. If, in the case of the Earth's atmosphere, we imagine that the air is contained in a stack of small cells stretching from the surface to beyond the stratosphere, then we know that the potential energy of the air in any particular cell will be the given by the mass, m, times the acceleration due to gravity, times the height, z, of the middle of the cell,

$$V = mgz.$$

We would anticipate that there might be more oxygen and nitrogen molecules in the lower energy cells close to the surface, and that the probability of finding air would decrease in some fashion, the further one would rise.

In the random walk, if the sidewalk is tilted down to the left, then a step to the left would require less energy than a step to the right, which would be uphill. We would expect our drunk to wander off to the left, and the mean displacement would no longer be zero. We could model this by setting $p = 0.4$ and $q = 0.6$ in the binomial distribution.

In the last section we showed that the most probable distribution of N particles among n States was solved by performing a constrained maximization of the probability of each arrangement of particles among States. The constraint we used was that of particle number (mass) conservation:

$$N = \sum_{i=1}^{n} N_i$$

that is, the total number of particles was constant, and the result we found was that $N = nN_n$. Now, what if we add another constraint, one based on the conservation of Energy? The total energy of a system is

$$E = \sum_{i=1}^{n} \varepsilon_i N_i$$

where ε_i is the energy of the ith State, with occupation number N_i. How does this added constraint change the equilibrium distribution of particles among states? In general, constrained maxima are only valid within their given constraints. The addition of a new constraint can therefore be expected to alter the details of the extrema.

Consider N particles distributed among n States. In the case of continuous, Classical Systems, these n States form a continuous distribution, but, as we shall see in Chap. 4, quantum systems can only exist in discrete states, which are often referred to as energy levels. However, our discussion of the distribution of particles among various energy states is completely general and does not change as one moves from Classical to quantum systems. Then, let there be N_i particles in the ith level with energy given by ε_i (Fig. 2.4).

What is the probability that the ith particle is in the jth state with energy ε_j, subject to our two constraints, the conservation of mass and the conservation of energy? We know that N particles can be arranged in $N!$ permutations, but permuting the N_i particles in the ith state does not produce a new arrangement, so $N!$ overestimates the number of unique arrangements by $N_i!$. Therefore, as we have seen in the last section, the number of distinct arrangements of N objects among n states is shown in Fig. 2.4.

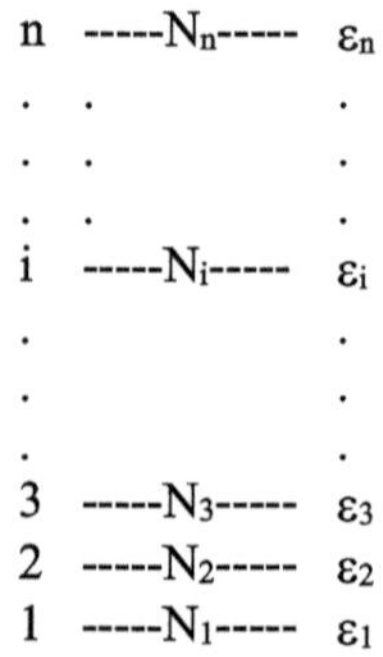

Fig. 2.4 A diagram illustrating the number of particles, N_i per energy level, ε_i

```
n   -----Nₙ-----   εₙ
.      .              .
.      .              .
.      .              .
i   -----Nᵢ-----   εᵢ
.      .              .
.      .              .
.      .              .
3   -----N₃-----   ε₃
2   -----N₂-----   ε₂
1   -----N₁-----   ε₁
```

$$W = \frac{N!}{N_1!N_2!N_3!\ldots N_n!}.$$

The most probable arrangement is then one in which W is once again a maximum, subject now to simultaneous particle (mass) and energy conservation.

We will solve this problem by setting $dW = 0$ and by using Lagrange multipliers. Again, we will use the logarithm of W because it varies more slowly than W itself, but it still has the same maximum as W.

$$\ln W = \ln N! - \sum_{i=1}^{n} \ln N_i!.$$

The maximum value of W is where $dW = 0$, or $d(\ln W) = 0$.

$$d(\ln W) = d(\ln N!) - \sum_{i=1}^{n} d(\ln N_i!) = 0.$$

Now $\ln N! = $ constant, so

$$d(\ln N!) = 0$$

and we are left with

$$d(\ln W) = - \sum_{i=1}^{n} d(\ln N_i!).$$

Again, we will use Stirlings's second approximation ($\ln N! = N \ln N - N$), to give

$$d(\ln W) = - \sum_{i=1}^{n} d(N_i \ln N_i - N_i)$$

$$d(\ln W) = - \sum_{i=1}^{n} d(N_i \ln N_i) + \sum_{i=1}^{n} dN_i$$

$$= - \sum_{i=1}^{n} d(N_i \ln N_i) + d \sum_{i=1}^{n} N_i$$

$$= - \sum_{i=1}^{n} d(N_i \ln N_i) + dN$$

but the first constraint equation states that the second sum on the right is N and $dN = 0$, and we are left with

$$d(\ln W) = - \sum_{i=1}^{n} d(N_i \ln N_i).$$

We can use the product rule from calculus ($d[f.g] = g\, df + f\, dg$) to compute the differential on the right to give

$$d(\ln W) = - \sum_{i=1}^{n} \ln N_i dN_i + \sum_{i=1}^{n} N_i d(\ln N_i)$$

However, $d(\ln N_i) = \frac{dN_i}{N_i}$, so the second term on the right is

$$\sum_{i=1}^{n} N_i d(\ln N_i) = \sum_{i=1}^{n} \frac{N_i}{N_i} dN_i = \sum_{i=1}^{n} dN_i = d \sum_{i=1}^{n} N_i = dN = 0$$

and we are left with

$$d(\ln W) = - \sum_{i=1}^{n} \ln N_i dN_i = 0.$$

Now, it is time to use Lagrange multipliers to incorporate the constraints of number and energy conservation. Let us start with the particle conservation constraint equation

$$N = \sum_{i=1}^{n} N_i, \text{ and } dN = d \sum_{i=1}^{n} N_i = 0$$

which we will multiply by the Lagrange multiplier α to give

$$\alpha dN = \alpha d \sum_{i=1}^{n} N_i = 0$$

and then use the energy constraint, noting that E is constant, so $dE = 0$,

$$E = \sum_{i=1}^{n} \varepsilon_i N_i, \text{ and } dE = \sum_{i=1}^{n} \varepsilon_i dN_i = 0.$$

We will multiply this equation by the Lagrange multiplier β to give

$$\beta dE = \beta \sum_{i=1}^{n} \varepsilon_i dN_i = 0.$$

Now we subtract the equations in α and β termwise to $d(\ln W)$ and find that

$$d(\ln W) - \alpha dN - \beta dE = -\sum_{i=1}^{n} \ln N_i dN_i - \alpha \sum_{i=1}^{n} dN_i - \beta \sum_{i=1}^{n} \varepsilon_i dN_i = 0$$

which we can rewrite by collecting all the terms on the right-hand side into a single sum

$$d(\ln W) - \alpha dN - \beta dE = -\sum_{i=1}^{n} (\ln N_i + \alpha + \beta \varepsilon_i) dN_i = 0$$

and then this implies that

$$-\sum_{i=1}^{n} (\ln N_i + \alpha + \beta \varepsilon_i) dN_i = 0.$$

In order for this equation to be true and the sum to be zero, each term in the sum must separately equal zero. The ith coefficient of dN_i must be zero, so

$$\text{Ln } N_i + \alpha + \beta \varepsilon_i = 0$$

$\forall$ i, or the number of particles in the ith state is given by

$$N_i = e^{-\alpha - \beta \varepsilon_i}$$

which is the **Boltzman distribution law**. It states that the N particles are distributed in a decreasing exponential manner among n states of increasing energy (Fig. 2.5).

However, what are the parameters, α and β? Direct our attention first to α. Our previous derivation of the Entropy maximum introduced us to the fact that the most

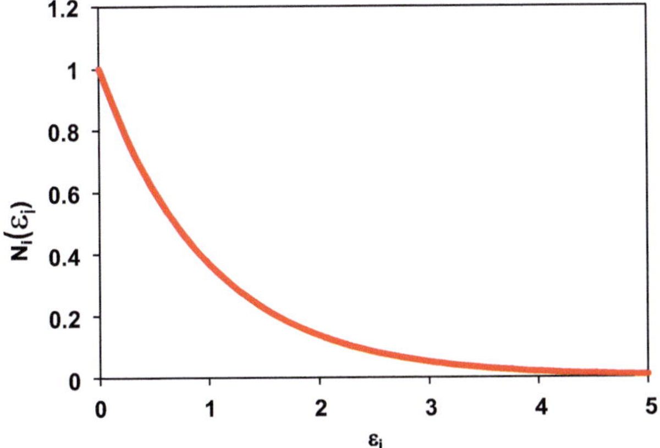

Fig. 2.5 An example of the Boltzmann (exponential) distribution of particles as a function of the energy of the states

probable distribution of N particles among n states places an equal number of particles per state, $N_n = N/n$. This was based on the constraint that the total number of particles was constant. That constraint is the same as the first constraint used in the derivation of Boltzmann's distribution, and it will lead to the same result, but we will revisit the derivation, this time using Lagrange multipliers.

The differential of the particle conservation equation is zero because N is a constant,

$$dN = \sum_{i=1}^{n} dN_i = 0, \tag{2.A}$$

and we have seen how maximizing the probability leads to

$$d(\ln W) = -\sum_{i=1}^{n} \ln N_i dN_i = 0. \tag{2.B}$$

Now, multiply (Eq. 2.A) by α to give

$$\alpha dN = \alpha \sum_{i=1}^{n} dN_i = 0$$

and subtract this from (Eq. 2.B)

$$d(\ln W) - \alpha dN = - \sum_{i=1}^{n} (\ln N_i + \alpha)dN_i = 0$$

Again, for the sum to equal zero, each of the coefficients must vanish identically. This results in an equation for N_i and α,

$$\text{Ln } N_i + \alpha = 0$$

or

$$N_i = e^{-\alpha}$$

which is the same constant we found earlier, $N = nN_i$, so that

$$e^{-\alpha} = \frac{N}{n}.$$

Thus, the first term in the Boltzmann distribution corresponds to the maximum entropy for N particles distributed among n states.

$$N_i = \frac{N}{n} e^{-\beta\varepsilon_i}$$

Now, turn our attention to the second term to find the value of β. In order to motivate this search, we note that the average energy per particle, $<\varepsilon>$, is defined using the mean value theorem for a probability distribution as

$$<\varepsilon> = \frac{E}{N} = \frac{\sum_{i=1}^{n} \varepsilon_i N_i}{\sum_{i=1}^{n} N_i},$$

where E is the total energy, $E = \sum_{i=1}^{n} \varepsilon_i N_i$, N is the number of particles, and we have just found N_i. Then, we can rewrite the average energy expression as

$$<\varepsilon> = \frac{\frac{N}{n} \sum_{i=1}^{n} \varepsilon_i e^{-\beta\varepsilon_i}}{\frac{N}{n} \sum_{i=1}^{n} e^{-\beta\varepsilon_i}} = \frac{\sum_{i=1}^{n} \varepsilon_i e^{-\beta\varepsilon_i}}{\sum_{i=1}^{n} e^{-\beta\varepsilon_i}}.$$

Remember that the entropy, $S = k \ln W$, with

$$W = \frac{N!}{\prod_{i=1}^{n} N_i}$$

so that

$$S = k \ln W = k \ln N! - k \sum_{i=1}^{n} \ln N_i!$$

Using Stirling's approximation again, we find that

$$S = Nk \ln N - Nk - k \sum_{i=1}^{n} [N_i \ln N_i - N_i].$$

However, $N_i = e^{-\alpha - \beta \varepsilon_i}$, $\ln N_i = -\alpha - \beta \varepsilon_i$, and we can write the equation for S as

$$S = Nk \ln N - Nk - k \sum_{i=1}^{n} [N_i(-\alpha - \beta \varepsilon_i) - N_i]$$

$$S = Nk \ln N - Nk + k \sum_{i=1}^{n} N_i(\alpha + \beta \varepsilon_i + 1)$$

$$S = k \ln N! + k \sum_{i=1}^{n} [N_i \alpha + N_i \beta \varepsilon_i + N_i]$$

where we have taken the minus sign inside the sum, and factored the N_i. The collection of terms in this last equation was accomplished by noting first that

$$Nk \ln N - Nk = k \ln N!$$

and then that

$$Nk\alpha = k\alpha \sum_{i=1}^{n} N_i$$

$$Nk = k \sum_{i=1}^{n} N_i$$

and

$$k\beta E = k\beta \sum_{i=1}^{n} \varepsilon_i N_i$$

The result of the term collection is

$$S = k \ln N! + Nk\alpha + k\beta E + Nk.$$

Now, the differential of S, keeping N and ε_i constant, is

$$dS = Nk \, d\alpha + kE \, d\beta + k\beta \, dE. \tag{2.C}$$

Remember: $N_i = e^{-\alpha - \beta\varepsilon_i}$, so that

$$N = \sum_{i-1}^{n} N_i = \sum_{i=1}^{n} e^{-\alpha - \beta\varepsilon_i} \tag{2.D}$$

The differentiation of (Eq. 2.D) produces

$$0 = dN = \sum_{i=1}^{n} (-d\alpha - \varepsilon_i d\beta) e^{-\alpha - \beta\varepsilon_i}.$$

The separate terms here are

$$-d\alpha \sum_{i=1}^{n} e^{-\alpha - \beta\varepsilon_i} = -d\alpha \sum_{i=1}^{n} N_i = -N d\alpha \tag{2.E}$$

$$-d\beta \sum_{i=1}^{n} \varepsilon_i e^{-\alpha - \beta\varepsilon_i} = -d\beta \sum_{i=1}^{n} \varepsilon_i N_i = -E d\beta \tag{2.F}$$

From the above equation in $dN = 0$, we have that

$$0 = -N \, d\alpha - E \, d\beta$$

or

$$N \, d\alpha = -E \, d\beta.$$

We can now insert this result into (Eq. 2.C),

$$dS = k(Nd\alpha) + k(Ed\beta) + k\beta dE,$$
$$dS = k(\text{---}Ed\beta) + k(Ed\beta) + k\beta dE,$$
$$\Rightarrow dS = k\beta dE$$

$$\frac{dS}{dE} = k\beta.$$

And, from thermodynamics, we know that $dS = dE/T$, where T is the absolute temperature. Then, we have an alternate expression for $\frac{dS}{dE} = \frac{1}{T}$, which gives us two equations for the same quantity

$$\frac{dS}{dE} = k\beta = \frac{1}{T},$$

so that we find that $\frac{1}{T} = k\beta$ or $\beta = \frac{1}{kT}$. Thus, the **_Boltzmann distribution_** is given by

$$N_i = \frac{N}{n}e^{-\frac{\varepsilon_i}{kT}} \ \forall i \in \{1, 2, 3, \ldots, n\}$$

That is, for a System of N particles, distributed among n States, where the energy of the ith State is ε_i, this gives the probability of finding N_i particles in the ith State. Notice that the Boltzmann distribution explicitly includes the entropy, N/n, as well as the energy.

2.3   The Chemical Potential

We have previously seen that the entropy is given by

$$S = k \ \ln N! + Nk\alpha + Nk\beta E + Nk$$

When this expression for the entropy was compared with experiment in the late nineteenth century, it was found that the measured entropy deviated from the theoretical value by $k \ln N!$. J. W. Gibbs in 1902 provided the necessary correction to this equation by pointing out that the prevailing definition of the thermodynamic probability, upon which the entropy was based, assumed that one could label each particle, implying that the particles were distinguishable, while quantum mechanics now tells us that it is impossible to put a label on an atom. In reality, one atom or molecule is the same as another; therefore, our above expression for the entropy overcounts states by $N!$, which is the number of distinguishable arrangements of N particles. In taking differentials, since N was a constant, one never confronted this discrepancy because $dN = 0$. However, in biochemistry, the number of particles is rarely conserved in a reaction and we can no longer rely on $dN = 0$. The consequences of this fact are that we must revisit our derivation of the state of maximum probability and allow for the variation in the number of particles, and that we must drop the term in the entropy that involves $k \ln N!$. Given these considerations, the correct entropy expression is

$$S = Nk\alpha + k\beta E + Nk$$

and we can proceed in finding the maximum in its differential to find the most probable state of the system, even in the face of a violation of particle number conservation. The total differential of the entropy, keeping ε_i constant, is then

$$dS = Nkd\alpha + k\alpha dN + kEd\beta + k\beta dE + kdN$$
$$dS = Nkd\alpha + kEd\beta + k\beta dE + (k\alpha + k)dN$$

and dN is no longer zero. How can we find dN? We can determine it from the equation for the total number of particles (Eq. 2.D)

$$N = \sum_{i=1}^{n} N_i = \sum_{i=1}^{n} e^{-\alpha - \beta\varepsilon_i} \tag{2.D}$$

and its differential, given by

$$dN = \sum_{i=1}^{n} d(\alpha - \varepsilon_i\beta)e^{-\alpha - \beta\varepsilon_i}$$

or, by computing the differential of the terms in parentheses, we have that

$$dN = \sum_{i=1}^{n} (-d\alpha - \varepsilon_i d\beta - \beta d\varepsilon_i)e^{-\alpha - \beta\varepsilon_i}$$

Now, we note that the exponential is just N_i, so that we can write dN as

$$dN = \sum_{i=1}^{n} (-d\alpha - \varepsilon_i d\beta - \beta d\varepsilon_i)N_i$$

which we can write, using Eqs. 2.E and 2.F as

$$dN = -Nd\alpha - Ed\beta - \beta \sum_{i=1}^{n} N_i d\varepsilon_i$$

Let us denote the last sum as $V = -\beta \sum_{i=1}^{n} N_i d\varepsilon_i$ and multiply this equation by k so that we can write $k\,dN$ as

$$k\,dN = -N\,k\,d\alpha - E\,k\,d\beta + Vk$$

and we can rearrange this to give

$$N\,k\,d\alpha + k\,E\,d\beta = V\,k - k\,dN$$

Note that the left-hand side of this equation occurs in the terms in brackets in dS

$$dS = (Nkd\alpha + kEd\beta) + k\beta dE + (k\alpha + k)dN$$

so that we can replace the terms in parenthesis by the right-hand side to give

$$dS = Vk - kdN + k\beta dE + (k\alpha + k)dN$$
$$dS = Vk - kdN + k\beta dE + k\alpha dN + kdN$$
$$dS = Vk + k\beta dE + k\alpha dN$$

and we can read off the values of $k\beta$ and $k\alpha$ by referring to the total differential of the entropy as a function of energy, E, and particle number, N, $S = S(E, N)$

$$dS = \frac{\partial S}{\partial E}dE + \frac{\partial S}{\partial N}dN$$

so that

$$\frac{\partial S}{\partial E} = k\beta = \frac{k}{kT} = \frac{1}{T}$$

and

$$\frac{\partial S}{\partial N} = k\alpha,$$

or

$$\frac{1}{k}\frac{\partial S}{\partial N} = \alpha.$$

However, from thermodynamic considerations, the temperature times the negative partial derivative of entropy with respect to particle number is defined to be the "chemical potential," μ, for a single particle

$$\frac{1}{k}\frac{\partial S}{\partial N} = \alpha = -\frac{\mu}{kT},$$

and for N particles,

$$N\alpha = -\frac{N\mu}{kT}.$$

Thus, we arrive at the **Gibbs** probability function for particles as a function of entropy and energy as

$$P(N_i, E_i) = e^{(N_i\mu - E_i)/kT}.$$

Note that this function explicitly includes both the entropy and the energy, and this form is due to Gibbs. If one assumes constant particle number, then one arrives at the Boltzmann probability function, which is only a function of energy:

$$P(E_i) = e^{-E_i/kT}.$$

How does one use such an expression? Notice that if two states differ in energy alone, then the probability of a particle occupying the higher of the two states is given by the ratio

$$\frac{P(E_2) = e^{-E_2/kT}}{P(E_1) = e^{-E_1/kT}} = e^{-\frac{\Delta E}{kT}},$$

where $\Delta E = E_2 - E_1$, and this probability function has amazingly broad applications in Biochemistry, Chemistry, and Physics.

An interesting example is to find the probability of finding a charged ion, such as a sodium ion, in an environment of low dielectric constant, such as the hydrophobic interior of a phospholipid bilayer membrane, or the interior of a protein. It is energetically unfavorable for a charged particle to reside in such an environment. A simple calculation will illustrate this effect.

If the self-energy of a sodium ion ($r_s = 0.5$ nm) is given by $E_s = q^2/2\varepsilon r_s$, what is the relative probability of finding this ion in the center of a protein where the dielectric constant, $\varepsilon_p = 3.5$, instead of in the bulk water surrounding the protein, where $\varepsilon_w = 80$? The units of the dielectric constants ε are Coulomb^2/ J m and dielectric constants are referred to that of the vacuum given by $\varepsilon_o = 1.112 \times 10^{-10}$ C^2/J m. The charge on the ion is that of a single electron, 1.6×10^{-19} Coulomb.

- The Boltzmann distribution gives the occupation of states as a function of energy as the probability of a state $P = e^{-\Delta E/kT}$, where $\Delta E = E_p - E_w$. We can write ΔE as

(continued)

$$\Delta E = \left(q^2/2r_s\right)\left\{1/\varepsilon_p - 1/\varepsilon_w\right\} = \left(q^2/2r_s\right)\left\{\varepsilon_w - \varepsilon_p\right\}/\varepsilon_p\varepsilon_w$$

$$\Delta E = \left(1.6\times 10^{-19}\text{Coulomb}\right)^2/2^*5\times 10^{-10}\text{m}\left\{\varepsilon_w - \varepsilon_p\right\}/\varepsilon_p\varepsilon_w$$

$$\Delta E = 2.56\times 10^{-29}\text{Coulomb}^2/\text{m}\left\{\varepsilon_w - \varepsilon_p\right\}/\varepsilon_p\varepsilon_w$$

Therefore, $\{\varepsilon_w - \varepsilon_p\}/\varepsilon_p\varepsilon_w = (80\text{--}3.5)/280 = 0.273\,/\,1.112\times 10^{-10}$ J m/C^2.
$= 2.455 \times 10^9$ J m/C^2.
Now $\Delta E = 2.56 \times 10^{-29}$ C^2/m $*$ 2.455×10^9 J m/C^2 = 6.28×10^{-20} J,
and it remains to compare this with kT, which is 4.27×10^{-21} Joule.
Therefore, $\Delta E/kT = 14.7$, so that $P = e^{-14.7} = 4.12 \times 10^{-7}$
or there is less than one chance in a million that the sodium ion would be
found in the hydrophobic center of a protein.

2.4 Entropy Calculation and Measurement

We have seen how Entropy is defined in terms of probability, and have found that its
maximum value is attained at equilibrium; we would now like to know how to
calculate the Entropy for a given process, and, furthermore, how would one go about
measuring such a State function? It is conceptually simpler to contemplate measuring energy; this can be accomplished by measuring the mass and temperature or
velocity of a System, or by letting the System do work on another System and
examining the States prior to and after the work is done. However, Entropy is
probability: how does one measure a probability? The essence of science is the
interplay between theory and experiment. How then, would one measure something
as abstract as the number of states in which a system exists.

One method, in the case of a System composed of a pair of dice, would be to
perform a series of identical experiments on the dice and to record the States in a
table. Then, one would have a sample of the empirical probability of each State and,
by using $S = Nk \ln W$, one could compute the entropy of each State. For example, for
two 6-sided dice, $W = 36$ and $S = 2k \ln(36) = 9.89 \times 10^{-23}$ J/K. This scheme works
fine for a small number of particles distributed over a small number of states, and
becomes better as the number observations increases. For Avogadro's number of
particles, however, simple counting becomes cumbersome, tedious, or ultimately
impossible. In this case we can exploit the fact that the States of a System become
better defined as the number of particles becomes large and we can use measurement
to determine the mean values of States in terms of their State variables. This
connection between the microscopic properties of Systems and the values of macroscopic State functions is the realm of thermodynamics, which we can use to
measure the entropy for many biological Systems of interest.

The calculation of the entropy of a biochemical System can be accomplished through the use of the definition of Entropy and that of the Boltzmann distribution. Once again, let us assume that we have N (a constant number) particles, n States, with energy per State of ε_i, $i \in \{1, 2, 3,\ldots,n\}$. First, we recast the Boltzmann distribution into terms involving the relative occupation of two States,

$$N_i = \frac{N}{n}e^{-\varepsilon_i/kT}, N_j = \frac{N}{n}e^{-\varepsilon_j/kT}$$

and we will take the ratio of these to find that

$$\frac{N_i}{N_j} = e^{-\varepsilon_i/kT}e^{\varepsilon_j/kT} = e^{-\left(\varepsilon_i - \varepsilon_j\right)/kT} = e^{-\Delta\varepsilon/kT} \tag{2.G}$$

where $\Delta\varepsilon = \varepsilon_j - \varepsilon_i$ is the energy difference between these two States. Also

$$S = k \ln W,$$

$$= k \ln \left[\frac{N!}{\prod\limits_{i=1}^{n} N_i!}\right]$$

We have seen that we can write

$$S = k \ln W = kN \ln N - k \sum_{i=1}^{n} N_i \ln N_i$$

and that

$$dS = kd(\ln W) = -k \sum_{i=1}^{n} \ln N_i dN_i.$$

We will now insert Boltzmann's equation for N_i, the relative occupation of the ith state

$$N_i = Ne^{-(\varepsilon_i - \varepsilon_N)/kT} = Ne^{-\Delta\varepsilon/kT}$$

$$=> \ln N_i = \ln N - \Delta\varepsilon/kT$$

into this expression for dS to give

$$dS = -k \sum_{i=1}^{n} \left\{ \ln N - \frac{\Delta\varepsilon}{kT} \right\} dN_i$$

$$= -k \left(\sum_{i=1}^{n} \ln N dN_i - \sum_{i=1}^{n} \frac{\Delta\varepsilon}{kT} dN_i \right)$$

We have also seen that the first sum on the right side is zero because it reduces to the differential of the total number of particles, N, so $dN = 0$, and we are left with

$$dS = k \sum_{i=1}^{n} \frac{\Delta\varepsilon}{kT} dN_i.$$

Now, we can move $1/kT$ out of the sum, and cancel the k's to give

$$dS = \frac{1}{T} \sum_{i=1}^{n} \Delta\varepsilon dN_i = \frac{1}{T} \sum_{i=1}^{n} (\varepsilon_i - \varepsilon_N) dN_i.$$

This sum is just the heat absorbed by the system, dQ, in going from state i to state N, so the expression for the differential entropy change becomes, simply

$$dS = \frac{dQ}{T},$$

and, for a reversible process, we can replace the differential by a finite difference

$$\Delta S = \frac{\Delta Q}{T}$$

where the quantities on the right-hand side are easily measured since they are just the energy absorbed and the temperature at which the process takes place. Thus, we have a thermodynamic method for measuring probabilities, and the units for entropy are clearly Joules/Kelvin. Note that we have already used this result in the derivation of the Boltzmann distribution.

We can now use this macroscopic definition to find the entropy change of a System as the temperature is varied,

$$\Delta S = \int_{S_1}^{S_2} dS = \int_{Q_1}^{Q_2} \frac{dQ}{T},$$

but it is not convenient or easy to measure the changes in energy implied by the rightmost integral. We can, therefore, convert this integral into one involving quantities easier to measure in the following way. The First Law of Thermodynamics (conservation of energy) states that the heat, dQ, exchanged during a process is the sum of the change in the internal energy and the work done, $dQ = dE + dW = dE + p\,dV = dH - V\,dP$, and if no pressure–volume work is done, then $dQ = dE = dH$. The total differential of H is

$$dH = \left(\frac{\partial H}{\partial T}\right)_p dT = dQ$$

and the partial derivative of the Enthalpy, H, with respect to temperature occurs so often that it is referred to as the heat capacity C_p (here at constant pressure) and is the amount of added heat needed to raise the temperature of a substance by a given amount, usually 1 K.

$$C_p = \left(\frac{\partial H}{\partial T}\right)_p$$

For example, the heat capacity of water is 75.3 Joule/(mole K) at $T = 298.15$ K. Now, we can replace dQ in the integral for the entropy change by $C_p\,dT$, to give

$$\Delta S = \int_{T_1}^{T_2} \left(\frac{\partial H}{\partial T}\right)_p \frac{dT}{T} = \int_{T_1}^{T_2} \frac{C_p}{T} dT,$$

and we have now found a simple method for measuring the number of accessible states for a system. The heat capacities for many substances are tabulated in standard reference works. For biomolecules, the heat capacity is often measured with the aid of a calorimeter.

In order to cement our understanding of the intimate relationship between the statistical (microscopic) and thermodynamic (macroscopic) views of State functions, let us again calculate the change in entropy during the isothermal, reversible, expansion of an ideal gas, this time from a thermodynamic viewpoint. Imagine the gas trapped inside a container fitted with a frictionless piston against which the gas can expand, pushing out the piston as it does so (Fig. 2.6). The volume prior to the expansion is V_1 and after it is V_2. The energy does not change during the expansion, $dE = dQ - dW = 0$, so $dQ = dW$. The work, dW, done by the gas on the piston is pressure–volume work, $p\,dV$.

$$dW = p\,dV.$$

The ideal gas equation, $pV = nRT$, then gives for $n = 1$ mole of gas,

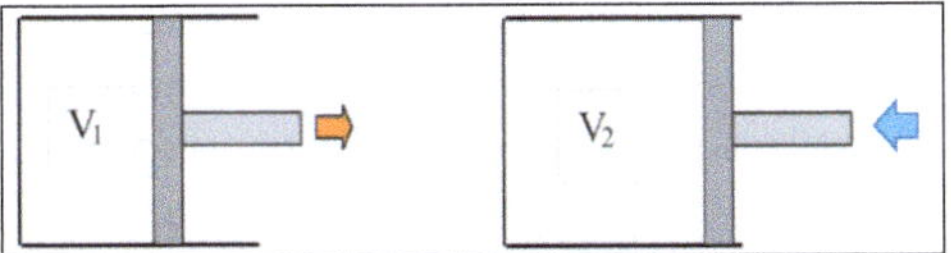

Fig. 2.6 A frictionless piston pushing against a volume of gas, V, in an insulated cylinder. The entropy-driven spontaneous expansion tends to move the piston to the right (orange arrow) but we can lower the entropy of the gas by doing work on the system by pushing on the piston (blue arrow) lowering the entropy of the gas

$$dQ = \frac{RT}{V}\,dV$$

and, with the differential entropy change,

$$dS = \frac{dQ}{T} = R\frac{dV}{V}$$

we can find the entropy difference by integration

$$\Delta S = \int_{S_1}^{S_2} dS = \frac{1}{T}\int_{Q_1}^{Q_2} dQ = R\int_{V_1}^{V_2}\frac{dV}{V}$$

which is

$$\Delta S = \frac{\Delta Q}{T} = R\ln\frac{V_2}{V_1}$$

and this is identical to the entropy change calculated from a statistical point of view before. We see that $\Delta S > 0$ for $V_2 > V_1$.

It is an obvious fact that living systems are not at equilibrium; that state of life is only attained after death. We are also not at a state of maximum entropy; in fact, we exist as humans and ants, as well as bacteria in a most improbable, highly organized, low entropy state of matter and energy. Thus, living systems must somehow manage their entropy and maintain its nonequilibrium state. The above simple example suggests a thermodynamic mechanism for accomplishing this necessary, but interesting feat. If we were to push on the piston, instead of letting the gas expand, we would compress it by doing work, ΔW, on it. The entropy change would then be

$$\Delta S = \frac{\Delta W}{T} = R\ln\frac{V_2}{V_1} < 0;$$

ΔS is negative because $V_2 < V_1$, and we have decreased the entropy of the gas by doing work on it. This is exactly how living systems maintain their highly ordered state; by coupling the reduction of entropy to the consumption of energy from their surroundings. In the case of biological systems, this energy ultimately comes from the solar energy stored in the biomolecules consumed.

2.5   The Gibbs Energy: Using Energy to Battle Entropy

Energy and entropy must therefore be coupled, and there must exist a thermodynamic State function that mathematically expresses this coupling, to explain how systems evolve toward equilibrium. Classical thermodynamic State functions depend on the State variables, pressure, temperature, volume, and chemical potential. Life occurs primarily at constant temperature and pressure, so the most useful State function will be one that does not depend upon T or P. These properties are inherent in the Gibbs energy, G, defined as

$$G = H - TS,$$

where H is the enthalpy ($H = E + PV$), T is the temperature, and S is the entropy. The differential of G obeys the product rule from calculus

$$\begin{aligned} dG &= dH - TdS - SdT \\ &= dE + VdP + PdV - TdS - SdT \end{aligned}$$

however, for a reversible process, $dE = TdS - PdV$, so

$$dG = \cancel{TdS} - \cancel{PdV} + VdP + \cancel{PdV} - SdT - \cancel{TdS}$$

which implies that we can write dG as

$$dG = VdP - SdT.$$

Now, if we require that T and P are held constant, we find for a reversible process that

$$dG = 0.$$

We have used similar equations previously to find, e.g., S_{max}, and we know that this equation implies that at equilibrium the Gibbs energy must attain an extremum; in this case it is a minimum. An examination of the change in the Gibbs energy can then be used to find the direction of evolution and positions of equilibrium in biochemical systems. If it is found that the change in the Gibbs energy during a biochemical process obeys

$$\Delta G = G_f - G_i < 0,$$

then this process will spontaneously proceed toward equilibrium.

The thermodynamic stability of proteins was left unexplained previously when only the entropy change on denaturation was taken into account. Let's revisit this problem in light of our new understanding of the coupling between entropy and energy. The entropy change on denaturation is

$$\Delta S = k \ln \frac{W_d}{W_n}$$

and since $W_d >> W_n$, i.e., the number of unfolded states, W_d, is much greater than the number of folded states, $W_n \sim 1$, $\Delta S > 0$. The key to protein stability, and that of life in general, is to use an input of energy to compensate for this entropy increase. In the case of biopolymers, this energy comes from the attractive potentials in the form of non-covalent bonds that stabilize the native state. These potentials will be examined in some detail later, but suffice it to say that among them are hydrogen bonds, Coulombic interactions, hydrophobic interactions with solvent water, and so on.

In general, the change in the Gibbs energy is given by

$$\Delta G = \Delta H - T\Delta S - S\Delta T$$

but at constant temperature, $\Delta T = 0$, so the change in the Gibbs energy becomes

$$\mathbf{\Delta G = \Delta H - T\Delta S}.$$

The stabilizing interactions among various parts of biomolecules then take the form $\Delta H > 0$. Thus, at biological temperatures, $\Delta H > T\Delta S$ and biomolecules are more stable in their native state than in their denatured state. The entropy term, however, increases as the temperature is raised, so that at some temperature the denatured state becomes more favorable. Energy, in the form of heat, must be added to the protein to denature it. Let us examine the temperature dependence of ΔG. Both ΔH and ΔS are >0, so as the temperature increases the entropy term in ΔG increases and overcomes the fixed stabilization energy imparted by ΔH. This is shown in Fig. 2.7 where we can distinguish two regions: (1) $\Delta G > 0$, or $\Delta H > T\Delta S$ and the native structure is stable, and (2) $\Delta G < 0$, or $\Delta H < T\Delta S$, and the denatured form of the protein is more stable. At the denaturation temperature, T_d, $\Delta G = 0$, and we have that $\Delta H = T_d \Delta S$, or

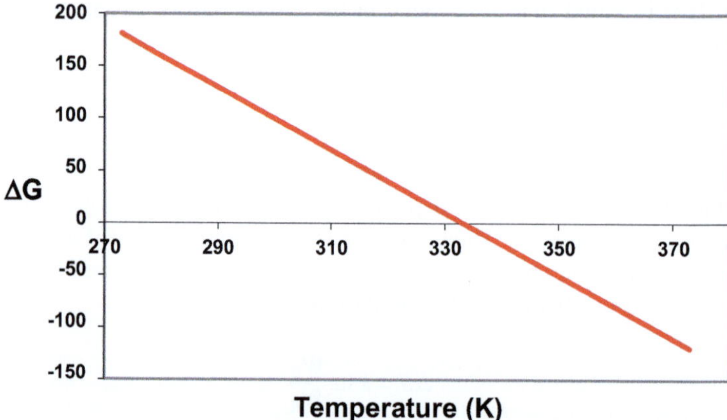

Fig. 2.7 The change in ΔG with temperature for a protein that denatures at 335 K

$$T_d = \Delta H / \Delta S$$

for the temperature at which the protein denatures (Fig. 2.7). In this hypothetical case, the protein is stable at temperatures below $T_d = 335$ K and denatures for $T > 335$ K.

2.6 Calorimetry to Measure the Number of States

One can measure the thermal properties of proteins, lipids, nuclei acids, and other biomolecules in a differential scanning calorimeter. Here, two samples contained in separate pressurized cells are heated at a constant rate. One cell contains a standard, usually only a buffer solution, while the other contains this buffer along with the molecule of interest. The cell's temperatures are measured and the heating power is controlled through a feedback circuit to keep both cells at the same temperature. The power needed to heat each cell is $P = IV$, i.e., the current times the voltage, so for cell, i, the power is $P_i = I_i V_i$. The energy E_i added to each cell is the power times the time, i.e., $E_i = P_i t$. Then the time-dependent difference in energy between the two cells is $\Delta E(t) = (I_2 V_2 - I_1 V_1)t$. The output of a scanning calorimeter is $\Delta E(t)$ versus t, or since we are heating at a constant rate, the time is proportional to the temperature, and the heat input is proportional to the difference in heat capacity at constant pressure, ΔCp. For example, Fig. 2.8 shows the denaturation of a protein at $T = 333$ K. Several proteins, such as ribonuclease and lysozyme, display reversible denaturation in that one can heat them, cool them, and reheat them and find that the

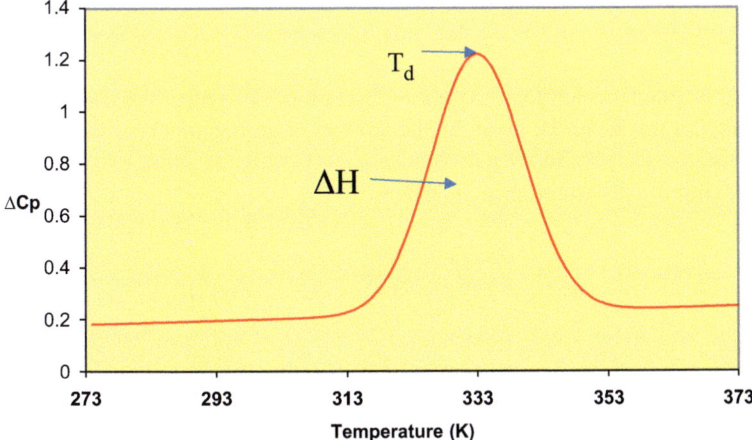

Fig. 2.8 The output of a differential scanning calorimeter used to measure the thermally induced unfolding of a protein. The area under the curve is ΔH and the peak is the temperature at which $T_d\, \Delta S = \Delta H$

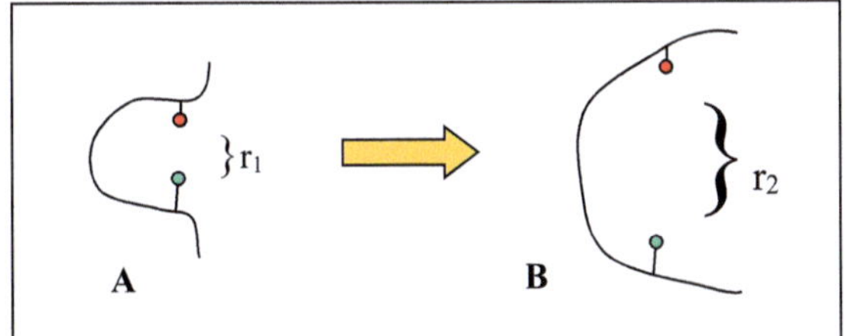

Fig. 2.9 The distance between a positive (red) and a negative (green) amino acid side chain in the native protein (**a**) is r_1, and in the denatured protein (**b**) is r_2

thermal properties do not depend on the sample history, the proteins will refold upon cooling; however, there are many other proteins whose denaturation is irreversible.

2.7 Protein Denaturation: Revisited

Boltzmann's distribution can be used to find the probability of the unfolding of a protein. The Gibbs energy for a protein can be schematically understood by referring to Fig. 2.9, where one can call the initial state that of the folded, or native protein, with energy E_n, and the final state one in which the protein is denatured, or unfolded, has energy E_d. The probability of unfolding is then

$$P = \frac{N_d}{N_n} = \frac{g_d}{g_n} e^{-(E_d - E_n)/kT},$$

where g is a degeneracy factor equal to the number of energy levels with identical energies. Let this factor be given by the number of arrangements of the protein in its various states, $g_i = W_i$. In the native state, $g_n = 1$, while there are $g_d = W_d \gg 1$ ways of arranging the denatured protein. Then

$$\frac{N_d}{N_n} = \frac{W_d}{W_n} e^{-\Delta E_n/kT},$$

or, since

$$e^{\ln\left(\frac{W_d}{W_n}\right)} = \frac{W_d}{W_n}$$

we can rewrite this as

$$\frac{N_d}{N_n} = e^{\ln\left(\frac{W_d}{W_n}\right)} e^{-\Delta E/kT} = e^{-\left(\Delta E - kT \ln\left(\frac{W_d}{W_n}\right)\right)/kT}.$$

On a per mole basis, we can make the replacements:

$$E_i = E/n_i,$$
$$N_i = N_o n_i,$$
$$N_o k = R,$$

where n_i is the number of moles of species, i. The left side of our probability equation now reads

$$\frac{n_d}{n_n} = e^{-\left(\Delta E - RT \ln\left(\frac{W_d}{W_n}\right)\right)/RT}.$$

The energy change, ΔE, can be replaced approximately by the enthalpy change, ΔH, and one recognizes the entropy change, $\Delta S = R \ln \frac{W_d}{W_n}$, so that one obtains

$$\frac{n_d}{n_n} = e^{-(\Delta H - T\Delta S)/RT}$$

and now the exponential term in parentheses is seen to be $\Delta G = \Delta H - T\Delta S$, giving

$$\frac{n_d}{n_n} = e^{-\Delta G/RT}.$$

The logarithm of the ratio of the number of moles of denatured protein, to that of the native protein, is ln K, where K is the equilibrium constant

$$\ln\left[\frac{n_d}{n_n}\right] = \ln K = -\frac{\Delta G^o}{RT}.$$

This equation, while of general validity, is an oversimplification of the real problem of protein denaturation, which implies that:

$$\mathbf{\Delta G = -RT\ \ln\ K}.$$

Note the very important point that our derivation of this equation did not actually use any specific chemical reaction or particular properties of proteins. Therefore, this equation is of general validity for any chemical reaction and is once again simply a formal expression for the probability of the initial and final states. The change in Gibbs energy is proportional to the logarithm of the ratio of the number of molecules in each state at equilibrium.

Even though, from an entropy point of view, proteins are more stable in their denatured state, we know from the fact that life exists, that proteins are indeed stable at $T = 310$ K. The stabilization energy for a macromolecule enters into ΔG in the form of the enthalpy, ΔH. One of the interactions which contributes to a positive ΔH is that of the Coulombic attraction among oppositely charged amino acid side chains, such as lysine (+) and glutamate (−). It is interesting to examine a model for this interaction in a protein.

Imagine two charged groups, separated from each other in the native structure by a distance r_1 (Fig. 2.9). We can find the equilibrium constant, K, the enthalpy change, ΔH, and the change in entropy, ΔS, from the potential energy change of the charges in the two configurations. The force between charges obeys an inverse square law: $F = -\frac{a}{\varepsilon r^2} = -\frac{\partial E}{\partial r}$, and can be written as the negative derivative of the potential energy, E. The constant $a = Z_1\,Z_2\,e^2$, where Z is the charge in units of the electron charge, e, and ε is the dielectric constant. The change in the Gibbs energy is equal to the work done in separating the charges; the work is the integral of the force, F, times the distance, Δr:

$$\Delta G = -\int_{r_1}^{r_2}\frac{a}{\varepsilon r^2}\,dr = -\left[\frac{a}{\varepsilon r}\right]_{r_1}^{r_2} = -\frac{Z_1 Z_2 e^2}{\varepsilon}\left[\frac{1}{r_2} - \frac{1}{r_1}\right].$$

The enthalpy change, ΔH, can be found from the thermodynamic relationship known as the Gibbs–Helmholtz equation:

$$\Delta H = - T^2 \frac{d}{dT}\left(\frac{\Delta G}{T}\right)$$

$$\Delta H = - T^2 \frac{d}{dT}\frac{a}{\varepsilon T}\left[\frac{1}{r_2} - \frac{1}{r_1}\right] = \frac{a}{\varepsilon}\left[\frac{1}{r_2} - \frac{1}{r_1}\right].$$

The entropy change can be found from $\Delta S = - \frac{d\Delta G}{dT} = 0$, since ΔG does not depend on temperature and the derivative of a constant is zero. We now have the Gibbs energy change because $\Delta G = \Delta H - T\Delta S$, and $\Delta S = 0$, so $\Delta G = \Delta H$.

The equilibrium constant is easily found because we know that

$$\Delta G = - RT \ln K :$$

$$\Delta G = - RT \ln K = \frac{a}{\varepsilon}\left[\frac{1}{r_2} - \frac{1}{r_1}\right]$$

$$\ln K = - \frac{a}{\varepsilon RT}\left[\frac{1}{r_2} - \frac{1}{r_1}\right]$$

so that

$$K = e^{- \frac{a}{\varepsilon RT}\left[\frac{1}{r_2} - \frac{1}{r_1}\right]}$$

Note here that the entire contribution to ΔG comes from the energy term, since $\Delta S = 0$ in this model. As long as $[-a < 0]$, the sign of the Gibbs energy change is positive, as can be found from the fact that $r_2 > r_1$, so the difference of the reciprocals is negative. In fact, the sign of the energy change depends solely on the signs of the individual charges, as shown in Table 2.1. The energy tends to decrease when like charges are separated, but it will increase when unlike charges are involved. This certainly fits with our understanding that like charges repel and unlike charges attract; for like charges the most probable state is one in which they are separated from one another, while the opposite is true for an unlike charge pair. The opposite charges on positive and negatively charged amino acid side chains of proteins then provide an electrostatic field which, along with other interactions, guides the folding, or self-assembly, of protein molecules as they emerge from the ribosome.

Table 2.1 The signs of the Gibbs energy changes associated with the unfolding of a protein containing charged amino acid residues of various signs

Z_1	Z_2	$- Z_1 Z_2 e^2$	ΔG
+	+	−	−
−	+	+	+
+	−	+	+
−	−	−	−

Table 2.2 Electromagnetic bonds that stabilize biological macromolecules

Bond type	Examples in proteins	Typical distance relation	Bond distance (nm)	Binding energy (kJ/mol)	Energy-to-noise ratio
Covalent	-C-C-C-C-	Nearest neighbor	0.15	356	143
Disulfide	-Cys-S-S-Cys-	Nearest neighbor	0.22	167	67
Salt bridge	- COO^-...H^+N-	Atoms <0.35 nm apart	0.28	12.5–30 depending on local ε	~8.5
Hydrogen	-C=O...H-N-	Atoms <0.35 nm apart	0.3	2–6 for neutral atoms; 12.5–21 if charged	~1.6 ~6.7
Electrostatic	$-COO^- H^+N$-	$\sim\frac{1}{\varepsilon r}$	Varies	Reduced in water, but strong in nonpolar regions	Variable; Very weak to very strong
Van der Waals	$-CH_3 CH_3$-	$\sim\frac{1}{r^6}$	0.35	4–17 in protein interior	~4.2
Hydrophobic effect	Water plus nonpolar	Bulk partners	Bulk partners	<40	
$\pi-\pi$ aromatic stacking	Trp, Tyr, His, Phe	$\sim\frac{1}{r^6}$	0.4	0–8	
Dipole–dipole	Protein backbone	$\sim\frac{1}{r^3}$	0.1–0.3	Biphasic, -6 to 6	

The major contributions to the stabilization of biological macromolecules arise from electrostatics and their magnitudes are summarized in Table 2.2. Note that we use SI (Systeme International) units (meters, kilograms, seconds) throughout this text, except in a few cases where natural units, such as electron volts, seem more appropriate.

Upon examination of the macromolecular stabilization energies given in Table 2.2, the first thing that we note is that compared with the thermal fluctuations in the aqueous environment of cells, which are of order $N_o kT = 2.5$ kJ/mol at 300 K, the only binding energies that are significantly larger than kT are for electron-sharing covalent interactions such as carbon–carbon, and sulfur–sulfur bonds (Table 2.2; column 6: Energy-to-noise ratio). For these bonds the probability of finding the partners in the free, unbound state is essentially zero ($\sim 10^{-30}$ to 10^{-60}). The remaining interactions, while much more comparable to $N_o kT$, still give probabilities of bond formation ranging from 0.91 to 0.99999. Several of these binding interactions vary in energy depending on the spatial orientation and separation of the partners, and the polarization properties of the intervening medium. The energy involved in the formation of a salt bridge, for example, depends on the local dielectric constant, ε. Water has a very high dielectric constant ($\varepsilon \sim 80$), while the interior of a protein consists of the buried hydrophobic chains of nonpolar amino acids, giving it a dielectric constant more like oil ($\varepsilon \sim 2$–5). Therefore, a salt bridge buried in the center of a protein can be significantly stronger than one on the surface of a protein. This fact is tempered by the energetic bias against transferring a charged amino acid into the center of a protein, with the result that salt bridges are mostly found at or near the surface of a protein.

The data in Table 2.2 indicate that hydrogen bonds are by far the weakest stable electromagnetic interactions found in biomolecules. Nevertheless, they still contribute greatly to the stabilization of, for example, DNA, where hundreds to thousands of hydrogen bonds form between the bases along the nucleic acid backbone and through their shear numbers act to stabilize DNA. Hydrogen bonds are responsible for the properties of water. Thus, hydrogen bonds are among the most important in the stabilization of biological macromolecules. Biomolecules also form hydrogen bonds with both the surface and interior water molecules; this leads to competition for hydrogen bonds as well as the possibility that water can act as a hydrogen bond bridge within a macromolecule. The interactions listed in Table 2.2 can be thought of as giving the strength with which the probabilities for biological self-assembly are biased by the energies involved in the interactions.

## 2.8	Gibbs Energy Changes for Biochemical Reactions

We have seen how the sign of the Gibbs energy change for a process determines the direction in which the process will spontaneously proceed to reach equilibrium. Living systems use the energy, ΔH, available from chemical reactions to establish and maintain order, i.e., to compensate for unfavorable entropy changes. How much energy can be derived from a particular reaction and how can this energy be captured and used, rather than dissipated as heat, and lost to the environment? As opposed to chemical reactions, such as the formation of NaCl, as represented by

$$Na + HCl = NaCl + H,$$

which occurs simply by the random collision of sodium and chloride ions in solution, biochemical reactions take place at the active sites of enzymes. All chemical reactions involve energy changes; in order to understand the amount of energy needed by or available from a reaction one requires knowledge of the energies of the reactants and products. Since reactions can occur under a wide variety of conditions of temperature, pressure, ionic strength, pH, metal ions, and so on, the energy changes are sensitive to these conditions present during measurements. For this reason, standard conditions have been specified as reference states for the reactants and products. These standard conditions for reporting the Gibbs energy changes in living systems are $T = 298$ K, reactants and products at a concentration of 1.0 molar, and at a pH of 7.0—parameters that are somewhat like the interior of a cell. These standard Gibbs energy changes are then denoted by ΔG° and are widely tabulated. The energy available from a reaction is the difference between ΔG° of the reactants and products. It should be noted that the standard conditions for biochemical reactions differ slightly from what might be found in physical chemistry texts or tables in that the standard conditions for the latter case involve the presence of all reactants and products at a concentration of 1.0 Molar. These standard conditions are the same for biochemical reactions, except when these involve the consumption or liberation of hydrogen ions. Biochemistry occurs largely at or near solution neutrality, pH 7, and here we certainly do not expect to have an entire mole of H^+ ions, for then the pH would be closer to 0. If H^+ ions are involved, then tabulated values of ΔG° would need to be adjusted in the following way.

(A) If the reaction involves H^+ ions as products, such as in the reaction

$$X + Y \rightarrow Z + nH^+$$

Then, the physical chemists value of $\Delta G_P = \Delta G^{\circ} + RT \ln\left(\frac{[Z][H^+ = 1]^n}{[X][Y]}\right)$,

while the biochemists value of $\Delta G_B = \Delta G^{\circ} + RT \ln\left(\frac{[Z][H^+ = 10^{-7}]^n}{[X][Y]}\right)$.

Or, since

$$\Delta G_P = \Delta G^{\circ} + RT(\ln\ [Z] + n\ \ln[H^+ = 1] - \ln\ [X] - \ln\ [Y])$$

and

$$\Delta G_B = \Delta G^{\circ} + RT(\ln\ [Z] + n\ \ln[H^+] - \ln\ [X] - \ln\ [Y])$$

So that we can subtract these two equations to find the difference between these two definitions. All the terms on the right side of both equations are the same except for the terms involving hydrogen ions, and the difference then becomes

$$\Delta G_P - \Delta G_B = nRT(\ln[H^+ = 1] - \ln[H^+]),$$

but $\ln(1) = 0$, and we are left with our sought-after correction factor of

$$\Delta G_P - \Delta G_B = -nRT\ \ln[H^+].$$

Now, at pH $= 7$, $[H^+] = 10^{-7}$, which gives rise to (for $n = 1$ and $T = 298$ K)

$$\Delta G_P - \Delta G_B = -nRT \ln\left(10^{-7}\right) = -8.31 \times 298 \times (-16.12)$$
$$\Delta G_P - \Delta G_B = +39.93 \text{ kJ mol}^{-1},$$
$$\Delta G_P = \Delta G_B + 39.93 \text{ kJ mol}^{-1},$$

and the reaction proceeds more readily to the right at pH 7 under standard state conditions.

(B) If the reaction under study involves H^+ ions as reactants, instead, as in

$$Z + nH^+ \rightarrow X + Y$$

then

$$\Delta G_P - \Delta G_B = -nRT \; \ln\left(10^7\right) = -39.93 \text{ kJ mol}^{-1},$$
$$\Delta G_P = \Delta G_B - 39.93 \text{ kJ mol}^{-1},$$

and the reaction proceeds more readily to the left at pH 7 under standard state conditions. In the case where the biochemical reaction does not consume or release hydrogen ions,

$$\Delta G_P = \Delta G_B.$$

As an example of an important biochemical reaction in which protons are involved, let us examine the reduction and oxidation of nicotinamide adenine dinucleotide (NAD^+/NADH), whose structure and reaction mechanism are shown in Figs. 2.10 and 2.11.

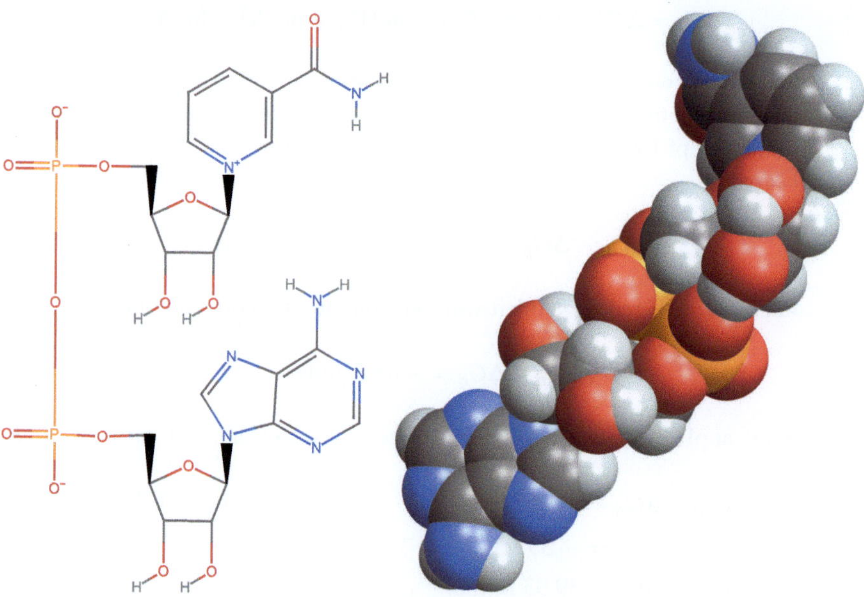

Fig. 2.10 A chemical and space-filling structure and for nicotinamide adenine dinucleotide. Nitrogen is blue, oxygen is red, carbon is dark gray, phosphorus is orange, and hydrogen is light gray

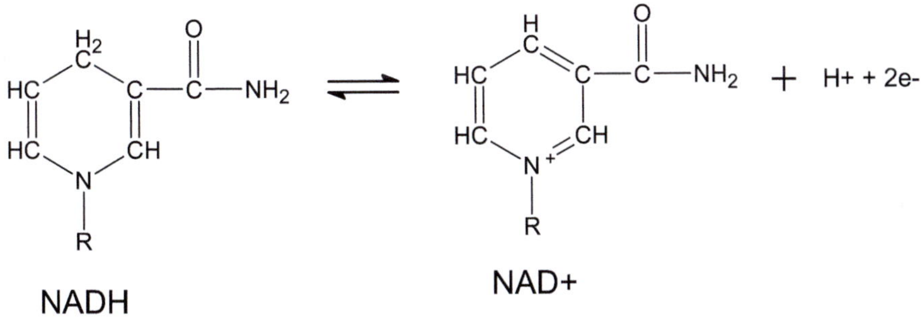

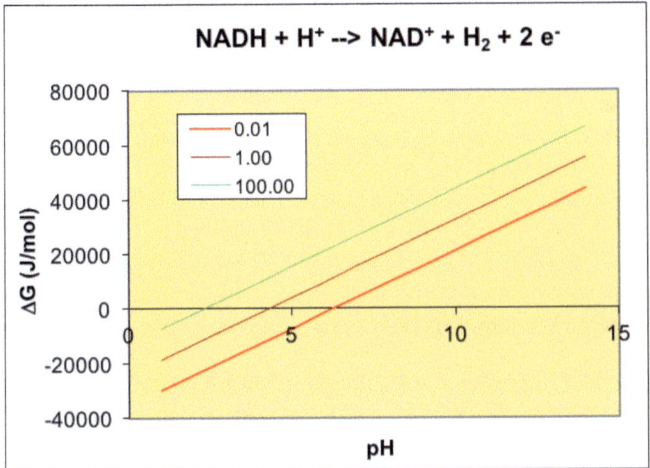

Fig. 2.11 Oxidation of NADH to form NAD$^+$

Fig. 2.12 The relationship between pH and the change in Gibbs energy for the reduction or oxidation of NADH. The three curves were computed for the values indicated in the example and for P_{H2} varying from 0.01 to 100 bar

One can use this relationship between ΔG and pH to examine the equilibrium as one changes the concentration of hydrogen ions. The results are shown in Fig. 2.12, where we observe that as the pH increases, [H$^+$] goes down and the reaction proceeds toward the left (reduction) where ΔG is positive, in the same manner as when one increases the partial pressure of hydrogen, P_{H2}, the reaction is strongly driven to the left, at a constant pH (Fig. 2.12).

The oxidation of NADH to form NAD^+ proceeds according to the reaction written as

$$NADH \rightarrow NAD^+ + 2\,e^- + H^+$$

in which protons enter the reaction as a product. The standard state change in the Gibbs energy is $\Delta G^o = -21.8$ kJ/mol at $T = 298$ K. What are ΔG_B and the equilibrium constant, k_p?

Protons appear on the right side of the reaction equation, so $\Delta G_P = \Delta G_B + 39.93$ kJ/mol. We then need to find the equilibrium constant, k_p, from

$$\Delta G^o = \Delta G_P = -21.8 \text{ kJ/mol} = -RT \ln k_p = -8.31 \text{ J/K/mol} \times 298 \text{ K } \ln k_p$$

$$\ln k_p = \frac{-21,800}{-8.31x298} = 8.8;$$

$$k_p = 6.6 \times 10^3$$

and we can find $\Delta G_B = \Delta G^o + 39.93$ kJ/mol $= -21.8$ kJ/mol $+ 39.93$ kJ/mol

$$\Delta G_B = 18.13 \text{ kJ/mol}$$

Now, what is the equilibrium constant, k_b, with respect to the Biochemist's standard state? We find it from

$$\Delta G_B = 18.13 \text{ kJ/mol} = -RT \ln k_b = -8.31 \text{ J/K/mol} \times 298 \text{ K } \ln k_b$$

$$\ln k_b = -7.32$$

$$k_b = 6.6 \times 10^{-4}$$

and the ratio of the two equilibrium constants is just 10^7, or the difference in the H^+ ion concentrations for the two different standard states.

How is the Gibbs energy change in this reaction different when we move away from the Biochemist's standard state? Let us assume that the following conditions hold: $T = 298$ K; $[NADH] = 12$ mM; $[H^+] = 3$ μM (pH $= 6.0$); $[NAD^+] = 4$ mM; and the partial pressure of H_2 is 0.02 bar. (Note that $RT = 8.31$ J/K/mol $\times 298$ K $= 2.48$ kJ/mol)

(continued)

$$\Delta G = \Delta G_B + RT \ \ln\left(\frac{[\text{NAD}^+][p_{H_2}]}{[\text{NADH}][H^+]}\right) = 18.13 \ \text{kJ/mol}$$

$$+2.48 \ \text{kJ/mol} \ \ln\left(\frac{(4 \ \text{mM})(0.02)]}{(12 \ \text{mM})(3 \ \mu\text{M}/0.1 \ \mu\text{M})}\right)$$

$\Delta G = -2.7$ kJ/mol. Note also that we compared the pH to the reference state of pH $= 7.0$, or $[H^+] = 0.1 \ \mu$M by dividing by this factor when we entered the $[H^+]$.

2.8.1   *The Enzyme Creatine Kinase*

An essential process for living systems is the capture of the energy from the hydrolysis of adenosine triphosphate (ATP). One of many ways cells accomplish this is through the activity of the enzyme, creatine kinase, which catalyzes the formation of phosphocreatine (PCr) from creatine and ATP

$$\text{PCr} + \text{ADP} < = > \text{Cr} + \text{ATP},$$

according to the scheme shown in Fig. 2.13. The net result of this reaction is either the formation of PCr or ATP depending on the equilibrium constant and the concentrations of each of the components in the cell at a given time.

We have just seen that the equilibrium constant, K, for a process is intimately related to the Gibbs energy change: $\Delta G^\circ = -RT \ln K$. The standard Gibbs energy changes for the hydrolysis of phosphocreatine and ATP (Table 2.3) are both very negative, indicating that the hydrolysis of both compounds is favored. The reaction catalyzed by creatine kinase is built up from the two reactions in Table 2.3 by reversing the ATP hydrolysis reaction and then adding the two as:

$$\text{PCr} + \text{H}_2\text{O} = \text{Cr} + \text{P} \qquad \Delta G^\circ = -43.51 \ \text{kJ/mole}$$

$$\text{ADP} + \text{P} = \text{ATP} + \text{H}_2\text{O} \qquad \Delta G^\circ = +39.75 \ \text{kJ/mole}$$

$$\text{PCr} + \text{ADP} = \text{Cr} + \text{ATP} \qquad \Delta G^\circ = -3.76 \ \text{kJ/mole}$$

The equilibrium constant (i.e., the probability of the reaction) is

$$K = 4.6 = \frac{[\text{Cr}][\text{ATP}]}{[\text{PCr}][\text{ADP}]} = e^{-\frac{\Delta G}{RT}} = e^{1.52}.$$

The creatine kinase reaction is close to equilibrium (where the equilibrium constant would be $K \sim 1$), but as the concentration of ADP is rather low in cells,

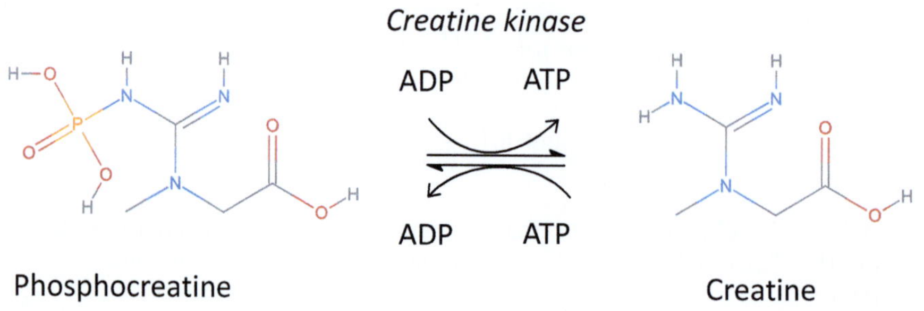

Fig. 2.13 The reaction catalyzed by the enzyme creatine kinase

Table 2.3 Standard ΔG^o of Hydrolysis (298 K, pMg = 4, I = 0.2, pH 7)

Reaction	ΔG^o (kJ/mole)
PCr + H_2O = Cr + P	−43.51
ATP + H_2O = ADP + P	−39.75

this reaction is quite sensitive to the ATP/ADP ratio and serves as a buffer for the ATP concentration in cells. Here, a reaction with a lower ΔG^o is used to drive a phosphorylation process with a less negative ΔG^o. The reaction takes place on the surface of the creatine kinase enzyme (Figs. 2.14 and 2.15). The transition state involves an in-line transfer of the phosphate group from ATP to the guanidine group nitrogen on creatine. In this manner, the large ΔG^o of the ADP~P bond is not lost to heat, as it would be if PCr were simply hydrolyzed and the products diffused away into the surrounding aqueous medium. Surface catalysis like this must have played a crucial role in the development of prebiotic chemistry. In addition, this reaction is a prime example of how living systems couple a thermodynamically unfavorable process to one that is thermodynamically spontaneous.

2.8.2 How Organisms Extract Energy from Their Environment

In order to maintain their existence in the face of continuous entropic pressure, living organisms must continuously extract energy from their environment. The three main extraction schemes are the metabolic pathways:

1. glycolysis (*sugar splitting*),
2. tricarboxylic acid cycle (**TCA,** citric acid, or *Krebs* cycle), and
3. oxidative phosphorylation.

Of these three pathways, glycolysis is universal and the most ancient, having been found in all living organisms, even in the earliest cyanophytes and prokaryotes.

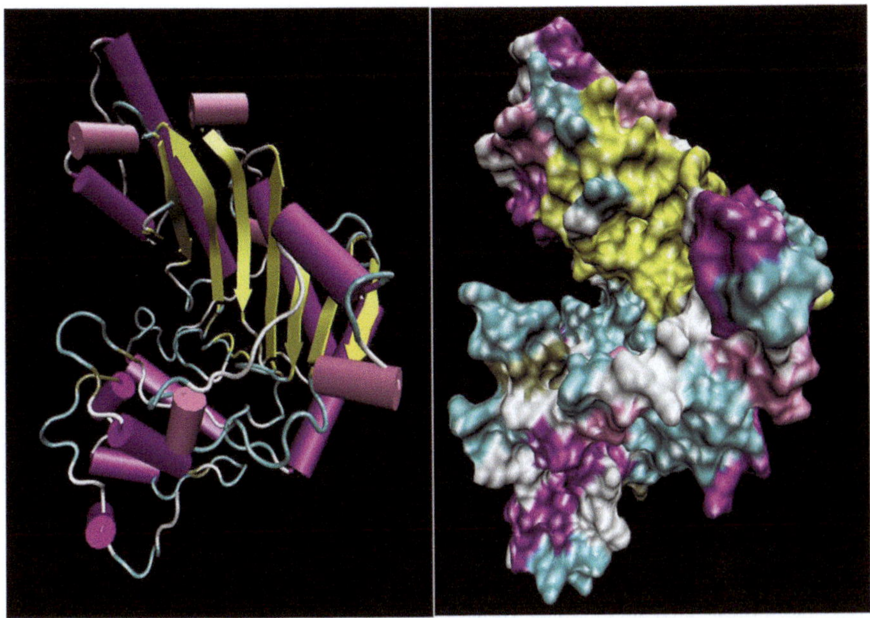

Fig. 2.14 The structure of the creatine kinase enzyme monomer drawn on the left as a cartoon showing the various structural motifs, with helices in purple and β-sheets in yellow. On the right is a space-filling model thresholded at the level of the Van der Waals radii of the constituent atoms colored as in the left frame according to secondary structure. (Rendered using VMD from the protein data bank file **1vrp.pdb** (Lahiri et al., Biochemistry **41**(2002)13861))

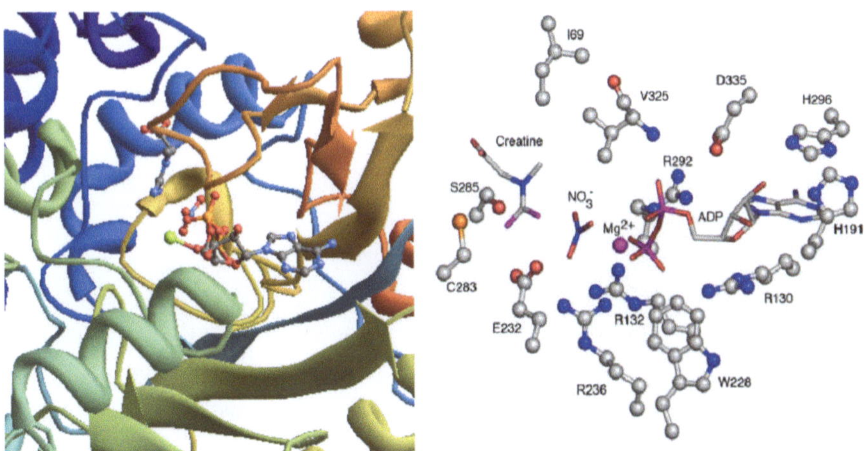

Fig. 2.15 The active site of the enzyme creatine kinase from *Torpedo californica* showing the transition state complex composed of creatine, NO_3^-, and MgADP. (Rendered using VMD from the coordinates in the protein data bank file **1vrp.pdb** (Lahiri et al., Biochemistry **41**(2002)13861)). Note that the reactants are held in a linear array by the Mg^{2+} and the protein side chains. The view on the right shows the important residues with the protein backbone removed for clarity

However, glycolysis is not very efficient at harvesting the energy stored from sunlight in the resulting simple, 6-carbon sugars, such as glucose, since it produces only two ATP molecules per glucose molecule consumed.

2.8.3 Energy Changes in Glycolysis

This primitive metabolic pathway (Fig. 2.16) is the first step in the catabolism (chemical disassembly) of sugars in all living cells. Glucose molecules, for example, are transported into cells by means of a glucose transporter, phosphorylated by hexokinase, trapping the neutral glucose within the cell as a charged molecule (step 1 in Fig. 2.16). The network of glycolytic enzymes converts glucose-6-phosphate (G6P) into pyruvate, which is cleaved by pyruvate dehydrogenase to form acetyl CoA that enters the tricarboxylic acid cycle (Fig. 2.17). The overall reaction of glycolysis is:

$$D\text{-glucose} + 2\,NAD^+ + 2\,ATP + 2\,P_i \rightarrow 2\,Pyruvate + 2\,NADH + 2\,H^+ + 2\,ATP + 2\,H_2O$$

and the Gibbs energy changes for each step shown in Fig. 2.16 are given in Table 2.4. The large overall change in Gibbs energy of -76 kJ/mol indicates that this biochemical network operates spontaneously.

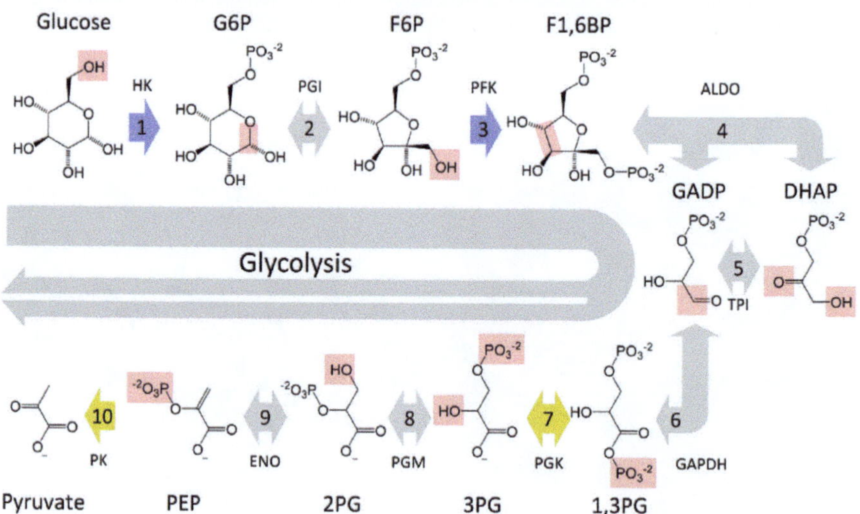

Fig. 2.16 An overview of glycolysis showing the entry (step 1) of a glucose molecule into the cell where it is trapped by phosphorylation by hexokinase. The subsequent reactions lead to pyruvate that is converted to acetyl CoA by pyruvate dehydrogenase for entry into the tricarboxylic acid cycle (Fig. 2.17). (Thomas Shafee / CC BY 4.0 / Wikimedia Commons)

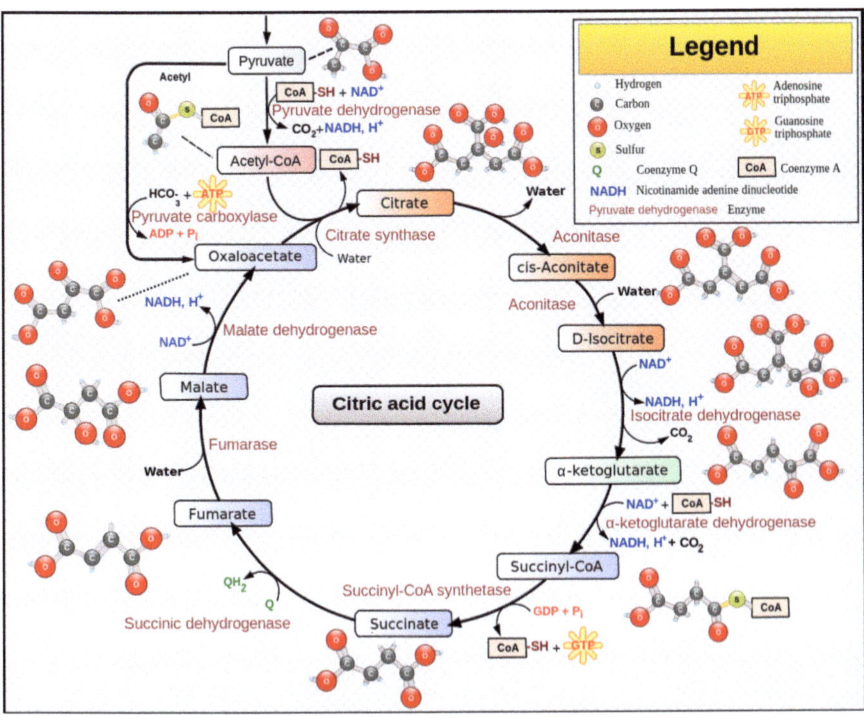

Fig. 2.17 The tricarboxylic acid (Krebs, or citric acid) metabolic cycle. Many of these enzymatic steps are so close to equilibrium that they are reversible (Fig. 2.15) such as the reactions catalyzed by aconitase, succinyl CoA synthetase, succinate dehydrogenase, and fumarase, so that this is actually only a logical rather than a physical cycle. We will observe in Chap. 8 that the flow of carbon goes in several directions simultaneously. (Narayanese, WikiUserPedia, YassineMrabet, TotoBaggins—http://biocyc.org/META/NEW-IMAGE?type=PATHWAY& object=TCA. Image adapted from Image:Citric acid cycle noi.svg CC BY-SA 3.0)

2.8.4 Energy Changes in the TCA Cycle

The tricarboxylic acid cycle is also highly conserved from the earliest to present organisms, but substantial variations have been found in the enzymes in the cycle among the different taxonomic classes of organisms (https://www.biocyc.org/ META/NEW-IMAGE?type=PATHWAY&object=TCA-VARIANTS). The tremendous increase in energy capture efficiency of the TCA cycle over glycolysis is manifest in the production of ATP per glucose molecule catabolized. Glycolysis generates two ATP molecules per glucose, while the TCA cycle generates approximately 30. The reactions in the TCA cycle are shown in Fig. 2.17 and the Gibbs energy changes at each step (Fig. 2.18) give an average $\Delta G \sim -61$ kJ/mol (which

Table 2.4 The Gibbs energy changes for each step in glycolysis

Step	Reaction	ΔG (kJ/mol)
1	Glucose + $ATP^{4-} \rightarrow$ Glucose-6-phosphate^{2-} + ADP^{3-} + H^+	−34.0
2	Glucose-6-phosphate$^{2-} \rightarrow$ Fructose-6-phosphate^{2-}	−2.90
3	Fructose-6-phosphate^{2-} + $ATP^{4-} \rightarrow$ Fructose-1,6-bisphosphate^{4-} + ADP^{3-} + H^+	−19.0
4	Fructose-1,6-bisphosphate$^{4-} \rightarrow$ Dihydroxyacetone phosphate^{2-} + Glyceraldehyde-3-phosphate^{2-}	−0.23
5	Dihydroxyacetone phosphate$^{2-} \rightarrow$ Glyceraldehyde-3-phosphate^{2-}	2.40
6	Glyceraldehyde-3-phosphate^{2-} + P_i^{2-} + $NAD^+ \rightarrow$ 1,3-Bisphosphoglycerate^{4-} + NADH + H^+	−1.29
7	1,3-Bisphosphoglycerate^{4-} + $ADP^{3-} \rightarrow$ 3-Phosphoglycerate^{3-} + ATP^{4-}	0.09
8	3-Phosphoglycerate$^{3-} \rightarrow$ 2-Phosphoglycerate^{3-}	0.83
9	2-Phosphoglycerate$^{3-} \rightarrow$ Phosphoenolpyruvate^{3-} + H_2O	1.10
10	Phosphoenolpyruvate^{3-} + ADP^{3-} + $H^+ \rightarrow$ Pyruvate$^-$ + ATP^{4-}	−23.0
	Sum =	**−76.0**

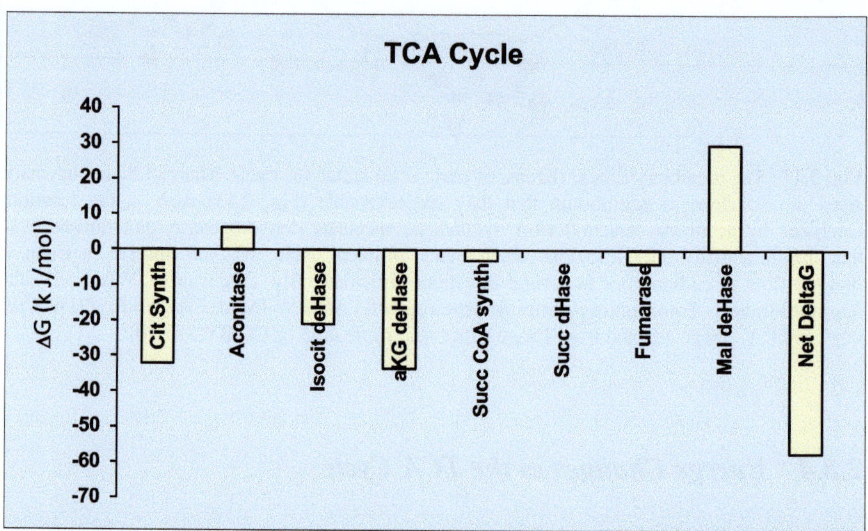

Fig. 2.18 The average Gibbs energy changes for biomolecules traversing the tricarboxylic acid (Krebs) metabolic cycle. The various enzymatic steps are shown with their individual ΔG values and the enzyme names. The complete cycle is shown in Fig. 2.17. The last bar shows the overall Gibbs energy gain from a single turn of the cycle

differs from that in Table 2.4 due to slight differences among the enzymes in a variety of organisms).

Before entering into the TCA cycle, the three-carbon molecule, pyruvate, is produced in several ways, through glycolysis as well as through the catabolism of

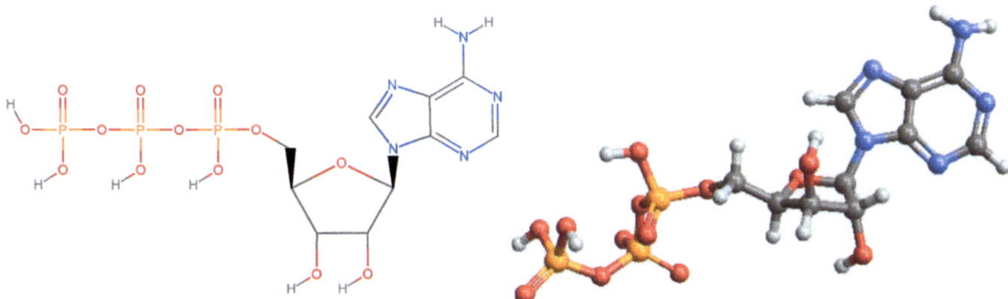

Fig. 2.19 The chemical structure (left) and a molecular model (right) of the ATP molecule. In cells, the three negatively charged phosphates chelate magnesium. The Gibbs energy from oxidative phosphorylation is captured by phosphorylating ADP and released by hydrolytic cleavage of the terminal phosphate

fats and proteins. The enzyme pyruvate dehydrogenase generates two-carbon acetate fragments in the form of acetyl-CoA which then enter the tricarboxylic acid cycle at the step of citrate synthase. Each turn of the cycle uses eight enzymes (Fig. 2.17) to completely oxidize the acetate moiety of acetyl-CoA to two molecules each of CO_2 and H_2O. In addition, the tricarboxylic acid cycle reactions reduce three NAD^+ molecules into NADH (Figs. 2.10 and 2.11) and one molecule of flavin adenine dinucleotide (FAD) into $FADH_2$. These latter molecules are then used in oxidative phosphorylation in the mitochondria of mammalian and other eukaryotic cells to generate what are basically the pennies of metabolism, the energy rich compound adenosine triphosphate (ATP; Fig. 2.19).

Beginning with glycolysis and ending with mitochondrial oxidative phosphorylation, the complete oxidation of a single glucose molecule can produce a theoretical maximum of 38 molecules of ATP, but transport and ATP synthetase inefficiencies reduce this number in practice to about 30. When the energy change from the hydrolysis of an ATP molecule is measured at the standard biochemical concentration of reactants and products of 1.0 M, the following reaction:

$$ATP + H_2O \rightarrow ADP + P_i^{\,-} + H^+$$

has a $\Delta G_o \sim -37$ kJ/mol, but the cellular concentration of ATP is about 5 mM and it is complexed with Mg_2^+ so that a more typical in vivo value is $\Delta G \sim -53$ kJ/mol.

The molecules that traverse the tricarboxylic acid cycle are not only used for energy production but also serve as precursors for a number of important structural components of cells. Citrate from the mitochondrion is transported out, cleaved by the enzyme ATP citrate lyase, and used for fatty acid and cholesterol synthesis, while the amino acids aspartate and glutamate are constructed from oxaloacetate and α-keto-glutarate, respectively, through transamination. These amino acids, along with carbons and nitrogens from other sources, form the pyrimidine bases needed for DNA and RNA synthesis. We will investigate further details of carbon metabolism in Chap. 8.

2.9   The Maxwell–Boltzmann Distribution of Molecular Speeds as a Source of Energy for Catalysis

Where does the energy come from to facilitate enzymatic reactions in living systems, to subtly alter the distribution of the electrons in molecules and lower the energy barrier from reactant to product? The short answer is that it comes from the random motions of the molecules in the environment such as water, because living systems operate at a finite, nonzero temperature. The molecules in cells are in thermal equilibrium with billions of other molecules at a temperature, $T \sim 300$ K, which provides the thermal kT background noise due to random fluctuations.

We can use the Boltzmann distribution to find the average energy of a molecule. The kinetic energy of a molecule of mass, m, is given by $E = \frac{1}{2} mv^2$. The total energy of a molecule is expressed as the Hamiltonian, $H = E + V$, where V is the potential energy of interaction between molecules. For a dilute, hard sphere gas the potential can be represented as shown in Fig. 2.20. The potential is zero for intermolecular distances larger than the radius of the molecules, $\mathbf{r_o}$, and infinite for distances less than $\mathbf{r_o}$. Collisions between such molecules are elastic, that is, energy is conserved and there is no dissipation of energy in the collisions. This potential does not describe the intermolecular interactions in a biological fluid such as water, where the intermolecular potential is attractive for small intermolecular distances. A more realistic potential $V(r)$ for aqueous solutions is that of Lenard–Jones (Fig. 2.21) which we will discuss for water in Chap. 11:

$$V_{LJ}(r) = \frac{A}{r^{12}} - \frac{B}{r^6} .$$

Many biomolecules engage in hydrogen bonds. For an ...O---H... hydrogen bond, $A = 1.03 \times 10^{-7}$ kJ-nm^{12}/mol and $B = 5.11 \times 10^{-4}$ kJ-nm^6/mol. The equilibrium O---H distance is $r_o = 0.27$ nm.

A plot of this potential (Fig. 2.21) shows that at distances around r_o the potential is attractive, while for $r < r_o$ the potential is strongly repulsive like the hard sphere case above. The potential goes to zero at $r \gg r_o$. This potential is a more realistic

Fig. 2.20 Diagram of a hard sphere potential. **V** $(\mathbf{r} > \mathbf{r_o}) = 0$; $V(\mathbf{r} \leq \mathbf{r_o}) = \infty$

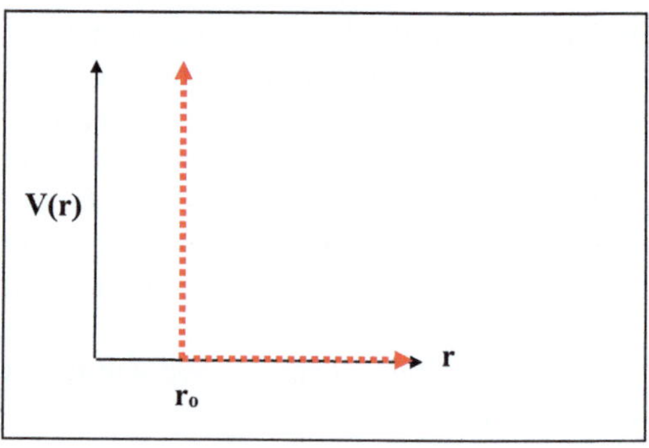

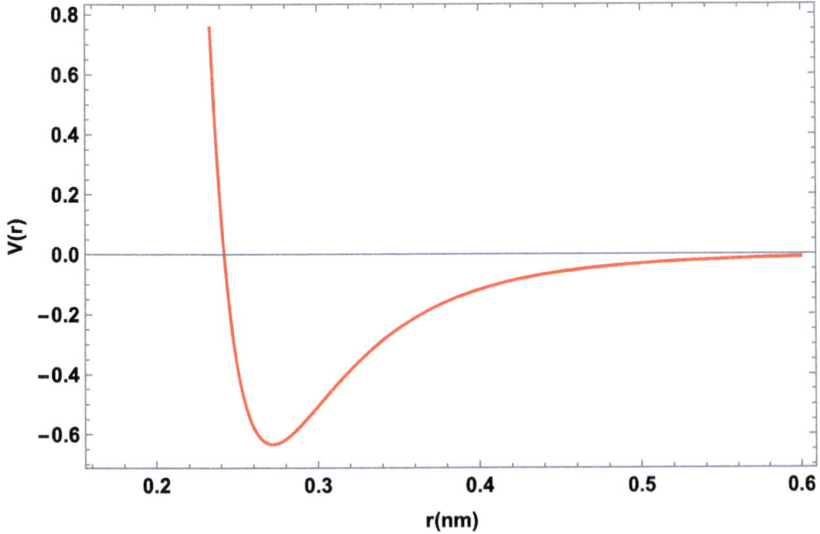

Fig. 2.21 Lenard–Jones potential for a hydrogen–oxygen H-bond

model of the water molecule. In order to develop a simple model of water, we can first examine an exactly soluble system, that of a hard-sphere ideal gas.

The Boltzmann distribution in one dimension gives the probability of the occupation of a level of a particular energy as: $P(E) = P_o e^{-E/kT}$; here, we can use the kinetic energy of a molecule

$$E = \frac{mv^2}{2}, \text{to give } P(E) = P_o e^{-mv^2/2kT},$$

which we recognize from Chap. 1 as a Gaussian function. We can find P_o from the fact that this is a probability distribution and that $P(E)$ must therefore be normalized:

$$\int_0^\infty P(E)dE = \int_0^\infty P_o e^{-E/kT} dE = 1.$$

Now, since P_o is a constant, we can take it out of the integral to obtain

$$P_o \int_{-\infty}^\infty e^{-\frac{mv^2}{2kT}} dv = 1 \Rightarrow P_o = \frac{1}{\int_{-\infty}^\infty e^{-\frac{mv^2}{2kT}} dv}.$$

The integral $\int_{-\infty}^{\infty} e^{-\frac{mv^2}{2kT}} dv$ is a Gaussian integral, which we have seen earlier. Its value is

$$\int_{-\infty}^{\infty} e^{-\frac{mv^2}{2kT}} dv = \sqrt{\frac{2\pi kT}{m}}$$

so that $P_0 = \sqrt{\frac{m}{2\pi kT}}$, and the normalized probability distribution of molecular speeds is

$$P(v) = \sqrt{\frac{m}{2\pi kT}} e^{-\frac{mv^2}{2kT}}.$$

The width (variance) of this distribution can be read from the defining equation for a Gaussian as

$$\sigma^2 = \frac{kT}{m}.$$

The full width at half maximum, Γ, of a Gaussian was found in Chap. 1 as

$$\Gamma = 2\sqrt{2\ \ln 2}\,\sigma = 2.355\,\sigma.$$

This speed distribution in one dimension (Fig. 2.22) is symmetric about $v = 0$; $P(v) = P(-v)$ and is a Gaussian with zero mean, which is easily shown from its definition: $<v> = \int vP(v)dv$. The integrand is an odd function of v so that the integral consists of equal negative and positive areas that cancel (Fig. 2.23). It is

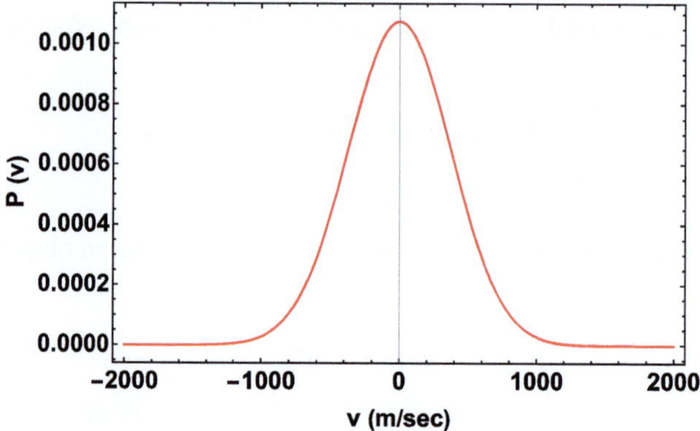

Fig. 2.22 One-dimensional distribution of speeds for an ideal, hard-sphere gas of H_2O molecules. The full width at half maximum is $\Gamma = 2.355\,\sigma = 874$ m/s

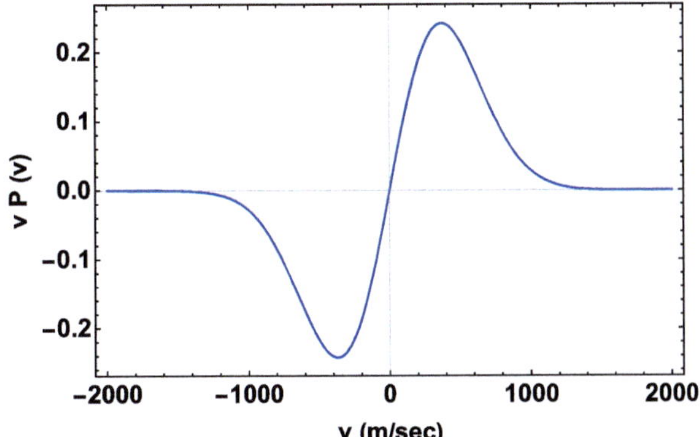

Fig. 2.23 The product, v $P(v)$, shows that it is an odd function about zero and that the average value of v would be zero

equally likely to have a positive or a negative speed. This makes sense since without a zero mean, the molecular speeds would have a nonnegative bias, which would tend to pile molecules against one side of a cell or other container and then the pressure on one side would exceed that on the other, a fact that is not observed. The motion of a molecule is a random walk known as self-diffusion.

In order to translate this result into three dimensions, we need to realize that we should be working with velocity, a vector, instead of speed, v, which is a scalar, $v = \sqrt{\vec{v} \cdot \vec{v}}$; however, the energy depends only on the average value of the speed

$$<E> = \left\langle \frac{1}{2}mv^2 \right\rangle = \frac{1}{2}m\langle v^2 \rangle,$$

since m is a constant. The velocity vector has three components:

$$\vec{v} = v_x\widehat{x} + v_y\widehat{y} + v_z\widehat{z}$$

where $\{\widehat{x}, \widehat{y}, \widehat{z}\}$ are a set of unit vectors along the x, y, and z directions. (Note that this set of vectors is a complete, orthonormal set, such that $\widehat{x}_i \cdot \widehat{x}_j = \delta_{ij}$, which is to be read $\delta_{ij} = 1$, if $i = j$, and $\delta_{ij} = 0$, if $i \neq j$.) Therefore, the square of the speed is $\vec{v} \cdot \vec{v} = v_x^2 + v_y^2 + v_z^2 = v^2$. Note that this latter equation is the equation of a sphere of radius v in $\{v_x, v_y, v_z\}$ space, reflecting the fact that there are many ways to combine the coordinates $\{v_x, v_y, v_z\}$ subject only to the constraint that the sum of their squares adds up to v^2. The number of ways to combine $\{v_x, v_y, v_z\}$ to produce v^2 is the *density of states*, proportional to the area of a sphere of radius v, and this is given by $4\pi v^2$. Then, to find the probability of a given speed in 3D, we must multiply the 1D probability by this factor, to produce:

$$P(v) = 4\pi \, v^2 \, P(v_x, v_y, v_z).$$

We know from our earlier work on probability that the simultaneous probability of obtaining speeds v_x, v_y, and v_z is given by the products of the individual probabilities

$$P(v_x, v_y, v_z) = P(v_x)\, P(v_y)\, P(v_z),$$

and, with $i \in \{x, y, z\}$,

$$P_i(v) = \sqrt{\frac{m}{2\pi kT}}\, e^{-\frac{mv_i^2}{2kT}}$$

so that

$$P(v) = 4\pi\, v^2\, P(v_x, v_y, v_z) = 4\pi v^2 \left[\frac{m}{2\pi kT}\right]^{3/2} e^{-\left(\frac{mv_x^2}{2kT} + \frac{mv_y^2}{2kT} + \frac{mv_z^2}{2kT}\right)}$$
$$= 4\pi v^2 \left[\frac{m}{2\pi kT}\right]^{3/2} e^{-\frac{m}{2kT}\left(v_x^2 + v_y^2 + v_z^2\right)}.$$

Or, since $v^2 = v_x^2 + v_y^2 + v_z^2$, we can rewrite this as

$$P(v) = 4\pi v^2 \left[\frac{m}{2\pi kT}\right]^{3/2} e^{-\left(\frac{mv^2}{2kT}\right)},$$

which is the Maxwell–Boltzmann distribution of speeds in an ideal gas. The shape of this distribution is shown in Fig. 2.24 for a water molecule at a temperature of 300 K. The most probable value for the speed of a water molecule is found by setting the derivative with respect to v to zero and solving for $v_{\max}$:

$$\frac{dP(v)}{dv} = \frac{\sqrt{\frac{2}{\pi}}}{m\left(\frac{kT}{m}\right)^{\frac{5}{2}}} v\left(2kT - mv^2\right) e^{-\frac{mv^2}{2kT}} = 0$$

$$\left(2kT - mv^2\right) e^{-\frac{mv^2}{2kT}} = 0.$$

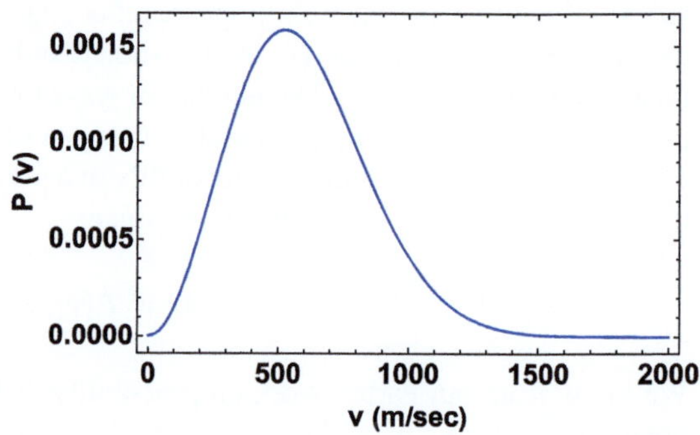

Fig. 2.24 The Maxwell–Boltzman probability distribution for the speeds of a gas of water molecules at $T = 300$ K

The solution is found from setting the term in parenthesis to zero:

$$(2kT - mv^2) = 0,$$

which gives rise to the simple relationship between the parameters of

$$v_{\max} = \sqrt{\frac{2kT}{m}},$$

and in our case the most probable speed for a water molecule in the gas phase at 300 K is 526 m/s, which is faster than a 0.22 caliber bullet. The derivative of the Maxwell–Boltzmann distribution is left as an exercise.

What is the average *energy* $<E>$ of a gas molecule? To answer this, we only need to calculate the average $<v^2>$ as indicated above, from its definition.

$$<E> = \frac{1}{2}m<v^2> = \frac{1}{2}m \int_0^\infty v^2 P(v)dv$$

$$<v^2> = \int_0^\infty 4\pi v^4 \left[\frac{m}{2\pi kT}\right]^{3/2} e^{-\left(\frac{mv^2}{2kT}\right)} dv = 4\pi \left[\frac{m}{2\pi kT}\right]^{3/2} \int_0^\infty v^4 e^{-\left(\frac{mv^2}{2kT}\right)} dv$$

This is another Gaussian integral, which can be evaluated with Mathematica to yield the important result that

$$<v^2> = 3\,k\,T/m,$$

or that the average energy is given by:

$$<E> = \left\langle \frac{1}{2}mv^2 \right\rangle = \frac{1}{2}m\langle v^2 \rangle = \frac{3}{2}kT.$$

Thus, the average energy is proportional to the temperature and the scale factor is kT. We note from the speed distribution (Fig. 2.22) that the molecules display a range of speeds. The width of the distribution gives the fluctuations in energy, and these fluctuations combine with the nonzero average energy to bombard enzymes in cells with water and other molecules.

This average energy of molecules in a gas also holds for molecules in liquids with certain modifications. Liquids are bound states of molecules in which the intermolecular potential is not one of hard spheres undergoing elastic collisions, but rather, a potential approximated by the Leonard–Jones form (Fig. 2.21) in which there is an attractive minimum. The total energy of a water molecule, for example, is then the sum of the kinetic energy $(3/2)kT$ and the potential energy, which is negative near the equilibrium intermolecular separation. Therefore, the average energy per molecule is lowered in a liquid with respect to the same molecules in a gas at the same temperature.

Many biochemical reactions involve positive changes in the Gibbs energy, and yet occur with a finite rate. The key to understanding this somewhat paradoxical behavior is to note that the Maxwell–Boltzmann distribution of energies has a long tail stretching to high temperatures so that enzymes can essentially borrow sufficient energy from these fluctuations in their environment to complete the reaction.

2.10   Enzymes Act as Probability Amplifiers for Biochemical Reactions

From this analysis of the Maxwell–Boltzmann distribution of energies, we can use transition state theory to examine the important role that enzymes play in altering the rates of biochemical reactions. Let's once again look at the progress graph (Fig. 2.1) for the approach to equilibrium for any spontaneous process, in which we see that the energy of a System tends to a minimum, E_f. The change in energy is $\Delta E = E_f - E_i < 0$. However, there exists an energy barrier of height E_b separating the lower energy state E_f of the products from the initial state of the reactants, such that the barrier energy $E_b > E_f$ or E_i, and molecules must have an energy at least as large as E_b in order to proceed over the barrier to reach the energy state E_f. The barrier height is referred to as the "transition state" and it is the rate-limiting step that must be overcome in order to reach the equilibrium product state.

We can use our knowledge of the Maxwell–Boltzmann distribution of molecular energies to show how enzymes alter the probability of biochemical reactions. The reaction progress graph (Fig. 2.1) illustrates how the energy barrier impedes the flow of molecules from the reactants to the products by only allowing those molecules with random energies greater than the barrier height to pass from the reactants to the products. If one knows the mass of the molecule and the temperature, then one can calculate the energy per molecule from $\frac{1}{2}mv^2 = 3kT/2$. The binding of the reactants to the surface of the enzyme allows the thermal fluctuations of the enzyme to couple to the reactants to add sufficient energy to lower the energy barrier and allow more molecules to flow down the energy gradient. For example, Fig. 2.25 shows the effect of the enzyme on the energies of water molecules; lowering the barrier height from 1000 to 800 m/s increases the probability that a molecule will overcome the barrier by a factor of 3. One can verify this by integrating the Maxwell–Boltzmann distribution where the integral from 1000 m/s gave a probability of 0.0652 (the purple region in Fig. 2.25), while decreasing the lower limit to 800 m/s increases the probability that a molecule will have an energy needed to overcome a 1000 m/s barrier of 0.202 (the purple plus the pink regions in Fig. 2.25). Another way of looking at this is that enzymes lower the effective temperature needed to cross the energy barrier of the transition state.

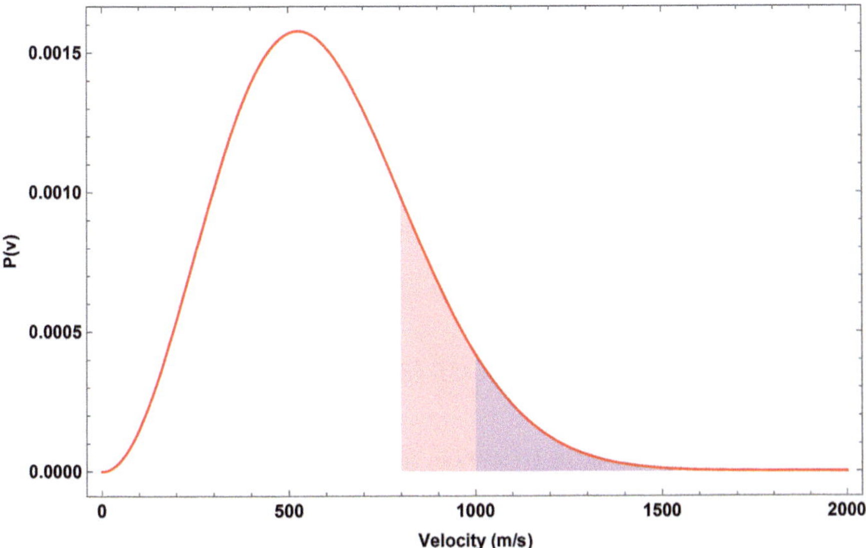

Fig. 2.25 The Maxwell–Boltzmann probability distribution of molecular speeds illustrating the change in probability brought about by the enzymatic lowering of the transition state barrier. The purple shaded part reflects the number of molecules with energy sufficient to cross over the barrier, while the pink shading shows the additional number of molecules that can cross over the barrier that has been lowered by binding to the surface of the enzyme

The width of the Maxwell–Boltzmann distribution also has a story to tell; it gives a measure of the magnitude of the fluctuations in the energies of the molecules. One can regard a plot like that of Fig. 2.25 as representing the spectrum of energy fluctuations of an object described by the Maxwell–Boltzmann distribution. The variance is the width of the spectral line in the thermal energy spectrum. Its value requires that we first calculate the mean of the Maxwell–Boltzmann distribution, say for a water molecule. This is given by the integral:

$$\langle v \rangle = \int_0^\infty v P(v) dv = \sqrt{\frac{8kT}{\pi m}}.$$

Inserting values gives us that the average speed is 594 m/s. Note that this is larger than the most probable value of the molecular speed of 526 m/s. The width of the Maxwell–Boltzmann distribution is then, from our investigation of probability distributions in Chap. 1, given by the variance

$$\sigma^2 = \langle v^2 \rangle - \langle v \rangle^2$$

$$\sigma^2 = \int_0^\infty v^2 P(v)dv - \left[\int_0^\infty vP(v)dv \right]^2,$$

$$\sigma^2 = \frac{kT}{m}\left(3 - \frac{16}{\pi}\right) - \frac{8kT}{\pi\, m},$$

giving the width as the square root of the variance as

$$\sigma = \sqrt{\frac{kT}{m}\left(\frac{3\pi - 8}{\pi}\right)}$$

and for our water molecule at $T = 300$ K, the fluctuation amplitude is ± 250 m/s.

2.11   Life at a Few Times kT: The Energy-to-Noise Ratio

Let's revisit Table 2.2 where we have listed some of the important energies that are involved in the processes of Life. The heading for the last column is the energy-to-noise ratio; what is this "noise"? Noise is thermal noise, the spectrum of energy fluctuations against which Life processes must achieve sufficient energy to, from a probabilistic perspective, form stable bound states, stable, that is with respect to the mean *and the variance* of the thermal background. One can use as a measure of the probability that a process will prevail against the random fluctuations of the environment the ratio of the energy to kT in the Boltzmann probability distribution. Or, one can use Shannon information theory to find the information excess of a process compared to that of a purely random thermal process. Life clearly succeeds in solving these puzzles. The subtlety of Life is captured in some quite interesting numbers. The hydrolysis of ATP in the cell produces about 53 kJ/mol. The thermal background at 310 K has an energy of ~2 kJ/mol, with fluctuations of ± 0.5 kJ/mol. The Gibbs energy released during ATP hydrolysis constitutes *pennies* in the currency of metabolism. These *pennies* are the smallest, but most important, amounts of energy used by living systems and their energy is only ~22-fold larger than that from random thermal background of the environment. Life operates at only a few times kT and yet it is stable. From an information theory perspective, Life operates with a basic information amount of $\log_2(22) \sim 5$ bits!

Problems

1. **Energy biases probability.** If the self-energy of a sodium ion ($r_s = 0.5$ nm) is given by $E_s = q^2/2\varepsilon r_s$, what is the relative probability of finding this ion in the center of a protein where the dielectric constant, $\varepsilon_p = 3.5$, instead of in the bulk water surrounding the protein, where $\varepsilon_w = 80$?
   Note that the units of the dielectric constants ε are Coulomb^2/J m and dielectric constants are referred to that of the vacuum given by $\varepsilon_o = 1.112 \times 10^{-10}$ C^2/J m. The charge on the ion is that of a single electron.

2. **Maximum entropy.** For N particles distributed over b states such that the entropy is a maximum given by

$$S_{\max} = k \ln W_o$$

   show that $S_{\max} = Nk \ln[b]$.

3. **Energy biases probability.** If we assume that the Earth's atmosphere is isothermal, with a temperature T, use Boltzmann's distribution to derive the variation of the atmospheric pressure $P(z)$ in terms of the height, z, above sea level, and the sea level pressure, P_o. Express your results in terms of the "molecular mass," M_o, of air, i.e., the number average of the masses of oxygen and nitrogen, and the gas constant, R.

4. **Entropy changes.** Find the entropy change,

$$\Delta S = k \ln W_2 - k \ln W_1,$$

   for the dialysis of a sucrose solution of N molecules of concentration C_1 out of a bag of volume V_1 into a volume $V_2 = V_1 + V_w$ with final concentration $C_2 = N/(N_o V_2)$. V_w is the volume of water into which the sucrose is dialyzing. Solve this by dividing up the volumes into n cells of volume V_o and then finding the number of ways of placing N molecules into n cells. N_o is Avogadro's number.

5. **Entropy changes.** Find ΔS for problem 4 if you start with 1.0 gram of sucrose in $V_1 = 100$ mL and that $V_2 = 1000$ mL.

6. **Energy biases the probability of nuclear spin states.** The difference in energy levels for the ground state and first excited state of different forms of spectroscopy are:

   NMR spectroscopy: $\Delta E = 1.19 \times 10^{-2}$ J/mol
   Rotational spectroscopy: $\Delta E = 11.9$ J/mol
   Electronic spectroscopy: $\Delta E = 119 \times 10^3$ J/mol

   Calculate the population ratio of the two states (in number of molecules) for each of the techniques:

   (a) At 298 K
   (b) At 4 K

(c) Plot the population ratio for electronic spectroscopy from 4 to 298 K

(d) Plot the population ratio for NMR spectroscopy from 4 to 298 K

(e) Which of the techniques is more sensitive and why?

7. **Protein states.** For ribonuclease, the enthalpy change on denaturation is $\Delta H = 15$ kJ/mol, and the denaturation temperature, $T_d = 335$ K (for all questions, assume ΔH does not change):

(a) What is the entropy change on denaturation, ΔS, in J/mol•K?

(b) Plot ΔS from $T = 0$ K to $T = 350$ K

(c) Above what temperature does the entropy become negative?

(d) Plot ΔG as ΔS decreases, assume $\Delta H = 15$ kJ/mol

(e) What happens to ΔG when ΔS is negative?

(f) Find ΔG when $\Delta S = -0.5$, 0, and 0.5 kJ/mol•K, still assume $\Delta H = 15$ kJ/mol

(g) How many accessible states does the ribonuclease protein have when $\Delta S = 0$?

8. **Thermal stability of proteins.** The temperature on primordial Earth was significantly hotter than it is now, and as a result required thermophilic bacteria that operated at high temperatures. One such bacterium, thought to have existed during this time, has been isolated from a deep sea hydrothermal vent and found to have an RNA polymerase of 342 amino acids long and a ΔG of -8230 J/mol.

(a) Using the very approximate equation for fraction of folded protein,

$$Pf = \frac{1}{1 + \partial e^{\frac{\Delta G}{RT}}}$$

plot the temperature denaturing plot assuming that the bonding coefficient (∂) for this RNA polymerase is 9.655.

(b) What is the melting temperature (T_M) for the protein?

(c) What is the entropy change upon protein unfolding?

(d) In a state of maximum entropy, $T \gg T_M$, each of the amino acids can exist in one of four different states. What is the approximate number of different configurations (W_d) the protein can exist under these conditions?

(e) How likely is it that one of these states is the correct folded protein?

9. **Helical proteins.** In some peptides there is an imaginary axis that amino acids wind around because of weak interactions between the residues.

(a) Using a `Graphics3D` Plot, simulate this axis using the `Cylinder` function from 0 to 2π example: $\{0, 0, 2\pi\}$. (Note: first, name the operation you are about to preform and then use)

```
Graphics3D[{{Opacity[0.2],Cylinder[{{0,0,0},{0,0,2π}},1]},
{Thickness[0.09],Line[{{0,0,0},{0,0,2π}}]}}]
```

Note that you can also change the axis color, make the box disappear, and so on.

(b) In a specific sequence of amino acids, the helical structure is determined to have the parameters {Cos[4t], Sin[4t], t}. Plot this using a Parametric 3D Plot from 0 to 2.

(c) Finally, show the alpha helix winding around the imaginary axis.

10. **Protein thermodynamics.** The peptide from problem **9** has 26 amino acid residues and upon denaturation of this peptide, the enthalpy change is measured to be 400,000 J/mol.

(a) If the entropy change of the molecule is 1200 J/mol*K what is the temperature at which this peptide denatures?

(b) What is ΔG when the temperature is 350 K?

(c) The protein above shows an abundance of glycine and proline residues. If it contained leucine and glutamate, would you expect to see an increase in the denaturation temperature?

11. **DNA sequence probability.** Four bases (A, C, G, and T) make up DNA. What is the probability of observing the following sequences? Assume that the appearance of each base is random.

(a) ACT

(b) AAA GGG

(c) ACT TAG CT

(d) CAT GAG TAC

12. **Molecular probability.** Consider an ideal gas of 1000 molecules in equilibrium within a container of 250 L (volume V_0). Let n be the number of molecules located in a 2.0 L sub-volume (V) of this container. The probability, p, that a given molecule is located within this sub-volume, V, is then given by $p = V/V_0$. **Hint:** You do not need to use any sums here.

(a) What is the mean number of molecules (n) located in the sub-volume V?

(b) What is the dispersion, σ^2, in the number of molecules within V?

(c) What is the ratio of fluctuations to the mean, $\sigma^2/<n>^2$?

(d) Plot the ratio in c, as $V \to V_0$

(e) What happens to the fluctuation as V approaches V_0? Why does this make sense?

13. **Proton/neutron ratio.** The probability that a particle will occupy a given state depends on the energy of that state. This probability distribution is given by the Boltzmann distribution. The ratio of the number of protons to neutrons during the first minute of the early Universe is

$$\frac{N_p}{N_n} = e^{-Q/kT}$$

where Q is the energy (mass) difference between the proton and neutron. This equilibrium favors the proton because the neutron is heavier by $Q = 2.07 \times 10^{-13}$ J.

(a) What is the proton/neutron ratio for $T = 10^{13}$ K?
(b) The proton/neutron ratio "freezes out" at a value of 1/6. What is the temperature of the Universe when this happens.

14. **The Maxwell–Boltzmann distribution.** The Maxwell–Boltzmann equation, which is based on the "kinetic theory of gases," gives the distribution of speeds for a gas at a certain temperature.

(a) Using the Maxwell–Boltzmann function, calculate the fraction of Argon gas molecules with a speed of 460 m/s at 640 K.
(b) Plot this function for v from 0 to 2000 m/s.
(c) If the system in part a has 0.34 moles of Argon gas, how many molecules have the speed of 460 m/s?
(d) What is the average energy of each of these molecules?

15. **Photosynthesis.** The photosynthetic reaction below makes glucose from carbon dioxide and water:

$$n \text{ Photons} + 6\, CO_2\ (g) + 6\, H_2O\ (l) < - - > C_6\, H_{12}\, O_6\ (s) + 6\, O_2\ (g)$$

This reaction requires the input of energy from light (photons) of a certain wavelength to proceed from the reactants to the products.

(a) Given the following standard enthalpies of formation, calculate the overall enthalpy change of this reaction in kcal:

$\Delta H°(CO_2\ (G)) = -393.5$ kJ/mol;
$\Delta H°(H_2O\ (l)) = -285.8$ kJ/mol;
$\Delta H°(C_6H_{12}O_6\ (s)) = -1274.5$ kJ/mol;
$\Delta H°(O_2\ (g)) = 0$ kJ/mol.

(b) How much energy is given off by red light? ($h = 6.626 \times 10^{-34}$ J s, $c = 3 \times 10^8$ m/s, and $\lambda = 700 \times 10^{-9}$ m)?
(c) How many photons of red light do we need to provide enough energy for the reaction to proceed?
(d) What if we use blue light ($\lambda = 400 \times 10^{-9}$ m)

16. **Glycogen phosphorylases.** Phosphorylases are a class of enzymes that catalyze the breakdown of glycogen (a polymer of glucose) in liver and muscle. Similar enzymes, occurring in plants, catalyze the hydrolysis of starch. The phosphorylase of skeletal muscle occurs in two forms: the active form, i.e., phosphorylase-a (PA) and a much less active form, i.e., phosphorylase-b (PB). These two forms are interconvertible. The enzyme PB requires the addition of AMP for activation:

$$PB + AMP \rightarrow PA : AMP$$

The K_{eq} for this reaction (the reverse of the reaction above) is of follows:

Temperature (K)	280	290	300	310	320
K_{eq} (10^5 M)	2.2	3.6	4.3	5.0	6.1

(a) Using Mathematica lists, plot ln K_{eq} vs. $1/T$. Plot the fitted function with the data and include axes labels.

(b) Use the Mathematica `Fit` function to obtain the linear regression line that fits the data in part a. This is a Van't Hoff plot, the slope of which is $-\Delta H/R$.

(c) Find the value of ΔH.

(d) What is the correlation coefficient (r^2) for this fit?

(e) Calculate ΔG (in J/mol) and ΔS (in J/mol*K) for all temperatures, assuming ΔH and ΔS are independent of temperature.

(f) Using lists that you have now made, show that $\ln[K] = \frac{-\Delta G}{RT}$, for ALL of the temperatures.

17. **Protein thermodynamics.** The enzyme PB from problem 16 contains 840 amino acid residues. When heated in a differential scanning calorimeter, PB is found to have an Enthalpy change, ΔH, of 450 kcal/mol and an Entropy change, ΔS, of 5000 J/mol•K, at denaturation.

(a) What is the enthalpy change in kJoule/mole?

(b) What is the denaturation temperature in Kelvin?

(c) Find ΔG at 0, 150, 310, 400, and 500 K (in kJoule/mol).

(d) Plot ΔG from 0 to 400 K. At what temperature (approximately) does the forward reaction occur spontaneously?

(e) What makes this reaction spontaneous at body temperature (310 K) and how might this work? Use the thermodynamic properties as the basis for your explanation.

18. **Protein thermodynamics.** Assume that phosphorylase from problem 16 exists in two folded states, PA and PB.

(a) Using ΔS on denaturation, estimate the number of unfolded states, W_d.

(b) Assuming each of the 840 amino acid residues in the denatured protein can access three states, and that the protein exists in a state of maximum entropy for $T > T_d$, then what is an estimate for the number of accessible states of the protein, W_d, for $T > T_d$?

(c) How does the answer in (**b**) compare with the results from calorimetry found in (**a**), that is, what are the actual number of accessible states from calorimetry?

(d) Since it is stated that PB is less active than PA, which would you expect to have a higher denaturation temperature? Which would you expect to have a lower ΔG at 310 K?

References

S.D. Lahiri, P.F. Wang, P.C. Babbitt, M.J. McLeish, G.L. Kenyon, K.N. Allen, The 2.1 A structure of Torpedo californica creatine kinase complexed with the ADP-Mg(2+)-NO(3)(−)-creatine transition-state analogue complex. Biochemistry **41**, 13861–7 (2002)

Bibliography

Y. Demirel, V. Vincent Gerbaud, *Nonequilibrium Thermodynamics: Transport and Rate Processes in Physical, Chemical and Biological Systems*, 4th edn. (Elsevier, Amsterdam, 2019)

E. Di Cera, *Thermodynamic Theory of Site-Specific Binding Processes in Biological Macromolecules* (Cambridge University Press, 1995)

D.T. Haynie, *Biological Thermodynamics*, 2nd edn. (Cambridge University Press, 2008)

G.G. Hammes, *Thermodynamics and Kinetics for the Biological Sciences* (Wiley-Interscience, New York, 2000)

M. Kurzynski, *The Thermodynamic Machinery of Life (The Frontiers Collection)* (Springer, 2006)

K. Michaelian, *Thermodynamic Dissipation Theory of the Origin and Evolution of Life: Salient Characteristics of RNA, DNA and Other Fundamental Molecules Suggest an Origin of Life Driven by UV-C Light* (CreateSpace, Mexico City, 2017)

P. Nelson, *Biological Physics Student Edition: Energy, Information, Life* (Chiliagon Science, Philadelphia, 2020)

Chapter 3
Space-Time Symmetry and Conservation Laws as Organizing Principles of Matter and Fields

"Symmetry, as wide or as narrow as you may define its meaning, is one idea by which man through the ages has tried to comprehend and create order, beauty and perfection."
Hermann Weyl (2015),
"Symmetry", Princeton University Press

In the first two chapters we developed the probability distributions for generic objects without specifying in detail just what those objects might be. Göthe's *Faust* sought to know what held the innermost parts of the world together; when closely examined, this question involves two separate concepts, that of the elementary constituents themselves and that of the forces binding them together. Perhaps the most profound insight into both of these realms was discovered in the twentieth century when the invariances of space-time were fully understood, for it was then found that the only objects allowed in space-time are the irreducible representations of the double cover of the Poincaré group. The eigenvalues of its two Casimir operators are mass and spin, with mass ≥ 0 and spin given by the sequence $j = \{0, 1/2, 1, 3/2, 2, \dots\}$ in units of $\hbar$. The integral spin irreducible representations are the *Bosonic* fields that bind the *Fermionic*, half-integral spin irreducible representations that are the matter particles.

It is quite impressive that the simple properties of the universe's space-time constrain its contents. By Occam's razor, the simplest hypotheses would appear to be the most applicable. Then, what is space-time; what are its invariances? Space-time is what we live in: a four-dimensional coordinate system in which we make measurements. The invariances of space-time are derived from how one can change coordinate systems and yet preserve the results of these measurements, leading naturally to the concept of symmetry. A symmetry is a transformation of the coordinate system (or, equivalently, of the object within the coordinate system) under which the object is the same before or after the transformation. A common way of describing this is to view an object, close your eyes, have someone perform a transformation on the object, open your eyes and if you cannot tell that any transformation occurred, then that was a symmetry transformation.

L. O. Sillerud, *Abiogenesis*, https://doi.org/10.1007/978-3-031-56687-5_3

3.1   Definition of a Symmetry

What do we mean by a symmetry? It is a property of the universe that is independent of your point of view. That is, a property that remains the same, or is invariant, as your point of view changes. Here, we define a point of view with respect to your *local inertial coordinate system*. What are ways that your point of view could change? You could measure the height of the page in this book today and do the same tomorrow with the firm expectation that you would get the same value. You could place the book on a table and rotate it from East to North and expect that the height would not change. You could take the book to Timbuktu and expect that the height would remain the same. These are examples of the extremely simple, yet also fundamentally important, transformations that leave a measured quantity invariant. One could also measure the book's height while in an airplane traveling on Mach 10, high in the sky, with the same expectation of invariance and you would have discovered the basis for the first four of the axioms about the universe mentioned at the beginning of this chapter; *Temporality*, *Isotropy*, *Homogeneity*, and *Physicality*.

OK, so much for symmetry; what then is space-time? In the pre-Einsteinian universe, space-time only consisted of space, with time as an external parameter quite distinct from space. This space was three-dimensional, ordinary Euclidean space described by a Cartesian coordinate system. In 1905, Einstein pointed out that this separation disturbed Maxwell's equations of electrodynamics and required that information could only be transmitted at a finite speed given by that of light. By incorporating time as a coordinate equal to space in a four-dimensional space-time manifold, the full structure of Minkowskian, 4D space-time was revealed.

Both Euclidean 3D space and 4D space-time are homogeneous, isotropic, and temporal, but 4D space-time must also obey the principle of relativity, which I call physicality. These simple properties serve to give rise to fundamental organizing principles at the foundation of Life. The symmetry transformations that take space-time into itself give us the bedrock conservation laws that govern the universe and constrain the properties of the objects that exist within it.

3.2   Symmetries of the Universe

We shall begin our study of the symmetries of the universe with an examination of the classical symmetries and their basis as organizing principles. Our goal is to show how the properties of space-time determine the types of objects from which the universe is constructed and only allow the existence of objects with those properties. We will derive the types of objects and their physical properties from the following set of axioms ("grand organizing principles") regarding space-time:

1. *Isotropy.* Space-time looks the same from all directions.
2. *Temporality.* Space-time looks the same yesterday, today, tomorrow, and ever onward.

3. *Homogeneity.* Every point in space-time is equivalent.
4. *Physicality.* There is a universal physical limit to the speed of information transfer between separated space-time points (i.e., special relativity).
5. *Measurability.* Knowledge of the contents of a volume in space-time can only be obtained through measurements conducted on that volume. Objects must be detectable. If you cannot measure it, it does not exist!
6. *Uncertainty.* There is a universal physical limit to the amount and type of knowledge we can obtain from a volume of space-time. A corollary is that space-time itself must be quantized; there must exist a smallest unit of space-time.
7. *Unitarity.* In Chap. 1, we examined the important properties of probability and later used these in our definition of entropy. For these to be invariant under changes in our point of view, we require that the probability of an event must be independent of our point of view. This is true if the transformations from one point of view to another are accomplished by probability-preserving or *unitary* transformations.
8. *Permuttivity.* At the level of elementary particles, the Universe should be invariant when two identical, indistinguishable objects are interchanged. This permutation symmetry is explicitly violated by the irreducible representations of the Poincaré group, which divide the objects in the Universe on the basis of an internal quantum number known as spin into Fermions, with half-integral spin, and Bosons with integral spin. The resulting spin-statistics theorem constitutes an essential grand organizing principle at the foundation of Life.
9. *Neutrality.* The Universe is electrically neutral. Charge is conserved due to the gauge invariance of the electromagnetic field.
10. *Antiparticality.* Although the Universe appears to be composed solely of matter, antimatter obeys the same physical laws as matter. Electrons and positrons only differ in their charge.

There is perhaps no greater insight into the properties of nature than that revealed by Emmy Noether in her 1918 classic paper (Noether 1918) on the study of the invariances of the physical laws under coordinate transformations. Now, mind you, the principles of nature have existed since the time of Decartes as mathematical objects that existed within an abstract, but very real and palpable, 3D space in which all physical processes take place. We have all seen this Cartesian space during our initial exposure to analytic geometry in our first calculus class. The space in which we live was first thought to be described by three mutually orthogonal axes: a basis for this vector space consisted of the three unit vectors $\{e_i\}$, $i \in \{1, 2, 3\}$ such that $e_i \cdot e_j = \delta_{ij}$ and with $|e_i| = 1$, $\forall\, i \in \{1, 2, 3\}$. These three unit vectors spanned the 3D linear vector space and formed a complete orthonormal basis set. The distance, s, between the origin and points in this 3D space was given by the Pythagorean theorem:

$$s^2 = x^2 + y^2 + z^2$$

Time, in this Cartesian view, was a distinct parameter that merely translated our 3D world from one instant to the next. In keeping with this worldview, time was measured in different units from space, but that will be taken up later in this chapter.

3.3 Classical Symmetries

The first conceptual breakthrough came when it was experimentally determined in the middle of the nineteenth century that several important physical quantities, such as mass, energy, momentum, and angular momentum, were constants of motion. These ideas formed the basis for statements, such as the conservation of energy, that were soon elevated to the status of laws of nature, but their origin was unknown. Then, Noether showed that one could derive these conservation laws from symmetry principles that described how the universe behaved when the underlying space-time was altered. For example, the simple fact that one should be able to do an experiment anytime and obtain the same result was shown to give rise to the conservation of energy. That is, the universe was invariant under time translation. In similar fashion, the invariance of space-time under coordinate translations directly led to the conservation of momentum, while the invariance of the world under rotations produced the conservation of angular momentum. Noether exposed the deep relationship between the structure of space-time and the properties of the objects within it. We will discover that the symmetries of 4D space-time only allow certain types of objects to exist, and of course, those objects that do exist have the special characteristics needed to form the basis for living systems by self-assembly.

3.4 Noether's Theorem: Continuous Symmetries of Space-Time

We are interested here in the symmetries of space-time because they serve as grand organizing principles by logically and importantly restricting the properties of matter and energy. In the seventeenth century, it was implicitly understood from *Newton's* laws of motion that energy and momentum were conserved. If we write these in terms of a potential energy function $V(x)$ then

$$\frac{dP}{dt} = F = -\frac{\partial V(x)}{\partial x}$$

In a potential-free region, there is no force acting on a particle and the time derivative of momentum is zero, which implies that momentum is a constant and conserved, i.e., invariant in time. One can use similar methods to show that energy is conserved, but it remains a puzzle that the conservation of energy was not experimentally verified until almost two centuries later, by Mayer, Joule, and Helmoltz in 1848, and it remained for Emmy Noether, in 1918, to show how continuous symmetries give rise to general conservation laws.

3.4.1   The Lagrangian and the Hamiltonian

In order to understand how this is possible, let us step back in time and explore the reformulation of Newton's laws by Lagrange and Hamilton in the late nineteenth century. We have already encountered Hamilton's expression for the total energy of a system as $H = T + V$ in Chap. 2. Through Hamilton's equations in 1D

$$\frac{dP}{dt} = -\frac{\partial H}{\partial x}$$

and

$$\frac{dx}{dt} = \frac{\partial H}{\partial P}$$

using the Hamiltonian for a particle of mass m in a potential $V(X)$

$$H = \frac{1}{2}mv^2 + V(x)$$

we can recover Newton's laws as

$$\frac{dP}{dt} = -\frac{\partial V(x)}{\partial x}$$

and

$$\frac{dx}{dt} = v.$$

Hamilton himself derived the function that bears his name as the Legendre-transform of an earlier function of the kinetic and potential energies due to *Lagrange* and named the Lagrangian, $L = T - V$, where L is the "free energy" of a system, much in the same way that in Thermodynamics we use the Gibb's "free energy" $G = H - TS$.

3.4.2 The Euler–Lagrange Equations of Motion

Just as there are Hamilton's equations that can be used to recover the dynamics of a system from the Hamiltonian, there are the Euler–Lagrange equations that do the same for the Lagrangian:

$$\frac{\partial L}{\partial x} - \frac{d}{dt}\left(\frac{\partial L}{\partial v}\right) = 0.$$

Here we have written the Euler–Lagrange equations again in only one spatial and one temporal dimension, but both Hamilton's and the Euler–Lagrange equations are equally valid in what is known as generalized coordinates, which are commonly given by (where the dot denotes the time derivative)

$$p_k = \frac{\partial L}{\partial \dot{q}_k}; \quad \dot{q}_k = \frac{dq_k}{dt} \quad \text{(Lagrange's equations)}$$

and

$$\dot{p}_k = -\frac{\partial H}{\partial q_k}; \quad \dot{q}_k = \frac{\partial H}{\partial p_k} \quad \text{(Hamilton's equations)}$$

as the generalized momenta, p_k, and the generalized positions, q_k. The Lagrangian is a function of these generalized coordinates and time, $L = L(q, \dot{q}, t)$. The action, S, is defined to be the time integral of the Lagrangian,

$$S = \int_{t_1}^{t_2} L(q, \dot{q}, t)dt$$

and Hamilton's principle states that the physical path taken by a system is the one for which infinitesimal variations in the path, such that the variation vanishes at the endpoints $\{t_1, t_2\}$, produce no change in the action and this gives rise to the Euler–Lagrange equations. The *action* is a type of mathematical structure called a functional, which is essentially a machine with an input slot (the Lagrangian) that performs an operation (time integral) on the input and produces an output (the action), which is a scalar (number). Other functionals arise in the theory of molecules embodied in quantum density functional theory. For ease of understanding in what follows, I will revert to the simple $\{x, t\}$ coordinates so that the structure of the equations is evident, with the knowledge that we could generalize the derivation to any number of coordinates if we wished. Any text on classical mechanics will provide the full treatment (e.g., Goldstein et al. 2002).

3.4.3   Noether's Theorem

Noether's Theorem states that any continuous symmetry of the action gives rise to a conservation law. Note that here conservation means *time-invariant*. If the Lagrangian is invariant under a symmetry transformation, then the action certainly is. This is one of the most beautiful relationships between the structure of the universe and the properties of the objects within it that I know of. Let's begin the proof of Noether's Theorem by considering a 1D Lagrangian $L(x, v, t)$ and transform this to new spatial coordinates by $x' = x + \delta x$, where δ is an infinitesimal variation that vanishes at the endpoints: $\delta x(t_1) = \delta x(t_2) = 0$. The invariance of the Lagrangian means that we seek to show that the Lagrangian remains the same after this transformation. We can write the variation in the Lagrangian as

$$\delta L = L\left(x, \frac{dx}{dt}, t\right) - L\left(x + \delta, \frac{d(x + \delta x)}{dt}, t\right)$$

The new velocity is given by

$$v' = \frac{d(x + \delta)}{dt} = \frac{dx}{dt} + \frac{d\delta x}{dt} = v + \delta v$$

so we suggest that $\delta L = 0$, or

$$\delta L = L\left(x, \frac{dx}{dt}, t\right) - L\left(x + \delta, \frac{dx}{dt} + \frac{d\delta x}{dt}, t\right) = 0,$$

but the requirement that the Lagrangian itself is invariant under this transformation is often too restrictive because it is the action that determines the dynamics, so we really only need the variation in the action to vanish.

$$\delta S = \int L\left(x, \frac{dx}{dt}, t\right) dt - \int L\left(x + \delta, \frac{dx}{dt} + \frac{d\delta x}{dt}, t\right) dt = \int \delta L \, dt = 0$$

and this will be true if $\delta L = 0$. The action can remain invariant even if the Lagrangian changes because we can always add the total time derivative of any function $F(x)$ to the Lagrangian without changing the action.

$$L' = L + \frac{dF(x)}{dt}$$

To see this, let's compute the variation in the new action $\delta S'$ as

$$\delta S' = \delta S + \int_{t_1}^{t_2} \frac{d}{dt} \delta F \, dt$$

Now, the time integral vanishes because $\delta F = \frac{dF}{dx} \delta x$ and

$$\int_{t_1}^{t_2} \frac{d}{dt} \delta F \, dt = \int_{t_1}^{t_2} \frac{d}{dt} \frac{dF}{dx} \delta x \, dt = \frac{dF}{dx} \delta x \Big]_{t_1}^{t_2} = \frac{dF}{dx} [\delta x(t_2) - \delta x(t_1)] = 0$$

which is zero because the variation vanishes at the endpoints. Therefore, we do not need to require that $\delta L = 0$. We have instead the relaxed condition that

$$\delta L = L' - L = \frac{dF}{dt}$$

the change in the Lagrangian does not change the action as long as we can write the change in the Lagrangian as the total time derivative of some function F. We can then write the infinitesimal variation in the Lagrangian as

$$\delta L = L' - L = L\left(x + \delta x, \frac{dx}{dt} + \frac{d\delta x}{dt}, t\right) - L\left(x, \frac{dx}{dt}, t\right) = \frac{dF}{dt}.$$

In order to explicitly calculate the variation, we can expand the first term in parentheses using a Taylor's series and keep only the lowest order terms

$$L' = L\left(x + \delta x, v + \delta v, t\right) = L(x, v, t) + \frac{\partial L}{\partial x} \delta x + \frac{\partial L}{\partial v} \delta v = \frac{dF}{dt}$$

where $\frac{d\delta x}{dt} = \delta \frac{dx}{dt} = \delta v$. Then the variation in the Lagrangian is

$$\delta L = L' - L = L(x, v, t) + \frac{\partial L}{\partial x} \delta x + \frac{\partial L}{\partial v} \delta v - L(x, v, t) = \frac{dF}{dt}$$

$$\delta L = \frac{\partial L}{\partial x} \delta x + \frac{\partial L}{\partial v} \delta v = \frac{dF}{dt}$$

Up to this point we have only used the fact that the action is invariant under the transformation $x' = x + \delta x$, $v' = v + \delta v$. If we also consider the fact that the action should be an extremum, we can use the Euler–Lagrange equations of motion

$$\frac{\partial L}{\partial x} = \frac{d}{dt}\left(\frac{\partial L}{\partial v}\right)$$

to rewrite the variation as

$$\delta L = \frac{d}{dt}\left(\frac{\partial L}{\partial v}\right)\delta x + \frac{\partial L}{\partial v}\delta v = \frac{dF}{dt}$$

or, since $\delta v = \frac{d\delta x}{dt} = \delta\frac{dx}{dt}$ this can be interpreted as the time derivative of the product $\frac{d}{dt}\left(\frac{\partial L}{\partial v}\delta x\right)$

$$\delta L = \frac{d}{dt}\left(\frac{\partial L}{\partial v}\right)\delta x + \frac{\partial L}{\partial v}\frac{d\delta x}{dt} = \frac{dF}{dt}$$

$$\delta L = \frac{d}{dt}\left(\frac{\partial L}{\partial v}\delta x\right) = \frac{dF}{dt}$$

or

$$\delta L = \frac{dF}{dt}$$

which leads directly to this conservation law

$$\frac{d}{dt}\left(\frac{\partial L}{\partial v}\delta x - F\right) = 0$$

that is one form of Noether's Theorem. The quantity in the parentheses is conserved in time and is often referred to as a Noether charge G

$$G = \frac{\partial L}{\partial v}\delta x - F$$

whose derivative is zero, $\frac{dG}{dt} = 0$, so that G is a constant with respect to time.

3.5   Momentum Conservation Results from the Homogeneity of Space

In order to apply Noether's Theorem to concrete examples, let's look at the Lagrangian for a free particle of mass m

$$L = T - V = T = \frac{1}{2}mv^2$$

and note that this does not depend on the position in space so that according to the Euler–Lagrange equations, the spatial derivative vanishes

$$\frac{\partial L}{\partial x} = \frac{d}{dt}\left(\frac{\partial L}{\partial v}\right) = 0$$

while $\frac{\partial L}{\partial v} = mv = P$ and we have that

$$\frac{dmv}{dt} = \frac{dP}{dt} = 0$$

and momentum is conserved (i.e., independent of time).

3.6 Temporality Gives Rise to Energy Conservation

That space-time looks the same yesterday, today, tomorrow, and ever onward gives rise to the conservation of energy. The results of an experiment should not depend on when it was performed. Therefore, time translation must be a symmetry of space-time. We can examine the variation of the Lagrangian under infinitesimal time translations, where $t' = t = \delta t$. As above, we can compute δL using a Taylor's expansion for L'

$$L' = L\left(x(t+\delta t), \frac{dx(t+\delta t)}{dt}, t+\delta t\right) = L(x, v, t) + \frac{\partial L}{\partial x}\frac{\partial x}{\partial t}\delta t + \frac{\partial L}{\partial v}\frac{\partial v}{\partial t}\delta t + \frac{\partial L}{\partial t}\delta t$$

and then $\delta L = \frac{\partial L}{\partial x}\frac{\partial x}{\partial t}\delta t + \frac{\partial L}{\partial v}\frac{\partial v}{\partial t}\delta t + \frac{\partial L}{\partial t}\delta t$ and we can factor out the total derivative of $L = L(x, v, t)$ with the result that the variation in L is then

$$\delta L = \left(\frac{\partial L}{\partial x}\frac{\partial x}{\partial t} + \frac{\partial L}{\partial v}\frac{\partial v}{\partial t} + \frac{\partial L}{\partial t}\right)\delta t = \frac{dL}{dt}\delta t$$

Now this means that the variation in L does not always vanish, but we can use $\delta L = \frac{dF}{dt}$ with $F = L$, so that $\delta L = \frac{dL}{dt}$ and the equation for the Noether charge is then

$$G = \frac{\partial L}{\partial v}\delta x - L$$

The conservation equation for the Noether charge then reads

$$\frac{d}{dt}\left(\frac{\partial L}{\partial v}\delta x - L\right) = 0$$

Let's compute this time derivative explicitly

$$\frac{dG}{dt} = \frac{d}{dt}\left(\frac{\partial L}{\partial v}\delta x - L\right) = 0$$

$$\frac{dG}{dt} = \frac{d}{dt}\left(\frac{\partial L}{\partial v}\delta x\right) - \frac{dL}{dt} = 0$$

and expanding the derivative of the product, we have that

$$A = \frac{d}{dt}\left(\frac{\partial L}{\partial v}\delta x\right) = \left(\frac{d}{dt}\frac{\partial L}{\partial v}\right)\delta x + \frac{\partial L}{\partial v}\frac{d\delta x}{dt}$$

now $\frac{d\delta x}{dt} = \delta v$ and by the Euler–Lagrange equations $\frac{d}{dt}\left(\frac{\partial L}{\partial v}\right) = \frac{\partial L}{\partial x}$ so that

$$A = \frac{\partial L}{\partial x}\delta x + \frac{\partial L}{\partial v}\,\delta v = \delta L = \frac{dL}{dt}$$

and we can expand the total time derivative of L as

$$B = \frac{dL}{dt} = \frac{\partial L}{\partial x}\frac{dx}{dt} + \frac{\partial L}{\partial v}\frac{dv}{dt} + \frac{\partial L}{\partial t}$$

$$B = \frac{d}{dt}\left(\frac{\partial L}{\partial v}v\right) + \frac{\partial L}{\partial t}$$

and

$$\frac{dG}{dt} = A - B = 0 = \frac{dL}{dt} - \frac{\partial L}{\partial t} - \frac{d}{dt}\left(\frac{\partial L}{\partial v}v\right)$$

which gives the important result that the Lagrangian is conserved if

$$-\frac{\partial L}{\partial t} = \frac{d}{dt}\left(\frac{\partial L}{\partial v}v - L\right) = 0$$

where the quantity in parentheses is the total energy of the system, or the Hamiltonian, H, which can be defined as

$$H = \frac{\partial L}{\partial v}v - L$$

and we have that

$$-\frac{\partial L}{\partial t} = \frac{dH}{dt}$$

energy is conserved if either the time derivative of the Hamiltonian $\frac{dH}{dt} = 0$ or if $\frac{\partial L}{\partial t} = 0$.

3.7 The Isotropy of Space-Time Leads to the Conservation of Angular Momentum

The axiom that space-time should look the same from all directions leads to the conservation of Angular momentum. We have been working in the simplest space-time up to now in that we only needed to consider one-dimensional problems in either position or time. As is often the case, one-dimensional problems are easiest to solve but there are features of the actual universe that do not make sense in only one dimension; an important example of this understanding is that of rotations. Rotations in one dimension are generated by the parity operator, π

$$\pi(x) = -x; \pi^2(x) = \pi(\pi(x)) = \pi(x)\pi(x) = (-x)(-x) = x$$

Rotations are a class of transformations that preserve the length of a vector, in this case the length of the angular momentum vector. Let $\vec{x} = x\vec{e}_1$, where $\vec{e}_1$ is an orthonormal unit vector. The rotation operator in one dimension preserves the length of $\vec{x}$. We can see this by finding the length of $\vec{x}$ before and after a 1D rotation

$$\text{Before: } \left|\vec{x}\right| = x\vec{e}_1 \cdot x\vec{e}_1 = x^2$$

$$\text{After: } \left|\pi\left(\vec{x}\right)\right| = -x\vec{e}_1 \cdot \left(-x\vec{e}_1\right) = x^2.$$

Rotations in more than one dimension mix coordinates and this is not possible when you only have a single coordinate. The universe is not one dimensional. Therefore, it is necessary to consider rotations in at least three spatial dimensions; later we will need to understand more about 4D rotations in our modest study of relativity.

Vectors in 3D need three coordinates, scalars that multiply the basis vectors which we can label with an index; the vector $\vec{x} = x_1\vec{e}_1 + x_2\vec{e}_2 + x_3\vec{e}_3$ has components x_i, $i \in \{1, 2, 3\}$. The completely antisymmetric, rank 3, *Levi-Civita* tensor ϵ_{ijk} allows the cross product of two vectors in three-dimensional Euclidean space, to be expressed in index notation. In what follows, we use the Einstein summation convention where there is an implied sum when any two indices are repeated. A rotation can then be written as

$$x'_i = x_i + \sum_{j=1}^{3} \sum_{k=1}^{3} \epsilon_{ijk} x_j a_k$$

or, more simply as $x'_i = x_i + \epsilon_{ijk} x_j a_k$ using the summation convention, from which we can derive an infinitesimal coordinate variation as

$$\delta x_i = x'_i - x_i = \epsilon_{ijk} x_j a_k.$$

where a_k is an infinitesimal rotation vector parallel to the axis of the rotation. Now the Lagrangian for a free particle $L = \frac{1}{2} m\left(\frac{dx}{dt}\right)^2$ does not depend on x, $(F = 0)$ and the conserved Noether charge is

$$G = \frac{\partial L}{\partial \dot{x}_i} \delta x_i = \frac{\partial L}{\partial \dot{x}_i} \epsilon_{ijk} x_j a_k$$

We can simplify the variation δx_i even further by introducing an antisymmetric infinitesimal rotation tensor a_{ij} defined as

$$a_{ij} = \epsilon_{ijk} a_j$$

and then the variation becomes

$$\delta x_i = -a_{ij} x_i.$$

The change in the Lagrangian as a result of this variation is

$$\delta L = L\left(x'_i, \dot{x}'_i, t\right) - L(x_i, \dot{x}_i, t) = \frac{\partial L}{\partial x_i} \delta x_i + \frac{\partial L}{\partial \dot{x}_i} \delta \dot{x}_i$$

which, after factoring the δ and replacing it by $\delta x_i = -a_{ij} x_i$, gives

$$\delta L = -\left(\frac{\partial L}{\partial x_i} x_i + \frac{\partial L}{\partial \dot{x}_i} \dot{x}_i \right) a_{ij} x_i$$

Rotations are length preserving, so if the Lagrangian contains only scalar terms involving lengths, such as $x_i \cdot x_i$, $\dot{x}_i \cdot \dot{x}_i$, or $x_i \cdot \dot{x}_i$, then the right side will vanish due to the antisymmetry of the a_{ij} tensor.

Note that we have not yet used the fact that the action must be an extremum. For that we can calculate the variation in the Lagrangian and use the Euler–Lagrange equations of motion to complete this proof of the conservation of angular momentum.

$$\delta L = \left(\frac{\partial L}{\partial x_i} - \frac{d}{dt} \frac{\partial L}{\partial \dot{x}_i} \right) \delta x_i + \frac{d}{dt} \left(\frac{\partial L}{\partial \dot{x}_i} \delta x_i \right)$$

$$\delta L = -\frac{d}{dt} \left(\frac{\partial L}{\partial \dot{x}_i} x_i \right) a_{ij}$$

$$\delta L = -\frac{1}{2} \frac{d}{dt} \left(x_i \frac{\partial L}{\partial \dot{x}_j} - x_j \frac{\partial L}{\partial \dot{x}_i} \right) a_{ij} = 0$$

and here we see the mixing of coordinates due to the rotation. Now we note that the momentum is obtained from the Lagrangian by $p_i = \frac{\partial L}{\partial \dot{x}_i}$ so the variation in the Lagrangian gives us the conserved Noether charges as the angular momentum for each antisymmetric pair $\{i, j\}$ as

$$J_{ij} = x_i \frac{\partial L}{\partial \dot{x}_j} - x_j \frac{\partial L}{\partial \dot{x}_i}$$

$$J_{ij} = x_i p_j - x_j p_i$$

and conservation of angular momentum results from the vanishing of the time derivative of the Noether charges

$$\delta L = -\frac{1}{2} \frac{d}{dt} \left(J_{ij} \right) a_{ij} = 0$$

The coordinate mixing here is typical of the cross product of vectors. One can write the angular momentum in the more familiar form of the cross product of position and momentum

$$J_k = \frac{1}{2} \epsilon_{ijk} J_{ij} = \left(\vec{x} \times \vec{p} \right)_k$$

In the transition from classical to quantum mechanics, the momentum becomes an operator, $p_i = -i\hbar \frac{\partial}{\partial x_i}$, leading to the quantum version of angular momentum operator as

$$J_{ij} = -i\hbar \left(x_i \frac{\partial}{\partial x_j} - x_j \frac{\partial}{\partial x_i} \right)$$

and the angular momentum operators have the commutators

$$\left[J_i, J_j \right] = i\hbar \epsilon_{ijk} J_k$$

which can also be written, using the definition of J_{ij}, as

$$\left[J_{ij}, J_{kl}\right] = -i\hbar\left(\delta_{ik}J_{jl} - \delta_{il}J_{jk} + \delta_{jl}J_{ik} - \delta_{jk}J_{il}\right)$$

3.8 Symmetries as Length Preserving Transformations of Vector Spaces

3.8.1 Euclidean 3D Space

The symmetries of the universe are based on both external and internal properties of the objects within it and also on the properties of space-time itself. An understanding of the properties of space-time is necessarily based on the study of coordinate systems, reference frames, that is, on its external properties. These external properties can be grasped by means of the transformations that change one frame of reference into another in physical space. Once the mathematical behavior of these external properties is elucidated, we can use them as models for the internal properties as well.

While the profound insights of Emmy Noether showed that an important class of the universe's grand organizing principles, the conservation laws, arose as a result of the *time-invariant* nature of the universe under simple transformations, they left open the question of the nature of space-time. What is space-time? We have been taught since the time of *Decartes* and *Newton* that objects existed within an abstract, mathematical, but very real, Euclidean (i.e., *flat*) 3D vector space in which all physical processes take place. This Cartesian space was presented as a differentiable manifold of real, three-tuples (coordinates or vector components). For now, let us study the properties of the above-mentioned, differentiable manifold of real, three-tuples, that was thought to describe all of the universe until 1905.

3.8.2 Vectors and Vector Spaces

Reference frames, or coordinate systems, are familiar to each and every one of us through our daily interactions with the environment. We all use rulers without giving them a second thought. Coordinate systems are characterized by their dimensionality. The ruler that we use so frequently is a one-dimensional coordinate system with its origin at the zero mark, but by using it to measure, not just the length of an object, but also its width, to obtain a second dimension, and its height for a third, we immediately grasp the fact that these measurements are easily associated with points in a one-, two-, or three-dimensional space, the space that Descartes. Another thing that we often do not ponder but is nevertheless interesting is the fact that this 3D Cartesian space comes equipped with a standard length; the ruler is calibrated in units, be they inches, meters, furlongs, or light years. Any distance can be expressed in terms of these standard units by simple multiplication of this standard unit by a real number. The sum of two or more distances can similarly be expressed as the sum of the real numbers times the standard length. These and more are the mathematical properties of a real, three-dimensional vector space.

3.8.2.1   What Is a Vector?

The definition of a vector depends on your level of mathematical sophistication. In elementary school, we learned that a vector was an arrow: an object with both length and direction. Later a vector might have been defined as an ordered n-tuple of numbers, while as our level of mathematical abstraction increased, we discovered that a vector is an object that obeys certain rules when you change the coordinate system. One of those rules is that the length of a vector is invariant; it does not depend on the coordinate system. What differs from one system to the next is the representation of the vector in terms of its components, the projections of a vector onto each of the coordinate axes. These components are unique for a given coordinate system, but change in interesting ways upon transformation to another system. In our example with the ruler, we noted that there was a standard length and one can similarly define a standard length vector, assign it unit length by dividing it by its length, and call it a unit vector. One of these unit vectors lies along each of the coordinate axes in the space. The set of unit vectors forms a basis for the space. The dimension of the space is given by counting the number of unique (i.e., linearly independent) unit vectors in the basis. Any vector in the space can be represented by multiples of the unit vectors, the coefficients are then the components of the vector in that space and form the representation of the vector in that space. We see that we already know quite a lot about vectors from our everyday experience with rulers and a coordinate system consisting of three spatial dimensions whose axes meet at right angles at the origin, but there are many more coordinate systems that arise in nature, some with some rather surprising features.

3.8.2.2   What Then Is a Vector Space?

Mathematicians like to boil ideas down to their core concepts, which they then codify as axioms, because the simpler a set of rules, the larger will be their range of applicability. We will take this axiomatic approach because its simplicity gives it great power. The price we pay is one of abstraction, but the rewards of learning about vector spaces will be great because we will see that the mathematics apply to a large number of cases where physical principles come to bear on the self-assembly of living systems. These include the external properties, symmetries such as special relativity, rotational and translational invariances, as well as internal properties of objects derived from the rich mathematical structure of quantum mechanics, such as spin and other purely quantum labels.

A vector space is a set of objects (vectors) that obey the following collection of axioms:

1. A vector space V is a set of objects (vectors) such that for any two vectors in the space, their sum (to be defined) is a third object (another vector) that is also within the space. Let a, b be two vectors in the space, then their sum, c, is also in the space:

$$\forall a, b \in V, a + b = c \in V.$$

   Think of measuring the length of a box by marking a point on the box between the two ends. The length of the box will be the sum of the lengths from one end to the mark, and from the mark to the other end.
2. This operation of addition is *associative* and *commutative*: for any two vectors a, b in the space, $a + b = b + a$, and for a third vector c in the space $(a + b) + c = a + (b + c)$. Addition also has an inverse: for any vector $a \in V$, there exists a unique inverse vector $-a$ such that $a - a = 0$.
3. For any vector $a \in V$ and for any number n, the vector na is also in V. In general, n can be either real, in which case we have a Real vector space, or n can be complex, such as in the Hilbert space of quantum mechanics. This multiplication is *associative* and *distributive*:

$$n'(na) = (n'n)a$$

$$n(a + b) = na + nb$$

$$(n' + n)a = n'a + na$$

4. There is a unique $\mathbf{0}$ vector such that for any vector a in V:

$$\mathbf{0} + a = a$$

$$\mathbf{0}\,a = \mathbf{0}$$

$$n\,\mathbf{0} = \mathbf{0}$$

5. A *normed* vector space is one on which the inner (scalar) product of two vectors is defined. For any two vectors a and b in the space, we denote the scalar product using the dot $(a \cdot b)$.
6. Many sets of objects satisfy these axioms and form vector spaces. For example, all the $n \times m$ matrices with fixed n and m form a vector space, and if m or $n = 1$, this space is the same as that formed by the space of n or m real numbers as n or m-tuples (vectors). Other objects that form vector spaces are familiar from quantum mechanics as the set of continuous functions defined on a closed interval $\{a, b\}$. In this case the inner product is defined for, say two functions $f(x)$ and $g(x)$ in the Hilbert space, by the integral

$$f \cdot g = \int_a^b f(x)g(x)dx$$

We also often will use Dirac's *bra-ket* notation as a shorthand for the scalar product of two functions $a(x)$, $b(x)$ as

$$<a \mid b> \equiv \int_c^d a(x)b(x)dx$$

which is linear

$$<a \mid (n\,a + n'\,a)> \; = \; <a \mid n\,a> + <a \mid n'a>$$

and positive definite, implying that the scalar product of a vector with itself is real and greater than zero

$$<a \mid a> \; > 0$$

as long as $a(x)$ is not the zero vector. The scalar product of real vectors is a real number, but for complex vectors the action of complex conjugation applied to the scalar product reverses the order of the vectors

$$<a \mid b> * \; = \; <b \mid a>.$$

Since this is true for any $<a \mid$, we conclude that bras and kets are complex conjugates of each other:

$$\mid b> * \; = \; <b \mid$$

Vector spaces can be finite dimensional, such as ordinary three-space, or infinite dimensional. Later, we will learn that the dimensions of a vector space often have important consequences for the vectors themselves. Note that we have not yet defined the addition or multiplication of vectors, because this requires us to explore the concept of a basis.

3.8.2.3   A Basis for a Vector Space

By axioms 1–3 above, any linear combination of vectors in V is also a vector in V; this gives rise to the concept of a basis for V. We can write any vector $\boldsymbol{a} \in V$ in an n-dimensional vector space V as

$$a = \sum_{i=1}^{n} a_i e_i$$

which is a linear combination of the set of n linearly independent basis vectors $\{e_i\}$ with scalar coefficients a_i (a set of n-tuples) that are called the *components* of the vector a in this basis. The basis vectors are linearly independent in that the above sum can only be zero if the a_i are all zero. If none are zero, then the set $\{e_i\}$ forms a basis because no member of this basis can be written as a linear combination of the other basis vectors. These basis vectors form a complete orthonormal set of unit vectors that *spans* the vector space, in the sense that any vector in the space can be formed as linear combination of the basis vectors: completeness means without needing any other vectors. The number of basis vectors is the dimension of the space.

3.8.2.4   Addition of Vectors

We often will work with a set of orthonormal basis vectors, meaning that the basis vectors are normalized (i.e., are unit vectors) and they are mutually orthogonal. In a normed vector space, one can form a unit vector by dividing a vector by its length, which is given by the square root of the inner product of the vector with itself.

We can now define the addition of two vectors, a and b, by making use of their definitions.

$$a = \sum_{i=1}^{n} a_i e_i$$

$$b = \sum_{i=1}^{n} b_i e_i$$

Then, their sum is given by

$$a + b = \sum_{i=1}^{n} a_i e_i + \sum_{i=1}^{n} b_i e_i$$

$$a + b = \sum_{i=1}^{n} (a_i + b_i) e_i$$

And we see that in order to add two vectors that have the same basis, we just need to add their components in that basis.

In ordinary Euclidean 3D space, we often denote the unit vectors as $\{e_1, e_2, e_3\}$ and express their orthonormality by their dot product

$$e_i \cdot e_j = \delta_{ij} \forall i,j \in \{1,2,3\}$$

where δ_{ij} is the Kronecker delta:

$$\delta_{ij} = 1 \text{ if } i = j, \text{ and } 0 \text{ if } i \neq j.$$

This equation states that the set of unit vectors $\{e_1, e_2, e_3\}$ form an orthonormal basis in 3D space:

$$\|e_i\| = 1, \, \forall i \in \{1,2,3\} \, .$$

It is often required that we transform our point of view from one coordinate system to another. In an n-dimensional Euclidean space, a coordinate system is a collection of ordered sets of n-tuples (a basis)

$$\{e_1, e_2, e_3, \ldots, e_n\}$$

and the kth coordinate axis as the set of points of the form

$$\{0, \ldots, 0, e_k, 0, \ldots, 0\}$$

where $\|e_k\| = 1$.

We can define objects in this coordinate system by how they transform under a change in the coordinate system, and we will see shortly that this transformation reduces to a change of the basis vectors.

3.8.2.5   Invariance of the Scalar Product

Because it is simply a number, a scalar is invariant under a change in the coordinate system. We are now ready to define the multiplication of vectors. The length of a vector is the scalar product of the vector with itself and therefore the length of a vector is invariant under a transformation of coordinates. Let us consider the vector a defined above as a linear combination of basis vectors; its length squared is then

$$a \cdot a = \left[\sum_{j=1}^{n} a_j e_j \right] \cdot \left[\sum_{i=1}^{n} a_i e_i \right]$$

$$a \cdot a = \sum_{i=1}^{n} \sum_{j=1}^{n} a_i a_j e_i \cdot e_j$$

but, since the unit vectors form an orthonormal basis, their inner products are the Kronecker delta

$$e_i \cdot e_j = \delta_{ij}$$

and the double sum collapses to a single sum for $i = j$ so that we find the form for the scalar product of a vector with itself as the sum of the squares of the components (the coefficients of the basis vectors):

$$a \cdot a = \sum_{i=1}^{n} a_i^2$$

In a similar fashion we can find the equation for the inner product of two different vectors, a and b, by writing each vector in terms of its linear combination of basis vectors, in the same basis as

$$a = \sum_{i=1}^{n} a_i e_i$$

$$b = \sum_{i=1}^{n} b_i e_i$$

$$a \cdot b = \sum_{i=1}^{n} \sum_{j=1}^{n} a_i b_j e_i \cdot e_j$$

$$a \cdot b = \sum_{i=1}^{n} \sum_{j=1}^{n} a_i b_j \delta_{ij} = \sum_{i=1}^{n} a_i b_i$$

By the way, the rank of a tensor is defined as the number of indices it carries, so that a scalar is a tensor of rank zero, while a vector is a tensor of rank one, and a general tensor has rank two or greater.

3.8.2.6   Transformations of Vectors

We can use the properties of an orthonormal basis to find out how a vector transforms between two coordinate systems, which is the same as transforming the vector space into itself. We will see that the transformation can be found by using the basis vectors themselves in the two coordinate systems, as follows. In two orthonormal bases, an n-dimensional vector has two different representations. Let us suppose that the two bases are $\{e_j\}$ and $\{e_j'\}$ for $j, j' \in \{1 \leq j, j' \leq n\}$. Then the vector a has two different representations in the vector space given by

$$a = \sum_{i=1}^{n} c_i e_i$$

and

$$a = \sum_{i=1}^{n} d_i e_i'$$

and we can find the relationship between the components c_i and d_i by computing the inner product of a with one of the vectors from each of the bases in turn with a separate equation for each value of $p \ni 1 \leq p \leq n$. In the unprimed basis,

$$a \cdot e_p = \sum_{i=1}^{n} c_i e_i \cdot e_p = \sum_{i=1}^{n} c_i \delta_{ip} = c_p$$

and in the primed basis we have an expression for the same scalar product as well:

$$c_p = a \cdot e_p = \sum_{i=1}^{n} d_i e_i' \cdot e_p$$

from which we conclude that the pth component of this inner product is

$$c_p = \sum_{i=1}^{n} d_i e_i' \cdot e_p$$

and we can write the vector transformation law as

$$c_p = \sum_{i=1}^{n} T_{ip} d_i$$

where the elements of the matrix T_{ip} (which is a tensor of rank 2) are given by scalar products of the basis vectors in the two systems:

$$T_{ip} = e_i' \cdot e_p$$

which we can rewrite as

$$e_i' \cdot e_p = T_{ip}$$

and by multiplying both sides by e_i',

$$\left(e_i' \cdot e_p\right)e_i' = T_{ip}e_i'$$

$$\left(e_i' \cdot e_i'\right) e_p = T_{ip}e_i'$$

Because $e_i' \cdot e_i' = 1$, we obtain the transformation law

$$e_p = T_{ip}e_i'$$

where we have again used the Einstein summation convention. This transformation law defines a vector in the abstract mathematical sense; any object v that transforms in this manner is defined to be a vector:

$$v_p' = T_{ip}v_i$$

Note here that the transformation matrix elements are defined in terms of inner products of the basis vectors and the transformation law appears as a change of basis.

We can derive the inverse transformation in an analogous fashion by computing the inner product of a with a vector from the primed basis:

$$a \cdot e_p' = \sum_{i=1}^{n} c_i e_i \cdot e_p' = \sum_{i=1}^{n} d_i e_i' \cdot e_p' = \sum_{i=1}^{n} d_i \delta_{ip} = d_p$$

or

$$d_p = \sum_{i=1}^{n} c_i e_i \cdot e_p'$$

and this can again be written in terms of a transformation matrix

$$T_{ip}' = e_i \cdot e_p' \quad \text{as}$$

$$d_p = \sum_{i=1}^{n} T_{ip}' c_i$$

We can write this as a matrix equation as $a' = T'a$.

As a simple example, let's compute the T matrix for the 3D transformation $x' = x$, $y' = y$ and $z' = z$. The orthonormal basis vectors in $\{\mathbf{x}, \mathbf{y}, \mathbf{z}\}$ and $\{\mathbf{x}', \mathbf{y}', \mathbf{z}'\}$ are both $\{e_1, e_2, e_3\}$, therefore T is given by:

$$T_{ij} = \begin{bmatrix} e_1 \cdot e_1 & e_1 \cdot e_2 & e_1 \cdot e_3 \\ e_2 \cdot e_1 & e_2 \cdot e_2 & e_2 \cdot e_3 \\ e_3 \cdot e_1 & e_3 \cdot e_2 & e_3 \cdot e_3 \end{bmatrix} = \delta_{ij} = \begin{bmatrix} 1 & 0 & 0 \\ 0 & 1 & 0 \\ 0 & 0 & 1 \end{bmatrix} = I,$$

the identity matrix, as expected, since this corresponds to the identity transformation. Although we will not need this concept here, we note that T defined here is called the *metric*, or metric tensor for 3D Euclidean (flat) space, and it is a tensor of rank 2, because it carries two indices.

For a general transformation from one basis to another, say from $\{e_1, e_2, e_3\}$ to $\{e_1{}', e_2{}', e_3{}'\}$, the transformation matrix will be given by

$$T_{ij} = \begin{bmatrix} e_1 \cdot e_1{}' & e_1 \cdot e_2{}' & e_1 \cdot e_3{}' \\ e_2 \cdot e_1{}' & e_2 \cdot e_2{}' & e_2 \cdot e_3{}' \\ e_3 \cdot e_1{}' & e_3 \cdot e_2{}' & e_3 \cdot e_3{}' \end{bmatrix}$$

which will not necessarily be the identity transformation.

We can still find general properties of the transformation matrix if the two different bases are orthonormal. The basis vectors can be written as a linear combination of basis vectors as any other vector in the vector space

$$e_j{}' = \sum_{i=1}^{n} c_i e_i$$

and by taking the inner product of this with a basis vector from the unprimed set

$$e_j{}' \cdot e_p = \sum_{i=1}^{n} c_i e_i \cdot e_p = \sum_{i=1}^{n} c_i \delta_{ip} = c_p$$

but this is also, by definition, one of the elements of our transformation matrix

$$T_{jp} = e_j{}' \cdot e_p = c_p$$

and we find that the basis vectors transform just as any other vector in the vector space

$$e_p{}' = \sum_{i=1}^{n} T_{pi} e_i$$

A similar calculation using the primed basis gives

$$e_q = \sum_{j=1}^{n} T_{qj}{}' e_j{}'$$

Now, inserting this equation for e_q into the above expression for $e_p{}'$ gives

$$e_p{'} = \sum_{j=1}^{n} \sum_{q=1}^{n} T_{pj} T_{jq}{'} e_q{'} = \sum_{q=1}^{n} \left(\sum_{j=1}^{n} T_{pj} T_{qj} \right) e_q{'}$$

This equation must be true for any p or q, and in particular for $p = q$, therefore the coefficients on the left and right sides must agree term by term and because the basis vectors are linearly independent, we cannot have that $e_p{'} = e_q{'}$ for arbitrary p and q, the sum in parenthesis must reduce to a Kronecker delta such that

$$\sum_{j=1}^{n} T_{pj} T_{qj} = \delta_{pq}$$

The left side of this equation is the inner product of two row vectors of the transformation matrix T. Written in matrix notation, the transpose of T (T^T) times T is the identity $T^T T = I$. We see therefore that two different row vectors of the transformation matrix are orthogonal since their inner product is zero, while the length of any row vector is one. The transformation matrix is an orthogonal matrix.

3.8.2.7   Orthogonal Transformations

The orthogonal transformation matrix T has the following properties:

1. T is an $n \times n$ matrix, and if nonsingular, is orthogonal if its inverse is equal to its transpose

$$T^{-1} = T^T$$

2. If T is an orthogonal matrix, its determinant equals 1.

$$\text{Det } T = \pm 1$$

3. If T is an orthogonal matrix, then its transpose is also an orthogonal matrix, giving the relationship

$$\sum_{j=1}^{n} T_{jp} T_{jq} = \delta_{pq}$$

which is equivalent to the matrix equation

$$T \, T^T = I$$

where the product of T with its transpose is equal to the identity matrix.

An important point to note is that this also states that orthogonal transformations are *unitary*, or that they conserve probability. The implication of this is that any two column vectors of an orthogonal matrix are orthogonal and that the norm of any column vector is 1. This is shown by multiplying the above equation by T^{-1}

$$T^{-1}\left(T\,T^{T}\right) = T^{-1}$$

For this equation to hold, $T\,T^{T} = I$ must hold. We can also show that Det $T = 1$ by using Det $T =$ Det T^{T} as follows:

$$T\,T^{T} = I$$

then

$$\det\left(T\,T^{T}\right) = \det I = 1$$

and

$$\det\left(T\,T^{T}\right) = \det(T)\,\det\left(T^{T}\right) = \det(T)^{2} = 1$$

or

$$\det(T) \pm 1.$$

Finally, to show that if T is an orthogonal transformation, then so it its transpose we note that

$$\left(T^{T}\right)^{-1} = \left(T^{-1}\right)^{T} = \left(T^{T}\right)^{T}$$

so that

$$\sum_{j=1}^{n} T_{pj}T_{qj} = \delta_{pq} = \sum_{j=1}^{n} T_{jp}T_{jq}$$

and we can use this as an alternative expression of orthogonality. The sign of the determinant above is $+1$ for a proper transformation, while is it -1 for an improper transformation (a reflection of the vector through the origin, also known as a parity transformation).

Now the length of a vector is a scalar and should not depend on our point of view, so let's examine how the inner product of a vector transforms under a general transformation, T. Using matrix notation, the inner product of a vector with itself transforms as

$$\boldsymbol{a} \cdot \boldsymbol{a} = (T\boldsymbol{a}') \cdot (T\boldsymbol{a}')$$

which we can explicitly expand using

$$\boldsymbol{a} = \sum_{p=1}^{n} c_p \boldsymbol{e}_p = \sum_{p=1}^{n} d_p \boldsymbol{e}_p'$$

and the components of $\boldsymbol{a}$ in the primed and unprimed bases

$$c_p = \sum_{i=1}^{n} T_{ip} d_i$$

and

$$d_p = \sum_{i=1}^{n} T_{ip}' c_i$$

to write a in terms of the transformed bases

$$\boldsymbol{a} = \sum_{p=1}^{n} \sum_{i=1}^{n} T_{ip} d_i \boldsymbol{e}_p = \sum_{p=1}^{n} \sum_{i=1}^{n} T_{ip}' c_i \boldsymbol{e}_p'$$

then, the length of $\boldsymbol{a}$ is the square root of the inner product of $\boldsymbol{a}$ with itself and it suffices to find the length squared, or the inner product

$$\boldsymbol{a} \cdot \boldsymbol{a} = \sum_{p=1}^{n} \sum_{i=1}^{n} T_{ip} d_i \boldsymbol{e}_p \cdot \sum_{q=1}^{n} \sum_{j=1}^{n} T_{qj} d_j \boldsymbol{e}_q = \sum_{p=1}^{n} \sum_{i=1}^{n} T_{ip}' c_i \boldsymbol{e}_p' \cdot \sum_{q=1}^{n} \sum_{j=1}^{n} T_{jq}' c_j \boldsymbol{e}_q'$$

Let's now rewrite the left-hand side of this as

$$\sum_{p=1}^{n} \sum_{i=1}^{n} T_{ip} d_i \boldsymbol{e}_p \cdot \sum_{q=1}^{n} \sum_{j=1}^{n} T_{qj} d_j \boldsymbol{e}_q = \sum_{p=1}^{n} \sum_{q=1}^{n} \sum_{i=1}^{n} \sum_{j=1}^{n} T_{ip} T_{qj} d_i d_j \boldsymbol{e}_p \cdot \boldsymbol{e}_q$$

and we use the fact that we are working in an orthonormal basis so that

$$\sum_{p=1}^{n} \sum_{q=1}^{n} \sum_{i=1}^{n} \sum_{j=1}^{n} T_{ip} T_{qj} d_i d_j \boldsymbol{e}_p \cdot \boldsymbol{e}_q = \sum_{p=1}^{n} \sum_{q=1}^{n} \sum_{i=1}^{n} \sum_{j=1}^{n} T_{ip} T_{qj} d_i d_j \delta_{pq} = \sum_{p=1}^{n} \sum_{i=1}^{n}$$

$$\times \sum_{j=1}^{n} T_{ip} T_{pj} d_i d_j$$

and also the T matrix is its own inverse so that $T_{pj} = T_{jp}$

$$= \sum_{p=1}^{n} \sum_{i=1}^{n} \sum_{j=1}^{n} T_{ip} T_{jp} d_i d_j$$

and that

$$\sum_{p=1}^{n} T_{ip} T_{jp} = \delta_{ij}$$

to give

$$\boldsymbol{a} \cdot \boldsymbol{a} = \sum_{p=1}^{n} \sum_{q=1}^{n} \sum_{i=1}^{n} \sum_{j=1}^{n} T_{ip} T_{qj} d_i d_j \boldsymbol{e}_p \cdot \boldsymbol{e}_q = \sum_{i=1}^{n} \sum_{j=1}^{n} d_i d_j \delta_{ij} = \sum_{i=1}^{n} d_i^2$$

which is the inner product $\boldsymbol{a} \cdot \boldsymbol{a} = \sum_{i=1}^{n} d_i^2$ that we sought. Now, let's turn our attention to the right-hand side of our transformed inner product equation

$$\boldsymbol{a} \cdot \boldsymbol{a} = \sum_{p=1}^{n} \sum_{i=1}^{n} T_{ip}' c_i \boldsymbol{e}_p' \cdot \sum_{q=1}^{n} \sum_{j=1}^{n} T_{jq}' c_j \boldsymbol{e}_q' = = \sum_{p=1}^{n} \sum_{q=1}^{n} \sum_{i=1}^{n} \sum_{j=1}^{n} T_{ip}' T_{jq}' c_i c_j \boldsymbol{e}_p' \cdot \boldsymbol{e}_q'$$

By an identical line of reasoning as used above we find that

$$\boldsymbol{a} \cdot \boldsymbol{a} = \sum_{p=1}^{n} \sum_{q=1}^{n} \sum_{i=1}^{n} \sum_{j=1}^{n} T_{ip}' T_{jq}' c_i c_j \boldsymbol{e}_p' \cdot \boldsymbol{e}_q' = \sum_{i=1}^{n} \sum_{j=1}^{n} c_i c \delta_{ij} = \sum_{i=1}^{n} c_i^2$$

and we observe that even though the components of $\boldsymbol{a}$ differ in the primed and unprimed representations of $\boldsymbol{a}$ in orthonormal bases, the inner product (i.e., length) is invariant under an orthogonal transformation:

$$\boldsymbol{a} \cdot \boldsymbol{a} = \sum_{i=1}^{n} c_i^2 = \sum_{i=1}^{n} d_i^2$$

3.8.2.8   Rotations Are Orthogonal Transformations That Also Preserve the Length of Vectors

There are many orthogonal transformations that change our point of view of the universe: translations in time, space, and velocity, to name but a few. Here, we shall be exclusively concerned with symmetry transformations that map a vector space unto itself, the most important of which are *rotations* of the axes. Rotations are the poster child of symmetry transformations and understanding their properties will be seen to provide a key constraint on the types of objects of which the universe can be constructed. The rotation operators will become the angular momentum operators of quantum mechanics.

A rotation in n dimensions changes the coordinates of a vector

$$x = \sum_{i=1}^{n} x_i e_i$$

by

$$x_i' = R_{ij} x_i$$

where R_{ij} is an orthogonal rotation matrix. This is the definition of a vector as a rank 1 tensor, i.e., an object described by a single index. A rotation mixes the coordinates as can be seen in 2D Euclidean space by explicitly writing down the rotation matrix

$$R_{ij}(\theta) = \begin{bmatrix} \cos\theta & -\sin\theta \\ \sin\theta & \cos\theta \end{bmatrix}$$

and computing its effect on an arbitrary 2D vector x_i gives

$$\begin{bmatrix} x_1{}' \\ x_2{}' \end{bmatrix} = R_{ij}(\theta) \begin{bmatrix} x_1 \\ x_2 \end{bmatrix} = \begin{bmatrix} \cos\theta & -\sin\theta \\ \sin\theta & \cos\theta \end{bmatrix} \begin{bmatrix} x_1 \\ x_2 \end{bmatrix} = \begin{bmatrix} x_1\cos\theta - x_2\sin\theta \\ x_2\sin\theta + x_2\cos\theta \end{bmatrix}$$

It is straightforward to find the metric for rotations in two dimensions and then to generalize this to three dimensions. Consider a vector $v = v_x\, e_1 + v_y\, e_2$, where $\{e_1, e_2\}$ are the standard unit vectors along the x and y axes in 2D Cartesian space and $\{v_x, v_y\}$ are the components of v along the x and y axes. If we rotate the x and y axes by the angle θ, which is conventionally defined to be positive in the counterclockwise (Fig. 3.1) direction, the T matrix for the transformation from the unprimed to the primed coordinates is the rotation matrix such that the basis unit vectors in the unprimed system transform to the primed system according to $v_x\, e_1 + v_y\, e_2$.

Fig. 3.1   A rotation of the vector v by the positive angle θ in the 2D Euclidean plane

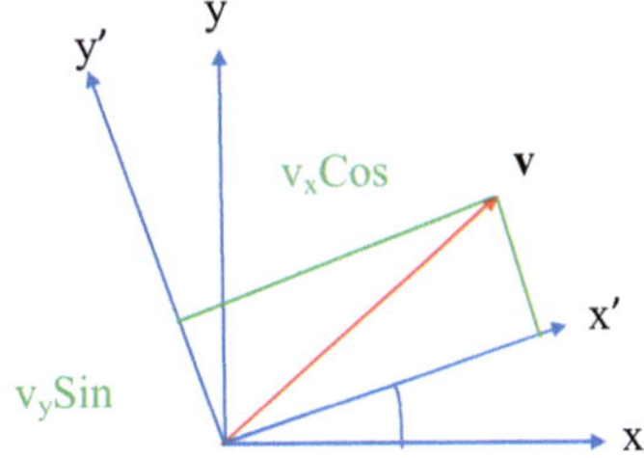

$$\begin{bmatrix} e_1{}' \\ e_2{}' \end{bmatrix} = T \begin{bmatrix} e_1 \\ e_2 \end{bmatrix}$$

where the T matrix is given by

$$\begin{bmatrix} e_1{}' \\ e_2{}' \end{bmatrix} = \begin{bmatrix} e_1{}' \cdot e_1 & e_1{}' \cdot e_2 \\ e_2{}' \cdot e_1 & e_2{}' \cdot e_2 \end{bmatrix} \begin{bmatrix} e_1 \\ e_2 \end{bmatrix} = \begin{bmatrix} \cos\theta & \sin\theta \\ -\sin\theta & \cos\theta \end{bmatrix} \begin{bmatrix} e_1 \\ e_2 \end{bmatrix}$$

which was derived from inner products such as $e_1{}' \cdot e_1 = |\,|e_1{}'|\,|\,|\,|e_1|\,|\,\cos\theta$ (which reduces simply to $\cos\theta$ because the basis vectors have unit length). The transformation matrix is orthogonal because the inner product of both row vectors is zero:

$$-\sin\theta\,\cos\theta + \cos\theta\,\sin\theta = 0, \text{etc.},$$

and the norm of each row and column vector is 1:

$$\sin^2\theta + \cos^2\theta = 1, \text{etc.}$$

We also see that $T^{-1} = T^T$ and that det $T = \sin^2\theta + \cos^2\theta = 1$. Therefore, our vector v becomes in the primed system

$$v' = v'_x e'_1 + v'_y e'_2 = v_x\,(e_1\,\cos\theta + e_2\,\sin\theta) + v_y\,(-e_1\,\sin\theta + e_2\,\cos\theta)$$

where it is important to note that, unlike reflections and translations in time and space, rotations mix the components of vectors. This is a major and very important distinction between these two types of transformations.

3.8.2.9   3D Coordinate Axes Rotation Matrices

Generalization of rotations to three dimensions is straightforward by the introduction of the components of a rotated vector using standard trigonometry. The matrices R for rotations by the angle θ around the z-axis, by the angle ϕ around the x-axis, and by the angle ψ around the y-axis are given by:

$$R_z(\theta) = \begin{bmatrix} \mathrm{Cos}[\theta] & \mathrm{Sin}[\theta] & 0 \\ -\mathrm{Sin}[\theta] & \mathrm{Cos}[\theta] & 0 \\ 0 & 0 & 1 \end{bmatrix},$$

$$R_x(\phi) = \begin{bmatrix} 1 & 0 & 0 \\ 0 & \text{Cos}[\phi] & \text{Sin}[\phi] \\ 0 & -\text{Sin}[\phi] & \text{Cos}[\phi] \end{bmatrix},$$

and

$$R_y(\psi) = \begin{bmatrix} \text{Cos}[\psi] & 0 & -\text{Sin}[\psi] \\ 0 & 1 & 0 \\ \text{Sin}[\psi] & 0 & \text{Cos}[\psi] \end{bmatrix}.$$

Mathematica can be used to show that, for example,

$$\det(R_z(\theta)) = \text{Cos}^2[\theta] + \text{Sin}^2[\theta] = 1,$$

and that

$$R_z(\theta = \pi/3).\ \text{Transpose}[R_z(\theta = \pi/3)] = I,$$

the identity matrix in 3D.

That proper rotations preserve the inner product (length of a vector) is guaranteed by the fact that the rotation matrix is orthogonal and has $\text{Det}(R) = 1$. Improper rotations (reflections through the origin) have $\text{Det}(R) = -1$ and correspond to parity operations. In its simplest but most general form, a 3D rotation matrix can be given in terms of the Euler angles (α, β, γ) between the unprimed axes and the new, primed axes. Then a rotation matrix, $R(\alpha, \beta, \gamma)$, can be written as the product of rotation matrices about each axis:

$$R(\alpha, \beta, \gamma) = R_z(\gamma)R_y(\beta)R_x(\alpha).$$

Rotations are symmetry operations since the rotated object is the same as the original object viewed from a different coordinate system. Rotations are a set of linear operators on a vector space that transforms the vector space into itself. The definition of unitarity is that the complex conjugate transpose of an operator is equal to the inverse of the operator itself: $R^\dagger = R^{-1}$, or if we multiply both sides by R, we have that $R^\dagger R = R^{-1}R = R^\dagger R = 1$. Now, the rotation operators are real, so that they are unitary because $R^\dagger = R^T$, and we have that $R^T R = 1$; rotations of a vector space preserve probability. The rotation operators also satisfy the axioms of group theory and therefore form a *group*. Further study of rotations, necessary for a complete understanding of their role in constraining the contents of the universe, therefore requires us to examine their group theoretical properties.

3.9 Group Theory

Like a vector space, a *group* is an abstract mathematical structure consisting of a set of elements that satisfy a small set of axioms. While the axioms are deceptively simple, the great power of Group theory comes from the fact that many sets of elements satisfying these axioms can be devised, but the mathematical relationships among the group elements remain fixed, independent of the exact nature of the elements. This is much like the abstraction of number or color from the environment; three red bricks have the same color and number properties as three red bacteria, but their underlying natures are vastly different. We will define a group first, and then look at some concrete examples that will illustrate this point. Later, we will restrict our attention to group elements that are transformation operators on space-time.

3.9.1 Definition of Groups

A *Group* G consists of a set of elements $\{g_i\}$ and single binary rule of combination (called the group operation), which we denote here by $\circ$, such that the group G consists of the elements along with the combination rule, $G = \{g_i; \circ\}$, where the following axioms are fulfilled:

1. The group G is *Closed* under the group operation. This means that

$$\forall (g_1, g_2) \in G$$

$$g_1 \circ g_2 = g_3 \in G$$

   that is, the combination of any two elements of G results in another element of G.
2. The group operation is *Associative:*

$$\forall (g_1, g_2, g_3) \in G$$

$$g_1 \circ (g_2 \circ g_3) = (g_1 \circ g_2) \circ g_3$$

3. There is an *Identity* element $\mathbf{e} \in G$ such that $\forall\, g_i \in G$ satisfies

$$\mathbf{e} \circ g_i = g_i$$

   The use of $\mathbf{e}$ for the identity comes from the German word *einheit*. Here $\mathbf{e}$ is not to be confused with an element of the basis for a vector space.
4. There is an *Inverse* $g_i^{-1} \in G$ for every $g_i \in G$, such that $g_i^{-1} \circ g_i = \mathbf{e}$.

Many collections of elements in nature satisfy these axioms and hence form a Group. The binary rule of combination (the group operation) is often referred to as a multiplication, but, more generally, can be addition, multiplication, or e.g., in physics, a construct called a commutator, which we encountered previously when we used Noether's Theorem to derive the conservation of angular momentum.

3.9.2 The Integers Form a Group

As an example of a group, the elements can consist of the set of integers, Z

$$G = \{\ldots, \ -3, \ -2, \ -1, \ 0, \ 1, \ 2, \ 3, \ \ldots \ ; \ +\} = \{Z; \ +\}$$

with addition of two group elements as the binary rule of combination. Closure implies that any two integers, a and b, added together, form another integer c, $a + b = c$. The addition of integers is associative in that if (a, b, c) are integers, then $(a + b) + c = a + (b + c)$. The Identity element here is zero because for any integer a, $a + 0 = a$. And, finally, the inverse of an integer a is $-a$ because $a - a = \mathbf{0}$, which is the identity element. This group is discrete and has an infinite number of elements because the set of integers is infinite.

3.9.3 The Rotation Operators Form a Group

Let's check that the rotation operators satisfy the axioms of group theory and therefore form a *group*.

Closure The product of two rotation operators is also a rotation operator.

$$R_1 \circ R_2 = R_3$$

Associativity If R_1, R_2, R_3 are rotation operators then

$$R_1 \circ (R_2 \circ R_3) = (R_1 \circ R_2) \circ R_3$$

Identity The identity element is a rotation operator that rotates a vector by zero.

$$e = R(0)$$

Inverse The inverse element is the inverse rotation operator, and

$$R^{-1} \circ R_1 = e$$

Therefore, rotation operators form a group. Note, however, that the group elements do not commute with each other since the order of two rotations determines the result and if the operations are reversed, a different rotation is obtained.

$$R_1 \circ R_2 \neq R_2 \circ R_1$$

Groups in which the elements commute are called Abelian, after George Abel, a Norwegian mathematician. The rotation group is an example of a non-Abelian group.

The astute reader will now be thinking that these above statements can be true for abstract rotations in general, but will be wondering how this works for the representation of rotation operators as matrices given above? Each rotation matrix is a function of a parameter, the rotation angle about the specified axis. Although we firmly expect that the product of two rotation matrices, each with arbitrary rotation angles about their axes, will produce another rotation matrix, it is not obvious how this will play out in terms of the group axioms. It is possible, however, to simplify this problem by inspecting the properties of the rotation matrices.

To motivate this, let's find the eigenvalues of $R_z(\theta)$, a task performed in *Mathematica* as

$$\text{Eigenvalues}[\text{Rz}[\theta]]//\text{MatrixForm} = \begin{bmatrix} 1 \\ \text{Cos}[\theta] - i\text{Sin}[\theta] \\ \text{Cos}[\theta] + i\text{Sin}[\theta] \end{bmatrix} = \begin{bmatrix} 1 \\ e^{-i\theta} \\ e^{i\theta} \end{bmatrix}$$

We used *Mathematica* here because it can save us a great deal of complicated algebra for the following task. Note that one of the eigenvalues of $R_z(\theta)$ is 1. That this is a general property of rotation matrices and their products is found by using *Mathematica* to compute the eigenvalues of all rotation matrices and products of rotation matrices, which I leave as an exercise. Since the rotation matrices defined above are all rotations about a single axis, the Euler angles for each rotation will consist simply of the angle about that axis. For example, for a rotation of $R_z(\pi/3)$, *Mathematica* tells us that

$$\text{EulerAngles}[R_z[\pi/3]] = \left\{ -\frac{\pi}{3}, 0, 0 \right\}$$

as expected. Euler generalized this to a theorem that states that any rotation can be written as a rotation by a single angle around a unit vector along the axis of rotation. In this case the unit vector is e_3, the z-axis itself. The eigenvectors of R_z show this explicitly

$$\text{Eigenvectors}[R_z[\theta]]//\text{MatrixForm} = \begin{bmatrix} 0 & 0 & 1 \\ i & 1 & 0 \\ -i & 1 & 0 \end{bmatrix}$$

where we observe $e_3 = \begin{bmatrix} 1 \\ 0 \\ 0 \end{bmatrix}$ in the last column; similarly, we can find the basis vectors e_1 and e_2 from the eigenvectors of R_x and R_y.

Writing a rotation as $R(\theta, \mathbf{n})$ where $\mathbf{n}$ is a unit vector about the axis of rotation and θ is the rotation angle, we can investigate the group properties of rotation matrices. The inverse rotation is

$$R^{-1}(\theta, \mathbf{n}) = R(-\theta, \mathbf{n})$$

Identity:

$$R(0, \mathbf{n}) = I$$

Closure:

$$R(\theta, \mathbf{n})R(\theta', \mathbf{n}) = R(\theta + \theta', \mathbf{n})$$

Associativity:

$$R(\theta, \mathbf{n})[R(\theta', \mathbf{n})R(\theta'', \mathbf{n})] = [R(\theta, \mathbf{n})R(\theta', \mathbf{n})]R(\theta'', \mathbf{n}) = R(\theta + \theta' + \theta'', \mathbf{n})$$

And therefore, we have shown explicitly that the rotation operators form a group. This group is named the *Special Orthogonal* group in three-dimensions and assigned the symbol SO(3): special, because it has determinant +1 and orthogonal, because it is represented by orthogonal operators.

3.9.4 *Discrete Versus Continuous Groups:* Lie Groups

We have encountered two different types of groups so far; discrete groups, such as the group of integers and now the rotation group, which is a continuous group since the group elements (sines and cosines) depend on the rotation angles and these can vary continuously. The theory of continuous groups was investigated by Sophus Lie, and for that reason they are referred to as Lie groups. Lie's brilliant insight into continuous groups is reminiscent of the foundations of calculus by Newton and Leibniz; the tangent to a curve at a point is found through a limiting process. One picks two points on the curve and then lets the distance between them shrink to zero; in this limit the derivative is found. Now, if we think of groups as being composed of a set of transformation operators that depend on a continuous set of parameters, we could envision picking two values of a parameter and finding the limit as the parameter difference was taken to zero. We would be somehow finding the tangent to a transformation.

In calculus, any two points on a curve will suffice because the limiting process brings them together; neither of the two points is privileged. However, a moment's reflection suggests that for Lie groups there indeed exists a single privileged point in the parameter space. The group axioms require the existence of this privileged transformation, the identity, the transformation that leaves each element unchanged. Lie's idea was to investigate the properties of the transformation operators for values of their parameters infinitesimally different from the identity.

In the case of rotations, one can see the effect of letting the rotation angle tend to zero in the rotation matrices

$$\lim_{\theta \to 0} R_z(\theta) = \lim_{\theta \to 0} \begin{bmatrix} \mathrm{Cos}[\theta] & \mathrm{Sin}[\theta] & 0 \\ -\mathrm{Sin}[\theta] & \mathrm{Cos}[\theta] & 0 \\ 0 & 0 & 1 \end{bmatrix} = \begin{bmatrix} 1 & \theta & 0 \\ -\theta & 1 & 0 \\ 0 & 0 & 1 \end{bmatrix}$$

This is true because $\mathrm{Cos}(\theta) \sim 1$ for $\theta \sim 0$ and $\mathrm{Sin}(\theta) \sim \theta$ for $\theta \sim 0$. The resulting matrix has a unit eigenvalue and eigenvector, indicating the axis of rotation, and its eigenvectors are identical to those of R_z.

The idea of the limit of the transformation operators as the transformation approaches the identity suggests that one should explore infinitesimal rotations. These can be defined as a series expansion in the parameters. Let's write a rotation group element as $r(\theta) = \lim_{N \to \infty} \left(I + \frac{\theta}{N} J \right)^N$ where I is the identity transformation and θ is an infinitesimal rotation angle. We can build up finite rotations by successively multiplying small ones. The limit is the definition of the exponential function

$$e^{\theta J} = \lim_{N \to \infty} \left(I + \frac{\theta}{N} J \right)^N$$

And we can express rotation operators as exponentials of group elements. This form is reminiscent of Euler's theorem; here J is called the basis for a set of generators of rotations, or simply the generator of rotations, and θ is the rotation angle, $R(\theta, \mathbf{n}) \sim e^{\theta J}$. It is comforting that *Mathematica* includes a proper function that produces the correct exponential of a matrix with its `MatrixExp[]`.

That exponentiation of the generator produces rotations when acting upon a vector can be seen in the example of rotations in the 2D complex plane about the origin. A vector z of unit length in the complex plane can be written in terms of its real components using Euler's formula as

$$z(\theta) = e^{i\theta} = \cos \theta \, \mathbf{e_1} + \sin \theta \mathbf{e_2}$$

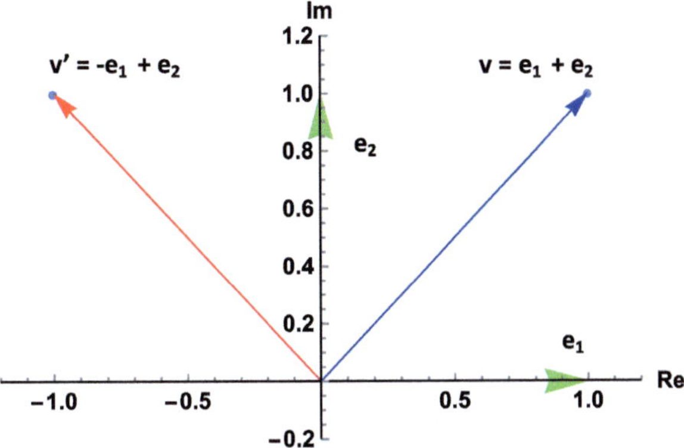

Fig. 3.2 A vector v (blue) in the complex plane can be rotated through an angle θ by multiplication with $e^{i\theta}$; in this case, the rotated vector (red) $v' = v\, e^{i\theta}$, where $\theta = \pi/2$. The basis vectors (green) are $\{e_1, e_2\} = \{\{1, 0\}, \{0, i\}\}$ and are given a two-dimensional representation by the 2×2 matrices in the text

with the basis vectors $\left\{ e_1 = \begin{bmatrix} 1 & 0 \\ 0 & 1 \end{bmatrix} = I, e_2 = \begin{bmatrix} 0 & -1 \\ 1 & 0 \end{bmatrix} = i \right\}$. Let's rotate a com-

plex vector $v = e_1 + e_2$ by $\frac{\pi}{2}$:

$$v' = e^{i\frac{\pi}{2}} v = \left(\cos\frac{\pi}{2}\, e_1 + \sin\frac{\pi}{2}\, e_2 \right) v$$

$$v' = (0\, e_1 + 1\, e_2)(e_1 + e_2)$$

$$v' = (e_2 \cdot e_1 + e_2 \cdot e_2)$$

For this last step, we need to find the metric in the complex plane. This, as you will remember from above, is the matrix T:

$$T = \begin{bmatrix} e_1 \cdot e_1 & e_2 \cdot e_1 \\ e_1 \cdot e_2 & e_2 \cdot e_2 \end{bmatrix} = \begin{bmatrix} e_1 & e_2 \\ e_2 & -e_1 \end{bmatrix} = \begin{bmatrix} 1 & i \\ i & -1 \end{bmatrix}$$

and we can read off the necessary dot products of the basis vectors, so that the rotated vector becomes

$$v' = (-e_1 + e_2)$$

which is indeed rotated 90° counterclockwise from v (Fig. 3.2).

3.9.4.1   The Transition from Rotations to Lie Groups and Their Lie Algebra

That the rotation operators in matrix form are a continuous function of the rotation angles means that one can differentiate these matrices with respect to their parameters. In particular, we can expand an element of the group of rotation matrices about the identity in a Taylor's series as

$$r(\theta) = I + \frac{dr(\theta)}{d\theta}\theta + \frac{d^2r(\theta)}{d\theta^2}\theta^2 + \frac{d^3r(\theta)}{d\theta^3}\theta^3 + \ldots = \sum_{n=0}^{\infty} \frac{d^n r(\theta)}{d\theta^n}\theta^n$$

where the derivatives are all evaluated at $\theta = 0$. This series on the right can be rewritten in terms of the exponential function by making note of the Taylor's series for e^x:

$$e^x = \sum_{n=0}^{\infty} \frac{x^n}{n!}$$

and by letting $x = \frac{dr(\theta)}{d\theta}\theta$, one can see that

$$r(\theta) = \sum_{n=0}^{\infty} \frac{d^n r(\theta)}{d\theta^n}\theta^n = e^{\frac{dr(\theta)}{d\theta}\theta}$$

$$r(\theta) = e^{\frac{dr(\theta)}{d\theta}\theta}$$

again evaluated at $\theta = 0$. We previously showed that $r(\theta) = e^{\theta J}$, so if we examine these two ways of writing a rotation in terms of group elements, we come to the conclusion that the basis generators J can be calculated from

$$e^{\theta J} = e^{\frac{dr(\theta)}{d\theta}\theta}$$

$$J = \lim_{\theta \to 0} \frac{dr(\theta)}{d\theta}$$

We will see from quantum mechanics in Chap. 4 that observables are represented by Hermitian operators on the complex function space of states known as Hilbert space. Hermitian operators obey $H^\dagger = H$, where the dagger means the complex conjugate transpose (adjoint) of the operator. Since we need to work with Hermitian operators, it is conventional in physics to introduce a factor of i so that the definition of the generators is modified to be

$$J = \frac{1}{i} \lim_{\theta \to 0} \frac{dr(\theta)}{d\theta}$$

With the explicit matrix representations of the rotation operators given above, we can find these generators by differentiation. The rotation matrices are

$$R_z(\theta) = \begin{bmatrix} \mathrm{Cos}[\theta] & \mathrm{Sin}[\theta] & 0 \\ -\mathrm{Sin}[\theta] & \mathrm{Cos}[\theta] & 0 \\ 0 & 0 & 1 \end{bmatrix},$$

$$R_x(\phi) = \begin{bmatrix} 1 & 0 & 0 \\ 0 & \text{Cos}[\phi] & \text{Sin}[\phi] \\ 0 & -\text{Sin}[\phi] & \text{Cos}[\phi] \end{bmatrix},$$

and

$$R_y(\psi) = \begin{bmatrix} \text{Cos}[\psi] & 0 & -\text{Sin}[\psi] \\ 0 & 1 & 0 \\ \text{Sin}[\psi] & 0 & \text{Cos}[\psi] \end{bmatrix}.$$

Evaluation of the generators in *Mathematica* requires two steps, the first is to find the derivatives and the second is to take the limits.

```
dRx[ϕ_] := (1/i) D[Rx(ϕ),(ϕ)]  (*etc. for Ry and Rz.*)
Jx = Limit[dRx[ϕ],ϕ→ 0//MatrixForm  (* etc. for ψ and θ. *)
```

This yields our set of basis generators for rotations about the three axes

$$J_x = \begin{bmatrix} 0 & 0 & 0 \\ 0 & 0 & -i \\ 0 & i & 0 \end{bmatrix}, \quad J_y = \begin{bmatrix} 0 & 0 & i \\ 0 & 0 & 0 \\ -i & 0 & 0 \end{bmatrix}, \quad \text{and } J_z = \begin{bmatrix} 0 & -i & 0 \\ i & 0 & 0 \\ 0 & 0 & 0 \end{bmatrix}.$$

as a set of traceless, antisymmetric matrices, as they must be in order to satisfy the conditions of orthogonality ($R^T R = 1$) with $\det(R) = 1$. We can show this by writing an element of the group as $R = e^{J\theta}$ in terms of the generators, then

$$R^T R = \left(e^{J\theta}\right)^T e^{J\theta} = e^{J^T\theta} e^{J\theta} = e^{\theta(J^T + J)} = 1$$

which is true if $J^T + J = 0$, so that $e^0 = 1$ and we have that $J^T = -J$: the generators must be antisymmetric. Looking at the condition that $\det(R) = 1$,

$$\det(R) = \det\left(e^{J\theta}\right) = e^{\theta Tr[J]} = 1$$

gives us that

$$Tr[J] = 0,$$

the matrices of the generators must have trace zero.

3.9.4.2 Lie Algebras and the Commutator

At this point, we come to a subtle but extremely important point that often puzzles students. We began with a set of rotation operators as real matrices. We showed that this set formed a Lie Group by explicitly calculating the group combination table using matrix multiplication as the combination rule ($\circ$). However, we could quickly discover, using *Mathematica*, that the set of *Js* we just found do not form a group at all if matrix multiplication is used as the combination rule:

$$j = \{Jx, Jy, Jz\};$$

$$\text{TableForm}[\text{Table}[\text{Table}[j[[i]].j[[k]], \{i, 1, 3\}], \{k, 1, 3\}],$$

$$\text{TableHeadings} \rightarrow \{\{"Jx", "Jy", "Jz"\}, \{"Jx", "Jy", "Jz"\}\}]$$

results in

	Jx	Jy	Jz
Jx	$\begin{bmatrix} 0 & 0 & 0 \\ 0 & 1 & 0 \\ 0 & 0 & 1 \end{bmatrix}$	$\begin{bmatrix} 0 & -1 & 0 \\ 0 & 0 & 0 \\ 0 & 0 & 0 \end{bmatrix}$	$\begin{bmatrix} 0 & 0 & -1 \\ 0 & 0 & 0 \\ 0 & 0 & 0 \end{bmatrix}$
Jy	$\begin{bmatrix} 0 & 0 & 0 \\ -1 & 0 & 0 \\ 0 & 0 & 0 \end{bmatrix}$	$\begin{bmatrix} 1 & 0 & 0 \\ 0 & 0 & 0 \\ 0 & 0 & 1 \end{bmatrix}$	$\begin{bmatrix} 0 & 0 & 0 \\ 0 & 0 & -1 \\ 0 & 0 & 0 \end{bmatrix}$
Jz	$\begin{bmatrix} 0 & 0 & 0 \\ 0 & 0 & 0 \\ -1 & 0 & 0 \end{bmatrix}$	$\begin{bmatrix} 0 & 0 & 0 \\ 0 & 0 & 0 \\ 0 & -1 & 0 \end{bmatrix}$	$\begin{bmatrix} 1 & 0 & 0 \\ 0 & 1 & 0 \\ 0 & 0 & 0 \end{bmatrix}$

None of these matrices are any of the *Js*, even if multiplied by *i*. Our set of generators do not form a *Lie Group* because we need a different role of combination. The generators form the basis for a linear vector space known as the *Lie algebra*. From our derivation of the generators as derivatives, it can be understood that they lie in the tangent space at the identity, just as the derivative of a continuous function gives the tangent to the function at any point. The rule of combination is that obtained from the Baker–Campbell–Hausdorff (BCH) theorem that connects the combination of *Lie Group* elements with combinations of elements of the corresponding *Lie algebra* (the generators). Let *j* and *h* be elements of the *Lie Group G*, $\{j, h\} \in G$, and their corresponding generators, *J* and *H* be elements of the associated *Lie algebra* **g**, $\{J, H\} \in$ **g**, then the combination of elements of the *Lie Group G* is written, using the BCH theorem, as

$$j \circ h = e^J \circ e^H = e^{J+H+\frac{1}{2}[J,H]+\frac{1}{12}[J,[J,H]] - \frac{1}{12}[H,[J,H]]+\dots}$$

where the left side is the combination of two elements of the *Lie Group*, while the right side contains a combination of two elements (generators) of the *Lie Algebra*. The new concept introduced here is that of the *Lie Bracket*, $[J, H]$ defined, for matrix *Lie Algebras*, as

$$[J, \ H] = JH - HJ$$

which in physics is called a *commutator*. It is interesting that although the matrix products JH and HJ of the generators of the *Lie Algebra* do not need to be in the *Lie Algebra*, their commutator must be.

So, let's see if our generators of the *Lie Algebra* for rotations are closed under commutation. We can define the commutator in *Mathematica* as

$$\mathrm{comm}[a\_, b\_] := a.b - b.a$$

where the dot signifies matrix multiplication, as usual.

$$\mathrm{TableForm}[\mathrm{Table}[\mathrm{Table}[i\ \mathrm{comm}[j[[l]], j[[k]]], \{k, 1, 3\}], \{l, 1, 3\}],$$

$$\mathrm{TableHeadings} \rightarrow \{\{"Jx", "Jy", "Jz"\}, \{"Jx", "Jy", "Jz"\}\}]$$

results in

	Jx	Jy	Jz
Jx	$\begin{bmatrix} 0 & 0 & 0 \\ 0 & 0 & 0 \\ 0 & 0 & 0 \end{bmatrix}$	$\begin{bmatrix} 0 & i & 0 \\ -i & 0 & 0 \\ 0 & 0 & 0 \end{bmatrix}$	$\begin{bmatrix} 0 & 0 & i \\ 0 & 0 & 0 \\ -i & 0 & 0 \end{bmatrix}$
Jy	$\begin{bmatrix} 0 & -i & 0 \\ i & 0 & 0 \\ 0 & 0 & 0 \end{bmatrix}$	$\begin{bmatrix} 0 & 0 & 0 \\ 0 & 0 & 0 \\ 0 & 0 & 0 \end{bmatrix}$	$\begin{bmatrix} 0 & 0 & 0 \\ 0 & 0 & i \\ 0 & -i & 0 \end{bmatrix}$
Jz	$\begin{bmatrix} 0 & 0 & -i \\ 0 & 0 & 0 \\ i & 0 & 0 \end{bmatrix}$	$\begin{bmatrix} 0 & 0 & 0 \\ 0 & 0 & -i \\ 0 & i & 0 \end{bmatrix}$	$\begin{bmatrix} 0 & 0 & 0 \\ 0 & 0 & 0 \\ 0 & 0 & 0 \end{bmatrix}$

Or, what this tells us is that the *Lie Algebra* (commutators) of the rotation generators is indeed closed when commutation is used as the rule of combination, for this set of matrices reduces to

$$\begin{bmatrix} 0 & -Jz & Jy \\ Jz & 0 & -Jx \\ -Jy & Jx & 0 \end{bmatrix}$$

The generators of the *Lie Algebra* form a group, with the identity as 0, and the inverse rotation as the negative of the generator. The commutator of any pair of generators is within the group. These J matrices form a basis for the set of generators of the special orthogonal group in that any generator belonging to SO(3) can be written as a linear combination of the basis generators. A shorthand for the generators is

$$(J_k)_{lm} = -\varepsilon_{klm}$$

where ε_{klm} is the Levi–Civita symbol (the completely antisymmetric tensor in three dimensions; *N.B.*: *Mathematica* recognizes this object).

We can summarize the commutation relations of the *Lie Algebra* generators in a manner we have already seen as

$$[J_k, J_l] = i\,\varepsilon_{klm}J_m$$

Since the generators of rotation are the angular momentum operators of quantum mechanics, they must also be Hermitian; i.e., we must have that

$$(J_k)^\dagger = (J_k), \forall k \in \{1, 2, 3\}$$

where the dagger symbol indicates the adjoint or complex-conjugate transpose of the matrix for the generator.

3.9.5   *The Special Unitary Group SU(2)*

In quantum mechanics (Chap. 4) and particularly with respect to spin magnetic resonance (Chap. 8), we will need to understand not only transformations such as rotations in external spaces but also rotations in internal spaces. We therefore need to introduce at this point a 2D representation of the special unitary group SU(2). SU(2) describes spin, whose smallest nonzero eigenvalue is $\pm\frac{1}{2}$. The group elements of SU(2) are a set of 2×2 special, Hermitian, traceless, unitary matrices with unit determinant that describe rotations in 2D.

Let $U \in$ SU(2) be an element of SU(2), then by unitarity we have that

$$U^\dagger U = 1$$

and by its special nature we know that the determinant is $+1$:

$$\det(U) = 1$$

A 2×2 matrix contains four complex elements or eight total elements. Then properties of Unitarity, Hermiticity, Tracelessness, and Specialness reduce this number to three independent elements.

We can find one of the matrices that serves as a generator for the Lie Algebra for SU(2) by differentiating the 2D rotation matrix to give σ_2:

$$R(\theta) = \begin{bmatrix} \mathrm{Cos}\theta & \mathrm{Sin}\theta \\ -\mathrm{Sin}\theta & \mathrm{Cos}\theta \end{bmatrix}$$

$$d\sigma_2[\theta\_] := (1/i)D[R[\theta], \theta]$$

$$\mathrm{Print}[''\sigma_2 = '', \mathrm{Limit}[d\sigma_2[\theta], \theta - > 0]//\mathrm{MatrixForm}$$

$$\sigma_2 = \begin{bmatrix} 0 & -i \\ i & 0 \end{bmatrix}$$

It remains to find σ_1 and σ_3. The eigenvectors of spin are $\{1, 0\}$ and $\{0, 1\}$ with eigenvalues $\{1, -1\}$ in a representation in which σ_3 is diagonal, therefore, the matrix for σ_3 is

$$\sigma_3 = \begin{bmatrix} 1 & 0 \\ 0 & -1 \end{bmatrix}$$

We can find the remaining σ_1 by the following reasoning. For any traceless 2×2 matrix u:

$$u = \begin{bmatrix} a & b \\ c & d \end{bmatrix}$$

we must have $a = -d$, so that $\mathrm{Trace}[u] = 0$, and

$$u = \begin{bmatrix} a & b \\ c & -a \end{bmatrix}$$

Hermiticity implies that $c = b^*$, and from Specialness and Hermiticity, the matrix elements can only be $\{0, \pm 1, \pm i\}$. Then, u should be

$$u = \begin{bmatrix} a & b \\ b^* & -a \end{bmatrix}$$

From $\mathrm{Det}[u] = 1$, we have that

$$\text{Det}[u] = \text{Det}\begin{bmatrix} a & b \\ b^* & -a \end{bmatrix} = \begin{bmatrix} aa^* + bb^* & -ab + ba^* \\ ab^* - a^*b^* & aa^* + bb^* \end{bmatrix} = \begin{bmatrix} 1 & 0 \\ 0 & 1 \end{bmatrix}$$

which gives $a = a*$, or that a is real, and if we let $a \in \{-1, 0, 1\}$, we find from

$$aa^* + bb^* = 1$$

that for $a = 0$, $bb* = 1$ so that b is either i or 1. However, with $b = i$, we get σ_2 again, and because these generators need to be linearly independent, the only remaining choice is $a = 0$, $b = 1$ and we have that

$$\sigma_1 = \begin{bmatrix} 0 & 1 \\ 1 & 0 \end{bmatrix}$$

The σ matrices are the famous Pauli matrices and form a set of basis generators for SU(2). Any 2×2 Hermitian traceless matrix can be written as linear combination of these matrices. We can also define a vector $\boldsymbol{\sigma} = \{\sigma_1, \sigma_2, \sigma_3\}$.

3.9.6 *The* Lie Algebra *of SU(2)*

The *Lie Algebra* of SU(2) is found by putting the Pauli matrices into the commutator and computing all combinations:

$$
\begin{array}{c|ccc}
 & \sigma_1 & \sigma_2 & \sigma_3 \\
\hline
\sigma_1 & \begin{bmatrix} 0 & 0 \\ 0 & 0 \end{bmatrix} & \begin{bmatrix} 2i & 0 \\ 0 & -2i \end{bmatrix} & \begin{bmatrix} 0 & -2 \\ 2 & 0 \end{bmatrix} \\
\sigma_2 & \begin{bmatrix} -2i & 0 \\ 0 & 2i \end{bmatrix} & \begin{bmatrix} 0 & 0 \\ 0 & 0 \end{bmatrix} & \begin{bmatrix} 0 & 2i \\ 2i & 0 \end{bmatrix} \\
\sigma_3 & \begin{bmatrix} 0 & 2 \\ -2 & 0 \end{bmatrix} & \begin{bmatrix} 0 & -2i \\ -2i & 0 \end{bmatrix} & \begin{bmatrix} 0 & 0 \\ 0 & 0 \end{bmatrix}
\end{array}
$$

Or, as a table of all the combinations:

$$
2i \begin{bmatrix} 0 & \sigma_3 & \sigma_2 \\ -\sigma_3 & 0 & \sigma_1 \\ -\sigma_2 & -\sigma_1 & 0 \end{bmatrix}
$$

which we can summarize as

$$[\sigma_k, \sigma_l] = 2i \, \varepsilon_{klm}\sigma_m$$

The generators of the *Lie Algebra* of SU(2) form a group, with the identity as 0, the inverse rotation as the negative of the generator, and the commutator of any pair of generators is within the group. The *Lie Algebra* of SU(2) is closed under commutation. Note that using the commutator as the combination rule for SU(2) results in the same *Lie Algebra* as that found for SO(3), except for an *important factor of 2*. One incorporates this factor of 2 into the definition of the generators of SU(2) and writes the generators as

$$J_k = \frac{1}{2}\sigma_k$$

so that the *Lie Algebra* becomes

$$[J_k, J_l] = i \, \varepsilon_{klm}J_m$$

which we found previously for the generators of SO(3). We conclude that SU(2) and SO(3) share the same *Lie Algebra*. The factor of 2 obtained when computing the commutators is related to the difference between rotations in 2D and 3D; rotations in 2D operate on *spinors*, rather than vectors, and SU(2) serves to describe half-integral spin objects.

3.9.7 *The* Lie Algebra *of SO(3)*

One might expect that SO(3) is a more important group in physics than SU(2), but that expectation is misleading, as can be seen from the fact that they share the same *Lie Algebra*; they are somehow related and the reason for the relationship should tell us which is most important. A theorem in Lie groups states that one can have only a single distinguished (simply-connected) *Lie group* that corresponds to a *Lie Algebra*, and this distinguished *Lie group* is referred to as the *covering group*. Distinguished (simply-connected) *Lie groups* are therefore the most fundamental.

In order to understand the meaning of simply-connected, let's once again consider the *Lie group* SO(3); the transformation operators are rotation matrices that depend on continuous parameters (the rotation angles). We can think of these rotation angles as a set of coordinates, but mathematicians prefer to call this set a *manifold*. We used the fact that trigonometric functions of these angles are continuously differentiable in order to find the generators for the *Lie Algebra*. The manifold of rotation angles is simply-connected in the sense that any closed path on this manifold can be smoothly shrunk to a point. The single distinguished, covering group of SO(3) is SU(2) and, in fact, SU(2) covers SO(3) twice. Hence, SU(2) is called the *double cover* of SO(3). We will find that SO(3) contains two representations of SU(2). What then is a representation of a group?

In quantum mechanics, a representation of a group, is a set of operators that obey the group rules; in our exploration of rotations in 2D and 3D, we came across the angular momentum operators and, by starting with rotation matrices, we showed how they combined. A theorem in Group theory states that any group can be represented by a set of matrices. The matrices form a representation of the group. The group representations (matrices) often contain one or more copies of smaller representations (matrices of lower dimension) within them, and when the matrices of lowest dimension are identified, it turns out that these form the bases for the representation. These bases are called *irreducible representations*. Any other representation can be formed using linear combinations of the *irreducible representations*. Objects that are allowed to exist in the universe must conform to the symmetries of space (and as later we will see, time as well). A most profound statement is that the objects that exist in space-time *are* the *irreducible representations* of the symmetries of the universe. How more fundamental can you get in your search for the answer's to Faust's quest? The poster child for this is the discovery that Quarks are point-like constituents of the proton through deep inelastic scattering experiments done by Taylor, Friedman, and Kendall at the Stanford Linear Accelerator Center around 50 years ago, work that was awarded the Noble Prize in physics in 1990 (https://www.nobelprize.org/prizes/physics/1990/summary/).

3.10   Symmetries of 4D Spacetime

We could now profitably proceed to a direct determination of the *irreducible representations* of SU(2) and SO(3), but, with the reader's indulgence, we will instead move on to expand our study of the symmetries of the universe with the derivation of the *Lie Algebra* of a larger group, the *Poincaré* group, which includes rotations in 4D, as well as velocity changes (boosts) and incorporates Einstein's special relativity. The *irreducible representations* of the rotation group in 3D will then be seen as a subgroup of the *Poincaré* group, while the *Lie Algebra* of the complete group will be seen to provide a natural constraint on the properties of the contents of the universe.

3.10.1   *The Only Objects Allowed in Space-Time Are Irreps of the Double Cover of the Poincaré Group*

Up to this point, we have dealt with symmetries that involved one, two, or three dimensions, discovering along the way that the simplest axiomatic properties of the Universe lead to profound logical constraints on it. Examination of these symmetries was then a prelude to a deeper study of the structure of space-time. We now need to investigate the full set of space-time symmetries that arise from the fact that the universe is not simply one or three dimensional. The work of Einstein and

Minkowski showed that spacetime (and even the use of this term itself hints at its 4D nature) is four dimensional. Furthermore, it is quite clear that the four dimensions must be treated on equal footing where space and time are the same thing. All of the properties of three spatial and one time dimension must therefore be embedded within the four-dimensional spacetime continuum (manifold). We might also expect to find additional properties in this unification that only emerge in a 4D universe and we will see that this expectation is fulfilled in a very sophisticated way.

3.10.2 Physicality

The unification of space-time would not have the interest that it does without the attendant discovery that there is a universal physical limit to the speed of information transfer between spatially separated points in spacetime. This limit is, of course, the speed of light. That there exists a finite information transfer velocity, in hindsight, should not have surprised anyone in science. Physical, as opposed to mathematical, infinities give rise to logical paradoxes. One of the lessons of the twentieth century is that there are only a finite set of means at our disposal with which to determine the properties of the universe. We can use only those tools that *exist* in the universe in order to measure its properties. Well, let's use them and find out what additional features of the universe arise from this unification of spacetime. Our simple study of symmetries has given us the expectation that we will find more conservation laws, and these we will find, but the bonus awaiting us is the discovery that the properties of 4D space-time tell us what can exist in the universe.

3.10.3 Special Relativity and the Poincaré Group

> There once was a man named Wright
> Who could travel faster than light.
> He departed one day,
> In a relative way,
> And returned the previous night.
> Source: Common knowledge among physics majors.

Spacetime is 4D and therefore points in space-time are represented by vectors with four components (commonly called 4-vectors, in order to distinguish them from vectors in 3D; i.e., 3-vectors). The symmetries in relativistic physics are then the set of operators in 4D that translate, rotate, and boost (increase the velocity in one frame to match that in another) a coordinate system that preserve the length of 4-vectors.

3.10.3.1   The Lorentz Invariance of Spacetime

In 1905, Einstein derived special relativity from two postulates:

1. *The Laws of Physics are the same in all inertial reference frames.*
2. *The velocity of light is constant in all inertial reference frames.*

An inertial frame is one in which the observer is at rest, although it can be moving with respect to another inertial frame, but it is not accelerated. Special relativity quickly became known as an invariance, or symmetry, principle arising from the more general statement:

2'. *There is a finite, physical limit to the speed with which information can be sent from one point in spacetime to another.* A corollary to this is:
There are **no** *physical infinities.*

A way to send information faster that the speed of light does not exist. We understand this now as a fundamental property of the universe. We will use this property to derive the types and properties of objects that can exist in spacetime. It might appear at first thought that 4D spacetime would have the same properties as 3D Euclidean space, simply with time as an added coordinate, but that is not what we will discover.

Our first task is then to find the metric of the transformations in 4D spacetime. We can investigate this using a famous *gedanken experiment* (thought experiment) of Einstein. We have seen that rotations in a vector space preserve the length of any vector. In 4D space-time, events are denoted by their coordinates $\{x, y, z, t\}$ and the distance between two events is given by the length of the vector connecting them. This distance should be the same in all inertial reference frames. Orthogonal transformations in 4D preserve the length of this vector.

In ordinary Euclidean 3D space, the squared length of a vector S with components $\{S_x, S_y, S_z\}$

$$S = \sum_{i=1}^{3} S_i e_i$$

is its inner product with itself

$$[\text{Length}(S)]^2 = S \cdot S = S_x^2\, e_x \cdot e_x + S_y^2\, e_y \cdot e_y + S_z^2\, e_z \cdot e_z$$

If we write the vector S as a matrix $\{S_x e_x,\ S_y e_y,\ S_z e_z\}$, then we can write its inner product with itself as

$$S \cdot S = S^T S = \begin{bmatrix} S_x e_x \\ S_y e_y \\ S_z e_z \end{bmatrix} \begin{bmatrix} S_x e_x & S_y e_y & S_z e_z \end{bmatrix} = \sum_{i=1}^{3} S_i^2 e_i \cdot e_i$$

Here, we add the squares of the components of S, and all the squares add with the same sign. Notice that we have introduced the transposition operation to transform S from a row vector to a column vector so that the matrix multiplication makes sense.

Another way to do this involves the properties of the orthonormal basis $\{e_x, e_y, e_z\}$ and the introduction of the transformation (T) matrix, which is the Euclidean metric, g_{ij}, $i, j \in \{1, 2, 3\}$, where

$$g_{ij} = \begin{bmatrix} e_x \cdot e_x & e_x \cdot e_y & e_x \cdot e_z \\ e_y \cdot e_x & e_y \cdot e_y & e_y \cdot e_z \\ e_z \cdot e_x & e_z \cdot e_y & e_z \cdot e_z \end{bmatrix} = \delta_{ij} = \begin{bmatrix} 1 & 0 & 0 \\ 0 & 1 & 0 \\ 0 & 0 & 1 \end{bmatrix}$$

and then

$$S \cdot S = \sum_{i=1}^{3} \sum_{j=1}^{3} S_i g_{ij} S_j = \begin{bmatrix} S_x & S_y & S_z \end{bmatrix} \begin{bmatrix} 1 & 0 & 0 \\ 0 & 1 & 0 \\ 0 & 0 & 1 \end{bmatrix} \begin{bmatrix} S_x \\ S_y \\ S_z \end{bmatrix} = S_x^2 + S_y^2 + S_z^2$$

In 3D Euclidean space, we note that the metric is just the identity 3×3 matrix that gives the correct signs for the sum of the components. 3D Euclidean space is flat because the metric is diagonal, i.e., the only nonzero matrix elements are arrayed along the diagonal. We could also incorporate time into this formalism by writing the position of an event in space-time as a 4D vector $\{x, y, z, t\}$ but here we notice that the units of time and space are different: time is measured in seconds, while distance is determined in meters. This makes about as much sense as measuring the length of a football field in meters and its width in furlongs. We then have to be careful about the meaning of the "distance" between two events in space-time. Einstein solved this problem by converting time into a distance using $x = ct$ so that space and time are equivalent.

We can understand this revolution in thought by considering in two dimensions: what happens when two observers move with respect to each other. The space-time distance between events in one reference frame must be the same in the other reference frame if the laws of physics are to be invariant under the transformation from one to the other. Let us represent an event in one frame by a vector, S, which in two dimensions has two components, $S = \{x, t\} = \{x e_1, t e_2\}$. An object stationary in this frame will trace out a path (called a "world line") characterized by a constant position, such as $x = 1$, as illustrated in Fig. 3.3 (red solid line) where the object rests at $x = 1$ for all time.

If we instead consider an object moving with velocity v such that $x(t) = vt$, we will find that the line describing the motion of the object tilts to the right with a slope v (green dashed line in Fig. 3.3). The transformation from the stationary frame to the moving frame is a rotation of the t axis by an angle θ, while $x = x'$. This transformation is a change of basis from unit vectors $\{e_1, e_2\} = \{\hat{x}, \hat{t}\}$ to $\{e_1', e_2'\} = \{\hat{x}', \hat{t}'\}$, where $v\hat{t}' = \hat{x}'$, and $\hat{t} = \hat{t}'$. The rotation angle, θ, can be obtained from the Pythagorean

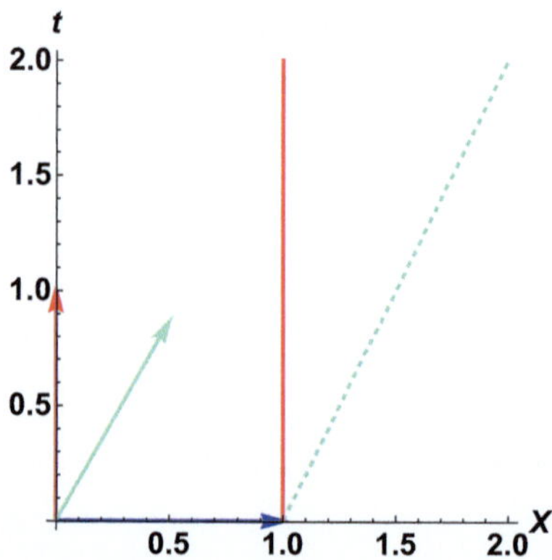

Fig. 3.3 The Galilean transformation from a stationary to a moving system. Shown are the world lines of objects in a 2D $\{x, t\}$ coordinate system at rest (red—solid line) for which the equation of motion is $x = 1$. For an object moving with velocity $v = 1/2$ (green — dashed line), the equation of motion is $x'(t) = t/2 + 1$. The unit vectors $\{e_1, e_2\} = \{\hat{x}, \hat{t}\}$ and $\{e_1', e_2'\} = \{\hat{x}' = \hat{x}, \hat{t}'\}$ in the x, t and x', t' dimensions are shown as red, blue, and green arrows along the axes. The transformation from the stationary frame to the moving frame is a rotation of only the t axis while $x = x'$

theorem as $\sin\theta = \frac{v}{\sqrt{v^2+1}}$ and $\cos\theta = \frac{1}{\sqrt{v^2+1}}$ (which gives $\tan\theta = v$). The relationship between the primed and unprimed unit vectors is

$$e_1 = e_1'$$

and

$$e_2' = \cos\theta\, e_2 + \sin\theta\, e_1.$$

Note that $\{e_1', e_2'\}$ are still unit vectors. The transformation matrix T_{ij} is the matrix whose elements are the pairwise inner products of the basis vectors:

$$T_{ij} = \begin{bmatrix} e_1 \cdot e_1' & e_1 \cdot e_2' \\ e_2 \cdot e_1' & e_2 \cdot e_2' \end{bmatrix} = \begin{bmatrix} 1 & \sin\theta \\ 0 & \cos\theta \end{bmatrix}$$

which is a rotation of the t axis in 2D about the origin by an angle θ, leaving the x-axis unchanged. This transformation does *not* mix the x and t coordinates. It corresponds to a Galilean *boost* of the coordinate system from zero to velocity v along the x-axis.

$$\begin{bmatrix} x' \\ t' \end{bmatrix} = \begin{bmatrix} 1 & -v \\ 0 & 1 \end{bmatrix} \begin{bmatrix} x \\ t \end{bmatrix}$$

As the velocity of the moving frame increases, the t-axis rotates into the x-axis: at infinite velocity the t- and x-axes are parallel (see Box 3.1). The negative t'-axis also makes an angle θ with the t-axis, thereby forming a cone of opening angle 2θ. This Galilean transformation is not described by orthogonal matrices:

$$T^T T = \begin{bmatrix} 1 & \mathrm{Sin}[\theta] \\ \mathrm{Sin}[\theta] & 1 \end{bmatrix}$$

and this is *not* the identity matrix. A *Mathematica* notebook that explores the transformation of the coordinate axes as the velocity increases from zero (Box 3.1) shows that as v increases the angle θ between the t- and t'-axes increases. Here we use the Manipulate wrapper function (see Appendix 1) so that we can continuously vary the relative velocity of the two frames. I urge the reader to use this notebook to observe how the time axis tilts into the space axis, while the space axis does not change. Such was the prevailing wisdom until Einstein found that the correct transformation rotated both axes onto the limiting line at $x = ct$.

Box 3.1 A Galilean Transformation of Space-Time
(* This is a Galilean transformation that only alters the x-axis. *)

```
gtp[v_] = 1/Sqrt[1 + v^2]
gxp[v_] = v/Sqrt[1 + v^2]


tphat[v_, t_, x_] = t gtp[v] + x gxp[v]
xphat[v_, t_, x_] = x gtp[v] + t gxp[v]


bag[v_] =
Graphics[{Blue, Arrow[{{0, 0}, {xphat[v, 0, 1], tphat[0,
0, 1]}}]},
Axes -> True, AxesLabel -> {x, t}]
rag[v_] =
Graphics[{Red, Arrow[{{0, 0}, {tphat[-v, 0, 1], xphat[0,
0, 1]}}]},
Axes -> True, AxesLabel -> {x, t},
LabelStyle -> Directive[Black, Medium, Bold]]


Manipulate[Show[
{rag[v],
ListPlot[Table[{tphat[-v, t, x], x}, {x, 0, 5, 1}, {t, 0, 5, 1}],
Joined -> True, PlotStyle -> Hue[0.7]], bag[v],
ListPlot[Table[{xphat[-v, t, x], t}, {x, 0, 5, 1}, {t, 0, 5, 1}],
Joined -> True, PlotStyle -> Hue[0]]}], {v, 0, -0.9}]
```

(continued)

Box 3.1 (continued)

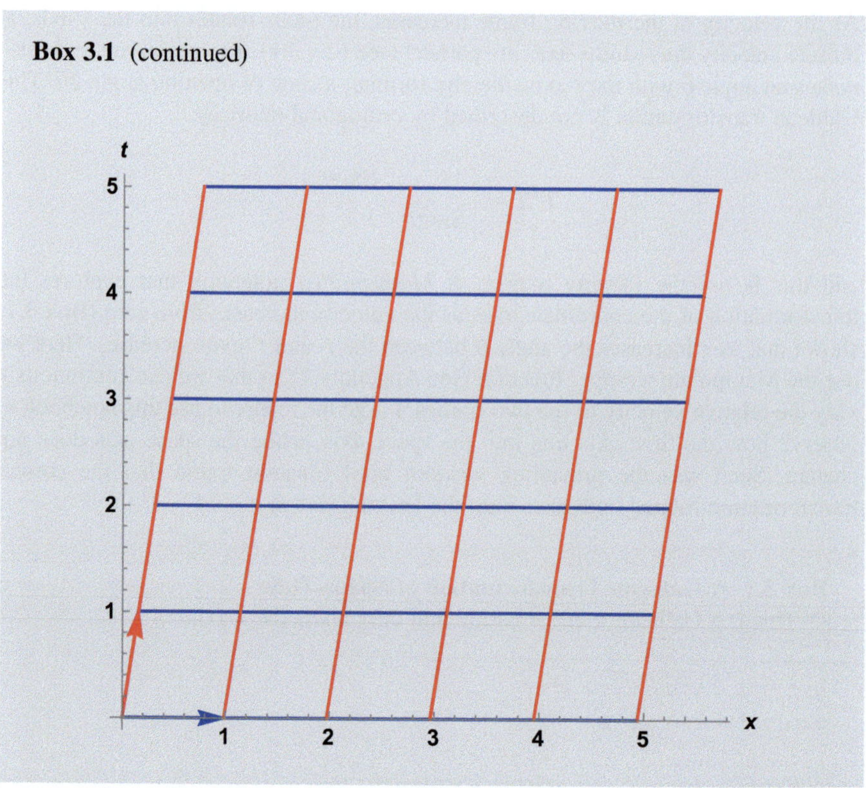

A change of basis cannot change the length of a vector, in this case, the space-time interval between two events. In the stationary frame the distance between two events $\{x_1, t_1\}$ and $\{x_2, t_2\}$ is the length of the vector S, where

$$S^2 = (x_2 - x_1)^2 + (t_2 - t_1)^2$$

while in the moving frame this same distance is

$$S'^2 = (x'_2 - x'_1)^2 + (t'_2 - t'_1)^2$$

Note the minus signs here.

3.10.3.2   Einstein's *Gedanken Experiment*

How do we compute the distance between events in this new 4D approach in which space and time are on equal footing? Let's proceed with Einstein's *gedanken experiment*. Imagine standing on a stationary platform and watching a train with speed u pass us that carries a mirror on an inside wall visible to you through a

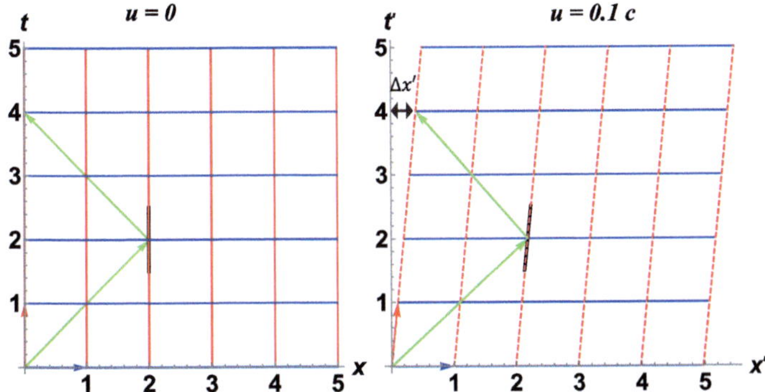

Fig. 3.4 Space-time diagrams ($c = 1$) showing the light pulse (green arrow) leaving the origin $(0, 0)$, striking the mirror at $(2, 2)$, and returning to the experimenter at $(0, 4)$. (*Left*) The path of light as observed within the train car by the experimenter. (*Right*) The path of the light pulse as seen by an observer standing on the stationary platform, as the train passes at a speed of $u = 0.1c$. Unit vectors along the space (blue) and time (red) axes are shown extending from the origin. The extra distance ($\Delta x'$) traveled by the light in the primed (observer's) coordinate system is indicated on the left

window. There is an experimenter in the train car (in an unprimed coordinate system) that uses a flashlamp at position x_1 to send a pulse of light across the car to the mirror at x_2 and he can measure the amount of time, Δt, that the light takes to leave his lamp, strike the mirror, and return to the position x_1 of the flashlamp. In Fig. 3.4 (left), we have illustrated this in a 2D space-time diagram. We can separate this action into three main events in the unprimed coordinates (where $u = 0$, and we have set $c = 1$, in order to have a square grid):

1. The light pulse leaves his lamp at position $(x_1, t_1) = (0, 0)$.
2. The light pulse strikes the mirror at $(x_2, t_2) = (2, 2)$ and is reflected.
3. The light pulse returns to the starting point $(x_3, t_3) = (x_1, t_3) = (0, 4)$.

First, let's determine the space-time distance between Events **1** and **3** in the experimenter's unprimed coordinate system. If the distance between the flash lamp and the mirror is $D = x_2 - x_1$, then the elapsed time between events **1** and **3** measured by our experimenter in the moving train car is

$$\Delta t = t_3 - t_1 = \frac{2D}{c}$$

where c is the speed of the signal (light) and the light returns to the same position, so $\Delta x = x_3 - x_1 = 0$.

For us, standing on the platform (in the primed coordinate system), the train car passes by at a velocity u. If the origin of our primed coordinate system coincides with that of the unprimed coordinate position of the flashlamp at t_1 ($x_1(t_1) = x_1{}'(t_1) = 0$), then we will observe that the light returns to a different position x_3' from the flashlamp at t_3. This difference is given by (see Fig. 3.4):

$$x_1'(t_1) = 0 \neq x_3' = u\Delta t' = \Delta x'$$

where

$$\Delta t' = t_3' - t_1' = \frac{2d}{c}$$

and the distance we observed the light pulse to travel is

$$2d = \Delta x' = x_3' - x_1' > 0$$

The Pythagorean theorem can be used to find the extra distance traveled as

$$d^2 = D^2 + \left(\frac{u\Delta t'}{2}\right)^2$$

or, using $d = \frac{c\Delta t'}{2}$ and $u\Delta t' = \Delta x'$

$$(c\Delta t')^2 = 4\left(D^2 + \left(\frac{\Delta x'}{2}\right)^2\right)$$

or

$$(c\Delta t')^2 - (\Delta x')^2 = 4D^2$$

Now, we remember that the elapsed time in the train car is $\Delta t = \frac{2D}{c}$ and we can substitute for D above to obtain $(c\Delta t)^2 = 4D^2$, giving us the suggestive result that

$$(c\Delta t')^2 - (\Delta x')^2 = (c\Delta t)^2 - (\Delta x)^2$$

where have added zero to the right-hand side, since $(\Delta x)^2 = 0$. In this simple 2D problem, we have found that the quantity $(\Delta s)^2$ is the same for all observers:

$$(\Delta s)^2 = (c\Delta t')^2 - (\Delta x')^2 = (c\Delta t)^2 - (\Delta x)^2$$

and is the invariant we were seeking.

Except for the minus sign, this expression looks like the distance between two space-time points represented by the vectors, s_1 and s_2:

$$s_1 = \sum_{i=1}^{2} s_{1i}\, e_i$$

$$s_2 = \sum_{j=1}^{2} s_{2j}\, e_j$$

$$\Delta s = s_2 - s_1 = \sum_{j=1}^{2} \left(s_{2j} - s_{1j} \right) e_j$$

where $\{e_i\}$ are the basis vectors in the spacetime. Then $(\Delta s)^2$ is given by

$$(\Delta s)^2 = \sum_{j=1}^{2} \left(s_{2j} - s_{1j} \right) e_j \cdot \sum_{k=1}^{2} \left(s_{2k} - s_{1k} \right) e_k$$

$$(\Delta s)^2 = \sum_{k=1}^{2} \sum_{j=1}^{2} \left(s_{2j} - s_{1j} \right) \left(s_{2k} - s_{1k} \right) e_j \cdot e_k$$

By comparing this with the earlier expression for $(\Delta s)^2$, we are forced to conclude that the transformation matrix $e_j \cdot e_k$ must be the space-time metric for relativity and has the properties that

$$e_j \cdot e_k = \begin{bmatrix} 1 & 0 \\ 0 & -1 \end{bmatrix}$$

And this is the same metric as was found for the complex numbers. All of the above arguments hold, with an extension from one to three spatial dimensions, for 4D spacetime as well and we can write the distance between space-time events as

$$(\Delta s)^2 = (c\Delta t)^2 - (\Delta x)^2 - (\Delta y)^2 - (\Delta z)^2$$

In this manner we arrive at the *Minkowski* metric for 4D space-time (*Minkowski* space) as

$$g_{ij} = e_j \cdot e_k = \begin{bmatrix} 1 & 0 & 0 & 0 \\ 0 & -1 & 0 & 0 \\ 0 & 0 & -1 & 0 \\ 0 & 0 & 0 & -1 \end{bmatrix}$$

This metric tells us the relative signs of the components of the inner product of any two 4-vectors. It also indicates that matter-free spacetime is flat since the metric is once again diagonal. The distance between spacetime events in 4D Minkowski space is a vector; its length is a scalar, and must be invariant because the separation between events cannot depend on our point of view.

3.10.3.3   The Separation Between Space-Time Events

The minus signs on the metric diagonal markedly differ from those for a Euclidean 3D vector space and give rise to different behavior of the spacetime interval between events in 4D. In particular, the separation between events can be zero even if the events do not occur at the same 4D point. To understand this, consider the fact that the length of a 4-vector connecting any two points in *Minkowski* space can be positive, negative, or zero, corresponding to three cases:

$$(\Delta s)^2 > 0; \text{if } (\Delta x)^2 + (\Delta y)^2 + (\Delta z)^2 > (c\Delta t)^2$$

$$(\Delta s)^2 = 0; \text{if } (\Delta x)^2 + (\Delta y)^2 + (\Delta z)^2 = (c\Delta t)^2$$

$$(\Delta s)^2 < 0; \text{if } (\Delta x)^2 + (\Delta y)^2 + (\Delta z)^2 < (c\Delta t)^2$$

Case (1) Spacelike. If the spacetime separation between two points is greater than zero, $(\Delta s)^2 > 0$, then the spatial distance between the points is larger than the light travel distance and the separation is *spacelike*.

Case (2) Lightlike. If the spacetime separation between two points is zero, $(\Delta s)^2 = 0$, then the spatial distance between the points is equal to the light travel distance and the separation is *lightlike* (or *null*). The hypersurface in 4D satisfying $(\Delta s)^2 = 0$ is called the *light cone* (or the *null* surface).

Case (3) Timelike. If the spacetime separation between two points is less than zero, $(\Delta s)^2 < 0$, then the spatial distance between the points is smaller than the light travel distance and the separation is *timelike*.

This hyperbolic separation of spacetime is illustrated in Fig. 3.5. One can use *Mathematica* to generate such a hyperbolic surface in $\{x, y, t\}$ with the code: ContourPlot3D[$t^2 - x^2 - y^2$], $\{t, -2, 2\}$, $\{x, -2, 2\}$, $\{y, -2, 2\}$, AxesLabel $\rightarrow$ $\{t, Y, X\}$ as is shown on the right in Fig. 3.5. A similar plot for rotations in 3D Euclidean space would be a 3-sphere in 4D.

3.10.3.4   The Speed of Light Limits the Rate of Information Transfer

Based on the knowledge that the metric of Minkowski space contains both minus and plus signs, we can understand that there will be situations where the interval between space-time points is a minimum; where $(\Delta s)^2 = 0$ and

$$(c\Delta t)^2 = (\Delta x)^2 + (\Delta y)^2 + (\Delta z)^2$$

Let's divide this expression by $(\Delta t)^2$ and take the infinitesimal limit $\lim\limits_{\Delta x_i \to 0} \Delta x_i = dx_i$, $i \in \{1, 4\}$, to arrive at

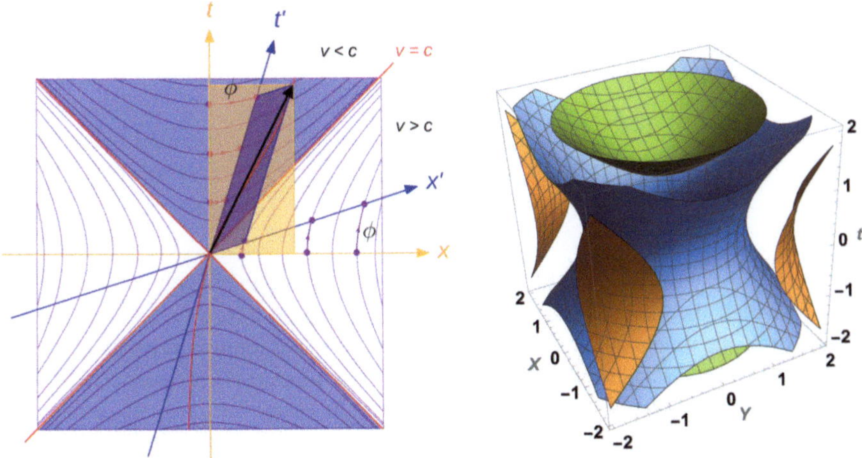

Fig. 3.5 (*Left*) A 2D contour plot in one time and one spatial dimension of Minkowski spacetime, using $c = 1$, with the observer at the origin. The hyperbolic contours are drawn at $(\Delta s)^2 = $ constant. Space-time is divided into three regions: (1) The *timelike*, causally connected purple region: $(\Delta s)^2 < 0$, $v < c$. When the coordinate frame is boosted to velocity, v, space-time points (events) on one hyperbola slide along that hyperbola, but cannot leave it. The light cone is inside the blue region bounded by the (2) *lightlike*, $(\Delta s)^2 = 0$, red lines at 45° passing through the origin for $v = c$ and $x = \pm t$. The present is the line $t = 0$, while the future and the past are in the regions where $t > 0$ and $t < 0$, respectively. For $0 < v < c$ the Lorentz transformation tilts both the space and time axes toward the lightcone by the same angle ϕ. (3) The white *spacelike* region is where $(\Delta s)^2 > 0$, $v > c$. (This image is in the public domain. https://commons.wikimedia.org/wiki/File:Minkowski_lightcone_lorentztransform.svg). (*Right*) A 3D plot in one time and two spatial dimensions of Minkowski space-time. The regions are divided into the future and past (green) and elsewhere (blue) with the observer again at the origin

$$c^2 = \frac{(dx)^2 + (dy)^2 + (dz)^2}{(dt)^2}$$

which is the square of a velocity and since this is computed from the minimum in the space-time separation between two events in Minkowski space, it corresponds to the maximum speed with which one region can communicate with another; this speed is that of light and its velocity is the maximum information transfer speed that exists. One can also derive the correct value for the speed of light from the fundamental constants that enter into Maxwell's equations. It was known in the late nineteenth century that electromagnetic waves propagated at the speed of light.

The fundamental symmetry of Minkowski space is embodied in the set of transformations of special relativity: the Lorentz transformations that leave the inner product invariant in all inertial reference frames. These are embodied in the function $\gamma(v)$ defined as

$$\gamma(v) = \frac{1}{\sqrt{1 - \frac{v^2}{c^2}}}$$

where c is the velocity of light. The difference between a Galilean space-time and that in which the speed of light is a constant is explored in Box 3.2. In Galilean space-time, the time axis rotates into the space axis as the relative velocity increases, whereas in "special relativity," both the space and time axes rotate into each other and at the speed of light, converge onto the light cone. The primary lesson that Einstein taught the world was that space and time are really the same thing and must be placed on equal footing. Once again, I suggest that the reader enjoy manipulating this notebook to see for themselves the marked difference between the Galilean and relativistic transformations of the coordinate system when moving from stationary to moving frames of reference.

Box 3.2

(* This notebook shows the relativistic transformation of 2D space-time as the velocity increases. In order to have a square space-time grid, the speed of light, c, is set to 1. *)

```
c = 1;
γ[v_] := 1/√(1 - v²/c²)
xp[v_, t_, x_] := γ[v](x − vt)
tp[v_, t_, x_] := γ[v](t − vx/c²)
ba[v_]       :=       Graphics[{Blue, Arrow[{{0, 0}, {xp[v, 0, 1], tp[v, 0, 1]}}]}
     Axes  →  True, AxesLabel  →  {x, t}, LabelStyle  →  Directive[Black,
     Medium, Bold]]
ra[v_] := Graphics[{Red, Arrow[{{0, 0}, {tp[v, 0, 1], xp[v, 0, 1]}}]}]
Manipulate[Show[{ba[v], ListPlot[Table[{xp[v, t, x], tp[v, t, x]}, {x, 0, 5, 1},
     {t, 0, 5, 1}], Joined
     → True, PlotStyle
     →Hue[0]], ra[v], ListPlot[Table[{tp[v, t, x], xp[v, t, x]}, {x, 0, 5, 1},
     {t, 0, 5, 1}], Joined
     → True, PlotStyle → Hue[0.7]]}], {v, 0, −0.9}]
```

(continued)

Box 3.2 (continued)

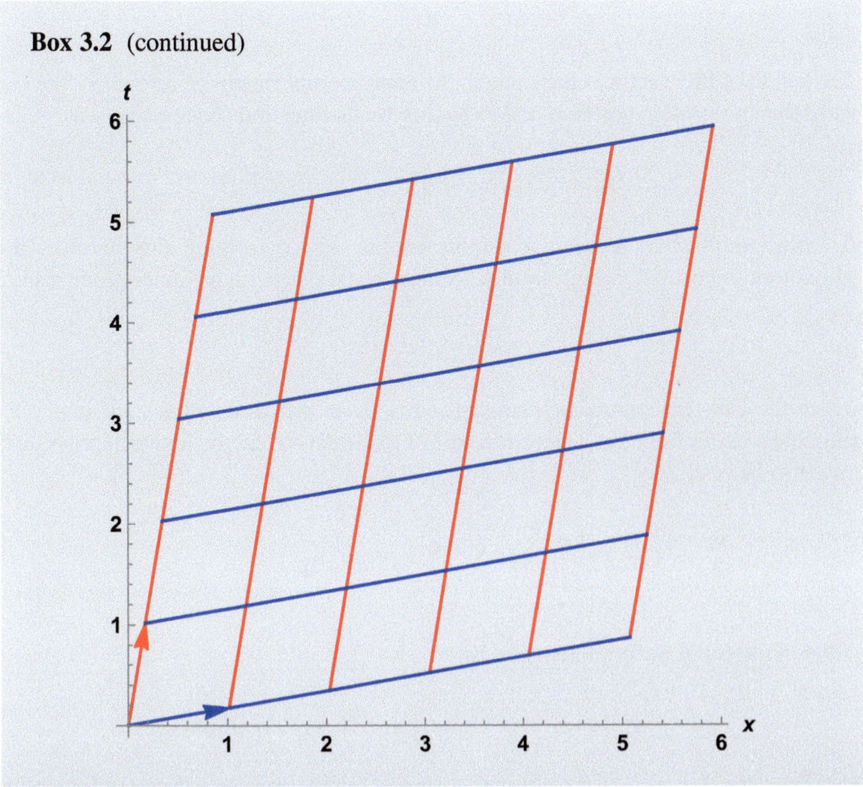

We have motivated our discussion by considering points in spacetime as represented by vectors with four components: $\{x, y, z, t\}$, but we also noticed that space and time in this formalism are measured using different units. We can convert these coordinates to a common system of units by either dividing the spatial coordinates by c or by multiplying the time by c; conventionally, the latter course is taken so that our coordinate manifold becomes $\{x, y, z, ct\}$. This still distinguishes the time from the other three coordinates. In order to treat all four coordinates equally, it is common to let $c = 1$ and then to use the same label for each as in $x_\mu = \{x_1, x_2, x_3, x_4\}^T$ to denote a four-dimensional vector (or 4-vector). (The transpose $[^T]$ is used to indicate that this is a column vector, to save space. The `Transpose` function in *Mathematica* can do the same for vectors, which it treats as lists.) We can use this notation to write the differential of a 4-vector as

$$dx_\mu = \{dx_1,\ dx_2,\ dx_3,\ dx_4\}^T$$

The Lorentz invariant is the length of the 4-vector connecting two points in Minkowski space, one of which can always be chosen to be the origin:

$$(ds)^2 = (cdt)^2 - (dx)^2 - (dy)^2 - (dz)^2$$

The length of this vector is the same in different inertial frames of reference. We can write this in our new notation that explicitly treats time and space equally as

$$(ds)^2 = (dx_1)^2 - (dx_2)^2 - (dx_3)^2 - (dx_4)^2$$

A more complicated looking, but quite useful, way of writing this involves the Minkowski metric $g^{\mu\nu}$, which simply accounts for the signs in the inner product sum:

$$ds^2 = g^{\mu\nu} dx_\mu dx_\nu$$

where the Einstein summation convention is used; repeated indices are implicitly summed over, in this case μ and ν run from 1 to 4 each so that the above expression is meant to be read as

$$ds^2 = \sum_{\mu=1}^{4} \sum_{\nu=1}^{4} g^{\mu\nu} dx_\mu dx_\nu$$

which written out for completeness is

$$ds^2 = g^{11}(dx_1)^2 + g^{22}(dx_2)^2 + g^{33}(dx_3)^2 + g^{44}(dx_4)^2$$

because there are only four nonzero elements of the metric and these lie on the diagonal. Putting in these four metric elements gives the same result we obtained above:

$$ds^2 = (dx_1)^2 - (dx_2)^2 - (dx_3)^2 - (dx_4)^2$$

In this manner we also can define the length of an arbitrary 4-vector as the inner product with itself as

$$x \cdot x = x^2 = x_\mu g^{\mu\nu} x_\nu$$

and in an analogous fashion the inner product for two distinct 4-vectors is given by

$$x \cdot y = x_\mu g^{\mu\nu} y_\nu$$

You may have noticed that we have placed the indices in these last expressions as both subscripts and superscripts. The reason for this has to do with a subtle point that arises from differential geometry and to understand this, we need to take a brief journey into the land of *tensors*.

3.11 Tensors, Vectors, and Scalars

We have already encountered scalars and vectors, but tensors are a new concept here. Most (perhaps all?) texts on tensor analysis begin with a sentence that reads something like "Tensors are objects defined in terms their transformation properties." Not very illuminating, to say the least. A more useful definition is that tensors are a generalized type of indexed objects whose simpler forms include scalars and vectors. A *scalar* has a single component without indices, a *vector*, such as A_i, has a single index and n components, where n is the dimension of the vector space, while a *tensor* can have any number of indices from zero to any arbitrary number. An example of a tensor is $A_{ij}{}^k$. Tensors are classified according to the number of indices: A scalar is a rank 0 tensor, a vector is a rank 1 tensor, and the tensor $A_{ij}{}^k$ is a rank 3 tensor. By now, we have encountered many rank 2 tensors as matrices, e.g., T_{ij}. A rank 3 tensor could be visualized as a 3D matrix, but for four or more dimensions, our ability to form mental images of objects fails. Is this why 4D spacetime was such a novel concept?

Turning our attention from objects to their transformations: an object that transforms as a rank 0 tensor under a change of basis is a scalar, an object that transforms as a rank 1 tensor is a vector, and object that transforms as a rank 2 tensor is a matrix (or simply a rank 2 tensor). It should be noted that the converse of these statements is not always true, e.g., not all scalars are rank 0 tensors, not all vectors are rank 1 tensors, and not all matrices are rank 2 tensors. The reason for this is that there is an additional property of tensors; they are also invariant under a change of basis, and therein lies the connection with transformations given in the standard textbook definition of tensors. Hidden within this last statement is the most important conclusion that, just as vectors are only completely defined once a set of basis vectors is specified, so it is with tensors; they are only completely defined once a basis is given. For example, let's multiply two vectors, not using dot or cross products, but using a new concept, a tensor product, denoted by the symbol $\otimes$. Given the vectors

$$a = a_1 e_1 + a_2 e_2 + \ldots + a_n e_n$$

and

$$b = b_1 e_1 + b_2 e_2 + \ldots + b_n e_n,$$

we define the tensor product as

$$a \otimes b = a_1 b_1 e_1 e_1 + a_1 b_2 e_1 e_2 + a_2 b_1 e_2 e_1 + a_2 b_2 e_2 e_2 + \ldots + a_n e_n b_n e_n$$

where we form all the n^2 combinations of the components *and* of their basis vectors. Each of the terms in the tensor product can be indexed by their position in the parent vectors. Let $u_{11} = a_1 b_1 e_1 e_1$, $u_{12} = a_1 b_2 e_1 e_2$, and so on. The products of the basis vectors are called *dyads*. Then, we can arrange the terms in $\boldsymbol{a} \otimes \boldsymbol{b}$ as a matrix U (a rank 2 tensor) as an $n \times n$ matrix:

$$U = \begin{bmatrix} u_{11} & u_{12} & \cdots & u_{1n} \\ u_{21} & u_{22} & \cdots & u_{11} \\ . & . & \cdots & . \\ u_{n1} & u_{n2} & \cdots & u_{nn} \end{bmatrix},$$

or we can denote this as a rank 2, n-dimensional covariant tensor U_{ij}.

Tensors bear a number of indices. For example, the rank 3 tensor $A_{ij}{}^k$ contained three indices, two (ij) as subscripts and one (k) as a superscript, and the superscripts are not arranged directly above the subscripts. The superscripted indices are called *contravariant* indices, while the subscripts are *covariant* indices. The difference is based on the concept that I call *dualism*.

3.11.1   Dualism

Mathematical objects and their duals play a significant role in physics, but there is unfortunately a branch of physics devoted to a loosely defined concept denoted "duality," which refers to equivalences among separate theories and is not related to what we are referring to as dualism here. In mathematics, the definition of duality depends on the branch of study one is pursuing and there is only general, but not specific, agreement on its meaning such that one is certain that the same concept is being referred to in each branch. Therefore, I am going to avoid the use of "duality" in favor of dualism to describe its various features and applications in physics here.

Objects, such as column vectors, obey well-defined transformation laws $\boldsymbol{x}' = T\boldsymbol{x}$ or in component form $x_i = T_{ij}x_j$, where T is a linear operator, a transformation matrix. One knows, from linear algebra, that in order to multiply two matrices together, say A and B, the matrix product AB is only defined if the number of columns of A is equal to the number of rows of B. For example, if A is an $n \times r$ matrix and B is an $r \times m$ matrix, then their matrix product $D = AB$ is an $n \times m$ matrix with elements d_{ij} defined by

$$d_{ij} = \sum_{q=1}^{r} a_{iq} \, b_{qj}$$

where the indices $\{i, j\}$ satisfy $i \in \{1, n\}$ and $j \in \{1, m\}$. The matrix product in reverse order BA is not defined.

An m-dimensional vector may be considered as a matrix consisting of a single row with m columns, or as a single column with m rows. If we let A be an n-dimensional row vector:

$$A = [a_1, a_2, \ldots, a_n]$$

and B be an n-dimensional column vector:

$$B = \begin{bmatrix} b_1 \\ b_2 \\ \cdot \\ \cdot \\ \cdot \\ b_n \end{bmatrix}$$

then their product $D = AB$ according to the above equation is

$$D = \sum_{i=1}^{n} a_i\, b_i = [a_1, a_2, \ldots, a_n] \begin{bmatrix} b_1 \\ b_2 \\ \cdot \\ \cdot \\ \cdot \\ b_n \end{bmatrix}$$

which is the same as the inner product of two vectors $\boldsymbol{A}$ and $\boldsymbol{B}$, and is a scalar:

$$\boldsymbol{A} \cdot \boldsymbol{B} = a_1 b_1 + a_2 b_2 + \ldots + a_n b_n$$

However, the matrix product in reverse order, BA, is an $n \times n$ matrix. This distinction between row and column vectors is an example of the difference between objects that are said to be "*duals*" of each other. A row vector A' is simply the transpose of the same column vector, A, $A' = A^T$, so this distinction does not appear to be very important. However, if we try to compute the length of A we discover, by using the definition of matrix multiplication given above, that in order to properly compute the inner product of a column vector A with itself, we must first transpose A to form A^T (a row vector) before we can form the inner product:

$$\boldsymbol{A} \cdot \boldsymbol{A} = A^T A = [a_1, a_2, \ldots, a_n] \begin{bmatrix} a_1 \\ a_2 \\ \cdot \\ \cdot \\ \cdot \\ a_n \end{bmatrix} = a_1 a_1 + a_2 a_2 + \ldots + a_n a_n$$

Thus, row and column vectors are not the same thing: they are *duals* of each other.

The use of dualism here is restricted to its definition with respect to the mathematical definition of a vector space and its associated dual space. A simple example of a vector space and its associated dual space comes from the study of the X-ray diffraction patterns from crystals. A crystal is a periodic structure (lattice) of atom(s) or molecule(s) formed by translating the basic asymmetric unit to other lattice sites by integral multiples of the primitive lattice vectors $\{a, b, c\}$ that describe the unit cell. X-rays coherently scatter (diffract) from various collections of lattice planes; the diffraction pattern is related to the Fourier transform of the lattice. Since the Fourier transform in this case takes physical space $\{x, y, z\}$ into momentum space where $p = \hbar k$ and the wavevector $k = 2\pi/\lambda$ has dimensions of $1/L$, we note that physical space is transformed (mapped) into its dual space, or what the crystallographers refer to as reciprocal space. Assume that the primitive lattice vectors $\{a, b, c\}$ form a basis for a 3D vector space, V, we can then form a basis $\{a^*, b^*, c^*\}$ for the dual space, V^*, by using the fact that the basis vectors of the dual basis must be orthogonal to $\{a, b, c\}$. We know that the cross product of two 3D vectors $\{a, b\}$ is orthogonal to the plane containing a and b so that $a^* = \frac{b \times c}{v}$ is a unit vector orthogonal to b and c and $v = a \cdot (b \times c)$ is the scalar volume of the unit cell and is used to normalize a^*. Similar equations hold for $b^* = \frac{a \times c}{v}$ and $c^* = \frac{a \times b}{v}$.

Now, let's return to the question of the upper and lower indices in

$$x \cdot y = x_\mu g^{\mu\nu} y_\nu$$

We can rewrite our derivation of the dual basis for reciprocal space in a crystal in terms of our standard Euclidean basis with orthonormal unit vectors written as covariant vectors with subscripts $\{e_1, e_2, e_3\}$ $(= \{a, b, c\})$, where $e_1 = (1, 0, 0)$, and so on. We can write the dual basis vectors as contravariant vectors with superscripts such that $e^1 = e_2 \times e_3$, with $v = e_1 \cdot (e_2 \times e_3) = 1$, $e^2 = e_1 \times e_3$, $e^3 = e_1 \times e_3$, and we have that the dual basis is orthogonal to the covariant basis:

$$e_i \cdot e^j = \delta_i^j$$

and the transformation matrix from one basis to the other is the 3D unit matrix. Note here that e^2 does not, in this context, mean $e \cdot e$, but rather the second contravariant unit vector $e^2 = (0, 1, 0)$: of course $e^1 = (1, 0, 0)$ and $e^3 = (0, 0, 1)$. In an orthogonal coordinate system, such as the flat Euclidean space considered here, the basis vectors for the dual space are proportional to (and in this case the same as) the basis vectors in the regular space, since $e_1 = e^1$, and so on, but this is not necessarily true for other spaces.

In the older literature (beginning with Einstein), vectors with superscripts are referred to as vectors with contravariant components (or often shortened to contravariant vectors), while those with subscripts are called vectors with covariant components (just as often shortened to covariant vectors). These terms arose from the early days of Riemannian and differential geometry. The more modern approach uses language derived from tensor analysis: a scalar is a zero-dimensional tensor, while a vector is a one-dimensional tensor; the dual space to a vector space contains objects now-called 1-forms and these also form a vector space called the dual space.

Examples of dual spaces from physics include the Hermitian adjoint of a quantum state vector, defined as the complex-conjugate transpose. We have already mentioned position and its dual, momentum; in general, x is a coordinate vector and its dual, momentum is the 1-form $\frac{d}{dx}$, which we will later encounter as the generator of spatial translations. Another important dual pair consists of time and its dual, energy, which is the 1-form $\frac{d}{dt}$, i.e., the generator of time translations. Dirac invented a very useful notation dealing with quantum state vectors $|n>$ as "kets" and their duals, the "bras" $<n|$ related by the adjoint operation, $<n|^{\dagger} = |n>$. The uncertainty principle is a statement about the commutation relations for objects and their duals.

3.11.2   Covariant and Contravariant Tensors

From this discussion we can infer that covariant and contravariant tensors are duals of each other. We will see that other duals arise in quantum mechanics: the dual for position is the derivative with respect to position, while the dual for time is the time derivative. At any point on a curve in a coordinate system, one can define a set of covariant basis vectors as the tangent vectors at that point. The Gram–Schmidt process may then be used to construct an orthonormal basis from this set. The coordinate surfaces passing through that same point define the dual space. Correspondingly, one can define a set of contravariant basis vectors in this dual space as the derivatives. The metric tensor is then used to convert between these two points of view.

3.12   The Poincaré Group

When one considers the symmetry of space-time, the operators that transform one point of view to another consist of rotations, translations, *reflections*, and boosts (changes in relative velocity). These operators form the basis for the Poincaré group, and since they depend on continuous parameters, form a *Lie* group. We need, therefore, to find the *Lie Algebra* corresponding to this Lie group structure and this will include the generators of all of these transformations. Of fundamental importance are the Casimir operators of the Poincaré group, the set of operators that commute with all of the group elements, whose eigenvalues label the irreducible representations (Irreps) of the Poincaré group. The universe is constructed from elementary particles that belong to one or another of these Irreps, which therefore constrain the properties of the objects allowed to exist within space-time. One can now appreciate why we have spent so much time and effort learning group theory; we are going to provide the answer of the first part (*Die Welt*, i.e., *the Universe*) of Faust's query about "*Was die Welt im innersten zusammenhält.*" We have purposely delayed an exploration of the Irreps of the rotation group until now because, as we will see, rotations form a subgroup of the Poincaré group.

The bases for Minkowski space-time are the sets of orthonormal covariant unit vectors $e_\mu = \{e_0, e_1, e_2, e_3\}$ and their contravariant duals $e^\nu = \{e^0, e^1, e^2, e^3\}$. The invariant scalar product of two contravariant 4-vectors x^μ, x^ν in two different inertial frames is given by

$$T_{\mu\nu}x^\mu x^\nu = T_{\mu\nu}x'^\mu x'^\nu$$

(using the Einstein summation convention over repeated upper and lower indices) where the covariant metric is the 4D transformation matrix:

$$T_{\mu\nu} = e_\mu \cdot e_\nu' = \begin{bmatrix} e_0 \cdot e'^0 & e_0 \cdot e'^1 & e_0 \cdot e'^2 & e_0 \cdot e'^3 \\ e_1 \cdot e'^0 & e_1 \cdot e'^1 & e_1 \cdot e'^2 & e_1 \cdot e'^3 \\ e_2 \cdot e'^0 & e_2 \cdot e'^1 & e_2 \cdot e'^2 & e_2 \cdot e'^3 \\ e_3 \cdot e'^0 & e_3 \cdot e'^1 & e_3 \cdot e'^2 & e_3 \cdot e'^3 \end{bmatrix}$$

The transformation metric is commonly written as

$$T_{\mu\nu} = \begin{bmatrix} 1 & 0 & 0 & 0 \\ 0 & -1 & 0 & 0 \\ 0 & 0 & -1 & 0 \\ 0 & 0 & 0 & -1 \end{bmatrix}$$

although there are other definitions, such as $T'_{\mu\nu} = -T_{\mu\nu}$. We have seen that this metric characterizes the hyperbolic nature of space-time (Fig. 3.5). Other forms of the metric can contain both upper and lower indices:

$$T_\mu{}^\nu = e_\mu \cdot e'^\nu = \begin{bmatrix} e_0 \cdot e'^0 & e_0 \cdot e'^1 & e_0 \cdot e'^2 & e_0 \cdot e'^3 \\ e_1 \cdot e'^0 & e_1 \cdot e'^1 & e_1 \cdot e'^2 & e_1 \cdot e'^3 \\ e_2 \cdot e'^0 & e_2 \cdot e'^1 & e_2 \cdot e'^2 & e_2 \cdot e'^3 \\ e_3 \cdot e'^0 & e_3 \cdot e'^1 & e_3 \cdot e'^2 & e_3 \cdot e'^3 \end{bmatrix}$$

Contraction of a contravariant 4-vector with the metric tensor produces the dual, covariant 4-vector:

$$x_\mu = T_{\mu\nu}x^\nu$$

and using this definition, the invariance of the scalar product of two 4-vectors becomes simply

$$x^\mu x_\mu = x'^\mu x'_\mu$$

The energy–momentum 4-vector P^μ unifies energy (a scalar, E, like the time component of a spacetime 4-vector) and the 3-vector momentum p into single contravariant object:

$$P^\mu = \{p^0, p^1, p^2, p^3\} = \{E, \boldsymbol{p}\}$$

with the relativistic relationship between the energy, rest mass, and momentum being $E^2 = m^2 + \boldsymbol{p} \cdot \boldsymbol{p}$. Since the length of this 4-vector (and the rest mass) must be the same in all inertial frames, we have that

$$T_{\mu\nu}\, P^\mu P^\nu = T_{\mu\nu}\, P'^\mu P'^\nu$$

which we can evaluate by introducing the covariant momentum $P_\mu = T_{\mu\nu}P^\nu$ through its transformation into its dual via the metric: note that $P_\mu = \{E, -\boldsymbol{p}\}$.

Then, the invariance of the length of the momentum 4-vector in the two reference frames becomes $P^\mu P_\mu = P'^\mu P'_\mu$. Let's explicitly find the length of the momentum 4-vector. Then, we have the very important results that the length of the momentum 4-vector is the rest mass

$$P^\mu P_\mu = \{E, \boldsymbol{p}\} \cdot \{E, -\boldsymbol{p}\} = E^2 - \boldsymbol{p} \cdot \boldsymbol{p} = m^2 + \boldsymbol{p} \cdot \boldsymbol{p} - \boldsymbol{p} \cdot \boldsymbol{p} = m^2.$$

3.13 The Lorentz Group

The Lorentz transformation between two inertial frames is a 4D orthogonal matrix $\Lambda^\mu{}_\nu$ that transforms a contravariant 4-vector according to

$$x'^\mu = \Lambda^\mu{}_\nu x^\nu$$

By including translations, a^μ, this becomes the most general relativistic relationship between the stationary and boosted systems: a Poincaré transformation where

$$x'^\mu = \Lambda^\mu{}_\nu x^\nu + a^\mu$$

For the time being, we will assume that one frame is just boosted and not translated with respect to the other and let $a^\mu = 0$ since if we incorporate translations then only the differences between vectors are invariant, rather than their lengths. We can learn several properties of the Lorentz transformation from its mathematical structure. The first is that the length of a 4-vector, x^ν, must be the same in each frame, which implies that its inner product remains the same in both frames

$$T_{\mu\nu}\, x^\mu x^\nu = T_{\mu\nu}\, x'^\mu x'^\nu$$

and, by inserting the Lorentz rank 2 tensor (matrix) for each of the right hand, primed, 4-vectors, and using dummy indices $x'^\nu = \Lambda^\nu{}_\rho x^\rho$ and $x'^\mu = \Lambda^\mu{}_\sigma x^\sigma$

$$T_{\mu\nu}\, x'^\mu x'^\nu = T_{\mu\nu}\Lambda^\mu{}_\sigma x^\sigma \Lambda^\nu{}_\rho x^\rho$$

Matrix transposition (the rows and columns of Λ) is the same as interchanging the indices of a rank 2 tensor. The transposition of Λ is accomplished with the aid of the metric according to

$$\left(\Lambda^T\right)_\mu{}^\sigma = T_{\mu\lambda}\left(\Lambda^T\right)^{\lambda\sigma} = T_{\mu\lambda}\Lambda^{\sigma\lambda} = \Lambda^{\sigma\lambda}T_{\lambda\mu} = \Lambda^\sigma{}_\mu.$$

Using this, we arrive at the expression for the length of a 4-vector in the primed frame as

$$T_{\mu\nu}\, x'^\mu x'^\nu = T_{\sigma\rho}\Lambda^\sigma{}_\mu\Lambda^\rho{}_\nu x^\mu x^\nu$$

This expression must hold for the inner product of any 4-vector and therefore

$$T_{\mu\nu} = \Lambda^\sigma{}_\mu T_{\sigma\rho}\Lambda^\rho{}_\nu.$$

In order words, the Lorentz transformation preserves the Minkowski metric. As a matrix equation, this is given by

$$T = \Lambda^T T\Lambda.$$

The second important property of Lorentz transformations is also found from the invariance of the metric as

$$\left(\Lambda^T\Lambda\right)^\nu{}_\mu = \left(\Lambda^T\right)^\nu{}_\sigma \Lambda^\sigma{}_\mu = \delta_\mu{}^\nu = \delta^\nu{}_\mu$$

which gives that the transpose of Λ is equal to its inverse, $\Lambda^T = \Lambda^{-1}$:

$$\left(\Lambda^T\right)^\nu{}_\sigma = \Lambda_\sigma{}^\nu = \left(\Lambda^{-1}\right)^\nu{}_\sigma$$

The demonstration that the Lorentz transformations satisfy the group axioms is an interesting exercise for the student to pursue, but we note that the product of two Lorentz transformations is another Lorentz transformation, the identity is $\Lambda_e = \delta^\nu{}_\sigma$, or the identity matrix. We have constructed the inverse transformation from $\Lambda^T = \Lambda^{-1}$, and furthermore, the determinant of Λ is found from

$$\mathrm{Det}\, T = \mathrm{Det}\, \Lambda^T\, \mathrm{Det}\, T\, \mathrm{Det}\, \Lambda$$

where the Det T cancels, and by means of Det $\Lambda^T = $ Det T, we find that

$$(\mathrm{Det}\, \Lambda)^2 = 1$$

or that Det $\Lambda = \pm 1$. Those Lorentz transformations with Det $\Lambda = +1$ are known as special, while we find from the T_{00} term in the metric that

Table 3.1 The Lorentz transformations four regions

Region	$\Lambda^0_{\;0}$	Det Λ	Meaning	Example $\Lambda^\mu_{\;\nu}$
(1)	>0	+1	Special orthochronus	$\Lambda_e = \delta^\mu_{\;\nu} = \begin{bmatrix} 1 & 0 & 0 & 0 \\ 0 & 1 & 0 & 0 \\ 0 & 0 & 1 & 0 \\ 0 & 0 & 0 & 1 \end{bmatrix}$
(2)	>0	−1	Space inversion	$\begin{bmatrix} 1 & 0 & 0 & 0 \\ 0 & -1 & 0 & 0 \\ 0 & 0 & -1 & 0 \\ 0 & 0 & 0 & -1 \end{bmatrix}$
(3)	<0	−1	Time inversion	$\begin{bmatrix} -1 & 0 & 0 & 0 \\ 0 & 1 & 0 & 0 \\ 0 & 0 & 1 & 0 \\ 0 & 0 & 0 & 1 \end{bmatrix}$
(4)	<0	+1	Space-time inversion	$\begin{bmatrix} -1 & 0 & 0 & 0 \\ 0 & -1 & 0 & 0 \\ 0 & 0 & -1 & 0 \\ 0 & 0 & 0 & -1 \end{bmatrix}$

$$T_{00} = 1 = \Lambda^\sigma_{\;0} T_{\sigma\rho} \Lambda^\rho_{\;0} = \left(\Lambda^0_{\;0}\right)^2 - \left(\Lambda^i_{\;0}\right)^2$$

or, that

$$\left(\Lambda^0_{\;0}\right)^2 = 1 + \left(\Lambda^i_{\;0}\right)^2 \geq 1$$

An *orthochronus* (*correct time* in Greek) Lorentz transformation is one with $\Lambda^0_{\;0} \geq 1$, preserving the direction of the time axis. We can classify Lorentz transformations L_i into four disjoint, non-connected, categories based on the signs of $\Lambda^0_{\;0}$ and Det Λ. Examples are given in Table 3.1.

3.13.1   **Lie Algebra** *of the Lorentz Group*

We next proceed to find the generators of the Lorentz Group. A Lorentz transformation from $\{x, y, z, ct\} \rightarrow \{x', y', z', ct'\}$ involves the well-known equations (for a boost along the x-axis):

$$x' = \gamma(x + vt)$$
$$y' = y$$
$$z' = z$$

$$t' = \gamma\left(t + \frac{vx}{c^2}\right)$$

where $\beta = \frac{v}{c}$ and

$$\gamma = \frac{1}{\sqrt{1 - \beta^2}}$$

or, in terms of the contravariant position 4-vector x'^{μ}:

$$x'^{\mu} = \left\{\gamma\left(x^0 + \beta x^1\right), \gamma\left(\beta x^0 + x^1\right), x^2, x^3\right\}.$$

By noticing that $\gamma^2 - \beta^2\gamma^2 = 1$, we can make the substitution in terms of the hyperbolic variables:

$$\gamma = \cosh\varphi$$

and

$$\gamma\beta = \sinh\varphi$$

To define the rapidity,

$$\psi = \tanh^{-1}\varphi$$

which enables us to write the Lorentz transformation tensor along the x-axis as

$$\Lambda_x = \begin{bmatrix} \cosh\varphi & \sinh\varphi & 0 & 0 \\ \sinh\varphi & \cosh\varphi & 0 & 0 \\ 0 & 0 & 1 & 0 \\ 0 & 0 & 0 & 1 \end{bmatrix}$$

with similar expressions for boosts along the y and z axes. Lie theory enables us to find the generator K_x of boosts in the x direction as

$$K_x = \lim_{\varphi \to 0} \frac{1}{i} \frac{\partial \Lambda_x}{\partial \varphi}$$

which, after a brief foray into *Mathematica*, yields the Hermitian matrix:

$$K_x = -i \begin{bmatrix} 0 & 1 & 0 & 0 \\ 1 & 0 & 0 & 0 \\ 0 & 0 & 0 & 0 \\ 0 & 0 & 0 & 0 \end{bmatrix}.$$

The other boost generators are found in a similar fashion from

$$\Lambda_y = \begin{bmatrix} \cosh\varphi & 0 & \sinh\varphi & 0 \\ 0 & 1 & 0 & 0 \\ \sinh\varphi & 0 & \cosh\varphi & 0 \\ 0 & 0 & 0 & 1 \end{bmatrix}$$

and

$$\Lambda_z = \begin{bmatrix} \cosh\varphi & 0 & 0 & \sinh\varphi \\ 0 & 1 & 0 & 0 \\ 0 & 0 & 1 & 0 \\ \sinh\varphi & 0 & 0 & \cosh\varphi \end{bmatrix}.$$

Taking the indicated derivatives gives that the boost matrices for K_y and K_z are

$$K_y = -i \begin{bmatrix} 0 & 0 & 1 & 0 \\ 0 & 0 & 0 & 0 \\ 1 & 0 & 0 & 0 \\ 0 & 0 & 0 & 0 \end{bmatrix}$$

and

$$K_z = -i \begin{bmatrix} 0 & 0 & 0 & 1 \\ 0 & 0 & 0 & 0 \\ 0 & 0 & 0 & 0 \\ 1 & 0 & 0 & 0 \end{bmatrix}.$$

Appropriate products of these boosts will provide boosts in any direction. The transition from the generators to finite transformations involves matrix exponentiation. For example, for Λ_x

$$\Lambda_x(\varphi) = e^{\varphi K_x}.$$

The commutation relations for the K_i are

$$[K_x, K_y] = \begin{bmatrix} 0 & 0 & 0 & 0 \\ 0 & 0 & 1 & 0 \\ 0 & -1 & 0 & 0 \\ 0 & 0 & 0 & 0 \end{bmatrix}$$

$$[K_y, K_z] = \begin{bmatrix} 0 & 0 & 0 & 0 \\ 0 & 0 & 0 & 0 \\ 0 & 0 & 0 & 1 \\ 0 & 0 & -1 & 0 \end{bmatrix}$$

and

$$[K_z, K_x] = \begin{bmatrix} 0 & 0 & 0 & 0 \\ 0 & 0 & 0 & -1 \\ 0 & 0 & 0 & 0 \\ 0 & 1 & 0 & 0 \end{bmatrix}.$$

Note that the Lorentz transformations involving boosts alone do not form a group because the algebra is not closed under commutation. The form of the commutator matrices is identical to those found for the rotation generators, if we remember that rotations do not act on the time component of 4-vectors, but only on the spatial portion. We can incorporate this knowledge into the rotation generators by expanding the matrices for the J_i by adding a fourth row and column of zeros to give the 4D versions, which we now rename as M_i, where

$$M_i = \{0, J_i\}, i \in \{1, 2, 3\}$$

such that

$$M_x = -i \begin{bmatrix} 0 & 0 & 0 & 0 \\ 0 & 0 & 0 & 0 \\ 0 & 0 & 0 & 1 \\ 0 & 0 & -1 & 0 \end{bmatrix} = [K_y, K_z],$$

$$M_y = -i \begin{bmatrix} 0 & 0 & 0 & 0 \\ 0 & 0 & 0 & -1 \\ 0 & 0 & 0 & 0 \\ 0 & 1 & 0 & 0 \end{bmatrix} = [K_z, K_x],$$

and

$$M_z = -i \begin{bmatrix} 0 & 0 & 0 & 0 \\ 0 & 0 & 1 & 0 \\ 0 & -1 & 0 & 0 \\ 0 & 0 & 0 & 0 \end{bmatrix} = [K_x, K_y].$$

We have now found that the commutators of the boost generators K_i give rotations $[K_x, K_y] = -iM_z$ and cyclic permutations.

We confirm that products of boosts involve not only changes in velocity but also rotations of the coordinate axes. The commutators of the rotation generators have been worked out already, and we repeat them here for convenience, along with those for the boost generators. The matrices defined above also enable us to find the commutators of boosts and rotations using *Mathematica* as:

$$\left[M_i, M_j\right] = i\epsilon_{ijk}M_k$$

$$\left[M_i, K_j\right] = i\epsilon_{ijk}K_k$$

$$\left[K_i, K_j\right] = -i\epsilon_{ijk}M_k$$

These six generators, three boosts and three rotations, comprise the most general Lorentz transformations and form a basis from which any Lorentz transformation may be obtained by forming linear combinations. The boost and rotation generators do not commute, but we can form new linear combinations that are closed under commutation and do commute. Let us define these new linear combinations $N_i^{\pm}$ such that

$$N_i^{\pm} = \frac{1}{2}(M_i \pm iK_i)$$

Computing these matrices, we find that for the sum of generators, we have that

$$N_x^{+} = \frac{1}{2}\begin{bmatrix} 0 & 1 & 0 & 0 \\ 1 & 0 & 0 & 0 \\ 0 & 0 & 0 & -i \\ 0 & 0 & i & 0 \end{bmatrix}$$

$$N_y^{+} = \frac{1}{2}\begin{bmatrix} 0 & 0 & 1 & 0 \\ 0 & 0 & 0 & i \\ 1 & 0 & 0 & 0 \\ 0 & -i & 0 & 0 \end{bmatrix}$$

$$N_z{}^+ = \frac{1}{2} \begin{bmatrix} 0 & 0 & 0 & 1 \\ 0 & 0 & -i & 0 \\ 0 & i & 0 & 0 \\ 1 & 0 & 0 & 0 \end{bmatrix}$$

while for the difference of the generators, we find another set

$$N_x{}^- = \frac{1}{2} \begin{bmatrix} 0 & -1 & 0 & 0 \\ -1 & 0 & 0 & 0 \\ 0 & 0 & 0 & -i \\ 0 & 0 & i & 0 \end{bmatrix}$$

$$N_y{}^- = \frac{1}{2} \begin{bmatrix} 0 & 0 & -1 & 0 \\ 0 & 0 & 0 & i \\ -1 & 0 & 0 & 0 \\ 0 & -i & 0 & 0 \end{bmatrix}$$

$$N_z{}^- = \frac{1}{2} \begin{bmatrix} 0 & 0 & 0 & -1 \\ 0 & 0 & -i & 0 \\ 0 & i & 0 & 0 \\ -1 & 0 & 0 & 0 \end{bmatrix}$$

with the commutation relations (which can be confirmed with *Mathematica*) showing that they form a closed algebra, and commute with one another:

$$\left[N_i{}^+, N_j{}^+\right] = i\epsilon_{ijk}N_k{}^+$$

$$\left[N_i{}^-, N_j{}^-\right] = i\epsilon_{ijk}N_k{}^-$$

$$\left[N_i{}^+, N_j{}^-\right] = 0$$

Rather than leave these pretty equations to themselves and move on to finding the Irreps of the Lorentz Group, let us reflect on what we have just accomplished. A very important point to note here is that the commutation relations for the $N_i{}^+$ and $N_i{}^-$ are each *individually* the same as those found previously for SU(2) and each forms its own, closed Lie algebra.

3.13.2 *SU(2) Is the Double Cover of the Lorentz Group*

We have found one representation of the Lorentz Group and shown that SU(2) is its *double cover*. The fact that the commutator of the $N_i{}^\pm$ is zero means, these operators have simultaneous eigenfunctions in the Hilbert space of states and bear two eigenvalues of angular momentum (j, j'), one quantum number from N_i^+ and the other from the N_i^- representation. One or the other can contain the spin zero state where there are two possibilities:

$$\text{Case (1) } N_j^- = 0 \Rightarrow M_j = iK_j$$

$$(j, j') = (j, 0)$$

$$\text{Case (2) } N_{j'}^+ = 0 \Rightarrow M_{j'} = -iK_{j'}$$

$$(j, j') = (0, j')$$

The Lie Algebra is then the tensor product $SU(2) \otimes SU(2)$. We can therefore use these eigenvalues to characterize the allowed states as we build up the eigenfunctions in the Hilbert space of states. Our main task is to count the number of eigenfunctions for this number will provide the dimension of the representation. The group characterized by the 2×2 matrices of SU(2) has as its basis a set of rank 1 tensors. Racah's theorem states that a group based on rank 1 tensors can have only a single Casimir operator, one that commutes with all the rest of the group operators. Schur's lemma states that any group element that commutes with all the rest of the group elements must be a multiple of the identity, i.e., a scalar. The Casimir operator for SU(2) must be an invariant, one that does not depend on the orientation of the axes of space-time. Such an invariant is therefore restricted to be a scalar, which in this case is the length S^2 of the spin vector, S:

$$S^2 = S_x{}^2 + S_y{}^2 + S_z{}^2$$

3.14   The Poincaré Group: Revisited

We stated at the beginning of this section that we would first consider the rotation group SO(3), then the Lorentz Group, and finally we come to the full group of relativistic space-time, the Poincaré group. The full symmetry of space-time involves boosts, rotations, reflections (in both space and time), and now we include translations so that the transformation equations read

$$x'^\mu = \Lambda^\mu{}_\nu x^\nu + a^\mu$$

where we have added an arbitrary 4-vector a^μ. This set still forms a group, which is available for the interested student to pursue in the exercises. The Nobel prize went to Eugene Wigner for his classic analysis of what was referred to as the "inhomogeneous Lorentz Group," inhomogeneous in the sense that it included a nonzero translation 4-vector (Wigner 1939). Lie theory tells us that we should study transformations that barely do anything, those close to the identity. In the case of Lorentz transformations, the identity is given by $\Lambda^{\mu}{}_{\nu} = \delta^{\mu}{}_{\nu}$ and $a^{\mu} = 0$. The transformation matrix is then written in terms of the infinitesimal parameters $\omega^{\mu}{}_{\nu}$ and $a^{\mu} = \varepsilon^{\mu}$:

$$\Lambda^{\mu}{}_{\nu} = \delta^{\mu}{}_{\nu} + \omega^{\mu}{}_{\nu}$$

Inserting this into the Lorentz invariance equation:

$$T_{\mu\nu} = \Lambda^{\sigma}{}_{\mu} T_{\sigma\rho} \Lambda^{\rho}{}_{\nu}$$

gives

$$T_{\sigma\rho} = \left(\delta^{\mu}{}_{\rho} + \omega^{\mu}{}_{\rho}\right) T_{\mu\nu} \left(\delta^{\mu}{}_{\sigma} + \omega^{\mu}{}_{\sigma}\right) = T_{\sigma\rho} + \omega_{\sigma\rho} + \omega_{\rho\sigma} + O\left(\omega^2\right) + O\left(\omega^3\right) + \ldots$$

Because we have assumed an infinitesimal Lorentz transformation, we neglect terms in ω higher than the first, and have that (using contractions with the metric)

$$T_{\sigma\rho} = T_{\sigma\rho} + \omega_{\sigma\rho} + \omega_{\rho\sigma}$$
$$T_{\sigma\rho} - T_{\sigma\rho} = 0 = \omega_{\sigma\rho} + \omega_{\rho\sigma}$$

so that

$$\omega_{\sigma\rho} + \omega_{\rho\sigma} = 0$$

the infinitesimal tensor $\omega_{\mu\nu}$ must be antisymmetric

$$\omega_{\sigma\rho} = -\omega_{\rho\sigma}$$

The inclusion of translations into the Lorentz Group to form the Poincaré group means that if the operators for translations commute with those of the existing Lorentz Group, then the Hilbert space of quantum states will contain functions that are simultaneous eigenfunctions of all of the commuting operators, with simultaneous eigenvalues. We can then characterize the states by their set of eigenvalues. Do rotations and translations commute? Let's see.

The Lie generator for 3D translations is the quantum momentum operator, P, the spatial derivative:

$$P_i = -i\hbar \frac{\partial}{\partial x^i}$$

while in Minkowski space-time, it is the energy–momentum 4-vector that generates space-time translations:

$$P_{\mu} = (E, -\boldsymbol{p})$$

which is then added to the group generators. The commutator of different momenta is

$$[P_\mu, P_\nu] = 0$$

because the order in which translations are carried out does not matter. The Lie algebra of the Lorentz Group is

$$[L_{\mu\nu}, L_{\rho\sigma}] = i\left(T_{\mu\rho}L_{\nu\sigma} - T_{\mu\sigma}L_{\nu\rho} - T_{\nu\rho}L_{\mu\sigma} + T_{\nu\sigma}L_{\mu\rho}\right)$$

where we have used the definition of a new operator $L_{\mu\nu}$ in terms of our rotation operator M and boost operator K as

$$\frac{1}{2}\epsilon_{ijk}L_{ij} = M_i$$

and

$$L_{0i} = K_i.$$

The commutator of the momentum operator with that for angular momentum reflects the mixing of coordinates commonly seen with rotations and gives that

$$[P_\mu, L_{\rho\sigma}] = i\left(T_{\mu\rho}P_\sigma - T_{\mu\sigma}P_\rho\right).$$

These commutators also indicate that P and L commute with $P_0 = H$, the Hamiltonian, the generator of time translations (i.e., the energy of the state).

The Poincaré group is characterized by rank 2 tensor operators and by Racah's theorem that possesses two Casimir operators whose eigenvalues are invariants. We already know, due to our examination of the role of rotations in the Lorentz Group, that one of the Casimir operators must be related to the angular momentum operator, and, since we have only boost and translation operators to consider, the other has to be the length of the momentum 4-vector whose eigenvalue is the rest mass:

$$\boldsymbol{P_\mu P^\mu = m^2}$$

Therefore, one of the eigenvalues we can use to define the states in the Hilbert space is the rest mass of the object. A Lorentz transformation preserves both the length p^2 and the sign of p^0 of the 4-momentum vector. There are six individual cases, as summarized in Table 3.2.

Class 1 corresponds physically massive objects, class 3 to massless photons with $E = \pm p$, and class 5 to the all-important quantum vacuum. Classes 2 and 4 do not describe physical objects, while class 6 is occupied in quantum field theory by virtual objects with spacelike momentum. These are the non-energy-conserving intermediate states that will be extensively discussed in Chap. 6 on the origin of the Universe.

Table 3.2 A classification of the 4-momentum P^μ based on its properties under Lorentz transformations

Class	p^2	p^0	Object
1	$m^2 > 0$	$p^0 > 0$	Physical massive
2	$m^2 > 0$	$p^0 < 0$	*Unphysical*
3	0	$p^0 > 0$	Physical massless
4	0	$p^0 < 0$	*Unphysical*
5	0	0	Vacuum
6	$p^2 < 0$		Virtual

In the transition from the Lorentz Group to the Poincaré group, we expanded the group to include another operator, the 4-momentum, the generator of translations in space-time. The irreducible representations of this larger Poincaré group are found with the aid of Schur's lemma: A representation is irreducible if the only group elements that commute with all the elements of the group are multiples of the identity operator, i.e., the group elements that commute with all the rest are the Casimir operators for the Poincaré group. The Hilbert space, within which the representation is defined, must be spanned by a set of eigenvectors corresponding to single eigenvalues of the Casimir operators. To each set of eigenvalues of the Casimir operators, there is one and only one irreducible representation. Then, finding the Irreps of the Poincaré group is reduced to determining the eigenvalues of the Casimir operators.

3.15 Casimir Operators for the Poincaré Group

We found that one of the Casimir operators for the Poincaré group was the scalar length of the momentum 4-vector, the rest mass,

$$P_\mu P^\mu = m^2$$

What then is the second Casimir operator for the Poincaré group? One is tempted to speculate that it must involve spin, and one would be correct, because we began our investigation of the Lorentz Group by examining invariances under rotations, which form a subgroup. However, we note that the second Casimir operator is not the square of the angular momentum as was the case for the rotation group in 3D. It is a tensor contraction of angular momentum and momentum first worked out by Pauli and Lubanski as the pseudovector W_μ defined as

$$W_\mu = -\frac{1}{2}\epsilon_{\mu\nu\rho\sigma}M^{\nu\rho}P^\sigma$$

where, again, $\epsilon_{\mu\nu\rho\sigma}$ is the Levi Civita totally antisymmetric symbol in 4D. W_μ is orthogonal to P^μ:

$$W_\mu P^\mu = 0.$$

The commutation relations for W_μ are

$$\left[M_{\mu\nu}, W_\rho\right] = i\left(T_{\nu\rho}W_\mu - T_{\mu\rho}W_\nu\right)$$

$$\left[W_\mu, P_\rho\right] = 0$$

$$\left[W_\mu, W_\nu\right] = i\epsilon_{\mu\nu\rho\sigma}W^\rho P^\sigma$$

Since the Pauli Lubanski pseudovector involves both momentum and angular momentum and it commutes with P^μ, one expects that as an operator product, its eigenvalues would be simultaneous products of each: we already know that the eigenvalue of momentum is m^2. The second Casimir operator is the square of the Pauli–Lubanski operator whose eigenvalues are given by the scalar:

$$W_\mu W^\mu = -m^2 s(s+1)$$

Therefore, objects (i.e., quantum states, Irreps) allowed in space-time are the $2s + 1$ simultaneous eigenfunctions of the rotation and translation operators. They can be either massless, as in the case of the photon, or massive. The spin can take on the values $\{0, 1/2, 1, 3/2, 2,\ldots\}$. These objects divide into two distinct and *mutually exclusive* families of objects in the Universe, the *Fermions*, with half-integral spin, and the *Bosons*, with integral spin: what we now recognize as matter and fields.

According to the spin-statistics theorem, *Fermions*, with spin $\{1/2, 3/2, 5/2,\ldots\}$ can only occupy states whose wave functions are antisymmetric (i.e., change sign) under the interchange on any two particles, while Bosons, with spin $\{0, 1, 2,\ldots\}$ are described by symmetric wave functions that do not change sign under particle interchange. The Pauli exclusion principle is therefore at the root of life, for without the antisymmetry of the Fermion wave function, all of the electrons in atoms would occupy the ground state and the periodic table would not exist. Fermions (matter) are the fundamental constituents of Faust's Universe (**Welt**) in his search for "*Was die Welt im innersten….*" We see that symmetry is not only interesting in its own right, but serves as a logical confinement for the contents of our observable world. The other part of Faust's search "*…im innersten zusammenhält.*" is satisfied by the Bosonic fields that hold the matter together.

The standard model of elementary particle physics (Fig. 3.6) directly displays the bosonic and fermionic structure of the Universe. In Chap. 6, we will examine how these players formed us and the world around us.

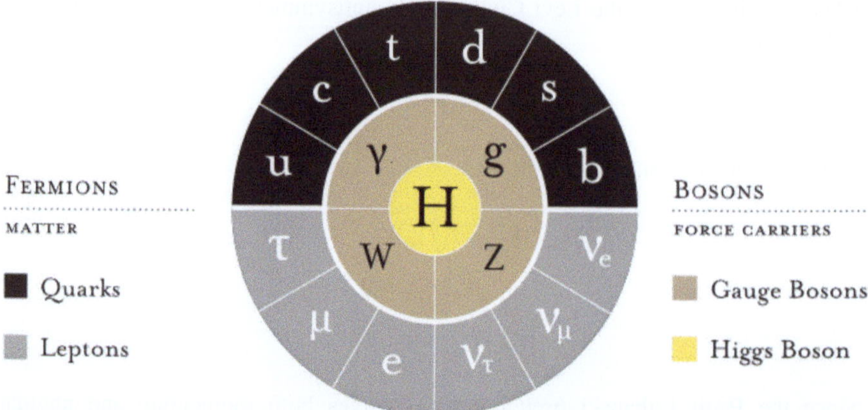

Fig. 3.6 A schematic illustration of the structure of the constituents of the Universe. Fermions exclusively make up matter, while Bosons are the carriers of the fundamental forces. (Source for Fig. 3.6: https://upload.wikimedia.org/wikipedia/commons/0/00/Standard_Model_of_Elementary_Particles.svg CC By 4.0)

Problems

1. *Basis.* Find the basis vectors e_1 and e_2 from the eigenvectors of R_x and R_y.
2. *Lorentz transformations.* Show that the Lorentz transformation forms a group.
3. *Poincaré transformations.* Show that the Poincaré transformation forms a group.
4. *Translation operator.* A translation by the distance a in 1D by the operator $U(a)$ acting on a wave function $\psi(x)$ generates the new function

$$U(a)\psi(x) = \psi(x+a).$$

   The generator of translations p_x is the momentum operator in quantum mechanics

$$p_x = -i\hbar \frac{d}{dx}.$$

   By expanding $\psi(x+a)$ in a Taylor's series, show that

$$U(a) = e^{i\,a\,p_x}.$$

5. *Pauli–Lubanski pseudovector.* Use the expressions given in the text to find an explicit representation of the Pauli–Lubanski pseudovector.
6. *Rotation matrices.* Use *Mathematica* to show that

$$\det(R_z(\theta)) = \text{Cos}^2[\theta] + \text{Sin}^2[\theta] = 1,$$

and that

$$R_z(\theta = \pi/3) \, \text{Transpose}\left[R_z\left(\theta = \frac{\pi}{3}\right)\right]] = I,$$

the identity matrix in 3D.

References

H. Goldstein, C. Poole, J. Safko, Classical Mechanics, 3rd Edition, Addison-Wesley (San Francisco CA: 2002)

D.E. Neuenschwander, *Emmy Noether's wonderful theorem* (Johns Hopkins University Press, 2011)

E.A. Noether, Invariante variationsprobleme. Nachr. Akad. Wiss. Gottingen, Math.-Phys. Kl. **II**, 235–257 (1918)

E. Wigner, On unitary representations of the inhomogeneous Lorentz group. Ann. Math. **40**, 149–203 (1939)

Bibliography

W. Greiner, B. Muller, *Quantum mechanics: symmetries*, 2nd edn. (Springer, New York, 1994)

M.P. Hobson, G. Efsthathiou, A.N. Lasenby, *General relativity: An introduction for physicists* (Cambridge University Press, 2006)

L.H. Ryder, *Quantum field theory*, 2nd edn. (Cambridge University Press, 1996)

S. Schweber, *An introduction to relativistic quantum field theory* (Dover, New York, 1989)

J. Schwichtenberg, *Physics from symmetry*, 2nd edn. (Springer, 2018)

M. Tinkham, *Group theory and quantum mechanics* (McGraw Hill, New York, 1961)

A. Zee, *Group theory in Nutshell for physicists* (Princeton University Press, Princeton, 2016)

Chapter 4
Quantum Mechanics: The Self-assembly of Atoms

> *"Never trust atoms, they make up everything!"*
>
> *Yours truly*

> *"A particle in a box,*
> *A fox in socks.*
> *Green eggs and ham,*
> *Spam I am.*
> *Schrödinger's cat in a Hat,*
> *Imagine that!"*
>
> *Yours truly, with apologies to Dr. Seuss*

Perhaps at the end of the nineteenth century, one would have expected that Faust's search for "...*was die Welt im innersten zussamenhält*" would have been satisfied by the energy-directed self-assembly of atoms from Fermionic electrons and nuclei. The principles organizing charged particles in the electromagnetic field mediated by Bosonic photons are described by Maxwell's equations, which were thought to describe all of nature. However, as the twentieth century dawned and measurements entered the atomic realm, these principles were forced to yield in unexpected ways as the limits of measurement became apparent and then paramount, limits that manifested themselves in both the physical, and as we have seen in the last chapter, the logical confinement of matter.

The physical confinement of atomic matter, electrons bound in the electromagnetic potential well in the field of a positively charged nucleus, produced one of the most impressive, predictive theories of the self-assembly of matter: *quantum mechanics*. The philosophical implications of this twentieth century revolution in knowledge are still a matter of active debate and perhaps, astonishment, even today. Nevertheless, if one is interested in a proper understanding of the nature and properties of objects as small as molecules, atoms, nuclei, or subnuclear particles, one needs to develop a facility with "quantum" concepts. Fermionic matter particles, such as electrons, behave as waves, light waves behave as Bosonic particles (photons); both of these paradoxes are intimately linked to the limits that quantum mechanics places on our ability to obtain information about such small systems. It will also be necessary to revise our notion of what constitutes a complete description

© The Author(s), under exclusive license to Springer Nature Switzerland AG 2024
L. O. Sillerud, *Abiogenesis*, https://doi.org/10.1007/978-3-031-56687-5_4

of nature. We will utilize the concepts from quantum mechanics in this and several subsequent chapters to elucidate the interaction of light with nuclei, atoms, and molecules, to understand atomic and molecular structure, to determine the magnetic properties of nuclei and electrons, and to examine the origin of the elements, particularly those involved with the evolution of Life. For the time being, we will concentrate our attention in this and the next chapter on those Fermions that constitute ordinary, neutral atoms and molecules, the electrons and protons, and ignore the rest of the particles and fields in the Standard Model until we discuss the Big Bang and the origin of the Universe where their presence is essential for its understanding. The electromagnetic interaction and quantum mechanics are sufficient to elucidate atomic and molecular structure.

That the single theory of quantum mechanics enables one to predict and calculate the behavior of a vast range of self-assembling microsystems is a remarkable accomplishment, and one is therefore tasked with providing its basic principles. Many expositions of quantum theory begin with an examination of the historical development of the failure of classical physics to explain the Black-Body problem, the photoelectric effect, and the optical spectra of atoms. (The interested reader is directed to the several references in the bibliography.) However, quantum mechanics has been with us for more than 100 years now, and its strength as a predictive foundation for understanding the properties of systems smaller than the wavelength of the probing radiation has been solidly established by a wealth of experimental data. Therefore, we will only provide an axiomatic approach to the fundamentals of nonrelativistic quantum mechanics. These certainly suffice for an understanding of the electronic and nuclear properties of biomolecules.

The foundation for quantum mechanics is the fact that the only means available for obtaining knowledge of the biophysical world is through utilization of the tools that physically exist; the main conclusion of twentieth century science is that *there are no other means for obtaining experimentally verifiable knowledge from nature other than through measurement*. We have achieved a profound understanding of the limits on knowledge imposed by the physical limits of measurement. One cannot make measurements more precise than allowed by the physical mechanisms available to you; we have reached the limit of those instruments, in the sense that the only means which we can utilize for knowledge is the scattering of photons or massive particles from objects of interest. What we find is that matter, confined to places smaller than the wavelength of the incident light used in a scattering study behaves like the air pressure in organ pipes, or the vibrations of a drum head, or the resonant tones of a guitar string. Indeed, when Werner Heisenberg spent the summer of 1925 in a cabin on the island of Helgoland thinking about the regular patterns observed in the energy levels of the anharmonic oscillator, the lines in the emission spectrum of hydrogen, and the other elements, he was led to the mathematical properties of matrices (Heisenberg 1925). An absorbing account was written by Carlo Rovelli (Rovelli,C., Helogoland. Riverhead Books, New York: 2021). Max Born, Heisenberg, and Pascual Jordan then quickly realized that matrices give rise to eigenvalue equations whose solutions were the observables from measurements (Born et al. 1925). That same summer, Erwin Schrödinger knew, from the properties of differential equations, that the spectrum of the hydrogen atom could be solved as another eigenvalue equation (Schrödinger 1926). The known fact that the resolution of an optical experiment is inversely proportional to the wavelength of the light eventually gave rise to Heisenberg's "*Ungenauigkeit prinzip*" or unexactness principle, what later became known as the uncertainty principle (Heisenberg 1927).

The fact that both differential equations and matrices display eigenvalues (e.g., resonant frequencies) inevitably led to the unification of both the matrix and the differential equation approach to quantum mechanics. The State variables in quantum mechanics turn out to be complex, rather than real, as in classical thermodynamics; this complexity is actually an advantage because it enables quantum mechanics to embrace the existence of antiparticles and to be described by first-order equations. However, in an examination of nature, one is interested in the results of measurements, which are always represented by real quantities. The way to produce a real quantity from a complex one is to perform the absolute square. The resulting real function was then interpreted as the probability of a particular object occupying a certain place at a certain time.

The implication of Heisenberg's uncertainty principle was that the size of a probing body (incident particle, wavelength) influences the knowledge we can obtain about an object. Although this appears as nonsense from a classical point of view, it is well illustrated by a *gedanken* experiment in which we, blindfolded, attempt to map the three-dimensional structure of an elephant with the aid of a tennis ball cannon. If we place the cannon near the elephant and shoot balls at it, some of the tennis balls will strike the elephant and bounce (scatter) off in different directions. In order to map the elephant, we would shoot balls and measure their direction and energy after they bounce off. Clearly, tennis balls would be suitable for this purpose, but beach balls would give us much less information because they are so large that they would average the elephant's shape and miss fine details such as the wrinkles in his skin. The resolution we would obtain would be even poorer if we were to use the scattering of hot-air balloons.

Quantum mechanics deals with exactly this problem, but our elephant is replaced by an atom, and the tennis balls by photons of a given wavelength. The energy of a photon is proportional to its frequency, ν, through the relationship $E = h\nu$, where h is Planck's constant, i.e., 6.626×10^{-34} Joule sec, and wavelength, λ, is inversely proportional to frequency, $\lambda = c/\nu$, where $c = 3 \times 10^8$ meters/sec is the speed of light. One would think that we could just throw something else at the atoms, and not use photons at all, or just use photons of shorter wavelengths, on the order of the size of the atom, or smaller. The revolution wrought by quantum mechanics is based on the fact that there is nothing else we can use, no other particles or waves, which would behave any differently in our atomic scattering experiment, and since the energy of a photon or particle increases with its frequency, we would reach the point where decreasing the wavelength of the particle led to such large energies that the collision event would knock the atom far out of its original position, and we would have no structural information at all. Imagine trying to map a mosquito by firing rifle bullets at it! Quantum mechanics, then, is the physical basis from which we can determine exactly what we *can* and, what is equally important, *cannot* know about tiny systems. And, as we saw in the exposition of thermodynamics from a statistical point of view in the previous chapter, we will see that quantum mechanics also provides only a statistical description of nature, and that, once again, this probabilistic description is nevertheless very precise.

4.1   The Basic Postulates of Quantum Mechanics

Group theory in the last chapter was based on a small set of logical axioms. In a similar vein, nonrelativistic quantum mechanics is founded on the following nine basic postulates:

1. **Wave function.** The State of a quantum system (electron, photon, proton, …) is completely determined by its _complex_ wave function, often written as $\psi(q,\ t)$, where q are the three spatial coordinates, $q \in \{x,\ y,\ z\}$, and t is the time. The shorthand $\psi(q,\ t)$ means $\psi(x,\ y,\ z,\ t)$. Here, we use the *Mathematica* notation for lists as objects contained within curly brackets: $\{\ldots\}$, so that q is the set $\{x, y, z\}$. Later we will condense this mathematical functional notation using a clever scheme developed by Paul Dirac (Dirac 1927) for eigenvalue equations.

2. **Probability.** The probability of finding the System in a State described by the wave function $\psi(q,\ t)$ is given by the absolute square of the wave function:

$$|\psi(q,t)|^2 = \psi^*(q,t)\ \psi(q,t),$$

   where $\psi^*(q, t)$ is the complex conjugate of $\psi(q, t)$, i.e., one where i is replaced by $-i$ in the functional form of $\psi(q, t)$. For example, if $\psi = e^{i\omega t}$, then $\psi^* = e^{-i\omega t}$.

3. **Normalization.** Since the absolute square of the wave function, $|\psi(q,t)|^2$, is a probability, its integral must be normalized to one:

$$\int \Psi^*(q,t)\Psi(q,t)dqdt = 1$$

4. **Observables.** Every observable is the eigenvalue of a linear, Hermitian operator. An operator, O, is Hermitian if $O^\dagger = O$, where $O^\dagger$ is the complex conjugate transpose (adjoint) of O. Operators can be either differential or algebraic. There are operators for energy, momentum, dipole moment, and so on. A Hermitian operator has real eigenvalues and therefore provides real values of observables.

5. **Measurements.** The average, or expectation, value, $\langle E \rangle$, of the measurement of an observable, such as energy, E, is given by

$$\langle E \rangle = \frac{\displaystyle\int \Psi^*(q,t)E\,\Psi(q,t)dqdt}{\displaystyle\int \Psi^*(q,t)\Psi(q,t)dqdt}$$

where E is the operator representing energy in the integral. Note that this is the same as the definition of the average value of a thermodynamic State function. Also, since by postulate (3) the wave function must be normalized, the denominator must be 1, and we can rewrite this as

$$\langle E \rangle = \int \Psi^*(q,t)E\,\Psi(q,t)dqdt.$$

6. **Eigenvalues.** Since the operators representing physical observables are Hermitian, they obey eigenvalue equations with real eigenvalues. For a defined State of the System, characterized by a value of an observable, p, which is the result of the measurement of the average value of the operator, P, we have this eigenvalue equation:

$$P\,\Psi(q,t) = p\Psi(q,t),$$

where p is the *eigenvalue* of the operator, P, and $\Psi(q,t)$ is the *eigenfunction* of P. Eigenvalues are just real numbers. The German word *eigen* was first used in this context by *David Hilbert* in 1904. "Eigen" can be translated as "own," or "characteristic." In English, the closest translation would be "characteristic," and some older references do use expressions such as "characteristic value" and "characteristic vector," or even "Eigenwert," which is the full German expression for eigenvalue.

As an example of an eigenvalue equation:

$$\text{Let } \Psi(x) = e^{2x}, \text{ and the differential operator, } P = \frac{d}{dx}, \text{ then}$$

$$P\Psi(x) = \frac{d\Psi(x)}{dx} = \frac{d}{dx}e^{2x} = 2e^{2x} = 2\Psi(x)$$

and we can write the eigenvalue equation as

$$P\Psi(x) = 2\Psi(x),$$

where the eigenvalue of P is $p = 2$.

7. **Eigenstates.** If $\Psi(q,t)$ is in a defined eigenstate of the operator P, then $\Psi(q,t)$ is an eigenfunction of P, and

Table 4.1 Relationship between classical variables and quantum operators

Classical variable	Quantum operator
Position, q	q
Momentum, p	$-i\hbar\frac{\partial}{\partial q}$
Time, t	t
Total energy, E	$-i\hbar\frac{\partial}{\partial t}$
Angular momentum, L	$i\hbar\frac{\partial}{\partial\theta}$
Potential energy, V	V
Kinetic energy, T	$\frac{\hbar^2}{2m}\nabla^2$

$$P\Psi(q,t) = p\Psi(q,t)$$

so that by (5) the average value of P is

$$\langle P\rangle = \int \Psi^*(q,t)P\,\Psi(q,t)dqdt$$

but, $P\,\Psi(q,t) = p\Psi(q,t)$, so we can take the constant, p, out of the integral

$$\langle P\rangle = \int \Psi^*(q,t)P\,\Psi(q,t)dqdt = p$$

because the remaining integral is unity by the normalization condition. Therefore, if the System is in an eigenstate of the operator, P, the average value of a measurement of the eigenvalue of P will be found to be p. If the system is not in an eigenstate of P one will measure an average value of P over an ensemble of States. Note the similarity here with thermodynamic average values over probability distributions from Chap. 2.

8. **_Classical variables become quantum operators._** The average values of classical quantities one might be interested in are replaced by the average, or expectation values of operators. Several operators for State variables of interest include those listed in Table 4.1:

9. **The _Schrödinger equation._** The wave function, $\Psi(q,t)$, is found as a solution of the time-dependent Schrödinger equation:

$$H\Psi(q,t) = i\hbar\frac{\partial}{\partial t}\,\Psi(q,t),$$

where H is the Hamiltonian operator corresponding to the classical Hamiltonian

$$H = T + V,$$

as the sum of the kinetic and potential energies of a System. Schrödinger's equation is the operator expression of the conservation of energy. In order to make this more explicit, we consider the classical expression for the kinetic energy

$$T = \tfrac{1}{2}\, m\, v^2$$

with

$$P = mv$$

therefore,

$$P^2 = m^2 v^2$$

so that

$$T = \frac{P^2}{2m}.$$

From our operator table (Table 4.1), we have that the operator for momentum, P, is

$$P = -i\hbar \frac{\partial}{\partial q}$$

and we can insert this into the above equation to give

$$T = \frac{P^2}{2m} = \frac{(-i\hbar)^2}{2m}\, \frac{\partial^2}{\partial q^2}.$$

The kinetic energy operator is the square of the reduced Planck's constant, $\hbar = \frac{h}{2\pi}$, divided by twice the mass of the particle, times the second partial derivative with respect to space. We have stated this result in only one dimension, q, such that $q =$ either x or y or z. In three dimensions, the kinetic energy operator uses the *Laplacian*,

$$\nabla^2 = \frac{\partial^2}{\partial x^2} + \frac{\partial^2}{\partial y^2} + \frac{\partial^2}{\partial z^2}$$

which is the dot product of the vector gradient operator, $\vec{\nabla}$, with itself

$$\nabla^2 = \vec{\nabla} \cdot \vec{\nabla} = \left(\hat{x} \frac{\partial}{\partial x} + \hat{y} \frac{\partial}{\partial y} + \hat{z} \frac{\partial}{\partial z} \right) \cdot \left(\hat{x} \frac{\partial}{\partial x} + \hat{y} \frac{\partial}{\partial y} + \hat{z} \frac{\partial}{\partial z} \right)$$

therefore, the kinetic energy operator in 3D is

$$T = \frac{\hbar^2}{2m} \left(\frac{\partial^2}{\partial x^2} + \frac{\partial^2}{\partial y^2} + \frac{\partial^2}{\partial z^2} \right) = \frac{\hbar^2}{2m} \nabla^2.$$

Note here that the $\hat{x}$, $\hat{y}$, and $\hat{z}$ are the orthonormal unit vectors along the x, y, and z axes, respectively. The time-dependent Schrödinger equation in three spatial dimensions is then

$$\frac{\hbar^2}{2m} \nabla^2 \Psi(q,t) + V\Psi(q,t) = i\hbar \frac{\partial \Psi(q,t)}{\partial t}$$

which is to be read as the sum of the kinetic and potential energies of a system, which is equal to the total energy. All solutions of Schrödinger's equation form a complete, orthonormal set, complete in the mathematical sense that one needs no other functions, the solutions form a basis for the Hilbert space, and orthogonal in that the inner product of two solutions is zero unless they are the same function,

$$\int \Psi_i^*(q,t) \Psi_j(q,t) dq dt = \delta_{ij}$$

where δ_{ij} is a Kronecker delta function.

10. **Dirac notation.** *P.A.M. Dirac* (1928) realized that the mathematical structure of quantum mechanics was compactly and elegantly based on that of complex linear vector spaces, in particular, on a variety known as a Hilbert space after *David Hilbert*. Dirac saw that wave functions were solely distinguished by their unique eigenvalues and not by their functional symbols or parameters (coordinates). Instead of wave functions, he introduced bras and kets, derived from the word bra-ket or "bracket." A ket is an object defined as $\Psi_n(x) = |n\rangle$, while its complex conjugate bra is $\Psi_n^*(x) = \langle n|$. When these are joined together, the implication is one of an inner product of vectors, or in terms of functions

$$\langle n \mid m \rangle = \int \Psi_n^*(x) \Psi_m(x) dx.$$

An eigenvalue equation for momentum

$$P\Psi(q,t) = p\Psi(q,t)$$

would be given as

$$P \,|\, p\rangle = p \,|\, p\rangle$$

where the wave function was an eigenfunction of the operator P with the eigenvalue p. The complex conjugate $\Psi^*(q,t)$ was then written as $\langle p|$ and the normalization condition (3) could be compactly written as

$$\langle p|p\rangle = \int \Psi_i^*(q,t)\Psi_j(q,t)\,dq\,dt = \delta_{ij}$$

The average, or expectation, value, $\langle E \rangle$, of the measurement of the observable energy, E, would be

$$\langle q|E|q\rangle = \frac{\displaystyle\int \Psi^*(q,t)E\,\Psi(q,t)\,dq\,dt}{\displaystyle\int \Psi^*(q,t)\Psi(q,t)\,dq\,dt}$$

where E is the operator representing energy in the integral.

4.2   Simple Solutions of *Schrödinger's Equation*

The modern exposition of quantum mechanics is to be understood in its Hilbert space description. Wave functions are state vectors in a Hilbert space that are acted upon by the quantum operators. From a biophysical vantage point, we would like to solve Schrödinger's equation for one or more problems of biophysical interest. Since electrons, nucleons, and atoms are all quantum systems, with dimensions much less than that of the wavelength of optical photons, and these particles constitute the basic components of biomolecules, an understanding of the quantum properties of these particles is our goal. In addition, since electrons, nucleons, and atoms are indeed all quantum systems, we might anticipate that a general approach to quantum systems would cover each specific case. Thus, we proceed on a strictly formal basis.

4.2.1   The Wave Function for a Free Particle

We can solve Schrödinger's equation if the Hamiltonian, $H(q,t) = T(q,t) + V(q,t)$, is explicitly independent of time:

$$H(q,t) = H(q) = T(q) + V(q),$$

$$H(q,t) \neq H(t).$$

In a potential-free region, where $V(q, t) = 0$, we have that $H(q, t) = H(q)$. Since there is no potential energy, the entire energy of the particle is kinetic; this will give us a solution of the Schrödinger equation for a free particle of mass m and energy E. To find the solution, we can use the technique of the separation of variables, commonly applied to partial differential equations.

Let the wave function, $\Psi(q, t)$, be written as the product of a function only of time, $\phi(t)$, and a function, $\chi(q)$, that only depends on space:

$$\Psi(q, t) = \chi(q)\ \phi(t).$$

Then, we inset this into Schrödinger's equation with the time-independent Hamiltonian

$$H\Psi(q, t) = i\hbar \frac{\partial}{\partial t}\ \Psi(q, t)$$

$$H\chi(q)\phi(t) = i\hbar \frac{\partial}{\partial t}\ \chi(q)\phi(t)$$

$$\phi(t)H\chi(q) = i\hbar\ \chi(q) \frac{\partial}{\partial t}\ \phi(t)$$

where the last two steps were possible because $\chi(q)$ depends only on spatial, and not temporal, coordinates, and the Hamiltonian is time-independent, so $\phi(t)$ is a constant. We now rearrange the terms to put the time-dependence on the right and the spatial dependence on the left:

$$\frac{H\chi(q)}{\chi(q)} = \frac{i\hbar}{\phi(t)} \frac{\partial}{\partial t}\ \phi(t) = C$$

In order for the left-hand side, which is only a function of space, to equal the right-hand side, which is only a temporal function, both sides can only be equal to a constant, C. We can determine this constant by noting that $\chi(q)$ must be a solution of the spatial-only, Schrödinger equation

$$H\chi(q) = E_n\chi(q)$$

The wave function, $\chi(q)$, is an eigenfunction of H, with the eigenvalue E_n; thus, the constant $C = E_n$.

$$\frac{H\chi(q)}{\chi(q)} = \frac{i\hbar}{\phi(t)} \frac{\partial}{\partial t}\ \phi(t) = E_n$$

We can solve the time-dependent part of this equation for $\phi(t)$ by means of a simple integration

$$i\hbar\,\frac{\partial}{\partial t}\,\phi(t) = E_n\phi(t)$$

Multiply by dt and divide by $\phi(t)$ to give

$$i\hbar\,\frac{d\phi(t)}{\phi(t)} = E_n dt$$

And now integrate this

$$i\hbar\int\frac{d\phi(t)}{\phi(t)} = \int E_n dt$$

noting that the eigenvalue, E_n, is a constant and the ϕ integral gives $\ln(\phi(t))$

$$i\hbar\int\frac{d\phi(t)}{\phi(t)} = E_n t$$

$$\ln\phi(t) = \frac{E_n t}{i\hbar}$$

and the solution for the wave function, $\phi(t)$, is that

$$\phi(t) = e^{\frac{-iE_n t}{\hbar}}.$$

This is the wave function for a free particle of energy, E_n. It is a wave oscillating with a frequency ω given by

$$\omega_n = \frac{E_n}{\hbar}.$$

4.2.2   Time Dependence

The general wave function $\Psi(q,\,t)$ is the product of $\phi(t)$ with the spatial function, $\chi(q)$,

$$\Psi(q,t) = \chi(q)\ \phi(t),$$

so we have that

$$\Psi(q,t) = \chi(q) \ e^{\frac{-iE_n t}{\hbar}}$$

The general wave function is an oscillating function of time.

Immediately, we see that we are confronted with a novel concept; the wave function for a free particle contains a complex number, i, where $i^2 = -1$. Physical experiments produce only real quantities as their results. Therefore, we must find a means to convert the complex wave function, Ψ, to the result of an actual measurement. The square of the absolute value of a function is always positive. The interpretation of the wave function is that its absolute square is the probability of locating a particle at the particular coordinates specified:

$$<\Psi \mid \Psi> \ = |\Psi|^2 = \Psi^* \ \Psi = \Psi^*(q,t) \ \Psi(q,t) = \chi^*(q) \ \chi(q) \ e^{\frac{iE_n t}{\hbar}} e^{\frac{-iE_n t}{\hbar}}$$

where Ψ^* is the function Ψ with i replaced by $-i$, i.e., Ψ^* is the complex conjugate of Ψ. Then, for the free particle, the time-dependent probability is

$$|\phi(t)|^2 = \phi(t) * \phi(t) = e^{\frac{iE_n t}{\hbar}} e^{\frac{-iE_n t}{\hbar}} = 1$$

and we see that the probability of locating the particle somewhere is 1. The wave function for a free particle is normalized if the spatial wave functions are also normalized:

$$\langle q|q \rangle = \int \chi^*(q)\chi(q)dq = 1.$$

4.2.3   Spatial Dependence

We have examined the behavior of Schrödinger's equation in the case when the Hamiltonian, H, was explicitly independent of time. Under these conditions, a biomolecule would be in a stationary state. In this case, the wave function, Ψ, would be an eigenfunction of the Hamiltonian, and Schrödinger's equation would read

$$H\Psi(q,t) = i\hbar \frac{\partial}{\partial t} \ \Psi(q,t)$$

Let us replace Ψ with the solution $\phi(t)$ of the free-particle Schrödinger equation

$$H\Psi = i\hbar\frac{\partial}{\partial t}\,\Psi$$

$$H\Psi = i\hbar\frac{\partial}{\partial t}\,e^{\frac{-iE_n t}{\hbar}}$$

$$H\Psi = i\hbar\left(\frac{-iE_n}{\hbar}\right)e^{\frac{-iE_n t}{\hbar}}$$

$$H\Psi = E_n\,e^{\frac{-iE_n t}{\hbar}}$$

$$H\Psi = E_n\,\Psi.$$

This is the important, *time-independent* Schrödinger equation. Notice that it is an eigenvalue equation of the operator, H, which is the total energy of the system, with an eigenvalue of E_n, which is the energy of the particle in the nth state, or energy level. In Dirac notation, we can write this as

$$H\,|\,n\rangle = n\,|\,n\rangle.$$

The separation of variables technique used above can also be applied to solve the time-independent Schrödinger equation for the spatial-dependent wave function of a free particle. We will demonstrate the solution in two distinct but related ways. The first is found by noticing that the wave function of a free particle must be an eigenfunction of the momentum operator, P, with eigenvalue, p:

$$P\,\Psi(q,\ t) = p\,\Psi(q,t),$$

or, with the definition of the momentum operator (Table 4.1)

$$P = -i\hbar\frac{\partial}{\partial q}$$

we have that

$$-i\hbar\frac{\partial}{\partial q}\Psi(q,t) = p\Psi(q,t)$$

Now, we again use the fact that we can rewrite $\Psi(q,\ t)$ as a product of a function of only time, $\phi(t)$, and a function of only space, $\chi(q)$,

$$\Psi(q,t) = \chi(q)\phi(t)$$

To give

$$-i\hbar\frac{\partial}{\partial q}(\chi(q)\phi(t)) = p\,\chi(q)\phi(t)$$

We can move $\phi(t)$ out from the derivative since it is not a function of q

$$-i\hbar\phi(t)\frac{\partial}{\partial q}\chi(q) = p\,\chi(q)\phi(t)$$

and now it appears on both sides as simply a product function, so we can cancel it and we are left with

$$-i\hbar\frac{\partial}{\partial q}\chi(q) = p\chi(q)$$

which can be integrated

$$-i\hbar\int\frac{d\chi(q)}{\chi(q)} = p\int dq$$

to give

$$\ln(\chi(q)) = \frac{ipq}{\hbar}$$

and if we have let $q = x$, for a one-dimensional solution, the spatially oscillating wave function is

$$\chi(x) = e^{\frac{ipx}{\hbar}}.$$

The wave vector or spatial frequency of a free particle is defined to be

$$k = p/\hbar = 2\pi/\lambda$$

where λ is the wavelength of the particle in an analogous fashion to the definition above of the frequency, ω, as

$$\omega = E/\hbar.$$

Then the spatial wave function for a free particle is given by

$$\chi(q) = e^{ikq}.$$

   The second method for showing this is to directly solve the time-independent Schrödinger equation

$$H\ \chi(q) = E\ \chi(q)$$

for $\chi(q)$. For a free particle, $H = T$ in a potential-free region of space. Then, inserting the kinetic energy operator from Table 4.1, we have that

$$-\frac{\hbar^2}{2m}\nabla^2\chi(q) = E\chi(q).$$

We can use the definition of k from above, along with $T = p^2/2m$ to express the energy as

$$-\hbar^2 k^2 = 2mE$$

to rewrite the time-independent Schrödinger equation as

$$\nabla^2\chi(q) + k^2\chi(q) = 0$$

which has solutions $\chi(q)$, which are the same as those found above

$$\chi(q) = e^{ikq}.$$

Now, the total wave function $\Psi(q, t)$ for a free particle of mass m is the product of the spatial and time-dependent functions

$$\Psi(q, t) = \chi(q)\ \phi(t)$$
$$\Psi(q, t) = e^{ikq}e^{-i\omega t}$$

and this is a traveling wave in both space and time with momentum, $p = \hbar k$, and energy, $E = \hbar\omega$.

As the basis functions for a Hilbert space, all solutions of Schrödinger's equation form a complete, orthonormal set. Completeness in mathematics means that no other functions are needed or necessary and orthonormal means that the basis functions are both orthogonal and normalized. For example, a periodic function of time, $F(t)$, may be rewritten as a sum of sines and cosines with frequencies, ω_i, and amplitudes, a_i, b_i as

$$F(t) = \sum_i f(\omega)[a_i\sin(\omega_i t) + b_i\cos(\omega_i t)].$$

In this case, the sines and cosines form the set of basis functions for rewriting the function, $f(\omega)$, of frequency, ω, into a function, $F(t)$, of time, t. The sines and cosines form a complete, orthonormal set of functions. We have already seen that the set of unit vectors, x_i, in three-space constitute a complete orthonormal set. Orthogonality is a property of functions where there is no overlap between one function and its immediate neighbor. Overlap is measured by the integral of the product of two functions. These properties of the solutions of Schrödinger's equation can be summarized by introducing the Kronecker delta, $\delta_{mn} = \{1, \text{if } m = n; 0, \text{if } m \neq n\}$.

For any two solutions of Schrödinger's equation, ψ_m and ψ_n, labeled by their eigenvalues, m and n, the overlap integral obeys

$$\langle m|n \rangle = \int \Psi_m^* \, \Psi_n dv = \delta_{mn} = \{1, \text{ if } m = n; \ = 0, \text{ if } m \neq n\}$$

where v is the set of coordinates over which the problem runs. Thus, the solutions of Schrödinger's equation comprise a complete, orthonormal set, and any other solution, ϕ, may be rewritten in terms of the solutions ψ_n as a linear combination of terms

$$\phi(x) = \sum_n a_n \Psi_n(x).$$

The coefficients, a_n, can be found by using the orthonormality of the ψ_n. We multiply this equation by ψ_m, and integrate over x:

$$\int \Psi_m^*(x)\phi(x)dx = \sum_n a_n \int \Psi_m^*(x)\Psi_n(x)dx = a_n \delta_{mn} = a_m$$

therefore,

$$a_m = \int \Psi_m^*(x)\phi(x)dx$$

where the integral for a_m measures the overlap between the $\psi_m(x)$ and $\phi(x)$ and we can find each of these coefficients by solving this set of integrals.

4.3   The Uncertainty Principle

Observations made over the last 100 years have shown that particles can act like waves, and that waves can act like particles. The classical double-slit diffraction experiment showed unambiguously that light is a wave with a given wavelength. On the other hand, Compton scattering, the elastic scattering of light from electrons, shows equally unambiguously that light consists of the particles know as photons. Einstein in 1905 attributed a momentum p to a photon with an energy E of

$$p = \frac{E}{c} = \frac{h\nu}{c} = \frac{h}{\lambda}.$$

The classical definition of momentum is $p = mv$, but the photon is a spin 0, massless Boson and always travels at the speed of light, c. The corollary that each particle can be described by a matter wave was proposed by de Broglie in 1924. He defined the energy and momentum of a particle in wave language (angular frequency, ω and wave vector, k) as

$$E = \hbar\omega \quad \text{and} \quad \vec{p} = \hbar\,\vec{k}$$

Therefore, waves share particle-like properties, and vice versa.

It had been known for a long time prior to the discovery of quantum mechanics that diffraction effects in optical microscopy became significant for objects near the wavelength of light. In keeping with the duality of waves and particles, the discovery of matter waves through the electron diffraction experiments of Davison and Germer in 1927 showed clearly that a revision was needed in our views with respect to how well the properties of small particles could be defined.

We can anticipate these results through a recollection of the wave function for a free particle. We have seen that the momentum operator in quantum mechanics is given by the spatial derivative of the wave function: $p\Psi \rightarrow -i\hbar\frac{\partial\Psi}{\partial x}$; therefore, momentum is related to $1/x$, where x is the spatial extent (wavelength) of the particle (represented as a wave packet). The relationship between momentum and position is that of conjugate Fourier pairs of variables $\{x,\, p\} = \left\{x,\, -i\hbar\frac{\partial}{\partial x}\right\} = \{x,\, \hbar k\}$. A Taylor's series represents a function in terms of the complete, orthonormal set of the derivatives of a function evaluated at a point. The derivative of a function is defined to be the tangent to the function, evaluated at that point, and therefore, by definition, the tangent is orthogonal to the function at every point.

The positional wave function $\Psi(x)$ of a particle is given by the Fourier transform of its momentum wave function

$$\Psi(x) = \int e^{ikx}\varphi(p)\,dp$$

where the term e^{ikx} is understood as a complete, orthonormal set of sines and cosines.

In momentum space, a free particle with well-defined momentum k_o is represented by a wave function, which is a delta function, $\delta(k_o) = 1$ if $k = k_o$; $= 0$ if $k \neq k_o$. Then the position wave function is the Fourier transform of this delta function,

$$\Psi(x) = \frac{\hbar}{\sqrt{2\pi}}\int e^{ikx}\delta(k_o)\,dk = \frac{\hbar}{\sqrt{2\pi}}$$

and this is a constant, independent of x, that is, independent of position and any spatial coordinates. The momentum of a particle described by a delta function is precisely defined as k_o.

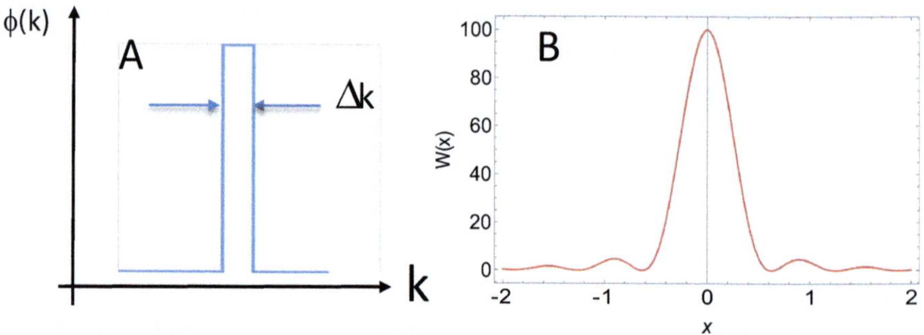

Fig. 4.1 (**a**) A square-wave packet $\phi(k)$ with width Δk defined in momentum space. (**b**) The Fourier transform $\psi(x)$ of the momentum distribution in (**a**)

What is this telling us about the location of a particle with a perfectly defined momentum? It says that if a particle's momentum is perfectly defined, then the position of the particle (which is given by the constant probability distribution) is completely undefined. The probability of locating the particle anywhere in any given volume in the Universe is a constant. Therefore, one cannot simultaneously know the position and momentum of a particle with arbitrary precision.

A more physically reasonable description of a real particle is that it has a wave function, $\phi(k)$, with a momentum defined by measurement to be within a small, but finite range, Δk, as shown in Fig. 4.1a. This is a square-wave function, whose Fourier transform is the well-known sinc function: $\psi(x) = \frac{\sin(\Delta k\, x)}{x}$. The spatial probability distribution, $|\psi^*(x)\psi(x)| = \frac{\sin^2(\Delta k\, x)}{x^2}$, is a peaked function (Fig. 4.1b) with ripples (zeroes) occurring at $\Delta k\, \Delta x = \pi$, or since $p = hk/2\pi$, we have that $\Delta k = 2\pi\, \Delta p/h$, and

$$2\pi\, \Delta p/h\ \Delta x = \pi,$$

therefore,

$$\Delta p\ \Delta x \sim h/2,$$

which is an approximate statement of *Heisenberg's Uncertainty Principle*. We will see in subsequent chapters that this principle constitutes the most profound physical limit on knowledge ever discovered.

The previous rationalization of the uncertainty principle can be put on more firm mathematical grounds based on the exposition in previous chapters of the characteristics of the moments of observables governed by probability distributions. Given a Fourier pair of conjugate variables, such as $\{x, p\}$, or $\{t, E\}$, we know that in one dimension the expectation value of, for example, the operator P is given by

$$\langle P \rangle = \int\limits_{-\infty}^{\infty} \Psi^*(x) P \Psi(x)\, dx$$

and for P^2, a similar expression holds

$$\langle P^2 \rangle = \int\limits_{-\infty}^{\infty} \Psi^*(x) P^2 \Psi(x)\, dx$$

with analogous expressions for $<x>$ and $<x^2>$. From our previous work on probability in Chap. 1, we can write the dispersions of x and p as

$$<\Delta x^2> = <x^2> - <x>^2$$

and

$$<\Delta p^2> = <p^2> - <p>^2.$$

Let us pick coordinate systems in which $<x> = <p> = 0$, such as we saw for a random walk that began at the origin. Then, we need only to calculate $<x^2>$ and $<p^2>$ in order to determine the dispersions in each variable.

Now, in order to examine the simultaneous relationship between position and momentum, we will need to calculate the root-mean-squared (rms) *product* of the dispersions of position and momentum,

$$\sqrt{\langle \Delta x^2 \rangle \langle \Delta p^2 \rangle}.$$

This will be remembered as the probability of simultaneously measuring the position within a range Δx and the momentum within a range Δp, given a probability distribution characterized by the absolute square of the wave function, $|\psi(x)|^2$. Let us begin by noting that the square of the 1D momentum operator, P, is given by

$$P^2 = \left(-i\hbar \frac{\partial}{\partial x} \right) \left(-i\hbar \frac{\partial}{\partial x} \right) = -\hbar^2 \frac{\partial^2}{\partial x^2}$$

and that the position operator is x. We can generate a positive definite form containing x^2 and p^2 by considering an expression such as $(x\psi + p\psi)^2 = x^2\psi^2 + 2xp\psi^2 + p^2\psi^2$. An evaluation of an integral of this form would allow us to compute the relationship between the dispersion in x and that in p. However, the wave function, ψ, is complex, so we need to take the absolute square, $|\psi|^2$, rather than the naïve square. This suggests that we write instead a form like

$$f(x, P) = |\alpha\, x\, \psi + P\, \psi|^2$$

which we note is an absolute square, and hence $f(x, P) \geq 0$, no matter what the values of the variables, as long as $\alpha \in \Re$, that is, α is contained in the set of Real numbers. The required dispersions are all contained within the integral of $f(x, P)$. Let us denote this integral by $I(\alpha)$:

$$I(\alpha) = \int_{-\infty}^{\infty} |\alpha\, x\, \psi + P\psi|^2 dx$$

$$I(\alpha) = \int_{-\infty}^{\infty} \left|\alpha\, x\, \psi - i\hbar \frac{\partial}{\partial x}\psi\right|^2 dx$$

When we expand the integrand, we find that we must compute the following three terms:

$$I(\alpha) = \alpha^2 \int_{-\infty}^{\infty} \psi^* x^2 \psi\, dx + \alpha\hbar^2 \int_{-\infty}^{\infty} x\left(\frac{\partial \psi^*}{\partial x}\psi + \psi^* \frac{\partial \psi}{\partial x}\right)dx + \hbar^2 \int_{-\infty}^{\infty} \frac{\partial \psi^*}{\partial x}\frac{\partial \psi}{\partial x}dx$$

call these terms A, B, and C, such that $I(\alpha) = A\alpha^2 + B\alpha + C$. First, we will examine

$$A = \int_{-\infty}^{\infty} \psi^* x^2 \psi\, dx$$

This is by definition equal to $<\Delta x^2>$, so that $I(\alpha) = <\Delta x^2 > \alpha^2 + B\alpha + C$. Next, we will examine B, which is given by

$$B = \hbar^2 \int_{-\infty}^{\infty} x\left(\frac{\partial \psi^*}{\partial x}\psi + \psi^* \frac{\partial \psi}{\partial x}\right)dx$$

which can be written using the product rule from calculus as

$$B = \hbar^2 \int_{-\infty}^{\infty} x\frac{d}{dx}(\psi\psi^*)dx$$

This integral can be computed by parts as follows. The differential of the product of two functions, u and v is

$$d(uv) = u\, dv + v\, du$$

then, the integral of $d(uv)$ is uv

$$\int d(uv) = \int u\,dv + \int v\,du = uv$$

We can use this to compute B as follows. Let $u = x$, $du = dx$, $dv = \psi^*\psi$, $v = \psi^*\psi$. We then have that

$$\int_{-\infty}^{\infty} d(uv) = -x\psi^*\psi = 0 = \int_{-\infty}^{\infty} x\frac{d}{dx}(\psi\psi^*)dx + \int_{-\infty}^{\infty} \psi\psi^*dx$$

and the left-hand side is zero because the wave function must vanish at infinity. The second integral on the right-hand side is just the normalization integral for the wave function, so must equal 1, and we have that

$$\int_{-\infty}^{\infty} x\frac{d}{dx}(\psi\psi^*)dx + 1 = 0 = B - 1$$

so that $B = 1$, and we have so far that

$$I(\alpha) = <\Delta x^2 > \alpha^2 + \alpha + C,$$

and it remains to calculate the integral C, which is given by

$$C = \hbar^2 \int_{-\infty}^{\infty} \frac{\partial\psi^*}{\partial x}\frac{\partial\psi}{\partial x}dx$$

This is again integrable by parts. Let $u = \psi^*$, $du = (d\psi^*/dx)$,

$$v = \frac{d\psi}{dx}dx, \text{ and } dv = \frac{d^2\psi}{dx^2}dx$$

Then

$$\int_{-\infty}^{\infty} \psi^*\frac{d\psi}{dx}dx = \int_{-\infty}^{\infty} \psi^*\frac{d^2\psi}{dx^2}dx + \int_{-\infty}^{\infty} \frac{d\psi^*}{dx}\frac{d\psi}{dx}dx$$

The integral on the left-hand side is $\psi^*\psi$ which, evaluated at the limits is zero, again because the wave function must vanish at infinity. This results in

$$0 = \frac{1}{\hbar^2} \int_{-\infty}^{\infty} \psi^* \frac{d\psi}{dx} dx - C$$

or that C is given by

$$C = \frac{1}{\hbar^2} \int_{-\infty}^{\infty} \psi^* \frac{d\psi}{dx} dx$$

Now, notice that this integral is $(2\pi/h^2)<p^2>$ due to the definition of $<p^2>$ given above. Finally, we have that

$$I(\alpha) = <x^2>\alpha^2 + \alpha + (2\pi/h^2)<p^2>.$$

In order for this form to be positive definite, the discriminant must be less than or equal to zero. In $I(\alpha) = <x^2>\alpha^2 + \alpha + (2\pi/h^2)<p^2>$, we have $I(\alpha) = A\alpha^2 + B\alpha + C$, with $A = <x^2>$, $B = 1$, and $C = (2\pi/h^2)<p^2>$, and the discriminant is

$$B^2 - 4AC \leq 0.$$

Inserting the above values for the coefficients, we find that this condition is satisfied if

$$1 \leq 4<x^2>(2\pi/h^2)<p^2>,$$

or

$$\sqrt{\langle \Delta x^2 \rangle \langle \Delta p^2 \rangle} = \Delta x \Delta p \geq \frac{h}{4\pi}$$

This gives the Heisenberg form of the "uncertainty principle," which is a quantitative understanding of the relationship between simultaneous measurements of Fourier conjugate variables:

$$\Delta x \Delta p \geq \frac{\hbar}{2}$$

We will show that this relationship has profound implications for many aspects of biology, including the origin of the Universe and of the elements of Life.

4.4   The Particle in a Box Model of a Simple Quantum System

One of the simplest, yet quite-interesting, analytically soluble problems in quantum mechanics is that of a free particle confined to a small region of space (a "box") in which the potential energy of the particle is zero. The walls of the box (Fig. 4.2) rise, essentially to infinity, so the particle cannot escape the box, no matter how large its energy might be. The Hamiltonian inside the box is $H = T + V(x)$; however, $V(x) = 0$, if $0 \leq x \leq a$, and $V(x) = \infty$, if $x < 0$ or $x > a$.

The Hamiltonian inside the box is then only the kinetic energy, expressed as an operator,

$$H = T = \frac{p^2}{2m} = \frac{\hbar^2}{2m}\frac{d^2}{dx^2}$$

Schrödinger's equation reads, in this case,

$$-\frac{\hbar^2}{2m}\frac{d^2}{dx^2}\Psi(x) = E_n\Psi(x)$$

where we have anticipated the discrete nature of the solutions by labeling them with the index (quantum number) n, and our task is to find the eigenvalues of energy, E_n. This is a second-order differential equation in x. Let us again substitute

$$k^2 = -\frac{2mE_n}{\hbar^2}$$

into Schrödinger's equation to simplify the algebra. Then the equation for the wave function becomes

$$\frac{d^2}{dx^2}\Psi(x) + k^2\Psi(x) = 0$$

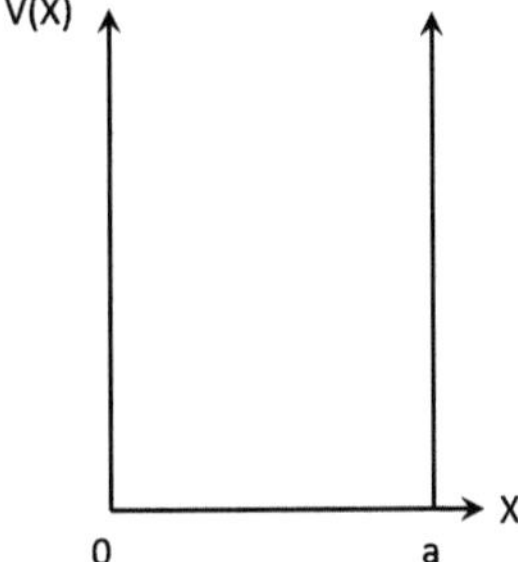

Fig. 4.2 A one-dimensional box, of width a, containing a particle of mass m

The solution to this is the wave function for a free particle, since $V = 0$ within the box,

$$\Psi(x) = e^{\pm ikx},$$

where the $\pm$ indicates that the particle could be traveling either to the left or the right. Any linear combination of solutions to a differential equation will also be a solution, so let us write the wave function as

$$\Psi(x) = A\, e^{ikx} + B\, e^{-ikx}.$$

Now, since $e^{ikx} = \cos(kx) + i\,\sin(kx)$, we can rewrite $\Psi(x)$ as

$$\Psi(x) = C\, \sin(kx) + D\, \cos(kx),$$

however, the cosine term does not satisfy the boundary conditions that $\Psi(0) = \Psi(a) = 0$, so it can be discarded, and we are left with

$$\Psi(x) = C\, \sin(kx).$$

An application of the boundary conditions gives both the trivial solution, $\Psi(0) = C\,\sin(0) = 0$, and the more interesting one, $\Psi(a) = C\,\sin(ka) = 0$. This equation is true whenever $ka = n\pi$.

Then the wave function is

$$\Psi(x) = C\, \sin(n\pi x/a).$$

We still have an arbitrary scale factor, C, to determine. The normalization condition can be used to find C by requiring that the integral of the absolute value of $\Psi(x)$ be 1.

$$\int \Psi^*(x)\Psi(x)dx = 1 = \int_0^a C^2 \sin^2\left(\frac{n\pi x}{a}\right)dx$$

$$\int_0^a \sin^2\left(\frac{n\pi x}{a}\right)dx = \frac{1}{C^2}$$

This integral has the value of $a/2$, so

$$\frac{1}{C^2} = \frac{a}{2}$$

or

$$C = \sqrt{\frac{2}{a}}$$

and the complete, normalized wave function is given by

$$\Psi_n(x) = \sqrt{\frac{2}{a}} \sin\left(\frac{n\pi x}{a}\right)$$

The important energy eigenvalues, E_n, can be found from the expression for k in the boundary value equation, $k = n\pi/a$, and

$$-k^2 = \frac{2mE_n}{\hbar^2} = \frac{n^2\pi^2}{a^2}$$

Solving this equation for E_n indicates that the energy levels for the particle in a box are given by

$$E_n = \frac{n^2 h^2}{2m\,a^2} \quad n \in \{1,\ 2,\ 3,\ \ldots,\ \infty\}.$$

Note that here we have absorbed a factor of π^2 into the expression for E_n, by replacing $\hbar$ by h.

The energy levels for a particle in a box increase as the square of n, and are shown schematically in Fig. 4.3a, for $1 \leq n \leq 4$. The first four eigenfunctions are also shown in Fig. 4.3b, and the probability, which is the absolute square of the wave function, is shown in Fig. 4.3c. Note that the wave function itself is both positive and negative, but the probability is always positive.

What are the actual values for the energies of the first few levels if the box is the size of a hydrogen atom, $a = 106$ pm, where m is the mass of the electron, $m_e = 9.10938188 \times 10^{-31}$ kg? These are shown in Table 4.2 for n ranging from 1 to 4. The electron Volt (eV) is a common unit of energy in the physics of particles; note that Einstein's $E = mc^2$ allows us to convert mass in kg to energy in Joules, and that 1 eV $= 1.602 \times 10^{-19}$ Joules. The wavelength corresponding to these energies is also included in the values in Table 4.2, and it is computed by means of Planck's formula for the energy of a quantum of wavelength, λ: $E = h\nu = hc/\lambda$, where h is Planck's constant and c is the speed of light, $c = 3 \times 10^8$ m/s.

The actual energy levels in a hydrogen atom are of lower energy due to the nonzero (negative) potential energy of an electron in the electric field from the proton. Transitions from one level to another involve the emission or absorption of photons of the wavelength of visible light, ~500 nm. We see, however, that the energy levels in a hydrogen atom modeled as an electron in a box are approximately of the correct order of magnitude; the ionization potential for the real hydrogen atom is 13.6 eV.

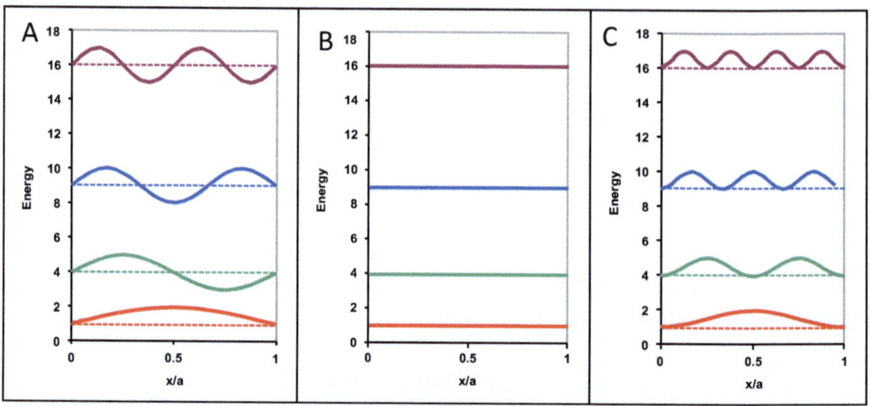

Fig. 4.3 (**a**) The first four wave functions for the particle in a box, plotted according to their energy levels normalized to the energy of the ground state. (**b**) The energy levels normalized to the energy of the ground state. (**c**) The probability, $P(x/a)$ of finding a particle at position x/a, where a is the width of the box. The wave functions and probabilities can be classified according to their parity as either even or odd, depending on their sign upon reflection about $a/2$, the center of the box

Table 4.2 Energies, and wavelengths for an electron in a 106 pm box	n	E_n (Joules)	λ_n (nm)	E_n (eV)
	1	5.36×10^{-18}	37.07	33.47
	2	2.14×10^{-17}	9.27	133.87
	3	4.83×10^{-17}	4.12	301.20
	4	8.58×10^{-17}	2.32	535.46

Will this model also work in an approximate manner for objects as small as the proton, which we will model as a mass, $m_p, = 1.6726 \times 10^{-27}$ kg, in a box the size of a typical atomic nucleus, $a \sim 10$ Fermi $= 1 \times 10^{-14}$ m? We find that for a box of the dimensions of a nucleon, the first energy level occurs at 2.05 MeV, the second at 8.20 MeV, and so on. These are certainly energies appropriate for nuclear transitions, where the mass of the proton is 938 MeV. That the particle in a box model for the confinement of matter should lead to numerically useful predictions is an astonishing feat for such a simple microscopic theory, but perhaps the most important result is the fact that *the physical confinement of matter leads to the generation of quantized energies.*

One feature of the particle in the box that we will find useful later is the force on the wall of the box. The force is the negative derivative of the energy as a function of distance,

$$F(x) = -\frac{d}{dx} E_n = \frac{n^2 h^2}{4m\, x^3}$$

Here we see that the force increases as the inverse cube of the linear dimension of the box.

The particle in a box model serves to illustrate many salient features of quantum mechanics. Using Dirac notation, one can succinctly express many features of quantum states; for example, the particle in a box states can then be written as:

$$\Psi_n(x) = \; \mid n> \; = \sqrt{\frac{2}{a}}\sin\left(\frac{n\pi x}{a}\right)$$

4.4.1   The Parity of the Particle in a Box States

Note from Fig. 4.3 that the wave functions for the particle in a box can be classified into two distinct groups: one group that peaks in the middle of the box and another that is zero in the box center. These two classes differ in an important quantum number, that of parity, or symmetry under reflections in a mirror. This is most easily seen if you think of the box shifted by $a/2$ to the left and then the origin would be in the center of the box. We can find the eigenvalues, λ, of the parity operator, Π, by noting that a reflection through the origin is denoted

$$\Pi\Psi_n(x) = \lambda\Psi_n(-x)$$

and if we apply the parity operator again, we regain the same wave function

$$\Pi\;(\Pi\Psi_n(x)) = \lambda^2\,\Psi_n(x) = \Psi_n(x)$$

therefore, $\lambda^2 = 1$, and $\lambda = \pm 1$ and we have that either

$$\Psi_n(x) = \Psi_n(-x) \text{ or}$$
$$\Psi_n(x) = -\Psi_n(-x).$$

Wave functions that do not change under reflection through the origin have *even* parity and those that do change sign have *odd* parity. Until 1956, it was thought that all interactions in nature conserved parity. It was predicted by Lee and Yang, as well as discovered experimentally by C.S. Wu, that the weak interactions violated parity conservation. No one would have expected that the mirror image of an object would be any different from the object itself, but the lesson from nature indicates that even supposedly bedrock symmetry principles must be tested experimentally. We now understand that the solution to this puzzle of parity nonconservation in the weak interactions has to do with the fact that nature only contains left-handed (left-chiral) leptons

Note that the Hamiltonian does not change if we change the box coordinates from $\{0, a\} \rightarrow \{-a/2, a/2\}$, but the eigenfunctions change from sines to cosines in order to satisfy the boundary conditions that now $\psi_n(-a/2) = \psi_n(a/2) = 0$. With this shift in coordinates, it is easy to see that the wave function is still an eigenfunction of the parity operator with eigenvalues $(-1)^n$. The Hamiltonian commutes with the parity operator so that the wave functions are simultaneous eigenfunctions of both operators and

$$[H, \ \Pi] = 0,$$

or that parity is conserved.

The product $\Pi \psi_n(x)$ is an eigenfunction of H with the same energy eigenvalue and we can write that

$$H\left(\Pi \, \psi_n(x)\right) = \Pi \ \left(H\psi_n(x)\right) = \Pi \ E_n\psi_n(x) = E_n\left(\Pi \ \psi_n(x)\right)$$

which states that the energy eigenfunctions are simultaneous eigenfunctions of the parity operator. We have seen above that the eigenvalues of Π are ± 1 and with $x \in \{-a/2, a/2\}$ the boundary conditions we have two types of solutions: one with even parity

$$| \, n+ > \, = \sqrt{\frac{2}{a}} \ \cos\left[\frac{(2n-1)\pi x}{a}\right] \ \text{with energy } E_n{}^+ = (2n-1)^2 \, \frac{\pi^2 \hbar^2}{2ma^2}$$

and the other with odd parity

$$| \, \mathrm{n}- > \, = \sqrt{\frac{2}{a}} \ \sin\left[\frac{2n\pi x}{a}\right] \ \text{with energy } E_n{}^+ = (2n)^2 \, \frac{\pi^2 \hbar^2}{2ma^2}$$

and these energies span the same series as n^2, viz, $\mathrm{n} \in \{1, 4, 9, 16, 25, \ldots\}$.

4.4.2   Selection Rules

As we will see in Chap. 10, when we study the interaction of light with biomolecules, the particle in box model also provides us with a prescription for predicting which transitions among energy levels are allowed and which are forbidden: these are known from the early studies of atomic spectra as selection rules.

4.4.3   Zero-Point Motion

Note that the lowest energy level for the particle in a box has the quantum number $n = 1$, and not zero. For $n = 0$ the wave function is identically zero everywhere, which implies that there is no particle in the box; the probability is zero for finding the particle in the box. For this reason, it is clear that any particle in the box must be in constant motion, even in the ground state. This is known as zero-point motion, and is a consequence of the uncertainty principle; if a particle were to have zero energy, it would also have zero momentum and its momentum would be perfectly defined since it would have zero momentum dispersion, $\Delta p = 0$, giving $\Delta x \Delta p = 0$, which violates the uncertainty principle.

We can use the probability theory developed in Chap. 1 to find the minimum momentum uncertainty for the particle in a box as follows. The average, or expectation value, of a quantity x governed by a probability distribution, $P(x)$ you will recall, is given by

$$\langle x \rangle = \int_{-\infty}^{\infty} x\, P_n(x) dx$$

where in quantum mechanics the probability distribution is given by the absolute square of the wave function

$$P_n(x) = \Psi_n^*(x)\ \Psi_n(x)$$

so that

$$\langle x \rangle = \int_{-\infty}^{\infty} \Psi_n^*(x) x\, \Psi_n(x) dx.$$

The variance was defined as

$$\sigma^2(n) = \int_{-\infty}^{\infty} (x - \langle x \rangle\,)^2\, P_n(x)\, dx$$

Then, let's compute the required means for x and p needed to find the dispersions and then calculate the product $\Delta x \Delta p$. The dispersion in x is given by

$$< \Delta x^2 > = <x^2> - <x>^2.$$

The average value of x is, in Dirac notation,

$$<x>^2 = \langle n|x|n \rangle = \int_{-\infty}^{\infty} \Psi_n^*(x) x\, \Psi_n(x) dx$$

$$\langle n|x|n \rangle = \frac{2}{a} \int_0^a \sin\left[\frac{2n\pi x}{a}\right] x \sin\left[\frac{2n\pi x}{a}\right] dx = \frac{a}{2}$$

which is the middle of the box, and this result holds for any value of $n \geq 1$, so

$$<x>^2 = \frac{a^2}{4}.$$

The average value of $<x^2>$ is, in Dirac notation,

$$<x^2> = \langle n|x^2|n\rangle = \int_{-\infty}^{\infty} \Psi_n^*(x)x^2\,\Psi_n(x)dx$$

It is left as a useful exercise (see "Problems") to show that the average value of $<x^2>$ is given by

$$<x^2> = \frac{a^2}{12}\left(4 - \frac{6}{\pi^2 n^2}\right)$$

which gives for the dispersion,

$$<\Delta x^2> = <x^2> - \frac{a^2}{4} = a^2\left(\frac{1}{12} - \frac{1}{2\pi^2 n^2}\right).$$

Now, we turn our attention to the momentum. The dispersion in p is given by

$$<\Delta p^2> = <p^2> - <p>^2.$$

The average value of p is, in Dirac notation,

$$<p> = \langle n|p|n\rangle = \frac{2}{a}\int_0^a \sin\left[\frac{2n\pi x}{a}\right]\left(-i\hbar\frac{d}{dx}\right)\sin\left[\frac{2n\pi x}{a}\right]dx$$

$$<p> = -\frac{i\hbar\,\mathrm{Sin}[n\pi]^2}{a}$$

but this is zero for any integer value of $n \geq 1$, because the probability of the particle moving to the left is equal to the probability of the particle moving to the right, and the average velocity is zero. The average value of p^2 is, in Dirac notation,

$$<p^2> = \langle n|p^2|n\rangle = \frac{2}{a}\int_0^a \sin\left[\frac{2n\pi x}{a}\right]\left(-\hbar^2\frac{d^2}{dx^2}\right)\sin\left[\frac{2n\pi x}{a}\right]dx$$

$$<p^2> = \hbar^2\left(\frac{n\pi(2n\pi - \mathrm{Sin}[2n\pi])}{2a^2}\right)$$

or, for integral values of $n \geq 1$, $<p^2> = \frac{n^2\,\pi^2\hbar^2}{a^2}$. It is left as an exercise to show that the minimum uncertainty in either x or p is obtained for the ground state, $n = 1$. We can find this uncertainty from the definition of the uncertainty principle:

Table 4.3 The minimum uncertainty product for a particle in a box

n	$\Delta x \Delta p$
1	0.568 $\hbar$
2	1.670 $\hbar$
3	2.627 $\hbar$
4	3.558 $\hbar$
5	4.479 $\hbar$

$$\Delta x \Delta p = \sqrt{\langle \Delta x^2 \rangle \langle \Delta p^2 \rangle} = \frac{\pi \hbar}{a} a \sqrt{\left(\frac{1}{12} - \frac{1}{2\pi^2} \right)},$$

in which the box length a cancels and we are left with the fact that the uncertainty product is

$$\Delta x \Delta p = \frac{\hbar}{2} \sqrt{\frac{1}{3} (\pi^2 - 6)}.$$

Evaluation of this expression results in a factor of 0.568 $\hbar$ and this is certainly greater than one-half $\hbar$.

Values for this expression are given in Table 4.3. The expression above for the uncertainty product does not depend on a, the size of the box, even though the uncertainty in position is proportional to the width of the box, but the uncertainty in momentum is inversely proportional to the box width, so they cancel.

4.5   The Momentum Wave Function

A new concept is that of the momentum wave function, or the particle wave function in momentum space. We alluded to the fact that position and momentum are essentially *duals*, conjugate variables where one is related to the reciprocal of the other; remember that momentum is inversely proportional to distance, $p = h/\lambda$, where λ is a distance, the wavelength of the particle. Position and momentum form a Fourier pair in which one can be determined from the other. What then is the wave function in momentum space? It is the Fourier transform of the position wave function. Let's denote the momentum wave function by $\phi(p)$, then

$$\phi(p) = \frac{1}{\sqrt{2\pi\hbar}} \int_{-\infty}^{\infty} \Psi_n(x) e^{-ipx/\hbar} dx$$

and, after inserting the wave function,

$$\phi(p) = \frac{1}{\sqrt{2\pi\hbar}} \sqrt{\frac{2}{a}} \int_{-\infty}^{\infty} \sin\left[\frac{n\pi x}{a} \right] e^{-ipx/\hbar} dx$$

we find that (with considerable help from *Mathematica*) the momentum wave function is given by:

$$\phi_n(p) = \sqrt{\frac{a}{\pi\hbar}} \left[\frac{n\pi}{n\pi + ap/\hbar}\right] \operatorname{sinc}\left[\frac{1}{2}\left(n\pi - \frac{pa}{\hbar}\right)\right] e^{-\frac{ipa}{2\hbar}}$$

This is, except for the normalization constants and the exponential phase factor, a sinc function as we found before when we examined the qualitative basis for the uncertainty principle (Fig. 4.1). It is instructive to compare the position wave functions (Fig. 4.3) with the momentum wave functions calculated above (Fig. 4.4). We observe that once again the wave function for $n = 0$ is identically zero, $\phi_0(p) = 0$, the probability is zero and there are no physical particles with zero momentum. Note also that all of momentum wave functions have a definite parity as do the spatial wave functions. As the momentum increases, the particle's wave function displays two prominent peaks, one at positive momentum and another negative peak symmetrically placed on the other side of zero with the opposite momentum corresponding to the particle going in the other direction. The momentum probability distributions (Fig. 4.5) indicate that the particle's momentum follows this pattern.

The uncertainty principle is one of the most important *grand organizing principles* discovered because it not only places physical limits on knowledge based on the fact that we can only use waves (light) to measure the properties of the Universe, but it also gives us a quantitative relationship between spatial and momentum fluctuations. There are other uncertainty principles: for continuous variables, such as time and energy, and for discrete variables, such as angular momentum. We will use these to understand the optical properties of colored biomolecules and to probe details of the Big Bang.

4.6   The Harmonic Oscillator: A Model for Light (Photons)

In nature, molecules, atoms, and elementary particles experience a wide variety of potentials in different environments. If the potentials are simple, several interesting, analytic solutions of Schrodinger's equation exist. We will examine the solutions in the quadratic potential experienced by particles undergoing simple harmonic motion and those for the electron in the central electrostatic Coulomb potential of the hydrogen atom.

The harmonic oscillator arises from a classical consideration of a mass suspended from the free end of a spring, whose other end is fixed. If we let the equilibrium position of the mass at rest be located at $x = 0$, and measure positions as positive when the mass is displaced by stretching the spring, then from Hooke's Law for a linear elastic spring, the force resisting the stretching of the spring is $F = -kx$, where k is the spring constant. The potential energy, $V(x)$, can be found from the integral of force times distance

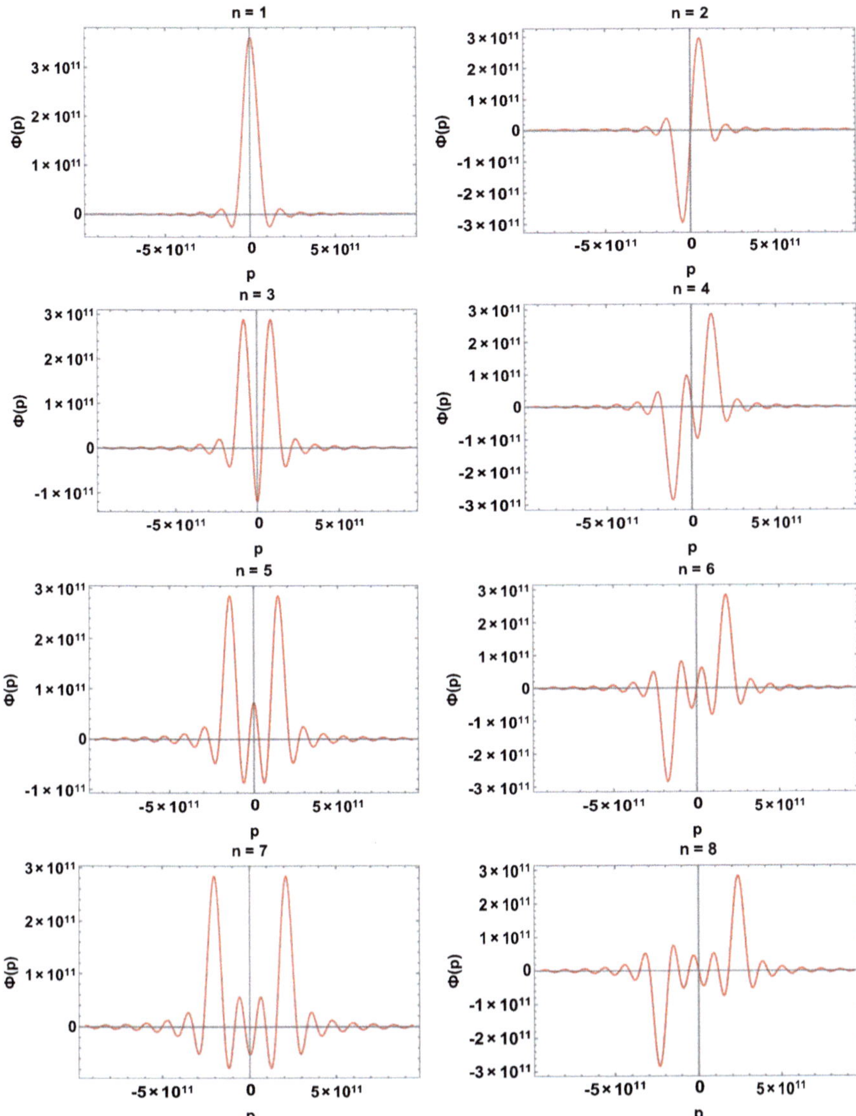

Fig. 4.4 The unnormalized momentum wave functions for a particle in a box the width of the hydrogen atom for quantum numbers $n = 1$–8. The momentum is in units of h/a, where a is the box width and h is Planck's constant

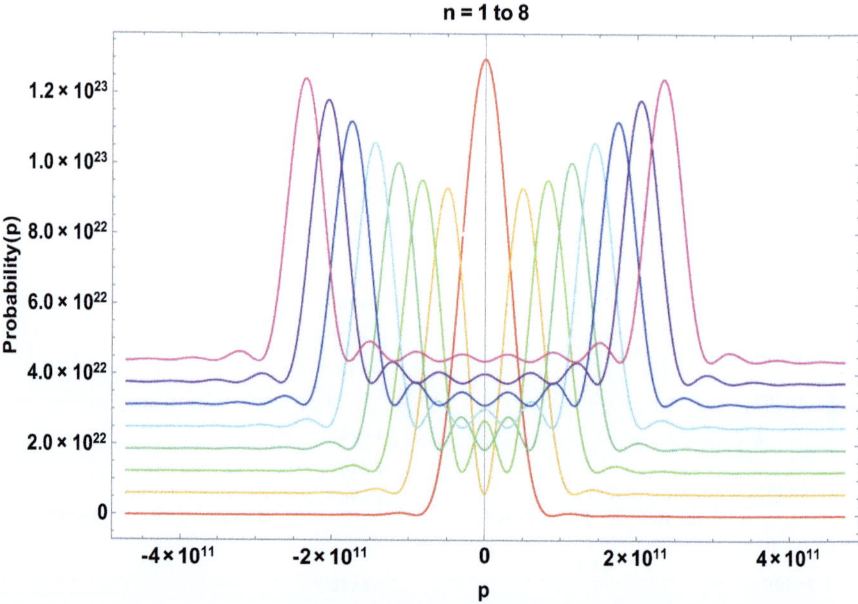

Fig. 4.5 The unnormalized momentum probability distribution for an electron in a box of width equal to the diameter of the hydrogen atom for quantum numbers $n = 1$–8

$$V(x) = - \int F(x)dx = \int kx\,dx = \frac{1}{2}kx^2$$

Newton's Law states that the force must equal the mass times the acceleration

$$F(x) = -kx = m\frac{d^2x}{dt^2}$$

which leads to the differential equation for the harmonic oscillator

$$m\frac{d^2x}{dt^2} + kx = 0$$

or, if we let $\omega = $ Sqrt(k/m), this gives the harmonic oscillator equation

$$\frac{d^2x}{dt^2} + \omega^2 x = 0$$

with its solution, $x(t) = \sin \omega t$. As we have seen above, this is the same as the wave function for a free particle, with frequency of oscillation ω.

The quantum mechanical harmonic oscillator solutions are of great importance in the study of the states of biomolecules because, even though the molecular potentials can be quite complicated, the behavior of the potentials for small oscillations can always be approximated by replacing the actual potential by that for the harmonic oscillator in the vicinity of the equilibrium position. We can see this more clearly if we expand a one-dimensional molecular potential, $V(x)$, in a Taylor's series about the equilibrium position, $x = a$:

$$V(x) = V(a) + V'(a)(x-a) + \frac{1}{2}V''(a)(x-a)^2 + \cdots$$

where $V'(a) = \left.\frac{dV(x)}{dx}\right|_{x=a}$, and so on. If the equilibrium is stable, then $V(x)$ has a minimum at $x = a$ and $V'(a) = 0$. We can choose the coordinate system so that $V(a) = 0$, and we are left with an approximate form for $V(x)$ as

$$V(x) = \frac{1}{2}V''(a)(x-a)^2,$$

which is indeed of the quadratic form of a harmonic oscillator potential.

The classical Hamiltonian for a particle of mass m oscillating with frequency ω is found from the kinetic and potential energies, $H = T + V(x) = \frac{p^2}{2m} + \frac{k}{2}x^2$, or if we rewrite this in terms of only mass and frequency, $k = \omega^2 m$, the classical Hamiltonian is

$$H = \frac{p^2}{2m} + \frac{m}{2}\omega^2 x^2.$$

From this it is straightforward to substitute in our expression for the quantum operator for T to obtain the quantum mechanical Hamiltonian

$$H = -\frac{\hbar^2}{2m}\frac{d^2}{dx^2} + \frac{m}{2}\omega^2 x^2.$$

The potential does not depend upon time so that the time-independent Schrodinger equation will give the stationary eigenfunctions $\psi_n(x)$ of the system and their corresponding energy eigenvalues, E_n, through $H\,\psi_n(x) = E_n\,\psi_n(x)$,

$$-\frac{\hbar^2}{2m}\frac{d^2}{dx^2}\psi(x) + \frac{m}{2}\omega^2 x^2\psi(x) = E\psi(x).$$

Let us replace the constants with

$$k^2 = \frac{2mE}{\hbar^2},$$

and

$$\lambda = \frac{m}{\hbar}\omega$$

to give Schrodinger's equation as

$$\frac{d^2}{dx^2}\psi(x) + \left(k^2 - \lambda^2 x^2\right)\psi(x) = 0$$

We state without proof that the behavior of solutions of this equation when $x \to \infty$ are of the form of a Gaussian envelope function times a polynomial $\varphi(x)$:

$$\psi(x) = e^{-\frac{\lambda x^2}{2}}\varphi(x)$$

where the Gaussian term makes intuitive sense due to the fact that the particle's wave function must vanish for $|x| \to \infty$. Solutions for $\varphi(x)$ are given in most texts on quantum mechanics and involve the Hermite polynomials $H_n(y)$, where $y = \sqrt{\lambda}x$.

$$\psi_n(x) = N_n \; e^{-\frac{\lambda x^2}{2}} H_n\left(\sqrt{\lambda}x\right)$$

as are shown in Fig. 4.6. The energy eigenvalues are

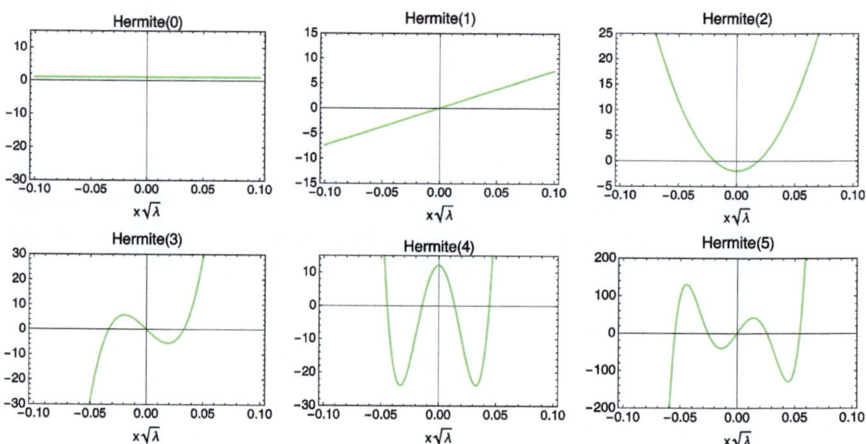

Fig. 4.6 The first six Hermite polynomials involved in the solutions to the quantum harmonic oscillator calculated for the mass of the electron. Note their orthogonality: integrals of products with different quantum numbers show no overlap

$$E_n = \left(n + \frac{1}{2}\right)\hbar\omega$$

For the quantum harmonic oscillator, the energies increase linearly with n, rather than as n^2 as for the particle in a box. The normalization constant, N_n is

$$N_n = \sqrt{\sqrt{\frac{\lambda}{\pi}}\frac{1}{2^n n!}}.$$

The Hermite polynomials can conveniently be obtained from a generating function

$$H_n(y) = (-1)^n\, e^{y^2}\,\frac{\partial^n}{\partial y^n}\,e^{-y^2}.$$

The first three normalized wave functions of the linear harmonic oscillator are then given by

$$\psi_0(x) = \sqrt[4]{\frac{\lambda}{\pi}}\;e^{-\frac{\lambda x^2}{2}}$$

$$\psi_1(x) = 2\sqrt{\frac{1}{2}}\sqrt{\frac{\lambda}{\pi}}\,e^{-\frac{\lambda x^2}{2}}\sqrt{\lambda}x$$

$$\psi_2(x) = \sqrt{\frac{1}{8}}\sqrt{\frac{\lambda}{\pi}}\,e^{-\frac{\lambda x^2}{2}}\left(4\lambda x^2 - 2\right)$$

And these are shown, together with their corresponding probability distributions in Fig. 4.7 for $n \in \{0, 5\}$. Note the interesting resemblance between the harmonic oscillator wave functions and those for the particle in a box (Fig. 4.3). Both solutions are bounded in that the wave functions go to zero outside of the limits of the potential (Fig. 4.8). We once again see that the physical confinement of matter in a potential leads to quantization of the energy levels into discrete states.

4.6.1   The Parity of the Harmonic Oscillator Wave Functions

The harmonic oscillator eigenfunctions $\psi_n(x)$ can be classified according to their parity as

$$\Pi\psi_n(x) = \psi_n(-x) = (-1)^n\psi_n(x)$$

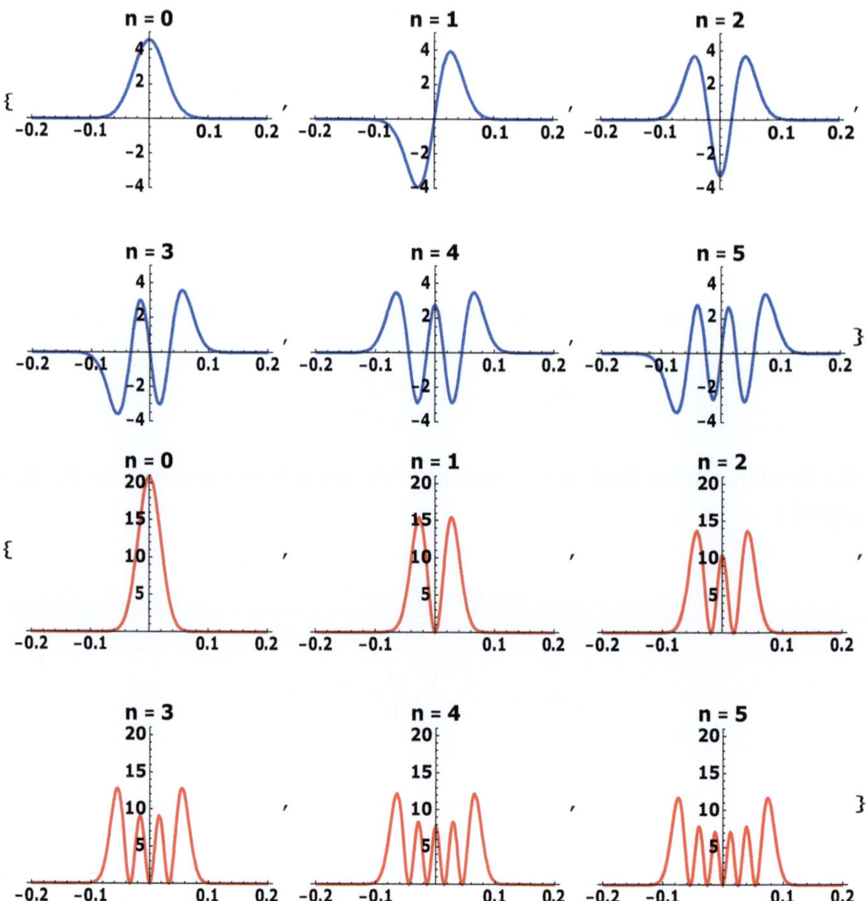

Fig. 4.7 The spatial dependence of the wave functions (top rows in blue) and the probabilities (bottom rows in red) for the harmonic oscillator for the first six states, $n \in \{0, 5\}$. These were calculated for a particle with the mass of the electron and for $\omega^2 = 1 = k/m$

where we have used Π as the parity operator to differentiate it from momentum. Here, if n is even then $\psi_n(x) = + \psi_n(-x)$ and the eigenvalue of Π is $+1$ and the parity is even, while if n is odd then $\psi_n(x) = -\psi_n(-x)$ and the eigenvalue of Π is -1 and the parity is odd. The Hamiltonian does not change under a parity transformation.

Even though we have now shown the solution to the differential equation that governs the quantum mechanical harmonic oscillator, there are several more insights to be gleaned by a closer examination of its quantum structure, and these require us to work directly with the quantum operators. The first is that the ground state energy E_0, for $n = 0$, is not zero, but $E_0 = \frac{1}{2}\hbar\omega$, the zero-point energy.

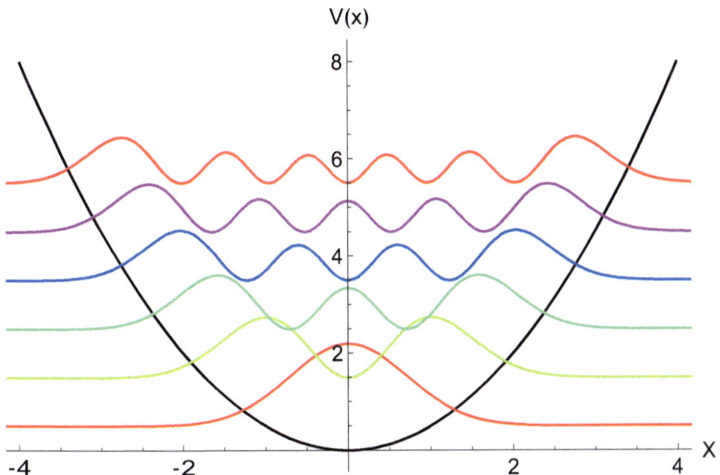

Fig. 4.8   The probability of finding a particle in the harmonic oscillator potential (black parabola) for the first six states, $n = \{0,5\}$ displaced according to their energies, which are a linear function of n: $E = (n + 1/2)\,\hbar\omega$

The second is that the energy eigenvalues are linearly spaced as $\left(n + \frac{1}{2}\right)\hbar\omega$, where $n \in \{0, 1, 2, 3, 4, \ldots\}$; in other words the energy occurs in single, discrete steps of $\hbar\omega$ separated by a constant amount. This suggests that we are dealing with unit excitations of a common object and that the energy levels correspond to the number of objects. Since the Hamiltonian is the total energy operator of the system, it must be that the Hamiltonian operating on the lowest energy state creates the higher energy states, and that these higher energy states are all the same and are composed of excitations of the same basic state; they only differ in the number of excitations of this basic state.

This more-elegant, and thus simpler, way to view the harmonic oscillator leads to a way of describing a large number of systems, such as quantized crystal excitations (phonons) as well the quantum electromagnetic field as a collection of photons (light quanta) and shows how to reinterpret the ground state of the harmonic oscillator as the vacuum, or the quantum state of nothing. Notice that this interpretation leaves the vacuum with finite, nonzero energy: the zero-point energy, and this is the third major point we wish to emphasize.

We have examined the quantum harmonic oscillator as a solution to a differential equation. However, a more elegant and consequently simpler way of looking at the same problem is via operators. In order to illustrate this reinterpretation, we can utilize the differential equation approach favored by Schrödinger to construct the operator approach of Heisenberg. These two ways of examining the same thing offer advantages that neither alone provides and enriches one's understanding of this system.

Instead of focusing on the energy of a state, E_n, let's think of the state as containing n objects, such as photons, and write the basis, eigenkets of the system as $|n>$ where n is an eigenvalue of a yet-to-be found occupation number operator, N, which we will define by its action on the basis states. We must have that $N|n> = n|n>$ and we can write the Hamiltonian in terms of this number operator such that $H|n> = (N + 1/2)\hbar\omega|n> = E_n|n> = (n + 1/2)\hbar\omega|n>$. Now, for the vacuum state, i.e., the state with no particles, $|0>$, we require that $N|0> = 0$.

Here, we can avoid wave functions and learn a new way of approaching quantum mechanics through algebra and commutation relations. We have motivated these ideas in Chap. 3 on "symmetry" and we can use these techniques to solve the harmonic oscillator in a very simple fashion that will provide us with the tools needed to later understand subtle and sophisticated phenomena related to the state of nothing and the origin of the Universe.

4.6.2   Group Theory, Commutators, and Symmetry

One of the many insights that we have found is that the commutator expresses the rule of combination for a group and that this provides another foundation for the uncertainty principle. For now, let it suffice that we will find a new, simpler way of understanding quantum systems by examining the commutator of their operators.

Using the Dirac bra-ket notation, we can write the position operator, **X,** eigenvalue equation as

$$\mathbf{X}|x> = x|x>$$

And the momentum operator, **P**, gives us

$$\mathbf{P}|x> = -i\hbar\frac{\partial}{\partial x}|x> = p|x>$$

We can write the Hamiltonian in a new form as

$$\mathbf{H} = \tfrac{1}{2}\left(\mathbf{P}^2 + \omega^2\mathbf{X}^2\right).$$

and the commutator of **P** and **X** is given by

$$[\mathbf{P},\ \mathbf{X}] = i\hbar$$

That the commutator of **P** and **X** is not zero is the basis for the uncertainty principle, and of primary importance for what follows. Although the form of this commutator may appear unfamiliar at first, let us remember that the position operator is simply a multiplicative operator, whereas the momentum operator, in this representation, is a derivative. It should be quite clear from an examination of these objects using elementary calculus that they cannot commute; position is orthogonal to momentum because derivatives of functions, together with the functions, form an orthogonal set, as is embodied in a Taylor's series expansions. Consider a function $f(x)$ and the operation of the commutator of $\mathbf{P}$ and $\mathbf{X}$ on it:

$$[\mathbf{P},\ \mathbf{X}]f(x) = P(Xf(x)) - X(Pf(x)) = -i\hbar\frac{\partial}{\partial x}\left(Xf(x)\right) - X\left(-i\hbar\frac{\partial}{\partial x}f(x)\right)$$

where the first term on the right is the derivative of a product and is certainly different from the product $X\left(-i\hbar\frac{\partial}{\partial x}f(x)\right)$.

Another, more physics-oriented way of looking at the uncertainty principle is to remember the difference between how position and momentum are measured. Position is measured with a ruler and requires only a single measurement of a point with respect to an origin, whereas momentum requires two position measurements separated by a time Δt to produce a velocity, $v = \Delta x/\Delta t$. In mathematics, one can take limits and define an instantaneous velocity as

$$v = \lim_{\Delta t \to 0}\frac{\Delta x}{\Delta t} = \frac{dx}{dt}$$

but in the real world, where measurements are limited by the speed of light, relativity tells us that you cannot make two position measurements simultaneously; they must always be separated by a finite interval time interval, Δt, which by $x = c\Delta t$ implies that by the time of the second position measurement the particle is no longer where your first measurement indicated. Consequently, by the time the second position measurement is completed, the particle has already moved. Note that unlike in mathematics where you can make Δt as small as you like, the real world is governed by physics, and, as Scotty says in Star Trek *"You cannot violate the laws of Physics!"*

The Hamiltonian above for the quantum harmonic oscillator

$$\mathbf{H} = \tfrac{1}{2}\left(\mathbf{P}^2 + \omega^2\mathbf{X}^2\right)$$

can be factored classically as if it were

$$a^2 + b^2 = (a + ib)(a - ib)$$

but if we reconstruct the Hamiltonian in this way as

$$\mathbf{H} = \tfrac{1}{2}\left(\mathbf{P} + i\omega\mathbf{X}\right)\left(\mathbf{P} - i\omega\mathbf{X}\right)$$

and keep track of the order of the X and P operators (knowing that they do not commute), we have that

$$\tfrac{1}{2}\,(\mathbf{P}+i\omega\mathbf{X})\,(\mathbf{P}-i\omega\mathbf{X}) = \tfrac{1}{2}\left(\mathbf{P}^2 + i\omega\mathbf{X}\mathbf{P} - i\omega\mathbf{P}\mathbf{X} - i^2\omega^2\mathbf{X}^2\right)$$

$$= \tfrac{1}{2}\left(\mathbf{P}^2 + i\omega[\mathbf{X},\ \mathbf{P}] - i^2\omega^2\mathbf{X}^2\right)$$

which we rewrite in terms of the commutator $[\mathbf{P},\mathbf{X}] = i\hbar$ so that the Hamiltonian becomes

$$\mathbf{H} = \frac{1}{2}\left(\mathbf{P}^2 + \omega^2\mathbf{X}^2\right) - \frac{1}{2}\,\hbar\omega$$

And this is smaller than our original Hamiltonian by $\tfrac{1}{2}\,\hbar\omega$, so we need to add this to offset the effect of this constant. The original Hamiltonian rewritten in these new terms is then

$$\mathbf{H} = \tfrac{1}{2}\left(\mathbf{P}^2 + \omega^2\mathbf{X}^2\right) + \tfrac{1}{2}\,\hbar\omega$$

Note that this includes the additive constant $\tfrac{1}{2}\,\hbar\omega$ that shifts the energy levels by this amount. Even for the lowest quantum state of the "harmonic oscillator," this zero-point energy mimics what we observed in the particle in the box solutions. If we interpret each quantum of the harmonic oscillator as an excitation of the electromagnetic field, then the field has nonzero energy even in the absence of excitations: the *zero-point energy.*

4.6.3   The Harmonic Oscillator Revisited: Creation and Annihilation Operators

Our goal here is to understand two features of quantum mechanics: the first is how to calculate the energy levels of the harmonic oscillator in this other formulation and the second is how to create something from nothing, or what is more correct in a quantum sense, how to excite the vacuum state to produce a physical object. Our factorization of the Hamiltonian gives us the clues about how to bring this about. We rewrote the Hamiltonian as a product of two terms $(\mathbf{P} + i\omega\mathbf{X})\,(\mathbf{P} - i\omega\mathbf{X})$ and it is useful to examine the action of each of these terms on the wave function as raising and lowering operators. First, look the lowering operator form of the right-hand term, which we denote a^-

$$a^- = (\mathbf{P} - i\omega\mathbf{X}) = \left(-i\hbar\frac{\partial}{\partial x} - i\omega x\right)$$

acting on a wave function

$$(\mathbf{P} - i\omega\mathbf{X})\,\Psi_0(x) = \left(-i\hbar\frac{\partial}{\partial x} - i\omega x\right)\Psi_0(x) = 0$$

gives a zero eigenvalue. A lowering operator acting on the ground state cannot produce a state with lower energy.

Remember that the wave function solution to the harmonic oscillator is

$$\psi_n(y) = \sqrt{\sqrt{\frac{\lambda}{\pi}\frac{1}{2^n n!}}}\ e^{-\frac{y^2}{2}} H_n(y)$$

and $H_n(y)$ are the Hermite polynomials where $y = \sqrt{\lambda}x$. These polynomials obey interesting recursion relations. We can easily derive one from the generating function as follows. The generating function is

$$H_n(y) = (-1)^n e^{y^2}\frac{\partial^n}{\partial y^n}e^{-y^2}.$$

And we can use this to show that we can write the derivative (which occurs in the momentum operator above) of the Hermite function as

$$\frac{\partial H_n(y)}{\partial y} = 2yH_n(y) - H_{n+1}(y).$$

To start, we take the derivative of both sides of the generating function equation

$$\frac{\partial H_n(y)}{\partial y} = \frac{\partial}{\partial y}\left((-1)^n e^{y^2}\frac{\partial^n}{\partial y^n}e^{-y^2}\right)$$

and use the chain rule on the product

$$= (-1)^n\left(\left[\frac{\partial}{\partial y}e^{y^2}\right]\left[\frac{\partial^n}{\partial y^n}e^{-y^2}\right] + \left[e^{y^2}\right]\frac{\partial}{\partial y}\left[\frac{\partial^n}{\partial y^n}e^{-y^2}\right]\right)$$

$$= (-1)^n\left(\left[2ye^{y^2}\right]\left[\frac{\partial^n}{\partial y^n}e^{-y^2}\right] + \left[e^{y^2}\right]\frac{\partial}{\partial y}\left[\frac{\partial^n}{\partial y^n}e^{-y^2}\right]\right)$$

$$= (-1)^n e^{y^2}\left(2y + \frac{\partial}{\partial y}\right)\left[\frac{\partial^n}{\partial y^n}e^{-y^2}\right]$$

$$= (-1)^n e^{y^2}2y\left[\frac{\partial^n}{\partial y^n}e^{-y^2}\right] + (-1)^n e^{y^2}\frac{\partial}{\partial y}\left[\frac{\partial^n}{\partial y^n}e^{-y^2}\right]$$

we see that we obtain two terms

$$= (-1)^n e^{y^2}2y\left[\frac{\partial^n}{\partial y^n}e^{-y^2}\right] + (-1)^{n+1}e^{y^2}\left[\frac{\partial^{n+1}}{\partial y^{n+1}}e^{-y^2}\right] = 2yH_n(y) - H_{n+1}(y)$$

and we recognize the first term as $2yH_n(y)$ while the second term is $H_{n+1}(y)$. Upon rearranging this we have an operator equation providing a method for creating the state with $n+1$ objects from the state with only n objects:

$$H_{n+1}(y) = 2y\,H_n(y) - \frac{\partial H_n(y)}{\partial y} = \left(2y - \frac{\partial}{\partial y}\right) H_n(y)$$

For future reference, note the form of the operator

$$A^\dagger = \left(2y - \frac{\partial}{\partial y}\right)$$

that we have factored out. This operator is a creation operator composed of two operators, familiar from our above treatment of quantum mechanics, that resemble the position and momentum operators:

$$H_{n+1}(y) = A^\dagger H_n(y).$$

The result is that the operator $A^\dagger$ creates a state with an energy $\hbar\omega$ higher than the state it operated on and is therefore called a "creation operator."

Using other recursion relations among the Hermite polynomials, we can also find a corresponding annihilation operator that lowers the number of objects by 1. First, we note that we can write

$$\frac{\partial H_n(y)}{\partial y} = 2nH_{n-1}(y)$$

and this is easily verified using *Mathematica* to compute $H_n(y)$, $2nH_{n-1}(y)$, and $\frac{\partial H_n(y)}{\partial y}$.

```
Table[{n,2n,HermiteH[n,x],2nHermiteH[n-1,x],D[HermiteH[n,x],]},{n,0,6}]
```

which produces the following table for $n \in \{0, 6\}$

n	$2n$	**HermiteH**$[ny]$	**2n HermiteH**$[n-1x]$	**D**$[$**HermiteH**$[ny]y]$
0	0	1	0	0
1	2	$2y$	2	2
2	4	$-2+4y^2$	$8y$	$8y$
3	6	$4y(-3+2y^2)$	$-12+24y^2$	$-12+24y^2$
4	8	$4(3-12y^2+4y^4)$	$32y(-3+2y^2)$	$32y(-3+2y^2)$
5	10	$8y(15-20y^2+4y^4)$	$40(3-12y^2+4y^4)$	$40(3-12y^2+4y^4)$
6	12	$8(-15+90y^2-60y^4+8y^6)$	$96y(15-20y^2+4y^4)$	$96y(15-20y^2+4y^4)$

The last two columns show that this relationship is valid for n up to 6, and we could have continued as far as we liked, but it is simpler to ask *Mathematica* if it is true for all n:

$$2n\text{HermiteH}[-1+n,y] == D[\text{HermiteH}[n,y],y]$$

True

Now we have our result that the expression for the state with one less object in it is given by

$$H_{n-1}(y) = \left(\frac{1}{2n}\right)\frac{\partial}{\partial y}H_n(y)$$

where the annihilation operator is

$$A^- = \left(\frac{1}{2n}\right)\frac{\partial}{\partial y}$$

$$A^- H_n(y) = H_{n-1}(y).$$

In like fashion

$$a^- = (\mathbf{P} - i\omega\mathbf{X}) = \left(-i\hbar\frac{\partial}{\partial x} - i\omega x\right)$$

is an annihilation operator that creates a state of energy $E - \hbar\omega$. This operator acting on the vacuum produces zero, $a^- \,|\,0> \,= 0$, or in terms of wave functions

$$a^- \Psi_0(x) = 0$$

or

$$\left(-i\hbar\frac{\partial}{\partial x} + i\omega x\right)\Psi_0(x) = 0$$

which is rearranged as

$$\frac{\partial}{\partial x}\Psi_0(x) = -(\omega x/\hbar)\,\Psi_0(x)$$

whose solution is

$$\Psi_0(x) = e^{-\frac{\omega x^2}{2\hbar}}$$

These creation and annihilation operators are the Hermitian conjugates of each other. Their commutation relation is

$$\left[a^-, a^\dagger\right] = 1$$

Now, we have seen that we can derive creation and annihilation operators that explicitly act on the Hermite polynomials to generate states that differ by unit excitations, let's develop similar concepts for the harmonic oscillator. When Paul Dirac was investigating both nonrelativistic and later relativistic quantum mechanics, he noticed that the momentum entered the Hamiltonian as the quadratic form P^2 and several times discovered how useful it was to factor the Hamiltonian into products of linear operators. We will pursue one of his many penetrating insights into the properties of matter by examining his beautiful restatement of the harmonic oscillator in terms of creation and annihilation operators. Remember that the Schrödinger equation for the harmonic oscillator is given by

$$-\frac{\hbar^2}{2m}\frac{d^2}{dx^2}\psi(x) + \frac{m}{2}\omega^2 x^2 \psi(x) = E\psi(x)$$

where both the position and momentum operators enter as their squares. Now, let's simplify this by changing variables to

$$x = \sqrt{\frac{\hbar}{m\omega}}q$$

with $x^2 = \frac{\hbar}{m\omega}q^2$ and $\frac{d}{dx} = \frac{dq}{dx}\frac{d}{dq} = \sqrt{\frac{m\omega}{\hbar}}\frac{d}{dq}$, which gives $\frac{d^2}{dx^2} = \frac{m\omega}{\hbar}\frac{d^2}{dq^2}$. Inserting this into the Schrödinger equation generates the simple form:

$$\frac{\hbar\omega}{2}\left(-\frac{d^2}{dq^2} + q^2\right)\psi_n(q) = E_n\psi_n(q)$$

in which we would have a perfect square $-a^2 - b^2 = (-a + b)(a + b)$ if a and b commuted, i.e., if $[a, b] = 0$. The term in front of the parenthesis on the left is $\hbar\omega = h\nu$, which is the energy of a photon of frequency ν. The derivative operator, however, does not commute with the position operator, and since we know we are dealing with position and momentum, we already know that $[p, x] = i\hbar$, and that this is the basis for the uncertainty principle. When we now factor the operator in parenthesis on the left, we must be careful to preserve the order of the operators

$$\left(-\frac{d^2}{dq^2} + q^2\right) = \left(-\frac{d}{dq} + q\right)\left(\frac{d}{dq} + q\right) + \frac{d}{dq}q - q\frac{d}{dq}$$

the last terms are the commutator $[p, q]$ where $p = -i\sqrt{\frac{m\omega}{\hbar}}\frac{d}{dq}$, but let's work in a representation, for the moment, in which $-i[p, q] = 1$ (in units in which $\hbar = 1$) and then we can insert the correct forms of the physics variables at the end. This makes the logic paramount and more transparent. Then we have that

$$\left(-\frac{d^2}{dq^2} + q^2\right) = \left(-\frac{d}{dq} + q\right)\left(\frac{d}{dq} + q\right) + 1$$

where we see that the factored operator now resembles what we found above,

$$A^\dagger = \left(2y - \frac{\partial}{\partial y}\right)$$

and we would anticipate that the other operator would just have the opposite sign. Let's now rewrite the Hamiltonian in terms of this factored product as

$$\frac{\hbar\omega}{2}\left[\left(-\frac{d}{dq} + q\right)\left(\frac{d}{dq} + q\right) + 1\right]\psi_n(q) = E_n\psi_n(q)$$

and we can distribute the factor of ½ and find that the Schrödinger equation becomes

$$\hbar\omega\left[\frac{1}{\sqrt{2}}\left(-\frac{d}{dq} + q\right)\frac{1}{\sqrt{2}}\left(\frac{d}{dq} + q\right) + \frac{1}{2}\right]\psi_n(q) = E_n\psi_n(q).$$

It was Dirac's insight that writing the Hamiltonian in this manner could be interpreted as

$$\hbar\omega\left(a^\dagger a + \frac{1}{2}\right)\psi_n(q) = E_n\psi_n(q)$$

where the creation, $a^\dagger$, and annihilation, a, operators are defined to be

$$a^\dagger = \frac{1}{\sqrt{2}}\left(-\frac{d}{dq} + q\right)$$

$$a = \frac{1}{\sqrt{2}}\left(\frac{d}{dq} + q\right)$$

and these are similar to, but more symmetric than, the raising and lowering operators we found for the Hermite polynomials, which of course makes sense because the Hermite polynomials form the basis for the energy eigenfunctions of the

Hamiltonian. Let's continue with these creation and annihilation operators by defining them in terms of position, q, and the dimensionless momentum,

$$p = -i\frac{d}{dq}, \text{ as}$$

$$a^\dagger = \frac{1}{\sqrt{2}}\left(q - \frac{d}{dq}\right) = \frac{1}{\sqrt{2}}(q - ip)$$

and

$$a = \frac{1}{\sqrt{2}}\left(q + \frac{d}{dq}\right) = \frac{1}{\sqrt{2}}(q + ip).$$

It is always useful to find the commutators of operators, so let's calculate the commutators of these creation and annihilation operators, noting from above that $[q, p] = i$ in these dimensionless quantities:

$$[a, a^\dagger] = \frac{1}{2}[(q + ip), (q - ip)] = \frac{1}{2}([q, -ip] + [ip, q]) = -\frac{i}{2}([q,p] + [q,p]) = 1$$

and since by the properties of commutators $[A, B] = -[B, A]$, we find that

$$[a^\dagger, a] = -1.$$

We can write the Hamiltonian using either order of the creation and annihilation operators

$$H = \hbar\omega\left(a^\dagger a + \frac{1}{2}\right) = \hbar\omega\left(aa^\dagger - \frac{1}{2}\right)$$

and use these to compute the commutation relations between it and the creation and annihilation operators

$$\begin{aligned}
[H, a] &= \left[\hbar\omega\left(a^\dagger a + \frac{1}{2}\right), a\right] \\
&= \hbar\omega(aa^\dagger a - aaa^\dagger) \\
&= \hbar\omega a(a^\dagger a - aa^\dagger) \\
&= \hbar\omega a[a^\dagger, a] \\
&= -\hbar\omega a
\end{aligned}$$

and the same holds for the other form of the Hamiltonian with $\left(aa^\dagger - \frac{1}{2}\right)$. We used here that $[a, a] = \left[a, \frac{1}{2}\right] = [a^\dagger, a^\dagger] = \left[a^\dagger, \frac{1}{2}\right] = 0$. The commutator of the Hamiltonian with the creation operator is

$$\begin{aligned}
\left[H, a^\dagger\right] &= \left[\hbar\omega\left(a^\dagger a + \frac{1}{2}\right), a^\dagger\right] \\
&= \hbar\omega\left(a^\dagger a a^\dagger - a^\dagger a^\dagger a\right) \\
&= \hbar\omega a^\dagger\left(a a^\dagger - a^\dagger a\right) \\
&= \hbar\omega a\left[a, a^\dagger\right] \\
&= \hbar\omega a^\dagger
\end{aligned}$$

We can now find all eigenstate energies using these commutators. Remember that

$$H\psi_n(q) = E_n\psi_n(q)$$

for the eigenfunctions and we can write this in bra-ket notation as

$$H \mid n> = E_n \mid n>$$

Then, applying the commutator of H with a to an eigenket of H gives

$$\begin{aligned}
[H, a]|n> &= Ha|n> - aH|n> = -\hbar\omega a|n> \\
&= Ha \mid n> - aE_n|n> = -\hbar\omega a|n> \\
&= H(a|n>) - E_n(a|n>) = -\hbar\omega(a|n>)
\end{aligned}$$

and we see that the term in parentheses ($a|n>$) is itself an eigenket of the Hamiltonian, with energy

$$H(a|n>) = (E_n - \hbar\omega)(a|n>)$$

of $(E_n - \hbar\omega)$, which is one quantum less than for level n. In a similar manner, we can find that

$$\begin{aligned}
\left[H, a^\dagger\right]|n> &= Ha^\dagger|n> - a^\dagger H|n> = \hbar\omega a^\dagger|n> \\
&= Ha^\dagger \mid n> - a^\dagger E_n|n> = \hbar\omega a^\dagger|n> \\
&= H(a^\dagger|n>) - E_n(a^\dagger|n>) = \hbar\omega(a^\dagger|n>) \\
&= H(a^\dagger|n>) = (E_n + \hbar\omega)(a^\dagger|n>)
\end{aligned}$$

the state ($a^\dagger|n>$) is an eigenket of H with energy eigenvalue ($E_n + \hbar\omega$), which is one quantum higher than that we started with. The difference in energy from one state to the next is $\hbar\omega$. The operators a and $a^\dagger$ annihilate and create states with this difference in energy. We can find the ground state wave function now by using the fact that when the annihilation operator acts upon it, the result is zero

$$a \mid 0> = 0$$

with $|0> \neq 0$. By using the definition of the annihilation operator, we have that

$$a\psi_0 = \frac{1}{\sqrt{2}}\left(q + \frac{d}{dq}\right)\psi_0 = 0$$

which we can rearrange and integrate as

$$\int \frac{d\psi_0(q)}{\psi_0(q)} = -\int q\,dq$$

and this gives $\ln(d\psi_0(q)) = -\frac{q^2}{2}$ so that our ground state wave function is

$$\psi_0(q) = Ce^{-\frac{q^2}{2}}$$

as we stated earlier without proof. The normalization constant C can be found from the requirement that the integral of the probability must equal 1.

$$\int\limits_{-\infty}^{\infty} \psi_0^*(q)\psi_0(q)\,dq = C^2 \int\limits_{-\infty}^{\infty} e^{-\frac{q^2}{2}}e^{-\frac{q^2}{2}}\,dq = C^2 \int\limits_{-\infty}^{\infty} e^{-q^2}\,dq = 1$$

or, since this integral evaluates to $\sqrt{\pi}$, we find that

$$C = \frac{1}{\sqrt[4]{\pi}}$$

and that the normalized ground state wave function is

$$|\,0> \,= \psi_0(q) = \frac{1}{\sqrt[4]{\pi}}e^{-\frac{q^2}{2}}.$$

One can find all the eigenfunctions of the harmonic oscillator by repeated application of $a^\dagger$ to $|0>$.

We have written this as $|0>$ to emphasize that this is not only the lowest energy state but that it contains no excitations, which is also a definition of the *vacuum* state. Note how the total energy operator (the Hamiltonian) creates objects by acting on the vacuum. The vacuum energy is found by operating on this vacuum state

$$\hbar\omega a^\dagger a|0> \,= \left(H - \frac{\hbar\omega}{2}\right)|0> \,= 0$$

and we note that the vacuum is an eigenket of the Hamiltonian with energy eigenvalue

$$E_0 = \frac{\hbar\omega}{2}$$

(this is known as zero-point energy) and that the entire spectrum of energies is now seen to be given by

$$E_n = \left(n + \frac{1}{2}\right)\hbar\omega.$$

The representation we are working with is known as the occupation number representation because the operator $a^\dagger a$ acting on a ket $|n>$ gives

$$a^\dagger a |n> = n|n>$$

the number of objects in the state, so we replace it with the number operator, $N = a^\dagger a$, and we can rewrite the Hamiltonian in the simple form:

$$H = \left(N + \frac{1}{2}\right)\hbar\omega.$$

What is the coefficient that results from the action of either a creation or an annihilation operator on a harmonic oscillator state? We know that these operators either raise or lower the energy of a state but what is the value of S in

$$a^\dagger \,|\,n> = S\,|\,n+1>\,?$$

We can find it using our operators by noting that

$$|S|^2 = <a^\dagger n\,|\,a^\dagger n> = <aa^\dagger n\,|\,n>$$
$$|S|^2 = <(a^\dagger a + [a, a^\dagger])n\,|\,n> = (n+1)<n\,|\,n>$$

and we find that

$$|S|^2 = (n+1)$$
$$|\,S\,| = \sqrt{n+1}$$

so

$$a^\dagger\,|\,n> = \sqrt{n+1}\,|\,n+1>$$

and, similarly, for $a|n> \,= R|n-1>$, we have that

$$|R|^2 = <a^\dagger a\, n\mid n>$$

$$|R|^2 = <N\, n\mid n>$$

$$|R|^2 = n<n\mid n> = n$$

so, $R = \sqrt{n}$ and we have that

$$a|n> = \sqrt{n}|n-1>$$

Note that we can make any state we wish from the vacuum and the creation operator as

$$|n> = \frac{1}{\sqrt{n!}}(a^\dagger)^n|0>.$$

The states in this occupation number representation are therefore normalized and since the number operator is Hermitian, the states |n> must be orthonormal

$$<n\mid m> = \delta_{n,m}.$$

It will be useful to find the inner products (matrix elements) of the creation and annihilation operators and then we can construct an explicit matrix representation of them.

$$
\begin{aligned}
<m|a|n> &= <m\left|a\frac{1}{\sqrt{n!}}(a^\dagger)^n\right|0> \\
&= \frac{1}{\sqrt{n!}}<m\left|n(a^\dagger)^{n-1}\right|0> \\
&= \frac{n\sqrt{(n-1)!}}{\sqrt{n!}}<m\mid n-1> \\
&= \sqrt{n}<m\mid n-1> \\
&= \sqrt{n}\,\delta_{n-1,m}
\end{aligned}
$$

which, by the same logic can also be written as $\sqrt{m+1}\,\delta_{n,m+1}$, and with

$$<m|a^\dagger|n> = <n|a|m>^*$$

we have that

$$<m|a^\dagger|n> = \sqrt{m}\,\delta_{n,m-1} = \sqrt{n+1}\,\delta_{m,n+1}$$

we leave it as an exercise to directly verify that $<n\mid m> = \delta_{n,\,m}$. The matrix elements consist of zeroes except where the Kronecker delta is 1, as it is for matrix entries 1 above or below the diagonal. For example, for $\{m,n\} \in \{0,5\}$

$$<m|a^{\dagger}|n> = \begin{bmatrix} 0 & 0 & 0 & 0 & 0 & 0 \\ 1 & 0 & 0 & 0 & 0 & 0 \\ 0 & \sqrt{2} & 0 & 0 & 0 & 0 \\ 0 & 0 & \sqrt{3} & 0 & 0 & 0 \\ 0 & 0 & 0 & 2 & 0 & 0 \\ 0 & 0 & 0 & 0 & \sqrt{5} & 0 \end{bmatrix}$$

4.7   The Hydrogen Atom

Although the harmonic oscillator potential serves as a good approximation for the bound states of biomolecules (see e.g., Problem 9), an exact solution for Schrodinger's equation is possible in the case of a single electron in the presence of the electric field due to the central, spherically symmetric Coulomb potential of a single proton. For a charged particle moving in an electromagnetic field, the Lorentz force vector on the particle is F, where

$$\vec{F} = e\left(\vec{E} + \frac{v}{c} \times \vec{B}\right)$$

is the force, e is the charge on the particle in Coulombs, E is the electric field, v is the particle's velocity, c is the speed of light, and B is the magnetic field strength. The electric and magnetic fields can be derived from their corresponding scalar, φ, and vector, A, potentials by

$$\vec{E} = -\vec{\nabla}\varphi - \frac{1}{c}\frac{\partial \vec{A}}{\partial t}$$

$$\vec{B} = \nabla \times \vec{A}$$

Here, $A(r, t)$ is the vector potential and $\varphi(r, t)$ is the Coulomb potential. The classical Hamiltonian for a particle in an electromagnetic field is

$$H = \frac{1}{2m}\left(\vec{p} - \frac{e}{c}\vec{A}\right)^2 + e\varphi$$

The transition to quantum mechanics involves the replacement of the momentum by its operator and we obtain the quantum Hamiltonian for a particle in an electromagnetic field

$$H = \frac{1}{2m}\left(\frac{\hbar}{i}\vec{\nabla} - \frac{e}{c}\vec{A}\right)^2 + e\varphi$$

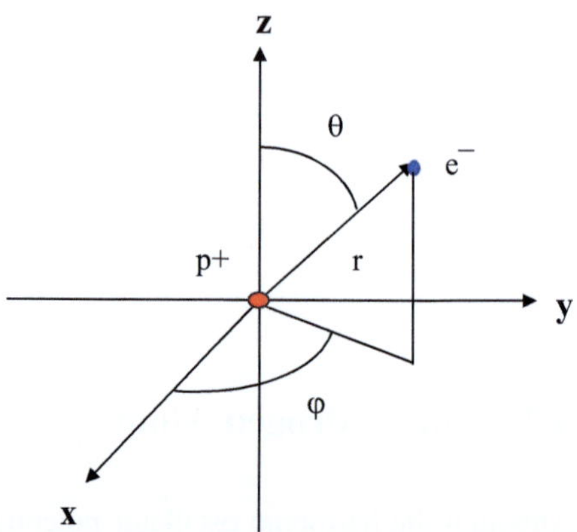

Fig. 4.9 Spherical polar coordinate system used to denote the position of the electron about the proton in the hydrogen atom

In expanding the quantity in brackets, it should be noted that the vector potential and the gradient operator do not in general commute. In a region of space free from magnetic fields, this Hamiltonian reduces to

$$H = \frac{-\hbar^2}{2m}\nabla^2 + e\,\varphi.$$

Electrons and protons attract each other with a force, $F = e^2/r^2$, resulting from a potential, $\varphi(r, t) = -e/r$, where r is the distance from the proton to the electron. An atom consisting of a single proton in the nucleus and a single, charge-balancing electron is the hydrogen atom. Then Schrodinger's equation for the hydrogen atom reads

$$\left(\frac{-\hbar^2}{2m}\nabla^2 + e\varphi\right)\psi = E\,\psi$$

or, with $\varphi(r) = -e/r$, this becomes

$$\left(\frac{-\hbar^2}{2m}\nabla^2 - \frac{e^2}{r}\right)\psi = E\,\psi.$$

The momentum operator, p, squared in spherical polar coordinates (Fig. 4.9), where

$$x = r\,\sin\,\theta\,\cos\,\varphi,$$
$$y = r\,\sin\,\theta\,\sin\,\varphi,$$
$$z = r\,\cos\,\theta,$$

can be expressed as

$$p^2 = -\hbar^2 \nabla^2 = -\hbar^2 \left(\frac{1}{r^2} \frac{\partial}{\partial r} r^2 \frac{\partial}{\partial r} + \frac{1}{r^2 \sin\theta} \frac{\partial}{\partial \theta} \left(\sin\theta \frac{\partial}{\partial \theta} \right) + \frac{1}{r^2 \sin^2\theta} - \frac{\partial^2}{\partial\varphi^2} \right)$$

which separates into a radial part

$$\frac{1}{r} \frac{\partial^2}{\partial r^2} r$$

and an angular part according to

$$L^2 = - \frac{1}{\sin\theta} \frac{\partial}{\partial \theta} \left(\sin\theta \frac{\partial}{\partial \theta} \right) + \frac{1}{\sin^2\theta} \frac{\partial^2}{\partial\varphi^2}$$

where L is the angular momentum operator. Schrodinger's equation then reads

$$\left(\frac{1}{r^2} \frac{\partial}{\partial r} r^2 \frac{\partial}{\partial r} - \frac{L^2}{\hbar^2 r^2} \right) \psi + \frac{2m}{\hbar^2} \left(E + \frac{e^2}{r} \right) \psi = 0$$

and we can use separation of variables, as seen above, to write the solution, $\psi(r, \theta, \varphi)$, as

$$\psi(r, \ \theta, \ \varphi) = \frac{R(r)}{r} Y(\theta, \varphi).$$

With this substitution, it is possible to separate ψ into a radial part, $R(r)$, and an angular part, $Y(\theta, \varphi)$.

$$\left(\frac{1}{r^2} \frac{\partial}{\partial r} r^2 \frac{\partial}{\partial r} - \frac{L^2}{\hbar^2 r^2} \right) \frac{R(r)}{r} Y(\theta, \ \varphi) + \frac{2m}{\hbar^2} \left(E + \frac{e^2}{r} \right) \frac{R(r)}{r} Y(\theta, \ \varphi) = 0$$

We note that

$$\left(\frac{1}{r^2} \frac{\partial}{\partial r} r^2 \frac{\partial}{\partial r} \frac{R(r)}{r} \right) = \frac{1}{r} \frac{\partial^2 R(r)}{\partial r^2}$$

$$\left(\frac{1}{r} \frac{\partial^2 R(r)}{\partial r^2} - \frac{R(r)}{r} \frac{L^2}{\hbar^2 r^2} \right) Y(\theta, \ \varphi) + \frac{2m}{\hbar^2} \left(E + \frac{e^2}{r} \right) \frac{R(r)}{r} Y(\theta, \ \varphi) = 0$$

We multiply this by r^3 to give

$$\left(r^2 \frac{\partial^2 R(r)}{\partial r^2} - R(r)\frac{L^2}{\hbar^2}\right) Y(\theta,\ \varphi) + Y(\theta,\ \varphi) r^2 \frac{2m}{\hbar^2}\left(E + \frac{e^2}{r}\right) R(r) = 0$$

and divide by $R(r)$:

$$\left(\frac{r^2}{R(r)} \frac{\partial^2 R(r)}{\partial r^2} - \frac{L^2}{\hbar^2}\right) Y(\theta,\ \varphi) + Y(\theta,\ \varphi) r^2 \frac{2m}{\hbar^2}\left(E + \frac{e^2}{r}\right) = 0$$

and divide by $Y(\theta, \varphi)$:

$$\frac{r^2}{R(r)} \frac{\partial^2 R(r)}{\partial r^2} - \frac{1}{Y(\theta,\ \varphi)} \frac{L^2}{\hbar^2} Y(\theta,\ \varphi) + r^2 \frac{2m}{\hbar^2}\left(E + \frac{e^2}{r}\right) = 0,$$

or, by bringing the angular-dependent function, $Y(\theta, \varphi)$, to the right-hand side, we have that

$$\frac{r^2}{R(r)} \frac{\partial^2 R(r)}{\partial r^2} + r^2 \frac{2m}{\hbar^2}\left(E + \frac{e^2}{r}\right) = \frac{1}{Y(\theta,\ \varphi)} \frac{L^2}{\hbar^2} Y(\theta,\ \varphi),$$

which by the standard separation of variables philosophy, the left-hand side, is only a function of the radial distance, r, and therefore can only be equal to the right-hand side, a function of the angular coordinates, if both sides are equal to a constant. We shall call this constant, $\hbar^2 l(l+1)$ and we then obtain two equations, one for the radial function $R(r)$ and another for the angular function $Y(\theta, \varphi)$:

$$\frac{\partial^2 R(r)}{\partial r^2} + \left[\frac{2m}{\hbar^2}\left(E + \frac{e^2}{r}\right) - \frac{\hbar^2 l(l+1)}{r^2}\right] R(r) = 0$$

and

$$\frac{1}{Y(\theta,\ \varphi)} \frac{L^2}{\hbar^2} Y(\theta,\ \varphi) = \hbar^2 l(l+1).$$

The angular equation is rewritten as

$$L^2 Y(\theta, \varphi) = \hbar^2 l(l+1) Y(\theta, \varphi),$$

which we immediately recognized as an eigenvalue equation for the angular function, $Y(\theta, \varphi)$, with the eigenvalues, $\hbar^2 l(l+1)$. To solve this, let us introduce another separation of variables, this time with $Y(\theta, \varphi) = P(\theta)\, Q(\varphi)$

$$L^2 Y(\theta, \varphi) = L^2 P(\theta) Q(\varphi) = \hbar^2 l(l+1) P(\theta) Q(\varphi)$$

or

$$\left(\frac{1}{\sin\theta}\frac{\partial}{\partial\theta}\left(\sin\theta\frac{\partial}{\partial\theta}\right)+\frac{1}{\sin^2\theta}\frac{\partial^2}{\partial\varphi^2}\right)P(\theta)Q(\varphi)=\hbar^2 l(l+1)P(\theta)Q(\varphi)$$

and if we expand the left-hand side:

$$Q(\varphi)\frac{1}{\sin\theta}\frac{\partial}{\partial\theta}\left(\sin\theta\frac{\partial}{\partial\theta}\right)P(\theta)+P(\theta)\frac{1}{\sin^2\theta}\frac{\partial^2}{\partial\varphi^2}Q(\varphi)=\hbar^2 l(l+1)P(\theta)Q(\varphi)$$

and divide by $P(\theta)\,Q(\varphi)$:

$$\frac{1}{\sin\theta}\frac{1}{P(\theta)}\frac{\partial}{\partial\theta}\left(\sin\theta\frac{\partial}{\partial\theta}\right)P(\theta)+\frac{1}{\sin^2\theta}\frac{1}{Q(\theta)}\frac{\partial^2}{\partial\varphi^2}Q(\varphi)=\hbar^2\,l(l+1)$$

and introduce another separation constant, $-m^2$, where

$$-m^2=\frac{1}{Q(\varphi)}\frac{\partial^2}{\partial\varphi^2}Q(\varphi)$$

so that the angular equation now reads

$$\frac{1}{\sin\theta}\frac{1}{P(\theta)}\frac{\partial}{\partial\theta}\left(\sin\theta\frac{\partial}{\partial\theta}\right)P(\theta)+\frac{1}{\sin^2\theta}m^2=\hbar^2\,l(l+1)$$

We find that this is independent of φ.

The equation for $Q(\varphi)$ is now familiar from the particle in a box problem (see §4):

$$\frac{\partial^2}{\partial\varphi^2}Q(\varphi)+m^2 Q(\varphi)=0$$

We know the solutions are $Q(\varphi)=C\,e^{im\varphi}$, and, because we require Q to satisfy the boundary condition that it is periodic with period 2π, $Q(\varphi)=Q(\varphi+2\pi)$, we find that $m=\{0,\pm1,\pm2,\pm3,\dots\}\hbar$ can only take on integral values. The equation for $Q(\varphi)$ is that of the z-component of angular momentum, which is quantized in units of Planck's constant (see Problem 8). Here we have an example of the introduction of quantization, not by physical, but by the logical confinement of an object to satisfy the boundary conditions.

Returning to the equation for $P(\theta)$, we make the substitution that

$$x=\cos\theta$$

and

Table 4.4 The associated Legendre polynomials $P_{lm}(x)$ for $\{l, m\} \in \{1, 4\}$

l	m	$P_{lm}(x)$
0	0	1
1	0	X
1	1	$-\sqrt{1-x^2}$
2	0	$\frac{1}{2}(3x^2 - 1)$
2	1	$-3x\sqrt{1-x^2}$
2	2	$-3(x^2 - 1)$
3	0	$\frac{1}{2}(5x^3 - 3x)$
3	1	$-\frac{3}{2}\sqrt{1-x^2}(5x^2 - 1)$
3	2	$-15x(x^2 - 1)$
3	3	$-15(1 - x^2)^{3/2}$
4	0	$\frac{1}{8}(35x^4 - 30x^2 + 3)$
4	1	$-\frac{5}{2}\sqrt{1-x^2}(7x^3 - 3x)$
4	2	$-\frac{15}{2}(x^2 - 1)(7x^2 - 1)$
4	3	$-105x(1 - x^2)^{3/2}$
4	4	$105(1 - x^2)^2$

$$\sin\theta = \sqrt{1 - x^2}$$

so that

$$\frac{\partial}{\partial\theta} = -\sqrt{1 - x^2}\,\frac{\partial}{\partial x}$$

and

$$\frac{\partial^2}{\partial\theta^2} = (1 - x^2)\frac{\partial^2}{\partial x^2} - x\frac{\partial}{\partial x}$$

Then, we can transform the equation for $P(x = \cos\theta)$ into the associated Legendre equation

$$(1 - x^2)\frac{\partial^2 P(x)}{\partial x^2} - x\frac{\partial P(x)}{\partial x} + \left(l(l+1) - \frac{m^2}{1 - x^2}\right)P(x) = 0$$

which has as solutions, the associated Legendre polynomials, $P_{lm}(x) = P_{lm}(\cos\theta)$. The first few associated Legendre polynomials $\mathbf{P}_{lm}(\mathbf{x})$ for $0 \leq l \leq 5$, $m \leq l$, are given in Table 4.4, where $x = \cos\theta$, and they are plotted in Fig. 4.10. One immediately notices that the associated Legendre polynomials can be divided into two groups, one with even and the other with odd parity. The parity of the state is determined by the quantum numbers $\{l, m\}$ because the associated Legendre polynomials are

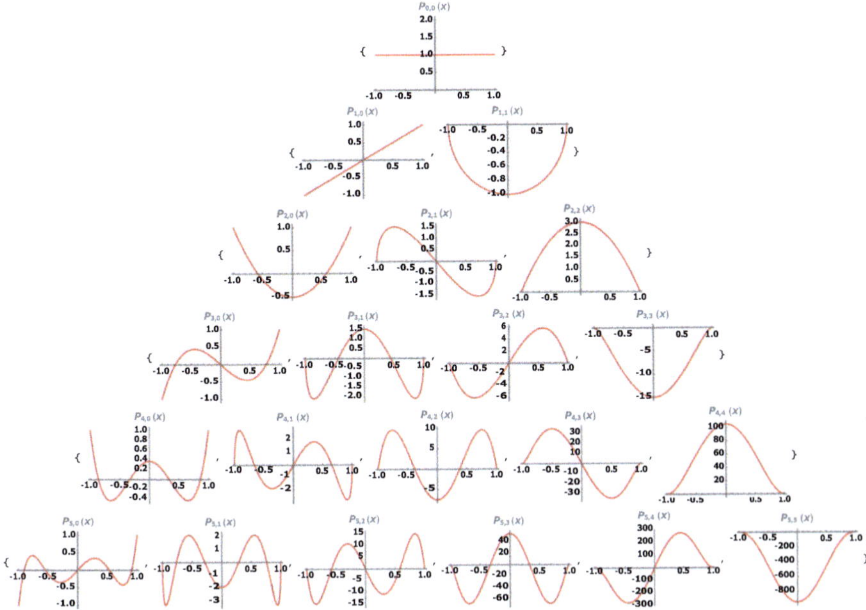

Fig. 4.10   The associated Legendre polynomials, $P_{lm}(x)$, for $0 \leq l \leq 5$, $0 \leq m \leq l$, where l increases from top to bottom and m increases from left to right

eigenfunctions of the parity operator, $\Pi P_{lm}(x) = (-1)^{l+m} P_{lm}(x)$, and this is evident in Fig. 4.10.

The associated Legendre polynomials are defined much like those for the Hermite polynomials as successive derivatives of a generating function:

$$P_{lm}(x) = \frac{1}{2^l \, l!} \left(1 - x^2\right)^{m/2} \frac{d^{l+m}}{dx^{l+m}} \left(x^2 - 1\right)^l$$

The solution for the angular functions is the spherical harmonics, $Y_{lm}(\theta, \varphi)$, defined in terms of the associated Legendre polynomials and the solution for $Q(\varphi)$ as

$$Y_{lm}(\theta, \ \varphi) = C_{lm} \ P_{lm}(\cos \ \theta) \ Q_m(\varphi) = C_{lm} \ P_{lm}(\cos \ \theta) \ e^{im\varphi},$$

where $m = \{0, \pm 1, \pm 2, \pm 3, \ldots, l\}$ and $l \geq |m|$. Note that m is bounded by l, the eigenvalue of the operator for the square of the angular momentum. The normalization constants, C_{lm}, are given by

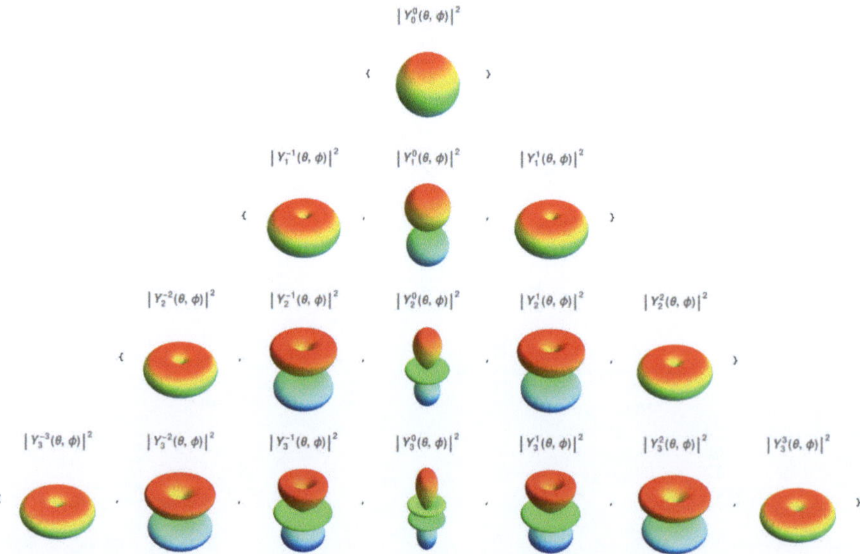

Fig. 4.11 3D plots of the absolute square of the spherical harmonics for $l = \{0,1,2,3\}$ and $m \leq 1$. The phase of the wave function is indicated by the color with red indicating positive and blue for negative

$$C_{lm} = \sqrt{\frac{2l+1}{4\pi}\frac{(l-m)!}{(l+m)!}}$$

and the angular wave function is then

$$Y_{lm}(\theta,\ \varphi) = \sqrt{\frac{2l+1}{4\pi}\frac{(l-m)!}{(l+m)!}}\ P_{lm}(\cos\ \theta)\ e^{im\varphi}.$$

The spherical harmonics form a complete, orthonormal set of functions. The shape in 3D of these functions as m and l are varied is shown in Fig. 4.11.

Let us turn our attention to the radial wave function equation.

$$\frac{d^2R(r)}{dr^2} + \left[\frac{2m}{\hbar^2}\left(E + \frac{e^2}{r}\right) - \frac{l(l+1)}{r^2}\right]R(r) = 0$$

Note that the energy of the electron depends only on the radial wave function, since E does not appear in the θ, or φ equations. We have converted this equation from a partial differential equation to a differential equation based on the fact that $R(r)$ is a function of the single variable, r. We can introduce dimensionless variables here by noting that the atomic unit of length can written as the Bohr radius,

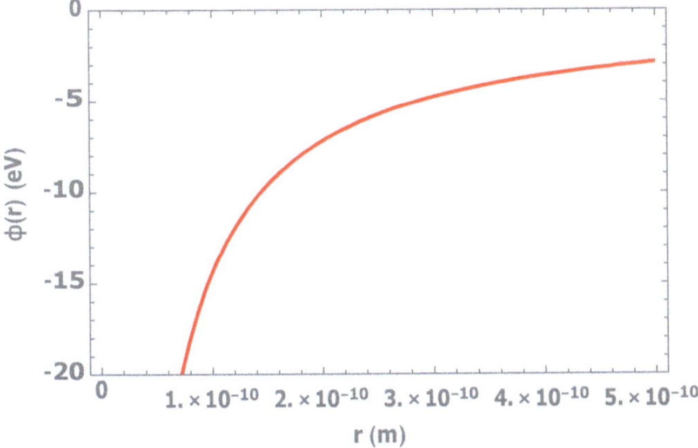

Fig. 4.12 The potential energy of an electron in the electric field of a proton as a function of the electron–proton separation

$$a_o = \frac{\hbar^2}{m e^2} = 0.529 \ \text{Å} = 52.9 \ \text{pm}.$$

The atomic unit of energy is

$$E_a = \ = \frac{e^2}{a_o} = \frac{m e^4}{\hbar^2} = 27.21 \ \text{eV} = 4.35 \ \times \ 10^{-18} \ \text{Joules}.$$

Now, we can replace E and r with the dimensionless quantities, ε and ρ

$$\varepsilon = E/E_a, \quad \rho = r/a_o,$$

And write the equation for the radial portion of the hydrogen atom wave function as

$$\left[\frac{d^2}{d\rho^2} + 2\varepsilon + \frac{2}{\rho} - \frac{l(l+1)}{\rho^2} \right] R(\rho) = 0$$

The potential energy due to the point charge at the position of the proton is $\varphi(\rho) = -e^2/\rho$, as shown in Fig. 4.12. This potential goes to zero as ρ goes to infinity and bound states of the electron correspond to negative total energy values, $\varepsilon < 0$. Let us therefore replace ε with the positive quantity, α

$$\alpha^2 = -2 \ \varepsilon > 0.$$

Then the radial equation becomes

$$\left[\frac{d^2}{d\rho^2} - \alpha^2 + \frac{2}{\rho} - \frac{l(l+1)}{\rho^2}\right] R(\rho) = 0$$

Exploration of this equation with *Mathematica* produces the limiting results that as $\rho \to \infty$, the solution for $R(\rho) \to e^{-\alpha\rho}$, and as $\rho \to 0$, $R(\rho) \to \rho^{l+1}$. The solution to this equation is stated without proof as the product of these two asymptotic functions:

$$R_l(\rho) = \rho^{l+1} \, e^{-\alpha\rho} \, F(\rho)$$

where the function $F(\rho)$ is a power series in ρ:

$$F(\rho) = \rho^\gamma \sum_{\nu=0}^{n_r} \beta_\nu \rho^\nu.$$

To determine the behavior of $F(\rho)$ as $\rho \to 0$, we substitute $F(\rho)$ into the radial equation and keep only the lowest power of $F(\rho)$, and find an equation for γ:

$$\gamma(\gamma - 1) = l(l+1)$$

which has the solutions, $\gamma = l + 1$ or $\gamma = -l$. Now, $R(\rho)$ must go to zero as $\rho \to 0$, so only the solution $\gamma = l + 1$ is finite at $\rho \to 0$. The solution of the radial equation satisfying the boundary conditions at $\rho \to 0$ is then

$$R_1(\rho) = e^{-\alpha\rho} \rho^{l+1} \sum_{\nu=0}^{n_r} \beta_\nu \rho^\nu.$$

We can find the coefficients, β_ν, by inserting this solution into the radial equation and by examining the values of β_ν for each power of ρ. This leads to the recurrence relation

$$\beta_{\nu+1} = \frac{2[\alpha(\nu + l + 1) - 1]}{(\nu + l + 2)(\nu + l + 1) - l(l+1)} \beta_\nu,$$

which we can use to find all the coefficients β_ν in terms of β_0, and this can be found from the normalization integral. The solution for the radial equation must remain finite, and this requires that the power series must terminate at some value of ν. Let us set this value of ν to the radial quantum number, n_r, so that $\nu = n_r$. By termination, we mean that for some value of $\nu = n_r$, the next coefficient $\beta_{\nu+1} = 0$. This implies that the numerator in the recurrence relationship must vanish:

$$2[\alpha(n_r + l + 1) - 1] = 0$$

or

$$\alpha = 1/(n_r + l + 1).$$

Now, let us return to the definition of $\alpha^2 = -2\varepsilon$ to find the energy eigenvalues, ε, as

$$\varepsilon = -\alpha^2/2$$

$$\varepsilon = -\frac{1}{2(n_r + l + 1)^2}.$$

The terms on the right-hand side are the eigenvalue of the angular momentum operator, l, and the radial quantum number, n_r. The quantity in parenthesis is the principal quantum number,

$$n = n_r + l + 1,$$

so called because it alone determines the energy of the bound states of the hydrogen atom. In atomic units, the energy levels for the hydrogen atom are then:

$$\varepsilon = -\frac{1}{2n^2}.$$

Note that since

$$n_r \in \{0, \ 1, \ 2, \ \ldots\},$$

and

$$l \in \{0, \ 1, \ 2, \ \ldots\},$$

the principal quantum number n takes on values of the positive integers, beginning with 1,

$$n \in \{1, \ 2, \ 3, \ldots\}.$$

It is also useful to write the energy eigenvalues in more traditional SI units as

$$E_n = -\frac{2\pi^2 m Z^2 e^4}{n^2 h^2}$$

where Z is the charge. We see that the energy is a function of the quantum numbers n and l and these serve to index the wave functions in terms of the quantized angular momentum and energy; for an energy level with a given value of n, there are n values of l:

$$0 \leq l \leq n - 1, \qquad l \in \{0, \ 1, \ 2, \ \ldots \ , \ n-1\}.$$

Radial functions $f(\rho) = R(\rho)/\rho$ for the hydrogen atom for the first $n = 4$ levels are given in Table 4.5, where $r = \rho a_0$ is measured in units of the Bohr radius, a_0. The

Table 4.5 The radial wave functions for the first few levels of the hydrogen atom

$R_{1,0}[r]$ $2e^{-r}$	""	""	""
$R_{2,0}[r]$ $\dfrac{e^{-r/2}\left(1-\dfrac{r}{2}\right)}{\sqrt{2}}$	$R_{2,1}[r]$ $\dfrac{e^{-r/2}r}{2\sqrt{6}}$	""	""
$R_{3,0}[r]$ $\dfrac{2e^{-r/3}\left(1-\dfrac{2r}{3}+\dfrac{2r^2}{27}\right)}{3\sqrt{3}}$	$R_{3,1}[r]$ $\dfrac{4}{27}\sqrt{\dfrac{2}{3}}e^{-r/3}\left(1-\dfrac{r}{6}\right)r$	$R_{3,2}[r]$ $\dfrac{2}{81}\sqrt{\dfrac{2}{15}}e^{-r/3}r^2$	""
$R_{4,0}[r]$ $\dfrac{1}{4}e^{-r/4}\left(1-\dfrac{3r}{4}+\dfrac{r^2}{8}-\dfrac{r^3}{192}\right)$	$R_{4,1}[r]$ $\dfrac{1}{16}\sqrt{\dfrac{5}{3}}e^{-r/4}r\left(1-\dfrac{r}{4}+\dfrac{r^2}{80}\right)$	$R_{4,2}[r]$ $\dfrac{e^{-r/4}\left(1-\dfrac{r}{12}\right)r^2}{64\sqrt{5}}$	$R_{4,3}[r]$ $\dfrac{e^{-r/4}r^3}{768\sqrt{35}}$

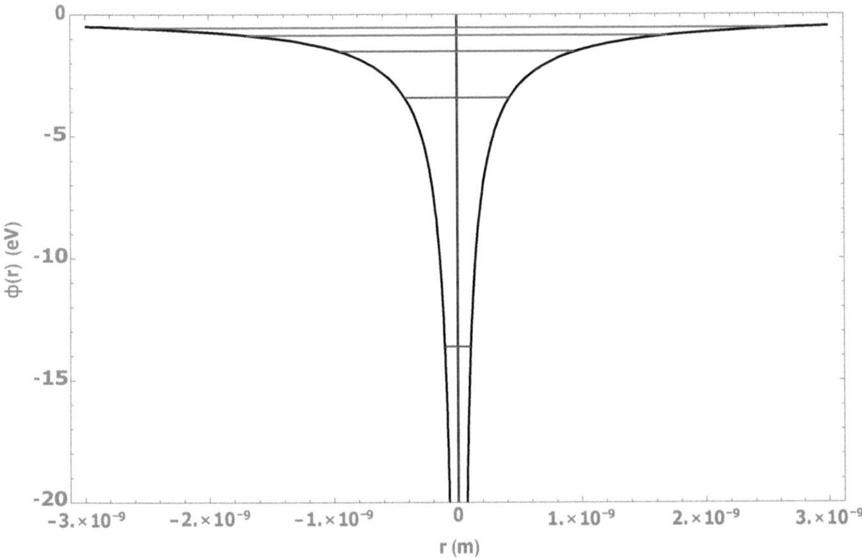

Fig. 4.13 The energy levels of the electron in the hydrogen atom shown within the Coulomb potential of the proton. The ground state is at -13.6 eV and the energies decrease as $1/n^2$

energy levels in the Coulomb potential are shown in Fig. 4.13; the ionization potential, which is the energy needed to completely remove the electron from the nuclear potential is given by the ground state ($n = 1$) energy of -13.6 eV. Note that once more, physical confinement of the electron in the Coulomb potential of the proton leads to quantization of the energy. This is a hallmark of quantum mechanics.

The radial functions are properly normalized by requiring that

$$\int_0^\infty [f(\rho)]^2 \, \rho^2 d\rho = 1$$

For the general state of the hydrogen atom, the normalized radial wave function is written in terms of the confluent hypergeometric function, $_1F_1\,(a,\ b,\ z) = \sum_{k=0}^{\infty} \frac{(a)_k}{(b)_k} \frac{z^k}{k!}$, where

$$(a)_k = a\ (a+1)(a+2)(a+3)\ldots(a+k-1)$$

and

$$(b)_k = b \ (b+1)(b+2)(b+3)\ldots(b+k-1)$$

$$f_{nl}(\rho) = N_{nl}\left(\frac{2\rho}{n}\right)^l {}_1F_1(l+1-n, \ 2l+2, \ 2\rho/n)\, e^{-\rho/n}$$

where the normalization constant, N_{nl} is given by

$$N_{nl} = \frac{1}{(2l+1)!}\sqrt{\frac{(n+l)!}{2n(n-l-1)!}}\left(\frac{2}{n}\right)^{3/2}$$

The radial wave functions are given in Table 4.5. The resulting radial wave functions and probability distributions are plotted as a function of the radius in Fig. 4.14.

The complete solution for the wave functions of the hydrogen atom then involves multiplying the radial wave function by the angular wave functions found earlier, $Y_{lm}(\theta, \varphi)$, where we remember that

$$Y_{lm}(\theta, \ \varphi) = \sqrt{\frac{2l+1}{4\pi}\frac{(l-m)!}{(l+m)!}}P_{lm}(\cos\ \theta)\ e^{im\varphi}.$$

Then we have the complete wave function as

$$\Psi_{nlm}(\rho, \ \theta, \ \varphi) = \sqrt{\frac{2l+1}{4\pi}\frac{(l-m)!}{(l+m)!}}P_{lm}(\cos\ \theta)\ e^{im\varphi}\, f_{nl}(\rho).$$

This is a function of three quantum numbers, n, l, and m. These three quantum numbers label the eigenfunctions with the eigenvalues of energy (n), the square of the angular momentum ($\hbar^2\, l(l+1)$), and the z-component of angular momentum (m). Thus, this wave function is a simultaneous eigenfunction of these three operators that commute with the Hamiltonian.

In the absence of an external electromagnetic field, the quantum states are degenerate with respect to the magnetic quantum number, m, because this eigenvalue is the projection of the angular momentum on the z-axis, and this axis can be oriented arbitrarily in space. When a field is applied to the hydrogen atom, this degeneracy is lifted and the states split into their m-distinct components. It is impressive that all of this structure of the hydrogen atom optical spectra is contained within the solutions of the Schrodinger equation for the electron in the central potential of the proton. The self-assembly of hydrogen proceeds from the Big Bang, as we shall see in Chap. 6. It is the most abundant element in the Universe and participates in that essential molecule, water, a necessary ingredient in the self-assembly of Life.

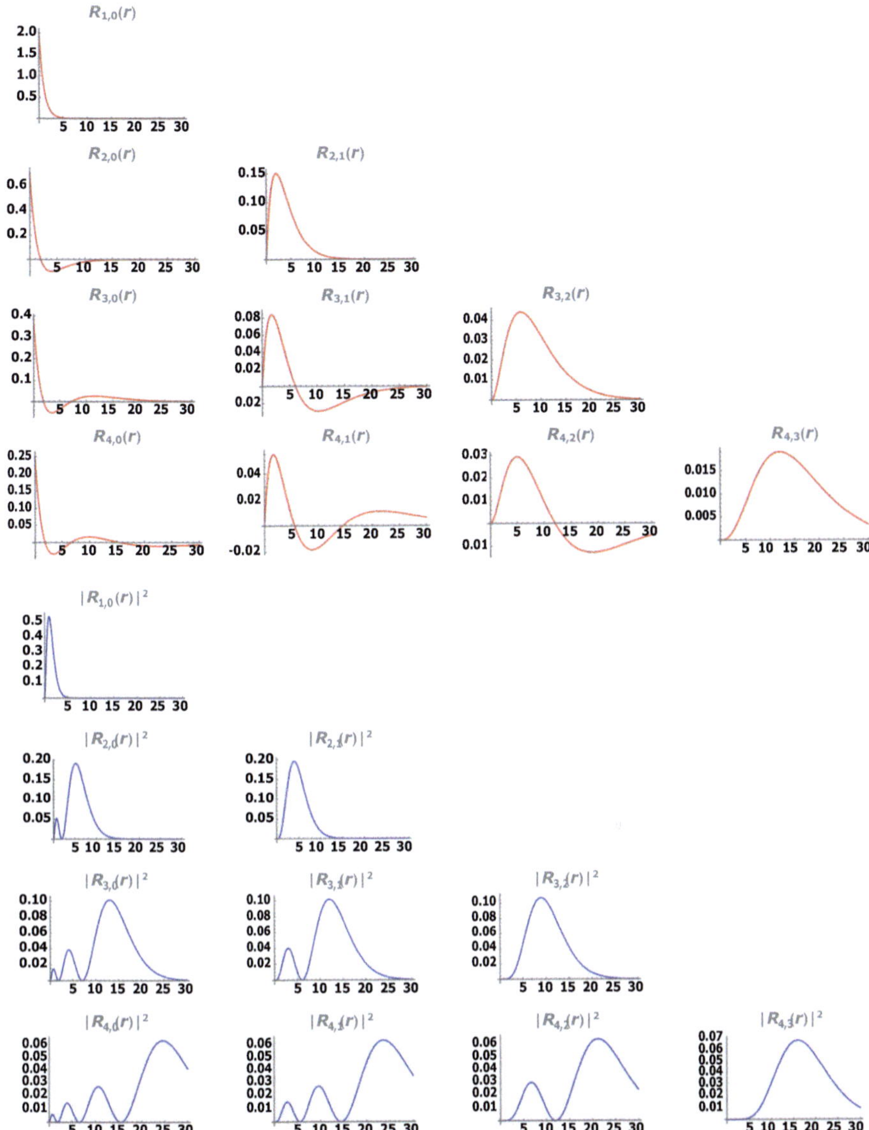

Fig. 4.14 Top (in red): the hydrogen atom radial wave functions versus distance (in Angstroms) from the proton for the $l = \{0, 4; m \leq l\}$ states. Bottom (in blue): the radial probability distributions for the same states

Problems

1. **Particle in a 3D box.** The energy levels $E(p,q,r)$ for a quantum particle in a three-dimensional box of equal sides given by $L = 0.4$ nm are:

$$E(p,q,r) = \frac{h^2}{8m}\left[\frac{p^2 + q^2 + r^2}{L^2}\right],$$

   where p, q, and r are quantum numbers for the x, y, and z dimensions, respectively, $\{p, q, r\} \in \{1,2,3,\ldots\}$.

   (a) For an electron, calculate $E(1,1,1)$ and $E(2,1,1)$ in Joules.
   (b) Calculate the same energy levels as in (a) but for a 9.11 kg chair in a cubic room with $L = 4.0$ m.
   (c) Compare the energy level spacing, $E(2,1,1) - E(1,1,1)$ for both (a) and (b).

2. **Planck's law.** Using Planck's law, $E = h\nu$, calculate:

   (a) the frequency, ν, in Hz,
   (b) the wavelength, λ, in nm, and
   (c) the energy as $1/\lambda$ in cm^{-1}
   for the electron in Problem 1 undergoing a transition from $E(1,1,1)$ to $E(2,1,1)$.

3. **Quantum dipole moment.** The dipole moment of an atom is like the quantum expectation value for the radius, but taken over two different quantum states. Calculate the dipole moment of the hydrogen atom in the transition between the 1 s and $2p_z$ orbitals given that the wave functions for these orbitals in spherical coordinates are:

$$\psi_{1s}(r,\theta,\varphi) = \frac{1}{\sqrt{\pi}}Z^{3/2}e^{-Zr};$$

$$\psi_{2pz}(r,\theta,\varphi) = \frac{1}{\sqrt{32\pi}}Z^{5/2}r\cos\theta e^{-Zr/2}$$

   where Z is the charge on the nucleus in units of the proton charge. That is, calculate the matrix element:

$$\vec{\mu} = \,<2p_z \mid \vec{r} \mid 1s>.$$

   Note that the volume element in spherical coordinates is given by:

   $dv = r^2 \sin\theta dr d\theta d\varphi$, and the limits of integration are $r \in \{0, \infty\}$, $\theta \in \{0, \pi\}$, and $\phi \in \{0, 2\pi\}$.

4. **Particle in a box wave functions.** Show by explicit calculation that for a particle in a one-dimensional box of width a, the wave functions given as

$\psi(x) = \sqrt{\frac{2}{a}}\sin\left(\frac{n\pi x}{a}\right)$ are:

(a) Orthogonal for $n = 1$, $n' = 2$, and
(b) Normalized.
(c) What is the most probable value of x for $n = 1$ and $n = 2$?
(d) Make a plot of the probability distributions for $n = 1, 2, 3$, and 4 in (c) in the box. What is unusual about the distribution, or different from what one might expect, for $n = 2$, or 4 compared with that for $n = 1$, or 3? The answer here involves the eigenvalues of the parity operator.
(e) What is $<n|x|n>$ the expectation, or average, value of x?
(f) Find the variance in x?

5. **Uncertainty principle.** Investigate the uncertainty principle by finding the spread in x space for a momentum eigenfunction of the free particle, $\phi(k) = e^{ikx}$. Assume the momentum spread is $\Delta k = \pm k_o/2$, and then compute the Fourier transform of a square wave with this spread in order to find the spatial wave function. From the zeroes of the probability, you can find Δx, the spread in x space. Make a plot of Δk versus Δx for various values of Δk from 0.05 to 50.

6. **Box energy levels.** For a quantum particle in a box, the energy levels are $E(n)$:

$$E(n) = \frac{h^2}{8m}\left[\frac{n^2}{L^2}\right],$$

where n is the quantum numbers for the x dimension, respectively, $\{n\} \in \{1,2,3,\ldots\}$.

(a) For an electron in a one-dimensional box of equal sides given by $L = 0.1$ nm, calculate $E(1)$ and $E(2)$ in Joules.
(b) Calculate the same energy levels as in (a) but for a proton in a nucleus with $L = 10^{-15}$ m.
(c) Calculate the same energy levels as in (a) but for a 10 g piece of chalk in a room with $L = 10$ m.

7. **Particles in the early Universe.** The probability that a particle will occupy a given state depends on the energy of that state. This probability distribution is given by the Boltzmann distribution. The ratio of the number of protons to neutrons during the first minute of the early Universe is

$$\frac{N_p}{N_n} = e^{-Q/kT}$$

where Q is the energy (mass) difference, given by $Q = (m_n - m_p)c^2$, where m_n and m_p are the masses of a neutron and proton, respectively, and c is the speed of light.

(a) What was the ratio of neutrons to protons when the temperature of the Universe was $5 \times 10^{13}, 5 \times 10^{12}, 5 \times 10^{11}, 5 \times 10^{10}$, and 5×10^{9} K, assuming thermal equilibrium and that an energy (mass) difference, given by $Q = 1.293$ MeV?

$$1 \text{ MeV} = 1 \times 10^{6} \text{ eV and } k = 8.617 \times 10^{-5} \text{ eV/K}$$

(b) The proton/neutron ratio "freezes out" at a value of 1/6. What is the temperature of the Universe when this happens?

(c) Plot the ratio of neutrons to protons from $T = 0$ K to $T = 5 \times 10^{13}$ K.

(d) What is the difference in mass between the neutron and proton (in kg)?

$$1 \text{ eV} = 1.609 \times 10^{-19} \text{ Joule and 1 Joule} = 1 \text{ kg*m}^2\text{/s}^2$$

(e) If a proton has a mass, mp $= 1.6726 \times 10^{-27}$ kg, using the mass difference found in (**d**), calculate the mass of a neutron (in kg).

8. **Eigenfunctions of angular momentum.** In spherical polar coordinates, we can write the operator for the z component of angular momentum as $\mathbf{L_z}$, where $L_z = -i\hbar \frac{\partial}{\partial \theta}$.

(a) What are the eigenfunctions $\psi(\theta)$ of $\mathbf{L_z}$? That is, solve $\mathbf{L_z \psi(\theta)} = \lambda \ \mathbf{\psi(\theta)}$ for $\psi(\theta)$.

(b) The eigenfunctions $\psi(\theta)$ found from (**a**) are not normalized, find the normalization constant.

(c) What is the normalized wave function?

(d) How would you calculate the average value of $\mathbf{L_z}$?

9. **Harmonic oscillator.** The quantum mechanical solutions for the simple harmonic oscillator can be used to estimate the strength of molecular bonds. You are studying a new protein with a reported spring constant, k, for a backbone cys-95 C=O peptide bond of $\mathbf{k = 1500 \ N/m}$. What wavelength (in $\boldsymbol{\mu}$**m**) of light will be absorbed? You may assume that the backbone carbon is fixed and that the oxygen nucleus vibrates in a C=O stretch mode. (1 amu $= 1.66 \times 10^{-27}$ kg and 1 N $= 1$ kg *m/s^2)

10. **Particle in a box.** For a particle in a box, if the kinetic energy was $\mathbf{2m^2v}$ instead of

(a) What would the Hamiltonian be? (Remember, inside the box, the potential $= 0$)

(b) What would Schrodinger's equation be?

(c) How would you calculate the probability of finding the particle in one of the quantum states?

11. **Uncertainty principle.** What is the uncertainty in position of a proton with a velocity of $3 \times 10^{8} \pm 1 \times 10^{6}$ m/s?

12. **Particle in a box.** Derive an equation for calculating the allowed energy levels for the particle in the box, starting the derivation with $\mathbf{E = KE + V}$, where KE is kinetic energy and V is potential energy.

13. **Orthonormality of particle in a box states.** Directly verify that $<n \mid m> = \delta_{n, m}$.
14. **Annihilation operator.** Find the matrix elements $<m|a|n>$ of the annihilation operator a, for $\{m, n\} \in \{0, 5\}$. How do they differ from the matrix elements of the creation operator, $a^\dagger$, for the same $\{m, n\}$?
15. **Harmonic oscillator.** Show that the harmonic oscillator wave functions are both orthogonal and normalized for $n = \{0, 5\}$.
16. **Harmonic oscillator.** Find the first five energy eigenvalues of the harmonic oscillator using $<n|E|n>$, where E is the energy operator.

References

M. Born, W. Heisenberg, P. Jordan, Zur Quantenmechanik II. Z. Phys. **35**, 557–615 (1925)

P. A. M. Dirac, The Physical Interpretation of Quantum Dynamics. Proc. Roy. Soc. A **113**, 621 (1927)

Dirac, P. A. M., The Quantum Theory of the Electron. Proc. Roy. Soc. A **117**, 610 (1928)

W. Heisenberg, Über quantentheoreticshe Umduetung kinematischer und mechanischer Beziehungen. Zeitschrift für Physik, **33**, 879–893 (1925)

W. Heisenberg, Über den anschaulichen Inhalt der quantentheoretischen Kinematik und Mechanik. Z. Phys. **43**, 172–198 (1927)

E. Schrödinger, An Undulatory Theory of the Mechanics of Atoms and Molecules. Phys. Rev. **28**, 1049 (1926)

Bibliography

P.A.M. Dirac, *The Principles of Quantum Mechanics* (Oxford University Press, 1930)

A. Messiah, *Quantum Mechanics volumes I and II* (Elsevier, Amsterdam, 1961)

J.J. Sakurai, J. Napolitano, *Modern Quantum Mechanics*, 3rd edn. (Cambridge University Press, 2021)

L. Susskind, A. Friedman, *Quantum Mechanics (The Theoretical Minimum)* (Basic Books, New York, 2014)

Chapter 5
The Self-assembly of Molecules: Molecular Quantum Mechanics

> *"… the theory of quantum mechanics could explain all of chemistry and the various properties of substances, it was a tremendous success."*
>
> *Richard P. Feynman*

We have now seen that self-assembly is driven by the biasing of probability (Entropy) by energy. Molecules form from atoms by forming bound states in which the energy of the molecule is lower than that of the constituent atoms, even though the atoms have fewer rather than more accessible states; this lowering of entropy is more than compensated by the energy decrease so that the final molecular states are more probable than are the separated atomic states. Applications of quantum mechanics to an elucidation of the electronic structures of atoms, while of intrinsic interest, are primarily useful in biochemistry as the basis for the study of biomolecules. Molecular quantum mechanics provides a detailed exposition of how matter self-assembles into the components of living systems. We will examine how the mobile electrons serve to dress the nuclei and generate bound states. In addition, all biochemistry takes place in water, therefore an understanding of the quantum mechanical properties of the water molecule cannot help but be of fundamental importance to the physical biochemist. Molecular quantum mechanics provides a physical description of one of the most fundamental of molecular properties, that of chemical bonding: How do you join atoms to form the molecules of life?

Biomolecules consist of a framework specified by the positions of the heavy backbone carbon, nitrogen, or oxygen nuclei whose positive charge is balanced by the negatively charged, ~1836-fold lighter, electrons. The goal of molecular quantum mechanics is then to determine how electrons on atoms combine to form chemical bonds that hold biomolecules together, that is, to form bound states of matter. The spectacular quantitative success of atomic structural analysis leads one to expect a similar level of understanding of molecular structure. We have seen how to apply Schrödinger's equation to simple, one-electron problems in our analysis of the energy levels of the hydrogen atom. It is hardly surprising then that a determination of the electronic structures of multi-electron atoms and molecules will not only require all that we have learned about such simple systems, but it will also

© The Author(s), under exclusive license to Springer Nature Switzerland AG 2024
L. O. Sillerud, *Abiogenesis*, https://doi.org/10.1007/978-3-031-56687-5_5

require new concepts that were not present in the solution of the simplest systems. Molecules are many-body structures and up to this point we have not learned how to write the wave functions for more than one electron at a time. However, we can proceed based on our knowledge of simple systems and seek to discover the quantum many-body laws as we need them. Along the way we will utilize the Born–Oppenheimer approximation and discover several new operators that commute with the Hamiltonian, one that describes the fact that all electrons are indistinguishable. These include the parity operator, P, and permutation symmetry arising from the spin-statistics theorem, otherwise known as the Pauli principle.

We will begin by studying the simplest molecules, for they already reveal necessary general features of molecular quantum mechanics. Just as the simplest atom is the hydrogen atom, the simplest molecule is that of the H_2^+ ion. This is a problem, which possesses exact solutions, of a single electron bound in the potential of two protons. We will then progress to an examination of the solutions of Schrodinger's equation for the hydrogen molecule, a two-electron problem. It would seem logical to examine the quantum energy levels and electronic structure of the water molecule next, given its central role in the evolution of life. The aromatic rings of Adenine, Guanine, Cytosine, and Thymine bear a strong relationship to that of the benzene molecule. Finally, of paramount importance for protein structure and absorption spectroscopy, we will seek to understand the electronic properties of the peptide bond. What we will discover is that the curve of binding energy for the simplest molecule, that of the H_2^+ molecular ion shows similarity with the already-presented Lennard–Jones potential, and serves as a prototype for understanding the bonding of atoms into molecules.

5.1   The Hydrogen Molecular Ion

The hydrogen molecular ion consists of a pair of positive protons about which a single negative electron is bound (Fig. 5.1) where the two protons are denoted by P_A^+ and P_B^+, with an internuclear distance, R, and the electron is located a distance

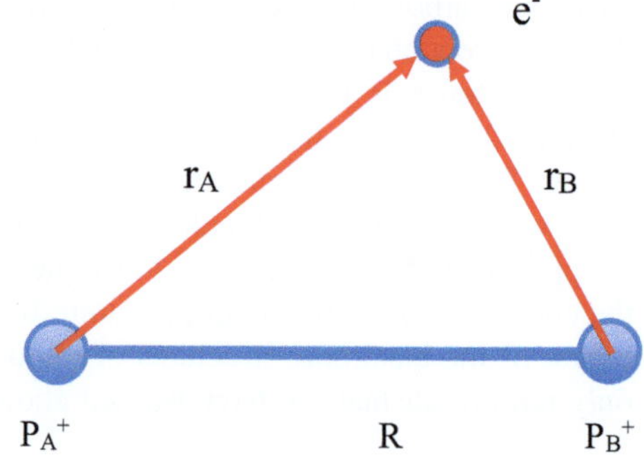

Fig. 5.1 The structural features of the hydrogen molecular ion in which the two protons (P⁺) are separated by a distance vector of length R originating at P_A^+, while the single electron e⁻ resides at vector distances r_A and r_B from the respective protons

r_A from proton A and r_B from proton B. The Hamiltonian for this system of three particles is then given by the sum of terms involving the kinetic energies of the electron and the two nuclei, and their interaction potential energies,

$$H = T_e + T_N + V(r_A, r_B, R).$$

The kinetic energy operators for each particle are given by

$$T_N = \frac{-\hbar^2}{2m_N} \nabla_N{}^2, \quad T_e = \frac{-\hbar^2}{2m_e} \nabla_e{}^2,$$

where m_N is the mass of the nucleon (the proton in this case) and m_e is the mass of the electron. In quantum mechanics, the potential is a multiplicative operator, which here is the Coulomb potential for a charged particle of charge e in the presence of another equally charged particle, $V(r) = e^2/r$. Therefore, in SI units we can write the Hamiltonian as

$$H = \frac{-\hbar^2}{2m_e} \nabla_e{}^2 - \sum_{i=1}^{2} \frac{-\hbar^2}{2M_i} \nabla_N{}^2 + \frac{e^2}{4\pi\varepsilon_0} \left[\frac{1}{R} - \frac{1}{r_A} - \frac{1}{r_B} \right]$$

where the first Laplacian $\nabla_e{}^2$ operates only on the electronic coordinates and involves the electron mass, m_e, while the second $\nabla_N{}^2$ operates only on the nuclear coordinates with the proton's mass, M. Note that the sign of the internuclear potential $(1/R)$ is positive, reflecting the fact that the nuclei, i.e., the two protons, repel each other, while the sign of the electron–proton interactions is negative as an indication that these two pairs of charged particles attract each other. Without this Coulomb attraction, molecules would not be stable and Life would not exist.

Schrodinger's equation for this system is written as

$$H \, \Psi(r_A, r_B, R) = E \, \Psi(r_A, r_B, R)$$

and, since we are dealing with a system consisting of two nuclei and an electron, we will attempt a solution in which the wave function separates into a product of electronic and nuclear wave functions. One strategy that suggests itself is to initially treat the nuclei as fixed at a distance R from each other and then to treat the internuclear distance as a variable parameter that we can alter later to find the minimum in the energy. In the spirit of the separation of variables as a method for the solution of partial differential equations, we suggest that we can separate the solution into terms involving only the nuclear, and only the electronic coordinates. Let the wave function then be written as a product

$$\Psi(r_A, r_B, R) = \Phi(r_A, r_B : R) \, \chi(R)$$

where $\Phi(r_A, r_B: R)$ is the electronic term, as a function of the distances of the electron from the nuclei, parameterized in terms of the internuclear distance, and $\chi(R)$ is the nuclear component of the wave function. The equilibrium distance between the nuclei will then occur at the place where the energy is a minimum.

5.1.1　The Born–Oppenheimer Approximation

What is the effect of applying the operator H to this wave function? Since the Hamiltonian is composed of both multiplicative (V) and second derivative (Laplacian) operators, we must be careful with respect to its application to the product wave function. Furthermore, the Hamiltonian contains terms that only operate on either the electronic or nuclear coordinates. Therefore,

$$H\,\Psi = H\,(\Phi\,\chi) = T_e\,(\Phi\,\chi) + T_N\,(\Phi\,\chi) + V\,(\Phi\,\chi) = E\,(\Phi\,\chi)$$

and the kinetic energy operators, T_e and T_N, contain second derivatives, so they must be treated using the product rule from calculus. The Hamiltonian applied to the wave function is

$$H\,\Psi = \chi\,T_e\,\Phi + \Phi\,T_e\,\chi + \Phi\,T_N\chi + \chi\,T_N\,\Phi + V\,\Phi\chi = E\,\Phi\chi.$$

Notice that T_e only operates on the electronic coordinates so that its effect on the nuclear wave function $\chi(R)$ is zero: $\Phi\,T_e\,\chi = 0$. Therefore, we are left with the following terms in the application of the Hamiltonian to the wave function:

$$H\,\Psi = \chi\,T_e\,\Phi + W + V\,\Phi\chi = E\,\Phi\chi.$$

The term W contains the operation of the nuclear Laplacian (a second derivative) on the product $\Phi\chi$, so we need to expand this in terms of the product rule again.

$$W = T_N\Phi\chi = -\sum_{i=1}^{2}\frac{-\hbar^2}{2M_i}\nabla_N^{\,2}(\Phi\chi)$$

We can write the nuclear Laplacian as the dot product of two gradient operators,

$$\nabla_N^{\,2} = \nabla_N \cdot \nabla_N$$

so that

$$\begin{aligned}
\nabla_N^2\left(\Phi_\chi\right) &= \nabla_N \cdot \left(\nabla_N \Phi_\chi\right) \\
&= \nabla_N \cdot \left(\chi \nabla_N \Phi + \Phi \nabla_N \chi\right) \\
&= \nabla_N \cdot \Phi \nabla_N \chi \\
&= \nabla_N \Phi \nabla_N \chi + \Phi \nabla_N^2 \chi
\end{aligned}$$

since $\chi \nabla_N \Phi = 0$ because Φ does not depend on the nuclear coordinates. Now, let us examine this last equation. Note that the nuclear momentum operator is proportional to ∇_N, ($p = -i\hbar \nabla_N$), and the nuclear kinetic energy operator is proportional to

$$\nabla^2{}_N, \left(T_N = \frac{-\hbar^2}{2M_i}\nabla_N{}^2\right),$$

so that the first term in W,

$$(\nabla_N \chi),$$

represents the nuclear momentum, while the second term,

$$(\nabla_N{}^2 \chi),$$

is the nuclear kinetic energy. These terms are referred to as vibronic components of the Hamiltonian by which is meant that they are due to molecular vibration-based oscillations of the internuclear distance, R. Note that each of these terms is divided by the proton mass, M_N, and is thus 1836-times smaller than the electronic terms.

The Born–Oppenheimer approximation treats the nuclei as fixed due to their much larger mass with respect to the electrons, and so we neglect W in the solution of the hydrogen molecular ion.

Schrodinger's equation for the hydrogen ion molecule is in this Born–Oppenheimer approximation given by

$$H\,\Psi = \chi\,T_e\,\Phi + \Phi\,T_N\,\chi + V\,\Phi\,\chi = E\,\Phi\,\chi.$$

Let us rewrite this as

$$\chi\,T_e\,\Phi + \Phi\,T_N\,\chi + V\,\Phi\,\chi = E\,\Phi\,\chi,$$

or

$$\Phi\,T_N\,\chi + \left(T_e\,\Phi + V\,\Phi\right)\chi = E\,\Phi\,\chi.$$

However, we have that

$$(T_e + V)\ \Phi = E_p\Phi(r_A, r_B : R)$$

and we can insert this into our equation for χ to give

$$T_N\ \chi + E_p\ \chi = E\ \chi,$$

Where the internuclear (molecular) potential energy is E_p and the total energy of the molecule is E in the Born–Oppenheimer approximation. In this manner, we have separated the nuclear and electronic wave functions.

To solve the hydrogen molecular ion, we then use for the electronic structure, the equation

$$T_e\Phi + V\ \Phi = E\ (r_A, r_B : R)\Phi.$$

where the Hamiltonian for the problem is $H = T_e + V$

$$H = \frac{-\hbar^2}{2m_e}\nabla_e^2 + \frac{e^2}{4\pi\varepsilon_0}\left[\frac{1}{R} - \frac{1}{r_A} - \frac{1}{r_B}\right]$$

which is usefully rewritten as

$$H = \left[\frac{-\hbar^2}{2m_e}\nabla_e^2 - \frac{e^2}{4\pi\varepsilon_0}\frac{1}{r_A}\right] + \frac{e^2}{4\pi\varepsilon_0}\left[\frac{1}{R} - \frac{1}{r_B}\right]$$

Or

$$H = H_o + \frac{e^2}{4\pi\varepsilon_0}\left[\frac{1}{R} - \frac{1}{r_B}\right]$$

where H_o gives the ground state energy of the hydrogen atom $E_o = \frac{e^2}{4\pi\varepsilon_0}\frac{1}{2a_o} = -13.6\,\text{eV}$, or in SI units, $E_o = 2.18 \times 10^{-18}\text{Joules}$:

$$H_o\,|\,A\rangle = \left[\frac{-\hbar^2}{2m_e}\nabla_e^2 - \frac{e^2}{4\pi\varepsilon_0}\frac{1}{r_A}\right]|A\rangle = E_o\,|\,A\rangle$$

and $|A\rangle$ is the ground state 1s wave function for the hydrogen atom, in Dirac notation (see Chap. 4). We remember that a wave function ψ_A is written as a ket $|A\rangle$ and its complex conjugate ψ^*_A is written as a bra $\langle A|$. A bra-ket is then the integral $\langle A|A\rangle = \int \psi^*_A\psi_A d\tau$ with the volume element, $d\tau$. In the calculations to follow, *it is important to note that we can express the constant $\frac{e^2}{4\pi\varepsilon_0}$ in the above equation as* $2a_oE_o$.

5.1.2   Ellipsoidal Coordinates

Schrodinger's equation for the hydrogen molecular ion using the above Hamiltonian is exactly soluble in ellipsoidal coordinates (μ, ν, ϕ) such that

$$\mu = (r_A + r_B)/R,$$

$$\nu = (r_A - r_B)/R,$$

although the solutions are in the form of series expansions and must be evaluated numerically and there is little physical insight provided by this approach even though it does provide accurate results (See: Grivet (2002)). More physical insight can be learned from simpler but solvable approximate methods that provide analytical solutions in the more conventional spherical polar coordinate system used for the hydrogen atom.

5.1.3   A Variational Method

Among the several useful approximate methods for the solutions of quantum mechanical problems is the variational approach that uses a trial wave function based on a linear combination of atomic states. We will show that the energies that result from this approach are always greater than or equal (vide infra) to the true energies, but often the calculations are considerably easier.

To begin, we note that the result of any quantum mechanical measurement is the expectation value of an operator (using Dirac's bra-ket notation (Dirac, 1939)). Let's examine the measurement of the energy, E, of a system described by the wave function, Ψ, in which case the measurement operator is H, the Hamiltonian:

$$E = \frac{\langle \Psi | H | \Psi \rangle}{\langle \Psi | \Psi \rangle} = \frac{\int \Psi^* H \Psi d\tau}{\int \Psi^* \Psi d\tau}$$

Suppose that we know the exact solution of Schrodinger's equation, $H \Psi_n = E_n \Psi_n$, for some related atomic system so that the wave functions $\{\Psi_n\}$ form a complete orthonormal set of functions (i.e., they span the Hilbert space of states), with

$$\langle \Psi_n | \Psi_m \rangle = \delta_{nm}.$$

We can then write a trial wave function, Φ, for our molecule as a linear expansion in this basis set $\{\Psi_i\}$ as

$$\Phi = \sum_i c_i \Psi_i$$

and note that if the set $\{\Psi_i\}$ is normalized, we have that

$$\langle \Phi | \Phi \rangle = \int \sum_i \sum_j c_i^* \Psi_i^* c_j \Psi_j d\tau = \sum_i |c_i|^2 \int \Psi_i^* \Psi_i d\tau = \sum_i |c_i|^2.$$

In order for our trial wave function to be normalized, we require that the sum of the squares of the linear coefficients must be 1.

$$\langle \Phi | \Phi \rangle = \sum_i |c_i|^2 = 1$$

We can use this normalization condition on the trial wave function to later find the coefficients c_i by means of:

$$\Phi' = \frac{\sum_i c_i \Psi_i}{\sqrt{\sum_i |c_i|^2}}$$

Let's assume we have already done that so that $\langle \Phi' | \Phi' \rangle = 1$ and have replaced Φ by this normalized function, and subsequently dropped the prime. Then let's calculate the expectation value of the energy using this trial wave function.

$$E = \langle \Phi | H | \Phi \rangle = \sum_i \sum_j \int c_i^* \Psi_i^* H c_j \Psi_j \, d\tau$$

$$E = \sum_i \sum_j c_i^* c_j E_i \int \Psi_i^* \Psi_j \, d\tau$$

$$E = \sum_i \sum_j c_i^* c_j E_i \delta_{ij}$$

$$E = E_i \sum_i c_i^* c_i = E_i$$

$$E = E_i \geq E_o$$

where E_o is the real ground state energy from $E_o = \langle \Psi_o | H | \Psi_o \rangle$ and the sum of the coefficients is 1. Here, one uses the fact that the energy of any state E_j $(j > 0)$ above the ground state is larger than that of the ground state. The variational principle thus states that the energy of any trial wave function is always greater than that of the true ground state wave function, and equal to the ground state energy if and only if the wave function is the true ground state wave function.

5.1.4   A Simple Variational Calculation

This can be explored using the normalized particle in a box ground state wave function (for $a = 1$ and $n = 1$; see Chap. 4):

$$\psi(x) = \sqrt{2}\,\sin(\pi x)$$

and a normalized parabolic trial wave function, φ, that is compared with ψ in Fig. 5.2:

$$\varphi(x) = \sqrt{30}(1 - x)$$

Note that the trial parabolic function is very similar to the true sine function on this interval. The difference is never greater than 6%. The Hamiltonian in this one-dimensional problem is

$$H = \frac{-\hbar^2}{2m}\nabla^2 = \frac{-\hbar}{2m}\frac{d^2}{dx^2}$$

since with $H = T + V$, and the potential V is zero within the box.

The expectation value of the true ground state energy E_o of the particle in a box is given by

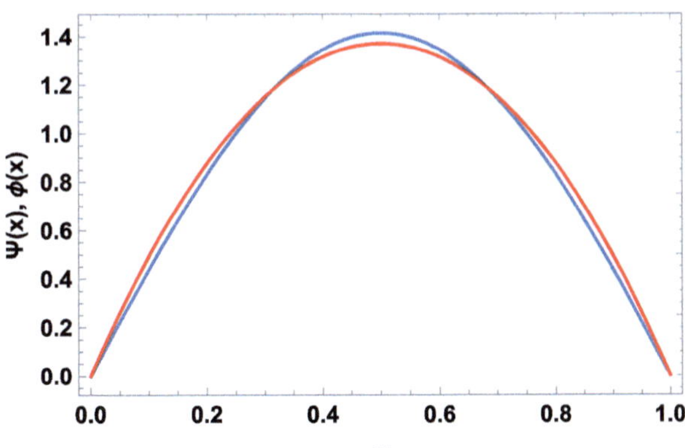

Fig. 5.2   The normalized ground state wave function $\psi(x)$ for the particle in a box with walls at $x = 0, 1$ (blue) and a normalized trial wave function $\varphi(x)$ (red) on the same interval

$$E_0 = \langle \Psi | H | \Psi \rangle = \frac{-\hbar^2}{2m} \int_0^1 \psi^*(x) \nabla^2 \psi(x) dx = \frac{-\hbar^2}{m} \int_0^1 \sin(\pi x) \frac{d^2}{dx^2} \sin(\pi x) dx$$

$$E_0 = \frac{2\pi^2 \, \hbar^2}{m} \int_0^1 \sin^2(\pi x) dx = \frac{\pi^2 \, \hbar^2}{m}$$

and the expectation value of the ground state energy of the particle in a box using our trial parabola is E', where

$$E' = \frac{-\hbar^2}{2m} \int_0^1 \varphi(x) \nabla^2 \varphi(x) dx = \frac{-30\hbar^2}{m} \int_0^1 x(1-x) \frac{d^2}{dx^2} x(1-x) dx$$

The value of the last integral on the right is $-1/3$, so

$$E' = \frac{10\hbar^2}{m}$$

And we have that $\frac{E'}{E} = \frac{10}{\pi^2} = 1.01321$, which is only ~1% different from the true ground state wave function but, congruent with the above proof, is still larger than the true ground state energy. Variational methods give robust upper bounds to the energies. We can also present these solutions in a manner that allows informative plots in *Mathematica* of the features of the wave functions that are most important for understanding the basis of molecular bonding and the self-assembly of molecules.

5.1.5   Linear Combination of Atomic States

We already have analytical solutions for the hydrogen atom, and we can expect that the wave functions for the hydrogen molecular ion would converge to these known states as the internuclear distance becomes large with respect to the Bohr radius and our system becomes an isolated proton and an isolated hydrogen atom. It, therefore, makes sense to use the normalized ground state hydrogen atom wave functions as the bases for our trial wave function. We begin by writing our trial solution φ as a sum of hydrogen atom 1s states. These are

$$\psi_A(r_A) = \frac{1}{\sqrt{\pi a_o^{3/2}}} e^{-r_A/a_o} = |A\rangle$$

for the electron's interaction with the proton at A, and a similar expression for the electron's interaction with the proton at B:

$$\psi_B(r_B) = \frac{1}{\sqrt{\pi a_o^{3/2}}} e^{-r_B/a_o} = \mid B\rangle$$

That these are normalized is shown by computing $\langle A|A\rangle = \langle B|B\rangle$ in *Mathematica*:

```
ψ[r_] = 1/√(πa₀³) e^(-r/a₀);
Integrate[ψ[r] ψ[r] r² Sin[θ], {θ,0,π}, {φ,0,2π}, {r,0,∞}]
Out[%] = 1
```

We write our trial wave function φ as

$$\varphi = c|A\rangle + d|B\rangle$$

We will need to accomplish two things: to normalize this sum wave function and to find the coefficients $\{c, d\}$, which we will find are like Lagrange multipliers as used in Chap. 1 to derive the Boltzmann distribution.

5.1.6   Solving for the Wave Function Coefficients

Inserting φ into our expression for the energy gives

$$E = \frac{\langle \varphi|H|\varphi\rangle}{\langle \varphi|\varphi\rangle}$$

with the Hamiltonian as

$$H = \frac{-\hbar^2}{2m_e}\nabla_e^2 + \frac{e^2}{4\pi\varepsilon_0}\left[\frac{1}{R} - \frac{1}{r_A} - \frac{1}{r_B}\right]$$

and we can subsequently drop the subscripts e because the Laplacian is now understood to operate only on the electronic coordinates and the mass is that of the electron. Then we need to evaluate the expression

$$E\langle \varphi|\varphi\rangle = \langle \varphi|H|\varphi\rangle \tag{5.1}$$

when we insert the trial wave function. The left-hand side of Eq. (5.1) then reads

$$E\langle\varphi|\varphi\rangle = E\left(c^2\langle A|A\rangle + cd\langle A|B\rangle + cd\langle B|A\rangle + d^2\langle B|B\rangle\right)$$

which reduces to

$$E\langle\varphi|\varphi\rangle = E\left(c^2 + 2\,c\,d\,S + d^2\right)$$

because

$$\langle A|A\rangle = \langle B|B\rangle = 1$$

and we let

$$S(R) = \langle A|B\rangle = \langle B|A\rangle = \frac{1}{\pi a_o{}^3}\int e^{-\frac{1}{a_o}(r_A + r_B)}\,d\tau.$$

$S(R)$ is a function of the nuclear separation, R, and is called the overlap integral because it measures the degree of overlap between the two wave functions.

Now, the right-hand side of Eq. (5.1) is

$$\langle\varphi|H|\varphi\rangle = c^2\langle A|H|A\rangle + cd\langle A|H|B\rangle + cd\langle B|H|A\rangle + d^2\langle B|H|B\rangle$$

or

$$\langle\varphi|H|\varphi\rangle = \left(c^2 + d^2\right)\langle A|H|A\rangle + 2cd\langle A|H|B\rangle$$

since $\langle A|H|A\rangle = \langle B|H|B\rangle$ and $\langle A|H|B\rangle = \langle B|H|A\rangle$ (as will be shown later). Now, equating both sides gives

$$E\left(c^2 + 2\,c\,d\,S + d^2\right) = \left(c^2 + d^2\right)\langle A|H|A\rangle + 2cd\langle A|H|B\rangle$$

which we can simplify to read as

$$E\left(c^2 + 2\,c\,d\,S + d^2\right) = \left(c^2 + d^2\right)K + 2cd\,J$$

where $K = \langle A|H|A\rangle$ and $J = \langle A|H|B\rangle$ are called the *Coulomb* and *Exchange* matrix elements, respectively. Reminiscent of the earlier usage of Lagrange multipliers, the minimum in the energy given by this trial wave function is found by taking the partial derivatives with respect to the coefficients $\{c,\,d\}$ and setting these to zero. This gives, first, for $\frac{\partial}{\partial c}$,

$$2\,c\,E + 2\,E\,d\,S = 2\,c\,K + 2\,d\,J$$

while the second, for $\frac{\partial}{\partial d}$, gives

$$2\,c\,E\,S + 2\,d\,E = 2\,d\,K + 2\,c\,J$$

Cancelling a common factor of 2 and rearranging, we find two equations in the unknown coefficients $\{c, d\}$

$$c\,(E - K) - d(J - ES) = 0$$

$$c\,(ES - J) - d(K - E) = 0$$

This set of equations has the trivial solution $c = d = 0$, and a nontrivial solution if the determinant of the coefficients is zero. *Mathematica* gives the two solutions to the quadratic equation

```
Solve[Det[| K-E   J-ES |]==0,E]
            | J-ES  K-E  |
```

as the energies, which we label with the German subscripts u (*ungerade*, or *un*straight, in English) for minus and g (*gerade*, or straight, in English) for plus, in terms of the integrals defined above:

$$E_{\mathrm{u}} = \frac{K - J}{1 - S} \quad \text{and} \quad E_{\mathrm{g}} = \frac{K + J}{1 + S}\,.$$

Inserting these into the above equations, we find that

```
Solve[{c(Eu - K)-d(J - Eu S)==0, c(Eu S - J)-d(K - Eu)==0}, c]
```

has the solution $c = -d$, giving

$$\varphi_{\mathrm{u}}(r) = c(|A\rangle - |B\rangle)$$

while for E_{g}, we find that

```
Solve[{c(Eg - K)-d(J - Eg S)==0, c(Eg S - J)-d(K - Eg)==0}, d]
```

has the solution $c = d$ giving

$$\varphi_{\mathrm{g}}(r) = d(|A\rangle + |B\rangle)$$

and the molecular wave functions are the sum and difference of the 1s atomic state functions. These are known as bonding, for $\varphi_{\mathrm{g}}(r)$, and antibonding, for $\varphi_{\mathrm{u}}(r)$, states, corresponding to the ground state and the first excited states, respectively. They differ in the fact that between the nuclei in the bonding state (Fig. 5.3a, c), there exists a large probability of finding an electron, while for the antibonding state (Fig. 5.3b, d), there is a markedly decreased probability of finding the electron between the two nuclei. The bonding state has no nodes between the two nuclei, while the antibonding state goes to zero (has a node) equidistant from the two protons.

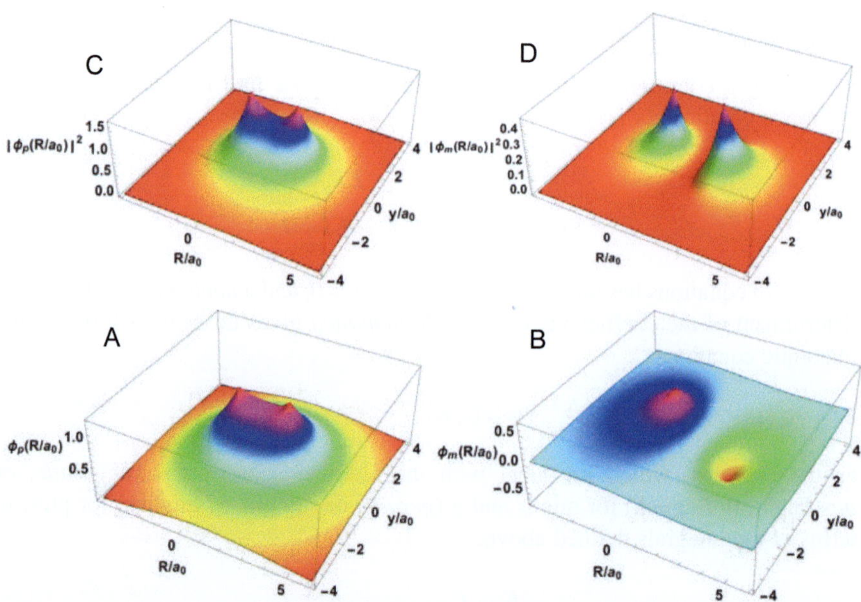

Fig. 5.3 Wave functions (**a, b**) and probabilities (**c, d**) for finding an electron in the region around the two protons in the linear combination of 1s atomic states approximation for the hydrogen molecule ion (H_2^+). The x and y axes are given in units of the Bohr radius, $a_o = 52.9$ pm; the nuclei in this figure are at a distance of $2.4a_o$ apart. Shown is the even (*gerade*) parity state φ_g on the left (**a, c**) and the odd (*ungerade*) parity state φ_u on the right (**b, d**)

5.1.7 *Symmetries of the Trial Wave Functions*

We have been introduced to group theory in Chap. 3 in the context of continuous groups and the symmetry of spacetime. An important subset of group theory involves the transformations, not of smooth functions, but of discrete points in 3D space, such as the assumed stable positions of nuclei in a molecule. Here, we anticipate that certain features of Point group theory (which we explore in some detail in Chap. 11) can be profitably applied to the hydrogen molecular ion.

The set of all point transformation operators that leave the Hamiltonian invariant form a group. If one orients the protons along one of the coordinate axes as in Fig. 5.4, the point group (see Chap. 11) containing the set of symmetry operators is that of all homonuclear diatomic molecules; labeled $D_{\infty h}$, whose operators consist of the set $U_i = \{e, I, \sigma_h, C_\infty, S_\infty, \sigma_v\}$ that operate on the, assumed fixed, coordinates of the heavy nuclei of the molecule. All of these operators can be written as matrices and are unitary (i.e., length and probability preserving) that commute with the Hamiltonian, $[H, U_i] = 0$, $\forall i \in U_i$ so that they have common eigenfunctions. Therefore, our trial wave function must transform according to the irreducible representations of this set.

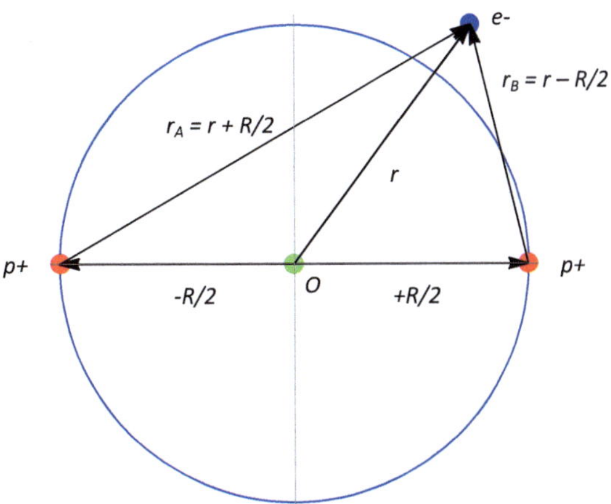

Fig. 5.4 A schematic diagram of the coordinate system used for the hydrogen molecule ion. The protons are shown in red, while the electron is shown in blue. The origin of the coordinate system is equidistant from the two protons. The distances from the two protons to the single electron are shown as r_A and r_B

Of particular interest in this case is the transformation under the operator i, the inversion through the origin (i.e., not the parity operator, which we will introduce later). The point group $D_{\infty h}$ is a non-Abelian (noncommutative) group in which the order of application of the operators matters; as we have seen in the uncertainty principle, it is not necessarily true that

$$U_i U_j H = U_j U_i H$$

for $U_i \in D_{\infty h}$. The character table for $D_{\infty h}$ gives the traces (characters) of the transformation matrices, and in this case shows that the character χ of the inversion, i, operator in each of the A_1 irreducible representations is either $\chi(A_{1g}) = +1$ or $\chi(A_{1u}) = -1$.

Although the electronic Hamiltonian is invariant under inversion through the origin, the trial wave functions are not necessarily invariant. Each of the hydrogen atom wave functions, $|A\rangle$ or $|B\rangle$, transform as the A_{1g} irreducible representation, Γ_A and Γ_B, and their product, $\langle A|B\rangle$, which is computed in the overlap integral, S, must transform as the tensor product $\Gamma_{AB} = \Gamma_A \otimes \Gamma_B$ of their symmetry species and Γ_{AB} must contain the irreducible representation A_{1g} of $D_{\infty h}$. In order for two quantum states (wave functions) to form a bound state, each must belong to the same irreducible representation of the symmetry group, with the same class and character, which here is the inversion operator, i. The even state φ_g has eigenvalue $+1$ under the inversion operator, while the odd state φ_u has eigenvalue -1. Our two trial wave functions are the pair

$$\varphi_g = d(|A\rangle + |B\rangle),$$

which transforms as A_{1g} and can form a bound state, and

$$\varphi_u = c(|A\rangle - |B\rangle),$$

which transforms as A_{1u} cannot form a bound state. These are called bonding and antibonding states. The overlap integral, S, must be nonzero in order to form a bound state. We will calculate the overlap integral $S = \langle A|B\rangle$ below and show that it is nonzero.

5.1.8 Normalization of the Trial Wave Function

We can find the normalization coefficients c and d of our trial wave function by means of the integral equations $\langle A|A\rangle = \langle B|B\rangle = 1$. The coefficients are real because the wave functions are real. From symmetry, if we interchange the positions of the two identical protons, we do not expect the wave function of the electron to change so that $\langle A|B\rangle = \langle B|A\rangle = S$, an integral that measures the overlap of the two, nonorthogonal atomic states. The normalization of φ then requires that $\langle \varphi_u \mid \varphi_u\rangle = \langle \varphi_g \mid \varphi_g\rangle = 1$, or for φ_g:

$$\langle \varphi_g|\varphi_g\rangle = d((\langle A| + \langle B|)d(|A\rangle + |B\rangle) = 1$$

$$\langle \varphi_g|\varphi_g\rangle = d^2(\langle A|A\rangle + 2\langle A|B\rangle + \langle B|B\rangle)$$

$$\langle \varphi_g|\varphi_g\rangle = d^2(2 + 2S) = 1$$

$$=> d = \frac{1}{\sqrt{2(1+S)}}$$

Note that $S \neq 0$ because the basis wave functions are not orthogonal, otherwise d would be $\pm\frac{1}{\sqrt{2}}$. Similarly, for φ_u:

$$\langle \varphi_u|\varphi_u\rangle = c((\langle A| - \langle B|)c(|A\rangle - |B\rangle) = 1$$

$$\langle \varphi_u|\varphi_u\rangle = c^2(\langle A|A\rangle - 2\langle A|B\rangle + \langle B|B\rangle)$$

$$\langle \varphi_u|\varphi_u\rangle = c^2(2 - 2S) = 1$$

$$=> c = \frac{1}{\sqrt{2(1-S)}}$$

and we have for our two normalized trial wave functions:

$$\varphi_g(r) = \frac{1}{\sqrt{2(1+S)}}(|A\rangle + |B\rangle)$$

and

$$\varphi_{\mathrm{u}}(r) = \frac{1}{\sqrt{2(1-S)}}(|A\rangle - |B\rangle).$$

5.1.9 Eigenfunctions of the Parity Operator

These two states also differ in their eigenvalues under the parity operator, Π. The electromagnetic interaction conserves parity, therefore the parity operator commutes with the Hamiltonian $[H, \Pi] = 0$, and the hydrogen molecule ion wave functions can be simultaneous eigenfunctions of both the Hamiltonian and the parity operator. The eigenvalues of Π are ± 1. If we place the origin of coordinates at the midpoint on the line joining the two nuclei, then the *gerade* wave function must correspond to even parity with eigenvalue $+1$, since interchanging the nuclei does not change the state, while interchanging the nuclei in the *ungerade* state changes the sign under the parity operator and therefore must have the -1 eigenvalue. The next task to address is to find the energies of the two states:

$$E_{\mathrm{u}} = \frac{J - K}{1 - S}$$

and

$$E_{\mathrm{g}} = \frac{J + K}{1 + S}.$$

For this we need to calculate three integrals:

$$S = \langle A|B\rangle,$$

$$K = \langle A|H|A\rangle,$$

and

$$J = \langle A|H|B\rangle;$$

the overlap integral (S), the Coulomb (K), and exchange (J) matrix elements, respectively.

5.1.10 Calculating the Overlap Integral

The overlap integral $S = \langle A|B\rangle$ in spherical polar coordinates $\{r, \theta, \phi\}$ is

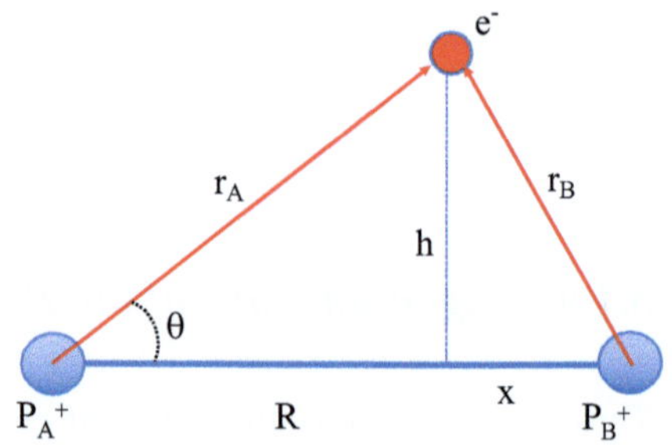

Fig. 5.5 Definition of the position vectors of the protons (blue) and the electron (red) in the coordinate system used for calculation of the hydrogen molecule ion wave functions

$$S = \frac{1}{\pi a_o^3} \int_0^{2\pi} \int_0^{\pi} \int_0^{\infty} e^{-\frac{r_A}{a_o}} e^{-\frac{r_B}{a_o}} \, r^2 \sin\theta \, dr d\theta d\varphi$$

As it stands, this seemingly simple integral cannot be performed as written, since it contains exponentials of two radii, r_A and r_B whose relationship is not yet defined. However, we can solve for r_B in terms of r_A and R using the triangle construction shown in Fig. 5.5.

Let the angle between the vectors R and r_A be θ, the height of the triangle be h, and the base of the smaller triangle, with sides h and r_B be x. Then we have that

$$r_B^2 = h^2 + x^2$$

$$r_A^2 = h^2 + (R-x)^2$$

$$x = R - r_A \cos\theta$$

$$h = r_A \sin\theta$$

$$\text{Then } r_B^2 = r_A^2 \sin^2\theta + (R - r_A \cos\theta)^2$$

$$r_B^2 = r_A^2 \sin^2\theta + R^2 - 2R\,r\,\cos\theta + r_A^2 \cos^2\theta$$

$$r_B^2 = r_A^2 (\sin^2\theta + \cos^2\theta) + R^2 - 2R\,r\,\cos\theta$$

$$r_B^2 = r_A^2 + R^2 - 2R\,r\,\cos\theta \text{ giving}$$

$$r_B = \sqrt{r_A^2 + R^2 - 2R\,r\,\cos\theta}$$

and we can eliminate r_B from the integral for S. Dropping the subscript A on r_A, the integral becomes

$$S = \frac{1}{\pi a_o^3} \int_0^{2\pi} \int_0^{\pi} \int_0^{\infty} e^{-\frac{r}{a_o}} e^{-\frac{\sqrt{r^2 + R^2 - 2R\,r\,\cos\theta}}{a_o}} \, r^2 \sin\theta \, dr d\theta d\varphi$$

The φ integration gives 2π. Then

$$S = \frac{2}{a_o{}^3} \int_0^\pi \int_0^\infty e^{-\frac{r}{a_o}} e^{-\frac{\sqrt{r^2+R^2-2R\,r\,\cos\theta}}{a_o}} r^2 \sin\theta \, dr d\theta$$

We rewrite the integral for S as an iterated integral

$$S = \frac{2}{a_o{}^3} \int_0^\infty e^{-\frac{r}{a_o}} r^2 dr \int_0^\pi e^{-\frac{\sqrt{r^2+R^2-2R\,r\,\cos\theta}}{a_o}} \sin\theta \, d\theta$$

This integral appears difficult due to the square root in the exponent, but *Mathematica* can directly perform the angular integration (it can take several minutes, so be patient):

$$j(r) = \int_0^\pi e^{-\frac{\sqrt{r^2+R^2-2R\,r\,\cos\theta}}{a_o}} \sin\theta d\theta =$$

$$= \text{Timing}\left[\text{Integrate}\left[e^{-\left(\frac{1}{a}\right)\sqrt{r^2+R^2-2rR\cos[\theta]}}\sin[\theta],\{\theta,\ 0,\ \pi\}\right]//\text{FullSimplify}\right]$$

$$= \left\{ 171.5 \text{ s}, \left[\frac{a_o\left(e^{-\frac{\sqrt{(r-R)^2}}{a_o}}\left(a_o+\sqrt{(r-R)^2}\right)-e^{-\frac{\sqrt{(r+R)^2}}{a_o}}\left(a_o+\sqrt{(r+R)^2}\right)\right)}{rR} \right] \right\}$$

The square roots of squared values give us a clue that the absolute values may be required. Now, we can insert $j(r)$ into the radial integral and attempt to solve it directly using *Mathematica*:

$$S = \frac{2}{a_o{}^3} \int_0^\infty j(r) \, e^{-\frac{r}{a_o}} r^2 dr$$

However, after 217 s, *Mathematica* failed to compute this integral; some reflection shows that the reason for the failure is the domain ambiguity associated with *Mathematica*'s use of $\sqrt{(r-R)^2}$ for what should really be the absolute value $|r-R|$ or, alternatively, the length of the vector $r-R$, so let's back up and use a trick from early calculus; that of variable substitution.

$$\text{Let } x = \frac{r}{a_o}, \quad X = \frac{R}{a_o}, \text{ and } y = \frac{r_B}{a_o} = \frac{1}{a_o}\sqrt{r_A{}^2+R^2-2R\,r\,\cos\theta}$$

so that

$$y^2 = x^2 + X^2 - 2\,X\,x\,\cos\theta$$

and

$$d\left(y^2\right) = 2y\,dy = 2\,x\,X\,\sin\theta\,d\theta$$

Or perhaps more transparently

$$\sin\theta\,d\theta = \frac{y\,dy}{x\,X}$$

The new limits on the integral are found from the substitution equations:

$$\text{For }\ \theta = 0,\ \ y^2 = x^2 + X^2 - 2\,X\,x = (x - X)^2,$$

$$\text{or }\ y = \sqrt{(x - X)^2} = |x - X|,$$

$$\text{while for }\ \theta = \pi,\ \ y^2 = x^2 + X^2 + 2\,X\,x = (x + X)^2$$

and

$$y = |x + X| = x + X,$$

from which we discover the reason for the need for absolute values. The angular integral is then transformed into

$$\int_0^\pi e^{-\frac{\sqrt{r^2 + R^2 - 2R\ r\ \cos\theta}}{a_o}}\sin\theta d\theta = \frac{1}{x\,X}\int_{|x - X|}^{x + X} y\,e^{-y}dy$$

which is done in *Mathematica* as

$$\text{Integrate}\left[\frac{1}{xX}ye^{-y},\{y,\text{Abs}[x - X],\ x + X\}\right]s = \frac{-e^{-x-X}(1 + x + X) + e^{-\text{Abs}[x - X]}(1 + \text{Abs}[x - X])}{xX}$$

or in conventional notation

$$\frac{1}{x\,X}\int_{|x - X|}^{x + X} y\,e^{-y}dy = \frac{-e^{-x-X}(1 + x + X) + e^{-|x - X|}(1 + |x - X|)}{xX}$$

Mathematica computed this integral using the correct absolute value Abs$[x - X]$ in the limits. Inserting this into our expression for the overlap integral, we have that

$$S = \frac{2}{a_o{}^3}\int_0^\infty e^{-\frac{r}{a_o}}r^2 dr\int_0^\pi e^{-\frac{\sqrt{r^2 + R^2 - 2R\ r\ \cos\theta}}{a_o}}\sin\theta\,d\theta$$

is reduced to

Fig. 5.6 The domain of $|x - X|$ used in the integration in the text for two values of the scaled radial coordinate, $x < X$ and $x > X$ (black arrows). The red circle represents the radius at which $x = X$, and the protons are represented by the blue dots

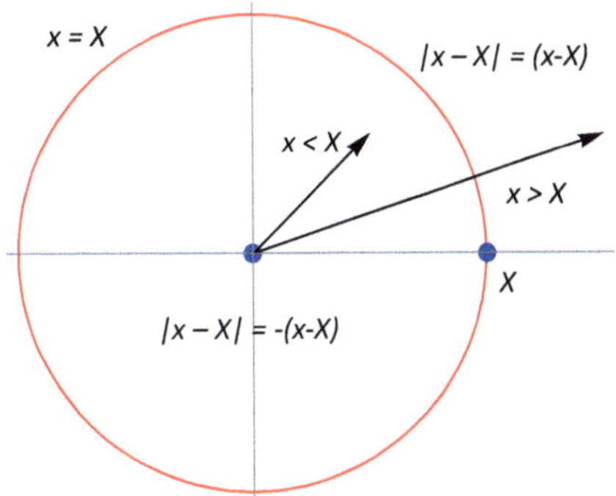

$$S = \frac{2}{a_o} \int_0^\infty e^{-\frac{r}{a_o}} r^2 \frac{-e^{-\frac{(r+R)}{a_o}}\left(1 + \frac{r+R}{a_o}\right) + e^{-\frac{|r-R|}{a_o}}\left(1 + \frac{|r-R|}{a_o}\right)}{rR} dr$$

or since $r = a_o x$ and $dr = a_o\, dx$; $R = a_o X$

$$S = \frac{2\, a_o}{X} \int_0^\infty x\, e^{-x}\left(-e^{-(x+X)}(1 + x + X) + e^{-|x-X|}(1 + |x - X|)\right) dx$$

Mathematica cannot directly perform this integral because the $|x - X|$ is now in the integrand, so we need to examine the domain of $|x - X|$ as shown in Fig. 5.5 and separately integrate two cases:

$$|x{-}X| = x{-}X \ if \ x \geq X \ or \ x - X \geq 0$$

$$|x{-}X| = -(x{-}X) \ if \ x < X \ or \ x - X < 0$$

It is useful to examine a plot (Fig. 5.6) of these variables in order to understand the domain of $|x - X|$; there are two regions, separated by the circular boundary at $x = X$, surrounding the position of proton A in Fig. 5.5, at the origin of the coordinate system. For radii less than the distance $X = R/a_o$ between the two protons, the integration needs to proceed from zero to X, while for $x = r/a_o \rangle R/a_o$, the integration proceeds from X to infinity. Then the two integrals are split into three and are easily done; *Mathematica* saves one from doing the algebra here:

$$S(X) = \frac{2}{X}\left[\int_0^\infty x\, e^{-x}\left(-e^{-(x+X)}(1 + x + X)\right) dx \right.$$
$$\left. + \int_0^X x\, e^{-x} e^{(x-X)}(1 - x + X)dx + \int_X^\infty x\, e^{-x} e^{-(x-X)}(1 + x - X)dx\right]$$

which *Mathematica* performs as:

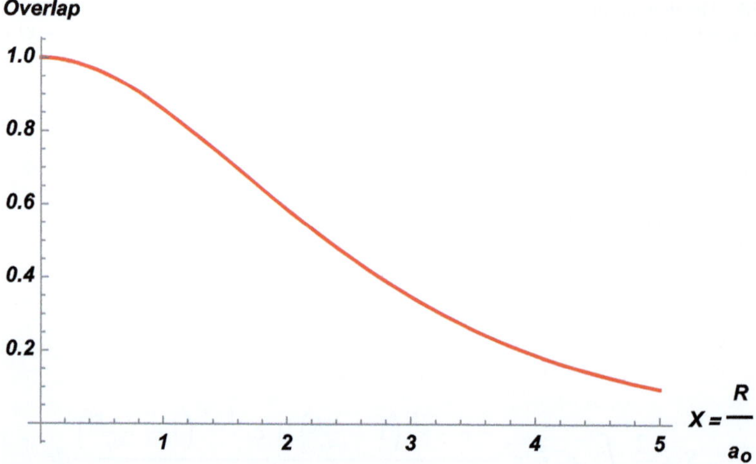

Fig. 5.7 The integral of the overlap $S(X)$ between the two hydrogen atom wave functions used in a linear combination to model the hydrogen molecule ion plotted versus X, the separation of the protons in units of the Bohr radius, a_o

$$
\begin{aligned}
2\,\Big(&\text{Integrate}\left[x^2 e^{-x}\,\frac{-e^{-(x+X)}(1+x+X)}{xX},\{x,\ 0,\ \infty\}\right] \\
&+ \text{Integrate}\left[x^2 e^{-x}\,\frac{e^{-(x-X)}(1+x-X)}{xX},\{x,\ X,\ \infty\}\right] \\
&+ \text{Integrate}\left[x^2 e^{-x}\,\frac{e^{(x-X)}(1-x+X)}{xX},\{x,\ 0,\ X\}\right]\Big)//\text{FullSimplify} \\
&= \tfrac{1}{3}e^{-X}(3+X(3+X))
\end{aligned}
$$

Giving the overlap function

$$
S(X) = e^{-X}\left(1 + X + X^2/3\right)
$$

Or, inserting the definition of $X = R/a_o$,

$$
S(R) = e^{-R/a_o}\left(1 + \frac{R}{a_o} + \frac{R^2}{3a_o{}^2}\right)
$$

we have the overlap integral expressed as a function of the distance between the nuclei. A plot of this (Fig. 5.7) shows that the overlap is maximal (unity) when the two nuclei are very close together and decreases to zero as they separate to infinity, as one would expect.

After first calculating this by the above method, I found that *Mathematica* can actually do this computation correctly in a very straightforward fashion, if one properly takes care of the separate integration domain boundaries using arguments to the `Simplify[]` and `FullSimplify[]` functions (I am indebted to M. S. Suzuki, http://bingweb.binghamton.edu/~suzuki/QuantumMechanicsII/8-4_Hydrogen_atom_H2+.pdf, for pointing this out):

This goes as follows:

$$\text{Clear}\left[\text{Global`} *\right];$$

$$p = \text{Sqrt}\left[r^2 - 2rR\text{Cos}[\theta] + R^2\right];$$

$$S1 = \text{Integrate}\left[2\frac{\text{Sin}[\theta]}{a_o^{\,3}}e^{-p/a_o}\{\theta, 0\pi\}\right]//\text{FullSimplify}\left[\#\left\{R>0, a_o>0, r>0(r-R)^2>0\right\}\right] \&$$

$$S1 = \frac{2\left(e^{-\frac{r+R}{a_o}}(-r-R-a_o) + e^{-\frac{\text{Abs}[r-R]}{a_o}}(\text{Abs}[r-R]+a_o)\right)}{rRa_o^2}$$

$$S2 = e^{-r/a_o}S1//\text{Simplify}[\#, r<R]\&$$

$$\frac{2e^{-\frac{r}{a_o}}\left(e^{-\frac{r+R}{a_o}}(-r-R-a_o) + e^{\frac{r-R}{a_o}}(-r+R+a_o)\right)}{rRa_o^2}$$

$$S3 = e^{-r/a_o}S1//\text{Simplify}[\#, r>R]\&$$

$$S3 = \frac{2e^{-\frac{r}{a_o}}\left(e^{-\frac{r+R}{a_o}}(-r-R-a_o) + e^{\frac{-r+R}{a_o}}(r-R+a_o)\right)}{rRa_o^2}$$

$$S4 = \text{Integrate}\left[r^2 S2, \{r, 0, R\}\right]//\text{Simplify}$$

$$S4 = \frac{e^{-\frac{3R}{a_o}}\left(3a_o(4R^2 + 5Ra_o + 2a_o^2) + e^{\frac{2R}{a_o}}(2R^3 + 6R^2a_o - 3a_o^2(R + 2a_o))\right)}{6Ra_o^2}$$

$$S5 = \text{Integrate}\left[r^2 S3, \{r, R, \infty\}\right]//\text{Simplify}[\#, a_o>0]\&$$

$$S5 = \frac{e^{-\frac{3R}{a_o}}\left(-4R^2 + a_o\left(-5R - 2a_o + e^{\frac{2R}{a_o}}(3R + 2a_o)\right)\right)}{2Ra_o}$$

$$S4 + S5//\text{FullSimplify}$$

which again gives the result for the overlap integral as

$$S(R) = \frac{e^{-\frac{R}{a_o}}\left(R^2 + 3a_0(R + a_0)\right)}{3a_o^2} = e^{-R/a_o}\left(1 + \frac{R}{a_o} + \frac{R^2}{3a_o{}^2}\right)$$

If we remember the solutions of the normalization equations, we found that normalization of the plus (*gerade*) and minus (*ungerade*) states differed, but was expressed in terms of the overlap as

$$\varphi_p(r) = \frac{1}{\sqrt{2(1+S)}}(|A\rangle + |B\rangle) = \frac{1}{\sqrt{2\left(1 + e^{-R/a_o}\left(1 + \frac{R}{a_o} + \frac{R^2}{3a_o{}^2}\right)\right)}}(|A\rangle + |B\rangle)$$

and

$$\varphi_m(r) = \frac{1}{\sqrt{2(1-S)}}(|A\rangle - |B\rangle) = \frac{1}{\sqrt{2\left(1 - e^{-R/a_o}\left(1 + \frac{R}{a_o} + \frac{R^2}{3a_o{}^2}\right)\right)}}(|A\rangle - |B\rangle)$$

These trial wave functions are orthogonal, as shown by computing the integrals $\langle \varphi_g | \varphi_u \rangle$ and $\langle \varphi_u | \varphi_g \rangle$

$$\langle \varphi_g | \varphi_u \rangle = c\,d\,((\langle A| + \langle B|)(|A\rangle - |B\rangle) = c\,d[\,\langle A|A\rangle - \langle A|B\rangle + \langle B|A\rangle - \langle B|B\rangle$$

$$\langle \varphi_g | \varphi_u \rangle = c\,d[1 - S + S - 1] = 0$$

and similarly for

$$\langle \varphi_g | \varphi_u \rangle = \langle \varphi_u | \varphi_g \rangle^* = 0.$$

The normalization is a function of the overlap between the wave functions and their parity. For the *gerade* state the normalization coefficient of the wave function has limits of

$$\lim_{X \to \infty} \frac{1}{\sqrt{2\left(1 + e^{-X}\left(1 + X + \frac{X^2}{3}\right)\right)}} = \frac{1}{\sqrt{2}}$$

and

$$\lim_{X \to 0} \frac{1}{\sqrt{2(1 + e^{-X}(1 + X + X^2/3))}} = \frac{1}{2}$$

while the *ungerade* state the normalization coefficient of the wave function has limits of

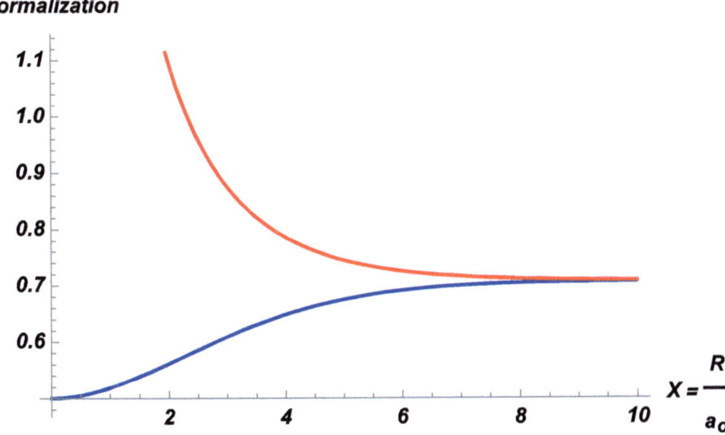

Fig. 5.8 The normalization coefficients for the *gerade* (blue) and *ungerade* (red) hydrogen atom wave functions used in a linear combination to model the hydrogen molecule ion plotted versus X, the separation of the protons in units of the Bohr radius, a_o

$$\lim_{X \to \infty} \frac{1}{\sqrt{2\left(1 - e^{-X}\left(1 + X + X^2/3\right)\right)}} = \frac{1}{\sqrt{2}}$$

and

$$\lim_{X \to 0} \frac{1}{\sqrt{2\left(1 - e^{-X}\left(1 + X + X^2/3\right)\right)}} = \infty$$

as a function of the distance between the protons. These relationships are displayed in Fig. 5.8. Note that the coefficients both tend to $1/\sqrt{2} = 0.707\ldots$, as the internuclear separation becomes infinite, and that this is the value of the coefficient if the wave functions had been orthogonal.

5.1.11   The Matrix Elements of the Hamiltonian

We have so far computed one of the three necessary integrals, that of the overlap of the atomic states. In order to continue, we need to compute the matrix elements of the Hamiltonian between the states; finding the energies of the states involves evaluating two more expressions, the Coulomb $K = \langle A| H| A \rangle$ and the exchange $J = \langle A |H| |B\rangle$ matrix elements. The techniques employed in the evaluation of the overlap integral will reappear here.

We begin by reminding the reader that the Hamiltonian used here is the Born–Oppenheimer approximate operator

$$H = \frac{-\hbar^2}{2m} \nabla^2 + \frac{e^2}{4\pi\varepsilon_0} \left[\frac{1}{R} - \frac{1}{r_A} - \frac{1}{r_B} \right]$$

And it is interpreted as giving the kinetic energy of the electron in the attractive Coulomb potential $(-1/r)$ of the two protons as well as a term $(1/R)$ reflecting the repulsion of the protons themselves. Or

$$H = H_o + \frac{e^2}{4\pi\varepsilon_0} \left[\frac{1}{R} - \frac{1}{r_B} \right]$$

where H_o gives the ground state energy of the hydrogen atom for each of the basis states.

Each of the ground state hydrogen atom wave functions $|i\rangle$, $i \in \{A, B\}$ obeys Schrödinger's equation with the hydrogen atom Hamiltonian, H_o and ground state energy E_o

$$H_o \mid i\rangle = \left[\frac{-\hbar^2}{2m} \nabla^2 - \frac{e^2}{4\pi\varepsilon_0 r_i} \right] \mid i\rangle = E_o \mid i\rangle$$

Or, more succinctly using Dirac notation, $H_o|A\rangle = E_o|A\rangle$ and $H_o|B\rangle = E_o|B\rangle$. I remind the reader that we can express the constant $\frac{e^2}{4\pi\varepsilon_0} = -2a_oE_o$

Since energies add linearly, we can separate the internuclear repulsion as the potential energy $E_p(R) = \frac{e^2}{4\pi\varepsilon_0 R} = \frac{-2a_oE_o}{R}$ from the interaction of the two protons, add it in later, and reduce the Hamiltonian to

$$H = \frac{-\hbar^2}{2m} \nabla^2 - \frac{e^2}{4\pi\varepsilon_0} \left[\frac{1}{r_A} + \frac{1}{r_B} \right] + E_p$$

$$H = \frac{-\hbar^2}{2m} \nabla^2 - \frac{e^2}{4\pi\varepsilon_0} \left[\frac{1}{r_A} + \frac{1}{r_B} \right] - \frac{2a_oE_o}{R}$$

The necessary energies from the definition of the expectation value of an operator are for the *gerade* state

$$E_g = \frac{\langle \varphi_g|H|\varphi_g \rangle}{\langle \varphi_g|\varphi_g \rangle} = d^2(\langle A|H|A\rangle + \langle A|H|B\rangle + \langle B|H|B\rangle + \langle B|H|A\rangle)$$

and in the same vein for the *ungerade* state

$$E_{\mathrm{u}} = \frac{\langle \varphi_{\mathrm{u}} | H | \varphi_{\mathrm{u}} \rangle}{\langle \varphi_{\mathrm{u}} | \varphi_{\mathrm{u}} \rangle} = c^2 (\langle A|H|A \rangle - \langle A|H|B \rangle + \langle B|H|B \rangle - \langle B|H|A \rangle)$$

5.1.12 Parity Conservation in the Electromagnetic Interaction

We know that the electromagnetic interaction conserves parity so that the wave functions must be eigenstates of the parity operator. The parity operator is also Hermitian so that $\Pi^{\dagger} = \Pi = \Pi^{-1}$ and it has real eigenvalues. If we move the origin of the coordinate system to a point equidistant between the two protons, at the center of mass, then our trial wave functions become

$$| A' \rangle = \frac{1}{\sqrt{\pi a_o{}^3}} e^{-\left(r+\frac{R}{2}\right)}$$

and

$$| B' \rangle = \frac{1}{\sqrt{\pi a_o{}^3}} e^{-\left(r-\frac{R}{2}\right)}$$

The parity operator Π acting on a function $f(R)$ gives $\Pi f(R) = f(-R)$ so that

$$\Pi \, | A' \rangle = \Pi \frac{1}{\sqrt{\pi a_o{}^3}} e^{-\left(r+\frac{R}{2}\right)} = \frac{1}{\sqrt{\pi a_o{}^3}} e^{-\left(r-\frac{R}{2}\right)} = \, | B' \rangle$$

and

$$\Pi \, | B' \rangle = \Pi \frac{1}{\sqrt{\pi a_o{}^3}} e^{-\left(r-\frac{R}{2}\right)} = \frac{1}{\sqrt{\pi a_o{}^3}} e^{-\left(r+\frac{R}{2}\right)} = \, | A' \rangle$$

and the parity operator acting on the expectation value of the Hamiltonian is

$$\langle A|H|B \rangle = \langle A|\Pi H \Pi|B \rangle = \langle B|H|A \rangle$$

and

$$\langle A|H|A \rangle = \langle A|\Pi H \Pi|A \rangle = \langle B|H|B \rangle$$

Therefore, we have just two types of matrix elements in the expectation value of the Hamiltonian, and these are referred to as Coulomb $\langle A|H|A \rangle = \langle B|H|B \rangle$ and exchange $\langle A|H|B \rangle = \langle B|H|A \rangle$ matrix elements.

$$E_g = \frac{\langle \varphi_g | H | \varphi_g \rangle}{\langle \varphi_g | \varphi_g \rangle} = 2d^2 (\langle A | H | A \rangle + \langle A | H | B \rangle)$$

$$E_g = \frac{\langle \varphi_g | H | \varphi_g \rangle}{\langle \varphi_g | \varphi_g \rangle} = \frac{2(\langle A | H | A \rangle + \langle A | H | B \rangle)}{2(S+1)}$$

$$E_g = \frac{(\langle A | H | A \rangle + \langle A | H | B \rangle)}{(S+1)} \quad \frac{K+J}{=1+S(R)}$$

Since $K = \langle A | H | A \rangle = \langle B | H | B \rangle$ and $J = \langle A | H | B \rangle = \langle B | H | A \rangle$ and in the same vein for the *ungerade* state

$$E_u = \frac{\langle A | H | A \rangle - \langle A | H | B \rangle}{1 - S(R)} = \frac{K - J}{1 - S(R)}$$

Therefore, the expectation of the Hamiltonian for the *gerade* state can be simplified as

$$\langle H \rangle_g = E_g = E_o + E_p - 2d^2 \frac{e^2}{4\pi\varepsilon_0} \left(\left\langle A \left| \frac{1}{r_A} \right| B \right\rangle + \left\langle A \left| \frac{1}{r_B} \right| A \right\rangle \right)$$

while for the *ungerade* state, the expectation value of the Hamiltonian reduces to

$$\langle H \rangle_u = E_u = E_o + E_p - 2c^2 \frac{e^2}{4\pi\varepsilon_0} \left(\left\langle A \left| \frac{1}{r_A} \right| B \right\rangle - \left\langle A \left| \frac{1}{r_B} \right| A \right\rangle \right)$$

These equations contain the constant $-\frac{e^2}{4\pi\varepsilon_0}$ that has the SI units of Joule-meter, which when multiplied by $\frac{1}{a_0}$ would give an energy in Joules. However, we can rewrite this constant in terms of E_o by noting that $E_o = -\frac{e^2}{4\pi\varepsilon_0} \frac{1}{2a_0}$, or we have an expression for the constant as $-2a_0 E_o = \frac{e^2}{4\pi\varepsilon_0}$ and this enables us to write the energies as

$$E_g = E_o + E_p + 4d^2 E_o a_0 \left(\left\langle A \left| \frac{1}{r_A} \right| B \right\rangle + \left\langle A \left| \frac{1}{r_B} \right| A \right\rangle \right)$$

$$E_g = E_o \left[1 + 4d^2 a_0 \left(\left\langle A \left| \frac{1}{r_A} \right| B \right\rangle + \left\langle A | \frac{1}{r_B} | A \right\rangle \right) \right] + E_p$$

with $d = \frac{1}{\sqrt{2(1+S)}}$ and $c = \frac{1}{\sqrt{2(1-S)}}$

$$E_g = E_o \left[1 + \frac{2}{(1+S)} a_0 \left(\left\langle A \left| \frac{1}{r_A} \right| B \right\rangle + \left\langle A \left| \frac{1}{r_B} \right| A \right\rangle \right) \right] + E_p$$

$$E_g(R) = E_o \left[1 + \frac{2a_0 (j(R) + k(R))}{(1 + S(R))} \right] + E_p(R)$$

and

$$E_u(R) = E_o \left[1 + \frac{2a_0 (j(R) - k(R))}{(1 - S(R))} \right] + E_p(R)$$

where the Coulomb integral is $k(R) = \langle A \mid \frac{1}{r_B} \mid A \rangle$ and the exchange integral is $j(R) = \langle A \left| \frac{1}{r_A} \right| B \rangle$. These integrals have units of m^{-1}, so when multiplied by the length a_0 become dimensionless coefficients that scale the energies. We can also express $E_p(R) = \frac{e^2}{4\pi\varepsilon_0} \frac{1}{R}$ in terms of E_o as $E_p(R) = \frac{-2a_0 E_o}{R}$ giving

$$E_g(R) = E_o \left[1 + 2a_0 \left(\frac{j(R) + k(R)}{1 + S(R)} - \frac{1}{R} \right) \right]$$

and

$$E_u(R) = E_o \left[1 + 2a_0 \left(\frac{j(R) - k(R)}{1 - S(R)} - \frac{1}{R} \right) \right]$$

5.1.13   Calculation of the Energies of the *gerade* and *ungerade* States

The expectation values of the energies E_g and E_u of the *gerade* and *ungerade* states from the variational principle are given by

$$E_g = \frac{(\langle A|H|A\rangle + \langle A|H|B\rangle)}{1 + S(R)}$$

and

$$E_u = \frac{\langle A|H|A\rangle - \langle A|H|B\rangle}{1 - S(R)}.$$

In order to calculate the energies of the states, we need to find the action of the molecular Hamiltonian on our trial *basis* wave functions. This can be computed as

$$H \mid A\rangle = \left(\frac{-\hbar^2}{2m} \nabla^2 - \frac{e^2}{4\pi\varepsilon_0} \left[\frac{1}{r_A} + \frac{1}{r_B} \right] + E_p \right) \mid A\rangle$$

$$H \mid A\rangle = \left(H_o + \frac{e^2}{4\pi\varepsilon_0} \left[\frac{1}{R} - \frac{1}{r_B} \right] \right) \mid A\rangle$$

$$H \mid A\rangle = E_o \mid A\rangle + \frac{e^2}{4\pi\varepsilon_0} \frac{1}{R} \mid A\rangle - \frac{e^2}{4\pi\varepsilon_0} \frac{1}{r_B} \mid A\rangle$$

$$H \mid A\rangle = E_o \mid A\rangle + E_p \mid A\rangle - \frac{e^2}{4\pi\varepsilon_0} \frac{1}{r_B} \mid A\rangle$$

and similarly for $|B\rangle$:

$$H \mid B\rangle = E_o \mid B\rangle + E_p \mid B\rangle - \frac{e^2}{4\pi\varepsilon_0} \frac{1}{r_A} \mid B\rangle$$

Then the matrix elements of the Hamiltonian are

$$K = \langle A \mid H \mid A\rangle = E_o \langle A \mid A\rangle + E_p \langle A \mid A\rangle - \frac{e^2}{4\pi\varepsilon_0} \langle A \mid \frac{1}{r_B} \mid A\rangle$$

$$K = E_o + E_p - \frac{e^2}{4\pi\varepsilon_0} \langle A \mid \frac{1}{r_B} \mid A\rangle$$

$$K = E_o + E_p - \frac{e^2}{4\pi\varepsilon_0} k(R)$$

Since $\langle A \mid A\rangle = 1$ and

$$J = \langle A \mid H \mid B\rangle = E_o \langle A \mid B\rangle + E_p \langle A \mid B\rangle - \frac{e^2}{4\pi\varepsilon_0} \langle A \mid \frac{1}{r_A} \mid B\rangle$$

$$J = S(E_o + E_p) - \frac{e^2}{4\pi\varepsilon_0} j(R)$$

where $S = \langle A \mid B\rangle$ and we need to calculate the Coulomb integral $k(R) = \langle A \mid \frac{1}{r_B} \mid A\rangle$ and the exchange integral $j(R) = \langle A \mid \frac{1}{r_A} \mid B\rangle$. Further progress then depends upon the calculation of the Coulomb and exchange integrals.

5.1.14   The Coulomb and Exchange Integrals

We now need to calculate the two integrals, the Coulomb integral $k = \langle A \mid \frac{1}{r_B} \mid A\rangle$ and the exchange integral $j = \langle A \mid \frac{1}{r_A} \mid B\rangle$. The energies are given by

$$\text{gerade:} \qquad \langle H \rangle = E_{\mathrm{o}} + E_{\mathrm{p}} - 2\,d^2\,\frac{e^2}{4\pi\varepsilon_0}\,(j+k)$$

$$\text{ungerade:} \qquad \langle H \rangle = E_{\mathrm{o}} + E_{\mathrm{p}} - 2\,c^2\,\frac{e^2}{4\pi\varepsilon_0}\,(j-k)$$

The calculations of these integrals will resemble those done above for the overlap integral, S.

5.1.14.1   The Coulomb Integral

The Coulomb integral is given by

$$k(R) = \left\langle A \left| \frac{1}{r_{\mathrm{B}}} \right| A \right\rangle = \left\langle B \left| \frac{1}{r_{\mathrm{A}}} \right| B \right\rangle = \frac{2}{a_{\mathrm{o}}^{3}} \int_0^\infty \frac{e^{-\frac{2p}{a_{\mathrm{o}}}}\,r^2}{r}\,dr \int_0^\pi \sin\theta\,d\theta = \frac{2}{a_{\mathrm{o}}^{3}} \int_0^\infty r\,e^{-\frac{2p}{a_{\mathrm{o}}}}\,dr \int_0^\pi \sin\theta\,d\theta$$

where $p = \sqrt{r^2 + R^2 - 2R\,r\,\cos\theta}$

Mathematica can be used to compute this integral by the following means:

$$\mathrm{Clear}\big[\mathrm{Global}^{`}\!*\big];$$

$$p = \mathrm{Sqrt}\big[r^2 - 2rR\cos[\theta] + R^2\big];$$

$$k1 = \mathrm{Integrate}\left[2\pi\,\frac{\mathrm{Sin}[\theta]}{a_{\mathrm{o}}^{3}\pi}\,e^{-2p/a_{\mathrm{o}}}\,\{\theta, 0\pi\}\right]//\mathrm{FullSimplify}\left[\#\big\{R>0, a_{\mathrm{o}}>0, r>0(r-R)^2>0\big\}\right]\,\&$$

$$k1 = \frac{e^{-\frac{2(r+R)}{a_{\mathrm{o}}}}(-a_{\mathrm{o}} - 2(r+R)) + e^{-\frac{2\mathrm{Abs}[r-R]}{a_{\mathrm{o}}}}(a_{\mathrm{o}} + 2\mathrm{Abs}[r-R])}{2a_{\mathrm{o}}^{2}rR}$$

$$k2 = r\,k1//\mathrm{Simplify}[\#, r<R]\,\&$$

$$k2 = \frac{e^{-\frac{2(r+R)}{a_{\mathrm{o}}}}(-2(r+R) - a_{\mathrm{o}}) + e^{\frac{2(r-R)}{a_{\mathrm{o}}}}(-2r + 2R + a_{\mathrm{o}})}{2Ra_{\mathrm{o}}^{2}}$$

$$k3 = r\,k1//\mathrm{Simplify}[\#, r>R]\,\&$$

$$k3 = \frac{e^{-\frac{2(r+R)}{a_{\mathrm{o}}}}(-2(r+R) - a_{\mathrm{o}}) + e^{-\frac{2(r-R)}{a_{\mathrm{o}}}}(2r - 2R + a_{\mathrm{o}})}{2Ra_{\mathrm{o}}^{2}}$$

$$k4 = \mathrm{Integrate}[\,k2, \{r, 0, R\}]//\mathrm{Simplify}[\#, \{r>0, R>0, a_{\mathrm{o}}>0\}]\,\&$$

$$k4 = \frac{e^{-\frac{4R}{a_{\mathrm{o}}}}\left(-1 + e^{\frac{2R}{a_{\mathrm{o}}}}\right)\left(-2R + \left(-1 + e^{\frac{2R}{a_{\mathrm{o}}}}\right)a_{\mathrm{o}}\right)}{2Ra_{\mathrm{o}}}$$

$$k5 = \mathrm{Integrate}[\,k3, \{r, R, \infty\}]//\mathrm{Simplify}[\#, a_{\mathrm{o}}>0]\,\&$$

$$k5 = \frac{a_0 - e^{-\frac{4R}{a_0}}(2R + a_0)}{2Ra_0}$$

$$k4 + k5 // \text{FullSimplify}$$

$$k(R) = \frac{1}{R}\left(1 - e^{-\frac{2R}{a_0}}(1 + R/a_0)\right)$$

Note that in the SI system the units of $k(R)$ are m^{-1}.

5.1.14.2   The Exchange Integral

Now, we turn to the calculation of the exchange integral, so called because it computes the energy when electron states are interchanged. This is written as

$$j = \langle A | \frac{1}{r_A} | B \rangle = \frac{2}{a_0{}^3} \int_0^\infty \frac{e^{-\frac{r}{a_0}} r^2}{r} dr \int_0^\pi e^{-\frac{p}{a_0}} \sin\theta \, d\theta = \frac{2}{a_0{}^3} \int_0^\infty r \, e^{-\frac{r}{a_0}} dr \int_0^\pi e^{-\frac{2p}{a_0}} \sin\theta \, d\theta,$$

which is done in *Mathematica* as

$$\text{Clear}\left[\text{Global`}*\right];$$

$$p = \text{Sqrt}\left[r^2 - 2rR\cos[\theta] + R^2\right];$$

$$j1 = \text{Integrate}\left[2\frac{\text{Sin}[\theta]}{a_0{}^3} e^{-p/a_0} \{\theta, 0\pi\}\right] // \text{FullSimplify}\left[\#\left\{R > 0, a_0 > 0, r > 0(r - R)^2 > 0\right\}\right] \&$$

$$j1 = \frac{2\left(e^{-\frac{r+R}{a_0}}(-r - R - a_0) + e^{-\frac{\text{Abs}[r-R]}{a_0}}(\text{Abs}[r - R] + a_0)\right)}{rRa_0^2}$$

$$j2 = re^{-r/a_0} J1 // \text{Simplify}[\#, r < R] \&$$

$$j2 = \frac{2e^{-\frac{r}{a_0}}\left(e^{-\frac{r+R}{a_0}}(-r - R - a_0) + e^{\frac{r-R}{a_0}}(-r + R + a_0)\right)}{Ra_0^2}$$

$$j3 = re^{-r/a_0} j1 // \text{Simplify}[\#, r > R] \&$$

$$j3 = \frac{2e^{-\frac{r}{a_0}}\left(e^{-\frac{r+R}{a_0}}(-r - R - a_0) + e^{\frac{-r+R}{a_0}}(r - R + a_0)\right)}{Ra_0^2}$$

$$J4 = \text{Integrate}[J2, \{r, 0, R\}] // \text{Simplify}[\#, \{r > 0, R > 0, a_0 > 0\}] \&$$

$$j4 = \frac{e^{-\frac{3R}{a_0}}\left(a_0(4R + 3a_0) + e^{\frac{2R}{a_0}}(2R^2 + 2Ra_0 - 3a_0^2)\right)}{2Ra_0^2}$$

$$j5 = \text{Integrate}[j3, \{r, R, \infty\}]//\text{Simplify}[\#, a_\text{o} > 0]\&$$

$$j5 = \frac{e^{-\frac{3R}{a_\text{o}}}\left(-4R + 3\left(-1 + e^{\frac{2R}{a_\text{o}}}\right)a_\text{o}\right)}{2Ra_\text{o}}$$

$$j4 + j5//\text{FullSimplify}$$

which gives for the *exchange* integral

$$j(R) = \frac{e^{-\frac{R}{a_\text{o}}}(a_\text{o} + R)}{a_\text{o}^2}$$

5.1.15 *Energies of the States*

From the variational principle, the energies E_p and E_m of the *gerade* and *ungerade* states are given by

$$E_\text{g}(R) = \frac{\langle\varphi_\text{p}|H|\varphi_\text{p}\rangle}{\langle\varphi_\text{p}|\varphi_\text{p}\rangle} = \frac{\langle A|H|A\rangle + \langle A|H|B\rangle}{1 + S(R)} = \frac{K + J}{1 + S}$$

and

$$E_\text{u}(R) = \frac{\langle\varphi_\text{m}|H|\varphi_\text{m}\rangle}{\langle\varphi_\text{m}|\varphi_\text{m}\rangle} = \frac{\langle A|H|A\rangle - \langle A|H|B\rangle}{1 - S(R)} = \frac{K - J}{1 - S}$$

The energies of the *gerade* and *ungerade* states are then

$$E_\text{g} = \frac{E_\text{o} + E_\text{p} - \frac{e^2}{4\pi\varepsilon_0}\frac{1}{R}\left(1 - e^{-\frac{2R}{a_\text{o}}}(1 + R/a_\text{o})\right) + S(E_\text{o} + E_\text{p}) - \frac{e^2}{4\pi\varepsilon_0}\left(\frac{e^{-\frac{R}{a_\text{o}}}(a_\text{o}+R)}{a_\text{o}^2}\right)}{(1 + S)}$$

$$= E_\text{o} + E_\text{p} - \frac{e^2}{4\pi\varepsilon_0(1 + S)}\left[\frac{1}{R}\left(1 - e^{-\frac{2R}{a_\text{o}}}(1 + R/a_\text{o})\right) + \left(\frac{e^{-\frac{R}{a_\text{o}}}(a_\text{o} + R)}{a_\text{o}^2}\right)\right]$$

and

$$E_\text{u} = \frac{E_\text{o} + E_\text{p} - \frac{e^2}{4\pi\varepsilon_0}\frac{1}{R}\left(1 - e^{-\frac{2R}{a_\text{o}}}(1 + R/a_\text{o})\right) - S(E_\text{o} + E_\text{p}) - \frac{e^2}{4\pi\varepsilon_0}\left(\frac{e^{-\frac{R}{a_\text{o}}}(a_\text{o}+R)}{a_\text{o}^2}\right)}{1 - S(R)}$$

$$E_{\mathrm{u}} = E_{\mathrm{o}} + E_{\mathrm{p}} - \frac{e^2}{4\pi\varepsilon_0(1 - S(R))} \left[\frac{1}{R}\left(1 - e^{-\frac{2R}{a_0}}(1 + R/a_0)\right) - \left(\frac{e^{-\frac{R}{a_0}}(a_0 + R)}{a_0{}^2}\right) \right]$$

In order to evaluate these expressions, we need to confront the units. For this purpose, in SI units, let us insert more factors of $\frac{e^2}{4\pi\varepsilon_0}$ from their definitions as E_{o} and E_{p} into the energy expressions for

$$\text{gerade:} \qquad E_{\mathrm{g}} = E_{\mathrm{o}} + E_{\mathrm{p}} - 2\,d^2\,\frac{e^2}{4\pi\varepsilon_0}\,(j + k)$$

and

$$\text{ungerade:} \qquad E_{\mathrm{u}} = E_{\mathrm{o}} + E_{\mathrm{p}} - 2\,c^2\,\frac{e^2}{4\pi\varepsilon_0}\,(j - k)$$

to give

$$E_{\mathrm{g}} = -\frac{e^2}{4\pi\varepsilon_0 2a_0} + \frac{e^2}{4\pi\varepsilon_0 R} - 2\,d^2\,\frac{e^2}{4\pi\varepsilon_0}\,(j + k)$$

$$E_{\mathrm{g}} = \frac{e^2}{4\pi\varepsilon_0}\left[-\frac{1}{2a_0} + \frac{1}{R} - 2\,d^2(j + k)\right]$$

$$E_{\mathrm{g}} = \frac{e^2}{4\pi\varepsilon_0}\left[-\frac{1}{2a_0} + \frac{1}{R} - \frac{(j + k)}{1 + S}\right]$$

$$E_{\mathrm{g}} = -2a_0 E_{\mathrm{o}}\left[-\frac{1}{2a_0} + \frac{1}{R} - \frac{(j + k)}{1 + S}\right]$$

$$E_{\mathrm{g}} = E_{\mathrm{o}}\left[1 + 2a_0\left(\frac{(k + j)}{1 + S} - \frac{1}{R}\right)\right]$$

and in a similar vein we find for the energy of the *ungerade* state, the value of

$$E_{\mathrm{u}} = E_{\mathrm{o}}\left[1 + 2a_0\left(\frac{(k - j)}{1 - S} - \frac{1}{R}\right)\right]$$

One finds from plots (Fig. 5.9) of the energies of the two states as a function of the interproton separation R that there are two different solutions for the energy. As we expected from our group theoretical analysis, only the *gerade* state displays a minimum in the energy and forms a bound state, while the *ungerade* state has no energy minimum below the energy of the isolated hydrogen atom, and is unbound, corresponding to the first excited state. The prediction from this simple linear combination model is that the equilibrium interproton separation occurs at $R = 2.49a_0$, (132 pm), while the measured value is $R = 106$ pm, which is the

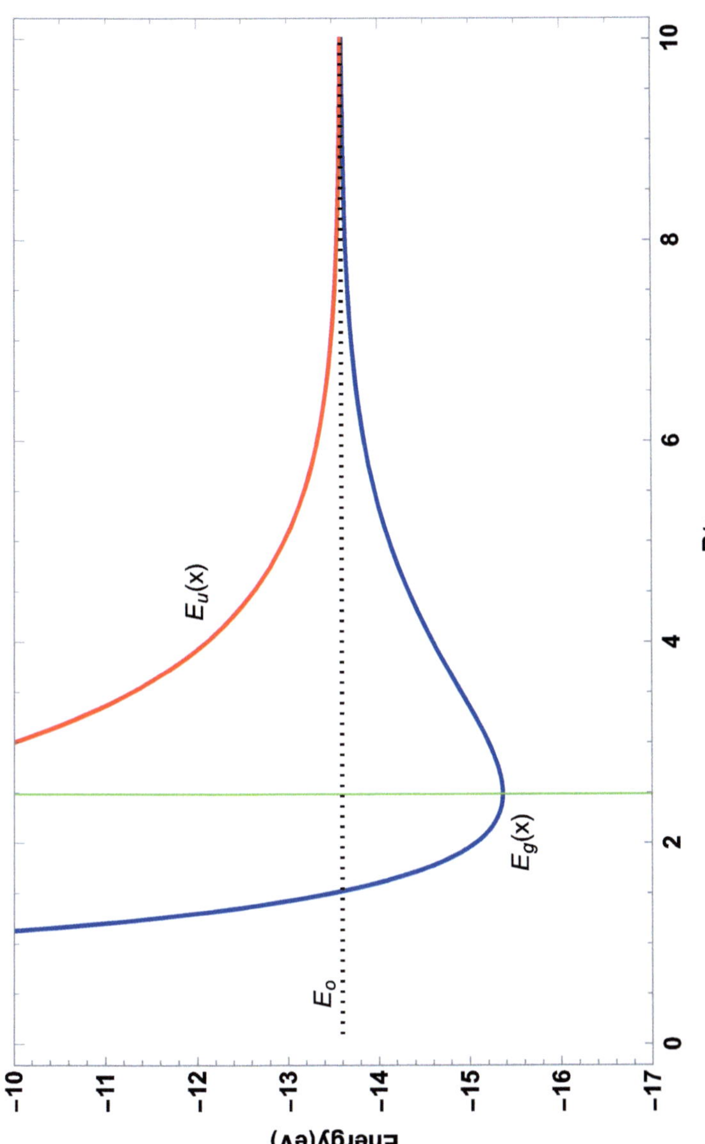

Fig. 5.9 The energy of the *gerade* (E_g) (red) and *ungerade* (E_u) (blue) states as a function of the internuclear separation in units of the Bohr radius for the hydrogen molecule ion. Only the *gerade* state shows an energy minimum, indicating the formation of a molecular bound state. The energy of an isolated hydrogen atom is E_o. The green line indicates the equilibrium distance ($x = 2.49a_o$) between the protons in this model

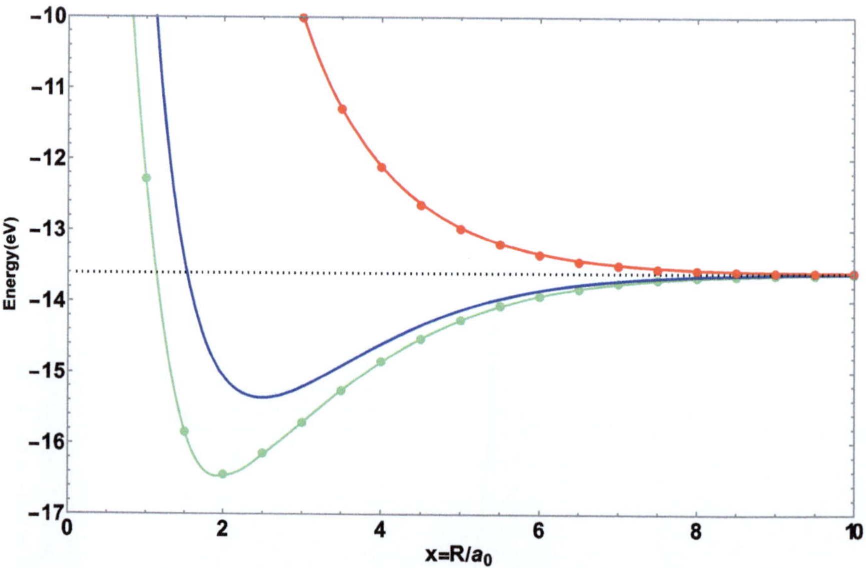

Fig. 5.10 The energies of the hydrogen molecule ion computed using the linear combination of 1s states for the E_g (blue line) and E_u (red line) wave functions compared with an exact calculation (Peek 1965) shown as the green points and curves and red points

value of $R = 2a_o$ that one might naively expect and which is found from an exact calculation (Peek 1965). The energy of the ground state from our approximate wave function is -15.36 eV, compared with the ground state of the hydrogen atom of -13.6 eV, indicating that the hydrogen molecule ion is -1.76 eV more stable than that of an isolated hydrogen atom. The measured ground state energy is found to be -16.393 eV, which is 2.793 eV below that of an isolated H atom, while an exact calculation gives -16.45 eV (Fig. 5.10). This is congruent with our use of a variational calculation that always gives energies greater than or equal to the actual ground state energies. One notes from Fig. 5.10 that the agreement between our simple approximate calculation agrees very well with the actual energy values for the *ungerade* state, but that the agreement is poorer for the *gerade* state (Feinberg et al. 1970).

5.2 The H_2^+ Ion as a Model for the Molecular Binding Potential

Molecules form when the energy of the molecular ground state is lower than that for the isolated atoms. A popular model for the potential energy of interaction between atoms and molecules is that of the Lennard–Jones potential. We have compared the potential energy of the hydrogen molecule ion with the Lennard–Jones potential

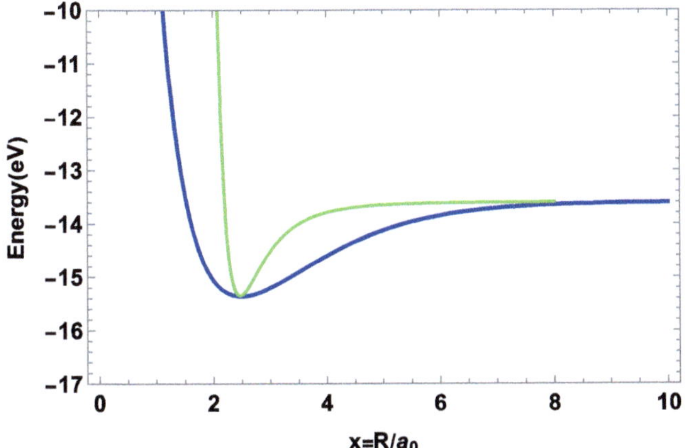

Fig. 5.11   A comparison between the potential energy curve for the hydrogen molecule ion (blue) and that of the Lennard–Jones potential (green) shows that the latter is more strongly peaked at its minimum

(Fig. 5.11). While the Lennard–Jones potential differs in detail from the results of this simple calculation using a linear combination of atomic states, the shape of the curve agrees with the idea that a hydrogen atom encountering a proton would self-assemble into a hydrogen molecule ion. The reverse process, that of an H_2 molecule losing an electron and becoming an ion, is actually quite common in interstellar space, where the collisions of cosmic rays (essentially high energy protons, among smaller numbers of heavier nuclei) with molecular hydrogen (H_2) produces hydrogen molecule ions by the reaction:

$$p^+ + H_2 \rightarrow H_2^+ + e^- + p^+ = H_2^+ + H,$$

leaving a hydrogen molecule ion and a hydrogen atom in the final state.

A final caveat with respect to this calculation of the energies of the hydrogen molecule ion is that it involves only a single electron, even though the predictions of the energies and equilibrium distances are not too far off from those measured. It does show several features that give insight into how the separate atoms in a molecule are bound together and these are present in every calculation of the energies of any molecule. The one important feature in multi-electron molecules that is missing from this single electron treatment, however, is that of the correlations among electrons, in particular the fact that the spin-statistics theorem (Pauli principle) requires that the wave function describing more than one electron must be antisymmetric under their interchange. Electrons are Fermions, i.e., half-integral spin particles and as such more than one electron cannot be in the same quantum state. This means that the wave function must include spin as well as spatial parts.

Without the Pauli principle, all electrons could reside in the ground state and the periodic table would not exist, and with it life would be impossible.

5.3   The Hydrogen Molecule

Let us next turn to the first real system that involves more than one electron, that of the hydrogen molecule. Here, we will have to confront, for the first time, the problem of writing a Hamiltonian and finding a wave function for more than one electron. The full Hamiltonian is

$$H = \frac{-\hbar^2}{2m_e} \sum_{i=1}^{2} \nabla_{ei}^{2} - \frac{\hbar^2}{2m_p} \sum_{i=1}^{2} \nabla_{i}^{2} + \frac{e^2}{4\pi\varepsilon_0} \left(\frac{1}{R} - \frac{1}{r_{A1}} - \frac{1}{r_{B1}} + \frac{1}{r_{12}} - \frac{1}{r_{A2}} - \frac{1}{r_{B2}} \right),$$

which represents the kinetic energy of the protons and electrons as well as their mutual Coulomb attraction and repulsion (Fig. 5.12). As we have seen with the hydrogen molecule ion, the solutions for the wave functions for multicenter molecules rapidly become so complex even within the Born–Oppenheimer scheme that we require additional approximations in order to make progress. With this in mind, we will drop the term in the Hamiltonian that gives the kinetic energy of the protons, $-\frac{\hbar^2}{2m_p} \sum_{i=1}^{2} \nabla_{i}^{2}$, and write the Hamiltonian as the sum of single electron terms with an interaction potential, $V_{12} = +\frac{e^2}{4\pi\varepsilon_0} \frac{1}{r_{12}}$, that is a function of the interelectron separation. We can then rewrite the Hamiltonian as

$$H = H_A(r_{A1}, r_{A2}) + H_B(r_{B1}, r_{B2}) + E_p + V_{12}$$

and the individual Hamiltonians H_A and H_B are functions only of their respective electron's coordinates:

$$H_A(r_{A1}, r_{A2}) = \frac{-\hbar^2}{2m} \nabla_A^2 - \frac{e^2}{4\pi\varepsilon_0} \left(\frac{1}{r_{A1}} + \frac{1}{r_{A2}} \right)$$

and

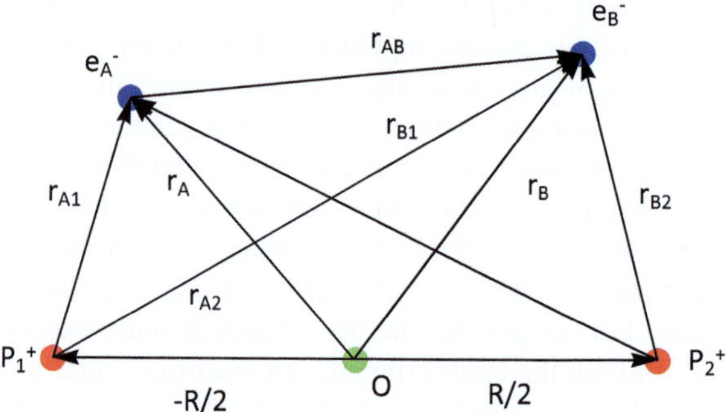

Fig. 5.12 Diagram defining the coordinates used in the discussion of the hydrogen molecule, H$_2$. The two electrons are denoted in blue, while the two protons are denoted in red. The distance between the two protons is R and the origin is green

$$H_B(r_{B1}, r_{B2}) = \frac{-\hbar^2}{2m} \nabla_B^{\ 2} - \frac{e^2}{4\pi\varepsilon_0}\left(\frac{1}{r_{B1}} + \frac{1}{r_{B2}}\right).$$

Here, E_p is the familiar $E_p = \frac{e^2}{4\pi\varepsilon_0 R}$ repulsive potential between the protons.

The properties of the hydrogen molecule cannot depend on which electron is labeled 1 and which is labeled 2: we could interchange the electrons and the protons and the molecule would remain invariant in all its properties because these Fermions are indistinguishable particles and their interchange is a permutation symmetry.

If we ignore, for the time being, the quantum states of the protons, the wave function for the two electrons can be symmetric or antisymmetric with respect to the interchange of the two identical electrons. According to the Pauli principle, the wave functions for two identical Fermions need to be antisymmetric with respect to the interchange of the two identical electrons. The wave functions can be written as products of spin and spatial components. The approximate "valence bond" wave function $\psi(A, B, \sigma)$ for this two-electron, two-proton problem is then written as a product of single electron states as $\psi(A, B, \sigma) = |AB\rangle|\sigma\rangle$, where A and B are shorthand for $\varphi(r_A)$ and $\varphi(r_B)$, respectively, r_A and r_B are the coordinates of the electrons, and $|\sigma\rangle$ is the spin part. Either the space or spin part can each be either symmetric or antisymmetric, but the product must be antisymmetric, so if the spatial part is symmetric, the spin component must be antisymmetric, i.e., the spins must be anti-parallel and the total state is then antisymmetric. This is reminiscent of the *gerade* and *ungerade* states of the hydrogen molecule ion.

For the ground state wave function of the hydrogen molecule, we take the product of 1s states centered on each proton (in an obvious notation):

$$\psi = (|A_1 B_2\rangle + |B_1 A_2\rangle)|\sigma\rangle$$

in which the spatial part is symmetric, and for the spin part we use the antisymmetric expression $|\sigma\rangle = \frac{1}{\sqrt{2}}(|\alpha_1\rangle|\beta_2\rangle - |\beta_1\rangle|\alpha_2\rangle)$. We also note that the spin part is already normalized, $\langle\sigma|\sigma\rangle = 1$, but the spatial part requires normalization. For this we need to compute the left-hand side of $\langle\psi|\psi\rangle = 1$.

$$\langle\psi|\psi\rangle = ((\langle A_1 B_2| + \langle B_1 A_2|)(|A_1 B_2\rangle + |B_1 A_2\rangle))$$
$$\langle\psi|\psi\rangle = ((\langle A_1 B_2|A_1 B_2\rangle + \langle B_1 A_2|A_1 B_2\rangle + \langle A_1 B_2|B_1 A_2\rangle + \langle B_1 A_2|B_1 A_2\rangle))$$

In this expression, we only have two different integrals, one equality is from the normalization of the basis functions

$$\langle A_1 B_2|A_1 B_2\rangle = \langle A_1|A_1\rangle\langle B_2|B_2\rangle = \langle B_1 A_2|B_1 A_2\rangle = \langle A_2|A_2\rangle\langle B_1|B_1\rangle = 1$$

and the other equality arises from the fact that the basis functions are symmetric under interchange of the nuclei and that electrons are indistinguishable so that the overlap integrals obey

$$S^2 = \langle A_1 B_2 | B_1 A_2 \rangle = \langle B_1 A_2 | A_1 B_2 \rangle$$

The normalization constant is therefore

$$\langle \psi | \psi \rangle = 2 + 2S^2.$$

The normalized wave function is then

$$|\psi\rangle = \frac{1}{\sqrt{2 + 2S^2}} (|A_1\rangle |B_2\rangle + |B_1\rangle |A_2\rangle) |\sigma\rangle.$$

5.3.1   The Matrix Elements of the Hamiltonian

The energies of the quantum states are found as the expectation values of the Hamiltonian operator. We can eliminate the spin part of the energies because the Hamiltonian is spin independent

$$\langle \psi(A, B, \sigma) | H | \psi(A, B, \sigma) \rangle = \langle \psi(A, B) | H | \psi(A, B) \rangle \langle \sigma | \sigma \rangle$$

$$\langle \psi(A, B, \sigma) | H | \psi(A, B, \sigma) \rangle = \langle \psi(A, B) | H | \psi(A, B) \rangle$$

The spin-statistics theorem is at work here in that the spatial wave function is symmetric and that is only possible with an antisymmetric spin state. Remembering that the Hamiltonian is

$$H = H_A + H_B + E_p + V_{12}$$

We will treat each term in succession to find its contribution of the energy of the hydrogen molecule. The energy is given by the matrix element

$$E = \langle \psi(A, B) | H | \psi(A, B) \rangle$$

$$E = \frac{1}{(1 + S^2)} [(\langle A_1 B_2 | + \langle B_1 A_2 |)(H_A + H_B + E_p + V_{12})(|A_1 B_2\rangle + |B_1 A_2\rangle)]$$

Multiplying out and collecting the myriad terms in this expression, and noting that the result cannot depend on which electron is labeled 1 or 2 (so that we can drop the indices on A and B) gives the energy as

$$E = \frac{h_{AA} + h_{BB} + S(h_{AB} + h_{BA}) + \langle AA|BB \rangle + \langle AB|AB \rangle}{(1 + S^2)} + E_p$$

and we have used the abbreviations

$$h_{AA} = E_A + \frac{e^2}{4\pi\varepsilon_0} \left\langle A \left| \frac{1}{r_B} \right| A \right\rangle \text{ and } h_{BB} = E_B + \frac{e^2}{4\pi\varepsilon_0} \left\langle B \left| \frac{1}{r_A} \right| B \right\rangle$$

along with

$$h_{BA} = E_A S + \frac{e^2}{4\pi\varepsilon_0} \left\langle A \left| \frac{1}{r_B} \right| B \right\rangle \text{ and } h_{AB} = E_B S + \frac{e^2}{4\pi\varepsilon_0} \left\langle B \left| \frac{1}{r_A} \right| A \right\rangle .$$

where E_A and E_B are the individual ground state energies of the separate hydrogen atoms and the second terms reflect the interatomic ground state energy. Writing out the expression for the ground state energy in detail, we have that

$$E = E_A + E_B + E_p + \frac{\frac{e^2}{4\pi\varepsilon_0}\left[\left\langle A \left|\frac{1}{r_B}\right| A\right\rangle + \left\langle B \left|\frac{1}{r_A}\right| B\right\rangle + S\left(\left\langle A \left|\frac{1}{r_B}\right| B\right\rangle + \left\langle B \left|\frac{1}{r_A}\right| A\right\rangle\right)\right] + \langle AA|BB \rangle + \langle AB|AB \rangle}{(1 + S^2)}$$

The six integrals shown here are themselves the subject of investigation and computation whose detailed exposition would take us too far afield from the question of how quantum mechanics demonstrates the self-assembly of molecules from their constituent atoms. Therefore, we will refer the reader to the literature (Hirschfelder and Linnett 1950) for the fine points in this and simply quote the results, as complex as they even are in summary. The required integrals encompass those from the study of the hydrogen molecular ion given in previous sections in this chapter.

5.3.2   A "Molecular Orbital" Approximation

An alternative linear combination of atomic states wave function $\psi(A, B, \sigma) = (a|A\rangle + b|B\rangle)|\sigma\rangle$ is referred to as a "molecular orbital" approximation and after operating on this with the Hamiltonian, and using $a = b = 1$, one obtains the energy in a form suitable for computation with *Mathematica*. We can write the expression for the energy of the ground state as

$$E = 2E_0 + E_p - 2a_0 E_0 \left[-\frac{2(j(R) + k(R))}{1 + S} + \frac{J + 2K + M + 4L}{2(1 + S)^2} \right]$$

where the Coulomb $k(R)$ and exchange integrals $j(R)$ are defined above as

$$j(R) = \left\langle A \left| \frac{1}{r_A} \right| B \right\rangle \text{ and } k(R) = \left\langle A \left| \frac{1}{r_B} \right| A \right\rangle$$

and $S(R)$ is the overlap integral defined above. The other integrals are

$$J = \left\langle AA \left| \frac{1}{r_{12}} \right| BB \right\rangle, K = \left\langle AB \left| \frac{1}{r_{12}} \right| AB \right\rangle, L = \left\langle AA \left| \frac{1}{r_{12}} \right| AB \right\rangle, M = \left\langle AA \left| \frac{1}{r_{12}} \right| AA \right\rangle$$

These integrals have been calculated (e.g., see Hirschfelder & Linnett *op cit.*, or Atkins and Friedman (2011), p. 528) and we will simply tabulate them here.

$$J(R) = \frac{1}{R} - \frac{1}{2a_o} \left[\frac{2}{R} + \frac{11}{4} + \frac{3}{2} R + \frac{1}{3} R^2 \right] e^{-2R}$$

$$K(R) = \frac{1}{5a_o} \left[\frac{6}{R} (\gamma + \ln R) S(R)^2 - E_1(4R) S(-R)^2 \right.$$
$$\left. + 2E_1(2R) S(-R) S(R) - \left\{ -\frac{25}{8} + \frac{23}{4} R + 3R^2 + \frac{1}{3} R^2 \right\} e^{-2R} \right]$$

where γ is the Euler gamma constant, $\gamma = 0.57722...$, and $E_1(x)$ is one version of the exponential integral $E_1(x) = \int_x^\infty \frac{e^{-z}}{z} dz$, which is found in *Mathematica* as

`ExpIntegralEi[x]`, and

$$S(-R) = \left[1 - R + \tfrac{1}{3} R^2 \right] e^R$$

$$M(R) = \frac{5}{8a_o}$$

$$L(R) = \frac{1}{2a_o} \left[\left(2R + \tfrac{1}{4} + \tfrac{5}{8R} \right) e^{-R} - \left(\tfrac{1}{4} + \tfrac{5}{8R} \right) e^{-3R} \right].$$

After evaluation of these expressions in *Mathematica* as a function of the internuclear separation R, we obtain the results shown in Fig. 5.13. The minimum in the energy plot is found at $R = 1.71a_o$ (90.4 pm), where the energy of the hydrogen molecule is -31.78 eV. The binding energy is -4.58 eV compared to the isolated hydrogen atoms. One should compare these calculated values with those found by Rydberg from measurements of the spectra of the excited states of the hydrogen molecule, which give a binding energy of -4.74 eV and a minimum of $R = 1.4a_o$ (74.1 pm). Nevertheless, this simple approximation shows that H_2 is a bound state; the two separate hydrogen atoms attract each other due to their van der Waals interaction energy and self-assemble into the hydrogen molecule. The binding energy of -4.74 eV requires an ultraviolet photon of wavelength 262 nm so that H_2 is stable in visible light, but in the presence of hot, blue stars, this molecule can be dissociated.

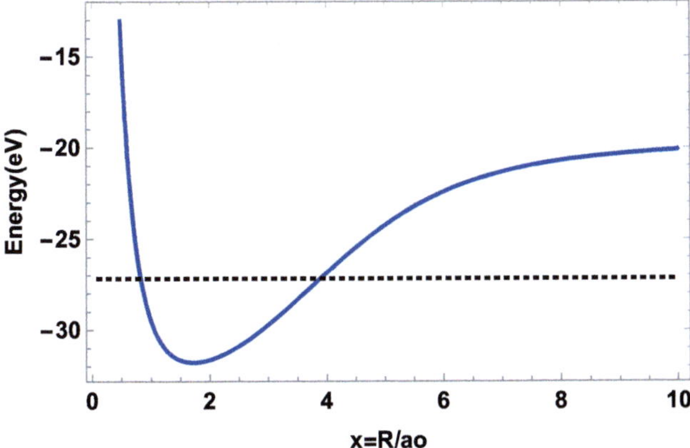

Fig. 5.13 The energy of the hydrogen molecule as a function of the interproton distance showing a bound state at $R = 1.71a_\mathrm{o}$ (90.4 pm) with a binding energy of -4.58 eV. This approximation is based on a linear combination of atomic 1s states and displays the known problem that the molecule does not have the correct ionization behavior at large R. The valence bond approximation gives the correct ionization. The dotted line is the enegry of two separated hydrogen atoms for comparison

5.4   Quantum Density Functional Theory

Although it is in principle possible to use the many-body Schrodinger equation to develop the wave functions of complex molecules, exact solutions are, as in all branches of physics, limited to two bodies or to three bodies if a single electron moves in the field of two, much heavier protons and by means of the Born–Oppenheimer approximation, one can ignore the motion of the nuclei and assume that they are fixed in space. Such methods are applicable to the helium ion, the H_2 ion, and so on but in building up the wave functions of more complex atoms, one must resort to further approximations and this leads to the realization that exact methods were not only available for more complex atoms, but the application of quantum mechanics to molecules required more advanced techniques as well. The Hartree–Fock wave functions for the atomic electrons were one way to enforce the antisymmetry that the Pauli Exclusion Principle required in the solutions for atoms larger than hydrogen. Since quantum mechanics is a linear theory, one can always use linear combinations of electronic wave functions in variational or perturbative solutions, although a major stumbling block is the rapid increase in the number of single electron wave functions needed for this program. Often this is met by using approximate wave functions, such as Gaussians and by using large digital computers to manipulate the various integrals needed. The limits on these methods were appreciated very quickly by the late 1950s in the history of molecular quantum mechanics and resulted in the search for an alternative method that could be applied more easily to the complex molecules found in living systems.

In 1964, Kohn (Kohn and Sham 1965; see also references in his Nobel lecture; Kohn 1999) recast molecular quantum mechanics by changing the focus from the complex wave function to that of the three-dimensional electron density, i.e., the spatial probability of locating an electron in the vicinity of the molecular backbone of a molecule as given by the square of the wave function. This breakthrough, embodied within modern computer programs, has enabled the accurate calculation of the properties of a vast array of molecular species. The method is referred to as quantum density functional theory and introduced a novel mathematical concept to molecular science, that of the function of a function, or a "functional."

Problems

1. *Point group of the hydrogen molecule.* To what point group does the H_2 molecule belong and how does this influence the possible choice of candidate wave functions?
2. *The hydrogen molecule in ellipsoidal coordinates.* The hydrogen ion molecule H_2^+ is exactly soluble in ellipsoidal coordinates (μ, ν, ϕ) such that

$$\mu = (r_A + r_B)/R,$$

$$\nu = (r_A - r_B)/R,$$

where the radial distances are shown in Fig. 5.1 and ϕ is the rotation angle around the axis joining the nuclei, which lie at the foci of ellipses of $\mu = $ constant. The volume element in this space is given by

$$dV = \frac{1}{8}R^3\left(\mu^2 - \nu^2\right)d\mu d\nu d\phi$$

and the variables are bounded by

$$1 \leq \mu \leq \infty$$

$$-1 \leq \nu \leq 1$$

$$0 \leq \phi \leq 2\pi.$$

Schrödinger's equation for the hydrogen molecular ion in the Born–Oppenheimer approximation uses the Hamiltonian

$$H = \frac{-\hbar^2}{2m_e}\nabla_e^2 + \frac{e^2}{4\pi\varepsilon_0}\left[\frac{1}{R} - \frac{1}{r_A} - \frac{1}{r_B}\right].$$

Given hydrogen atom wave functions centered on each nucleus of the form

$$\psi(r) = \sqrt{\frac{1}{\pi a_0{}^3}}\, e^{-r/a_0}$$

begin by expressing the overlap integral, $S(R) = \langle A| B\rangle$, in terms of the product $\psi_A(r)\psi_B(r)$.

(a) Evaluate S(R) and plot it for $0 \le R/a_0 \le 10$.
(b) Explain the behavior of $S(R)$ over this range of distances.
(c) Find the Coulomb integral $C(R)$ by noting that $\mu^2 - \nu^2 = (\mu - \nu)(\mu + \nu)$ from

$$C(R) = \frac{e^2}{4\pi\varepsilon_0} \int \frac{\psi_A{}^2(r)}{r_B}\, dV$$

and plot $C(R)$ for $0 \le R/a_0 \le 10$.
(d) Explain the behavior of $C(R)$ over this range of distances.
(e) Evaluate the nonclassical exchange integral $N(R) = \langle A| H| B\rangle$, which represents the charge density that overlaps the region between the two nuclei.
(f) Plot $N(R)$ for $0 \le R/a_0 \le 10$.
(g) Explain the behavior of N(R) over this range of distances.

3. *Electron density*. The probability of finding an electron at a particular point is given by the absolute square of the wave function at that point.

(a) Using

$$\psi(r) = \psi_A(r) + \psi_B(r)$$

and computer mathematics software (perhaps *Mathematica*), find the electron density in the middle of the bond between the two protons in the $H_2{}^+$ ion.
(b) Plot it versus R.

4. *Molecular vibrations*. To a first approximation, molecules vibrate about the equilibrium positions of their nuclei according to a potential of a simple harmonic oscillator for small nuclear displacements. The frequency of vibration is

$$\omega = \sqrt{\frac{k}{M}}$$

where the force constant k is the second derivative of the energy with respect to distance:

$$k = \frac{\partial^2 E}{\partial R^2}$$

and the effective mass M for a diatomic molecule whose atomic masses are m_1 and m_2 is

$$M = \frac{m_1\, m_2}{m_1 + m_2}.$$

For the hydrogen molecular ion, $m_1 = m_2 = m$, so that $M = m/2$.

(a) Find the vibrational frequency of the hydrogen molecular ion for $R = 130$ pm.

(b) The actual value for ω is found from the infrared spectrum of the hydrogen molecular ion. Look this up and compare your result with the data.

References

P.A.M. Dirac, A new notation for quantum mechanics. Math. Proc. Camb. Philos. Soc. **35**, 416–418 (1939)

M.J. Feinberg, K. Ruedenberg, E.L. Mehler, The origin of binding and antibinding in the hydrogen molecule ion. Adv. Quantum Chem. **5**, 27–98 (1970)

J.-P. Grivet, The hydrogen molecular ion revisited. J. Chem. Educ. **79**, 127–132 (2002)

J.O. Hirschfelder, J.W. Linnett, The energy of interaction between two hydrogen atoms. J. Chem. Phys. **18**, 130 (1950)

W. Kohn, Nobel Lecture: Electronic structure of matter—Wave functions and density functionals. Rev. Mod. Phys. **71**, 1253 (1999)

W. Kohn, L.J. Sham, Self-consistent equations including exchange and correlation effects. Phys. Rev. **140**, A1133–A1138 (1965)

J.M. Peek, Eigenparameters for the 1sσg and 2pσu Orbitals of H2+. J. Chem. Phys. **43**, 3004 (1965)

Bibliography

P.W. Atkins, R.S. Friedman, *Molecular Quantum Mechanics*, 5th edn. (Oxford University Press, New York, 2011)

P.A.M. Dirac, *The Principles of Quantum Mechanics* (Oxford University Press, Oxford, 1930)

V. Magnasco, *Elementary Molecular Quantum Mechanics: Mathematical Methods and Applications*, 2nd edn. (Elsevier, 2013)

A. Messiah, *Quantum Mechanics*, vol I and II (Elsevier, Amsterdam, 1961)

J.J. Sakurai, J. Napolitano, *Modern Quantum Mechanics*, 3rd edn. (Cambridge University Press, Cambridge, 2021)

L. Susskind, A. Friedman, *Quantum Mechanics (The Theoretical Minimum)* (Basic Books, New York, 2014)

Chapter 6
The Self Assembly of the Universe and the Elements of the Periodic Table

"The evolution of the world can be compared to a display of fireworks that has just ended; some few red wisps, ashes and smoke. Standing on a cooled cinder, we see the slow fading of the suns, and we try to recall the vanishing brilliance of the origin of the worlds."

George Lemaitre.

"I believe a leaf of grass is no less than the journeywork of the stars."

Walt Whitman, Leaves of Grass.

"We are Stardust, we are Golden, …"

Joni Mitchell, Woodstock

In our search to understand the physical basis for the origin of Life, we have encountered probability, energy, entropy, the symmetries of space-time, and quantum mechanics and have seen how they direct the self-assembly of matter into its various present forms. What needs to be developed next is the origin of matter and the elements that comprise living systems. Here quantum mechanics, and in particular, the uncertainty principle, will play a vital role because the universe is constructed from quantum objects. We therefore need to establish certain facts about the universe that will guide us. To begin, we must ask if the universe has a finite age.

6.1 How Old Is the Universe?

The universe cannot be younger than the oldest object in it and it must be at least as old as this object. Since measurement is the basis for all human knowledge, the age of the universe has been determined by measuring the ages of the objects in the universe in several different ways. These include determining the abundances of elements using the known properties of radioactive decays, finding the ages of the oldest stars as white dwarfs and in star clusters, and measuring the properties of the

© The Author(s), under exclusive license to Springer Nature Switzerland AG 2024
L. O. Sillerud, *Abiogenesis*, https://doi.org/10.1007/978-3-031-56687-5_6

cosmic microwave background radiation. All of these methods agree on a single important fact: The universe is *not* infinitely old. We can also understand this from first principles; if the universe were infinitely old, all matter and energy would now be at equilibrium at the same temperature and stars and life would have perished eons ago. We would like to know how old the universe is because, t_o, the age of the universe sets the ultimate time scale for subsequent events relative to the origin of life. It is useful therefore to briefly examine how this age was determined.

Not long after Becquerel discovered radioactivity in the 1890s, it was realized that the decay of unstable elements could be used as a clock with which to date the Earth, the solar system, and the universe if radioactive elements with suitably long half-lives could be found. As the twentieth century progressed, the decay series involving radium, uranium, and lead were examined and the ratios of the isotopes of these elements were successfully developed for use as chronometers. Alongside these developments in nuclear physics, the maturation of stellar spectroscopy allowed the measurement of isotopic abundances in the cosmos. Modern applications of these methods have refined the age of the universe so that now, we have an excellent understanding of when it was formed. For example, the half-lives of the radioactive isotopes ^{238}uranium and ^{232}thorium are sufficiently long ($t_{1/2}$ [^{238}U] = 4.468 Gyr and $t_{1/2}$ [^{232}T] = 14.05 Gyr) that Dauphas (2005) was able to combine measurements of the ratio of ^{238}U to ^{232}Th in the solar system with determinations of this ratio in the oldest stars to derive an age for the universe of $t_o = 14.5$ Gyr with an uncertainty of ~2.5 Gyr. Similarly, Cowan et al. (1999) used the ^{232}Th/^{63}Eu ratio in old stars to calculate an age of $t_o = 15.6 \pm 4.6$ Gyr, while the decay of ^{238}U in another old star gave an age of $t_o = 12.5 \pm 3$ Gyr (Cayrel et al. 2001).

A white dwarf is a star, made up of degenerate matter, that is at the end of its lifetime. These stars have masses about equal to that of our sun, but possess radii approximately the size of the Earth, giving them a density of ~10^9 Kg/m^3, or a million times the density of water. These collapsed stars no longer fuse nuclei to produce heat to support their mass through photon pressure; they simply shine as black bodies because they are very hot. They radiate away their heat, cooling over time, so that the oldest white dwarfs are the coolest and the faintest. With the aid of the Hubble Space Telescope, Hansen et al. (2004) used the cooling of white dwarfs to find that the globular cluster *Messier 4* had an age of $t_o = 12.1 \pm 0.9$ Gyr, which, combined with the time after the Big Bang required for the cluster to form, resulted in an age of the universe of $t_o = 12.8 \pm 1.1$ Gyr. The age of the universe can also be determined from the brightness of the most luminous star(s) in a globular star cluster. This method was applied by Chaboyer (1996) to find an age of $t_o = 14.6 \pm 1.7$ Gyr. Even though the oldest of these age determinations was published more than 20 years ago, the results agree with each other within their respective quoted errors. The unweighted average of these measurements is $<t_o> = 14.0 \pm 1.3$ Gyr, with a coefficient of variation of 9.4%. Such a large uncertainty in the universe's age remained until the advent of space-based methods using satellites. Data from the Planck satellite give an age of the universe of $t_o = 13.801 \times 10^9$ years (Aghanim 2020).

6.2   How Old Are the Sun and the Solar System?

The age of the sun can be obtained from measurements of the ratios of the decay products of ^{238}uranium and ^{235}uranium. Both of these uranium nuclei decay to lead; ^{238}U decays to ^{206}Pb, while ^{235}U decays to ^{207}Pb. Mass spectrometric measurements of the ratio ^{207}Pb/^{206}Pb, which were pioneered by Nier in 1939 (Nier 1939), give a meteoritic age of the sun of $t_{Sun} = 4.566 \pm 0.005$ Gyr (Bahcall 1995). The sun's age can also be directly measured from helioseismology where the oscillations in the solar plasma constrain the equation of state of the hydrogen and helium. Using these techniques Bonanno et al. (2002) determined an age of the sun of $t_{Sun} = 4.57 \pm 0.11$ Gyr in agreement with its age determined from meteorites.

The age of the solar system needs to be defined with respect to a particular event in the process of condensation from the primordial solar nebula. Let us pick the age as the amount of time that has elapsed from the formation of the first solid grains in the disk of material orbiting the proto-sun. We can determine this age by measuring the ^{207}Pb/^{206}Pb ratios in inclusions within meteorites that are rich in calcium and aluminum. These inclusions range in size up to about a centimeter and are thought to have been formed in the protoplanetary disk by high-temperature condensation, making them candidates for the earliest materials to form. Measurements of their ages can then be used to constrain the age of the solar system. Such data from an inclusion in the Northwest Africa 2364 CV3-group chondritic meteorite gave an age of $t_{SS} = 4.5682$ Gyr representing the oldest solar system material yet found (Bouvier 2010).

6.3   How Old Is the Earth?

Rock dating by examining the ratios of certain isotopes, such as ^{87}Sr/^{86}Rb, was one of the early applications of mass spectroscopy. When this technique was first used on rocks from the Earth's crust, it was found that the oldest rocks were about 3.8 Gyr of age. Plate tectonics have scrambled the Earth's crust, however, so that subduction and crustal recycling have produced surface rocks that are much younger than the expected age of the Earth. Recent studies of zirconium silicate crystals (zircons), such as those from the Jack Hills in Western Australia, have found ages of 4.382 Gyr, indicating that the Earth is at least that old (Valley et al. 2014). Using analysis of lead isotopes in meteorites, Patterson established the age of the Earth as $t_E = 4.55 \pm 0.07$ Gyr (Patterson 1956). The return of rocks from our moon by the Apollo astronauts enabled dating of the moon to an age of $t_M = 4.35 - 4.42$ Gyr (Carlson 2014) slightly younger than the Earth and supporting an origin for the moon in a giant impact after the Earth was formed.

This small sample of age measurements support the concept that even though the universe, the sun, the solar system, and the Earth are billions of years old, there must have been a beginning; the universe must have had a birthday, which we have referred to as t_o. We can observe the results of this *date de naissance* by looking around us at the Earth and the sun and particularly on a clear, dark night at the diffuse clouds of light from the Milky Way and the myriad bright, point-like stars. In Earth's distant past, a prescientific philosopher might have gazed up in wonder, night after night, noticing that the celestial sphere rotated in concert with the seasons, reappearing unchanged year after year, and that the groupings of the bright stars remained apparently the same with the passage of human time. This apparent constancy of the firmament was enshrined in the naming of the constellations. However, such a static picture was abruptly shattered in the late nineteenth century when the positions of the stars in the night sky were measured over time using photographic techniques, and it was discovered that some of the "fixed stars" actually moved by small, but significant amounts within a person's lifetime. Even more surprising was the discovery in the early twentieth century that galaxies, extended disk-shaped collections of stars, were not only at great distances, but were also moving along our line of sight, either towards or away from the Earth at great speeds. The universe ceased to be an eternal, static structure and took on, instead, the mantle of a dramatic rush through space and time.

Natural curiosity also led philosophers to speculate upon the origin and perhaps even the constitution of these brilliant, shining diamonds and foggy patches of light. Among the many fantastic creation myths, some imagined that the stars were simply thrown up into the sky, but any child who has thrown a rock up into the air knows that it will just fall back to Earth. On the other hand, in 1687, the year of the publication of Newton's *Principia*, it was known that if the rock were given a large enough initial velocity, it could leave the Earth's gravity, but then of course, it would soon leave one's sight and disappear forever. Other creation myths postulated that the universe arose from nothing: from an eternal, formless void. Of these myths, none is closer to our modern, predictive model for the origin of the universe and the life within it: *the Big Bang*.

6.4   The Universe from Nothing: Tests of the Big Bang

The birthday of the universe happened $t_o = 13.8$ billion years ago, when it self-assembled from a hot, dense, quantum vacuum fluctuation that rapidly grew by inflation and generated our present spacetime. This model, *the Big Bang* first proposed in the 1940s by George Gamov (1946), was jokingly given its moniker by Fred Hoyle, and the name has stuck. The succeeding years have witnessed its subsequent development and refinement into a robust quantum-mechanical theory that makes accurate, testable predictions of key features of the temporal and spatial evolution of the past, and the current, observable universe. At its foundation is the idea that the universe arose from fluctuations of the quantum vacuum state, the state of "nothing." We have previously encountered the vacuum in our study of the harmonic oscillator where we discovered that there exists zero-point energy even in the absence of field excitations. We can also write the quantum Hamiltonian in terms of creation and annihilation operators that generate these field excitations by operating on the vacuum. Heisenberg's uncertainty principle allows quantum fluctuations to produce the mass of the universe as long as these fluctuations occur for such a short time that no measurement can detect this seemingly obvious violation of the conservation of energy. We will discuss these vacuum fluctuation in more detail in a later section.

Once generated, the universe underwent a period of breathtakingly rapid inflation (Guth 1981) and three distinct epochs of nucleosynthesis. This Big Bang model quantitatively accounts for many key observations, such as the expansion of the universe, the origin and properties of the cosmic microwave background radiation, the resulting formation and abundance of the nuclei comprising the periodic table and the chemical elements of which living systems are composed, and the formation of galactic and large-scale structure.

The measured age of the universe of $t_{o} = 13.801 \times 10^{9}$ years (Planck 2108) tells us *when* the Big Bang occurred, but it leaves open the vital question of *how* it happened. In order to understand the *how*, we would like to travel backwards in time to the epoch at which $t \sim t_{o}$, when spacetime was first generated. However, just as we cannot replay the Earth's Cambrian explosion, we are permanently barred by the passage of time from directly examining the Big Bang. Modern accelerator measurements at multi-TeV energies reveal the state of matter around $t_{o} + 100$ picoseconds, but no closer than this to t_{o} itself. Nevertheless, like the events of the far distant past on Earth, the Big Bang left traces in the fossil record, so to speak. For, just as an examination of the Cambrian stratum laid down as the Burgess shale reveals the solidified exo- and endoskeletons of 550-million-year-old phyla, so it is with the Big Bang. There are indeed physical fossils from the distant past that survive to the present and have left their imprint on the current universe. In the case of the Big Bang, these imprints are revealed by measurement, as always, guided by sound physical theory. The Big Bang, like bigfoot, would be chased to its lair through observations.

The first of these observations was the discovery of the expansion of the universe, which, while not really a fossil, was certainly an imprint, a trace fossil as it were, like a dinosaur trackway, of the origin of spacetime and matter. More to the point was the discovery of the cosmic microwave background radiation and the important finding that its spectrum matched that from a blackbody. This was closely followed by measurements of the temperature anisotropies across the sky that showed the amplitude of the primordial vacuum fluctuations that subsequently, through inflation, gave rise to the stars and galaxies so vivid and shining in the sky of a dark night. Another fossil, albeit a statistical beast, was the image of the shock wave in the sky of the decoupling of radiation from matter at a redshift of $z \sim 1089$. Last, but of course of utmost importance for the origin of life in the universe, was nucleosynthesis, initially during the Big Bang and later in stars. The fossils from this epoch are nothing less than the elements of the periodic table and of living systems: ^{1}H, ^{2}H, ^{4}He, ^{7}Li, ^{12}C, ^{14}N, ^{16}O, etc. Of great irony is the fact that the Big Bang was so simple in its origin and yet so rich in its ramifications; the known principles of physics provide testable explanations as far back in time as the primordial hot quark-gluon plasma. What then is the scientific evidence (the measurements) supporting the origin of the universe and the elements of life in a Big Bang?

The evidence for the Big Bang has been drawn from many independent measurements. It is incontrovertible that the universe has *not* existed for an infinite time in the past; the fact that we exist refutes such an assertion. Measurements of the age of the universe and its contents clearly demonstrate a finite existence. The universe is not static, but expands with time, each galaxy receding from its neighbors at the velocity of the Hubble flow; the implication is that the universe was previously much, much smaller than its present awesome dimensions. The current volume of the universe, approximated as a sphere of radius 13.8 billion light-years centered on the Earth, is 10^{70} km^3. The Big Bang was hot and radiated photons with an energy spectrum described by the simplest of thermal objects, that of a blackbody, an object whose characteristics were quantitatively elucidated more than a century ago by Max Planck. Anisotropies in the distribution of these fossil photons across the dome of the sky reveal the seeds of present-day accumulations of matter in the form of galaxies. The decoupling of radiation and matter (protons) as the primordial universe expanded left its imprint on the distribution of present-day galaxies as the baryon acoustic oscillation signal. All of these phenomena are a direct result of quantum fluctuations of the vacuum and we will use the insights of Richard Feynman to show how this happens.

6.4.1   The Expansion of the Universe

Observations showing that the universe was not static were made by E. E. Barnard in 1916 (Barnard 1916) of a red dwarf star in the constellation Ophiuchus just 5.86 light-years from Earth whose proper motion across the sky was found to be 10.3 arc seconds/year, a value significantly different from zero (an arcsecond is 2.78×10^{-4} degrees of arc). Even more startling was the subsequent discovery by Edwin Hubble (1929) that the universe is expanding, which he made by correlating his photometric determinations of the distance to 22 galaxies, using Cepheid variable stars, with spectroscopic velocity measurements of these galaxies made by Vesto Slipher (1917). Slipher determined the galactic velocities from the redshifts of spectral lines in these distant galaxies.

6.4.1.1   Redshift

The concept of redshift is central to our understanding of the origin and structure of the universe. Slipher assumed that the redshift was a Doppler shift of the light waves, much like the decrease in frequency of the sound waves from a train whistle as it passes the observer and recedes into the distance. The Doppler parameter, z, is the ratio of the velocity of the source, v, to the speed of information transfer through the medium, or through space. In the case of the classical Doppler effect, as applied to train whistles, the information transfer velocity is the speed of sound in air, but for the astronomical redshift, the information transfer velocity is the fastest thing in the universe, i.e., the speed of light, c, (note that the speed of light is *not* infinite). Therefore, for optical measurements, the redshift is $z = v/c$, and the velocity is found from

$$z = \frac{v}{c} = \frac{\lambda}{\lambda_o} - 1$$

where λ is the measured wavelength of a spectral line in a distant object, and λ_o is the wavelength of the same line measured in the rest frame of the observer. The redshift of the present epoch is zero ($\lambda = \lambda_o$) and increases as one contemplates earlier and earlier events.

Hubble observed that the recession velocity was directly proportional to the distance, expressed as Hubble's law:

$$v = H_o d$$

where d is the distance to a galaxy and H_o is known as the current Hubble parameter. In this view, nearby galaxies recede from us at a constant rate per unit distance, $v/d = H_o$; the farther an object is from Earth, the faster it recedes. Dividing this equation by the speed of light gives

$$z = v/c = H_o d/c$$

which shows that the redshift is also proportional to the distance. The distance is also simply given by $d = vt$, so that we have the remarkable fact that the Hubble time is given by $t_H = 1/H_o$; this is the time since t_o, or the amount of time that the universe has been expanding. Friedmann (1922) and LeMaitre (1927) solved Einstein's general relativity equations (Einstein 1916) and found that the redshift was due to the expansion of the intervening space-time between the galaxies.

We encountered the symmetries of space-time in Chap. 3 and found that an examination of their properties led to profound insights about the allowed constituents of the universe. From an astronomical perspective, the two major symmetries of the universe are that it is *homogeneous* and *isotropic*. Homogeneity means that the distribution of matter (the density) does not depend on the origin of the coordinate system used and must therefore be associated with the conservation of momentum and its position independence. Isotropy implies that the universe looks the same in all directions and must therefore be associated with the conservation of angular momentum. The assumptions of cosmological isotropy and homogeneity are the correlates of the symmetries of space-time that we examined earlier, invariance with respect to position, time, or direction which directly led to these conservation laws. The Hubble-LeMaitre relationship between redshift and distance is directly the result of the homogeneity of the universe. There is no preferred origin for a coordinate system, any will do, and the distance-velocity relation holds for any observer. We will see that statistical isotropy can be measured from the distribution of matter (galaxies) but that it is best inferred from the observations of the microwave background radiation (vide infra).

6.4.1.2   The Hubble Parameter

The magnitude of the current expansion rate is given by the current value of the Hubble parameter; an unweighted average of 36 modern measurements from 2001 to 2020 (https://en.wikipedia.org/wiki/Hubble%27s_law) using the Hubble Space Telescope, the WMAP and Planck satellites, and other means, indicates that $H_o = 71.6 \pm 2.7$ (km/s)/Mpc, or 22.0 km/(second million light-years), although there remains an active controversy in which different astronomical techniques give slightly, but consistently, different values for H_o (Castelvecchi 2020). For example, data from the Planck satellite (Aghanim 2020) give a value of $H_o = 67.37 \pm 0.54$ (km/s)/Mpc, which agrees with that found by the Atacama Cosmology Telescope (Choi 2020) of $H_o = 67.6 \pm 1.5$ (km/s)/Mpc. Astronomers continue to refine the Hubble parameter; its uncertainty has led to the common usage of a dimensionless parameter, unfortunately known as h (not to be confused with Planck's constant of the same symbol) that is defined as $h = \dfrac{H0}{[100 \ (km/s)/Mpc]}$, where H_0 is the current value of the Hubble constant; in these terms the average value of the current Hubble parameter is $h = 0.716 \pm 0.027$.

The vast distance scale for the universe was dramatically revealed when, in 1963, Maarten Schmidt found that the spectral line for the hydrogen Balmer series H_β transition appeared in the spectrum of the Quasar 3C273 at $\lambda = 563.2$ nm, while at rest, this transition was found to have a wavelength of $\lambda_o = 486.1$ nm (Schmidt 1963). Using the value of the Hubble parameter given above, this redshift of $z = 0.159$ implied that this quasar was at a distance of 2.4×10^9 light-years from Earth (vide infra). A light-year (lyr) is the distance that light travels in an Earth year:

$$1 \ \mathrm{lyr} = 3 \times 10^5 \ \mathrm{km/s^*} \ 3.17 \times 10^7 \ \mathrm{s/year} = 9.46 \times 10^{12} \ \mathrm{km}$$

3C273 is therefore 2.27×10^{22} km from Earth, and yet it is so bright (magnitude 12.9) that amateur astronomers, such as myself, have easily observed its starlike appearance in small telescopes.

The measured, finite value of the Hubble constant is another, independent datum that shows that the universe is not infinitely old, but has existed for a finite time: the Hubble time, t_H. Although astronomers use obsolete, and obscure, classical cgs units, in this text we prefer standard SI units, in which H_o is a frequency (in Hz $= s^{-1}$) or, more insightfully, an expansion rate:

$$H_o = 1/t_H.$$

Therefore, it is instructive to convert the value of H_o to SI units:

For example, if $H_o = 71.6$ (km/s)/Mpc,
And 1 light-year $= 9.46 \times 10^{15}$ m,
1 Parsec $= 3.26$ light-years $= 3.0857 \times 10^{16}$ m,
therefore, 1 Mpc $= 3.0857 \times 10^{22}$ m,
so $H_o = 71.6$ (km/s)/Mpc $= 71.6 \times 10^3 / (3.0857 \times 10^{22})$
$\Rightarrow \mathbf{H_o = 2.19 \times 10^{-18}\ Hz}$
and

$$t_H = 1/H_o = 1.366 \times 10^{10}\ \text{years}$$

The reason for this conversion to SI units is to expose the simple fact that the Hubble time $t_H = 1/H_o$ in these standard units gives the age of the universe, if the expansion of the universe was strictly constant and under certain other assumptions with respect to the background cosmology. The most precise current measurement of the age of the universe (Aghanim et al. 2020) is that $t_o = 13.801 \pm 0.024$ Gyr (where 1 Gyr $= 10^9$ years).

The Hubble time is less than t_o because the expansion rate has been decreasing as the universe ages. General relativity indicates that the rate of expansion of the universe is governed by the energy density through Friedmann's equation for $H(z)$. There is only a single, dimensionless parameter, the scale factor $a(t)$, that specifies the expansion of space-time in the Friedmann-Lemaitre-Roberston-Walker metric for an expanding, flat ($\Omega_k = 0$) homogeneous, isotropic universe given by

$$ds^2 = c^2 dt^2 - a^2(t)\left(d\theta^2 + \sin^2\theta\, d\varphi^2\right)$$

The Hubble parameter is

$$H(t) = \frac{1}{a(t)} \frac{da(t)}{dt}$$

which is Hubble's law relating the expansion rate to the velocity divided by the distance. The scale factor is directly related to the redshift by

$$z(t) = \frac{a(t_o)}{a(t)} - 1$$

The absolute value of the scale factor is unknown; only the ratio at two different times has any physical meaning so we pick a value at a specific time, $a(t_o) = a_o = 1$ and find for the redshift

$$z(t) = \frac{1}{a(t)} - 1$$

or that

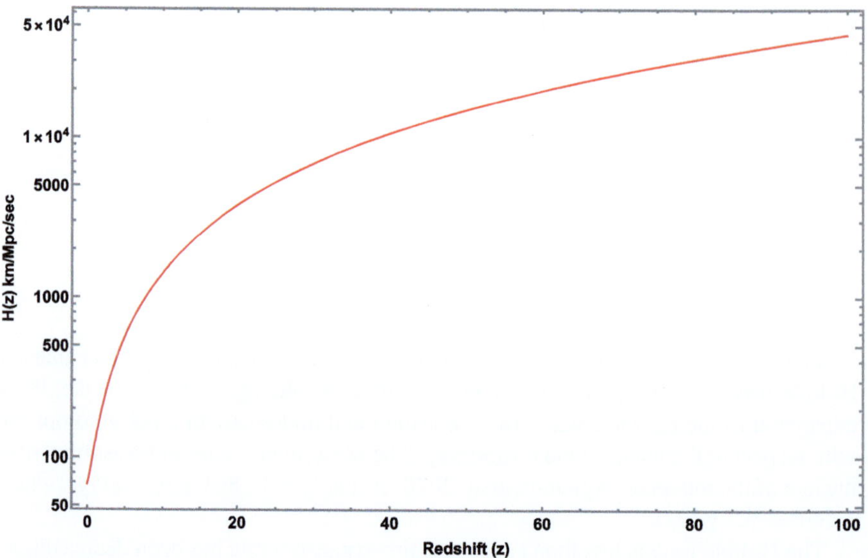

Fig. 6.1 The value of the Hubble parameter, $H(z)$, as a function of redshift using the Planck 2108 cosmological parameters from Table 6.1

$$a(z) = \frac{1}{z+1}$$

The expansion rate at a redshift of z is then found from Friedmann's equation:

$$H^2(z) = H_o{}^2 \left[\frac{\Omega_r}{a^4(z)} + \frac{\Omega_M}{a^3(z)} + \frac{\Omega_k}{a^2(z)} + \Omega_\Lambda \right]$$

And this is shown in Fig. 6.1 for the cosmological parameters from Table 6.1 (*q. v.*).

We can also use the cosmological parameters from the Planck satellite mission (Table 6.1) to find the age of the universe for a given value of the Hubble constant from

$$t_o(\Omega_M) = \frac{2}{3H_o} \left[\frac{1}{\sqrt{1-\Omega_M}} \mathrm{Log} \left[\frac{1 + \sqrt{1-\Omega_M}}{\sqrt{\Omega_M}} \right] \right]$$

which assumes that $\Omega_k = \Omega_r = 0$ and that $\Omega_M + \Omega_\Lambda = 1$ in Friedmann's equation. One can check that this gives the correct lifetime of $t_o = 1.38 \times 10^{10}$ years, for the measured value of $\Omega_M = 0.3147$ (Table 6.1). The finite, nonzero value of H_o indicates that the universe originated ~13.8 billion years ago, rather than infinitely far back in the past. Therefore, the universe has not existed forever, but had a beginning at a finite time in the past, and since it has been expanding, *it must have*

Table 6.1 Parameters of the universe determined by fitting the background radiation anisotropy spectrum determined by the Planck satellite to the ΛCDM cosmological model (Anghanim et al. 2020)

Parameter	Value	Value for $h = 0.6737$	Units
$\Omega_b h^2$	0.02233 ± 0.00015	0.04920	–
$\Omega_c h^2$	0.1198 ± 0.0012	0.2640	–
$\Omega_\Lambda h^2$	0.3107	0.6846	–
Ω_m	0.3147 ± 0.0074	0.3147	–
Ω_k	0	0	–
n_s	0.9652 ± 0.0042	0.9652	
A_s	$2.126 \pm 0.044 \times 10^{-9}$	2.126×10^{-9}	K^2
τ	0.0540 ± 0.0074	0.0540	–
H_o	67.37 ± 0.54	67.37	km/(s Mpc)
D_H	4.44	4.44	Gpc
m_ν	0.06	0.06	eV
r_{dec}	147.18 ± 0.29	147.18	Mpc
t_o	13.801 ± 0.024	13.801	Gyr

been extremely small and dense at its beginning. This observation of the expansion of the universe directly led to the development of *the Big Bang model*.

The discovery of the expansion of the universe also resolves Olber's paradox. In an infinite universe, any radial ray which leaves the Earth and proceeds outwards without limit would encounter a star. Let the luminosity of a star be L with a constant density of stars given by n. The flux from a star at a distance R is then $F = \frac{L}{4\pi R^2}$. If we imagine a series of shells of thickness dR extending out to infinity, the power from the light reaching the surface of the Earth from a shell at a distance R is then

$$dP = \frac{nL}{4\pi} dR$$

The total power from all the shells is the integral

$$P = \int dP = \int_0^\infty \frac{nL}{4\pi} dR = \infty$$

and this sum diverges for an infinite universe that is not expanding. The night sky should be blazing with light as bright as the sun instead of dark with visible stars. The finite travel time for light to reach us from the farthest portions of the visible universe means that the extrapolation to infinity for the shell radius is not physically meaningful. The longest distance a photon could travel since the Big Bang is the Hubble distance, $D_H = c/H_o$, which replaces the upper limit and the integral becomes finite. The actual density of stars as determined from surveys of galaxies is $n \sim 10^8/$ Mpc^3; therefore, with the Hubble distance $D_H = (3 \times 10^5 \text{ km/s})/67.37 \text{ km/s/}$

Mpc = 4.45 Gpc, we have that the total power from the stars is $P = 3.54 \times 10^{10}$ L/ Mpc^2. If we take our sun as an average star with a luminosity of $L = 3.8 \times 10^{26}$ watts, we can compare this to the power radiated per square meter by the stars (1 Mpc$^2 = 9.5 \times 10^{44}$ m^2) as $P = 1.4 \times 10^{-8}$ watts/m^2. The Earth receives ~300 watts/m^2 from the sun; therefore, the stars contribute only 4.7×10^{-11} as much power as our sun. The night sky is indeed very dark!

6.4.2   The Mass of the Universe

Measurements by the Planck satellite (Table 6.1; Aghanim et al. 2020) determined that the total mass/energy content of the present observable universe consisted of significant contributions from four types of objects: baryonic matter (4.8%), neutrinos (0.1%), cold dark matter (26.8%), and an object given the peculiar name "dark energy" (68.5%) with an insignificant contribution from radiation. It is estimated that the baryonic mass of the observable universe is ~10^{53} kg, giving a total mass/energy content of ~10^{54} kg, or using $E = mc^2$, this corresponds to about 10^{70} joules. The uncertainty principle, $\Delta E \Delta t = \hbar/2$, then implies that the Big Bang occurred in about 10^{-104} s. This time is the amount of time required for light ($c = 3 \times 10^8$ m/s) to cross an object with a diameter of ~10^{-96} m and is beyond physical meaning since it is unmeasurable. We must compare these times and distances to the smallest physically meaningful quantities, the Planck time, $t_P = \sqrt{\hbar G/c^5} = 5.39 \times 10^{-44}$ s, and the Planck length, $l_P = \sqrt{\hbar G/c^3} = 1.62 \times 10^{-35}$ m, where G is the universal gravitational constant. The Planck time is the amount of time needed for light to cross a black hole containing an amount of mass (the Planck mass, $m_P = \sqrt{\hbar G/c} = 2.176 \times 10^{-8}$ kg) whose size (Schwarzschild radius) equals its quantum mechanical wavelength (Compton wavelength, $\lambda_C = \frac{h}{mc}$).

In order to put these various parameters of the Big Bang into perspective, let us reflect on the quantum properties of space-time. Loop quantum gravity theories (Rovelli & Smolin 1995; Rovelli 2004, Rovelli & Vidotto 2014), among others, posit that space-time is quantized in units of the Planck length and time and that smaller dimensions have no physical meaning. This is logical since only measurable properties of an object can serve as the basis for our knowledge of the physical world. We know that a quantum description of the universe is provided by the eigenvalues of a complete set of operators that commute with the Hamiltonian and that physical observables are the eigenvalues of these operators. Spacetime must be an observable. We know that we can measure lengths and times. Then what are the quantum operators for spacetime? They are the area and volume operators, in loop quantum gravity, and the spacetime wave function must be an eigenfunction of these, with eigenvalues whose magnitude is given by the Planck length and time.

A volume element in space-time is a tetrahedron with vertices denoted by the set of space-time points $\{x_1, x_2, x_3, x_4\}$. The area operator is then constructed as the cross product of two vectors that join a single vertex with two other vertices; this cross product gives the area of a face of a tetrahedron. Therefore, the commutation relations of this operator must reflect its *vector* nature and turn out to be the same as those for angular momentum. Since volume is a *scalar*, the volume operator must commute with the area operator, and both volume and area can be simultaneous eigenvalues when their respective operators act on the spacetime wave function.

Spacetime is then a network of tetrahedra that share two common vertices. The eigenvalues of the area and volume operators (Rovelli and Smolin 1995) are given for a surface S, and volume R, of spacetime in terms of the Planck length by

$$A = \frac{l_P^2}{2} \sum_j \sqrt{p_j(p_j + 2)}$$

$$V = \frac{l_P^3}{4} \sum_i \sqrt{a_i b_i c_i + a_i b_i + a_i c_i + b_i c_i}$$

The index j labels the links of the spin network that crosses the surface S, while $\{a_i, b_i, c_i\}$ are integers defined by $a_i + b_i = p_i$, $b_i + c_i = q_i$, and $c_i + a_i = r_i$, with $\{p_i, q_i, r_i\}$ defined as the colors of the links connecting the ith vertex of the tetrahedral spin network and the sum runs over all the vertices contained in the volume R. Note that the area operator has eigenvalues of angular momentum of a spin ½ object, and hence, its states are irreducible representations of the group SU(2), the group of rotations, hence the name spin networks. The spectrum of eigenvalues of the area and volume operators consists of discrete values. The area operator in particular has eigenvalues much like spin $[s(s + 1)]$.

We have found that the universe contains a large mass and that this mass originated in a very short time and in a tiny volume. These volumes and times are much, much smaller than the Planck scale, which is reasonably viewed as the smallest measurable and physically meaningful physical dimension, and the Planck time as the smallest unit of time which has physical meaning. Grand unified theories in which all the four forces of nature were equivalent in strength in the distant past give values of this unification time as 10^{-43} s, certainly consistent with this naïve view of the origin of the universe. Thus, the early universe could be considered a very massive particle in a very tiny box, with an enormous zero-point energy generated by vacuum fluctuations. These quantum effects would produce an enormous pressure and temperature in such a system, forcing the rapid expansion of the universe. Inflationary effects would release this energy in a phase transition, much like the sudden warming that occurs during crystallization of a supersaturated solution, that would cause the expansion of spacetime (Guth 1981).

The standard model of particle physics predicts that the initial state of the universe, which can currently be traced back to $\sim 10^{-10}$ s, was one of an extremely hot, quark-gluon plasma, which rapidly cooled and expanded (Peebles et al. 1991). At the earliest meaningful time, t_P, the universe was at a temperature of more than 10^{32} Kelvin. Such a hot object radiates copious quantities of blackbody radiation as photons with a wavelength of $\lambda = hc/kT = 3.65 \times 10^{-36}$ m. Notice that this is again on the order of the Planck length. The high temperature of the early universe meant that the initial cosmic fireball emitted photons at a rate which was ten billion times that of heavy matter (baryons). We now detect this photon flux after it has been redshifted due to the expansion of the universe as the cosmic microwave background at a present temperature of only $T_o = 2.7$ K. The expansion of the universe stretched the metric of spacetime which resulted in the cooling of the Big Bang radiation according to the relation

$$T(z) = T_o(1 + z)$$

where T_o is the present temperature.

The expansion of the universe was an expansion of spacetime, the distance between spacetime points increased, and the distance between peaks and troughs of light waves increased by the same amount, linking the redshift of light to the expansion factor. We can then unambiguously characterize time periods in the past by their redshifts, z. For example, after $t_o + 3$ min, baryons were conserved and the average number of baryons per unit volume at a redshift of z is

$$n(z) = n_o(1 + z)^3$$

where n_o is the present baryon density. Extrapolation of this to large z (and concomitantly, the far distant past) implies that the density of the early universe was far, far greater than it is today.

The significance of the expanding universe is that when the expansion is run backward in time, it leads to a condensed state of matter and radiation. The temperature of this state was very, very high and can be qualitatively understood if we treat the universe as an ideal gas. The temperature of such a gas is inversely proportional to its volume; the smaller the volume, the higher the temperature. These effects argue for a hot Big Bang as the initial state of the universe. In the Big Bang model the universe expanded from a hot, dense plasma that was in thermal equilibrium. Any body at a finite temperature will radiate photons in a manner given by Planck's blackbody spectrum. The temperature of the universe decreased as it expanded. Therefore, one of the predictions of the Big Bang model is the presence of this cosmic background radiation that began, at large z, at a very high temperature and subsequently cooled (shifted to longer wavelengths) due to the expansion of the universe.

6.4.3   *The Cosmic Microwave Background Radiation*

The universe at the Planck time was certainly extremely hot. At first thought one supposes that the Big Bang produced an immediate burst of high-energy photons that filled the expanding universe and that is what we observe as remnants of the primordial fireball. A little reflection will suffice to show how that scenario cannot explain it, for then the universe would consist solely of boson (light) light and no fermions (matter). We must therefore also account for the synthesis of baryons in the early universe. We will examine Big Bang nucleosynthesis in the next section, but let us anticipate baryogenesis and state for our purposes here that after the first ~20 min the early universe consisted almost entirely of protons, alpha particles, electrons, and photons. There was indeed an immediate burst of photons, but these rapidly lost energy as the universe expanded and cooled. Baryons and other known elementary particles that comprise the standard model (Fig. 3.6) became stable when the temperature dropped to the point where baryons were stable against photon-induced decay. At this point, it was still too hot ($E_\gamma > 100$ eV) for electrons to combine with protons to form hydrogen (ionization potential 13.6 eV) and with alpha particles to form helium which has the highest ionization potential of any known gas at 24.6 eV. The large numerical excess of these energetic photons, with respect to the other particles, rapidly ionized any hydrogen or helium atoms that were formed from the electrons, protons, and alpha particles:

$$H + \gamma \rightarrow e^- + p^+$$

As the expansion of space-time continued, the wavelength of the photons increased, and their energies decreased until, at a certain temperature, the photons were no longer energetic enough to ionize the nascent atoms, and first electrons assembled onto alpha particles to form helium atoms, followed by the formation of hydrogen, deuterium, and tritium with their lower ionization potentials. Prior to this time, the photons strongly scattered from the free electrons by Thomson scattering and their mean free path was so short that light could not escape from this expanding plasma; rather than originating in a Big Flash, the early universe was opaque to light. When the temperature dropped to the point where neutral atoms were stable, the scattering of photons from free electrons ceased and radiation *decoupled* from matter leaving the universe transparent to light, allowing the photons to freely propagate, and which we now observe as the cosmic microwave background radiation.

[As an aside, astronomers refer to this epoch of decoupling as "recombination," but since protons and electrons were never combined prior to the epoch of decoupling, the "re" seems to stretch the point. Here we will refer to this as the epoch of decoupling of radiation from matter, or simply as *decoupling*. Nevertheless, "recombination" is almost universally used in the literature of the early universe but the reader should be aware of its true meaning.]

Of the two main products of Big Bang nucleosynthesis, helium and hydrogen, helium has the highest ionization potential; however, the removal of the first electron to form the He$^+$ ion deshields the second electron from the +2 charge on the nucleus, and since the Coulomb potential of the nucleus is proportional to the square of its charge, this results in a very large, i.e., four-fold, increase in the ionization potential of the second electron, compared with hydrogen to a value of $13.6^2 = 54.4$ eV. For this reason, the newly formed helium stabilized earlier, at a higher temperature, while hydrogen, with its lower ionization potential, was the last to remain ionized until the expansion of space-time reduced the energy of the photons.

We now observe the freely propagating photons after decoupling with satellites and high-frequency antennae as the surface of last scattering. We can calculate the decoupling temperature at which the universe became transparent to radiation by examining the ionization of atoms as a function of temperature. Gamow (1946) suggested that the current universe should then be filled with these background photons with a temperature around 5 K that was lowered by the expansion of space-time from its much higher initial value at decoupling.

In 1965 Penzias and Wilson (1965) discovered this cosmic background radiation and reported its temperature of 3.5 $\pm$ 1.0 K from measurements at a single micro-wave frequency of 4.08 GHz, which was on the low-frequency side of an expected peak at 160 GHz in the blackbody spectrum. The antenna that Penzias and Wilson used was originally designed for communication experiments during the Echo project in which a large aluminized mylar balloon was launched into orbit in August 1960 and used to bounce radio signals from California to New Jersey. The frequency chosen was, by necessity, within a frequency window where the Earth's atmosphere is transparent to radiofrequency waves. At higher frequencies, above ~30 GHz, Earth's atmosphere becomes opaque. Measurements of the cosmic micro-wave background radiation at these higher frequencies were needed to ascertain the detailed shape of the spectrum, but could not be performed from the surface of the Earth. The Cosmic Background Explorer (COBE) satellite was put into polar orbit in 1989 around the Earth in order to measure the spectrum of the background radiation and to map its anisotropies across the sky. The measurements by the COBE satellite (Fixen et al. 1994) showed that the spectrum of the cosmic microwave background matched that of a blackbody (Fig. 6.2) indicating that the primordial baryons were in equilibrium with the radiation and obeyed a Maxwellian distribution of energies (Chap. 2). Both of these probability distributions are characterized by a single parameter, the temperature, which sets the energy scale for their properties.

The measurements performed by the Far-InfraRed Absolute Spectrophotometer (FIRAS) also aboard COBE, reported in 1994 (Mather et al. 1994), indicated that the background radiation exactly followed the spectrum of a blackbody with a temper-ature of 2.726 $\pm$ 0.01 K. In generating the plots for Fig. 6.2 in Mathematica, I used the data points from (Fixen et al. 1994) but computed the blackbody spectrum from Planck's law and then plotted one on top of the other. The red line in Fig. 6.2 is the theoretical blackbody spectrum for a temperature of 2.726 K; there are no other adjustable parameters. One can see that the data points match the theoretical blackbody spectrum extremely well.

The latest temperature measurements by the Planck satellite (Aghanim 2018) give a present temperature of $T_0 = 2.72548 \pm 0.00057$ K. The temperature of the early universe was much, much higher than at present. This has been tested to epochs as far back as $z \sim 10^{10}$, where the temperature was ~2.7×10^{10} K, or around 27 billion K, a value greater than the temperature at the center of our sun (15 million K) where nuclear fusion reactions currently occur and have for the last 4.5 billion years. We will shortly take up this next aspect of the self-assembly of the universe.

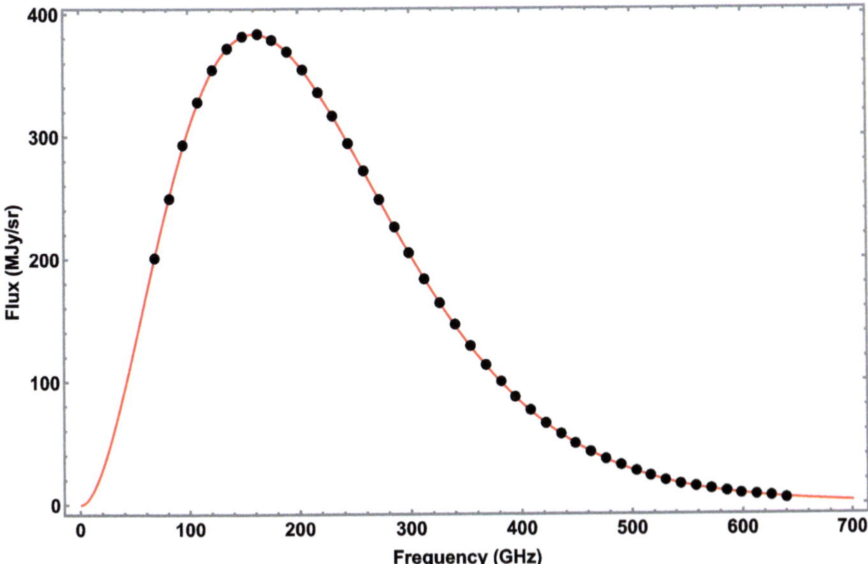

Fig. 6.2 The measurements (black circles) of the flux of the cosmic microwave background dipole radiation from the COBE satellite fit a blackbody spectrum (red line) with a temperature of 2.726 K and a peak at 160.23 GHz, corresponding to a photon energy of 6.626×10^{-4} eV. The flux of photons is reported in the units of megaJanskys per steradian. The unit Jansky is named after Karl Jansky, a pioneer in radio astronomy, and is defined as 10^{-26} Watt/(m^2 Hz); the maximum in the blackbody dipole spectrum is 383.478 Jy/sr

It was expected, and the COBE data also revealed for the first time, that the background radiation was not completely uniformly distributed across the sky, but was slightly lumpy and varied by tiny amounts (parts per million) from place to place (Smoot et al. 1992). The photon flux is very nearly isotropic, but the tiny (~10–20 µK) residual anisotropies found by COBE, and mapped later in much greater detail by the Wilkinson Microwave Anisotropy Probe (WMAP; Fig. 6.5. *q.v.*) and the Planck satellites, display a unique pattern, like that expected from turbulence in an otherwise uniformly distributed hot gas that has expanded and cooled after the Big Bang to the present size of the universe.

Detailed measurements of the anisotropies in the flux of photons across the sky made by the WMAP (Bennett et al. 2013) and Planck satellites (Aghanim 2020) and by the Atacama Cosmology Telescope (Choi et al. 2020) match expectations for quantum fluctuations of matter previously confined in a very small space that had expanded to the current dimensions of the observable universe. The general form of the blackbody spectrum and these additional features of the background radiation agree with the Big Bang model and have yet to be satisfactorily explained by any other theory. For this reason, the Big Bang model for the origin of the universe has supplanted all others as the primary explanation for the cosmic microwave background radiation. We will see (vide infra) that the Big Bang model also accurately predicts several other important features of the present universe. It is clear that the universe (and space-time itself) was once confined to an extremely tiny dense, hot region.

An examination of the detailed properties of blackbody radiation led Max Planck (1901) to introduce the concept of the quantization of energy through the now famous equation, $E = h\nu$, where E is the energy, h is Planck's constant, and ν is the frequency. The number of photons ρ per unit volume (i.e., the photon density) emitted by a blackbody at a photon frequency between ν and $\nu + d\nu$ is given by the Planck probability distribution as

$$\rho(\nu)d\nu = \frac{8\pi h\, \nu^2}{c^3 \left(e^{\frac{h\nu}{kT}} - 1 \right)} d\nu$$

where one will recognize the Bose-Einstein distribution from Chap. 2. This gives the probability of a photon with energy $h\nu$ at a temperature of T. Once the temperature is specified, there are no free parameters in the Planck distribution. The -1 appears in the denominator to account for the statistics of photons as bosons; for fermions, this would be $+1$, while for particles without statistical constraints, this constant would be 0, and we would have our usual Boltzmann term as $e^{-\frac{h\nu}{kT}}$ in the numerator.

The fact that the cosmic microwave background radiation displayed a blackbody spectrum to great precision has far-ranging implications for cosmology, beyond that of simply providing a shape for the spectrum. By measuring the temperature of the radiation one also obtains the present ($z = 0$) number density of photons. Let us recall that given a probability distribution, we can compute mean values, and in this case we seek the energy density in the radiation. Using $E = h\nu$ we can integrate the Planck distribution to find the energy density:

$$\rho(E) = \int_0^\infty \frac{h\nu\, 8\pi\nu^2}{c^3 \left(e^{\frac{h\nu}{kT}} - 1 \right)} d\nu$$

This integral gives (in Mathematica, or by hand)

$$\rho(E) = \frac{8\,\pi^5\, k^4}{15\, c^3 h^3} T^4 = aT^4$$

The combination of constants preceding the temperature is called the radiation constant, a, with the value of

$$a = \frac{8\,\pi^5\, k^4}{15\, c^3 h^3} = 7.56573 \times 10^{-16} \frac{\text{Joules}}{\text{Kelvins}^4\, \text{m}^3}$$

Inserting $T_o = 2.72548$ K informs us that the background radiation presently contains an energy density of

$$\rho_o(E) = 4.17 \times 10^{-14} \frac{\text{Joules}}{\text{m}^3}$$

$$\text{or } 261 \frac{\text{KeV}}{\text{m}^3.}$$

A quantity that is of utmost importance when we examine the predictions of Big Bang nucleosynthesis (vide infra) will be the ratio of radiation to matter (i.e., photons to baryons) in the early Big Bang. For this we need to compute the present number, n, of photons per unit volume in the background radiation. We can find this by integrating the density

$$n(T) = \int_0^\infty \rho(\nu)d\nu = \int_0^\infty \frac{8\pi\nu^2}{c^3\left(e^{\frac{h\nu}{kT}} - 1\right)}d\nu$$

This integral, in Mathematica, becomes

$$n(T) = \frac{16\,\pi\,\zeta(3)\,k^3}{c^3\,h^2}T^3 = \frac{30\,\zeta(3)\,a}{\pi^4 k}T^3$$

where $\zeta(3) \sim 1.2026$ is the Riemann zeta function evaluated at a real value of 3. By evaluating the constants we find that the number of photons per cubic meter is given by

$$n(T) = 2.02868 \times 10^7 \frac{T^3}{\text{Kelvin}^3\,\text{m}^3}$$

and at $T_o = 2.72548$ K we have that

$$n(T_o) = 4.107 \times 10^8 \text{ photons}/\text{m}^3.$$

In the section on Big Bang nucleosynthesis this will be compared with the baryon density to set the time scale of nucleosynthesis in the early universe.

We can calculate the temperature of the early universe when decoupling happened using our good friend probability theory as embodied in the Maxwell-Boltzmann distribution. One certainly would be correct in assuming that the photon's energies would need to bear some relation to the ionization potentials of hydrogen and helium. For example, if we convert the ionization potential of hydrogen to temperature, using $E = kT = 13.6$ eV, we might expect that $T \sim 158{,}000$ K was the temperature at decoupling. We will shortly observe that this is a rather gross overestimate of the decoupling temperature and that the universe was substantially cooler when the radiation finally broke free. As an initial estimate, let us use the Saha equation which relates the ionization of a gas of atoms to the temperature using the Maxwell-Boltzmann probability distribution. The ratio of n_p, the density of ionized hydrogen atoms (i.e., protons) to those in the ground state, n_H, is given by the ratio of their probabilities

$$\frac{n_p}{n_H} = \frac{1}{n_e} \left(\frac{2\pi \, m_e \, kT}{h^2} \right)^{3/2} e^{-\frac{B}{kT}}$$

where n_e is the electron density, B is the binding energy, and m_e is the mass of the electron. The term in parentheses is the reciprocal of the de Broglie wavelength of the electron and represents the volume element in phase space. We recognize the right-hand side of this equation from our studies of the Maxwell-Boltzmann probability distribution since it represents the probability of finding a particle with an energy of B at a temperature of T. The early universe was electrically neutral so that charge conservation implies that $Z_p n_p + Z_e n_e = 0$ where Z is the charge on each particle: $Z_p = +1$, while $Z_e = -1$, which leads to $n_p = n_e$. Then, if there are n hydrogen atoms and protons in each volume, we can write the number density of electrons using charge conservation as

$$n_e = \frac{n_p}{\left(n_H + n_p \right)} n$$

Inserting this into the Saha equation yields

$$\frac{n_p}{n_H} = \frac{\left(n_H + n_p \right)}{n_p n} \left(\frac{2\pi \, m_e \, kT}{h^2} \right)^{3/2} e^{-\frac{B}{kT}}$$

Now, if we let $N = n_p + n_H$, we can eliminate $n_H = N - n_p$ and achieve

$$\frac{n_p}{N \left(N - n_p \right)} = \frac{1}{n} \left(\frac{2\pi \, m_e \, kT}{h^2} \right)^{3/2} e^{-\frac{B}{kT}}$$

By introducing the fractional ionization, $\chi(T) = n_p/N$, this becomes the quadratic equation

$$\frac{\chi^2}{1 - \chi} = \frac{1}{n} \left(\frac{2\pi \, m_e \, kT}{h^2} \right)^{3/2} e^{-\frac{B}{kT}}$$

Let the right-hand side be replaced by

$$M = \frac{1}{n}\left(\frac{2\pi\, m_\mathrm{e}\, kT}{h^2}\right)^{3/2} e^{-\frac{B}{kT}}$$

so that

$$\frac{\chi^2}{1-\chi} = M,\ \text{or}\ \chi^2 + M\chi - M = 0$$

which has solutions

$$\chi = \frac{1}{2}\left(-M \pm \sqrt{M(M+4)}\right)$$

giving the fractional ionization as a function of temperature and baryon density as

$$\chi(n,T) = \frac{1}{2}\left(-\frac{1}{n}\left(\frac{2\pi\, m_\mathrm{e}\, kT}{h^2}\right)^{3/2} e^{-\frac{B}{kT}} \pm \sqrt{\frac{1}{n}\left(\frac{2\pi\, m_\mathrm{e}\, kT}{h^2}\right)^{3/2} e^{-\frac{B}{kT}}\left[\frac{1}{n}\left(\frac{2\pi\, m_\mathrm{e}\, kT}{h^2}\right)^{3/2} e^{-\frac{B}{kT}} + 4\right]}\right)$$

The physical solution uses the plus sign. The number density of baryons in the early universe is about 300,000 per cubic meter; a plot of the fractional ionization as a function of temperature (Fig. 6.3) shows that a cloud of hydrogen would become transparent to photons when the temperature dropped to about 3000 K, a value quite different from the naïve expectation of $T \sim 158{,}000$ K suggested just from the ionization potential of the hydrogen atom. The reason for this is that the phase space factor and the baryon density combination rapidly rise at temperatures much

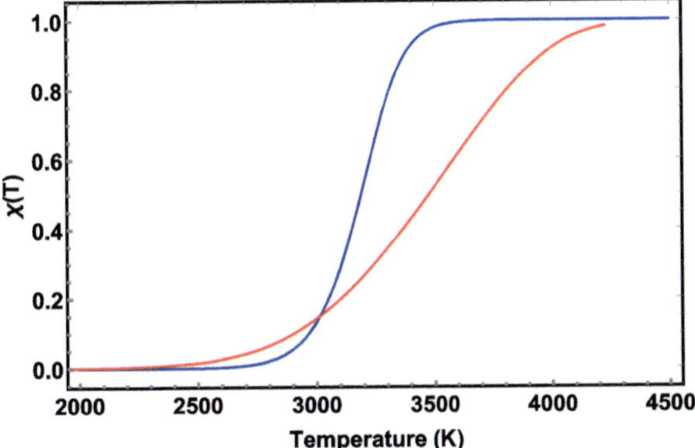

Fig. 6.3 The fractional ionization of hydrogen (blue line) as a function of the temperature of the photons in the expanding plasma of the early universe, as estimated from the Saha equation. A more accurate calculation (Weinberg 2009; red line) shows a broader transition as a function of temperature. Nevertheless, at a temperature of ~3000 K the universe became transparent to light and radiation decoupled from matter

lower compared with the exponential at the ionization potential of hydrogen. The photon density at decoupling is given by using

$$n(T) = 2.02868 \times 10^7 \; \frac{T^3}{\text{Kelvin}^3 \; \text{m}^3}$$

and at $T_{\text{dec}} = 3000$ K we have that

$$n(T_{\text{dec}}) = 5.477 \times 10^{17} \; \text{photons}/\text{m}^3.$$

We computed the ionization using a hydrogen plus proton density at decoupling of 3×10^5 atoms/m^3 indicating that there were about 10^{12} photons per hydrogen atom. This number is higher than the photon/baryon ratio since ~24% of the atoms were helium (^{4}He), which has 4 baryons/atom, and other baryonic contributions come from deuterium (^{2}H), tritium (^{3}H), ^{3}He, ^{6}Li, ^{7}Li, ^{7}Be, and ^{10}B.

Another clue that the CMB photons that we observe arise from the epoch of decoupling of radiation from hydrogen is found when we consider that the only other abundant baryonic component of the primordial fireball at times greater than ~20 min was helium. Its larger ionization potential of 24.4 eV gives a decoupling temperature of ~5100 K indicating that helium atoms were already stable by the time that the radiation broke free from the cooling plasma and that the hydrogen cloud formed the last scattering surface. The second ionization potential of helium is 54.4 eV, so much larger than the first that we only need to consider the first ionization in computing the decoupling temperature for helium.

What is the redshift and the time at which decoupling took place for helium and hydrogen? We can find this from the relation between redshift and temperature:

$$T(z) = T_\text{o}(1 + z)$$

And between temperature and time (prior to decoupling, when the universe was radiation dominated)

$$T(t) = 2.10 \times 10^{10} \; t^{-1/2}$$

for time in years. Using $T_\text{o} = 2.725$ K, we find that the decoupling of radiation from hydrogen (for $T = 3000$ K) took place at a redshift of $z_\text{H} = 1100$, while decoupling for helium occurred much earlier at a redshift of $z_{\text{He}} = 1907$. The time after the Big Bang that decoupling occurred was $t_{\text{dec}} = 380,000$ years for hydrogen and a shorter, $t_{\text{dec}} = 132,000$ years for helium. After decoupling the universe became matter dominated and the relationship between temperature and time changed to

$$T(t) = 1.48 \times 10^{11} \; t^{-2/3}$$

Our use of the Saha equation is only a first approximation. The hydrogen atoms were not actually in equilibrium with the radiation and the Saha approach neglects the fact that there are many other energy levels in the hydrogen atom that can be more easily ionized than electrons in the ground state and that these require only lower-energy photons. A more accurate calculation (Weinberg 2009) shows a broader transition as a function of temperature (Fig. 6.3) but the overall message remains unchanged; at a temperature of ~3000 K the universe became transparent to light and radiation decoupled from baryons. The surface of last scattering of photons reveals the distribution of hydrogen at decoupling. The fact that we have now generated an image of the Big Bang using photons that were emitted ~13.8 billion years ago lends great credence to our understanding of the current hot Big Bang model of the universe. But this does not exhaust the evidence in its favor, as we will see.

6.4.4   Anisotropies in the Cosmic Microwave Background Radiation

The discovery of the microwave background radiation in 1965 spurred a race to understand its detailed properties. The data from the COBE satellite in the early 1990s showed that its spectrum was indeed that of a blackbody to very high accuracy. However, the data also showed, to a first approximation, that the radiation was smoothly distributed across the sky, and this presented a puzzle. If this primordial radiation was so smooth, then how could the obvious clumps of galaxies arise? What was needed was a detailed mapping of the temperature variations across the sky. Before we examine the results of this endeavor, let us confront just what is meant by isotropy.

The Copernican or cosmological principle is postulated as a physical symmetry in which the universe is homogeneous and isotropic at the largest scales. It assumes that there is no preferred position or direction in space. The universe appears the same from every vantage point independent of where one looks in the sky. Any observer in the universe would then see the microwave background photons in the same way that we do here on Earth. One of the strongest lines of evidence for this isotropy is actually found from the observed uniformity of the cosmic microwave background radiation across the sky (Saadeh et al. 2016). Statistical isotropy is also supported by number count maps from the National Radio Astronomy Observatory Very Large Array Sky Survey (NVSS) radio source catalogue (Bengaly et al. 2019). Further evidence for isotropy is based on galaxy counts as a function of redshift from the Sloan Digital Sky Survey (Marinoni et al. 2012) where the scale above which isotropy is statistically robust is ~150 h^{-1} MPc ~100 Mpc. Given the average distance between galaxies of ~1 Mpc, there would be on the order of one million galaxies in this volume.

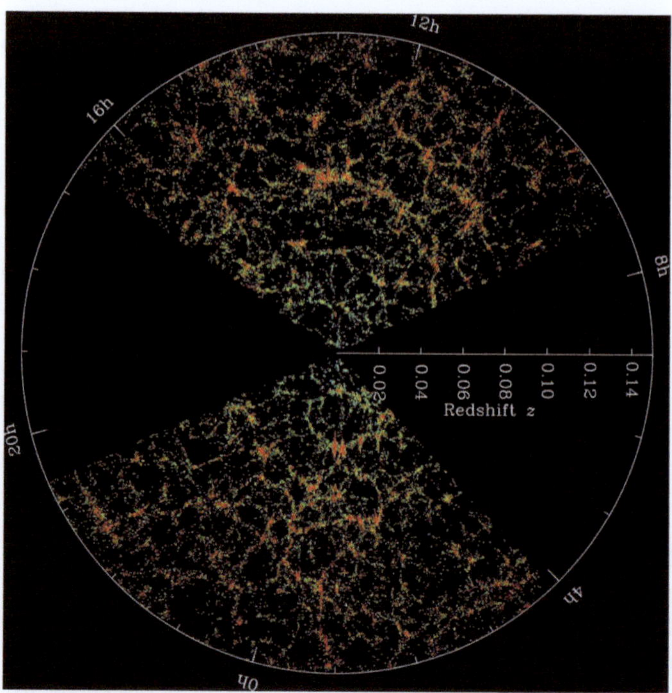

Fig. 6.4 A map of galaxies up to redshifts of $z \sim 0.15$ ($d \sim 2 \times 10^9$ light-years) shows that their distribution is lumpy, with clusters, filaments, and voids whose origin is reflected in the anisotropies present in the cosmic microwave background radiation. (Courtesy of the Sloan Digital Sky Survey, https://dev.sdss.org/wp-content/uploads/2014/06/orangepie.jpg, CC BY 4.0)

A cursory examination of the night sky shows that the actual distribution of matter in the universe is not very homogeneous and isotropic at smaller scales on the order of 10^7 light-years. We can estimate the matter density using our eyes. For example, on a clear, dark night the unaided human eye can easily observe Messier 31, the massive ($\sim 4 \times 10^{11}$ solar mass) Andromeda galaxy, behind the field of nearby foreground Milky Way stars, even though it is 2.537 million light-years (0.778 Mpc) from Earth. M31 is the largest of the galaxies in our local group; the other 53 are much smaller and are distributed within a radius of ~ 5 million light-years around the Milky Way. Most of the volume surrounding the local group is background space without stars. The density of stars is then about $4 \times 10^{11}/500$ Mpc^3 or $n \sim 8 \times 10^8/$ Mpc^3. The distance between galaxies is on the order of what we observe in our local group, about one Mpc. Large clumps of matter in the universe are present in the form of galaxies, clusters, filaments, and voids (Fig. 6.4). These structures, which grew during inflation from primordial quantum fluctuations, formed by gravitational instability from small initial density variations.

When the results from the COBE satellite were first presented, the point to point temperature variation was found to be very small, less than 20 μK. This smoothness presented an immediate puzzle; if the radiation was so uniform, how could stars and galaxies form? It was expected that the distribution of the flux across the sky had to be lumpy at some level. These lumps would trace the regions where matter clumped into higher-density regions; at the center of these regions stars and galaxies would have originated. These seeds of structure formation are imprinted on the CMB as subtle temperature variations from point to point. For this reason, the study of the anisotropies in the CMB has yielded perhaps the most important quantitative data with respect to the origin and composition of the universe. All-sky images of the anisotropies in the CMB (Fig. 6.5) dramatically display these lumps and literally provide a breathtaking view of our nascent universe only ~380,000 years after the Big Bang.

A closer examination of the position-dependent photon flux showed these lumps as regions where the temperature was some ~18 μK higher or lower than the average,

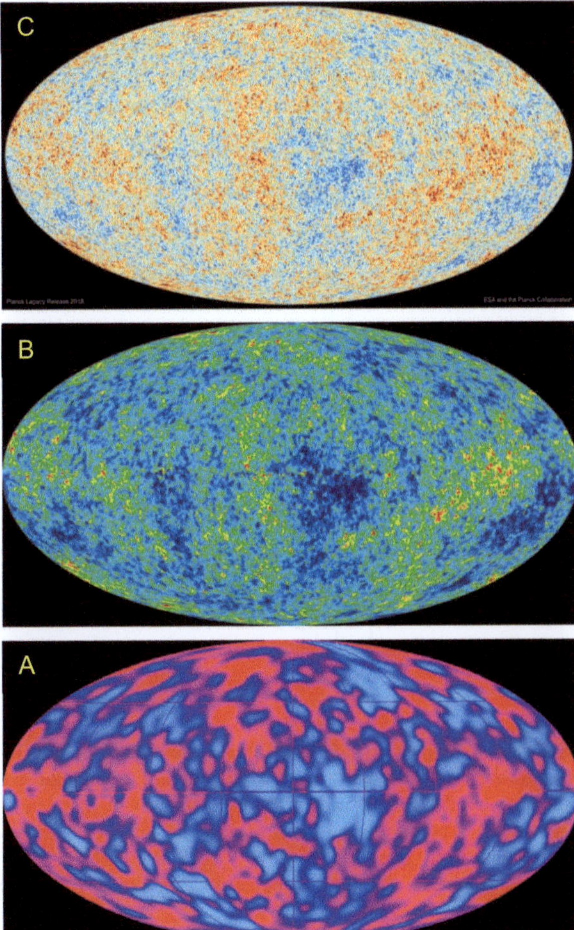

Fig. 6.5 All-sky maps of the anisotropies in the cosmic microwave background radiation 380,000 years after the Big Bang measured by the (**a**) COBE (in 1992); (**b**) WMAP (in 2003); and (**c**) Planck (Planck Collaboration 2013) satellites. The time-dependent differences in these maps reflect the large increase in resolution of the detectors as one progresses from **a** to **c**. Here, the large monopole and dipole signals have been subtracted to reveal the spatial dependence of the higher multipoles. These anisotropies arose as quantum fluctuations in the vacuum that were greatly amplified by an initial period of exponential inflation to become the seeds for the formation of galaxies. (Courtesy of NASA, the ESA, and the Planck Collaboration, used by permission)

i.e., deviations on the order of parts per million (Smoot et al. 1977). The images of the cosmic microwave background radiation seen in Fig. 6.5 show these anisotropies because the average temperature T_o has been subtracted from this map and one can then see the lumpiness of the radiation that gave rise to the clumps of matter that form the present structure of the universe. So much about the early and present universe can be gleaned from its study. The cosmological model, the density fluctuation spectrum, and the amount of dark matter in the universe can all be determined from the anisotropies in the cosmic microwave background radiation. It is no wonder then that considerable scientific effort has been directed at a detailed understanding of its features. The fundamental importance of the discovery and mapping of this radiation was recognized in several Nobel Prizes, in 1978 to Penzias and Wilson, for its discovery and in 2006 to Smoot and Mather "for their discovery of the blackbody form and anisotropy of the cosmic microwave background radiation."

The distribution of the amount of microwave power in these lumps as a function of the angular size of the patch of sky interrogated (Fig. 6.6) reveals that these seeds of structure obey a pattern of spatial oscillations, with a main peak at ~1° and an obvious set of ripples at smaller angular scales. A common method used to describe the anisotropies is to expand the difference in temperature readings across the two-dimensional surface of the sky in terms of spherical harmonics, which we first

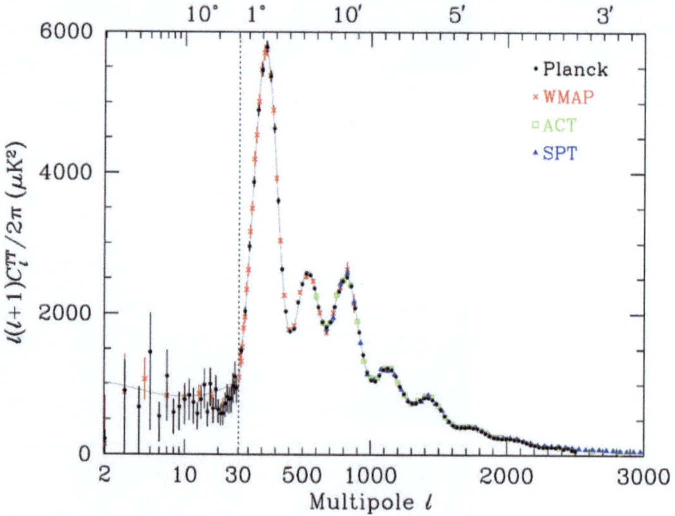

Fig. 6.6 The spectrum of fluctuations (the variance in μK^2) in the temperature of the cosmic microwave background radiation as a function of the angular size ($\theta° = 180/l$) of the region of sky sampled by the Planck and WMAP satellites and the Atacama Cosmology and South Polar Telescopes. The grey line is a fit to the data based on the current Planck satellite values of the ΛCDM cosmology model. (From Workman et al. (2022) and 2023 update. Content of the 2023 Review of Particle Physics is licensed under a Creative Commons 4.0 International (CC BY-NC 4.0) license)

encountered in our solution Schrodinger's equation for the hydrogen atom wave functions, as

$$\Delta T(\theta, \varphi) = T(\theta, \varphi) - T_o = \sum_{l=0}^{\infty} \sum_{m=-l}^{l} a_{lm} Y_{lm}(\theta, \varphi)$$

where the a_{lm} are the coefficients in the expansion that weight each of the spherical harmonics, $-l \leq m \leq l$, and l ranges over the positive integers. The average present temperature T_o is the temperature integrated over the whole sky:

$$T_o = \frac{1}{4\pi} \int T(\theta, \varphi) d\theta d\varphi$$

Because there exists no preferred direction in space, and there are only very weak phase correlations of the radiation at different spatial points, the variance in the temperature is independent of m and we can use the addition theorem for spherical harmonics to sum over m:

$$(\Delta T(l))^2 = \sum_m |a_{lm}|^2 = \frac{(2l+1)|a_{lm}|^2}{4\pi} = \frac{(2l+1)C_l}{4\pi}$$

$$C_l = |a_{lm}|^2$$

leaving the variance in the temperature distribution solely a function of l, the multipole moment. For $l = 0$, the monopole moment, a_{oo}, i.e., the average temperature, $T_o = 2.725$ K, is isotropic across the sky. This temperature corresponds to an energy of 3.76×10^{-23} Joules, or 235 μeV. For $l = 1$, $m = 0$, for a_{10}, the dipole moment of the temperature distribution becomes a simple cosine function (Smoot et al. 1997) with an amplitude of 3.3621 ± 0.0010 mK. This dipole, the largest anisotropy in the temperature distribution, arises as a Doppler shift of the monopole from the motion of the sun and the Earth at 368 ± 2 km/s with respect to the reference frame of the background radiation. There is an annual modulation of the radiation, a redshift as Earth's motion carries it away from the radiation and a blueshift as the Earth moves towards it. Higher-order anisotropies (multipoles with $l \geq 2$) are due to turbulence, or density fluctuations, in the early universe prior to the epoch of decoupling of radiation from matter ($z \sim 1100$) when electrons and protons combined into hydrogen atoms. These density variations were amplified by inflation and produced the regions where matter (and dark matter) clumped under the influence of gravity and collapsed to form the stars and galaxies that we observe in the present night sky.

Because measurements of the a_{lm} provide direct information about the universe's early physical properties, any model of the early universe must be able to reproduce the spectrum of anisotropies that we currently observe (Fig. 6.6). The temperature variance data in Fig. 6.6 are presented as a power spectrum, where all the peaks are positive; the variance is the square of the standard deviation (see Chap. 1) and is computed from the maps, like that shown in Fig. 6.5, by Fourier transformation. From our studies of NMR in Chap. 7, we will become familiar with this process and find the oscillating nature of the anisotropies is suggestive of the free induction decay of magnetic nuclei after a radiofrequency pulse, or of the decaying sound from that a bell makes after being struck.

Just as one can Fourier transform the carrier-subtracted audio signal from an NMR experiment and listen to it as a set of audio tones, so can one play the spectrum of anisotropies as a sound file. Dr. John Cramer, from the Department of Physics at the University of Washington in Seattle, has written a Mathematica program that uses the anisotropy data from the European Space Agency's Planck satellite to generate a sound file that reproduces, in a greatly compressed time frame, an audio representation of the Big Bang. This can be obtained from Dr. Cramer's website: https://faculty.washington.edu/jcramer/BBSound.html. He notes that he had to speed up this by 10^{24} in order to transform the frequencies into the range of human hearing.

One can readily understand that the early universe could be likened to a particle in a 3D box and that its wave function would display characteristic oscillations whose amplitude and wavelength would need to satisfy time-dependent boundary conditions. The spectrum of fluctuations in the temperature of the cosmic microwave background radiation as a function of the angular size of the region of observation shows oscillations that are sound waves that arose from the ringing of the early universe. These oscillations result from the competition between the photon pressure pushing matter, baryons and electrons, outward against the gravitational attraction of the baryons and the cold dark matter. The first peak is often called *the* acoustic peak and results from a wave that contracted once since decoupling. The second peak comes from fluctuations that contracted and expanded once since decoupling. It will come as no surprise by now that the third peak is from fluctuations that contracted twice and expanded once, and so forth. Combined data up to 2014 from the WMAP and Planck satellites (Aghanim et al. 2020) along with measurements by the Atacama Cosmology and South Polar Telescopes have traced at least nine of these acoustic peaks so far.

Any model of the early universe must be able to reproduce the spectrum of anisotropies that we currently observe (Fig. 6.6). As early as the 1990s, basic physics calculations (summarized by Hu and Sugiyama 1995) were able to reproduce the first peak at $l \sim 200$, as well as the reduced amplitudes of the second and third peaks at $l \sim 550$ and 850, and the subsequent decaying tail for $l > 1000$. The most recent work by the Planck and Atacama Cosmology Telescope teams (Choi et al. 2020 Arxiv.org/2007.072892; Aghanim et al. 2020) has produced anisotropy data out to $l \sim 8000$ and given precise measurements of the age of the universe, the density of baryons, and dark matter and supports the requirement for both dark energy (the cosmological constant) and dark matter. The spectrum of CMB anisotropies can be fitted with just 6 parameters that describe the universe. These, along with other measurements, are given in Table 6.1. Astronomers have still not settled on a value of the present Hubble parameter, H_o, so they hedge their bets by parameterizing it as

$100h = H_o$ and then quote measured values, such as the baryon density, Ω_b, as $\Omega_b h^2$ so that when there is agreement on H_o in the future, the actual value of Ω_b can be specified.

Table 6.1 shows both the parameterized and actual values for several constants used in the calculations of the CMB temperature anisotropy power spectrum, using a value for the Hubble parameter recently produced by both the Planck and ACT teams of $H_o = 67.37$ km/(s Mpc). The parameters in Table 6.1 are described as follows: The Ω_s are the densities; Ω_B, of baryons; Ω_C, of cold dark matter; Ω_Λ, of the cosmological constant; Ω_m, of total matter; and Ω_k, of curvature. The optical depth of the universe is τ; due to reionization of the intergalactic medium at $z \sim 8$ from the newly formed stars. Astronomers use the term optical depth instead of absorbance because the attenuation of optical photons can be due to scattering and/or reflection, or other physical phenomena in addition to absorption. In biochemistry, the absorbance is defined in Beer's law (Chap. 10) in terms of the removal of photons from the optical path, independent of the physical reason for their disappearance and includes scattering. The optical depth τ is defined as $\tau = -lnT$, where T is the transmittance, while the absorbance, a, is given by $a = \tau/ln10 = 0.4343\tau$. The transmittance of the universe is then $T = e^{-\tau} = 0.947$, which implies that ~95% of cosmic background radiation photons reach us on Earth from the last scattering surface only 380,000 years after the Big Bang. The absorbance of the universe is then $a = 0.02345$, a rather small value that indicates that the universe is currently ionized; otherwise, we would not be able to observe photons from high redshift ($z \sim 2.4$) quasars because these objects emit copious numbers of X-ray and ultraviolet photons with wavelengths $\lambda < 91$ nm (Lusso et al. 2015) that have sufficient energy (>13.6 eV) to ionize hydrogen in the intergalactic medium and would be absorbed and we would not be able to observe these quasars due to absorption by a neutral hydrogen fog. In Table 6.1 the upper limit to the neutrino mass is m_ν. That neutrinos are not massless is supported by the discovery that the three neutrino types (ν_e, ν_μ, ν_τ) oscillate, changing from one type to another, and this is only possible for massive neutrino eigenstates (Fukuda et al. 1998; Ahmad et al. 2001). The radius of the sound horizon at decoupling (about which we will have more to say later) is r_{dec}, and t_o is the age of the universe.

The Friedmann equation derived from Einstein's general relativity describes the space-time geometry of a homogeneous, isotropic universe in terms of its matter and energy density. If the density of matter and energy is above a critical value, ρ_c, then the geometry is positively curved, or closed in the sense that all light rays (geodesics) meet at infinity, or close in on themselves and the universe will eventually contract into a Big Crunch. If, on the other hand, the matter and energy density are below this critical value, all light rays diverge and the universe will expand at an accelerating rate. For a flat, i.e., Euclidian universe, in which parallel light rays remain parallel and never meet or diverge, the critical matter and energy density is

$$\rho_c = \frac{3H_o{}^2}{8\pi G}$$

where H_o is the current value of the Hubble parameter and G is the universal gravitation constant, $G = 6.674 \times 10^{-11}$ m^3/(kg s^2). Remembering that in SI units $H_o = 2.189 \times 10^{-18}$ Hz, the critical density is $\rho_c = 8.57 \times 10^{-27}$ kg/m^3. A single proton has a mass of 1.673×10^{-27} kg; the critical density corresponds to only 5.1 protons/m^3, a value that would be considered almost a perfect vacuum by any physicist. That the actual baryon density is only ~5% of this value (Table 6.1) or 0.25 protons/m^3 is the astonishing finding from detailed studies of the cosmic microwave background radiation. This means that currently observable matter in the form of baryonic particles in the universe accounts for a mere 5% of the total mass energy present.

It is common in the astronomical literature to compare the measured matter and energy densities to ρ_c using the ratio $\Omega = \frac{\rho}{\rho_c}$, and the measured baryon density Ω_b is ~5% of the critical density. For a flat universe, the densities add to the critical density. We can check the measurements, $\Omega_b + \Omega_c + \Omega_\Lambda + \Omega_k = 1.004$, which is unity within the errors of the measurements. The model we are using is called ΛCDM for cold dark matter with a cosmological constant, Λ. One notes from Table 6.1 that baryonic and cold dark matter only account for 32% of the critical density; the remaining density comes as the cosmological constant Ω_Λ. We will examine the origin of this baryonic matter in a later section.

6.4.5   Baryon Acoustic Oscillations

Imagine dropping a lighted firecracker into a pond of water and, as it explodes, watching the circular waves of light and fluid propagate outwards. The light flash would go on forever, but what if the water motion suddenly stopped. Such behavior would not be expected for water, which is a single component fluid, not coupled to the light. The early universe, on the other hand, was composed of several interacting components: electrons, baryons, photons, neutrinos, and dark matter. At a time after the Big Bang of around $t = t_o + \sim 380{,}000$ years, at a redshift of $z \sim 1089$, the universe was still opaque to radiation, the mean free path for a photon did not extend beyond that of the expanding cloud of ionized matter. Then, around $z \sim 1020$ as the expansion dropped the temperature to ~3000 K, below that needed to ionize hydrogen, protons and electrons combined to form neutral hydrogen. With no more free electrons to scatter the photons the mean free path for photons became essentially infinite and the universe became transparent to radiation. The photons propagated freely from then until now and were detected by the COBE, WMAP, and Planck satellites as the cosmic microwave background radiation.

Before this time, however, Thomson scattering of photons by electrons rendered the universe opaque to radiation, strongly coupling the pressure of the photons to the electrons and baryons, forcing the expansion of both matter and radiation. This drove the matter in the universe outwards in a spherical wave. At decoupling the photon pressure ceased to force the matter cloud outwards, and the velocity of the matter cloud abruptly dropped, while the photons continued to propagate unhindered. The outgoing matter wave stalled. It was as if the water wave from our pebble suddenly froze on the surface of the pond.

At decoupling acoustic wave propagation abruptly ceased, but these sound waves left a characteristic imprint both on the cosmic microwave background radiation, which we have seen as the anisotropies mentioned above, and on the distribution of matter in the universe at a radius of ~450,000 light-years. Rescaling this distance to compensate for the difference in redshift between then and now brings this distance to about 500 million light-years, or ~150 megaparsecs in the present universe. The predicted behavior of the photon-matter propagation is shown in Fig. 6.7 for several redshifts. Today one should observe a peak at this separation of galaxies when one measures the correlation function $\xi(x)$ of the matter density, $\delta(x) = \frac{\rho(x) - \rho}{\rho}$, compared with the average density, ρ. The correlation function is

$$\xi(s) = \int_0^\infty \delta(x)\delta(x + s)dx$$

The oscillations seen in the angular power spectrum of the cosmic microwave background radiation (Fig. 6.6) arise from the baryon acoustic oscillations in the early universe and are also observable in the spatial correlation function for galaxies. The Sloan Digital Sky Survey (SDSS; York et al. 2000) and the 2-degree Field Galaxy Redshift Survey (2dFGRS; Colless et al. 2001) were the first redshift surveys to have directly detected the BAO signal. Subsequent data from the SDSS (Eisenstein 2005; Anderson 2014) have resulted in a robust detection of this feature (Fig. 6.8) in a huge spatial volume of 6 Gpc^3.

The baryon acoustic oscillation signal in the correlation function for the distribution of galaxies measured by the Sloan Digital Sky Survey (Fig. 6.8a) shows two features, a peak at a distance of ~150 Mpc and a decaying exponential background reflecting the decrease in the spatial correlation of galactic matter with distance. The peak is (Fig. 6.8b) well modeled by a gaussian of width 15 Mpc, a value that represents the spatial extent of the stalled wave of baryons (the width of the waves in our pond when their expansion stopped). High-precision measurements of the baryon acoustic oscillation signal performed by the Planck satellite (Table 6.1) show that the radius of the universe at the epoch of decoupling was 147.18 ± 0.29 Mpc. The spatial correlation of galactic matter is described by a random exponential probability distribution with a correlation length of 68.63 Mpc, which is much smaller than the radius of the baryon acoustic oscillation signal. Note also that this background galactic correlation function decays to zero around 200 Mpc. The baryon acoustic oscillation signal is another physical fossil left over from the Big Bang.

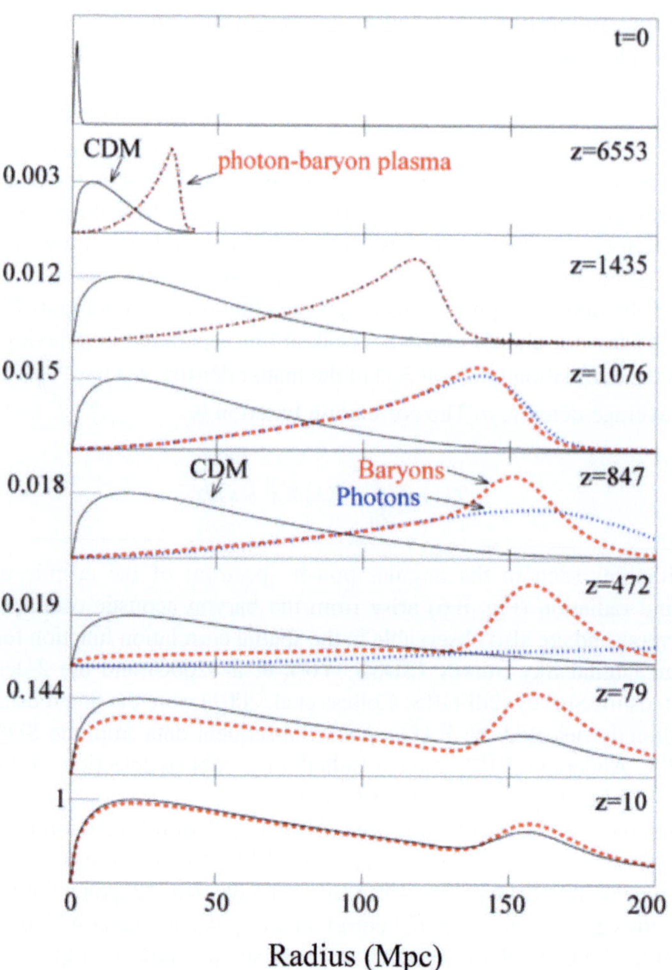

Fig. 6.7 Simulations of the expansion of the universe showing the spatial relationship between the cold dark matter (black), the baryons (dotted red), and the photons (dotted blue) as the universe expanded and cooled through the decoupling transition at $z = 1089$. Before decoupling photon pressure pushed the baryons, but afterwards the photons propagated freely, leaving the baryon wave stalled at ~150 Mpc. The gravitational attraction of the baryons also dragged the cold dark matter causing a matter clump at this distance. (Courtesy of Martin White: http://mwhite.berkeley.edu. Used with permission)

6.4.6   The Universe from Nothing

Either the universe has always existed in its present form, an idea that invokes some vague, unphysical form of infinite time and which is observationally ruled out by the measured expansion of space-time and the finite age of the universe, or based on our

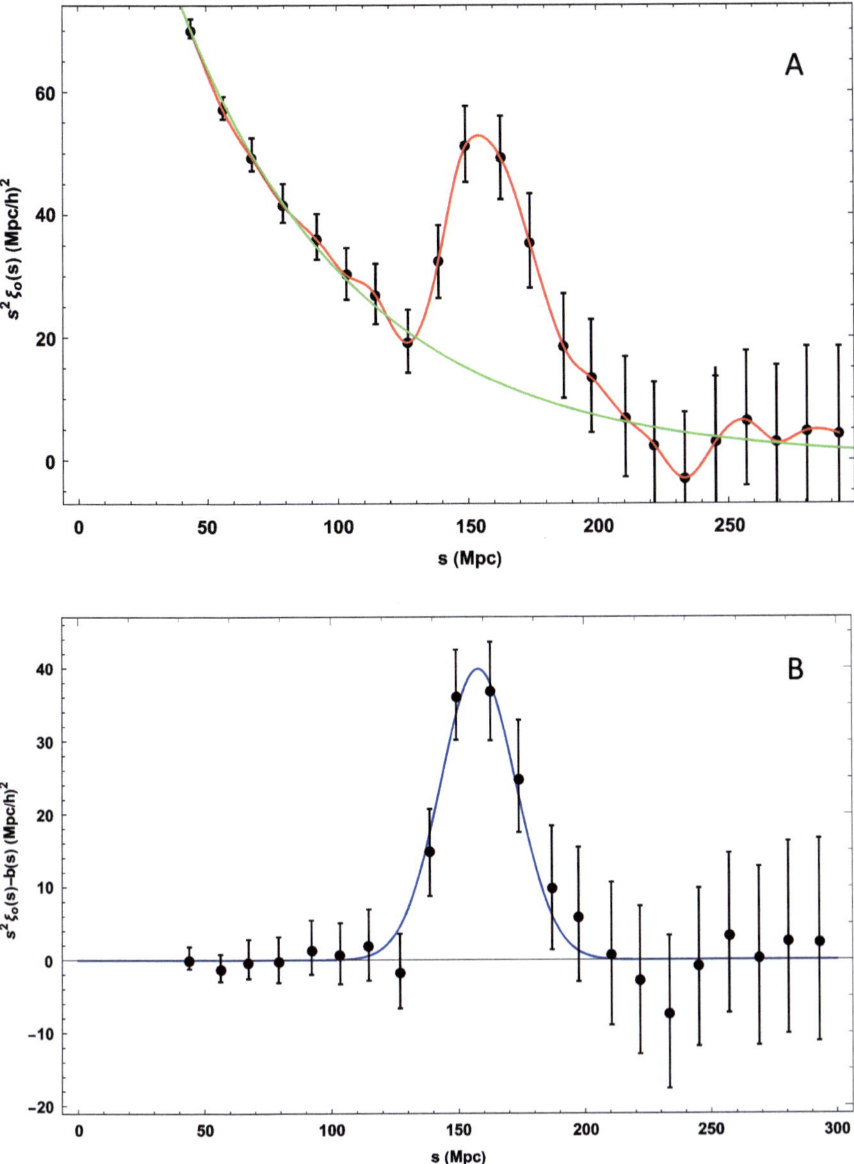

Fig. 6.8 The baryon acoustic oscillation signal in the correlation function for the distribution of galaxies measured by the Sloan Digital Sky Survey (Anderson et al. 2014). (**a**) The complete correlation function of the monopole is denoted by the black data points, with their errors; the red line connects the points to guide the eye. The green line is an exponential fit to the correlation function without the peak: $s^2\xi_o(s) = Ae^{-b\,s}$ with $A = 132.92$ $(\text{Mpc/h})^2$ and $b = 0.01457$ $(\text{Mpc})^{-1}$. (**b**) Subtraction of the exponential background reveals the baryon acoustic oscillation signal as a large peak at a distance of ~150 Mpc. The blue line is a Gaussian with a width of 15 Mpc. Here we have used $H_o = 67.6$ km/s/Mpc

measurements, we find that it arose as a tiny, hot, dense, initial state some 13.8 billion years ago. What preceded its formation? The simplest answer, and the one favored by Occam's razor, is that the universe arose from nothing. In quantum mechanics, the state of *nothing*, the vacuum, $|0>$ is a very special state. If I gave you a box and claimed that "nothing" was in the box, how could you know if I was telling the truth? The answer is so simple that even a young child could tell you; she would chime, "Open the box and look," and that is precisely how you would proceed. There are two important things to note here, one is that "looking" is a *measurement* process, subject to the rules of quantum mechanics, and the other is that we need to agree on a definition of nothing. What is *nothing*?

In order to understand the vacuum state, i.e., to perform measurements on it, we know that the uncertainty principle will appear from the depths, much like the Water Gnome in Dvorak's *Rusalka*, to remind us that human knowledge is different from wishful thinking. We need to examine two slightly different ways to understand the uncertainty principle. The first applies to discrete quantum states, such as spin, subject to logical confinement, while the second concerns continuous states like momentum and position, subject to physical confinement.

The uncertainty principle results from the fact that certain pairs of quantum operators, whose eigenvalues are the observables, do not commute. In the discrete case, it tells us that the wave function cannot simultaneously be an eigenfunction of a pair of similar operators, such as the spin operators S_x and S_y, but it does not constitute a limit on knowledge in the sense that we can measure, say S_x, precisely and obtain an unambiguous result, while an attempt to simultaneously determine S_y will produce random results. A measurement of the component of the spin in the x-direction will always produce a discrete result, either $+1/2$ or $-1/2$, as it were. In the continuous case, we find that the wave function also cannot be a simultaneous eigenfunction of, for example, position and momentum, but position and momentum can assume arbitrary values, so that a measurement will not produce discrete values such as spin up or down, but continuous values. Here the value of commutator cannot be less than $\frac{\hbar}{2}$, and this does indeed place limits on our knowledge of the system. It is this second property of the uncertainty principle that we invoke here.

Let us examine the uncertainty principle for energy and time: $\Delta E \Delta t \geq \hbar/2$. We contend that this must place limits on our knowledge of the energies of quantum systems, though, in nonrelativistic quantum mechanics, there is an operator for energy, but no operator for time, unlike that for the conjugate variables space and momentum. Nevertheless, by embracing relativistic quantum mechanics, which unifies space and time into spacetime, it is well understood from nuclear and elementary particle studies that the lifetime of quantum states can be determined through the application of the uncertainty principle for energy and time. One measures, ΔE, the dispersion of the energy probability distribution and uses the minimum version of the uncertainty principle, $\Delta E \Delta t = \hbar/2$, to find the lifetime of the state. Excited nuclear states are a great example. We can measure the size of the nucleus ($d \sim 10^{-15}$ m) and by knowing that the maximum information transfer velocity is, c, the speed of light, use $t = d/c$ to predict that the widths of nuclear resonances cannot be less than ~100 MeV. Massive numbers of experiments routinely confirm this result.

Let us look into the box we were just given and ask how much energy is in the box. The key here is measurement; it is the only way we can gain knowledge. We open the box, look inside, and then close it. However, if we open it again and ask if the amount of energy has changed, we expect that, by the conservation of energy, we would find exactly the same amount. In order for us to verify that energy is conserved in a process or an event, we must be able to measure the energy of a system before and after the event; this measurement requires a certain amount of time, Δt. If energy is not conserved for a time less than Δt, no measurement can reveal it.

What is more surprising, perhaps, in this context, is that the obverse is true; energy need not be conserved for a time less than Δt. Here, the role of measurement, or an observer, is crucial to our understanding of nature. Let us return to our gift box containing "nothing." In order to verify that the box indeed contained "nothing," we looked into it. This involved two processes; the first was to make a measurement, and here we will use an example of shining light into the box as the means of measurement. The second is to understand more fully the vacuum, i.e., the quantum state with zero excitations, $|0\rangle$. Let us first inquire about the interaction of light with the interior of the box.

Richard Feynman (1949) developed a novel, but powerful, way of visualizing the quantum field theory of the interaction light with matter, as embodied in quantum electrodynamics, through the use of graphs, which have now received the honorific of Feynman diagrams. In a classic paper entitled "A Space-Time Approach to Quantum Electrodynamics," he provided a direct correspondence between the underlying mathematical terms in an expression for the scattering of light (photons) from electrons (matter) and a set of simple metasymbols summarized as spacetime diagrams. These diagrams are graphs (Fig. 6.9) that represent the interactions between particles (Fermions) and fields (Bosons) and serve as a shorthand for remembering the various Green's and delta functions that are combined in order to give the amplitude—whose absolute square is the probability—for the described

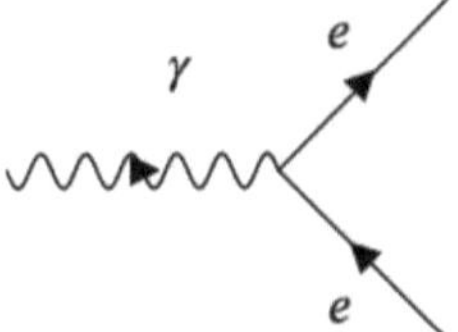

Fig. 6.9 Feynman diagram illustrating the conceptual interaction between a photon, γ, and an electron, e, in one spatial and one temporal dimension. The lines represent Green's functions, or "propagators" for each of the objects, and are meant to denote the logical, rather than the physical trajectories. The arrows on the electron line show the sense of time. This interaction is the basis for human sight. When we "look" at an object in an ordinary sense, we are enjoying the large probability of the scattering of an optical photon by a molecular electron

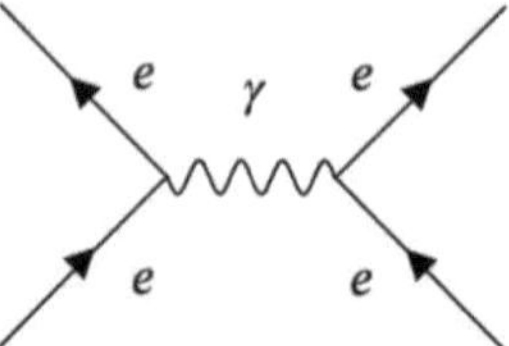

Fig. 6.10 A Feynman diagram illustrating the repulsive Coulomb interaction between two electrons mediated by a *virtual* photon, γ. The lines that enter and leave represent observable electrons, but the line for the photon is internal and hence virtual. All electromagnetic interactions can be written as a variation of diagrams such as this and Figs. 6.11 and 6.12

process. He described these as two-dimensional spacetime diagrams with a spatial dimension, x, and an orthogonal time dimension, ct. A simple Feynman diagram for the interaction between a photon and an electron graphically illustrates (Fig. 6.9) the scattering interaction and momentum transfer that occurs. This diagram describes how we can see; electrons (e) in an object scatter photons (γ) which travel to our eyes and interact there with the electrons in the rhodopsin molecules in our retinas (Chap. 10). In this case the exchanged photon is real and comes from a light source such as the sun. The progress of the electron's wave function, but not its physical trajectory, since this is not measurable by the uncertainty principle, is represented by a solid, directed, line, labeled e, while a photon, γ, is indicated by a wavy line. The intersection at a vertex represents the absorption or emission of a photon by an electron. At each vertex charge, lepton number, baryon number, and all the other conservation laws must hold.

The photon is the bosonic particle exchanged in the interaction between a pair of electrons; it is the exchange of this particle that mediates the electromagnetic interaction. The Feynman diagram (Fig. 6.9) graphically illustrates the absorption of a photon by an electron, and of course, the opposite diagram, one with the photon arrow reversed, would represent the emission of a photon by an electron. By combining these two diagrams we can arrive at the diagram for the Coulomb interaction, the inverse-square force between two electrons (Fig. 6.10).

In these diagrams, observable particles are indicated by external lines, ones that enter or leave the diagram, such as the two electron lines in Fig. 6.10. On the other hand, internal lines, ones that do not leave the diagram, represent virtual particles, in this case a virtual photon. This diagram also describes electron-electron scattering, as occurred in a particle accelerator, like the Stanford Positron Electron Accumulator Ring (SPEAR). In that case, the exchanged photon would have a lifetime too short to be measured and would be called a virtual photon. Here we see the crucial role played by measurement again. In order for a particle to have a physical existence, it must live long enough to be measured. Energy is conserved if quantum states live long enough to measure the energy. If states are too evanescent for measurement, then they can violate conservation of energy. In the above Feynman diagram energy is not conserved because electrons do not simply emit photons at random; they would have to give up mass in order to do so. Rather, virtual photons are emitted and

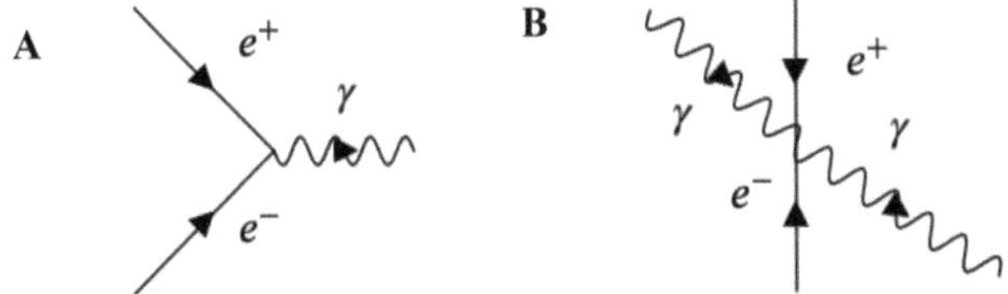

Fig. 6.11 (**a**) The Feynman diagram for electron-positron annihilation. Note that the positron is represented as an electron going backwards in time. This diagram conserves energy, but not momentum. (**b**) In this diagram the e^+e^- pair produces two colinear photons and thus conserves both momentum and energy

absorbed according to the uncertainty principle, for times too short to measure. The photon in this diagram acts as the carrier for the electromagnetic interaction so that there is no action at a distance. Other field-carrying particles (bosons) are exchanged in the weak and strong forces. We have, for simplicity, chosen to use Feynman diagrams illustrating quantum electrodynamics, but it should not be forgotten that we could also have used such diagrams for the weak and strong interactions as well; the same ideas hold.

Feynman's replacement of the complex, underlying, mathematical expressions for scattering by an easy-to-understand, diagrammatic approach is in keeping with our initial expectation that the symbolic manipulation of classical mathematics can be embedded in a higher language in which advanced metasymbols represent the lower machinery of actual calculations. Thus, Feynman's brilliance was revealed.

So what do we do with these metasymbols? Our task is to understand the origin of life. There will be many way stations on the way when we proceed to climb this mountain. A first step on this most pleasant journey is to find out how the universe came into existence, and this, it turns out, is the easy part because we can use Feynman diagrams and the uncertainty principle as our guide.

Now what about these virtual photons? In order to answer that question we need to expand our knowledge of the composition of the universe to include objects besides ordinary matter. In 1928 Paul Dirac unified quantum mechanics with Einstein's special relativity to produce the first relativistic equations to correctly describe Fermionic matter (in this case electrons). In doing so, he naturally predicted the existence of a positive electron (a positron or antielectron) which would annihilate an ordinary electron upon meeting one, as shown in the Feynman diagram (Fig. 6.11a). When matter meets antimatter, the sparks fly!

Positrons, as they were called, were soon discovered by Carl Anderson in the early 1930s. It turns out that in order to conserve both energy and momentum in the positron annihilation diagram, two photons must be produced and these photons must leave the interaction region in exactly opposite directions (Fig. 6.11b). This is now used as the basis for the medical imaging procedure positron emission tomography, or PET. Through the use of collimated detectors operating in coincidence, one can observe the two photons produced and can calculate their origin within the object and thus produce images of the emitting region. A common use of this

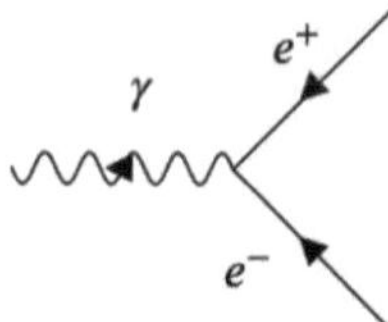

Fig. 6.12 A Feynman diagram for pair production of a pair of electrons and positrons through the disintegration of a photon of energy greater than or equal to the rest mass of the lepton pair, 1.022 MeV

technique is to label 2-deoxy-glucose, a non-metabolizable sugar, with a positron-emitting isotope, such as ^{18}F, and then to inject this labeled glucose into the blood. As the glucose is taken up by cells in the brain or in a tumor, images of the body will reveal the location of the emitting region.

When one can incorporate antimatter into Feynman diagrams, additional processes can occur (Fig. 6.12). A photon of high enough energy (greater than the rest masses of the two electrons, 1.022. MeV) can disintegrate into an electron-positron pair. This process is observed in nature and is called pair production; it is the opposite of electron-positron annihilation.

Let us return to the virtual photons shown in these diagrams. In order to understand their role we have to invoke the uncertainty principle in its quantitative form:

$$\Delta E \, \Delta t \geq \hbar/2$$

This states that for any two conjugate (orthogonal) variables, like energy and time, or position and momentum, the product of the accuracy in one variable and the other can never be greater than the reduced Planck's constant divided by 2. This is one of the profound limits on knowledge we alluded to earlier and has been verified many times. It correctly predicts, as we have seen, the widths of nuclear states.

But, you say, Noether's theorem tells us that the conservation of energy arises from a mathematical symmetry of space-time—time-translation invariance. However, the difference between mathematics and physical reality, as Einstein so forcefully argued in his essay "Geometry or Reality?," is that our knowledge of physical reality has no other basis than measurement, while mathematics is based solely on logic. Are these two knowledge systems at odds? The answer, which formed the quantum-mechanical foundation for the twentieth-century revolution in understanding, is a resounding yes. Our knowledge of the biophysical world is based on measurement, with its uncertainty principle limits, while logic admits no such limits and assumes infinite precision in its statements. For example, the logical statement

$$A = B$$

asserts that A and B are indistinguishable in the sense that if A is an apple, and B is an apple, then both A and B are apples; they are instances drawn from the same set of objects. But, as we depart from the world of logic and enter the physical world, it becomes more interesting to compare A with B; for this we need to make a measurement, and then uncertainties come into play. We now must decide if the following statement is true:

$$A \pm \Delta A = B \pm \Delta B$$

where ΔA and ΔB are the uncertainties (errors) in the measured values of A and B, and these uncertainties cannot be reduced to below values given by the uncertainty principle. This situation is of immense physical significance since, as our example of elephant mapping in Chap. 4 illustrated, the accuracy of measurement depends on the measuring apparatus, and there are only physically limited methods with which to do the measurement. We often use light waves (photons) to make measurements of the positions of matter particles in our environment, but we cannot increase the resolution of our measurements arbitrarily with light waves because the energy increases as the wavelength decreases; it is like trying to improve our map of the elephant by scattering bulldozers from him.

In order to ascertain if indeed $A = B$, we must instead answer the question: Does the measured value of A with its uncertainty agree with the measured value of B including its uncertainty? And this must be performed under the constraint that ΔA and ΔB cannot be reduced to zero because measurement is subject to the laws of quantum mechanics. These are the fundamental limits on human knowledge imposed by quantum mechanics.

What has all this to do with virtual photons? Conservation of energy is a logical proposition, based on a symmetry of nature, time translation invariance, which has in turn been amply verified through more than 170 years of measurements. Thus, the logical statement that energy is conserved is absolutely true because time translational invariance is a fundamental property of the Hamiltonian and is a spacetime symmetry of the universe. However, the conservation of energy as a physical law can only be true subject to our ability to measure it, and that is described by the uncertainty principle. It takes a certain amount of time Δt to measure whether energy has been conserved in an interaction, and the measurement of the energy is subject to another uncertainty, ΔE; their product must obey $\Delta E \, \Delta t \geq \hbar/2$. The implication of this fact is that energy conservation can be violated if it occurs for such a short time that no measurement can detect the nonconservation. And here enter the virtual photons seen in many Feynman diagrams; their existence is so fleeting that we cannot detect the fact that an electron must have given up rest mass in order to create them. The properties of the quantum world are seen to diverge in a fundamental way from those of the macroscopic world of our ordinary experience. Virtual particles can erupt and decay at will as long as they do not exist for long enough for us to measure their explicit violation of energy conservation. This is a fact based on the limits of measurement. And yet, it should be emphasized that these statements lead to testable predictions which when tested are found to agree with quantum mechanics and not with our naïve expectations.

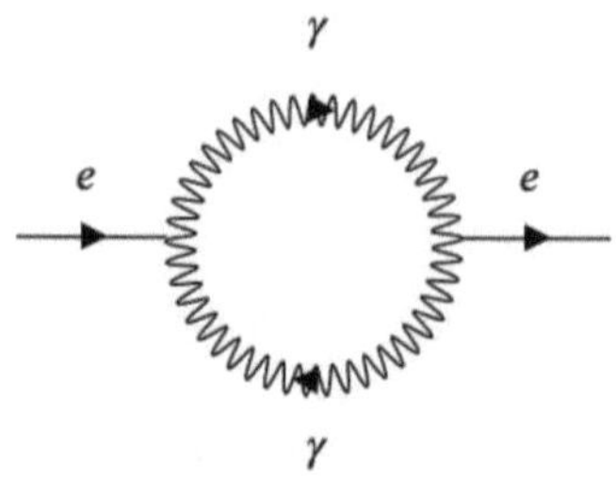

Fig. 6.13 A Feynman diagram for the spontaneous decay of an electron into a pair of photons, which then recombine to reproduce the electron. This diagram explicitly violates the conservation of energy because the energy of the virtual photons can assume any value

A major triumph of Feynman's quantum electrodynamics was to explicitly include virtual photons into diagrams such as Fig. 6.13, which represents the travel of an electron from left to right. By the uncertainty principle an electron can spontaneously decay into a pair of photons which can then recombine to reconstitute the electron for such a short time that you cannot measure the energy nonconservation. But more importantly, is this actually a measurable event?

The short answer is yes. This diagram (Fig. 6.13) is the basis for the Lamb shift in the energy levels of the $2s_{1/2}$ ($n = 2$, $I = 0$, $j = \frac{1}{2}$) $\rightarrow 2p_{1/2}$ $(2, 1, \frac{1}{2})$ transition in the hydrogen atom, whose measurement resulted in a Nobel Prize. Detailed quantum electrodynamic calculations, using Feynman diagrams, predicted that this level shift would be at the microwave frequency of 1058 MHz, a value that agreed very well with the measured value of 1057.9 MHz found in the late 1940s (Lamb and Retherford 1947).

Why does this diagram (Fig. 6.13) imply a level shift? Because when the negative electron decays into a neutral photon pair, there is a violation of charge conservation. The resulting neutral photon pair is not electrically bound to the positively charged nucleus so that its energy increases. A similar diagram with the roles of the photons and electrons reversed is shown in Fig. 6.14a in which a photon decays into an electron-positron pair, which then recombine to form the photon. This effect is called vacuum polarization. These quantum effects seem counterintuitive, but have testable and confirmed predictions and have a profound relationship with the origin of the universe.

In order to understand this let us examine the vacuum in the light of what we now understand about the limits of measurement and virtual particles. The fact that the vacuum contains "nothing" must now be understood with respect to measurement; it takes two things to make a measurement of "nothing": (1) sufficient time to provide an upper limit on the energy density of the vacuum and (2) someone to make the measurement, i.e., an observer.

We stated earlier the rule for Feynman diagrams, one of which was that *real*, physically observable, particles are represented by external lines. Internal lines (Fig. 6.14b, c) correspond to virtual particles, that is, particles whose existence

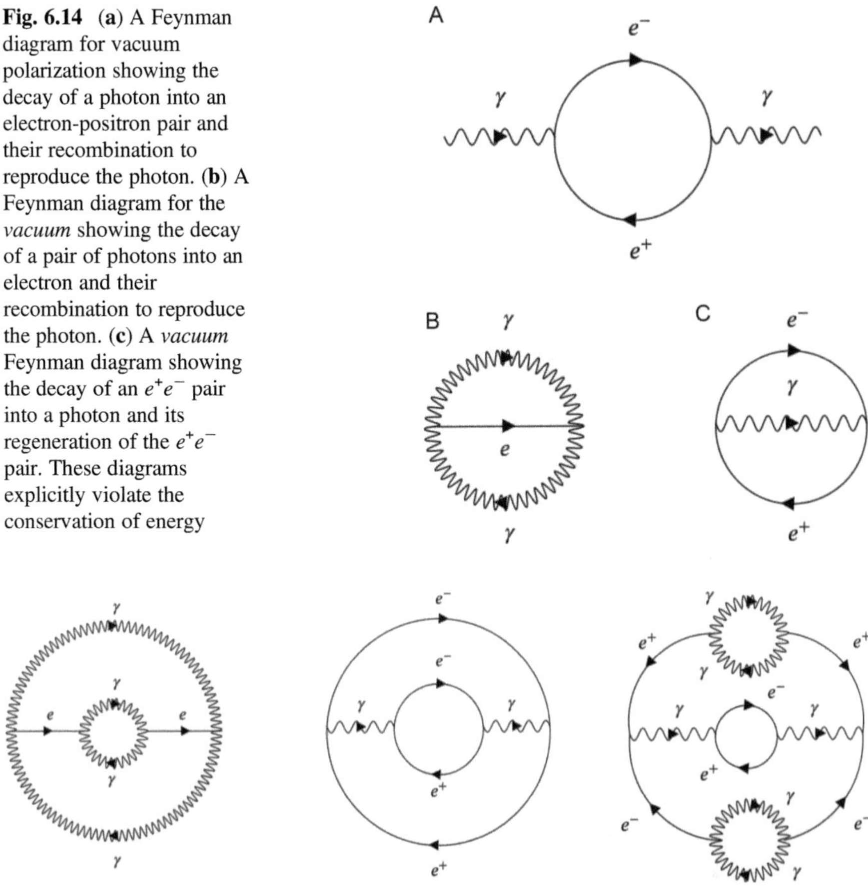

Fig. 6.14 (a) A Feynman diagram for vacuum polarization showing the decay of a photon into an electron-positron pair and their recombination to reproduce the photon. (b) A Feynman diagram for the *vacuum* showing the decay of a pair of photons into an electron and their recombination to reproduce the photon. (c) A *vacuum* Feynman diagram showing the decay of an e^+e^- pair into a photon and its regeneration of the e^+e^- pair. These diagrams explicitly violate the conservation of energy

Fig. 6.15 Some Feynman diagrams with no external lines illustrating the quantum electrodynamic *vacuum*. Of course, there are an infinity of ever more complicated such diagrams in quantum electrodynamics alone, as well as for the weak and strong interactions. The vacuum is more interesting than we thought!

violates the conservation of energy, but this violation can exist, by the uncertainty principle, as long as it is not measurable.

Since the vacuum is a state in which there are no particles in the initial state and none in the final state, what kinds of diagrams can we write which satisfy this requirement? This is easy since we can write many diagrams that only contain internal lines, such as shown in Figs. 6.14b, c and 6.15. Now, note that we can insert any number of electron-positron pairs into any photon line and any number of pairs of photons into any electron line, so that the actual number of vacuum diagrams is infinite. One can immediately grasp the idea that, because there are well-understood physical limits on measurement, and hence our knowledge, an infinite myriad of virtual events can occur *under the radar* so to speak, in the vacuum state, and that the existence of these events must inform our concept of *nothing*.

Another demonstration of these vacuum fluctuations is the Casimir effect in which quantum fluctuations generate a measurable physical force acting on two conducting metal plates separated by a vacuum. Hendrik Casimir predicted this force in 1948 (Casimir 1948) and it was quantitatively measured by Lamoreaux in 1997 (Lamoreaux 1997). Such effects are now understood to be the basis for the quantum levitation of small objects (Munday et al. 2009). It is now clear that the uncertainty principle both places limits on knowledge and allows events to occur in the vacuum which would otherwise seem fantastical. We have only shown vacuum diagrams that involve the electromagnetic interaction, but a myriad of Feynman diagrams can be similarly constructed to illustrate the vacuum events that occur in the weak and strong interactions among leptons and hadrons as well.

The most spectacular of these events is the origin of the universe. How is this possible? The mass of the universe is roughly 10^{54} Kg. If we were to "borrow" this much energy 10^{70} joules ($=$ mass) from a vacuum fluctuation, how long would the uncertainty principle allow us to do this before we would have to repay this energy debt? That is, how much time can a vacuum fluctuation of this magnitude survive before one can measure the violation of the conservation of energy? The answer is: not long, but "not long" corresponds to $\sim 10^{-104}$ s. This is an unmeasurably small amount of time. The smallest amount of time which is physically meaningful is the Planck time, or the time it takes for light to traverse a black hole whose radius is equal to its Compton wavelength, $t_P \sim 10^{-44}$ s. It is useful to note that the origin of the universe in the Big Bang did not actually have to violate the conservation of energy like this because the kinetic energy of the mass developed is positive, while the total energy stored in the gravitational attraction among the mass particles is negative; gravity is an attractive force. These two forms of energy are equal and opposite and their sum is zero; the energy density of the universe has been measured to exactly correspond to the critical density, Ω_c. Thus, if there was no energy prior to the Big Bang, there was equally zero total energy after it.

We conclude that not only do vacuum fluctuations actually occur, but that one must explicitly take them into account in quantum field theory to have theory agree with experiment. The vacuum state contains a hurly-burly of quantum processes (virtual particles) which flit in and out of existence faster than we can measure them, that is, for times allowed by the uncertainty principle. The universe arose from such a set of vacuum fluctuations; these could grow in amplitude through inflation to produce the mass of the known universe from within a volume of $\sim 10^{-33}$ m^3. As we have seen for the particle in the box in quantum mechanics, the pressure on the walls of a box so small, containing so much mass, would have been enormous (10^{58} Pascals), and by the ideal gas law, the temperature would have been in excess of 10^{20} K. It is therefore hardly surprising that this region of newly generated space-time would have rapidly expanded, such conditions are literally explosive and have given rise to the nomenclature referring to this epoch as that of George Gamov's "Big Bang."

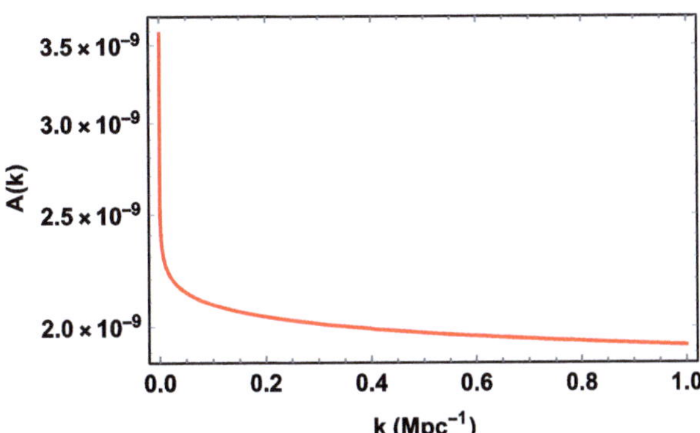

Fig. 6.16 The primordial comoving curvature power spectrum for quantum vacuum fluctuations in the spherical box containing the early universe during its expansion prior to decoupling (using $A_s(k)$ and n_s determined by the Planck satellite; Aghanim 2020)

6.4.7   Quantum Vacuum Fluctuations

Can the spectrum of primordial quantum vacuum fluctuations be measured? The anisotropies in the background radiation offer a way to determine both the strength of the early vacuum fluctuations and their variation with wavelength, $\lambda = 2\pi/k$. Measurements by COBE, WMAP, Planck, and several other experiments in the last few decades have found that the slope of the primordial vacuum energy fluctuation spectrum $A(k)$ is n_s, where

$$A(k) = A_s(k_o)\,(k/k_o)^{n_s - 1}$$

For a scale-independent spectrum of perturbations, $n_s = 1$, and $A(k) \propto k$. The measurements (Table 6.1) favor a value for n_s slightly less than one. A_s is the primordial comoving curvature power spectrum amplitude defined at the pivot scale wavenumber $k_o = 0.05$ Mpc^{-1} (Fig. 6.16). One can estimate the value of A_s from the mean value of the quadrupole temperature variance measured by COBE for $10 \leq l \leq 40$ of $Q = 18$ μK which gives $A_s \sim Q^2 \sim 10^{-10}$ K^2. The measured value from Planck (Table 6.1) is 2.14×10^{-9} K^2, using the estimated comoving wavelength from decoupling of ~126 Mpc, which is slightly smaller than the actual measured value (Table 6.1) of $r_{dec} = 147$ Mpc. These quantum fluctuations were amplified by inflation and produced the inhomogeneities in the distribution of both baryonic and dark matter that eventually coalesced into the presently observed (Fig. 6.4) cosmic web of galaxies.

6.5   The Early Universe

What did the early universe consist of? So far we have found, through measurement, that the universe arose from a tiny, hot, dense vacuum fluctuation. It certainly had to somehow evolve into the current constituents of the presently observed world. We

Standard Model of Elementary Particles

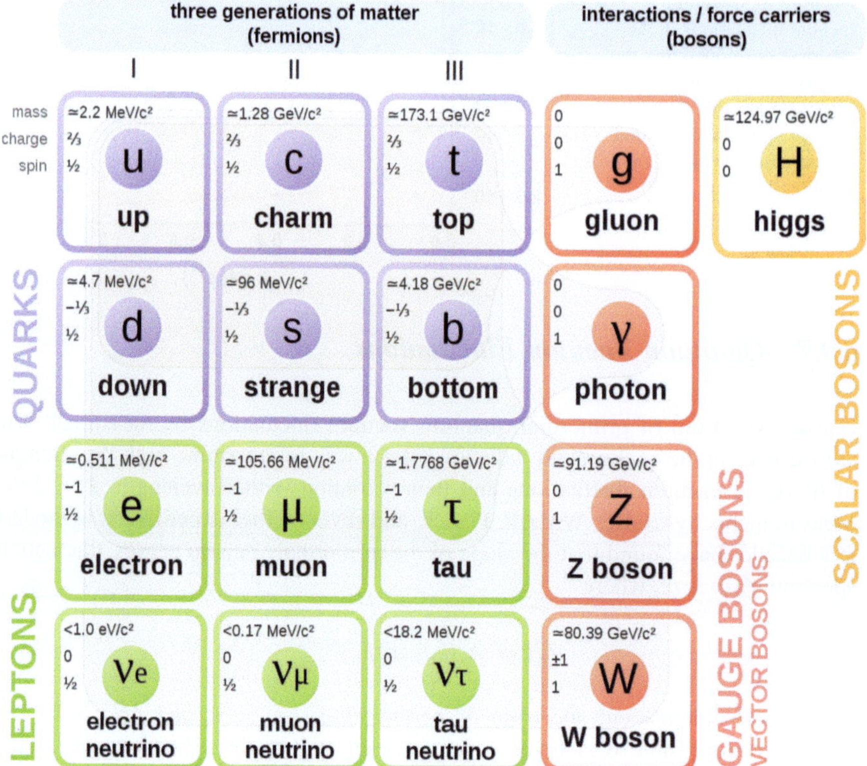

Fig. 6.17 The constituents of the universe embodied in a schematic of the standard model of modern particle physics divided into fermions and bosons, quarks and leptons. The mass, spin, and charges are shown, although the neutrino mass limits are perhaps a bit fanciful. The Planck 2018 results place an upper limit of 60 meV on the neutrino mass. (Content of the 2023 Review of Particle Physics is under a Creative Commons 4.0 International (CC BY-NC 4.0) license)

know from our examination of the irreducible representations of the Poincare group (Chap. 3) that the only objects that can exist are *fermions* with half-integral spin (intrinsic angular momentum I in units of $\hbar$, $I \in \{1/2, 3/2, 5/2, \ldots\}$ and *bosons* with integral spin, $I \in \{0, 1, 2, 3, \ldots\}$. In order to understand the world, the only thing we can do is to go out and measure it. As the result of more than a century of measurement, we have developed the *standard model* of elementary particle physics (Fig. 6.17) where the fermions, the quarks and leptons, make up matter, while the bosons, the gluons, photons, $W^{\pm}$, Z^0, and Higgs, are the carriers of the fields. The constituents of the present low-energy (low-temperature) universe are these bound states of the quantum fields.

Table 6.2 The number of relativistic species as a function of temperature after the Big Bang

Temperature (K)	Mass (eV)	Threshold particles	$N(T)$
5.93E+09	5.11E+05	γ's + ν's	7.25
1.24E+12	1.07E+08	$e^{\pm}$	10.75
1.61E+12	1.39E+08	$\mu^{\pm}$	14.25
1.74E+12[a]	1.50E+08	π's	17.25
1.17E+12	1.01E+08	π's + $u, \bar{u}$ + $d, \bar{d}$ + gluons	51.25
1.17E+12	1.01E+08	$s, \bar{s}$	61.75
1.47E+13	1.27E+09	$c, \bar{c}$	72.25
2.06E+13	1.78E+09	$\tau^{\pm}$	75.75
5.22E+13	4.50E+09	$b, \bar{b}$	86.25
9.63E+14	8.30E+10	$W^{\pm}, Z$	95.25
1.45E+15	1.25E+11	H°	96.25
2.00E+15	1.72E+11	$t, \bar{t}$	106.75

[a]The Hagedorn temperature

6.5.1   Standard Model of Elementary Particle Physics

Fermions interact by exchanging bosons whose properties provide the four known fundamental forces with their characteristic strengths and ranges given, along with their field quanta, in Table 6.2. We have already encountered the bosonic quanta of the electromagnetic field as the photons whose exchange mediates the Coulombic interaction of charged particles such as electrons and protons. We will discuss the fundamental forces in the order of their strength, beginning with the, perhaps unfamiliar to biologists, *strong* force, whose short-range attractive power surpasses the Coulombic repulsion of protons and binds them, along with neutrons, in nuclear matter.

*6.5.2   The **Strong** Force*

The *strong* force is responsible for the binding of quarks into protons and neutrons (and other, heavier, hadrons) through the exchange of spin 1 gluons. The massless gluons are confined within the proton and neutron (and other, heavier, hadrons) and therefore do not directly participate in the binding of nuclei. The neutrons and protons are bound into nuclei by pions (π-mesons) which are quark-antiquark $(q, \bar{q})$ pairs, Yukawa particles which are themselves bound together by gluons. The strong force has a range about the diameter of a small nucleus, $\sim 10^{-15}$ m determined by the large mass (140 MeV) of the exchanged pions. By the uncertainty principle, this is about the lightest particle that can bind nuclear matter. The smallest nucleus is the single proton in the hydrogen atom. The latest measurements (Grinin et al. 2020) of the proton diameter give a value of 1.6964×10^{-15} m. It is illuminating to calculate the strength of the Coulombic repulsion of two protons separated by the distance found in a helium nucleus, $r = 3.8 \times 10^{-15}$ m, from Coulomb's force law:

$$F = \frac{ke^2}{r^2}$$

with $k = 9 \times 10^9$ N m^2/Coulomb^2; $e = 1.6 \times 10^{-19}$ Coulomb. The force is an astonishing 15.96 N, for two tiny objects that we cannot even see. Yet the strong force is able to overcome this electrostatic repulsion as witnessed by proton-proton fusion in stars.

6.5.3 *The* Electromagnetic *Force*

The *electromagnetic* force is the interaction between two or more charges of like or unlike sign and is mediated by the exchange of massless, spin 1 photons, as shown in the Feynman diagrams (Fig. 6.10). Life and chemistry are the realm of the electromagnetic force between the electrons in atoms and molecules. Because the photon is massless, the electromagnetic interaction has infinite range, although it can easily be attenuated by electrically conductive materials, as in a Faraday cage.

6.5.4 *The* Weak *Interaction*

The *weak* interaction is responsible for β-decay of free neutrons and nuclei in which a neutron changes into a proton with the ejection of an electron and an antineutrino (Fig. 6.18). An example of β-decay that is relevant for biology is the decay of ^{14}C that is used to date carbonaceous materials. Radioactive ^{14}C has a half-life of about 5730 years so that all the ^{14}C on the Earth would have long since decayed by conversion to ^{14}N in the β-decay scheme:

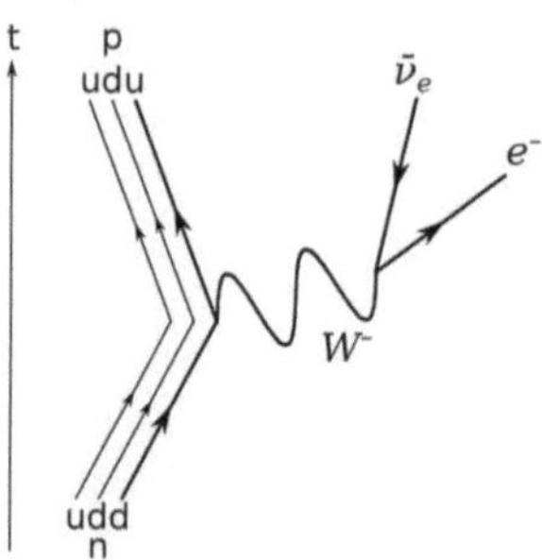

Fig. 6.18 β-Decay of the neutron

$$\,^{14}_{6}\mathrm{C} \rightarrow \,^{14}_{7}\mathrm{N} + e + \nu$$

However, the large flux of high-energy cosmic rays striking the Earth's atmosphere produces neutrons which are captured by the nucleus of the abundant atmospheric ^{14}N, constantly generating ^{14}C by neutron capture:

$$\mathrm{n} + \,^{14}_{7}\mathrm{N} \rightarrow \,^{14}_{6}\mathrm{C} + \mathrm{p}$$

The ^{14}C rapidly reacts with atmospheric oxygen and the resulting CO_2 is taken up by plants, or absorbed in the oceans and winds up in the tissues of plants and animals. By counting the amount of ^{14}C in biological materials, one can measure the age of the object. Note that in β-decay the original nuclide is transmuted to a new chemical element with the same mass number A, but with an atomic number $Z + 1$. Chemistry is determined by Z.

The weak force is carried by intermediate particles consisting of the three massive ($m \sim 80$–90 GeV) vector, spin 1, bosons, W^{+}, W^{-}, and Z^{o}, with charges, $+1$, -1, and 0, respectively. These particles are so massive that the weak force is very short range. In β-decay of the unstable free neutron (Fig. 6.18), a *down* quark in the neutron (u,d,d) transforms into an *up* quark in the proton (u,u,d), with the emission of a W^{-} boson which decays into an electron (in order to conserve overall charge) and an electron antineutrino:

$$\mathrm{n} \rightarrow \mathrm{p} + \mathrm{W}^{-} \rightarrow \mathrm{p} + \mathrm{e} + \bar{\nu}$$

The exact lifetime of the free neutron is something of a puzzle (reminiscent of the Hubble paramter) in that different types of measurements give slightly, but consistently different results: beam methods yield $\tau_{\mathrm{n}} \sim 887.7 \pm 2.2$ s, while trap methods give $\tau_{\mathrm{n}} \sim 878.5 \pm 0.8$ s (Pattie et al. 2018); nevertheless, these times are much longer than typical characteristic lifetimes of $\sim 10^{-10}$ s for decays mediated by the weak force and make it clear that neutron decay, while very slow for a weak decay, will still lead to the depletion of free neutrons after the Big Bang unless there exists an alternative pathway for their preservation. The process that is opposite to β-decay is electron capture, which occurs when an orbital electron is captured by a proton in the nucleus, transforming it into a neutron, and an electron neutrino is released in what is inverse β-decay.

6.5.5   *The Gravitational Interaction*

We are all familiar with the extremely weak, uniformly attractive, but infinite range, *gravitational* force that keeps us bound to the Earth and is responsible for the formation of large-scale structures in the universe. The proposed carrier of the gravitational interaction, the *graviton,* must be massless, because gravity is an infinite range force, and it must have a spin $I = 2$, in order to be attractive. Gravity is such a weak force that the *graviton* has never been observed. Quantum field theories of gravity (Rovelli and Smolin 1995; Rovelli 2004) would be distinguishable from Einstein's general relativity only in the limit of large masses, such as found at the center of black holes, or at the initial Planck time scale of the Big Bang. Nevertheless, since gravity only requires mass, not electrical charge, it is the ultimate organizer of the universe, pulling matter ineluctably closer until it causes the condensation of primordial hydrogen into stars which resist collapse by fusing it into helium and heavier nuclei. But, as stars exhaust their nuclear fuel, gravity eventually overcomes all other forces and stars collapse into white dwarfs and then, finally, into black holes. The latest data (Oct 2019) from LIGO and VIRGO (https://www.ligo.org/science/Publication-O3aCatalog/) indicate that at least 48 black hole mergers have been detected over the last few years. The farthest was at $z = 0.8$, and since gravitational waves travel at the speed of light, these ripples in space-time originated in a black hole merger from 7 billion years ago, when the universe was half its present age.

6.5.6   *The Thermal History of the Universe*

The universe was much, much hotter in the past. From studies of the cosmic microwave background radiation, we have found that the universe, which was at a temperature of 3000 K at $t = t_o + 380,000$ years, has cooled to a current temperature of only 2.725 K at $t = t_o + 13.8$ billion years. How hot was the universe at t_o? Our particle in a box estimate (vide supra) is that it was in excess of 10^{20} K. The observation that the universe has markedly cooled since t_o suggests that a physical perspective on the evolution of the Big Bang could be gained by examining its thermal history. Through $E = kT$ we can at the same time make a catalog of standard model processes and events whose thresholds were energy (temperature) dependent.

For these purposes, it is useful to revert to the non-SI units for energy of electron-volts (eV) while at the same time providing the same quantities in their equivalent SI units (joules). The rest masses of elementary particles that make up the standard model are reported in Fig. 6.17 in eV, rather than in Kg. One can use $E = mc^2$ to convert from one to the other. In SI units these energies are in joules and $1 \ \mathrm{eV} = 1.602 \times 10^{-19}$ joule. For example, the proton rest mass is $m_\mathrm{p} = 1.6726 \times 10^{-27}$ Kg which becomes

$$E_\mathrm{p} = m_\mathrm{p}c^2 = \frac{1.6726 \times 10^{-27} \ \mathrm{Kg} \ \left(2.9979 \times 10^8 \mathrm{m}\right)^2}{1.602 \times 10^{-19} \ \mathrm{s}^2 \ \mathrm{Joule}} \mathrm{eV} = 9.38272 \times 10^8 \mathrm{eV} = 938.272 \ \mathrm{MeV}$$

in energy units. We can then convert this to a temperature using Boltzmann's constant $k = 8.62 \times 10^{-5}$ eV/K and find that this corresponds to

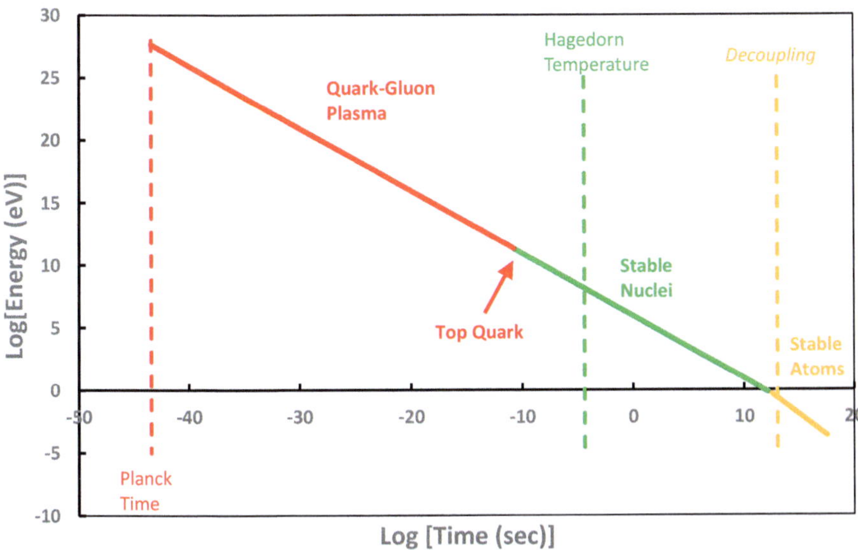

Fig. 6.19 The energy of the universe as a function of time after the Big Bang showing the epoch of radiation domination from the Planck time to the epoch of decoupling, after which the universe was matter dominated and the cooling rate increased. Indicated are the energies at which the quarks and gluons cooled from the plasma to form QCD bound states and at the Hagedorn temperature, stable baryons and nuclei. After decoupling the universe cooled sufficiently to form stable atoms of helium and hydrogen

$T = 3.47895 \times 10^{12}$ K. Knowledge of the temperature of the early universe therefore serves as a guide to the existence of particles whose mass can be generated as pair production from the energy present in the thermal fluctuations of the quantum vacuum state. The thermal history of the universe beginning at the Planck time and progressing to the present is therefore a map (Fig. 6.19) of its generation from the fluctuations of the vacuum i.e., from *nothing*. This has been put on a firm mathematical footing by using analytical solutions of the Wheeler-DeWitt equation (He et al. 2014) where it is shown that once a quantum vacuum fluctuation arises, it grows exponentially producing the early universe.

The early universe was dominated by radiation; the number of photons was a billion times the number of hadrons and the temperature decreased as the universe expanded according to $T(z) = T_o(1 + z)$. Between the Planck time and decoupling at 380,000 years, the temperature of the radiation-dominated universe decreased as $T(t) = 2.20 \times 10^{10} \, t^{-1/2}$, with t in s.

At the earliest physical time, on the order of the Planck time, the only state of matter currently known (measured) to exist was a simple, hot quark-gluon plasma. The temperature was so high that any quantum states for ordinary matter were not stable and were quickly converted back into free particle plasma states. The extremely high temperature caused pair production of particles, quarks and gluons, from the vacuum. This quark-gluon plasma has been detected at both CERN (Heinz and Jacob 2000; Heinz 2001) and RHIC (Arsene et al. 2005; Back et al. 2005; Adams et al. 2005; Adcox et al. 2005) and explored in depth in the intervening years (Aidala et al. 2019).

The most massive particle yet discovered is the top quark with a mass of 173.1 GeV, corresponding to a temperature of the early universe of 2×10^{15} K. Could the universe ever have been so hot? Let us use more quantum theory to estimate the temperature of the primordial universe.

At the earliest physical time, that of the Planck time, we can estimate the strength of the quantum fluctuations in the vacuum by querying the quantum mechanics of the harmonic oscillator. Its zero-point energy is $E = \frac{1}{2}\hbar\omega = \frac{1}{2}h\nu$. The oscillator frequency ν is the reciprocal of the Planck time t_P:

$$\nu = \frac{1}{t_P} = \frac{1}{5.39 \times 10^{-44}\,s}$$

or $\nu = 1.86 \times 10^{43}$ Hz, corresponding to an energy of 1.23×10^{10} joules (7.83×10^{28} eV) which is markedly above the threshold for generating a top quark (Fig. 6.18). The temperature would have been $T = 8.90 \times 10^{32}$ K, slightly higher than the Planck temperature, $T_P = 1.42 \times 10^{32}$ K (7.67×10^{28} eV), and greatly in excess of the temperature needed to generate any known standard model particle from the vacuum, leading to the conclusion that the early universe consisted of a simple quark-gluon plasma.

The very early phase of the Big Bang, at the Planck time, $t = t_o + 10^{-43}$ s, was extremely hot, with temperatures in excess of 10^{31} K, considerably above the Hagedorn temperature, T_H, which is the deconfinement temperature (energy) of the strong force. Above T_H hadrons are no longer stable bound states of quarks and gluons that normally bind them at lower temperatures, but dissociate into free charged quarks and gluons, a plasma. The amount of energy available from the thermal environment is above that needed to generate quark-antiquark pairs from the vacuum. As the universe cooled, it underwent a quantum chromodynamic phase transition in which the quark-gluon plasma was converted into hadrons, essentially freezing the baryonic components of the universe into their present state, beginning with the top quark and, as the cooling progressed, to the Hagedorn temperature, T_H. The lowest-mass object that can be formed from these quark-antiquark pairs is the pion, with a mass ~ 150 MeV.

We can understand this QCD phase transition from the quark-gluon plasma (Shuryak 2017) to ordinary baryonic matter by counting the number of relativistic species (particles, spin states, antiparticles, etc.) as a function of their mass. Table 6.2 shows the particles, their masses, and the temperature below which they would be stable. The state counting is left as an exercise for the student. In order to estimate the temperatures for each of these states, we assumed that the energy threshold for particle-antiparticle pair production from the vacuum was the rest mass of each of the species known in the standard model, beginning with the heaviest, currently known particle, the top quark, at 173.1 GeV, with a threshold temperature

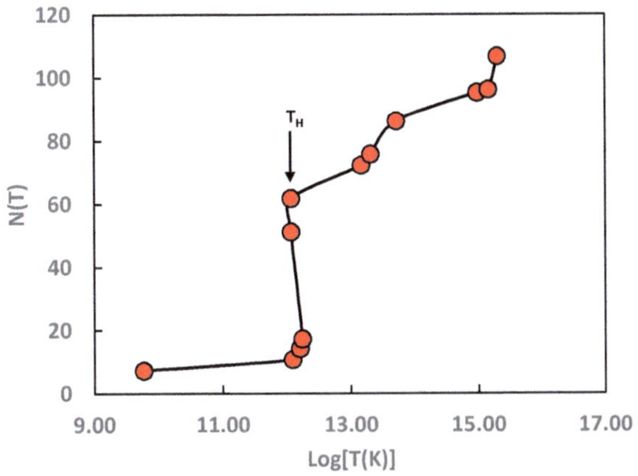

Fig. 6.20 The number of degrees of freedom (particles, spin and polarization states, etc.) of the relativistic species present in the Big Bang at various temperatures. As the plasma of quarks and gluons cooled through the confinement transition (T_H), hadrons emerged as bound states

(at t_o + 35 ps) of 2×10^{15} K, etc. The transition from a quark-gluon plasma occurred at t_o + 40 μs for $z = 6.4 \times 10^{11}$ when $T_H \sim 1.74 \times 10^{12}$ K (Fig. 6.20) below which quarks and gluons condense into mainly protons and neutrons (and their antiparticles), in a process of baryogenesis. The primary constituents of the universe then also included photons, neutrinos, electrons, and positrons, along with unstable, rapidly decaying muons and pions, but the baryons could not yet form hadronic bound states like the deuteron; the photon flux and energy were above the photodissociation threshold for the deuteron (binding energy = 2.2 MeV) of 2.6×10^{10} K, preventing the onset of Big Bang nucleosynthesis until the universe cooled still further, a stage it reached at t_o + 165 ms.

The cooling of the universe from the primordial quark-gluon plasma illustrates a fundamental feature of the origin of life; the flow of energy through a system organizes that system. As the universe cooled, it condensed into known bound states of matter, much like water flowing down a mountain stream collecting in pools dammed by fallen rocks. The water is temporarily held in the thrall of the gravitational well, but still flows downhill until it encounters the next rest stop, placid pool, etc. Cooling is likened to downhill flow. One might think of the pools as phase transitions, from one form, quarks and gluons as it were, to another, protons and neutrons, and on to atoms and molecules, ever downward in a cascade of energy. We now know the actors in the drama, the photons, electrons, protons, neutrons, and lovely little neutrinos that were the stations at which the universe hesitated on its headlong descent to life.

6.6 Nucleosynthesis

How do we make the present universe from the hot relic left after the Big Bang? We have a giant expanding ball of light and matter that is getting colder by the attosecond. One of the strongest lines of evidence for the probabilistic self-assembly of the universe in the Big Bang is its prediction of the abundances of the light elements resulting from Big Bang nucleosynthesis (or primordial nucleosynthesis) which refers to the production of heavier nuclei by the cascade of thermonuclear fusion from the neutrons and protons present during the $t_o + 165^+$ ms of the early phase of the universe, shortly after the Big Bang.

6.6.1 Primordial Nucleosynthesis in the Early Universe

Because it is based on standard model physics that have been well-tested in accelerator experiments, Big Bang nucleosynthesis gives us a very detailed way of probing deep into the early universe and quantitatively comparing our theories with measurements. Big Bang nucleosynthesis is responsible for the formation of the lightest elements in the periodic table, mainly hydrogen, its isotope deuterium, the helium isotopes, and lithium. The mathematical theory of Big Bang nucleosynthesis gives detailed, quantitative predictions for the production of these light elements formed when the universe was ~1 s old, at a temperature of 10^{10} K (~ 1 MeV) after the primordial quark-gluon plasma froze out to form protons and neutrons (Fig. 6.20).

These predictions are quite sensitive to the physical state of the early universe. Because of the very short period in which Big Bang nucleosynthesis occurred before being stopped by the universe's expansion and cooling, no elements heavier than lithium could be formed. The abundances predicted by these caculations have been extensively compared with observations and excellent agreement has been found, over a range of abundance ratios that spans a factor of ~10^9, from $n_{4He}/n_{1H} \sim 0.24$ to $n_{7Li}/n_{1H} \sim 10^{-10}$ (Cyburt et al. 2016). This is perhaps the most critical test of the Big Bang generation of the particles of life.

6.6.2 Nomenclature for Isotopes

The elements in the periodic table are composed solely of neutrons, protons, and electrons and are denoted by their chemical symbol, X (such as $X = Br$ for bromine), surrounded by superscripts and subscripts: $^A_Z X$. The number of neutrons plus protons is the mass number $A = N + Z$, and Z is the number of protons, or atomic number. Z determines the number of attendant electrons and hence the chemical identity of the element. One would write, for example, the symbol for zirconium as $^{90}_{40} Zr$, which informs us that its nuclear charge is $Z = 40$ protons, and in order to preserve charge neutrality, this is also the number of electrons that define the element's chemistry as that of zirconium. The number of neutrons in the $^{90}_{40} Zr$ nucleus is $N = A - Z = 90 - 40 = 50$. Note that there are two additional Zr isotopes with $A = 92, 94$, which are also stable: $^{92}_{40} Zr$ and $^{94}_{40} Zr$.

Since the charge on the nucleus is the sole determinant of the number and arrangement of electrons in an atom, and hence its chemical properties, it is reasonable to ask, where did the elements (i.e., nuclei) come from? The periodic table contains more than 100 elements, each with a number of isotopes that have constant Z, but vary in N. The number of stable, nonradioactive isotopes, increases with Z, reaching a maximum at ten with $_{50}$Sn. Electrons come from beta decay of the Big Bang neutrons, so we have {p, n, e, γ} as constituents of the primordial universe. All evidence so far indicates that the electron is not a composite particle and is therefore elementary. The proton and the neutron are composites and consist of bound states of the quark triplets p = (u, u, d) and n = (u, d, d). Massless, spin 1 bosons, called gluons (g) (Fig. 6.17), carry the strong force that binds the quarks together within the proton and neutron. Therefore, at the Planck time, the universe consisted of a plasma (a gas of charged particles) of quarks and gluons at a temperature in excess of 10^{32} K. By knowing the probability distribution of particles their over energy states, one can calculate the mean properties of the early universe and the amounts of heavier nuclei produced. One feature of Big Bang nucleosynthesis (BBN) is that the physical laws and constants that govern the behavior of matter at these energies are very well understood, and hence, BBN lacks some of the speculative uncertainties that characterize earlier periods in the life of the universe. Another feature is that the process of nucleosynthesis is determined by conditions at the start of this phase of the life of the universe, making what happened before t_o + 1 s irrelevant.

6.6.3   Big Bang Nucleosynthesis

We can use the properties of probability once again to calculate the composition of the universe as it cooled from the Big Bang. The probability density for a nonrelativistic particle species of mass number A, charge Z, and mass m_A, whose energy is much less than its rest mass, is given by the inverse of the Maxwell-Boltzmann distribution:

$$P(A) = \frac{g_A}{n_T} \left(\frac{m_A T}{2\pi}\right)^{3/2} e^{\left(\frac{\mu_A - m_A}{kT}\right)}$$

where g_A is the number of degrees of freedom, n_T is the total number of particles summed over species s,

$$n_T = n_\mathrm{p} + n_\mathrm{n} + \sum_{i=1}^{s} [An_A]_i$$

and μ_A is the chemical potential, essentially the number density. This equation holds equally for nuclides as well as protons and neutrons. In what follows we ignore the slight differences between the proton and neutron masses and m_A in the exponential and prefactors and replace them with a single *nucleon mass*, m_N. One second after the Big Bang, the temperature of the universe was roughly 10^{10} K and it was filled with neutrons, protons, electrons, anti-electrons (positrons), photons, and neutrinos. As the universe cooled, the neutrons either decayed into protons and electrons or combined with protons to make deuterium (an isotope of hydrogen, $_1^2H$).

Nuclear reactions that produce an element with mass number A from $N = A - Z$ neutrons and Z protons at a rate that is larger than the expansion rate (the Hubble parameter) will be in chemical equilibrium so that the chemical potential of A can be found by counting the species:

$$\mu_A = Z\,\mu_p + (A-Z)\mu_n$$

The binding energy $B(A)$ of a nuclide $A(Z)$ can be determined from the mass difference between the reactants and products as

$$B(A) = Zm_p + (A - Z)m_n - m_A$$

The probability density for a nuclide A can be written in terms of the proton and neutron probability densities as

$$e^{\frac{\mu_A}{kT}} = e^{\left[\frac{(Z\,\mu_p + (A-Z)\mu_n)}{kT}\right]}$$

$$= \frac{n_p{}^Z n_n{}^{A-Z}}{2^A}\left(\frac{2\pi}{m_N T}\right)^{3A/2} e^{\left[\frac{(Z\,m_p + (A-Z)m_n)}{kT}\right]}$$

And by replacing the exponential by the expression for the binding energy, we have that the probability density for nuclear species $A(Z)$ is

$$P(A) = \frac{g_A A^{3/2} n_p{}^Z n_n{}^{A-Z}}{2^A n_T}\left(\frac{2\pi}{m_N T}\right)^{3(A-1)/2} e^{\left(\frac{B(A)}{kT}\right)}$$

This can be converted to the measured mass fraction $\chi(A)$ of nuclide A:

$$\chi(A) = \frac{A\,n_A}{n_T}$$

by noting that the probability density for species A is

$$P(A) = \frac{n_A}{n_T}$$

so that

Table 6.3 Characteristics and binding energies of light nuclei

A	Z	Nucleus (^{A}Z)	$B(A)$ (MeV)	g_A
2	1	^{2}H	2.22	3
3	1	^{3}H	6.92	2
3	2	^{3}He	7.72	2
4	2	^{4}He	28.3	1
12	6	^{12}C	92.2	1

$$\chi(A) = AP(A)$$

or

$$\chi(A) = g_A A^{\frac{5}{2}} \left[\zeta(3)^{A-1} \pi^{\frac{1-A}{2}} 2^{\frac{3A-5}{2}} \right] \left(\frac{kT}{m_N} \right)^{\frac{3(A-1)}{2}} \eta^{A-1} \chi_p^{Z} \chi_n^{A-Z} e^{\left(\frac{B(A)}{kT} \right)}$$

where $\zeta(3)$ is the Riemann zeta function evaluated when its argument is 3, $\zeta(3) = 1.20206$, and η is the current value of the baryon to photon ratio, which the Planck satellite data indicates has a value of

$$\eta = \frac{n_T}{n_\gamma} = 2.73 \times 10^{-8} \Omega_B h^2$$

and with the Planck value for $\Omega_B\, h^2 = 0.02233$, we find that the baryon to photon ratio is 6.1×10^{-10}. The binding energies of some the light nuclei are shown in Table 6.3 and are seen to vary from ~1 to 7.7 MeV per nucleon.

The following stages occur during the first few minutes of the universe: At temperatures on the order of 10 MeV (~10^{11} K) and around a time of $t = t_o + 10$ ms, the interconversion of neutrons and protons was maintained by the weak interactions among the leptons and baryons through the reactions (all neutrinos are electron neutrinos here):

$$n \leftrightarrow p + e^- + \bar{\nu}$$

$$\nu + n \leftrightarrow p + e^-$$

$$e^+ + n \leftrightarrow p + \bar{\nu}$$

Since the free neutron is unstable, but its lifetime is much longer than 10 ms, these processes equalized the number of neutrons and protons: $\chi_p = \chi_n = 0.5$. The rates of these weak reactions were much faster than the Hubble expansion rate so that these reactions were at equilibrium but the mass fractions calculated from $\chi(A)$ were very small (Table 6.4) because η is such a small number. There must have been a process to incorporate neutrons into bound states in order to preserve them for the future universe. We will see (vide infra) that most of the neutrons end up in $^{4}_{2}$He nuclei.

As the temperature fell to ~1 MeV (~10^{10} K) nucleosynthesis began around $t = t_o + 1$ s, when the universe had cooled enough to form stable protons and

Table 6.4 Light nuclear mass fractions at given temperatures

Nucleus (^{A}Z)	10 MeV	1 MeV	0.3 MeV	0.2 MeV	0.1 MeV
^{2}H	3.43×10^{-12}	3.92×10^{-13}	1.14×10^{-11}	2.52×10^{-10}	5.89×10^{-6}
^{3}H	6.52×10^{-24}	4.62×10^{-25}	1.28×10^{-19}	3.88×10^{-15}	3.88×10^{-15}
^{3}He	7.06×10^{-24}	6.17×10^{-24}	1.11×10^{-17}	1.27×10^{-12}	1.27×10^{-12}
^{4}He	3.67×10^{-35}	3.21×10^{-29}	6.78×10^{-3}	1.20×10^{-1}	2.40×10^{-1}
^{12}C	1.05×10^{-128}	5.00×10^{-111}	3.18×10^{-26}	3.18×10^{-26}	3.18×10^{-26}

neutrons. From the Boltzmann probability distribution, one can calculate the fraction of protons and neutrons based on the temperature at this point. $N_p/N_n = e^{-Q/kT}$, where Q is the mass difference between the neutron and the proton, $Q = 1.293$ MeV. Note that $Q = 2.07 \times 10^{-13}$ joules. This distribution is in favor of protons, because the higher mass of the neutron results in the spontaneous decay of neutrons to protons with a half-life of about 882 s. This point is called neutron freeze out ($E = 0.664$ MeV) and the neutron/proton ratio stabilized at ~1/6. After 1 s, the only reaction that appreciably changes the number of neutrons is neutron decay:

$$n \rightarrow p + e^- + \bar{\nu}$$

Without further reactions to preserve neutrons within stable nuclei, the universe would be pure hydrogen.

The first reaction that preserves the neutrons is the formation of the deuteron, the nucleus of deuterium:

$$p + n \leftrightarrow d + \gamma$$

This reaction is exothermic with an energy difference of 2.2 MeV, but since photons are a billion times more numerous than protons, the reaction does not proceed until the temperature of the universe falls to 1 billion K or $T = 0.1$ MeV, about 100 s after the Big Bang. At this time, the neutron/proton ratio is about 1:7. Mass fraction calculations using $\chi(A)$ (Table 6.4) indicate that the light nuclei still had tiny mass fractions.

From 1 to 3 min after t_o, the temperature had dropped to 0.1–0.3 MeV (~10^9 K) and most of the deuterium had combined to make helium. Trace amounts of lithium were also produced at this time. The mass fraction in various isotopes versus time is shown in Fig. 6.21. Deuterium peaks around 100 s after the Big Bang and is then rapidly swept up into helium nuclei. A very few helium nuclei combine into heavier nuclei giving a small abundance of Li^7 coming from the Big Bang. Note that tritium, the radioactive form of hydrogen ^{3}H, decays into ^{3}He with a 12-year half-life so no tritium survives to the present, and ^{7}Be decays into ^{7}Li with a 53-day half-life and also does not survive.

Once deuteron formation has occurred, further reactions proceed to make helium nuclei. Both light helium ^{3}He and normal helium ^{4}He are made, along with tritium ^{3}H. These reactions can be photoreactions as shown here:

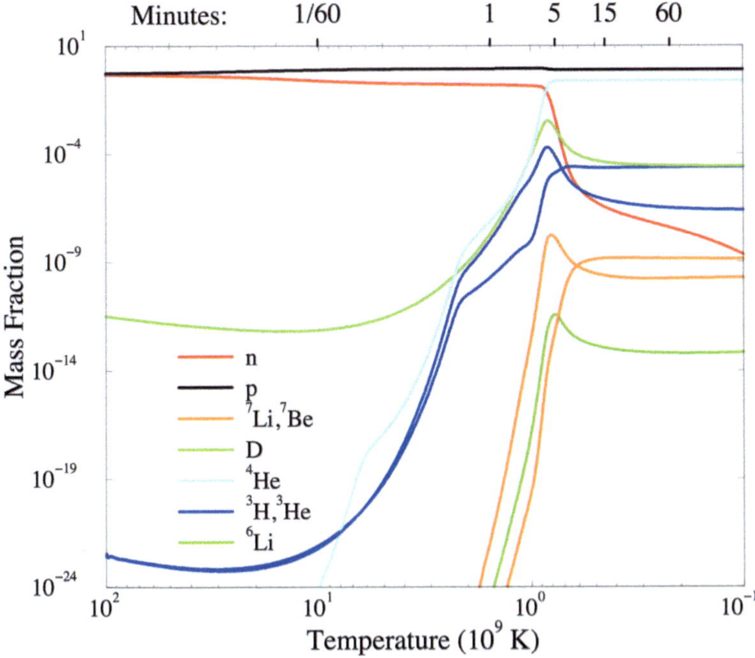

Fig. 6.21 Big Bang nucleosynthesis showing the fraction of the total mass of the universe contributed by each of the indicated nuclear fusion reactions as a function of both time after the Big Bang and the temperature of the early universe. Nucleosynthesis occurs only for the first few minutes after the Big Bang, producing nuclei up to ^{7}Li and ^{7}Be before the temperature and density drop at ~5 min below the fusion threshold. Neutrons are unstable and decay into protons with a lifetime of ~880 s. By 10^4 s, the constitution of the universe was essentially fixed at ~75% ^{1}H, ~24% ^{4}He, and small amounts of ^{2}H, ^{3}He, ^{7}Be, ^{7}Li, and ^{6}Li. Although neutrons still appear at $t = t_0 + 10^5$ s, they all eventually decay into protons, leaving no free neutrons in the present universe. (Courtesy of Michael Turner, used with permission from Burles et al. (1999))

$$d + n \leftrightarrow {}^3H + \gamma$$

$$^3H + p \leftrightarrow {}^4He + \gamma$$

$$d + p \leftrightarrow {}^3He + \gamma$$

$$^3He + n \leftrightarrow {}^4He + \gamma$$

Because the helium nucleus is 28 MeV more bound than the deuteron, and the temperature has already fallen so far that $T = 0.1$ MeV, these reactions only go one way. The reactions below also produce helium and usually go faster since they do not involve the relatively slow electromagnetic process of photon emission:

$$d + d \leftrightarrow {}^{3}\text{He} + n$$

$$d + d \leftrightarrow {}^{3}\text{H} + p$$

$${}^{3}\text{H} + d \leftrightarrow {}^{4}\text{He} + n$$

$${}^{3}\text{He} + d \leftrightarrow {}^{4}\text{He} + p$$

The net effect is $d + d \leftrightarrow {}^{4}\text{He} + \gamma$. Eventually the temperature gets so low that the electrostatic repulsion of the deuterons causes the reaction to stop at a deuteron/ proton ratio of only $\sim 5 \times 10^{-6}$ (Table 6.4) and essentially inversely proportional to the total density in protons and neutrons. Almost all the neutrons in the universe end up in normal helium ${}^{4}\text{He}$ nuclei. For a neutron/proton ratio of 1:7 at the time of deuteron formation, 25% of the mass ends up in helium, which has the highest binding energy of the light nuclei (7.1 Mev; Table 6.3). The mass fraction Y_p of ${}^{4}\text{He}$ can be estimated if all of the neutrons are captured in helium:

$$Y_{\text{p}} = \chi(4) = \frac{4P(4)}{n_{\text{T}}} = \frac{4\left(\frac{n_{\text{n}}}{2}\right)}{n_{\text{p}} + n_{\text{n}}} = \frac{2(n/p)}{1 + n/p} = \frac{2/7}{1 + 1/7} = 0.25$$

The predicted abundance of deuterium, helium, and lithium depends on the density of ordinary matter in the early universe. The results indicate that the yield of helium is relatively insensitive to the abundance of ordinary matter, above a certain threshold. We generically expect *and* do observe that about 24% of the ordinary (baryonic) matter in the universe is helium produced in the Big Bang. This is another major triumph for the Big Bang theory.

The results of detailed calculations of Big Bang nucleosynthesis (Fig. 6.22) using the latest computer codes give a very complete picture of the thermal evolution of the generation of the light nuclei. It is of substantial interest to compare these predictions with measurements of their abundances. One will notice that in the equation for the mass fraction $\chi(A)$, there are only three free parameters, the proton and neutron mass fractions, which can be estimated quite well (vide supra) and the baryon to photon ratio, which has been determined from measurements of the cosmic microwave background radiation. The measurements are in excellent agreement with the detailed predictions (Fig. 6.21) given the value of $\Omega_{\text{B}}h^2 = 0.02233$ measured by the Planck satellite (Table 6.1) at a baryon to photon ratio of 6.1×10^{-10}.

6.6.4 Stellar Nucleosynthesis

Gamow proposed the hot Big Bang as a means to produce all of the elements (Gamov 1946) in the periodic table. However, Big Bang nucleosynthesis stops at $t = t_{\text{o}} + 3$ min as the temperature falls below that needed for the efficient fusion of nucleons. The lack of stable nuclei with atomic weights of $A = 5$ (e.g., ${}^{5}_{3}\text{Li}$) or 8 (e.g.,

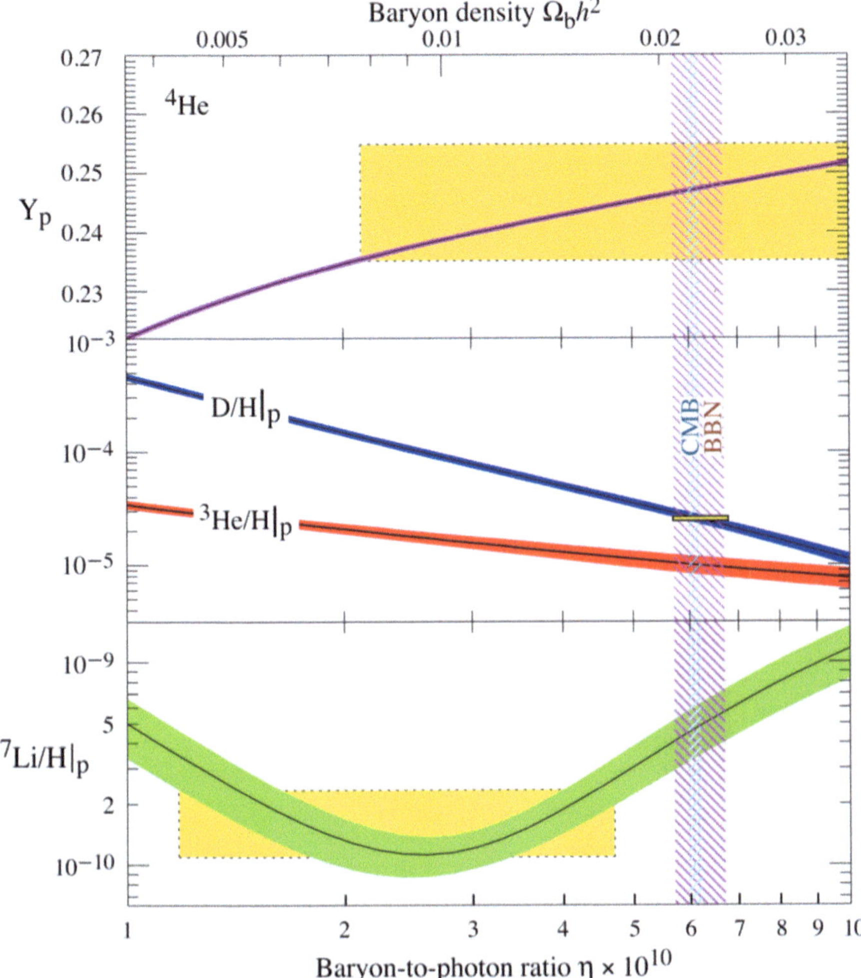

Fig. 6.22 The predicted number fraction of light nuclei synthesized during the first ~220 s after the Big Bang, shown as a function of the baryon to photon ratio while the universe was radiation dominated. The pink cross-hatching shows the range of measurements, while the light blue vertical band gives the value of η measured by the Planck satellite (Table 6.1). (Data from the Particle Data Group (Fields 2020). Content of the Review of Particle Physics is under a Creative Commons 4.0 International (CC BY-NC 4.0) license)

$^8_4$Be) limited the Big Bang to producing hydrogen and helium. Burbidge et al. (1957) worked out the nucleosynthesis processes that go on in stars, where the much greater nuclear densities and the longer time scales allow the triple-alpha process (^{4}He + ^{4}He + ^{4}He $\rightarrow$ ^{12}C) to proceed and make the elements heavier than helium. But the calculations by Burbidge et al. produced amounts of helium that were too small to acount for its presence as ~25% of the baryonic matter in the universe. How then did the heavier elements originate, particularly those intimately involved with the evolution of life: oxygen, carbon, sulfur, nitrogen, and phosphorus?

Now we know that both processes occured: Most helium was produced in the Big Bang but carbon and everything heavier is produced in stars. Notice that the carbon mass fraction from the Big Bang calculated in Table 6.4 is so low that its mass fraction does not appear on the plot in Fig. 6.21. Big Bang nucleosyntheis produced no elements heavier than beryllium. Most beryllium, and lithium, is produced by cosmic ray collisions breaking up some of the carbon produced in stars. The lack of a stable nucleus with 8 nucleons created a bottleneck in Big Bang nucleosynthesis that stopped the process. In stars, this bottleneck is passed by collisions of three ^{4}He nuclei, but this triple-alpha process takes tens of thousands of years to convert a significant amount of helium to carbon, and therefore, the Big Bang was unable to convert any significant amount of helium to carbon in the 3 min when its temperature was still high enough.

The Big Bang involved the strongest of the four fundamental forces, the strong, weak, and electromagnetic interactions, but in the formation of stars the weakest, gravity, played an essential role. The hydrogen generated in the Big Bang condensed under the influence of gravity and formed the stars, in which the fusion of hydrogen to helium begins the process of stellar nucleosynthesis to produce C, N, O, ..., up to Fe. An important clue to the processes for the formation of, for example, carbon and oxygen lies in the fact that both nuclides have even numbers of nucleons and that the number of protons equals the number of neutrons; even-even nuclei are well known by nuclear physicists to be relatively more stable than nuclei with odd numbers of nucleons (see also Chap. 11 on water). Most carbon in nature is ^{12}C, while oxygen is predominantly ^{16}O. The atomic numbers for these nuclides are divisible by 4, strongly suggesting that they were stable states generated through the aggregation of objects containing 4 nucleons, such as α particles—^{4}He nuclei. Oxygen (^{16}O), neon (^{20}Ne), nitrogen (^{14}N), and carbon (^{12}C), in that order, are the most abundant nuclides in the universe after hydrogen and helium. Note again that neon's atomic number of 20 suggests that it was formed from the aggregation of 5 objects containing 4 nucleons (α−particles).

Stellar nucleosynthesis occurs in stars (Fig. 6.23) during the process of stellar evolution where young stars first fuse hydrogen to helium (Fig. 6.24) in a process that supplies sufficient photon pressure to support the gases against gravitational collapse. We earlier saw that this coupling of photons to ionized matter drove the Big Bang outward until the epoch of decoupling. As stars evolve and age much of the available hydrogen is converted to helium, but helium fusion to heavier nuclei also releases energy and photons and continues to support the star's gaseous envelope. During further stellar evolution, nuclear fusion processes are responsible for generating the elements from carbon to calcium (Fig. 6.23). Of particular importance is carbon, because its formation from ^{4}He is a bottleneck in the entire process. Carbon is also the main element used in the production of free neutrons within the stars, giving rise to the s (slow) process which involves the slow absorption of neutrons to produce elements heavier than iron and nickel (^{57}Fe and ^{62}Ni). Carbon and other

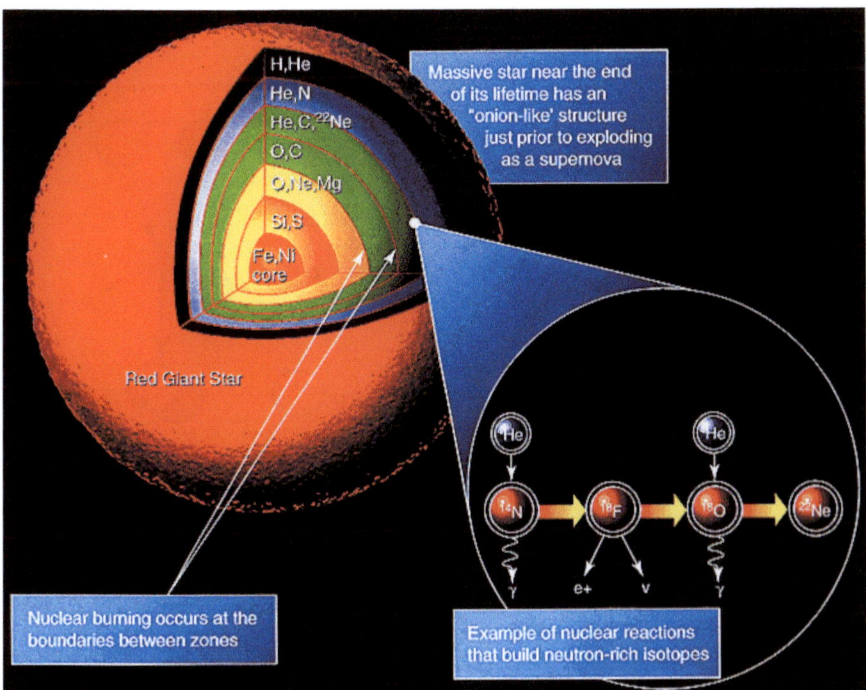

Fig. 6.23 A massive star such as Betelgeuse near the end of its lifetime has a structure consisting of layers characterized by the temperature that the elements require for thermonuclear fusion, with the lightest ^{1}H and ^{4}He close to the surface and the heavier elements towards the Fe, Ni core. (From https://terraforming.fandom.com/wiki/Category:Stars_and_other_hosting_celestial_bodies. Licensed under CC-BY-SA)

elements formed by this process are also fundamental to life. Red giant stars, such as Betelgeuse and Antares, are at the ends of their lives (Fig. 6.23) and contain all the fusion products up to iron and nickel. These products of stellar nucleosynthesis are generally distributed into the universe's interstellar medium as planetary nebulae when dying stars cast off their outer envelopes, or through stellar winds.

The first direct proof that nucleosynthesis occurs in stars was the detection of technetium $_{43}$Tc in the atmosphere of a red giant in the early 1950s by Merrill (1952) using the telescopes at Mts. Wilson and Palomar. The detection of spectral lines from technetium was a surprise because all of the isotopes of technetium are unstable (Table 6.5). In fact, when Mendeleev developed the periodic table the space for element 43 was blank. It took the discovery of radioactivity to resolve the puzzle of why, even after many decades of searching, no technetium was ever found on the Earth. We now know that technetium is the lightest nucleus in the periodic table that possesses no stable isotopes. Because all the isotopes of technetium are radioactive, with half-lives (Table 6.5) much less than the ages of the stars, its detection must reflect its creation within that star.

$$^2\text{H} + {}^1\text{H} \rightarrow {}^3\text{He} + \gamma + 5.49 \text{ MeV}$$

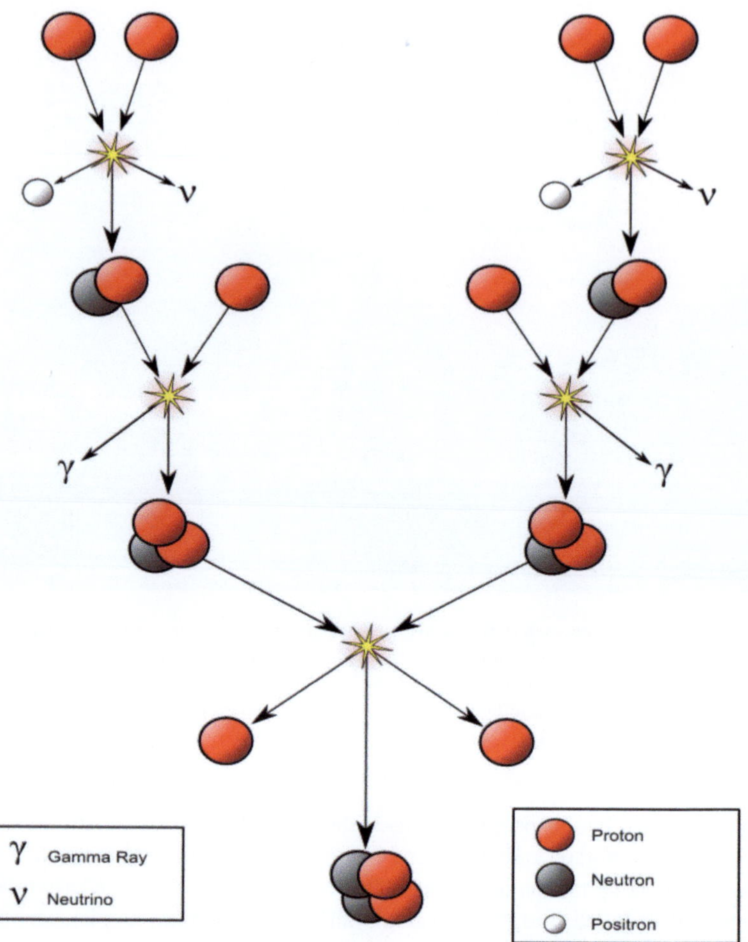

Fig. 6.24 The proton-proton chain reaction in which two protons fuse to ^{2}H, which reacts with another proton to form ^{3}He. Two ^{3}He nuclei then fuse to make ^{4}He with the release of two protons, which are free to reenter the chain reaction. (From https://www.sciencelearn.org.nz/images/244-proton-proton-chain-reaction. Borb Creative Commons Attribution-ShareAlike 3.0 Unported)

Less dramatic, but equally convincing evidence of stellar nucleosynthesis comes from the large overabundances of specific stable elements in stellar atmospheres. A historically important case was the observation that old stars possess barium abundances some 20–50 times greater than in unevolved younger stars, which is evidence for the operation of the s process within that star. Many modern examples of stellar

Table 6.5 Half-lives of several isotopes of technetium $_{43}$Tc

A	$t_{1/2}$	Units
93	2.73	Hours
94	4.88	Hours
95	20	Hours
96	4.3	Days
97	4.21×10^6	Years
98	4.2×10^6	Years
99	2.11×10^5	Years

nucleosynthesis arise from the examination of the isotopic composition of stardust. Stardust is one component of cosmic dust, found and extracted from meteorites as solid grains, that condensed from the gases of individual stars that were distributed into the interstellar medium. The measured isotopic compositions demonstrate nucleosynthesis within the stars from which the stardust grains condensed.

6.6.4.1   The Hydrogen Fusion Chain Reaction Initiates Stellar Nucleosynthesis

The all-important first step in stellar nucleosynthesis is that of the fusion of hydrogen in an energy-producing chain reaction to produce the heavier nuclides deuterium, tritium, and helium (Fig. 6.24). When the gravitational collapse of the primordial hydrogen cloud compressed the gas in the center to temperatures around 10^7 K, high enough to fuse two protons, stars ignited and first begin to shine. The energy at the center of our sun ($T = 1.6 \times 10^7$ K) is only 1.2 KeV, much lower than the classical energy needed to surmount the Coulomb barrier at the proton radius ($\sim 10^{-15}$ m), but proton-proton fusion nevertheless occurs for two reasons. One proton can tunnel through the Coulomb barrier and fuse with its neighbor, and the long high-energy tail of the Maxwell-Boltzmann distribution gives some small fraction of protons considerably greater energy than this average value. At these temperatures the hydrogen atom is not a bound state of an electron and a proton. A naïve estimation of the ionization temperature (vide supra) is that all atoms exist as dissociated ions at any temperature in excess of 158,000 K, but the use of the Saha equation (Fig. 6.3) lowers this estimate to only ~ 3000 K. Therefore, hydrogen at these temperatures consists solely of protons in a hot, dense plasma.

The density of the Big Bang at decoupling, $t = t_0 + 380,000$ years, was about 10^{18} protons/m^3. The average density of our sun is $\sim 10^{30}$ protons/m^3 and in the central core the proton density increases a hundredfold to $\sim 10^{32}$ protons/m^3, or, from a chemical point of view, a concentration of 10^6 moles/l. Gravitational attraction provided a powerful means of concentrating hydrogen into a self-assembling structure that through the eons has sent photons out into space to warm the planetary remnants of the primordial dusty disks surrounding these beacons we see in the night sky.

Proton fusion begins the *proton-proton* chain of nuclear reactions, which has three main branches, denoted *pp I*, *pp II*, and *pp III*. The first reaction in the proton-proton chain (Fig. 6.23) involves the fusion of two protons into deuterium ^{2}H, releasing a positron and an (electron) neutrino as one proton changes into a neutron through inverse β-decay (whose Feynman diagram is shown in Fig. 6.18):

$$^1\text{H} + {}^1\text{H} \rightarrow {}^2\text{H} + e^+ + \nu_e$$

with the neutrinos released in this step carrying energies up to 0.42 MeV. This first step is extremely slow, because it depends on the weak interaction to convert one proton into a neutron. In fact this is the limiting step in the pp chain, with a proton in our sun waiting an average of 10^9 years before fusing into deuterium. The positron emitted in this reaction immediately annihilates with an electron, and their mass energy is carried off by two gamma ray photons in a process whose Feynman diagram is shown in Fig 6.11b:

$$e^+ + e^- \rightarrow 2\gamma + 1.02\,\text{MeV}$$

After this, the deuterium produced in the first stage can fuse with another hydrogen to produce a light isotope of helium, ^{3}He:

$$^2\text{H} + {}^1\text{H} \rightarrow {}^3\text{He} + \gamma + 5.49\,\text{MeV}$$

From here there are three possible paths to generate the helium isotope ^{4}He, whose nucleus is the α-particle. As we have seen using the Saha equation and its subsequent modifications, helium would also be completely ionized and would participate in these reactions as an α particle.

6.6.4.2   The pp I Branch

In *pp I* helium-4 comes from fusing two of the helium-3 nuclei just produced (Fig. 6.24):

$$^3\text{He} + {}^3\text{He} \rightarrow {}^4\text{He} + {}^1\text{H} + {}^1\text{H} + 12.86\,\text{MeV}$$

The complete *pp I* chain reaction releases a net energy of 26.7 MeV. The pp I branch is dominant at temperatures of 10 to 14 megaKelvins (MK). Below 10 MK, the PP chain does not produce much ^{4}He. This is the dominant reaction in our sun with a frequency of 86%.

6.6.4.3 The pp II Branch

The *pp II* branch fuses ^{3}He with a preexisting ^{4}He to make beryllium-7:

$$^3\text{He} + {}^4\text{He} \rightarrow {}^7\text{Be} + \gamma$$

$$^7\text{Be} + \text{e}^- \rightarrow {}^7\text{Li} + \nu_\text{e}$$

$$^7\text{Li} + {}^1\text{H} \rightarrow {}^4\text{He} + {}^4\text{He}$$

The *pp II* branch is dominant at temperatures of 14–23 MK. Ninety percent of the neutrinos produced in the reaction 7Be(e$^-$,ν_e)7Li* carry an energy of 0.861 MeV, while the remaining 10% carry 0.383 MeV (depending on whether lithium-7 is in the ground state or an excited state, respectively). In our sun the *pp II* branch occurs 14% of the time.

6.6.4.4 The pp III Branch

The *pp III* branch fuses ^{3}He with a preexisting ^{4}He to make beryllium-7 and ^{4}He:

$$^3\text{He} + {}^4\text{He} \rightarrow {}^7\text{Be} + \gamma$$

$$^7\text{Be} + {}^1\text{H} \rightarrow {}^8\text{B} + \gamma$$

$$^8\text{B} \rightarrow {}^8\text{Be} + \text{e}^+ + \nu_\text{e}$$

$$^8\text{Be} \Leftrightarrow {}^4\text{He} + {}^4\text{He}$$

The *pp III* chain is dominant if the temperature exceeds 23 MK. The pp III chain is not a major source of energy in the sun (only 0.11%), but was very important in the solar neutrino problem because it generates very-high-energy neutrinos (up to 14.06 MeV).

6.6.4.5 Energy Release

It is very important to note that the *proton-proton* chain just described is a reaction in which the entering quantum states leave the reactions regenerated, ready to continue the chain reaction, but giving a net energy output. Comparing the mass of the final helium-4 atom with the masses of the input four protons reveals that 0.007 or 0.7% of the mass of the original protons has been lost. This mass has been converted into energy, in the form of gamma rays and neutrinos released during each of the individual reactions. The total energy output from one whole chain is 26.73 MeV. Only the energy released as gamma rays will interact with electrons and protons to heat the interior of the sun. This photon flux supports the sun and prevents it from collapsing under its own weight. Neutrinos do not interact significantly with matter and do not help support the sun against gravitational collapse. The neutrinos in the *pp I*, *pp II*, and *pp III* chains carry away the 2.0%, 4.0%, and 28.3% of the energy, respectively.

6.6.4.6   The CNO Cycle

In addition to the proton-proton chain reaction, stars can convert hydrogen to helium in another fusion reaction, the CNO cycle (for carbon-nitrogen-oxygen; Fig. 6.25), often referred to as the Bethe-Weizsäcker cycle after its discoverers. The proton-proton chain is more important in stars whose mass is less than that of our sun. Only

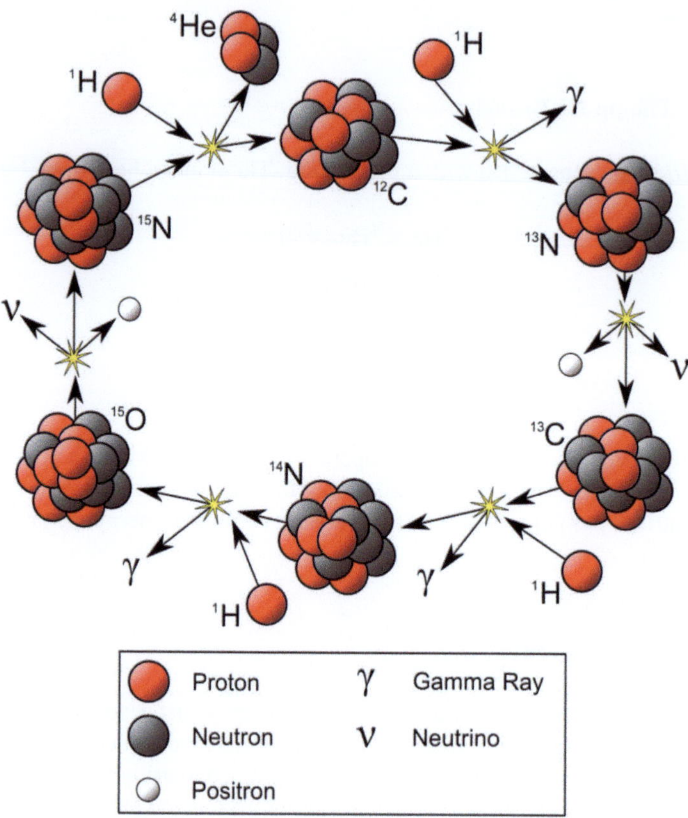

Fig. 6.25   The CNO cycle (for carbon-nitrogen-oxygen) is another fusion reaction by which stars heavier than our sun convert hydrogen to helium. Note the production of the stable magnetic nuclei ^{13}C and ^{15}N that are natural constituents of living systems and are detectable in vivo by nuclear magnetic resonance techniques. (https://commons.wikimedia.org/w/index.php?curid=691760, CC BY-SA 3.0)

1.7% of ^{4}He nuclei produced in the sun are born in the CNO cycle. However, theoretical models show that the CNO cycle is the dominant source of energy in heavier stars. The CNO process was proposed in 1939 by Hans Bethe (1939).

The reactions of the CNO cycle are as follows:

$$^{12}\text{C} + {}^{1}\text{H} \rightarrow {}^{13}\text{N} + \gamma + 1.95 \text{ MeV}$$

$$^{13}\text{N} \rightarrow {}^{13}\text{C} + \text{e} + \; + \nu_\text{e} + 2.22 \text{ MeV}$$

$$^{13}\text{C} + {}^{1}\text{H} \rightarrow {}^{14}\text{N} + \gamma + 7.54 \text{ MeV}$$

$$^{14}\text{N} + {}^{1}\text{H} \rightarrow {}^{15}\text{O} + \gamma + 7.35 \text{ MeV}$$

$$^{15}\text{O} \rightarrow {}^{15}\text{N} + \text{e}^+ + \nu_\text{e} + 2.75 \text{ MeV}$$

$$^{15}\text{N} + {}^{1}\text{H} \rightarrow {}^{12}\text{C} + {}^{4}\text{He} + 4.96 \text{ MeV}$$

The net result of the cycle is to fuse four protons into an alpha particle plus two positrons (annihilating with surrounding electrons and releasing energy in the form of 1.022 MeV gamma rays) plus two neutrinos which escape from the star with a portion of the energy. The carbon, oxygen, and nitrogen nuclei serve as catalysts and are regenerated.

In a minor branch of the reaction, occurring in the sun core just 0.04% of the time, the final reaction shown above does not produce ^{12}C and ^{4}He, but instead produces ^{16}O and a photon and continues as follows:

$$^{15}\text{N} + {}^{1}\text{H} \rightarrow {}^{16}\text{O} + \gamma + 12.13 \text{ MeV}$$

$$^{16}\text{O} + {}^{1}\text{H} \rightarrow {}^{17}\text{F} + \gamma + 0.60 \text{ MeV}$$

$$^{17}\text{F} \rightarrow {}^{17}\text{O} + \text{e}^+ + \nu_\text{e} + 2.76 \text{ MeV}$$

$$^{17}\text{O} + {}^{1}\text{H} \rightarrow {}^{14}\text{N} + {}^{4}\text{He} + 1.19 \text{ MeV}$$

Like the carbon, nitrogen, and oxygen involved in the main branch, the fluorine produced in the minor branch is merely catalytic and, at steady state, does not accumulate in the star.

The main branch of the CNO cycle is known as CNO-I, the minor branch as CNO-II. There exist also two subdominant branches of CNO-III and CNO-IV which are significant only for heavy stars. They are started when the last reaction in CNO-II results in oxygen-18 and gamma instead of nitrogen-14 and alpha:

$$^{17}\text{O} + {}^{1}\text{H} \rightarrow {}^{18}\text{O} + \gamma.$$

While the total number of "catalytic" CNO nuclei is conserved in the cycle, in stellar evolution the relative proportions of the nuclei are altered. When the cycle is run to equilibrium, the ratio of the ^{12}C/^{13}C nuclei is driven to 3.5, and ^{14}N becomes the most numerous nucleus, regardless of initial composition. During a star's evolution, convective mixing episodes bring material in which the CNO cycle has operated from the star's interior to the surface, altering the observed composition of the star. Red giant stars are observed to have lower $^{12}C/^{13}C$ and $^{12}C/^{14}N$ ratios than main sequence stars, which is considered to be proof of nuclear energy generation in stars by hydrogen fusion.

6.6.4.7   The Triple-Alpha Process

The CNO cycle, which is the main energy source in heavier stars, relies on the nucleosynthesis of carbon. Three helium nuclei (α-particles) are transformed into carbon via the triple-alpha process. Due to the much larger Coulomb repulsion of α-particles compared with protons, this nuclear fusion reaction can occur rapidly only at temperatures above 10^9 K and in stellar interiors having a high helium abundance, conditions found in older stars, where helium produced by the proton-proton chain and the carbon nitrogen oxygen cycle has accumulated in the center of the star (Fig. 6.21). After the completion of hydrogen burning in the stellar core, the core will collapse until the central temperature rises to the point where helium burning occurs:

$$^4He + {}^4He \rightarrow {}^8Be$$

$$^8Be + {}^4He \rightarrow {}^{12}C + \gamma + 7.367 \text{ MeV}$$

The net energy release of the process is 7.275 MeV. The ^8Be produced in the first step is unstable and decays back into two helium nuclei in 2.6×10^{-16} s. However, under the conditions of helium burning a small equilibrium abundance of ^8Be is formed; capture of another alpha particle then leads to ^{12}C. This conversion of three alpha particles to ^{12}C is called the triple-alpha process (Fig. 6.26).

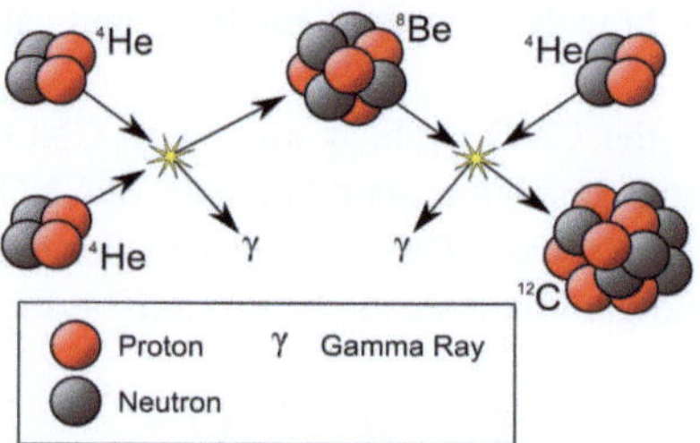

Fig. 6.26 The triple alpha process in which three ^{4}He nuclei (a particles) fuse to produce stable ^{12}C, one of the most important nuclei for living systems. (https://upload.wikimedia.org/wikipedia/commons/8/8d/Triple Alpha_Process.png?20060411182044. Creative Commons Attribution-Share Alike 3.0 Unported)

Ordinarily, the probability of the triple-alpha process would be extremely small. Because the triple-alpha process is unlikely, it requires a long period of time to produce carbon. We have already seen that one consequence of this is that no carbon was produced in the Big Bang because within minutes after the Big Bang, the temperature fell below that necessary for nuclear fusion. The triple-alpha process is highly dependent on ^{12}C having a resonance with the same energy as ^{4}He nuclei (α-particles) and ^{8}Be and before 1952 no such energy level was known. It was astrophysicist Fred Hoyle who used the fact that ^{12}C is so abundant in the universe (and that our existence depends upon it), as evidence for the existence of the ^{12}C resonance. Hoyle suggested the idea to nuclear physicist Willy Fowler, who conceded that it was possible that this energy level had been missed in previous work on ^{12}C. After a brief undertaking by his research group at the Kellogg Radiation Laboratory at the California Institute of Technology, they discovered a resonance near 7.65 MeV. The ground state of the ^{8}Be nucleus has almost exactly the energy of two alpha particles. In the second step in the triple-alpha process, ^{8}Be + ^{4}He has almost exactly the energy of an excited state of ^{12}C. These resonances greatly increase the probability that an incoming alpha particle will combine with ^{8}Be to form carbon. As a side effect of the process, some carbon nuclei can fuse with additional helium to produce a stable isotope of oxygen ^{16}O and release energy:

$$^{12}\text{C} + {}^4\text{He} \rightarrow {}^{16}\text{O} + \gamma$$

The next step of the chain in which oxygen combines with an alpha particle to form neon turns out to be more difficult because of nuclear spin rules, and as a result heavier elements cannot easily be formed in stellar nucleosynthesis. This creates a situation in which stellar nucleosynthesis produces large amounts of carbon and oxygen, elements of life, but only a small fraction of these elements is converted into neon and heavier elements. Both oxygen and carbon make up the ash of helium burning, but these intricacies of nuclear physics mean that stellar nucleosynthesis produced large amounts of the elements of life in a process of nuclear self-assembly.

6.6.4.8 Stellar Evolution and the Nuclear Reaction Rate

The triple-alpha process is strongly dependent on the temperature and density of the stellar material. The energy released by the reaction is approximately proportional to the temperature to the 30th power and the density squared. Contrast this to the proton-proton chain which produces energy at a rate proportional to the fourth power of temperature and directly with density. This strong temperature dependence of the relative reaction rates has consequences for the late stage of stellar evolution, the red giant stage. For lower-mass stars, the helium accumulating in the core is prevented from further collapse only by electron degeneracy (the spin-statistics theorem). The volume of the stellar core is thus dependent only on density and not on pressure. A consequence of this is that once a smaller star begins burning using the triple-alpha process, the core temperature can only increase, which results in the reaction rate increasing further still and becoming a runaway reaction. This runaway reaction, known as the helium flash, lasts only for minutes but burns 60–80% of the helium in the core and produces prodigious quantities of energy.

For higher-mass stars, the helium burning occurs in a shell surrounding a degenerate carbon core. Since the helium shell is not degenerate, the energy released by helium burning increases temperature and causes the star to expand. The expansion cools the helium layer and shuts off the reaction, and the star contracts again. This cyclical process causes the star to become strongly variable and results in it blowing off material from its outer layers.

Of fundamental importance for the understanding of stellar nucleosynthesis is the curve of binding energy (Fig. 6.27) where the average binding energy per nucleon is given as a function of A the number of nucleons in a given nucleus. We can calculate the nuclear binding energies from the masses of atoms using Einstein's $E = mc^2$. The

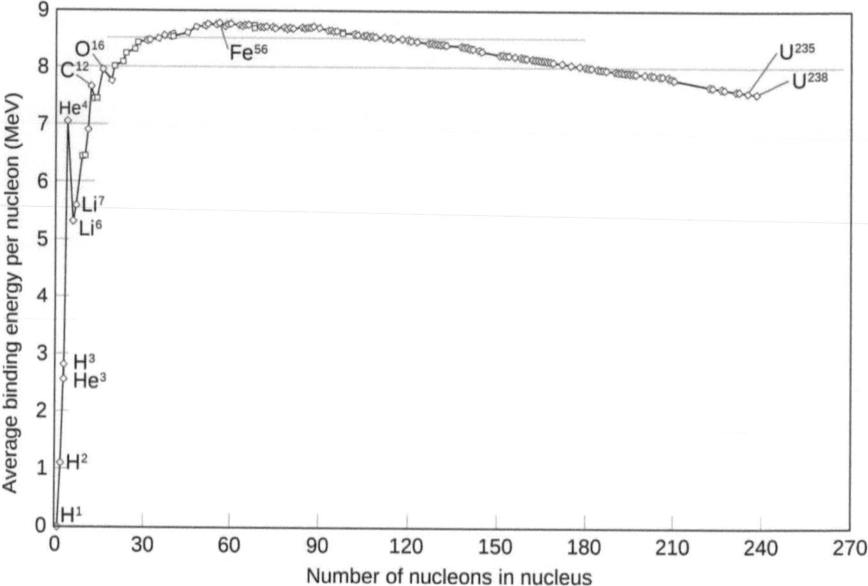

Fig. 6.27 The curve of nuclear binding energy per nucleon. This is the amount of energy released by the fusion of lighter nuclei into heavier nuclei in the periodic table. Note that the binding energy per nucleon for ^{4}He is higher than that for the several adjacent low-mass isotopes, reflecting its doubly magic status as an even-even nucleus. The curve then reaches additional local maxima at the even-even nuclei ^{12}C and ^{16}O, the elements of life. The global maximum in the binding energy is found at ^{56}Fe and then decreases indicating that the synthesis by fusion of elements heavier than ^{56}Fe requires an energy input. Therefore, the synthesis of elements heavier than ^{56}Fe can only occur if an additional source of energy is available, such as when a star explodes as a supernova. (This image is in the public domain: https://upload.wikimedia.org/wikipedia/commons/thumb/5/53/ Binding_energy_curve_-_common_isotopes.svg/2560px-Binding_energy_curve_-_common_iso topes.svg.png. Using data from the Atomic Mass Data Center, http://www.nndc.bnl.gov/amdc/web/ nubase_en.html)

masses of atoms can now be easily determined with a precision of 10^{-9} by ion cyclotron magnetic resonance.

The mass m_j of an atom j of atomic weight W_j is currently defined as $W_j = m_j/M_u$, where M_u is the molar mass unit, $M_u = m_{12}/12$. The molar mass unit M_u is given by m_{12} the mass per mole of ^{12}C divided by 12 to obtain the standard for a single mass unit. We can express the atomic weight in terms of the mass of the hydrogen atom and that of the neutron as

$$W_j = \left(Z_j m_H + (A_j - Z_j) m_n - \frac{B_j}{c^2} \right) / M_u$$

where B_j is the nuclear binding energy of species j. By using the atomic weight for ^{12}C $= W_{12} = 12 = A_{12}$, we can find its binding energy as

$$B_{12} = \frac{m_{12} c^2 (W_H - W_n)}{2} - 1$$

and by inserting this into the above equation, we find that the atomic weight of species j is given by

$$W_j = A_j + (A_j/M_u) \left(\frac{B_{12}}{12} - \frac{B_j}{A_j} \right) + (Z_j - A_j/2)(W_H - W_n)$$

from which we can compute the nuclear binding energy just from a knowledge of atomic weights and this is plotted versus the number of nucleons in Fig. 6.26.

Several important features stand out with respect to this curve of binding energy. The first is that for the isotopes with small numbers of nucleons (^{1}H, ^{2}H, ^{3}H, ^{3}He), the binding energy per nucleon is much smaller than that for nuclei with $A \geq 6$ implying that these nuclides were the easiest to make in the Big Bang, but are less stable than heavier nuclei. The second is that the the binding energy per nucleon for ^{4}He is considerably larger than for any nuclide lighter than ^{12}C as a result of the fact that the ^{4}He nucleus is the first of the nuclei with *magic* quantum numbers.

The stability of a nucleus is determined by the actual numbers of neutrons and protons. The shell model of the nucleus, which owes its great predictive power to a nuclear recapitulation of atomic physics and the shell model of the atom, indicates that nuclei with certain numbers of neutrons or protons are more stable than other nuclei. Those with even numbers of neutrons or protons are more stable than those with odd numbers due to the favorable dependence on spin pairing of the effective nuclear potential. Certain occupation numbers, called *magic*, correspond to closed shells where all of the quantum states with a given angular momentum are filled with spin-paired nucleons. These *magic* numbers are $N, Z = 2, 8, 20, 28, 50, 82$ or $N = 126$. ^{4}He is doubly magic, since both N and $Z = 2$, the first magic number. The binding energy of the ^{2}H nucleus is only 2.2 MeV, but this is higher than the energy difference between the neutron and the proton and the electron (0.78 MeV) when a free neutron decays:

$$n \leftrightarrow p + e^- + \bar{\nu}$$

so that the neutron in the deuterium nucleus is stable against β-decay. The binding energy of the ^{4}He nucleus, on the other hand, is 28.3 MeV giving it great stability.

After dropping from its high value at ^{4}He to the smaller value at ^{6}Li, the curve in Fig. 6.26 rises to reach ^{12}C and ^{16}O, both even-even nuclei, while for ^{16}O, $Z = N = 8$, making it a doubly magic nucleus. Notice that ^{14}N as an odd-odd nucleus is almost as stable as carbon or oxygen; these three elements form the basis for all living systems.

The curve of binding energy continues to increase through ^{20}Ne, another magic nucleus, until it reaches a maximum at ^{56}Fe where the addition of more neutrons or protons reduces the binding energy per nucleon. This means that in a nuclear fusion reaction within a star, energy is released by the fusion of nuclei lighter than ^{56}Fe, but the fusion of elements with $A > 56$ becomes an endothermic process and would actually cool a star, although the fission of such nuclei would again release energy (Lodders 2003). Here stellar nucleosynthesis stops with the prediction that the ashes of stellar nucleosynthesis consist of iron. Another prediction is that the abundance of nuclei heavier than iron should be smaller than that for the $A \sim 56$ nuclei. The core of the Earth contains large amounts of iron! We know that the periodic table continues beyond $A = 56$, so how do we make the heavy nuclei, like ^{238}U?

6.6.5   *Explosive Nucleosynthesis*

Stars come in all different sizes, i.e., masses, and their mass really determines their fate. Very massive stars are very hot for simple thermodynamic reasons. Their large mass compresses their gas to higher temperatures which increases the rate of nuclear fusion and increases their energy output. The most massive stars have the shortest lives and often explode as supernovae whose remnants decorate the night sky (Fig. 6.28). The resulting shock wave travels into the stellar core increasing its temperature to $\sim 10^{11}$ K triggering a massive wave of nucleosynthesis producing the elements heavier than iron by an intense burst of nuclear reactions that typically last only a few seconds. In the explosive environments of supernovae, the elements between silicon and nickel are synthesized by fast fusion. Also, in supernovae further nucleosynthetic processes can occur, such as the r (rapid) process, in which the most neutron-rich isotopes of elements heavier than nickel are produced by rapid absorption of free neutrons released during the explosions.

Explosive supernova nucleosynthesis is responsible for the natural cohort of radioactive elements on the Earth, such as uranium and thorium, as well as the most neutron-rich isotopes of each heavy element. In the Big Bang the fireball cooled so quickly that fusion reactions from the initial, simple quark-gluon plasma stalled at the lightest nuclei. For supernovae the stellar core contains a great accumulation of nuclear species that were never formed from the primordial

Fig. 6.28 Supernova! An image taken by NASA's Hubble Space Telescope of the Crab Nebula, a six-light-year-wide expanding remnant of a star's supernova explosion, 6500 light-years away. Japanese and Chinese astronomers recorded this violent event years ago in 1054, as did Native Americans, as reflected by the famous supernova petroglyph in Chaco Canyon, New Mexico. The orange filaments are the tattered remains of the star and consist mostly of hydrogen. The rapidly spinning neutron star embedded in the center of the nebula is the dynamo powering the nebula's eerie interior bluish glow. The blue light is synchrotron radiation from electrons whirling at nearly the speed of light around magnetic field lines from the neutron star. The colors in the image indicate the different elements that were expelled during the explosion. Blue in the filaments in the outer part of the nebula represents neutral oxygen, green is singly ionized sulfur, and red indicates doubly ionized oxygen. (Courtesy of the Space Telescope Science Institute under NASA Contract NAS5-26555. This image is in the public domain)

quark-gluon plasma. Both Big Bang nucleosynthesis and supernova nucleosynthesis are severely time-limited. The core collapse of a supernova occurs too rapidly ($\sim$0.25–10 s) for slow radioactive β-decay to increase the number of neutrons, so that many abundant isotopes having equal *even* numbers of protons and neutrons are synthesized in an r process. These include ^{44}Ti, ^{48}Cr, ^{52}Fe, and ^{56}Ni, all of which decay after the explosion to create abundant stable isobars at each atomic weight.

Many such decays are accompanied by the emission of gamma-ray lines capable of identifying the isotope that has just been created in the explosion. This nuclear astronomy was predicted in 1969 (Clayton et al. 1969) as a way to confirm explosive nucleosynthesis of the elements and that prediction played an important role in the planning for NASA's successful Compton Gamma-Ray Observatory.

The most convincing proof of explosive nucleosynthesis in supernovae occurred in 1987 when these gamma-ray lines were detected emerging from supernova 1987A. Gamma ray lines identifying ^{56}Co and ^{57}Co, whose radioactive half-lives limit their age to about a year, proved that ^{56}Fe and ^{57}Fe were created by radioactive parents. Other proofs of explosive nucleosynthesis are found within the stardust grains that condensed within the interiors of supernova as they expanded and cooled. In particular, radioactive ^{44}Ti was measured to be very abundant within supernova stardust grains at the time they condensed during the supernova expansion, confirming a 1975 prediction for identifying supernova stardust. Other unusual isotopic ratios within those grains reveal specific aspects of explosive nucleosynthesis.

The fossil record of the universe gives up its secrets once again; for now we can understand not only the origin of the light elements in Big Bang nucleosynthesis, but the subsequent stellar and explosive nucleosyntheses that also left their observable footprints on the universe. The details of the cosmic microwave background radiation have cemented a standard cosmological model of the Big Bang. Now come the stars and their nuclear fusion processes; the fossils of nucleosynthesis are the nuclei themselves, for their isotopic abundances are a record of the paths used in their production.

6.6.6   Nuclear Abundances

What are nuclear abundances? They are nothing less than a record of the details of the self-assembly of the universe, fossils recording the results of Big Bang, stellar, and supernova nucleosynthesis. Just as a paleontologist can understand the trajectory of evolution by examining fossils from a variety of creatures that lived in the past, we can determine the path matter has taken through time since the Big Bang by measuring the nuclear abundances of the isotopes. We can envision counting all the nuclei in a sample, say a meteorite, and assigning an abundance X_j as N_j the number of nuclei of species j normalized to n_T the total number of nuclei in a sample, $\chi_j = \frac{N_j}{n_T}$. With this beginning we can ask about the nature of samples. Clearly, we need not bother sampling air or water, for most material in the periodic table are not volatile, or water soluble; our best bet would be to sample something solid, such as a rock, but lacking representative rocks on the Earth we can turn to rocks from space. Meteorites often strike the atmosphere and the largest survive their blazing journey to land on the Earth's surface. The ages of these visitors from space are the same as the Earth's age, ~4.6 billion years, indicating that they were part of the pre-solar

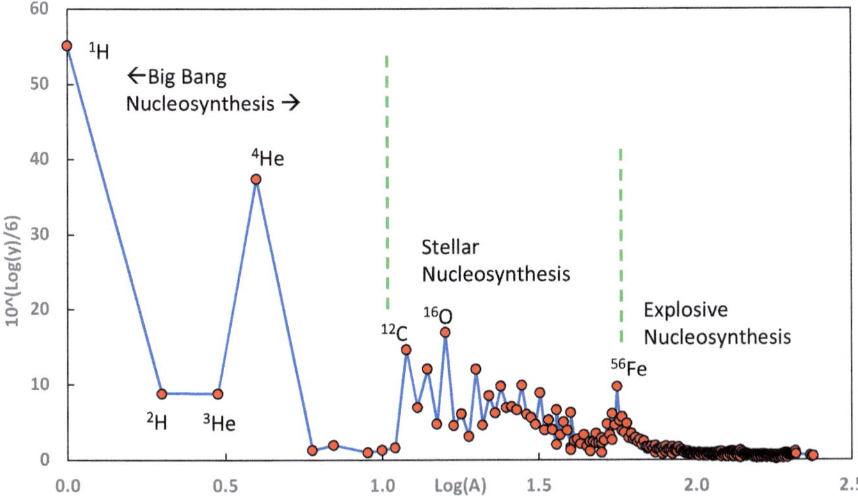

Fig. 6.29 The abundance of nuclei in the solar system and, by inference, the universe. The data are the mole fractions $Y_j = 10^{\wedge}(Log(y_j)/6)$ scaled to emphasize the three regions of nucleosynthesis, from the Big Bang, from nucleosynthesis in stars, and in supernovae. The x-axis is $Log(A)$ where A is the number of nucleons in the nuclide. Note that the three most abundant nuclei made in stars from hydrogen and helium are ^{12}C, ^{14}N, and ^{16}O, the elements of life

nebula and are likely to provide a good sample from which to measure the abundances of the nuclei. Another extraterrestrial source to examine is not a rock, but starlight, which, through spectroscopy, has provided a rich trove of abundances. The sun is the closest star and the one with the greatest photon flux at the Earth. Solar abundances were measured first, prior to the advent of the ultrahigh-resolution mass spectrometers used today to measure the abundances in meteorites. The detailed and excellent agreement between solar and meteoric abundances lends strong support to the universality of nuclear abundances and to the current view of the three stages of nucleosynthesis (Fig. 6.29) as arising from (1) the Big Bang, (2) nucleosynthesis in stars terminating with ^{56}Fe due to the shape of the curve of binding energy (Fig. 6.27), and (3) explosive nucleosynthesis (Fig. 6.28).

Nuclear abundances are quantitative. Their measurement requires that we determine not only the number of nuclei of a given isotope, but also the size of the sample, be it the mass, the number of nucleons, or the volume. Therefore, it is useful to begin with density, which we all know is the amount of something in some volume V. The mass density, ρ_m, is the total mass per unit volume, or

$$\rho_m = \frac{m_T}{V}$$

For a mixture containing s species of nuclei, the total mass can be decomposed into the sum of its parts:

$$\rho_{\mathrm{m}} = \frac{1}{N_{\mathrm{A}}} \sum_{j=1}^{s} N_j W_j$$

where N_j is the number of nucleons of species j per unit volume, W_j is the atomic weight of species j, and Avogadro's number is N_{A}. Avogadro's number is $6.02214076 \times 10^{23}$ and is the number of particles (nuclei, atoms, or molecules) that constitute a mole of particles and is defined as the number of, say atoms, of species j present in a sample of W_j grams.

The atomic weights are defined in terms of a standard; for these purposes, we have already presented the IUPAC standard of 1/12th of the weight of a single, neutral ^{12}C atom, which is given the name dalton, Da, or atomic mass unit, m_{u}:

$$1\,\mathrm{Da} = 1\,m_{\mathrm{u}} = \frac{M_{\mathrm{u}}}{N_{\mathrm{A}}} = \frac{M_{12}}{12N_{\mathrm{A}}} = 1.66053906660 \times 10^{-27}\,\mathrm{kg}$$

in which M_{u} is the molar mass unit (g/mol) and M_{12} is the molar mass of ^{12}C. Using this we find that the molar mass of ^{12}C is

$$M_{12} = 12\,N_{\mathrm{A}} \times 1.66053906660 \times 10^{-27}\,\frac{\mathrm{kg}}{\mathrm{mol}}$$

$$= 12 \times 6.02214076 \times 10^{23} \times 1.66053906660 \times 10^{-27}\,\frac{\mathrm{kg}}{\mathrm{mol}}$$

$$M_{12} = 0.012\,\frac{\mathrm{kg}}{\mathrm{mol}} = 12\,\frac{\mathrm{g}}{\mathrm{mol}}$$

That is, 12 g of ^{12}C contain exactly Avogadro's number (1 mole) of carbon atoms, and neglecting units, we have that $W_j = M_{12} = A_{12} = 12$. The atomic weights quoted in tables are not in SI mass units (kg), but are actually weight ratios, i.e., ratios of the weight to 1 m_{u}. To convert these to actual weights one must multiply by m_{u}. For example, ^{23}Na has a tabulated atomic weight of $W_{23} = 22.989769$ Da, which means that the actual weight of a ^{23}Na atom is 22.989769 Da $\times$ 1.661 $\times$ 10^{-27} kg/Da $= 3.818 \times 10^{-26}$ kg.

Atomic weights defined in this way are conveniently almost equal to the number of nucleons in species j, $W_j \sim A_j$, and it is therefore straightforward to recast mass density as *nucleon density*:

$$\rho = \frac{1}{N_{\mathrm{A}}} \sum_{j=1}^{s} N_j A_j$$

And we can use this to compute the nucleon fraction X_j for species j in a sample of material as

$$X_j = \frac{N_j A_j}{\rho N_{\mathrm{A}}} = \frac{N_j A_j}{n_{\mathrm{T}}}$$

where ρN_A is the total number of nucleons in the sample. The sum of the nuclear fractions must equal one, so

$$\sum_{j=1}^{s} X_j = \sum_{j=1}^{s} \frac{N_j A_j}{\rho N_A} = \sum_{j=1}^{s} \frac{N_j A_j}{n_T} = \frac{1}{n_T} \sum_{j=1}^{s} N_j A_j = \frac{n_T}{n_T} = 1$$

We observe here that the nucleon fraction X_j is the probability of finding species j in the sample. What we have shown in Fig. 6.28 is the mole fraction Y_j of species j as a function of the number of nucleons, and this is given by Y_j:

$$Y_j = \frac{X_j}{A_j} = \frac{N_j}{\rho N_A} = \frac{N_j}{n_T}$$

which is what we set out to show. This sets the scale for abundances as the total number of nucleons. We already expect from our study of Big Bang nucleosynthesis that $Y_H > Y_{He} \gg Y_{j \, > \, He}$ the abundances of hydrogen and helium are far larger than for any heavier nuclides. Dividing the abundance of heavier nuclides by the total means that small errors in the abundances of hydrogen and helium will introduce large errors in the heavier nuclides.

Just as atomic weights are defined with respect to a weight standard, ^{12}C, nuclear mole fractions are defined with respect to Si in the meteorics community and to ^{1}H in the solar community in the following manner. Taking the logarithms of Y_j gives

$$\mathrm{Log}(N_j) = \mathrm{Log}(n_T) + \mathrm{Log}(Y_j)$$

Or for Si

$$\mathrm{Log}(N_{Si}) = \mathrm{Log}(n_T) + \mathrm{Log}(Y_{Si})$$

And the standard is that $\mathrm{Log}(N_{Si}) = 6$. The actual mole fraction of silicon nuclei is $Y_{Si} = 2.529 \times 10^{-5}$ and $\mathrm{Log}(Y_{Si}) = -4.597$ so that the normalization constant $\mathrm{Log}(n_{Si}) = 10.597$.

We rescale the abundances from Y_j to y_j which are then given by

$$\mathrm{Log}(y_j) = 10.597 + \mathrm{Log}(Y_j)$$

And these are the rescaled abundances shown in Fig. 6.28 as $Y_j = 10^{\wedge}(\mathrm{Log}(y_j)/6)$ where we have compressed the large dynamic range of $\sim 10^{12}$ in abundances by dividing the $\mathrm{Log}(y_j)$ values by 6 prior to exponentiation. This rescaling serves to emphasize the three regions of nucleosynthesis, light nuclides arose from the Big Bang, intermediate-mass nuclides were generated from nucleosynthesis in stars, and the nuclides heavier than iron were formed in the explosions of supernovae.

Other features of the abundances (Fig. 6.28) are germane to the origin of life. In particular, the three most abundant nuclei made in stars from hydrogen and helium are ^{12}C, ^{14}N, and ^{16}O, the elements of life. The *even-even* carbon and oxygen nuclei

Table 6.6 Comparison of the abundances of certain elements important for living systems in the universe with that found in organisms

Element	In organisms	In universe
Hydrogen	80–250	10,000,000
Carbon	1000	1000
Nitrogen	60–300	1600
Oxygen	500–800	5000
Sodium	10–20	12
Magnesium	2–8	200
Phosphorus	8–50	3
Sulfur	4–20	80
Potassium	6–40	0.6
Calcium	25–50	10
Manganese	0.25–0.8	1.6
Iron	0.25–0.8	100
Zinc	0.1–0.4	0.12

have nucleon numbers divisible by 4, as do ^{20}Ne, ^{24}Mg, ^{28}Si, ^{32}S, ^{36}Ar, and ^{40}Ca, and are called alpha nuclides because alpha particles (^{4}He nuclei) are a particularly stable arrangement of two protons and two neutrons (*magic numbers* again). The highly abundant ^{14}N nucleus is an *odd-odd* nuclide formed easily from hydrogen fusion. There is also a high abundance of *doubly magic* nuclei, ^{4}He with $Z = N = 2$, ^{16}O with $Z = N = 8$, and ^{40}Ca with $Z = N = 20$. The next most abundant nuclide is ^{56}Fe formed as the end of the energy generation process from the fusion of lighter nuclei, but this iron isotope is not the most tightly bound of the $Z = N$ nuclides; that honor falls to ^{56}Ni which is radioactive with a half-life of 6.075 days. This unstable nucleus rapidly decays to Co and then to ^{56}Fe. The production and decay of ^{56}Ni dominates the light curve of supernovae.

The abundances of certain elements that are fundamental to living systems are compared with their present abundances in the universe in Table 6.6. There is a good, general correspondence between the amounts of these elements in the biosphere and in the universe at large, indicating that living systems utilized the chemical species available to them, an idea that supports the universal nature of life in the universe. The linear correlation coefficient for a power law fit $O(U) \sim 3.3\sqrt[3]{U}$ is $r^2 \sim 0.4$. It seems very likely that any alien species we discover in the future would resemble us in terms of its structure and metabolism because the same elements we find in the solar system were made across the universe and we can be certain that the same periodic table, and hence chemistry, would also prevail across the endless leagues of space.

6.6.7   Why the Earth Is Mostly Fe/Ni

One of the main classes of meteorites consists of those lumps from space that are primarily made up of iron and nickel. The core of the Earth is also chiefly constructed from these elements. Why are iron and nickel so abundant in the planetesimals and planets of the universe? As we earlier stated, iron and nickel are the ashes of nuclear fusion. They are made in the cores of massive stars and cast out into the interstellar medium when these short-lived stars explode as supernovae. Gravity then sweeps up this stellar debris, along with fresh hydrogen, into a protoplanetary disk and begins the stellar and planetary formation process all over again.

Problems

1. *Baryon density of the universe.* Given the present known mass of the observable universe, $\sim 3 \times 10^{52}$ kg, and the size of 13.8×10^9 light-years, estimate the present baryon density, and use it to predict the baryon density at distant epochs back to $z \sim 10^{10}$ in steps of 10. At what value of z does this density approach the uncertainty principle estimate for the universe in a box of dimension given by the Planck length 1.62×10^{-35} m?

2. *Particles in the early universe.* The probability that a particle will occupy a given state depends on the energy of that state. This probability distribution is given by the Boltzmann distribution. The ratio of the number of protons to neutrons during the first minute of the early universe is

$$\frac{N_\text{p}}{N_\text{n}} = e^{-Q/kT}$$

where Q is the energy (mass) difference between the proton and neutron. This equilibrium favors the proton because the neutron is heavier by $Q = 2.07 \times 10^{-13}$ joules.

(a) What is the proton/neutron ratio for $T = 10^{13}$ Kelvins?

(b) The proton/neutron ratio "freezes out" at a value of 1/6. What is the temperature of the universe when this happens?

3. *Isotopes of carbon.* Carbon has four main isotopes: ^{11}C, ^{12}C, ^{13}C, and ^{14}C.

(a) Which of these are stable, and which are unstable, and for what reasons?

(b) At what stage of nucleosynthesis would these isotopes form?

(c) How many neutrons and protons are contained in each of these nuclei, and hence, what would a likely spin state be for each of them?

(d) Which of these isotopes would be useful for dating material derived from living systems?

3(a) *Solutions for isotopes of carbon*

Isotope	Stability	A	N	Z	Reason	Spin (I)
^{11}C	Unstable	11	5	6	$Z > N$	½
^{12}C	Stable	12	6	6	$Z \sim N$	0
^{13}C	Stable	13	7	6	$Z \sim N$	½
^{14}C	Unstable	14	8	6	$N > Z$	0

(b) These are formed during stellar nucleosynthesis.

(c) The spin states are *estimated* from the fact that nucleons (n, p) are fermions with spin $\frac{1}{2}$, and the Pauli principle only allows a maximum of 2 fermions per energy level if they spin pair. Therefore, for example, for ^{11}C,

Protons:
$$I_p(^{11}\text{C}) = (\tfrac{1}{2} - \tfrac{1}{2})_p + (\tfrac{1}{2} - \tfrac{1}{2})_p + (\tfrac{1}{2} - \tfrac{1}{2})_p = 0$$

Neutrons:
$$I_n(^{11}\text{C}) = (\tfrac{1}{2} - \tfrac{1}{2})_n + (\tfrac{1}{2} - \tfrac{1}{2})_n + (\tfrac{1}{2})_n = \tfrac{1}{2}$$
$$\text{And} \quad I = I_p + I_n = \tfrac{1}{2}.$$

while for ^{12}C,

Protons:
$$I_p(^{12}\text{C}) = (\tfrac{1}{2} - \tfrac{1}{2})_p + (\tfrac{1}{2} - \tfrac{1}{2})_p + (\tfrac{1}{2} - \tfrac{1}{2})_p = 0$$

Neutrons:
$$I_n(^{12}\text{C}) = (\tfrac{1}{2} - \tfrac{1}{2})_n + (\tfrac{1}{2} - \tfrac{1}{2})_n + (\tfrac{1}{2} - \tfrac{1}{2})_n = 0$$
$$\text{And} \quad I = I_p + I_n = 0.$$

The nuclides ^{13}C and ^{14}C behave like ^{11}C and ^{12}C, respectively, except that for ^{13}C, the spin $\frac{1}{2}$ comes from the spin of the extra neutron, where in ^{11}C the spin $\frac{1}{2}$ comes from the extra proton. These spin values are an approximation which ignores quantum systematics; the spins for ^{12}C, ^{13}C, and ^{14}C are correctly predicted, but the actual spin for ^{11}C is 3/2 rather than $\frac{1}{2}$.

(d) To be useful for dating formerly living systems, the two stable nuclides would be useless because their amounts do not change with time. Therefore, only the radioactive ^{11}C and ^{14}C would be candidates. An examination of the half-lives for these gives:

$$t_{\frac{1}{2}}(^{11}\text{C}) = 20 \text{ min, and } t_{\frac{1}{2}}(^{14}\text{C}) = 5700 \text{ years.}$$

The half-life for ^{11}C is too short; all the ^{11}C ever formed would have decayed by now. However, the lifetime for ^{14}C is long enough so that if ^{14}C uptake ceases at death, then one could date organic matter back to ~15,000 years. Indeed, many of the accurate dates for the Anasazi sites in the southwest (~1 kya) were derived from ^{14}C dating.

4. *Stability of isotopes.* Use the data found on http://www.nndc.bnl.gov/chart/ to make plots of the half-life (in seconds) of carbon (^{8}C $\rightarrow$ ^{22}C), nitrogen (^{10}N $\rightarrow$ ^{25}N), and oxygen (^{12}O $\rightarrow$ ^{28}O) versus $N = A - Z$, i.e., the number of neutrons in the nucleus. What conclusions can you draw with respect to the number of neutrons versus the number of protons in the nucleus with respect to the stability of a given nuclide?

Here is a table of the data.

Carbon	N	n/p	$t\ 1/2$ (s)	Nitrogen	N	n/p	$t\ 1/2$ (s)	Oxygen	N	n/p	$t\ 1/2$ (s)
8	2	0.33	1.00E−21								
9	3	0.50	0.127								
10	4	0.67	19.3	10	3	0.429	2.00E−22				
11	5	0.83	1200	11	4	0.571	4.00E−22				
12	6	1.00	4.42E+17	12	5	0.714	0.011	12	4	0.5	1.65E−21
13	7	1.17	4.42E+17	13	6	0.857	598.2	13	5	0.625	0.00858
14	8	1.33	1.798E+11	14	7	1.000	4.42E+17	14	6	0.75	70.6
15	9	1.50	2.4	15	8	1.143	4.42E+17	15	7	0.875	122
16	10	1.67	0.75	16	9	1.286	7.13	16	8	1	4.42E+17
17	11	1.83	0.193	17	10	1.429	4.17	17	9	1.125	4.42E+17
18	12	2.00	0.092	18	11	1.571	0.624	18	10	1.25	4.42E+17
19	13	2.17	0.049	19	12	1.714	0.271	19	11	1.375	27
20	14	2.33	0.014	20	13	1.857	0.13	20	12	1.5	13.5
21	15	2.50	3.00E−08	21	14	2.000	0.085	21	13	1.625	3.42
22	16	2.67	0.0061	22	15	2.143	0.024	22	14	1.75	2.25
				23	16	2.286	0.0141	23	15	1.875	0.082
				24	17	2.429	5.20E−08	24	16	2	0.065
				25	18	2.571	2.60E−07	25	17	2.125	5.00E−08
								26	18	2.25	4.00E−08
								27	19	2.375	2.60E−07
								28	20	2.5	1.00E−07
A	**N**	**n/p**	**t 1/2** (s)	**A**	**N**	**n/p**	**t 1/2** (s)	**A**	**N**	**n/p**	**t 1/2** (s)

Let us plot this using Excel, but note that the range of half-lives encompasses 38 orders of magnitude, so a Log plot would be a good idea. Also note that the stable nuclides can have a half-life assigned on the order of the age of the universe, ~14 Gyr. Finally, some half-lives are reported in energy units, keV or MeV. To convert these to seconds, use the uncertainty principle, $\Delta E \Delta t = \hbar$, where the reported value of $t_{1/2}$ is ΔE, and one must convert this energy to joules (1 eV = 1.6×10^{-19} joules). Then the lifetime of the nucleus is $t_{1/2} = \hbar/\Delta E$ in seconds.

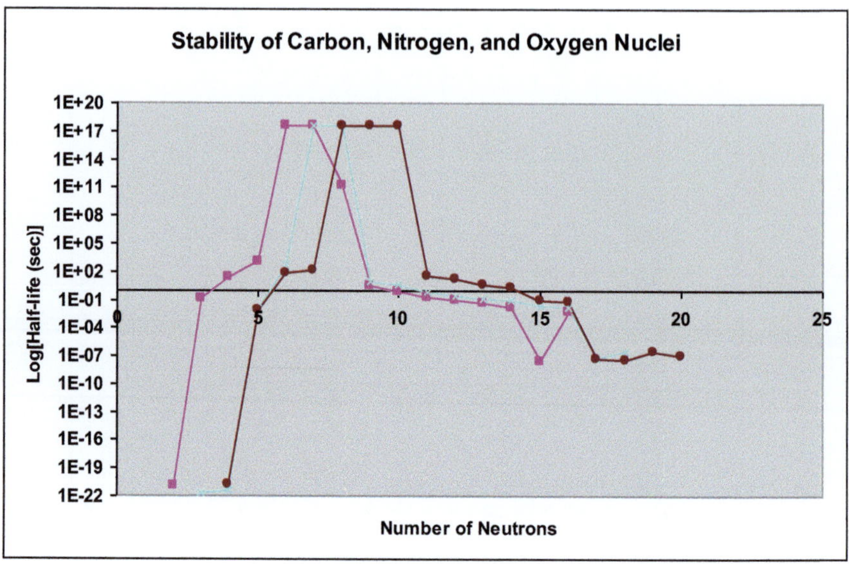

One way to plot this data is as I suggested $t_{1/2}$ versus N: Here we see that the most stable nuclides have $N \sim Z$, i.e., the number of neutrons equals the number of protons for each of these elements.

However, it is more suggestive to plot the lifetime versus the neutron to proton ratio as below:

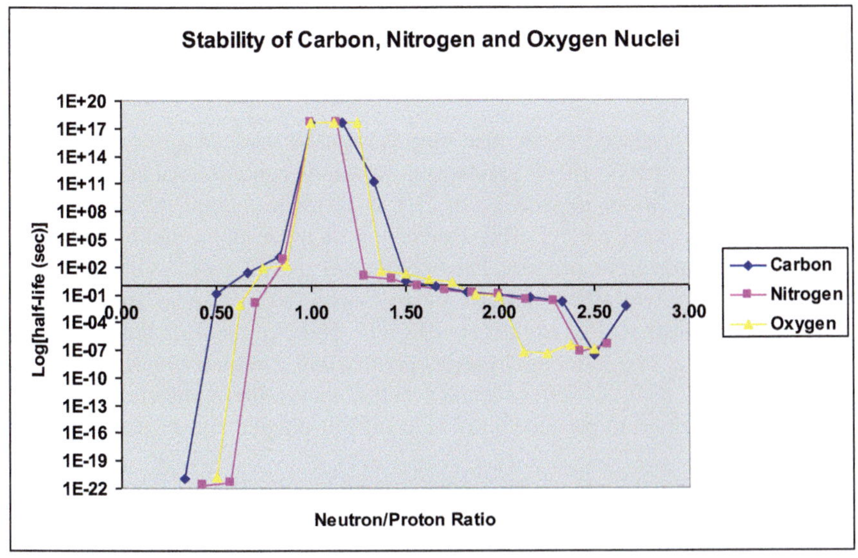

Here we note two things.

1. Nuclear stability requires a neutron/proton ratio ~ 1.
2. One or two extra neutrons appear to be well tolerated, particularly in the case of oxygen (^{16}O, ^{17}O, and ^{18}O are all stable).

    The addition of more neutrons up to an n/p ratio of ~ 2.0 has only a small effect on the lifetimes. On the other hand, a deficit of neutrons, even by a single one, immediately implies instability (^{11}C, ^{13}N, and ^{15}O are all *unstable*), and there is a strong effect on stability if $n/p < 1.0$.

5. *Isotopes that make up biomolecules.* Biomolecules, like proteins, are made up primarily of carbon (^{12}C), nitrogen (^{14}N), and oxygen (^{16}O), which are all light elements synthesized in the process of stellar nucleosynthesis. Explain from the viewpoint of stellar nucleosynthesis why these nuclides are so abundant, while lithium (^{7}Li), beryllium (^{10}Be), and neon (^{20}Ne) are not found as constituents of cells, even though these too are light elements.

6. *Stellar nucleosynthesis.* The elements most common in life are also the most abundantly synthesized during stellar nucleosynthesis. The most abundant nuclides are oxygen (^{16}O), neon (^{20}Ne), nitrogen (^{14}N), and carbon (^{12}C), in that order. Carbon-12 and oxygen-16 contain 3 and 4 α-particles, respectively. Alpha particles consist of 2 protons and 2 neutrons and are very stable: These are the nucleus of the noble gas, helium. Note again that neon's atomic number of 20 suggests that it was formed from the aggregation of 5 objects containing 4 nucleons (α-particles again). After the completion of hydrogen burning in the stellar core, the core will collapse until the central temperature rises to the point where helium burning occurs:

$$^4\text{He} + {}^4\text{He} \rightarrow {}^8\text{Be}$$

$$^8\text{Be} + {}^4\text{He} \rightarrow {}^{12}\text{C} + \gamma$$

The ^{8}Be produced in the first step is unstable and decays back into two helium nuclei in 2.6×10^{-16} s. However, under the conditions of helium burning a small equilibrium abundance of ^{8}Be is formed; capture of another alpha particle then leads to ^{12}C. This conversion of three alpha particles to ^{12}C is called the triple-alpha process. The beryllium-8 ground state has almost exactly the energy of two alpha particles. In the second step, ^{8}Be + ^{4}He has almost exactly the energy of an excited state of ^{12}C. These resonances greatly increase the probability that an incoming alpha particle will combine with beryllium-8 to form carbon. As a side effect of the process, some carbon nuclei can fuse with additional helium to produce a stable isotope of oxygen and release energy:

$$^{12}\text{C} + {}^4\text{He} \rightarrow {}^{16}\text{O} + \gamma$$

The next step of the chain in which oxygen combines with an alpha particle to form neon turns out to be more difficult because of nuclear spin rules, and as a result heavier elements cannot easily be formed in stellar nucleosynthesis. This creates a situation in which stellar nucleosynthesis produces large amounts of carbon and oxygen but only a small fraction of these elements is converted into neon and heavier elements.

Beryllium and lithium are underproduced in both the Big Bang and stellar nucleosynthesis because there is no stable nuclide with $A = 8$ nucleons. Neon is almost as abundant as carbon, but forms no chemical compounds due to its filled electron shells, which make it a noble gas. Lithium and beryllium only form ionic bonds, not covalent bonds, and Be is extremely toxic to living systems.

7. *Opacity of the early universe.* Find the temperature at which the opacity (i.e., ionization) of clouds of the elements generated by Big Bang nucleosynthesis would have dropped to 1%. Use the primary and secondary ionization potentials of ^{1}H, ^{2}H, ^{3}H, ^{3}He, ^{4}He, ^{6}Li, ^{7}Li, and ^{7}Be.

8. *The early universe as a particle in a box.* As we have seen with the particle in a box model of an atom, the energy levels of an electron are of the correct order of magnitude. The same is true for the energy levels of a nucleus. The wave functions of the simple harmonic oscillator display a nonzero energy eigenvalue. Put a particle in a box the size of the Planck length and compute the lowest-energy eigenvalue for a rectangular potential and the simple harmonic oscillator potential for a particle with the mass of:

   (a) An electron
   (b) A proton
   (c) A Higgs boson
   (d) A top quark
   (e) A Planck mass

(f) The mass of the observed universe

From the energies, compute the minimum temperature of the system. What conclusions can you draw with respect to the origin of the universe?

9. *The early universe as a particle in a box.* Think of the early universe as a particle in a spherical box, with expanding walls. The expansion rate is given by the Hubble parameter at that redshift, z. The eigenfunctions would be the spherical Bessel functions, $j_l(n,x)$ where n is an integer and x is a continuous parameter. Would these be eigenfunctions of a perturbation operator? The particle in a spherical box is a standard problem for the motivation of the hydrogen atom solutions for the wave function. How do the energies ($E = \frac{\hbar^2 k^2}{2m}$) scale as the boundary of the box, called the sound horizon, moves outwards?

$$a_1{}^2 = \frac{\pi\,\Omega_o{}^2 H_o{}^4 a_o{}^4}{D_o{}^2} \int_0^\infty \left(j_1(k\,r_m)\right)^2 \frac{P(k)}{k^2}\,dk$$

where $j_l(x)$ is in Mathematica a `SphericalBesselFunction[l,x]`. For a primordial power spectrum of vacuum fluctuations, $P(k) = bk$, where b is a constant.

10. *Temperature dependence of the degrees of freedom of the early universe.* Find the number of states $N(T)$ in Table 6.2 by counting degrees of freedom for the constituents of the universe as a function of temperature. Degrees of freedom are unique states and include particle (antiparticle) identity, spin, charge, etc. as identifiers.

References

N. Aghanim et al., Planck 2018 results. VI. Cosmological parameters. Astron. Astrophys. **641**, A6 (2020)

Q.R. Ahmad et al., Measurement of the rate of $\nu_e+d{\to}p+p+e^-$ interactions produced by ^{8}B solar neutrinos at the Sudbury Neutrino Observatory. Phys. Rev. Lett. **87**, 071301 (2001)

L. Anderson et al., The clustering of galaxies in the SDSS-III Baryon Oscillation Spectroscopic Survey: Baryon acoustic oscillations in the Data Releases 10 and 11 Galaxy samples. Mon. Not. R. Astron. Soc. **441**, 24–62 (2014)

J.N. Bahcall, M.H. Pinsonneault, G.J. Wasserburg, Solar models with helium and heavy-element diffusion. Rev. Mod. Phys. **67**, 781 (1995)

C.A.P. Bengaly, R. Maartens, N. Randriamiarinarivo, A. Baloyi, Testing the Cosmological Principle in the radio sky. J. Cosmol. Astropart. Phys. **9**, 25 (2019)

C.L. Bennett et al., Nine-year Wilkinson Microwave Anisotropy Probe (WMAP) observations: Final maps and results. Astrophys. J. Suppl. **208**, 20 (2013)

H. Bethe, Energy production in stars. Phys. Rev. **55**, 434–456 (1939)

A. Bonanno, H. Schlattl, L. Paterno, The age of the Sun and the relativistic corrections in the EOS. Astron. Astrophys. **390**, 1115 (2002)

A. Bouvier, M. Wadhwa, The age of the Solar System redefined by the oldest Pb–Pb age of a meteoritic inclusion. Nat. Geosci. **3**, 637 (2010)

E.M. Burbidge, G.R. Burbidge, W.A. Fowler, F. Hoyle, Synthesis of the elements in stars. Rev. Mod. Phys. **29**, 547–650 (1957)

S. Burles, K.M. Nollett, M.S. Turner, Big-Bang nucleosynthesis: Linking inner space and outer space. arXiv:astro-ph/9903300 (1999)

R.W. Carlson, L.E. Borg, A.M. Gaffney, M. Boyet, Rb-Sr, Sm-Nd and Lu-Hf isotope systematics of the lunar Mg-suite: The age of the lunar crust and its relation to the time of Moon formation. Philos. Trans. A. Math. Phys. Eng. Sci. **372**, 20130246 (2014)

H.B.G. Casimir, On the attraction between two perfectly conducting plates. Proc. R. Neth. Acad. Arts Sci. **51**, 793–795 (1948)

D. Castelvecchi, Mystery over Universe's expansion deepens with fresh data. Nature **583**, 500–501 (2020)

Cayrel et al., Measurement of stellar age from uranium decay. Nature **409**, 691–692 (2001)

B. Chaboyer, The age of the Universe. Nucl. Phys. Proc. Suppl. **51B**, 10–19 (1996)

S.K. Choi et al., The Atacama Cosmology Telescope: A measurement of the Cosmic Microwave Background power spectra at 98 and 150 GHz. J. Cosmol. Astropart. Phys. **12**, 45 (2020)

D.D. Clayton, S.A. Colgate, G.J. Fishman, Gamma ray lines from young supernova remnants. Astrophys. J. **155**, 75 (1969)

M. Colless et al., The 2dF Galaxy Redshift Survey: Spectra and redshifts. Mon. Not. R. Astron. Soc. **328**, 1039 (2001)

J.J. Cowan, B. Pfeiffer, K.-L. Kratz, F.-K. Thielemann, C. Sneden, S. Burles, D. Tytler, T.C. Beers, R-Process abundances and chronometers in metal-poor stars. Astrophys. J. **521**, 194 (1999)

R.H. Cyburt, B.D. Fields, K.A. Olive, T.-H. Yeh, Big bang nucleosynthesis: Present status. Rev. Mod. Phys. **88**, 015004 (2016)

N. Dauphas, The U/Th production ratio and the age of the Milky Way from meteorites and Galactic halo stars. Nature **435**, 1203 (2005)

A. Einstein, The foundation of the general theory of relativity. Ann. Phys. **354**, 769 (1916)

D.J. Eisenstein et al., Detection of the baryon acoustic peak in the large-scale correlation function of SDSS luminous red galaxies. Astrophys. J. **633**, 560–574 (2005)

R.P. Feynman, Space-time approach to quantum electrodynamics. Phys. Rev. **76**, 769–789 (1949)

B.D. Fields, P. Molaro, S. Sarkar., http://pdg.lbl.gov/2020/reviews/rpp2020-rev-bbang-nucleosynthesis.pdf. Chapter 24: Big-Bang Nucleosynthesis (2020)

D.J. Fixsen, E.S. Cheng, D.A. Cottingham, R.E. Eplee Jr., R.B. Isaacman, J.C. Mather, S.S. Meyer, P.D. Noerdlinger, R.A. Shafer, R. Weiss, E.L. Wright, C.L. Bennett, N.W. Boggess, T. Kelsall, S.H. Moseley, R.F. Silverberg, G.F. Smoot, D.T. Wilkinson, Cosmic microwave background dipole spectrum measured by the COBE FIRAS instrument. Astrophys. J. **420**, 445–449 (1994)

A. Friedman, Über die Krümmung des Raumes. Z. Phys. **10**, 377–386 (1922)

Y. Fukuda et al., Evidence for oscillation of atmospheric neutrinos. Phys. Rev. Lett. **81**, 1562 (1998)

G. Gamow, Expanding universe and the origin of elements. Phys. Rev. **70**, 572 (1946)

A. Grinin et al., Two-photon frequency comb spectroscopy of atomic hydrogen. Science **370**, 1061–1066 (2020)

A.H. Guth, Inflationary universe: A possible solution to the horizon and flatness problems. Phys. Rev. D **23**, 347–356 (1981)

B.M.S. Hansen, H.B. Richer, G.G. Fahlman, P.B. Stetson, J. Brewer, T.G. Currie, B.K. Gibson, R. Ibata, M. Rich, M.M. Shara, Hubble space telescope observations of the white dwarf cooling sequence of M4. Astrophys. J. Suppl. Ser. **155**, 551–576 (2004)

D. He, D. Gao, Q. Cai, Spontaneous creation of the universe from nothing. ArXiv: 1404.1207 (2014)

E. Hubble, A relation between distance and radial velocity among extra-galactic nebulae. Proc. Natl. Acad. Sci. USA **15**, 168–173 (1929)

W.E. Lamb Jr., R.C. Retherford, Fine structure of the hydrogen atom by a microwave method. Phys. Rev. **72**, 241 (1947)

S.K. Lamoreaux, Demonstration of the Casimir Force in the 0.6 to 6 μm range. Phys. Rev. Lett. **78**, 5–8 (1997)

G. Lemaître, Un univers homogène de masse constante et de rayon croissant rendant compte de la vitesse radiale des nébuleuses extra-galactiques. Ann. Soc. Sci. Bruxelles A. **47**, 49–59 (1927)

K. Lodders, Solar system abundances and condensation temperatures of the elements. Astrophys. J. **591**, 1220–1247 (2003)

E. Lusso, G. Worseck, J.F. Hennawi, J.X. Prochaska, C. Vignali, J. Stern, J.M. O'Meara, The first ultraviolet quasar-stacked spectrum at z ≃ 2.4 from WFC3. Mon. Not. R. Astron. Soc. **449**, 4204 (2015)

C. Marinoni, J. Bel, A. Buzzi, The scale of cosmic isotropy. J. Cosmol. Astropart. Phys. **10**, 036 (2012)

J.C. Mather et al., Measurement of the cosmic microwave background spectrum by the COBE FIRAS instrument. Astrophys. J. **420**, 439 (1994)

J.N. Munday, F. Capasso, V. Adrian Parsegian, Measured long-range repulsive Casimir–Lifshitz forces. Nature **457**, 170–173 (2009)

A.O. Nier, The isotopic constitution of radiogenic leads and the measurement of geological time. II. Phys. Rev. **55**, 153 (1939)

C. Patterson, Age of meteorites and the Earth. Geochim. Cosmochim. Acta **10**, 230–237 (1956)

R.W. Pattie Jr. et al., Measurement of the neutron lifetime using a magneto-gravitational trap and *in situ* detection. Science **360**, 627–632 (2018)

P.J.E. Peebles, D.N. Schramm, E.L. Turner, R.G. Kron, The case for the relativistic hot Big Bang cosmology. Nature **352**, 769 (1991)

A.A. Penzias, R.W. Wilson, A measurement of excess antenna temperature at 4080 Mc/s. Astrophys. J. **142**, 419–421 (1965)

M. Planck, Ueber das Gesetz der Energieverteilung im Normalspectrum. Ann. Phys. **309**, 553–563 (1901)

Planck Collaboration, Planck 2013 results. XVI. Cosmological parameters. Astron. Astrophys. **571**, A16 (2013)

C. Rovelli, *Quantum Gravity* (Cambridge University Press, Cambridge, 2004)

C. Rovelli, L. Smolin, Discreteness of area and volume in quantum gravity. Nucl. Phys. **B442**, 593–622 (1995)

C. Rovelli, F. Vidotto, Covariant Loop Quantum Gravity. 2015 Cambridge University Press (2014)

D. Saadeh, S.M. Feeney, A. Pontzen, H.V. Peiris, J.D. McEwen, How isotropic is the Universe? Phys. Rev. Lett. **117**, 131302 (2016)

M. Schmidt, 3*C* 273: A star-like object with large red-shift. Nature **197**, 1040 (1963)

E. Shuryak, Strongly coupled quark-gluon plasma in heavy ion collisions. Rev. Mod. Phys. **89**, 1 (2017)

V.M. Slipher, Radial velocity observations of spiral nebulae. The Observatory **40**, 304–306 (1917)

G.F. Smoot, M.V. Gorenstein, R.A. Muller, Detection of anisotropy in the cosmic microwave background radiation. Phys. Rev. Lett. **39**, 898 (1977)

G.F. Smoot, C.L. Bennett, A. Kogut, E.L. Wright, J. Aymon, N.W. Boggess, et al., Structure in the COBE differential microwave radiometer first-year maps. Astrophys. J. **396**, L1–L5 (1992)

J.W. Valley, A.J. Cavosie, T. Ushikubo, D.A. Reinhard, D.F. Lawrence, D.J. Larson, P.H. Clifton, T.F. Kelly, S.A. Wilde, D.E. Moser, M.J. Spicuzza, Hadean age for a post-magma-ocean zircon confirmed by atom-probe tomography. Nat. Geosci. **7**, 219 (2014)

S. Weinberg, *Cosmology* (Cambridge University Press, Cambridge, 2009)

R.L. Workman et al. (Particle Data Group), Prog. Theor. Exp. Phys. **2022**, 083C01 (2022)

D.G. York et al., (SDSS Collaboration), The Sloan digital sky survey: Technical summary. Astrophys. J. **120**, 1579 (2000)

Bibliography

G. Börner, *The Early Universe*, 4th edn. (Springer, New York, 2003)
E.W. Kolb, M.S. Turner, *The Early Universe* (CRC Press, Boca Raton, 2018)
J.V. Narlikar, *An Introduction to Cosmology*, 3rd edn. (Cambridge University Press, Cambridge, 2002)
P.J.E. Peebles, *Principles of Physical Cosmology* (Princeton University Press, Princeton, 1993)
C. Rovelli, *Quantum Gravity* (Cambridge University Press, Cambridge, 2004)
J. Silk, *A Short History of the Universe* (W. H. Freeman, New York, 1994)
J. Silk, *The Big Bang*, 3rd edn. (W. H. Freeman, New York, 2001)
G. Smoot, K. Davidson, *Wrinkles in Time* (Avon Books, New York, 1994)
S. Weinberg, *The First Three Minutes* (Basic Books, New York, 1993)
S. Weinberg, *Cosmology* (Cambridge University Press, 2009)

Chapter 7
Quantum Mechanical Spin Magnetic Resonance Can Determine the Structure of Self-Assembled Molecules

"Therefore, we can draw the conclusion that long-term exposure to high magnetic fields has no known harmful physical effect."

In Chap. 3 we found that the irreducible representations of the Poincaré group are characterized by two fundamental quantities, mass and spin. Mass was understood quite early in physics, by Galileo, Newton, and Einstein, as a *scalar*, and its coupling to fields depended solely on the quantity of mass. Newton's law of universal gravitation gives the force between two masses m_1 and m_2 as $F = G\frac{m_1 m_2}{r^2}$ where the amount of mass enters only as a multiplicative constant. Spin, on the other hand, is quite another matter. It was the first *internal* quantum state discovered (vide infra).

---

[1] ENC cartoon courtesy of Jürgen Schulte http://nmr.binghamton.edu/comics/ NoHarmfulPhysicalEffect.jpg. Used with permission.

L. O. Sillerud, *Abiogenesis*, https://doi.org/10.1007/978-3-031-56687-5_7

Spin is a *vector*, with the commutation relations of angular momentum, and hence, its interaction with an applied field depends not only on its magnitude, but importantly also on its orientation. Both bosons and fermions possess spin. The number of quantum states for spin S is $2S + 1$. Spins couple to fields as vectors. Of the four fields found in nature, the electromagnetic field is perhaps the easiest to manipulate in the laboratory. Exploration of the energy eigenvalues of spin operators in the presence of electromagnetic fields has given us a rich window into the structures that resulted from the self-assembly of the universe.

The previous chapter introduced the self-assembly of the current matter component of the universe from electrons, photons, protons, and neutrons. Atoms consist of electrons and nuclei formed, in the Big Bang and in stars, from the fermions, protons, and neutrons. All electrons are the same, but nuclei differ by their charges. In nuclei, the spin-statistics theorem separately restricts the wave functions of the proton and the neutron sector. The overall spin of a nucleus is the sum of the spins of each sector, giving nuclei small net spins of 0, ½, 1, 3/2, 2, 5/2, . . ., etc. Any spin greater than zero will couple to the magnetic field across the nucleus. These simple properties of quantum mechanical spin belie the fact that the last 70 years has witnessed a marvelous application of spin physics to abiogenesis, through which we can not only observe the small organic molecules of life, but also monitor their chemical alterations in metabolism. In addition, the biological functions of macromolecules are predicated on their three-dimensional structure and this spatial arrangement of atoms can be determined by exploiting the properties of spin. Therefore, it behooves us to understand the basic quantum properties of spin.

7.1   The Discovery of Spin

Almost as soon as Schrödinger had solved the hydrogen atom, showing how to find the energy levels for the electron in the Coulomb field of the proton, it was noticed by Goudsmit and Uhlenbeck, and Wolfgang Pauli, that there was additional fine structure in the optical spectra of hydrogen. It was observed that each of the spectral lines was split into a close doublet, even in the absence of an applied electromagnetic field. The optical spectra of other atoms displayed similar doublets. Sodium, for example, contains a single valence electron whose transition from the first excited state to the ground state ($2p \rightarrow 1s$) was observed to split into two spectral lines at 589.05 and 589.55 nm. It was obvious that another degree of freedom, beyond the three already discovered, was needed to explain these observations and the purely quantum phenomenon of intrinsic spin was introduced.

The extra spectral lines in the spectra of hydrogen and other atoms always came in pairs, suggesting that the operator representing this new degree of freedom required only two eigenvalues. Perhaps another form of angular momentum had been somehow overlooked by Schrödinger? A simple reflection on the properties of angular momentum operators showed that this was very unlikely because we have seen that in the hydrogen atom, angular momentum occurs in only integral values, l, and that there are $2l + 1$ projections along any defined axis. The smallest, nonzero value of angular momentum, $l = 1$, would then produce three, rather than two additional states.

One feature known to have been ignored in Schrödinger's solution for the hydrogen atom was that of the potential interaction of an electron with a magnetic field. We can use the classical electromagnetic theory to estimate the magnetic field experienced by an electron moving in an orbit around the proton. A moving charge produces a current, which in turn results in a magnetic field. The resulting field is 5.8×10^4 G (5.8 T). From the splitting in the sodium optical spectrum quoted above, one can determine the energy difference between the levels involved as

$$\Delta E = hc\left(\frac{1}{\lambda_2} - \frac{1}{\lambda_1}\right) = 3 \times 10^8 \text{ m/s} \times 6.626 \times 10^{-27} \text{ ergs}\left(\frac{1}{589.0} - \frac{1}{589.6}\right)10^9/\text{m}$$
$$= 3.3196 \times 10^{-15} \text{ erg}$$

The magnetic moment of the electron was known to be given by the Bohr magneton:

$$\mu_\mathrm{B} = \frac{e\hbar}{2mc} = 9.27 \times 10^{-21} \text{ erg/G}$$

The energy of interaction of a magnetic moment with a magnetic field, B, was also known from classical electromagnetic theory to be $\mathbf{U} = -\mu_\mathrm{B} B$. The ratio of the energies determined from the optical splitting to that from the interaction of a magnetic moment with a magnetic field will then give the magnitude of any magnetic moment in units of $\hbar$. This ratio is $\sigma = \frac{\Delta E mc}{4\pi e\hbar} = 0.500$, or the surprising value of ½. Clearly, if this results from an interaction with an additional degree of freedom of the electron, the eigenvalue of ½ is at odds with the eigenvalue of any known angular momentum operator based on the prescription suggested above of replacing classical observables with quantum operators. The young Goudsmit and Uhlenbeck proposed the concept of electron spin to account for this new degree of freedom. The eigenvalues of the purely quantum spin operator would then be $\pm\frac{1}{2}\hbar$. Pauli did not find this reasoning compelling because others had also suggested the concept of spin. The classical picture of the electron as a charged sphere required that a point on the surface of the electron would be traveling in excess of the speed of light in order to account for the observed magnitude of the spin angular momentum. When Goudsmit and Uhlenbeck persisted in their desire to publish their theory, Pauli is reputed to have responded with the *soubriquet*, "Go ahead and publish. You are young enough to be wrong once," or with words to that effect. The modern view of the electron is one of a point charge, so this classical objection now vanishes. Nevertheless, spin was the first purely quantum-mechanical, internal degree of freedom discovered; many more have been discovered over the years.

7.2 Spin Angular Momentum

The classical concept of angular momentum, L, is based on the rotation with a velocity, v, at a radius, r, of a mass, m, about an axis. L is a vector $L = r \times p$, where all the quantities in this equation are vectors, and $\times$ is the vector cross product. The classical magnetic moment due to the electron's orbital angular momentum, L, about the nucleus of an atom is M:

$$M = eL/2mc$$

The magnitude of the electron's magnetic moment has been found experimentally to be a multiple of the Bohr magneton, μ_B, given by (in SI units)

$$\mu_B = \frac{e\hbar}{2mc} = 9.2732 \times 10^{-24} \text{ J/T}$$

A comparison of the electron's orbital angular momentum with the Bohr magneton indicates that its magnetic moment is proportional to its angular momentum: $\mu_B = L\hbar$. The z-component of orbital angular momentum, L_z, is quantized in units of $m_l \in \{0, 1, 2, \ldots, l\}$ and there are $(2l + 1)$ values of m_l. In the absence of a preferred direction in space-time, these states are degenerate and all have the same energy. When a magnetic field is applied, a direction is defined and the m_l components split into $2l + 1$ states with discrete energies. The magnetic moment is proportional to the z-component of angular momentum, $M_z = \mu_b \, m_l$. If we replace orbital angular momentum with spin $S = \frac{1}{2}$, only two values of m_l are observed:

$$(2S + 1) = 2$$

and the intrinsic angular momentum of the electron is then $S = \hbar/2$.

In quantum mechanics (see exercises for Chap. 4) angular momentum is represented by an operator. When quantum mechanics was discovered, both Werner Heisenberg and Erwin Schrödinger arrived at a formalism simultaneously, but neither formalism agreed with the other, at first sight. However, it was shortly evident that Heisenberg's matrix mechanics was formally equivalent to Schrödinger's wave mechanics. Whereas Schrödinger emphasized an approach based on differential equations, Heisenberg used matrices to compute the energy levels of quantum systems. Both of these systems utilize operators, be they differential or matrix, and each displays the same eigenvalues, but often the choice of which to use in a modern calculation depends on the ease of manipulation afforded in a given case. Differential operators offer the advantage that one can use the familiar territory of calculus and immediately grasp the relationship between derivatives and operators. However, in the case of angular momentum, one quickly finds that the quantization rules lend themselves most naturally to the use of matrix operations. And this is the time-honored course which we will take here.

Spin is not a classical concept; electrons do not rotate about their axes in any physical sense. Spin obeys the rules of angular momentum; in particular, spin, S, like L, is a vector with components along the three spatial dimensions.

$$\vec{S} = S_x \hat{x} + S_y \hat{y} + S_z \hat{z}$$

In quantum mechanics, spin is associated with a spin operator, σ, which has spin eigenfunctions and spin eigenvalues.

7.3   Pauli Matrices and Spin, SU(2)

Although the spin vector of a particle can point in any arbitrary direction in the absence of a preferred direction in space, when the underlying rotational symmetry of space-time is broken by, for example, the application of a magnetic or electric field, a direction is defined specified by the field vector. From a quantum mechanical perspective, the magnetic moment of a particle is proportional to its spin. The magnetic moment operator, M, is then given by $M = -\mu\sigma$, with σ as the spin operator. Nonzero spin states couple to magnetic fields, altering the energy of the states.

The potential energy associated with the interaction of spin with a magnetic field, B, is then the dot product of these two vectors:

$$U = -M \cdot B = \mu\sigma \cdot B$$

where the lowest energy state is when the spins are parallel to the field. The spin Hamiltonian for a quantum particle is

$$H_s = \mu\sigma \cdot B$$

and the average value for the magnetization is

$$\langle M \rangle = -\mu \int \psi \times \sigma\psi \, d\tau$$

In a magnetic field the vector for spin ½ can be oriented in only one of two ways; it can align parallel or antiparallel to the defined axis. There are only two eigenvalues of the spin operator for a spin ½ wave function; we call these spin "up" and spin "down," that is, "up" is parallel to the field, and "down" is the antiparallel state. A convenient representation for a wave function, which obeys such rules, is a 2 × 1 matrix. Let us call the spin wave functions $\chi_\pm$, where

$$\chi_- = \begin{bmatrix} 0 \\ 1 \end{bmatrix}, \quad \text{and} \quad \chi_+ = \begin{bmatrix} 1 \\ 0 \end{bmatrix}$$

Then, if we let the axis of symmetry breaking be the z-axis, a simple representation of a suitable spin operator is given by the following set of Pauli matrices:

$$\sigma_x = \frac{\hbar}{2}\begin{bmatrix} 0 & 1 \\ 1 & 0 \end{bmatrix}, \quad \sigma_y = \frac{\hbar}{2}\begin{bmatrix} 0 & -i \\ i & 0 \end{bmatrix}, \quad \sigma_z = \frac{\hbar}{2}\begin{bmatrix} 1 & 0 \\ 0 & -1 \end{bmatrix}$$

and the spin vector operator is $\vec{\sigma} = \sigma_x \hat{x} + \sigma_y \hat{y} + \sigma_z \hat{z}$. If the spin wave functions, χ, are eigenfunctions of σ, then they must satisfy an eigenvalue equation. Let us check this statement by computing the effect of σ_z on χ_+:

$$\sigma_z \chi_+ = \frac{\hbar}{2}\begin{bmatrix} 1 & 0 \\ 0 & -1 \end{bmatrix}\begin{bmatrix} 1 \\ 0 \end{bmatrix} = \frac{\hbar}{2}\begin{bmatrix} 1 \\ 0 \end{bmatrix}$$

and we see by explicit computation that $\sigma_z \chi_+ = \frac{\hbar}{2}\chi_+$, that is, χ_+ is an eigenfunction of the z-component of the spin operator, σ_z, with eigenvalue $\frac{\hbar}{2}$. By a similar argument we can find that χ_- is an eigenfunction σ_z of with eigenvalue $-\frac{\hbar}{2}$. Notice that neither χ_+ nor χ_- are eigenfunctions of σ_x or σ_y in this representation because

$$\sigma_x \chi_+ = \chi_-$$
$$\sigma_x \chi_- = \chi_+$$
$$\sigma_y \chi_+ = i\chi_-$$
$$\sigma_y \chi_- = -i\chi_+$$

In general, only one component of a 3-dimensional angular momentum vector can be measured at a time as a direct consequence of the uncertainty principle and the angular momentum commutation relations; the three components of the spin operator do not commute with one another:

$$\left[\sigma_i, \sigma_j\right] = i\hbar\sigma_k, \quad \text{for } \{i,j,k\} \in \{x,y,z\}$$

What is the average value of the spin? We can find this by using the definition of the average value of an operator if we know the eigenfunctions of the spin operator. In this case we can use the matrix eigenfunctions:

$$\langle \vec{\sigma} \rangle = \sum_{i=1}^{3} \chi^* \sigma_i \chi \hat{x}_i$$

where $\hat{x}_i \in \{\hat{x}, \hat{y}, \hat{z}\}$ are the unit vectors along the x-, y-, and z-axes, for the spinors, χ defined above:

$$\chi_+^{\dagger} \sigma_z \chi_+ = \frac{\hbar}{2}[1 \ 0]\begin{bmatrix} 1 & 0 \\ 0 & -1 \end{bmatrix}\begin{bmatrix} 1 \\ 0 \end{bmatrix} = \frac{\hbar}{2}$$

and

$$\chi_-{}^{\dagger}\sigma_z\chi_- = \frac{\hbar}{2}\begin{bmatrix} 1 & 0 \end{bmatrix}\begin{bmatrix} 1 & 0 \\ 0 & -1 \end{bmatrix}\begin{bmatrix} 0 \\ 1 \end{bmatrix} = -\frac{\hbar}{2}$$

Note that here (†) means the Hermitian conjugate, which is the complex conjugate transpose of the matrix. We see that the average value of the z-component of spin in the "spin up" eigenfunction is $+\frac{1}{2}$ in units of Planck's constant divided by 2π, and for the "spin down" eigenfunction, it is $-\frac{1}{2}$.

7.4   Magnetic Moments of Electrons and Nucleons

One of the most important consequences of spin in biophysics, in addition to the spin-statistics theorem (the Pauli principle), is that spin angular momentum gives rise to a magnetic moment and that this moment couples to magnetic fields. Atoms are composed of electrons, protons, and neutrons, all of which are spin $\frac{1}{2}$ fermions. The value of the electron spin is identical to that of the protons and neutrons. By the Pauli principle essentially all atomic and molecular electrons are spin paired into states with $s = 0$ and consequently do not possess a net magnetic moment. On the other hand, the rules of spin angular momentum in quantum mechanics allow nuclei with an odd number of nucleons (protons or neutrons) to possess a net spin and hence a magnetic moment proportional to the spin; nuclei with nonzero spin are magnetic.

The magnetic dipole moment, μ, of a nucleon is proportional to its spin:

$$\mu = -\mu_N\sigma$$

where μ_N is called the nuclear magneton,

$$\mu_N = \frac{e}{2m_N c}$$

and m_N is the mass of a nucleon (a proton or a neutron; the mass difference between these two particles, while finite, is quite small). It is very important to note here that the magnetic moment of a nucleus is inversely proportional to its mass and that the masses of nuclei differ greatly as one progresses through the periodic table. All electrons have identical magnetic moments, but that is not true for nuclei. The electron is a point particle, while protons and neutrons are extended objects, even though they are very small. The root mean square charge radius of the proton was recently measured (Beyer et al. 2017) to be 0.834 fm.

The magnetic moment of a quantum particle is its charge to mass ratio: The moment of the electron is therefore $\mu_e = \frac{e\hbar}{2m_e}$, where e is the charge, $\hbar$ is the reduced Planck's constant, and m_e is the electron rest mass, giving a value of

$$\mu_e = -9.285 \times 10^{-24} \text{ J/T}$$

negative because the electron's charge is negative. The charge on the proton only differs in sign from that of the electron. For an electron modeled as a macroscopic sphere with a radius of 2.82 fm (the classical electron radius), the rotational speed at the surface would exceed that of light. This was known in the late nineteenth century and served as an early indication that the application of classical electrodynamics to the realm of the ultrasmall produced predictions at odds with measurements.

The measurement of magnetic moments can help us to understand the important structural differences between complex nucleons and structureless point particles, like the electron. The simplest nucleus is that of the hydrogen atom, a single proton with spin angular momentum $\frac{\hbar}{2}$. In 1927 Dennison suggested that his results on the hydrogen molecule would correspond with theory if "...the nuclear spin is taken equal to that of the electron...," a clear indication that the spin of the proton was ½, making it also a fermion (Dennison 1927). Using the above equation, one would then predict that the magnetic moment of the proton would be smaller than the electron's by the ratio of their masses:

$$\mu_p = \mu_e \frac{m_e}{m_p} = \frac{\mu_e}{1836} = +5.05 \times 10^{-27} \text{ J/T}$$

However, a measurement of the proton's magnetic moment by Esterman and Stern (1933) gave the unexpected value of

$$\mu_p = +1.41 \times 10^{-26} \text{ J/T} = +2.79 \frac{e\hbar}{2m_p} = +2.79 \, \mu_N$$

which is almost threefold higher than that suggested by its mass. It was therefore clear by this time that the structure of the proton was not that of a point particle, but that it was a composite particle, made up of smaller constituents.

Another important clue that there were fundamental differences between nuclear matter and electrons came from estimates (Tamm & Altshuler 1934) and measurements (Esterman and Stern 1934) of the neutron magnetic moment. One would expect that since the neutron was uncharged, $e = 0$, the magnetic moment would be identically zero. However, when these investigators subtracted the proton's moment (2.79 μ_N) from that of the deuteron's (0.85647 μ_N) (Rabi et al. 1934) it was found that the neutron's moment (Breit & Rabi 1934)

$$\mu_n = \mu_D - \mu_p = 0.85647 \, \mu_N - 2.79 \, \mu_N = -1.93 \, \mu_N$$

was not only different from zero, in contradiction to its zero charge, but surprisingly, the neutron's moment was actually negative! (Direct measurements in 1940 gave the same value (Alvarez & Bloch 1940). Later direct measurements refined this value to the presently accepted $\mu_n = -1.91 \, \mu_N$. This nonvanishing magnetic moment of the neutron was another suggestion that nuclear matter differed in a fundamental way from electronic matter. One could also conclude that in order to give a negative moment, the constituents of the neutron must have opposite charges from those in the proton; the negative constituents must also be displaced from the center and surround the positive core, and their charges must cancel, electrically, to endow the neutron with zero charge. An alternative possibility, that the neutron was a boson with spin one, was ruled out when the spin of the neutron was directly determined (Byrne 2011) to be ½, identical to that of the proton, implying that the proton was a fermion and obeyed Fermi-Dirac statistics. As is often the case in science, when a measurement radically disagrees with preconceptions, it is the preconceptions that require reexamination.

To explain these results, perhaps another degree of freedom was required, another internal quantum number, the eigenvalues of yet another operator. Heisenberg had previously suggested that the proton and the neutron were states that differed in terms of a two-state, spin-like quantum number that he dubbed isospin, since it would obey the same algebra as spin; the proton had isospin +1/2, while the neutron's isospin would be $-1/2$. That isospin symmetry appeared to be correct was reflected in the tiny mass difference between the proton and the neutron of 0.28%. Clearly, the quantum states of the proton and neutron differed through the influence of the isospin operator on these particles wave functions, but how this happened remained an open question.

A resolution of this problem came almost 30 years later, when, with the evidence presented by large datasets derived from measurements of the myriad particles generated in high-energy collisions at accelerators, Gell-Mann, Zweig, and Ne'eman (Gell-Mann 1964; Zweig 1964; Ne'eman 1961) gave the group structure of the strong interaction its first correct embodiment. The bases for the irreducible representations of its internal SU(3) symmetries were called "quarks," by Gell-Mann, after a phrase in James Joyce's *Finnegan's Wake* where a call is made in a pub for "Three quarks for Muster Mark!," which Gell-Mann, prior to his demise, confessed to the author in a physics colloquium was a coded version of "Three quarts of ale, for mister Mark!"

The measured transformation properties of the excited nuclear states fitted into a group consisting of operators whose basis was a tensor product of spin, isospin, and a new operator, hypercharge; the three basis vectors, $\{u, d, s\}$, of the irreducible representations were identified with the quarks (up, down, strange) of Gell-Mann. Models of elementary particles as combinations of quarks soon flooded the literature. The proton was proposed to be composed of two *up* quarks plus a *down* quark, while the neutron was made of two *down* quarks plus an *up* quark. The individual quarks were expected to possess fractional positive and negative charges (multiples of 1/3) so that the total charge on the proton was divided among the three quarks. The charges also were required to sum to zero for the neutron, but, and a crucial point about this model is that the quarks were not assumed to be uniformly distributed within the nucleons.

The magnetic moment of a composite object, such as a collection of bar magnets of random orientation, would be expected to be a linear superposition of the individual moments due to the linearity of electromagnetism. This was apparently true even for tiny subatomic particles as indicated by the agreement between the neutron's moment and the difference between the proton's and deuteron's moments. Decomposing the magnetic moments of the proton (neutron) into the moments of the quarks should then reveal properties of the quarks themselves.

A simple model of the proton and neutron would be quarks orbiting a common center of mass. If we assume that the three quarks are approximately of equal mass, with $m_q \sim m_p/3$, and the moment of a single quark is given by

$$\mu_q = \frac{e\hbar}{2m_q}$$

and that the moment of the proton is

$$\mu_p = 3\mu_q$$

which we find to be approximately correct because the proton moment is almost three nuclear magnetons:

$$\mu_p = 2.79\frac{e\hbar}{2m_p} \sim 3\mu_N$$

What about the neutron? Its moment μ_n is negative:

$$\mu_n = -1.91\frac{e\hbar}{2m_n} = -1.91\,\mu_N$$

suggesting that the configuration of the charged quarks is somehow reversed in it with respect to the proton. A point to remember for later is that the ratio of the proton and neutron moments

$$\frac{\mu_p}{\mu_n} = -1.46$$

is a value very close to $-3/2$. We can understand this value with the aid of a simple quark model for the nucleons.

Imagine the proton $|u\,u\,d\rangle$ as consisting of a negatively charged down quark and two positively charged up quarks orbiting their center of mass. [Oh my, it sounds like the Bohr model of the hydrogen atom all over again!] The neutron $|d\,d\,u\rangle$ would correspond to a pair of negative down quarks surrounding a positive up quark. Each nucleon would consist of three quarks whose fractional charges would sum to either zero for the neutron or one for the proton. A solution to this puzzle was that the up quark had charge $+2/3\ e$, while the down quark charge is $-1/3\ e$. The proton charge is then $C_p = c(\text{up}) + c(\text{up}) + c(\text{down}) = 2/3\ e + 2/3\ e - 1/3\ e = 1\ e$, and the neutron charge is $C_n = c(\text{up}) + c(\text{down}) + c(\text{down}) = 2/3\ e - 1/3\ e - 1/3\ e = 0$.

Quarks were assumed to be spin ½ Fermions, consistent with their occurrence as matter fields allowed by Poincare invariance, and their wave functions would therefore be subject to the spin-statistics theorem and must be antisymmetric under quark interchange. No two quarks could occupy the same quantum state. This is satisfied for the proton or neutron wave functions because the two up (down) quarks in the proton (neutron) can pair with opposite spins to produce a spin 0 state, leaving the spin of the nucleon determined by the spin ½ of the "left over" quark.

However, there exist more massive, excited states of the nucleon, such as the Δ^{++} with mass 1232 MeV, spin 3/2, and charge $+2e$ that consists of three spin $+1/2$ up quarks (total spin $J = +1/2 + 1/2 + 1/2 = 3/2$). The wave function must therefore be symmetric under quark exchange, and this would violate the Pauli principle. However, the Pauli principle only applies to *indistinguishable* fermions. Another quark degree of freedom was required, and thus, color charge was proposed. One could have three up quarks, all with spin up, in the Δ^{++} if each quark could be distinguished by a different color, red, green, or blue, much like shown in Fig. 7.1, and one could write their wave function as an antisymmetric linear combination such as

$$\frac{1}{\sqrt{6}}(\text{rgb} + \text{brg} + \text{gbr} - \text{rbg} - \text{bgr} - \text{grb})$$

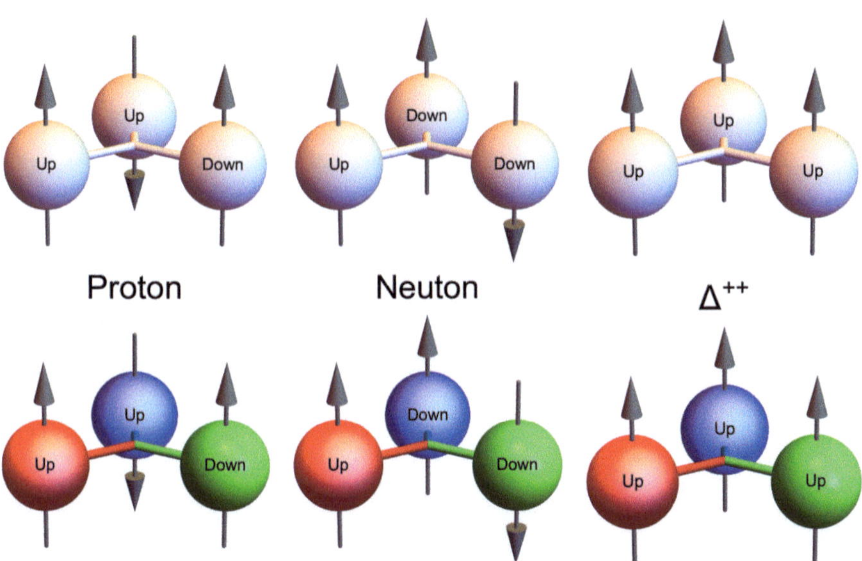

Fig. 7.1 In the top row is shown a naïve view of the quark and spin $|q\ s\rangle$ states making up the proton $(|u + \frac{1}{2}\rangle, |u - \frac{1}{2}\rangle, |d + \frac{1}{2}\rangle)$, the neutron $(|u + \frac{1}{2}\rangle, |d + \frac{1}{2}\rangle, |d - \frac{1}{2}\rangle)$, and the Δ^{++} excitation $(|u + \frac{1}{2}\rangle, |u + \frac{1}{2}\rangle, |u + \frac{1}{2}\rangle)$. In the proton, the two up quarks are spin paired, giving $I_u = s_{u1} + s_{u2} = 0$ and leaving the spin of the proton to be determined by the spin ½ of the down quark. In the neutron, the spins of the two down quarks are paired, giving $I_d = s_{d1} + s_{d2} = 0$ and leaving the spin of the neutron to be determined by the spin ½ of the up quark. The up quarks have charge $+2e/3$ each and the charge on the down quark is $-e/3$; these sum to a total charge of $+e$ for the proton and to zero for the neutron. The Δ^{++} state consists of all up quarks, with spin 3/2 and charge $2e$. The Pauli principle would forbid this entirely symmetric configuration. The bottom row shows a solution which introduces an eigenvalue of the additional color charge operator; individual quarks can be either red, green, or blue. The total quark wave function is colorless, with red + green + blue = white (colorless) and now the Δ^{++} is an allowed state because the color charge allows the total wave function to be antisymmetric

The nucleon magnetic moments are the matrix elements of the magnetic moment operator:

$$\mu_{\mathrm{N}} = \left\langle N\frac{1}{2}\frac{1}{2} \Big| \sum_{i=1}^{3} \frac{q_i S_{zi}}{2m_i} \Big| N\frac{1}{2}\frac{1}{2} \right\rangle$$

where $N = \{n, p\}$ is either the neutron or proton, q is the quark electrical charge operator, m is the quark mass, and S_z is the spin operator. The nucleon wave function must also have the correct spin, isospin, color, and exchange quantum properties. If we assume that the quarks are equally massive, then the moments add as the quark charges:

$$\mu_{\mathrm{p}} = 2\mu_{\mathrm{u}} + \mu_{\mathrm{d}}$$

$$\mu_{\mathrm{n}} = 2\mu_{\mathrm{d}} + \mu_{\mathrm{u}}$$

Combining these we have a simple result that the magnetic moment of the proton is

$$\mu_{\mathrm{p}} = \left(\frac{4}{3}\right)\mu_{\mathrm{u}} - \left(\frac{1}{3}\right)\mu_{\mathrm{d}}$$

And that of the neutron is

$$\mu_{\mathrm{n}} = \left(\frac{4}{3}\right)\mu_{\mathrm{d}} - \left(\frac{1}{3}\right)\mu_{\mathrm{u}}$$

We can solve these for the magnetic moments of the up and down quarks, using the experimental values for the proton and neutron moments:

$$\mu_{\mathrm{d}} = \left(\frac{1}{5}\right)\left[\mu_{\mathrm{p}} + 4\mu_{\mathrm{n}}\right]$$

and

$$\mu_{\mathrm{u}} = \left(\frac{1}{5}\right)\left[4\mu_{\mathrm{p}} + \mu_{\mathrm{n}}\right]$$

Putting in the measured proton and neutron values then gives us that

$$\mu_{\mathrm{d}} = \mu_{\mathrm{N}}[4(-1.91) + 2.79] = -0.97\,\mu_{\mathrm{N}}$$

$$\mu_{\mathrm{u}} = \mu_{\mathrm{N}}[4(2.79 - 1.91] = +1.85\,\mu_{\mathrm{N}}$$

The ratio $\frac{\mu_{\mathrm{u}}}{\mu_{\mathrm{d}}} = -1.90$ is very close to the ratio of charges for the up and down quarks, $\frac{e_{\mathrm{u}}}{e_{\mathrm{d}}} = \frac{2/3}{-1/3} = -2$, and this suggests that $\mu_{\mathrm{d}} = -\mu_{\mathrm{N}}$ and $\mu_{\mathrm{u}} = 2\mu_{\mathrm{N}}$. Using measurements (Dothan 1982) of the magnetic moments of a number of excited states of quarks, these magnetic moments were found to be

$$\mu_u = +2.08\,\mu_N$$

$$\mu_d = -1.31\,\mu_N$$

Under these assumptions, the calculated magnetic moment of the $|u\,u\,u\rangle$ Δ^{++} state would be $3\mu_u = 6.24\,\mu_N$, while the measured moment for this state was found to be $6.14\,\mu_N$. The ratio of the magnetic moments of the proton and the neutron would be given by

$$\frac{\mu_p}{\mu_n} = \frac{\left(\frac{4}{3}\right)\mu_u - \left(\frac{1}{3}\right)\mu_d}{\left(\frac{4}{3}\right)\mu_d - \left(\frac{1}{3}\right)\mu_u} = \frac{2.91}{-1.94} = -1.5$$

The measured ratio of the magnetic moments of the proton and the neutron is

$$\frac{\mu_p}{\mu_n} = \frac{2.79}{-1.91} = -1.46$$

The agreement between this simplistic quark model and the subsequent measurements of the magnetic moments of the *many* excited states of nuclear matter (Dothan 1982) was considered a major early triumph of the constituent quark model. Subsequent calculations using the MIT bag model of quarks have continued to produce results of magnetic moments that agree with experiment to a surprising degree (Zhang et al. 2021). However, the reader should not be inappropriately impressed with these results; one needs to be acutely aware that the spin structure of the nucleons and their excited states continues to be a significant source of investigation as the quark model has evolved. Our introduction to QED and the uncertainty principle led us to expect that virtual particles will play a very important role in the strong interactions. An introduction to some of the subtleties of proton structure is reviewed in Roberts (1990) and for the relativistic quark model in Bass (2005). Nevertheless, we cannot ignore the fact that protons are diamagnetic and that this has spawned entire industries devoted to molecular structural analysis, investigations of metabolism, and biomedical imaging.

7.5   Spin Precession in a Magnetic Field

The potential energy, U, of a nucleus of moment μ_n, when placed into a magnetic field, B, is

$$U = -\mu \cdot B = \mu_n \sigma \cdot B$$

When a charged, magnetic nucleus of mass m_n moves through a region of a nonzero, homogeneous magnetic field, the particle's spin vector rotates around the magnetic field vector with a frequency $\omega_L = -\frac{e}{2m_n c} B$, called the Larmor frequency, named after Sir Joseph Larmor, who first developed this concept. This precession of the spin moment about the field axis can be understood from the fact that at equilibrium the magnetic force on the particle must be balanced by the mass times the acceleration: $-e\,B\,v/c = m_n\,r\,\omega^2$, or $-e\,B/m_n c = r\,\omega^2/v$. However, since $2v/r = \omega$, we have that

$$\omega_L = -\frac{e}{2m_n c} B$$

This is the fundamental relationship between the nuclear magnetic resonance frequency and the magnetic field. We can rewrite this in even simpler terms as

$$\omega_L = \gamma B$$

where $\gamma = -\frac{e}{2m_n c}$ is called the magnetogyric ratio. This value is a constant for a given nucleus, but differs greatly among nuclei due to their markedly differing masses. The angular frequency, ω, which has units of radians per second, is rarely used in NMR today; rather one uses the linear frequency, $\nu = \omega/2\pi$. The value of ν for the proton is 42.3 MHz/T. Modern NMR spectrometers operate at fields up to ~23 T, with frequencies up to ~1 GHz.

7.6 The Spin Hamiltonian

The Hamiltonian of a charged particle with spin is the sum of the spin-independent terms, which describe the kinetic and potential energies, and the spin-dependent terms. Let us write this as $H = H_o + H_s$, where H_s is the spin portion. The potential energy, U, associated with nuclear spin in a magnetic field is the vector dot product of the magnetic moment, μ, with that of the field, $U = \mu_N\,\sigma \cdot B$. Then the Hamiltonian becomes

$$H = H_o + \mu_N\,\sigma \cdot B$$

Since the magnetic energy levels are much smaller than those due to kinetic energy and the electrostatic potential, we will ignore H_o and work only with the spin Hamiltonian in order to illustrate the motion of a nuclear spin in a magnetic field. We can also deal only with the spin wave functions because we can always decompose the total wave function into a product of spin and spatial functions, $\Psi(q, s) = \phi(q)\,\chi(s)$.

The time-dependent spin wave function of our system of nuclei is described by a linear combination of spin up and spin down functions

$$\chi(t) = a(t)\chi_+ + b(t)\chi_-$$

with time-dependent coefficients $a(t)$ and $b(t)$. At time zero $a(0) = a_{\mathrm{o}}$ and $b(0) = b_{\mathrm{o}}$, so that

$$\chi(0) = a_{\mathrm{o}}\chi_+ + b_{\mathrm{o}}\chi_-$$

If we let the magnetic field point along the z-axis, $B = B_{\mathrm{o}}\,\hat{z}$, we can calculate the expectation value of the spin as a function of time. We need to solve the time-dependent Schrödinger equation:

$$i\hbar\frac{\partial\chi}{\partial t} = H_{\mathrm{s}}\chi = \mu_{\mathrm{N}}\sigma\cdot B\chi = -\frac{e\hbar}{2mc}\sigma_z B_{\mathrm{o}}\chi$$

which we can rewrite using the Larmor frequency, $\omega_{\mathrm{L}} = -eB_{\mathrm{o}}/2mc$, as

$$i\hbar\frac{\partial\chi}{\partial t} = \hbar\omega_{\mathrm{L}}\sigma_z\chi$$

We now need to insert our spin wave functions and solve the resulting equation for the time evolution of the spin. Remember that our spin wave functions were column vectors, so that

$$\chi(t) = a(t)\chi_+ + b(t)\chi_- = a(t)\begin{bmatrix}1\\0\end{bmatrix} + b(t)\begin{bmatrix}0\\1\end{bmatrix} = \begin{bmatrix}a(t)\\b(t)\end{bmatrix}$$

and, inserting this expression into our time-dependent Schrodinger equation, we find that

$$i\frac{\partial}{\partial t}\begin{bmatrix}a(t)\\b(t)\end{bmatrix} = \omega_{\mathrm{L}}\begin{bmatrix}1 & 0\\0 & -1\end{bmatrix}\begin{bmatrix}a(t)\\b(t)\end{bmatrix} = \omega_{\mathrm{L}}\begin{bmatrix}a(t)\\-b(t)\end{bmatrix}$$

and this leads, after multiplying out the matrices, to the following two equations:

$$\frac{da(t)}{dt} = -i\omega_{\mathrm{L}}a(t),\text{ and}$$

$$\frac{db(t)}{dt} = i\omega_L b(t)$$

which we can integrate

$$\int_0^b \frac{db(t)}{b(t)} = i\omega_L \int_0^t dt = i\omega_L t$$

to give

$$b(t) = e^{i\omega_L t}$$

and, similarly for $a(t)$,

$$a(t) = e^{-i\omega_L t}$$

Our time-dependent spin wave function becomes

$$\chi(t) = \begin{bmatrix} e^{-i\omega_L t} \\ e^{i\omega_L t} \end{bmatrix}$$

$$\chi^{\dagger}(t) = \begin{bmatrix} e^{i\omega_L t} & e^{-i\omega_L t} \end{bmatrix}$$

This can now be used to compute the expectation value of the spin through

$$\left\langle \vec{S} \right\rangle = \frac{\hbar}{2}\langle \vec{\sigma} \rangle = \frac{\hbar}{2}\chi^{\dagger}\vec{\sigma}\chi = \frac{\hbar}{2}\left(\chi^{\dagger}\sigma_x\chi\hat{x} + \chi^{\dagger}\sigma_y\chi\hat{y} + \chi^{\dagger}\sigma_z\chi\hat{z} \right)$$

The result is obtained by inserting our spin wave functions and the Pauli spin operator matrices and noting that $2\sin\omega_L t = e^{i\omega_L t} - e^{-i\omega_L t}$, and $2\cos\omega_L t = e^{i\omega_L t} + e^{-i\omega_L t}$:

$$\langle S \rangle = \frac{\hbar}{2}\left[\cos(2\omega_L t)\hat{x} + \sin(2\omega_L t)\hat{y} + \hat{z} \right]$$

where we see that the component of the spin angular momentum vector in the direction of the magnetic field, S_z, is constant, with the value $\frac{\hbar}{2}$, and that the x and y components precess around the z-axis with a frequency that is twice the Larmor frequency, $2\omega_L = -eB/mc$. The reason why the spin precesses about the z-axis with twice the Larmor frequency has to do with the fact that the ratio of the spin angular momentum ($s = 1/2, 3/2, 5/2, \ldots$) to the mass, the so-called g-factor, or gyromagnetic ratio, is approximately 2. For orbital angular momentum ($L = 0, 1, 2, \ldots$), this ratio is one so that orbital moments precess at the Larmor frequency.

7.7   Spin Magnetic Resonance

By virtue of their spin angular momentum, which we now call $\boldsymbol{I}$, electrons, protons, and neutrons have a magnetic moment, $\boldsymbol{\mu}$. These two quantities are parallel to each other, and in fact, the magnetic moment is proportional to the spin angular momentum:

$$\boldsymbol{\mu} = \gamma \hbar \boldsymbol{I}$$

The constant of proportionality, γ, which is different for each nucleus, but the same for each electron, is referred to as the magnetogyric ratio because we can rewrite this equation as

$$\gamma = \frac{\mu}{\hbar I}$$

and we see that this is the ratio of the magnetic moment to the angular momentum (spin I). Since the magnetic moment of a particle is inversely proportional to its mass, the magnetogyric ratio for the electron is 1836 times larger than that for the proton and is found to be 2.80 MHz/G $=$ 28 GHz/T. As we have seen in the discussion above with respect to the discovery of spin, this quantum degree of freedom interacts with a magnetic field, B, because there is a coupling between the spin and the field which removes the m-degeneracy and splits the states into $2I + 1$ states of differing energy (Fig. 7.2). For spin ½, we have $2(1/2) + 1 = 2$ states. This coupling with the field leads to an interaction energy, U, given by

$$U = -\boldsymbol{\mu} \cdot \mathbf{B}$$

Since the applied magnetic field defines a preferred direction (axis) in space, this interaction energy is characterized by the projection quantum numbers along

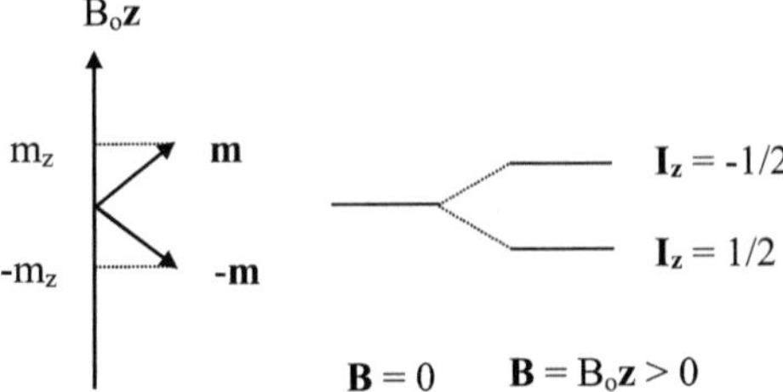

Fig. 7.2   On the left is shown the splitting of the m-degenerate quantum spin states into sublevels characterized by their projection rsquantum number along the field direction. On the right a spin system is placed in a magnetic field which lifts the degeneracy and splits the states of the spin ½ system into sublevels according to their energies

(parallel to) the applied field, $\mathbf{B}$. If we apply the magnetic field along the z-direction, $\mathbf{B} = B_o\mathbf{z}$, then the energy is

$$U = -\mu_z \cdot \mathbf{B} = -\gamma\hbar B_o I_z$$

The allowed values of I_z are $m_I \in \{I, I - 1, \ldots, -I\}$, from which we find that the energy is given by

$$U = -m_I\gamma\hbar B_o$$

The m-degenerate quantum spin states split into sublevels characterized by their projection quantum number along the field direction (Fig. 7.2). The z-component of the spin angular momentum, m_z, is given by $m_z = \gamma\hbar m$. Since the interaction energy is found from the dot product of the magnetic moment and the applied field, we can write that

$$U = -mB_o \cos\theta$$

Since $\theta \in \{0, \text{ or, } \pi\}$, $\cos\theta = \pm 1$, and for a spin parallel to the applied field, $\theta = 0$, and $\cos\theta = 1$, the energy is $U = -m\,B_o$, while for $\theta = \pi$, $U = m\,B_o$, and we see that the parallel state with $\theta = 0$ is of lower energy. In a magnetic field, a nucleus with spin $I = \frac{1}{2}$ will split into two energy levels. The spins tend to align parallel to the applied field. The important energy difference between these two spin orientations is $\Delta E = 2\mu B_o = \gamma\hbar B_o$. We can express this as a frequency as $\Delta E = \hbar\omega = \gamma\hbar B_o$, or

$$\omega = \gamma B_o$$

which provides the fundamental equation for resonant photon absorption of a spin system in a magnetic field. We recognize this as stating that the absorption of energy by a quantum system only occurs if the frequency of the applied photon field corresponds to the difference in energy between the levels. Of course, when we notice that there is an energy difference between states, we immediately wonder about the difference in populations for particles among these states. This is given by the Boltzmann distribution:

where the energy difference is

$$\Delta E = 2\mu B_o = \gamma\hbar B_o$$

For a system of N spins, we denote the number of spins in the upper energy state as n^- and the number in the lower energy state as n^+ (Fig. 7.3). The total number of spins, N, must equal the numbers in the various states, so that $N = n^+ + n^-$; the number of spins in the upper energy state is

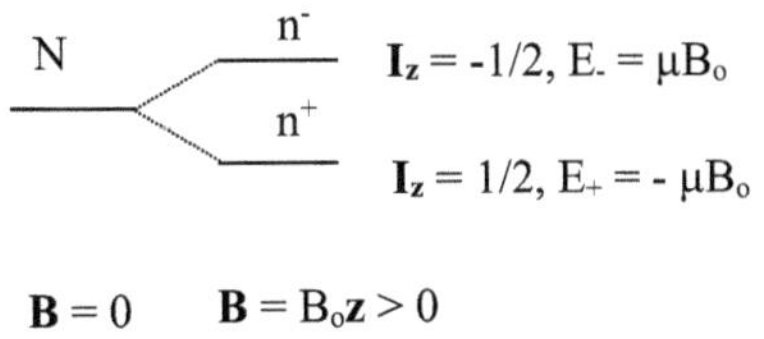

Fig. 7.3 The energy difference between spin states for a particle with a magnetic moment μ in a magnetic field B

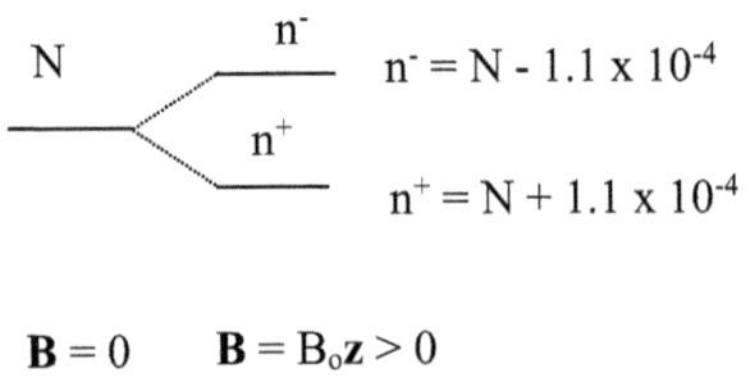

$$N \qquad \begin{array}{l} \text{n}^- \quad I_z = -1/2, E_- = \mu B_o \\ \text{n}^+ \quad I_z = 1/2, E_+ = -\mu B_o \end{array}$$

$$\mathbf{B} = 0 \qquad \mathbf{B} = B_o \mathbf{z} > 0$$

Fig. 7.4 The population differences for a proton in a field of 11.7 T at 300 K

$$N \qquad \begin{array}{l} \text{n}^- \quad \text{n}^- = N - 1.1 \times 10^{-4} \\ \text{n}^+ \quad \text{n}^+ = N + 1.1 \times 10^{-4} \end{array}$$

$$\mathbf{B} = 0 \qquad \mathbf{B} = B_o \mathbf{z} > 0$$

$$n^- = N e^{-\Delta E/kT} = (n^+ + n^-) e^{-\Delta E/kT}$$
$$= N e^{-2\mu B_o/kT}.$$

The magnetic moment of the proton, μ_P, is a multiple of the nuclear magneton, μ_N:

$$\mu_P = 2.79 \, \mu_N$$

where $\mu_N = 1.41 \times 10^{-26}$ J/T. At a temperature of $T = 300$ K and in a magnetic field of $B_o = 11.7$ T (this corresponds to a 500-MHz NMR system) (Fig. 7.4), the excess number of spins in the upper energy level is e^{-x}, where

$$x = \Delta E/kT = \frac{2 \times 1.41 \times 10^{-26} \times 11.7}{3 \times 10^2 \times 1.38 \times 10^{-23}} = 1.1 \times 10^{-4}$$

We note that this quantity is much less than one, so that we can expand the exponential in a
   Taylor's series as

$$e^{-x} = 1 - x + \cdots$$

to show that the ratio of up to down spins in a 11.7-T field is only 1.1×10^{-4}. It is for this reason that NMR is such an insensitive spectroscopic method. Only a small excess of spins are in the lower-energy state and this small excess gives rise to an induced magnetic moment, M_o, along the z-axis. We can calculate this magnetic moment from our work on quantum theory and thermodynamics.

7.8 Curie's Law

The average value of any thermodynamic quantity Q is given by

$$\overline{Q} = \frac{\sum\limits_{n} Q_n P_n(Q)}{\sum\limits_{n} P_n(Q)}$$

where $P_n(Q)$ is the probability that the system is in the nth state, whose value is given by Q_n. For nuclei in a magnetic field the probability that the system is in the nth state is given by the Boltzmann distribution:

$$P_n(E) = e^{-E_n/kT}$$

where E_n is the energy of the nth state. The energy of a spin in a field of B_o is $E = -\mu \cdot B_o$; the magnetic moment is $\mu = \gamma\hbar m$, where m is the spin of the particle; and this quantum number ranges from $-I_z$ to $+I_z$. For N particles, the magnetic moment is $M = N\mu$. The average value of the magnetic moment is then given by

$$M = N<\mu> = N\gamma\hbar <m>,$$

And we need to compute the average value of the spin quantum number, m. The magnetic moment is then given by

$$M = \frac{N\gamma\hbar \sum\limits_{m=-I_z}^{I_z} m e^{\frac{\gamma\hbar m B_o}{kT}}}{\sum\limits_{m=-I_z}^{I_z} e^{\frac{\gamma\hbar m B_o}{kT}}}$$

Let us look at the denominator first. If the temperature is high, i.e., greater than about 1 K, then the exponent is very small. Let $x = \gamma\hbar B_o/kT$, then we can approximate the exponential by $e^{-x} = 1 - x + \cdots$, so that

$$e^{\frac{\gamma\hbar m B}{kT}} \cong 1 + \frac{\gamma\hbar m B_o}{kT}$$

and

$$\sum_{m=-I}^{I} e^{\frac{\gamma\hbar m B_o}{kT}} = \sum_{m=-I}^{I} e^{mx} = e^{-I_x} + \cdots + e^{I_x} = 1 - I_x + \cdots + 1 + I_x = 2I + 1,$$

There are $2I + 1$ values of m, the projection of spin I on the z-axis. Because there are an equal number of $-I_x$ and $+I_x$ terms, these cancel, and we are left with $2I + 1$ terms of 1. We find then that

$$\sum_{m=-I}^{I} e^{\frac{\gamma \hbar m B_0}{kT}} = 2I + 1$$

So now our calculation reduces to

$$M = N\gamma\hbar <m> = N\gamma\hbar \frac{\sum\limits_{m=-I}^{m=I} m e^{\frac{\gamma \hbar m B_0}{kT}}}{2I+1}$$

The numerator can be addressed by expanding the exponential to give

$$\sum_{m=-I}^{I} m e^{\frac{\gamma \hbar m B_0}{kT}} = \sum_{m=-I}^{I} m(1 + mx) = \sum_{m=-I}^{I} \left(m + m^2 x\right) = \sum_{m=-I}^{I} m + \sum_{m=-I}^{I} m^2 x$$

$$= x \sum_{m=-I}^{I} m^2,$$

since $\sum\limits_{m=-I}^{I} m = 0$. Now we need to know the sum of m^2. This is given by

$$\sum_{m=-I}^{I} m^2 = \frac{1}{3}I(I+1)(2I+1)$$

and we put this back into our equation for M_o to obtain

$$M = \frac{N\gamma^2\hbar^2}{3kT}I(I+1)B_o$$

which is known as *Curie's law*, after Pierre Curie. It predicts that the magnetic moment, M_o, is proportional to the applied field, B_o, and is inversely proportional to the temperature. If we write that

$$M_o = \chi_o B_o$$

then we see that the magnetic susceptibility, χ_o, is inversely proportional to the absolute temperature, T, where

$$\chi_{\mathrm{o}} = \frac{N\gamma^2\hbar^2 I(I+1)}{3kT}$$

The direct measurement of a magnetic moment is not particularly easy and can be done only to a limited accuracy by electromechanical means. The present discussion suggests that there may be a superior method based on magnetic resonance.

7.9 Nuclear Magnetic Resonance

In order to induce transitions between different spin states, one needs to irradiate the sample with photons tuned to the frequency corresponding to this energy difference. What is this frequency? It is found from the Planck equation, $E = h\nu$. We have found the energy difference above, so that

$$\Delta E = h(\nu_2 - \nu_1) = \gamma\hbar\,(I_2 - I_1)\,B_{\mathrm{o}} = \gamma\hbar\,[1/2 - (-1/2)]B_{\mathrm{o}} = \gamma\hbar B_{\mathrm{o}}$$

The absorption frequency, ν, is then given by

$$\nu = \frac{\gamma h B_{\mathrm{o}}}{2\pi h} = \frac{\gamma B_{\mathrm{o}}}{2\pi}$$

or, if we rewrite this in terms of the angular frequency, ω, we have the NMR equation:

$$\omega = 2\pi\nu = \gamma B_{\mathrm{o}}$$

The frequency of photon absorption, ω, is linearly proportional to the applied magnetic field, B_{o}. As a practical matter, the most accurately measured quantities in science are time and frequency; one can determine frequency or time to one part in 10^{12} with only moderate effort. Therefore, the measurement of magnetic moments is most accurately accomplished through the measurement of NMR frequencies which are distinguished from each other by their dependence on the masses of the various nuclei with nonzero spin. This has been done for all of the known nuclear isotopes, and the results are given in terms of NMR frequency tables. One can find these on the web at, for example, http://web.mit.edu/speclab/www/nmrfreq.html and their frequencies, nuclear spins, and natural isotopic abundances are given in Appendix 2, Table A.2. One notes that the NMR frequencies vary markedly from nucleus to nucleus. What causes this variation? The magnetic moment is proportional to the spin and an inverse function of the mass. We then expect that the NMR frequency of a nucleus would decrease as we move up the periodic table from hydrogen, and that is indeed what is observed (Fig. 7.5).

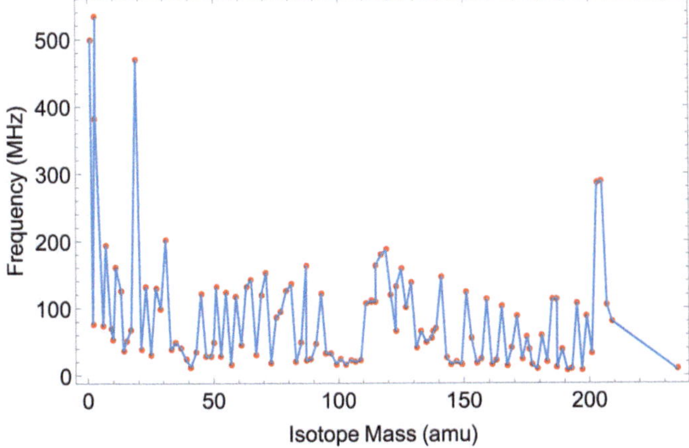

Fig. 7.5 NMR frequencies in a magnetic field of 11.74 T as a function of the mass of the nucleus. The highest frequencies are for the light spin ½ nuclei: ^{1}H, ^{3}H, ^{3}He, and ^{19}F

7.10 The Chemical Shift

If all of the nuclei in a sample experienced the same local magnetic field, then all the resonance frequencies for a large molecule would coincide and NMR would not be a very interesting form of spectroscopy. Fortunately, the nuclei in atoms are surrounded by electrons, which, we have seen, have magnetic moments almost 2000-fold larger than the nucleons. Therefore, even a small probability that an electron's wave function has a finite value at the position of the nucleus $|\Psi(0)|^2 > 0$ will lead to a local magnetic field at the nucleus which adds to or subtracts from the applied field, and since the resonance frequency for a nuclear spin is proportional to the local magnetic field, the NMR signal will be shifted slightly to higher or lower field (frequency). The distribution of electrons in an atom or a molecule is determined by the chemical bonding pattern. The chemical nature of an atom in a molecule will influence the NMR frequency of the nucleus of that atom. This is the important link between chemical structure and NMR frequency which has given rise to the extreme usefulness of this form of spectroscopy in, for example, organic chemistry.

It is said (an urban legend) that the physicists who discovered NMR in the late 1940s (Bloch, Pound, and Purcell) first examined the NMR signal from water, which gives a single peak. Then, ethanol was studied, with the result that three peaks were found at slightly different frequencies. Since the difference between water and ethanol is due to chemistry, the frequency shift was referred to as a "chemical shift." The resonance frequency, ν, for a nucleus in a molecule can then be written as

$$\nu = \nu_0(1-\sigma)$$

where

$$\nu_0 = \frac{\gamma B_0}{2\pi}$$

corresponds to the resonance frequency of a standard compound in the field, B_0. The actual resonance frequency for a nucleus in an atom or molecule differs by the quantity σ (not the spin operator) from that of a standard. This quantity σ is the shielding constant and is proportional to the probability of finding an electron in the nucleus, which is given by $\sigma \propto \Psi^* (0) \Psi (0)$, where Ψ is the electron's wave function. Calculations of the chemical shifts of nuclei in various molecules are then concerned with computations of this shielding constant.

Although it is possible to directly measure the absolute resonance frequency for a nucleus in a given magnet, upon doing so, one quickly finds that such measurements produce results which depend on the temperature, on the presence of nearby mobile magnetic material, and even on the solvent used to prepare the sample. For this reason, it was found that measurements of the relative NMR frequency of a compound with respect to a standard material produced more reliable and reproducible results. The standards used as references in NMR vary according to the observed nucleus; for protons two standards (Fig. 7.6) are now in almost universal use, (1) tetramethylsilane (TMS) for molecules soluble in organic solvents and (2) sodium 2,2-methyl-2-sila-d_6-pentane-5-sulfonate (DSS, also known as TSP, [D_6]-trimethylsilyl-sodium-propionate) for water-soluble compounds, proteins, peptides, and nucleic acids. These molecules are chosen as standards because the bonding of the methyl groups to the silicon atom shifts electron density onto the protons and shields them from the applied field to such a large extent that the protons

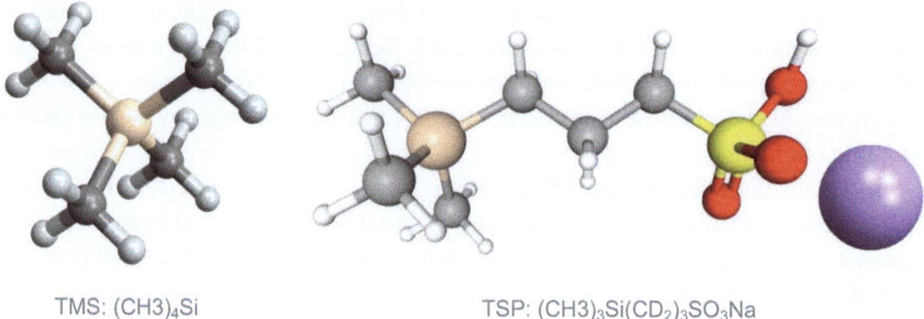

Fig. 7.6 The chemical structures for two compounds commonly used as standards for proton NMR spectroscopy. (*Left*) Tetramethylsilane (TMS), (*right*) sodium 2,2-methyl-2-sila-d_6-pentane-5-sulfonate (DSS, or TSP)

resonate at frequencies higher than that for most other molecules and the resonance from the standard does not interfere with the observation of the resonances from the molecules of interest.

In practice, the absolute resonance frequencies for a given nucleus can easily be measured with an accuracy of 1 part per billion, but, instead of reporting the absolute resonance frequency for a nucleus, one determines its frequency shift from the standard resonance. This **chemical shift** is given the symbol δ and is determined from the frequencies of the reference standard, ν_r; the resonance frequency of the nucleus of interest, ν; and the spectrometer frequency, $\nu_o = \frac{\gamma B_o}{2\pi}$, by

$$\delta = \frac{\nu_r - \nu}{\nu_o}$$

For a proton resonance frequency, $\nu_o = 500$ MHz, the shifts from a standard like TMS or DSS are only a few kHz. Therefore, the chemical shifts are reported in the dimensionless units of parts per million, or ppm. For example, water protons at $T = 295$ K resonate at +2421.5 Hz with respect to DSS (Table 11.16) which gives a chemical shift of water, δ, as

$$\delta = \frac{5 \times 10^8 - \left(5 \times 10^8 + 2421.5\right)}{5 \times 10^8} = 4.843 \times 10^{-6} = 4.843 \text{ ppm}$$

Note that the chemical shift scale increases from right to left, opposite to that for many other scales, and that the chemical shift increases as the shielding of a nucleus decreases. These are simply conventions embedded in NMR spectroscopy, much as other spectroscopic domains report their measurements in, for example, wavenumbers, or m/z.

The atoms which account for the bulk molecular composition of living systems consist of isotopes of hydrogen, carbon, nitrogen, oxygen, sodium, phosphorus, potassium, magnesium, calcium, iron, and chlorine; each of these nuclides has an isotope with a nonzero spin (Appendix Table A.2). An isotope is an atom with a nucleus containing a given number of protons, but with a variable number of neutrons. Since atoms are electrically neutral, chemistry is solely determined by the charge, Z, on the nucleus, which is equal to the number of protons alone. Therefore, the number of neutrons, N, determines the atomic mass and the atomic number, $A = N + Z$. But neutrons do more than simply this for the NMR spectroscopist because both neutrons and protons have spin $\frac{1}{2}$. Therefore, the total spin on a nucleus is determined by the quantum mechanical addition of the spins of both constituents of the nucleus. For example, the spin of the proton, i.e., the $Z = A = 1$ nucleus of the simplest atom in the periodic table, hydrogen, is $\frac{1}{2}$. The addition of a neutron to hydrogen produces the heavier isotope, deuterium, with $A = N + Z = 1 + 1 = 2$, which obeys essentially all of the chemistry that one observes for hydrogen. The spin of the deuteron can be either 0 or 1 corresponding to the antiparallel or parallel addition of the neutron and proton's spins. The spin of the deuteron was found to be one, and consequently, one can perform NMR experiments on deuterium. Since the deuteron is more massive than the proton, we find from Table A.2 that the NMR frequency for deuterium is lower than that for hydrogen. Adding another neutron to the nucleus of deuterium produces tritium, an unstable (radioactive) nuclide with $A = N + Z = 2 + 1 = 3$. By the Pauli principle, the two identical neutrons can occupy the same energy state if their spins add to zero, or the spins can add to one if they are in different states. We predict that the spin of the triton is then either ½ or 3/2 because the sum of the neutron spins must be added to that of the proton (1/2); a spin of ½ is actually found. The nomenclature of nuclides is defined such that A, the atomic number, or the sum of the neutrons and protons in a nucleus is entered to the upper left of the chemical symbol of the element. For example, ordinary hydrogen is denoted ^{1}H; deuterium, ^{2}H; tritium, ^{3}H; etc. With this in mind we can amplify our list of biochemically interesting nuclides to include ^{1}H, ^{2}H, ^{3}H, ^{13}C, ^{14}N, ^{15}N, ^{17}O, ^{23}Na, ^{25}Mg, ^{31}P, ^{35}Cl, ^{39}K, and ^{57}Fe.

The applications of NMR in biochemistry encompass a wide variety of areas, ranging from the structures of small molecules, metabolites, peptides, proteins, and nucleic acids, anatomical and functional imaging, to noninvasive determinations of the metabolism of cells, tissues, and organs. When one is contemplating the application of NMR to a particular biochemical problem, more factors than simply the resonance frequency must be considered, although a close examination of the data presented in Table A.2 indicates that the NMR frequencies for each nuclide do not overlap. Therefore, one can be assured that a study of hydrogen, for example, will not be contaminated by NMR signals from phosphorus. The chemical shift ranges for each isotope are also narrow enough to guarantee this fact.

The ease of detecting a given isotope of interest depends on its natural abundance (Table A.2) and its magnetogyric ratio. From our discussion of the isotopes of hydrogen given above, one can understand both of these factors. Normal hydrogen, ^{1}H, is present as 99.98% of all of the hydrogen on the earth, while its more massive isotope, deuterium (^{2}H), accounts for the remaining 0.02%, and tritium, ^{3}H, is not present naturally because it is not stable. The NMR spectrum of a sample of water would obviously be most-profitably obtained at the NMR frequency of ^{1}H, and although modern NMR spectrometers can detect ^{2}H in water at natural abundance, one would normally wish to employ enrichment techniques in order to increase the amount of deuterium in a sample so the acquisition of useful data would not occupy an impractical amount of instrument time. There exists an industry dedicated to providing useful isotopes at much higher than natural abundance levels. This began with the separation of deuterated water from protonated water by electrolysis by the Norwegians at Vemork in 1935.

Besides knowing the natural abundance of an isotope when planning an NMR study, it is also important to understand an isotope's relative sensitivity, which is intimately related to its magnetogyric ratio. We have seen that the energy difference between nuclear spin states induced by placing the sample in a magnetic field depends on the product of the nuclear moment and the field as $E = -\mu \cdot B_0$ and the magnetic moment, μ, is $\mu = \gamma \hbar m$, so that

$$\Delta E = -\gamma \hbar m \cdot B_o$$

Therefore, the energy difference is directly related to the size of the magnetic moment, and this depends on γ. The sensitivity of an NMR experiment is determined by the difference in populations between the nuclear ground state and the excited state(s). We can see then that low γ nuclei do not produce as much NMR signal because the population differences between states are much smaller than those for higher γ nuclei, like protons. The sensitivity for the detection of a given isotope, x, relative to hydrogen is $(\gamma_x/\gamma_H)^3$ for an equal number of nuclei. The absolute sensitivity is this relative sensitivity times the natural abundance of the isotope.

7.11   The NMR Experiment

How does one actually perform an NMR study? We have seen from our examination of Curie's law that a sample, when placed into a magnetic field, acquires a magnetization, or magnetic moment, which is parallel to the applied field. Even though the NMR phenomenon is purely quantum mechanical, we can understand some of its basic features by resorting, initially, to a classical electromagnetic approach. A sample placed into a magnetic field acquires a magnetic moment, M, which is a vector parallel to the applied field, B_o. This sample magnetization, M, precesses about the applied field with at the Larmor frequency ω_L given by the NMR equation:

$$\omega_L = \gamma B_o$$

The net magnetization parallel to the z-axis is then M_o, and this is independent of time (Fig. 7.7). The energy of this magnetization vector is

$$E = -M \cdot B_o$$

the dot product of the moment with the applied field. This magnetic moment vector experiences a torque, $T = M \times B_o$, which is equal to the time derivative of the angular momentum, I:

Fig. 7.7 The precession of a magnetic moment, M, about the magnetic field, B, and the resulting magnetization, M_o

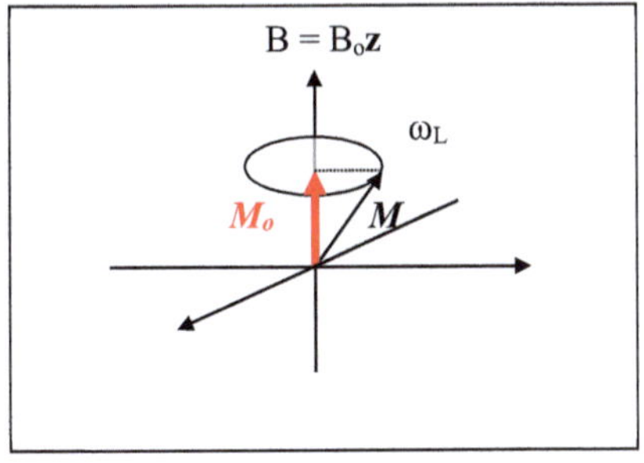

$$T = \frac{d\hbar I}{dt} = M \times B_{\mathrm{o}}$$

if we multiply both sides of this equation by γ, we obtain

$$\gamma T = \gamma \frac{d\hbar I}{dt} = \gamma M \times B_{\mathrm{o}} = \frac{d}{dt} M$$

or

$$\frac{d}{dt} M = M \times \gamma B_{\mathrm{o}}$$

and here we have the mathematical equivalent of the statement above that the moment precesses around $\boldsymbol{B_o}$ with a frequency $\omega_{\mathrm{L}} = \gamma \, \boldsymbol{B_o}$. It is simpler to understand many NMR processes if we transform to a coordinate system rotating at the frequency ω_{L}. The relationship between this rotating frame and the frame fixed in the laboratory is

$$\left(\frac{d}{dt} M \right)_{\mathrm{rot}} = \left(\frac{d}{dt} M \right)_{\mathrm{lab}} + M \times \omega_{\mathrm{L}}$$

but, since we just saw that the time derivative of the magnetization in the lab frame is $\frac{d}{dt} M = \gamma M \times B_{\mathrm{o}}$, we can replace the derivative on the right to give

$$\left(\frac{d}{dt} M \right)_{\mathrm{rot}} = \gamma M \times B_{\mathrm{o}} + M \times \omega_{\mathrm{L}}$$

which we can rewrite by collecting terms as

$$\left(\frac{d}{dt} M \right)_{\mathrm{rot}} = M \times (\gamma B_{\mathrm{o}} + \omega_{\mathrm{L}})$$

This has the same form as $\frac{d}{dt} M = M \times \gamma B_{\mathrm{o}}$, but with the applied field, B_{o}, replaced by an effective field: $B = B_{\mathrm{o}} + \omega_{\mathrm{L}}/\gamma$. If we chose the frequency of the rotating frame to be $\omega_{\mathrm{L}} = -\gamma \, B_{\mathrm{o}}$, then we see that the effective field vanishes and that the moment, M, is time-independent in this rotating frame. However, if we look at M from the perspective of the laboratory frame, it is seen to be rotating about the z-axis (B_{o}) with the frequency vector given by the Larmor frequency: $\omega_{\mathrm{L}} = -\gamma B_{\mathrm{o}}$. Let me reiterate, in the rotating frame, the magnetization vector M does not rotate, but appears to be a static vector parallel to the z-axis (the direction of the applied field). As long as the magnetization remains parallel to the z-axis in a steady state, there is no NMR signal from the nuclei. One can understand this from a classical viewpoint as well. Consider a long cylindrical magnet polarized such that one end is

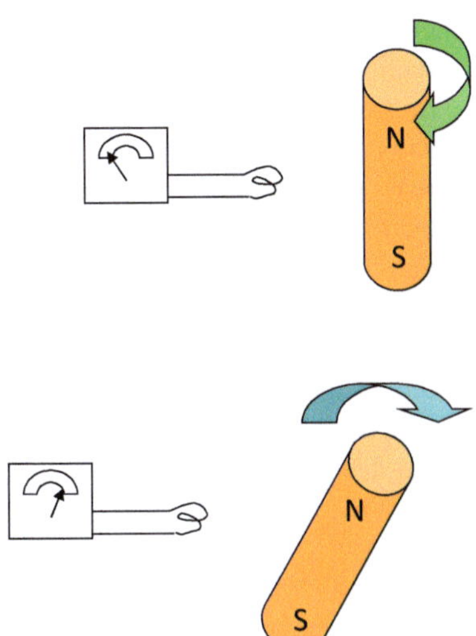

Fig. 7.8 A cylindrical magnet rotating about an axis parallel to a cylindrical coil, by symmetry, induces no voltage in the coil windings

Fig. 7.9 A cylindrical magnet rotating about an axis perpendicular to a cylindrical coil breaks symmetry and induces a voltage in the coil windings

the north and the other end is the south pole. If this magnet spins around its long axis, from elementary symmetry considerations, there is no change in the magnetic field experienced by an observer anywhere around the magnet, because it is rotationally invariant about its symmetry axis (Fig. 7.8).

However, if we take this same cylindrical magnet and rotate it about an axis perpendicular to its long axis, an observer will see an induced current in a pickup coil placed nearby (Fig. 7.9).

This simple demonstration indicates how we would go about detecting the NMR signal from a sample placed in a magnetic field. We just need to arrange to tip the magnetization vector away from the z-axis and into the x–y plane, where we have placed a pickup coil. We will then detect an alternating voltage in the pickup coil from the precession of magnetization vector at the Larmor frequency. One immediately recognizes this as the basic physics of a generator.

We can use a spinning gyroscope as a model of a magnetic moment in an applied magnetic field. As long as the axis of a spinning gyroscope remains strictly parallel to the Earth's gravitational field vector, the gyroscope axis (which is parallel to its angular momentum) will continue to point in the same direction in three dimensions. Now, however, if we place the spinning gyroscope on a stand at an angle with respect to the Earth's gravitational field vector, g, a torque, T, will be exerted on the gyroscope given by the vector cross product of the angular momentum vector, L, and Earth's gravitational field vector:

$$T = L \times g$$

This torque will cause the gyroscope axis to precess around the Earth's gravitational field vector. By the conservation of angular momentum, the gyroscope's spin axis is observed to be difficult to redirect. One needs to apply a large transient (with a big hammer, say) to the axis to cause it to move substantially away from its initial direction. The conservation of angular momentum will also cause the spin axis of the gyroscope to return to its original direction. If we were to make the gyroscope's spin axis out of a permanent magnet and to detect its precession after an impulse with a nearby pickup coil, we would have a good classical model of an NMR experiment. Note that the precession frequency of a gyroscope in the Earth's gravitational field is given by known quantities, just as the Larmor frequency for a nuclear spin is given by known physical quantities.

What can we use as a big hammer for the NMR experiment? Since a magnetic moment interacts with an applied field, we can certainly expect to use a second magnetic field orthogonal to the applied field. It is also necessary to tip all of the nuclear spins in a sample even though from the chemical shift variation of the resonance frequencies we know that the various nuclei in a real sample resonate over a band of frequencies. The nuclear resonance frequencies also are known from the basic NMR equation to be in the 100–1000 MHz region, and this means that we require a magnetic "hammer" which is a radiofrequency field in this frequency range.

In our search for a suitable source of a magnetic impulse, we can take a clue from the uncertainty principle, which states that the product of the uncertainty for two conjugate variables can be no smaller than the reduced Planck's constant, $h/2\pi$. Let us examine the uncertainty principle for energy and time:

$$\Delta E \, \Delta t = h/2\pi$$

if we now write the energy as $h\nu$, the appearance of Planck's constant on both sides of this equation cancels and we are left with

$$\Delta \nu \, \Delta t = 1/2\pi$$

in other words, if we apply a short pulse of suitable radiofrequency photons at the Larmor frequency, $\omega_L = 2\pi\nu$, for a duration, Δt, we find that this pulse will cover a frequency bandwidth of

$$2\pi \, \Delta \nu = 1/\Delta t$$

Now, if the bandwidth needed for covering the, say, proton spectrum at 500 MHz is 12 ppm, or 6000 Hz, we would need to have a pulse with a duration no longer than

$$\Delta t = 1/(2\pi \, \Delta \nu) = 1/(6.26 \times 6000) = 26.5 \, \mu s$$

Such pulses are easy to generate with modern electronics. The following Mathematica notebook:

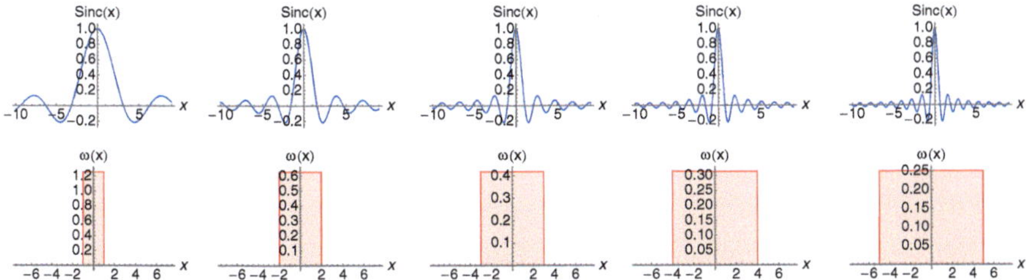

Fig. 7.10 The top row of plots shows the sinc function (blue) whose Fourier transforms are the square waves (red) shown just below. Note that as the pulse gets narrower, the sinc function broadens, and vice versa

```
sinc[x_] = Sin[x]/x
tbsinc = Table[Plot[sinc[n x],{x,-10,10},PlotRange -> All,
 PlotStyle -> Hue[0.65]], {n, 5}]
tbft = Table[Plot[Abs[FourierTransform[sinc[n x],x,ω]],
 {ω,-10, 10}, PlotStyle -> Hue[0]], {n, 5}]
```

shows the inverse relationship between the duration of a square wave pulse and the width of the resulting frequency domain coverage (Fig. 7.10), i.e., the bandwidth of the pulse. The relationship between the first zeroes on the sinc function and the width of the square wave is $\Delta x \Delta \omega \sim \pi$, which is again a manifestation of the uncertainty principle (see also Fig. 4.1).

The NMR signal generated in the pickup coil is clearly a sine or cosine wave; exciting the sample with a short radiofrequency photon pulse at the Larmor frequency will bring all the nuclei into resonance at once. If we remember from the basic NMR equation, the frequency (energy; $E = h\nu$) needed to cause transitions between spin states is equal to the energy difference between those states, and then a short pulse of radiofrequency photons will contain all the frequencies needed to induce energy absorption by every nucleus in our sample. The result is that the sine or cosine modulated voltage induced in our detector coil will contain the sum of all the resonance frequencies in our sample. We therefore require some means of sorting out the amplitudes and frequencies for all the various nuclei into a useable NMR spectrum. Again the uncertainty principle can offer us a clue as to how to proceed. We saw that pulsed NMR was motivated from a desire to apply an excitation which covered the range of frequencies present in our sample. This was based on the fact that frequency and time are conjugate variables; frequency has the units of 1/time, and vice versa. Such variables are also known as Fourier transform pairs.

What is a Fourier transform? We have seen from quantum mechanics that the solutions of Schrödinger's equation form a complete orthonormal set of functions and that any other function can be written as an infinite sum of these functions with coefficients determined by means of the orthogonality condition. One such complete orthonormal set of functions was the solution for the particle in the box, which was a set of sines and cosines. We can express sines and cosines in the compact exponential notation, $e^{i\omega t} = \cos \omega t + i \sin \omega t$. A Fourier transform is simply a way of expressing a function of a given variable, either time or frequency, in terms of a sum (integral) over a complete orthonormal set of sines and cosines of another function of the opposite conjugate variable.

To be explicit, if the time-dependent voltage induced by the x–y precession of the nuclear magnetic moments is denoted $s(t)$, then the NMR spectrum is its Fourier transform given by $S(\omega)$, where

$$S(\omega) = \int_{-\infty}^{\infty} s(t)\, e^{i\omega t} dt$$

Here we see the expansion of $S(\omega)$ in a complete orthonormal set of sines and cosines, whose coefficients are the measured voltages in the NMR coil, $s(t)$. It is then to the structure of $s(t)$ we next turn.

Just as a spinning gyroscope struck by a hammer will precess about the Earth's gravitational field vector and return to assume the initial direction of its spin angular momentum, a nuclear moment will precess about the direction of the applied field after the application of a suitable radiofrequency pulse and then return to its initial orientation parallel to B_o. Therefore, the NMR signal, $s(t)$, will contain a decaying function of time, $d(t)$, which we can write as

$$s(t) = A\, d(t)$$

where A is a constant representing the amplitude of the NMR signal at $t = 0$. Many decaying processes in nature obey exponential kinetics, so we will write $d(t) = e^{-gt}$, where g is a decay time constant. Furthermore, the NMR signal will consist of the sum of the resonance frequencies of all the nuclei in the sample; for a single frequency, a suitable oscillating function is $e^{i\omega t}$, where ω is the Larmor frequency. To represent NMR frequencies different from the Larmor frequency, we can add another frequency, ω_o, where this is the resonance frequency of, for example, a water proton. A simple model time domain NMR signal is then represented by

$$s(t) = A\, e^{-gt}\, e^{i(\omega - \omega_o)t}$$

According to our line of reasoning above, the actual NMR spectrum is then the Fourier transform of this time domain signal. Before we embark on its solution, however, it is useful to examine the behavior of $s(t)$. Notice first that the NMR signal is complex; it contains the imaginary number $i^2 = -1$, representing the fact that we could have used two orthogonal detector coils in our NMR instrument, one along the x-axis and another along the y-axis which would independently sample the sine and cosine terms of $s(t)$. A plot of the NMR signal (Fig. 7.11) shows that it consists of exponentially decaying sines and cosines. For historical reasons, the NMR signal, $s(t)$, is called the free-induction decay, or FID.

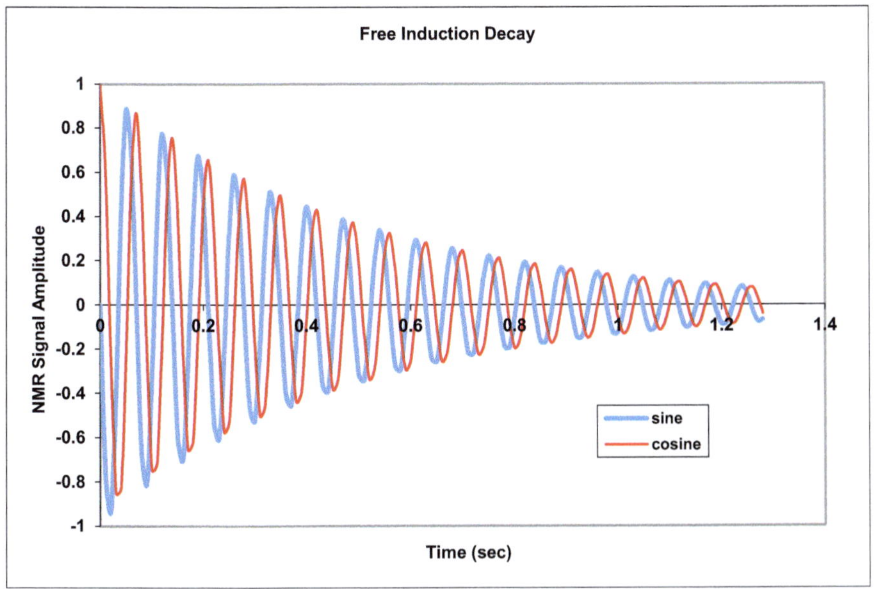

Fig. 7.11   A plot of $s(t)$ for $A = 1$, $g = 2$ Hz, $\omega = 10$ Hz, and $\omega_o = 100$ Hz

7.12   A Mathematica Notebook Simulating the NMR Experiment

We will next turn our attention to the Fourier transform of this model FID. A simple way to do this Fourier transform is to use a Mathematica notebook and the following notebook illustrates a number of interesting features of the Fourier transform NMR experiment. The first is that the analytical form of the NMR signal, $S(t)$, given as an exponentially damped sum of sines and cosines produces the correct line shape function; this function is known as a *Lorentzian*, named after Hendrik Lorentz, a contemporary of Einstein. It is useful to show that the Fourier transform of an exponential is a *Lorentzian*. Let $S(t)$ be our decaying function, and the width of the *Lorentzian* is proportional to the inverse of the decay time g of the exponential:

$$S(t, g) = e^{-g|t|}$$

which is shown in Fig. 7.12 (Top) for $g = 0.25$ Hz. The Fourier transform of $S(t,g)$ is (Fig. 7.12, bottom) indeed a Lorentzian function whose full width at half maximum is g.

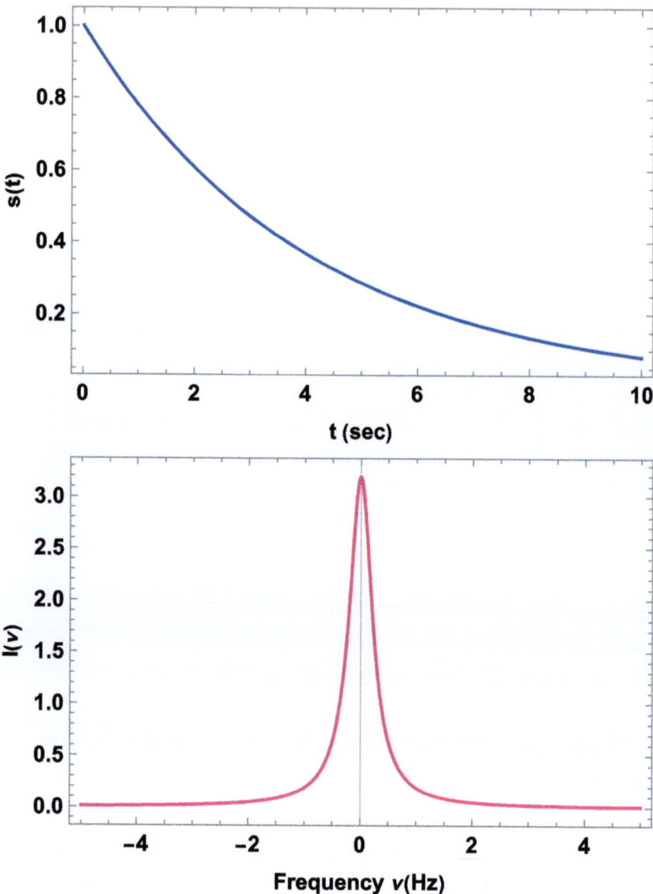

Fig. 7.12 (*Top*) The exponential decay of the FID for a decay constant of 0.25 Hz. (*Bottom*) The Fourier transform of the FID showing a Lorentzian of half width 0.25 Hz

$$F[\omega, g] = \text{Fourier transform}\,[f[t, g], t, \omega] = \sqrt{\frac{2}{\pi}}\,\frac{g}{\omega^2 + g^2}$$

```
Plot [F[ω, 0.25 ], {ω, -5, 5}, PlotRange -> All, PlotStyle -> Hue[0.9]]
```

Here, we have seen that the Fourier transform of an exponential is a Lorentzian, whose width (as defined as the full width at half maximum, FWHM) is the inverse of the exponential decay constant, g. The time domain NMR signal is a damped exponential containing all the resonance frequencies of the sample. Let us use Mathematica to examine how this comes about.

This notebook calculates the NMR spectrum from a continuous time-domain
signal by means of a filtered Fourier transform.

s(t) is the time-domain NMR signal, with a decay constant, g = 0.25 Hz, and
a frequency, ω_o = 12.5 Hz.

```
s[t_] = Exp[-g t] Exp[-I (ω - ωₒ) t]
```

```
Plot[Re[s[t]], {t, 0, 10}, PlotRange -> {-1, 1},
 PlotStyle -> Hue[0], Frame -> True,
 FrameLabel->{"Time (sec)", "NMR Signal"}]
```

The real part of this signal is shown in Fig. 7.13.

Fourier transform this to produce the NMR spectrum, S(ω).

```
Integrate[s[t], {t, 0, Infinity}, Assumptions-> 1/(g + I (ω - ωₒ)), Im[ωₒ]
+ Re[g] > Im[ω]]
```

```
S[ω_] = ComplexExpand[1/(g + I (ω - ωₒ))]
```

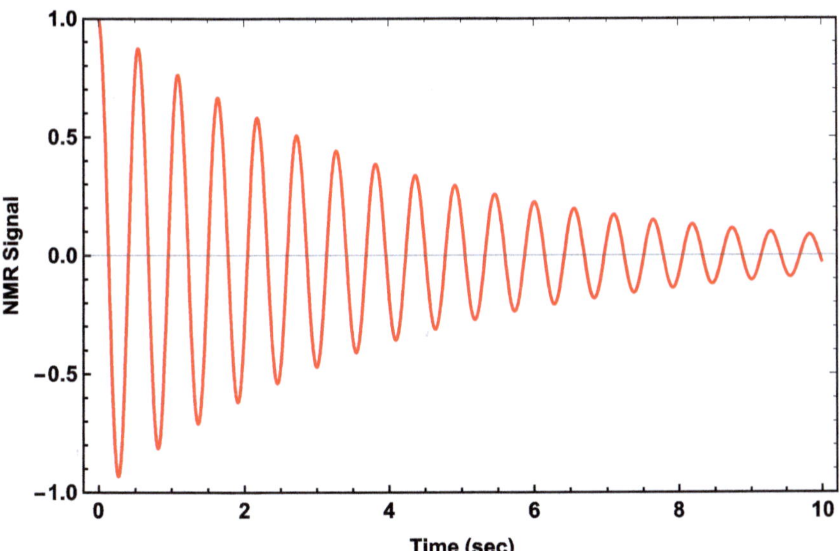

Fig. 7.13 The real part of the time domain NMR signal (free induction decay) after a
radiofrequency pulse

$$S[\omega] = \frac{g}{g^2 + (\omega - \omega_{\mathrm{o}})^2} + i\left(\frac{\omega_{\mathrm{o}} - \omega}{g^2 + (\omega - \omega_{\mathrm{o}})^2}\right)$$

```
Plot[Re[S[ω]], {ω, 0, 25}, PlotRange -> {0, 1.2 Table[Re[S[w]], {w, 0, 25,
0.01}]},
PlotStyle -> RGBColor[0, 1, 0], Frame -> True,
FrameLabel -> {"Frequency (Hz)"}]
```

The real part of the NMR absorption signal (Fig. 7.14) is a *Lorentzian* which peaks at a frequency of $\omega_{\mathrm{o}} = 12.5$ Hz.

The imaginary part of the NMR absorption signal (Fig. 7.15) is a dispersion signal.

```
Plot[Im[S[ω]], {ω, 0, 25}, PlotStyle -> RGBColor[0, 0, 1], Frame -> True,
FrameLabel -> {"Frequency (Hz)", "NMR Signal (Imaginary Part)"}]
```

While the above correctly illustrates the NMR experiment, it uses analytical continuous functions. Actual NMR signals are discrete digital samples of the FIDs and therefore contain noise from the sample and from the receiver electronics. We can sample the FID at a number of discrete points in the time dimension by using Mathematica's Table function and we can add noise to the signal using the Random [] function.

Digital sampling of the FID needs to conform to the Nyquist theorem. The sampling rate must be at least twice the highest frequency in the FID in order to

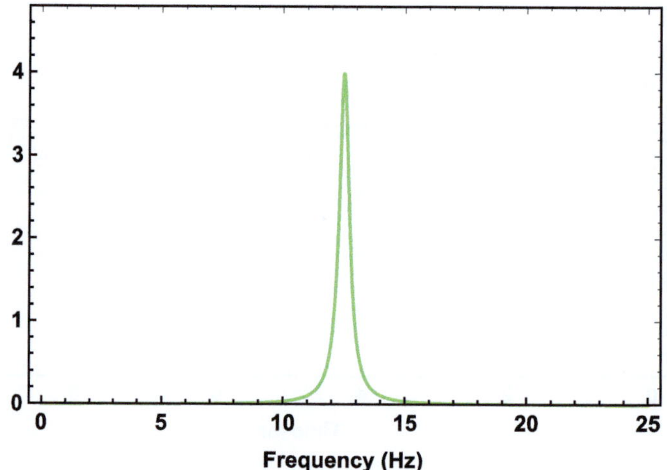

Fig. 7.14 The real part of the NMR absorption signal

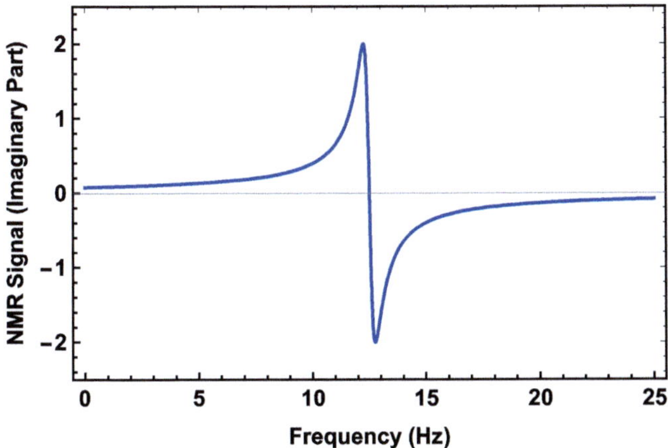

Fig. 7.15 The imaginary part of the NMR absorption signal

correctly determine both the frequency and the phase of the signal. Parameters common to the digital acquisition of NMR signals include (1) the sampling, or acquisition time, aq, in seconds; (2) the sampling rate, or spectral width, sw, in Hz; and (3) the number of points sampled, dp. These are related by the Nyquist theorem as $sw = \frac{dp}{2aq}$, so, for example, if one samples the FID for 10 s and uses 4096 data points, the spectral width is 204.8 Hz.

The following Mathematica code illustrates digital sampling of an NMR signal that peaks at peak = 12.5 Hz, with a full width at half maximum of g = 2.5 Hz, sampled for aq =10 s into dp = 4096 data points. The dwell time dw = aq/(2dp) is the interval between sampling points in the time domain.

Let the FID be given by the complex free induction decay, i.e., the NMR time domain signal:

$$\mathrm{fid}[t\_, w\_, g\_] := e^{-gt}e^{iwt}$$

$w = 2$ aq peak (∗The position of the signal maximum in the aq window.∗)

Generate the time and frequency axes for the plots:

tim = Table[t, {t, 0, aq, aq/(2dp)}]; (∗Generate the time axis. ∗)
fq = Table[1.25(sw − sw$i$/(2dp)), {i, 0, 2dp − 1}]; (∗Generate the frequency axis. ∗)
gfid = Table[fid[t, w, g], {t, 0, aq, aq/(2dp)}]; (∗Sample the FID eachdw/2∗)
regfid = Re [gfid]; (∗Take the real part of the FID∗)

Then plot the real part of the FID:

ListLinePlot[Table[{tim[[i]], regfid[[i]]}, {i, 1, Length[tim]}], PlotRange $\rightarrow$ All, PlotStyle $\rightarrow$ Hue[0.4], Frame $\rightarrow$ True, FrameLabel $\rightarrow$ {"t (sec)", "s(t)"}, FrameStyle $\rightarrow$ Medium, LabelStyle $\rightarrow$ Directive[Black, Bold]]

to give Fig. 7.16 (top). The imaginary part can also be obtained as Im[gfid] and plotted similarly. Fourier transforming the digitally sampled FID results in the NMR spectrum (Fig. 7.16, bottom) which shows the signal at a position of 12.5 Hz. Note that in Fig. 7.16 (bottom) the frequency axis has been truncated from its total width of sw = 204.8 Hz in order to better display the width of the resonance.

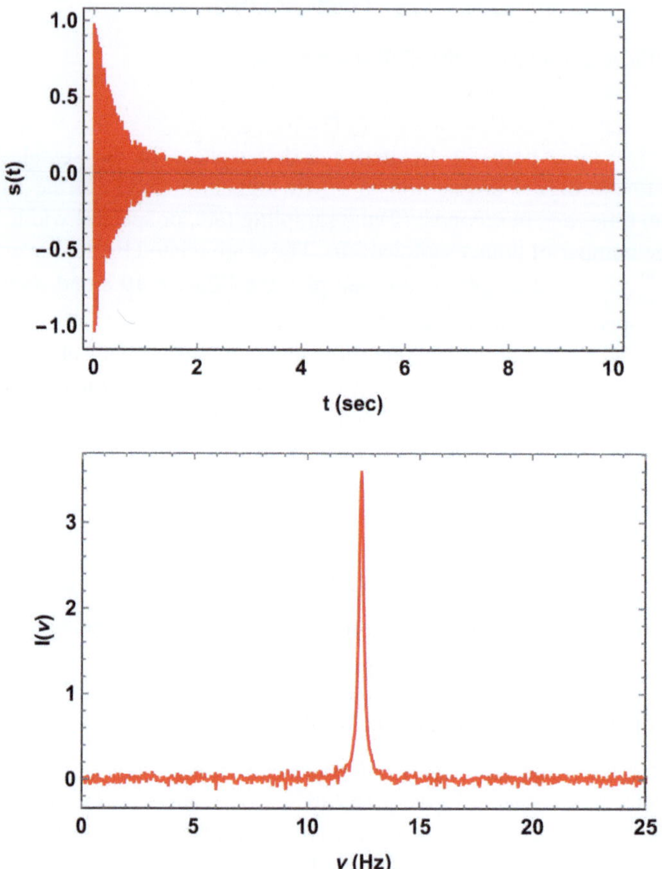

Fig. 7.16 (*Top*) The free induction decay (FID) from a resonance that peaks at 12.5 Hz, with a linewidth of 2.5 Hz, sampled with 4096 data points in 10 s, with 10% added noise. (*Bottom*) The NMR signal resulting from the Fourier transformation of the noisy FID

```
gspec = Fourier[gfid];
grespec =  Re [gspec];
gnrespec = Table[{fq[[i]], grespec[[i]]}, {i, 0, 2dp − 1}];
ListLinePlot[gnrespec, PlotRange → {{0, 25}, All}, PlotStyle → Hue[0.3]]
```

Adding 10% noise to the FID with the aid of the Random[] function in Mathematica provides an FID that exhibits a large, random variation in signal intensity (Fig. 7.16, top).

```
sfid  =   Table[fid[t, w, g]  +  noise  RandomReal[{−1, 1}], {t, 0, aq, aq/(2dp)}];
   (∗Add noise to the FID∗)
refid =  Re [sfid]; (∗Take the real part∗)
ListLinePlot[Table[{tim[[i]], refid[[i]]}, {i, 1, Length[tim]}], PlotRange   →   All,
   PlotStyle   →   Hue[0], Frame   →   True, FrameLabel   →   {"t  (sec)", "s(t)"},
   FrameStyle → Medium, LabelStyle → Directive[Black, Bold]]
```

The Fourier transform of these noisy data results in a noisy spectrum (Fig. 7.16, bottom).

```
noispec = Fourier[sfid];
renoispec =  Re [noispec];
nrenoispec = Table[{fq[[i]], renoispec[[i]]}, {i, 0, 2dp − 1}];
ListLinePlot[nrenoispec, PlotRange → {{0, 25}, All}, PlotStyle → Hue[0]]
```

7.13   The Convolution Theorem

By sampling the FID in the time domain and Fourier transforming the data into the frequency domain, one can use time domain filtering to improve either the signal to noise ratio or the spectral resolution of an NMR data set. The basis for time domain filtering lies in the mathematics of the convolution theorem, which we now prove.

Let us define two functions of time, $S(t)$, which is the NMR signal, and $F(t)$, which is a time domain filter function. We assume that both of these functions possess Fourier transforms, given by $s(\omega)$ and $f(\omega)$, where $s(\omega)$ is the NMR spectrum, and $f(\omega)$ is the frequency bandwidth of the filter function:

$$s(\omega) = \frac{1}{\sqrt{2\pi}} \int_{-\infty}^{\infty} S(t)e^{-i\omega t} dt$$

$$f(\omega) = \frac{1}{\sqrt{2\pi}} \int_{-\infty}^{\infty} F(t)e^{-i\omega t} dt$$

Note that the Fourier transform, $\mathcal{F}$, is linear and possesses an inverse, $\mathcal{F}^{-1}$, so that

$$\mathcal{F}(A) + \mathcal{F}(B) = \mathcal{F}(A + B)$$

$$\mathcal{F}^{-1}\mathcal{F} = 1$$

The **convolution theorem** then states that the Fourier transform of a convolution in the frequency domain is equal to the product of two functions in the time domain. The convolution of two functions, f and s, is denoted by * and is defined by this integral transform:

$$s(\omega) * f(\omega) = \frac{1}{\sqrt{2\pi}} \int_{-\infty}^{\infty} s(\omega_o) f(\omega_o - \omega) d\omega_o$$

Now, we insert the value of $f(\omega_o - \omega)$ defined in terms of the Fourier transform of $F(t)$:

$$f(\omega_o - \omega) = \frac{1}{\sqrt{2\pi}} \int_{-\infty}^{\infty} F(t) e^{i(\omega_o - \omega)t} dt$$

into the definition of the convolution

$$s(\omega) * f(\omega) = \frac{1}{\sqrt{2\pi}} \int_{-\infty}^{\infty} s(\omega_o) \frac{1}{\sqrt{2\pi}} \int_{-\infty}^{\infty} F(t) e^{i(\omega_o - \omega)t} dt \, d\omega_o$$

and rearrange the terms under the integrals by distributing the frequency difference in the exponential

$$s(\omega) * f(\omega) = \frac{1}{2\pi} \int_{-\infty}^{\infty} s(\omega_o) e^{-i\omega_o t} d\omega_o \int_{-\infty}^{\infty} F(t) e^{-i\omega t} dt$$

Now, we note that first integral on the right side is $S(t)$, which is just the Fourier transform of $s(\omega)$ given by

$$S(t) = \frac{1}{\sqrt{2\pi}} \int_{-\infty}^{\infty} s(\omega_o) e^{-i\omega_o t} d\omega_o$$

and so we have that

$$s(\omega) * f(\omega) = \frac{1}{\sqrt{2\pi}} \int_{-\infty}^{\infty} S(t) F(t) e^{-i\omega t} dt$$

Now, if we take the inverse Fourier transform of both sides of this, we obtain a statement of the **convolution theorem:**

$$\sqrt{2\pi}\int_{-\infty}^{\infty} s(\omega) * f(\omega)e^{i\omega t}\,dt = S(t)F(t)$$

On the left-hand side of this equation, we have a complicated Fourier transform of a convolution of two functions, a signal and a filter, both in the frequency domain, while on the right we simply have the product of a signal and a filter function in the time domain. Therefore, we can filter the frequency domain NMR spectrum by multiplying the time domain NMR signal by a filter function and then Fourier transforming the resulting product. This is an extremely important result because it forms the basis for most digital signal processing theory.

There are two basic reasons to filter an NMR signal; the first is to improve the signal to noise ratio, while the second is to increase spectral resolution. Signal to noise improvement is most often accomplished by using a simple exponential filter, whose effect on the FID is to essentially chop off the noise at the long time end of the FID (see Fig. 7.17, top). Resolution enhancement uses a filter which peaks at a certain time during the FID later than $t = 0$, because the broadest lines decay quickly and occur for short times.

Let us illustrate time domain filtration using our digitally sampled FID from Fig. 7.16. The convolution theorem allows us to multiply the FID by a decaying exponential, with a time constant of `filter` $= 5$ Hz, which has no effect at the first point ($t = 0$) in the acquisition window because $e^0 = 1$, but increasingly truncates the noise at subsequent time points.

Let our filtered FID be

filfid $=$ Table[sfid[[n]]$e^{-\text{filter } n/\text{dp}}$, $\{n, 1, \text{Length}[\text{sfid}]\}$];
refilfid $=$ Re [filfid];

and then we plot it (Fig. 7.17, top). Note how the noise is attenuated for times greater than around 1 s.

ListLinePlot[Table[$\{$tim[[i]], refilfid[[i]]$\}$, $\{i, 1, \text{Length}[\text{tim}]\}$], PlotRange $\rightarrow$ All,
    PlotStyle $\rightarrow$ Hue[0.7], Frame $\rightarrow$ True, FrameLabel $\rightarrow$ $\{$"t (sec)", "s(t)"$\}$,
    FrameStyle $\rightarrow$ Medium, LabelStyle $\rightarrow$ Directive[Black, Bold]]

When this FID is Fourier transformed, we retrieve our spectrum, but with much better signal to noise ratio (Fig. 7.17, bottom) compared with the unfiltered data (Fig. 7.16, bottom).

filspec $=$ Fourier[filfid];
refilspec $=$ Re [filspec];
ListLinePlot[Table[$\{$fq[[i]], refilspec[[i]]$\}$, $\{i, 1, \text{Length}[\text{tim}] - 1\}$], PlotRange $\rightarrow$
    $\{\{0, 25\}, \text{All}\}$, PlotStyle $\rightarrow$ Hue[0.7], Frame $\rightarrow$ True, FrameLabel $\rightarrow$
    $\{$"ν (Hz)", "I(ν)"$\}$, FrameStyle $\rightarrow$ Medium, LabelStyle $\rightarrow$ Directive[Black, Bold]]

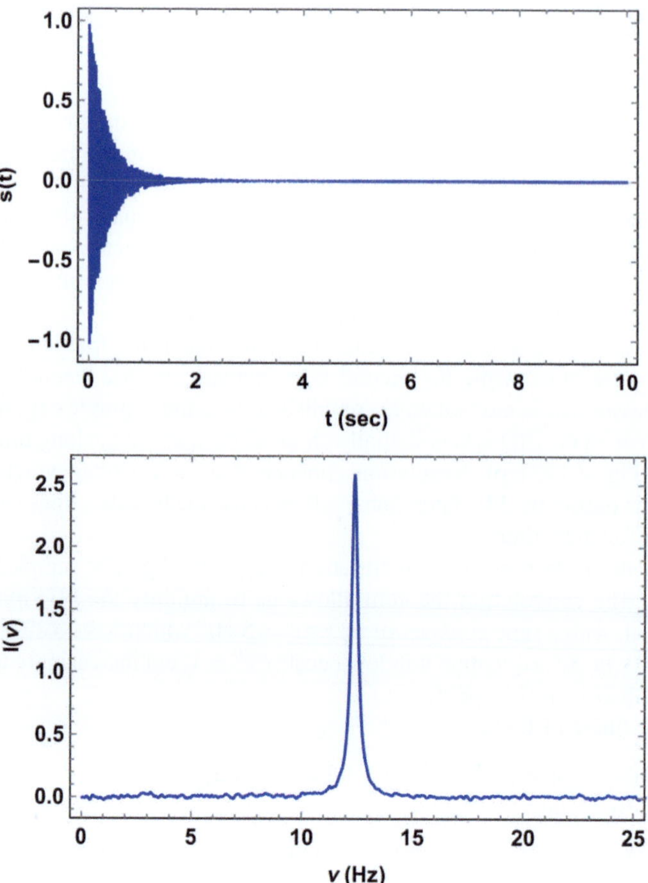

Fig. 7.17 (*Top*) The free induction decay (FID) from a resonance that peaks at 12.5 Hz, with a linewidth of 2.5 Hz, sampled with 4096 data points in 10 s, filtered with a decaying exponential with a time constant of 5 Hz. (*Bottom*) The NMR signal resulting from the Fourier transformation of the filtered, noisy FID

By designing time domain filter functions that are small at $t = 0$, rise to a maximum of ~1 at a particular nonzero time (and therefore emphasizing certain frequencies) and then fall to a small value for larger times, one can develop filters that both improve the signal to noise ratio and improve the resolution of the spectrum. The construction of such functions is left as an exercise for the interested reader, but the study by Daniel Traficante (Traficante and Rajabzadeh 2000) will provide several useful points of departure.

7.14 Protein Structure in Solution by NMR

Proteins, as enzymes and as structural components of cells, are the ultimate product of the *Central Dogma* of molecular biology, where information flows from DNA to RNA to proteins. Proteins range in molecular mass from a few thousand daltons to large multi-subunit enzymes, such as fatty acid synthetase, and with an increasing number of atoms, to the proteins in the nuclear pore complex and the ribosome. When the NMR structure of a protein is desired, it is most often the structure of the smallest functional unit or domain which is sought first because the width of the nuclear resonances increases as the mass of the monomer increases. The basic features of protein NMR are based on an appreciation of the physical parameters which influence spectral resolution.

One of the most important of these is the rate of relaxation of the nuclear spins back to their equilibrium magnetization after an excitation. When we motivated the physical principles behind the NMR experiment, we indicated that we used a magnetic pulse to tip the sample magnetic moment into the x–y plane. We have seen that the NMR signal does not last forever; the sample magnetic moment, once transiently polarized, relaxes from the x–y plane back to the z-axis as the magnetization returns to thermal equilibrium and the FID decays into the noise. One of the more interesting aspects of NMR is the weakness of the coupling between nuclei and their environment. While electrons forget their magnetization after a few nanoseconds at most, nuclei exist in a calm wonderland of the framework of molecules awaiting their next resonant photon. If the proton at the center of a hydrogen atom was the sun, the nearest electron would be beyond the orbit of Pluto. The time required for a polarized nucleus to forget its orientation once tipped into the x–y plane axis is measured in times that range from seconds to months, extremely long times for subatomic processes.

The cylindrical symmetry of all modern solenoidal magnets, in which the static field is directed along the z-axis, has the result that only two, rather than three, parameters are needed to understand the return to thermal equilibrium of the sample magnetization. The rate of decay of the sample polarization can be, and is, different in the x–y plane from along the z-axis. The decay of x–y magnetization obeys a time constant called T_2, while the recovery of the thermal equilibrium magnetization along the z-axis obeys a time constant called T_1. Other names applied to T_1 include spin-lattice, or longitudinal relaxation, and T_2 is often referred to as spin-spin, or transverse relaxation.

Both of these relaxation time constants are a strong function of the dynamics of molecular motion; T_1 and T_2 are related directly to the rotational correlation time, τ_c. The simplest relationship between molecular parameters and the rotational correlation time is given by *Stokes' law*:

$$\tau_c = \frac{4\pi\eta r^3}{3kT}$$

which can be rewritten, by noticing that $4\pi r^3/3$ is the volume, V, of the molecule, so that

$$\tau_{\rm c} = \frac{V\eta}{kT}$$

where η is the viscosity of the solvent, and r is the hydrodynamic radius of the molecule. We may assume for proteins that the hydrodynamic radius of the molecule is approximated by

$$r = \left[\frac{3\bar{v}M}{4\pi N_{\rm o}}\right]^{1/3} + r_{\rm w}$$

Here, $r_{\rm w}$ is the radius of the water hydration shell, 0.16–0.32 nm, corresponding to either one or two water molecules in thickness, and $\bar{v}$ is the partial specific volume of the protein in units of cm^3/g. The partial specific volume for most amino acids does not differ significantly from 0.74 cm^3/g. For example, the protein ubiquitin ($M_{\rm r} = 8564$ Da) has a hydrodynamic radius of $r = 1.65$ nm, with a rotational correlation time given by Stokes' law of 3.8 ns at 300 K. This compares favorably with a measured value of $\tau_{\rm c} = 4.1$ ns.

7.15 Nuclear Spin Relaxation

Relaxation is the return to equilibrium of the nuclear magnetization. Although it makes perfect sense that the nuclear magnetization should decay back to its equilibrium value after a pulse has perturbed it, what is more interesting is that we can develop a quantitative formalism for predicting the nuclear relaxation times, T_1 and T_2, based on the rotational correlation time of a nucleus within a molecular framework. The causes for relaxation are molecular motions at the Larmor frequency which are sources of photons that lead to stimulated emission of radiation, as developed by Einstein (see Chap. 10).

For a nuclear spin transition involving an energy difference ΔE between levels, nuclear resonance is achieved through the absorption of a photon of frequency $\nu = \Delta E/h$ which raises the energy of the spin from the ground state. When this spin is in an excited state, the irradiation of the nucleus with photons of frequency $\nu = \Delta E/h$ will cause the excited nuclear spin to relax back to the ground state in a process known as stimulated emission of radiation. Where do the photons come from to cause relaxation? They come from the surroundings of the nucleus of interest.

We can understand the origin of these photons if we consider our nuclear spin to resemble an ice skater who is executing a spin. She will see the surrounding spectators moving around her at a frequency proportional to her spin frequency. If each of the spectators carried a magnet, then our spinning skater would observe an oscillating magnetic field arising from her surroundings at a frequency proportional to her spin frequency. For nuclear spins, we can call the spectator spins the "lattice" in reference to a crystal lattice in solids. In solutions, the concept of the "lattice" must be generalized to include the surrounding nuclei; the magnets held by the spectators are really other magnetic nuclei. For example, for protons in water, relaxation is due to the fluctuating magnetic fields generated by protons from neighboring water molecules. Nuclear relaxation is not due to the larger magnetic moments of surrounding electrons, except in magnetic materials, because essentially all electrons in objects are spin paired and have no net magnetic moment.

We can calculate the relaxation properties of a molecule by finding the amount of lattice radiation generated by the motion of the molecular framework containing the nuclei of interest. Relaxation is caused by this molecular motion which is characterized by the molecular rotational correlation time, τ_c. This motion gives rise to frequency components at the Larmor frequency. Molecular motion, while often described by a single average rotational correlation time, will, nevertheless, be distributed over a range of frequencies, or equivalently, times described by a correlation function. Let us denote the correlation function by $g(\tau)$, where

$$g(\tau) = \langle f(t) f(t + \tau) \rangle$$

and the brackets mean that we have performed an ensemble average (sum or integration) over all molecules. Note that the correlation function is very similar to a convolution; the difference is a reversal of the sign of the running variable, τ. We can motivate the form of the correlation function by considering its limits for small and large τ values. At $\tau = 0$ the correlation function must become one since all motions are frozen in time, $g(0) = 1$, while for long times the correlation function must become zero, $g(\infty) = 0$, as all memory of previous motions is forgotten. One function obeying these limits is an exponential such as

$$g(\tau) = e^{-\tau/\tau_c}$$

where the correlation time, τ_c, is the time it takes for the correlation function to decay to $1/e$ of its initial value. We can find the frequency components, $J(\omega)$, of molecular motion described by a correlation function by computing the Fourier transform of $g(\tau)$:

$$J(\omega) = \int_0^\infty g(\tau) e^{i\omega\tau} d\tau$$

$J(\omega)$ are known as the spectral density functions because they provide information about the range of lattice frequencies found in each spectral band. We have seen from prior computations that the Fourier transform of an exponential is a Lorentzian:

$$J(\omega) = \frac{2\tau_c}{1 + \omega^2 \tau_c^2}$$

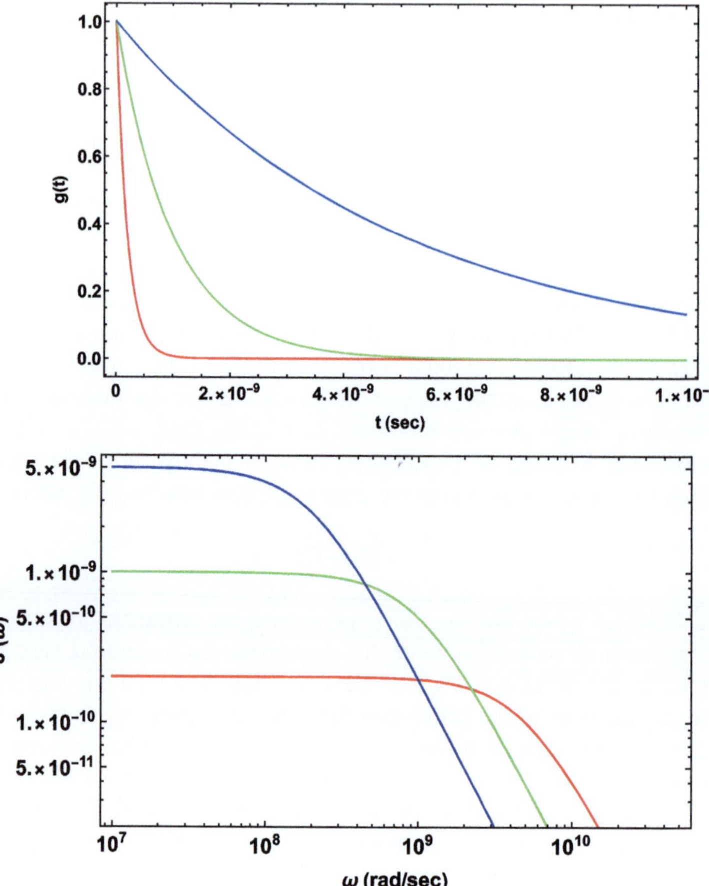

Fig. 7.18 (*Top*) Correlation functions, $g(\tau)$, for $\tau = 0.2$ ns (red), $\tau = 1.0$ ns (green), and $\tau = 5.0$ ns (blue). (*Bottom*) Spectral density functions, $J(\omega)$, for $\tau = 0.2$ ns (red), $\tau = 1.0$ ns (green), and $\tau = 5.0$ ns (blue). *Note that the J(ω) are plotted on a log-log scale*

In this case the width of the Lorentzian is $1/\tau_c$, and unlike the case for a spectral line where the width is on the order of Hz, here the width is on the order of 100 s of MHz. The correlation functions (Fig. 7.18, top) and spectral densities (Fig. 7.18, bottom) for $\tau_c = \{0.2, 2, 5 \text{ ns}\}$ support our expectation that for short correlation times, the correlation function decays quickly, and since the spectral density is the Fourier transform involving 1/time, the spectral density is spread out over a large region of frequency space. The opposite is true for longer correlation times and their spectral densities: The correlation function decays more slowly, and the spectral density is spread out over a smaller region of frequency space.

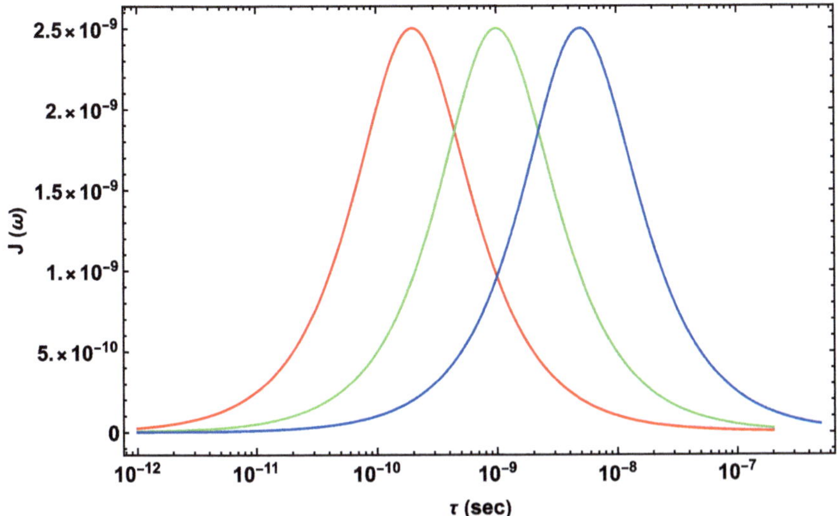

Fig. 7.19 Spectral density functions, $J(\omega)$, for (red) $\tau_c = 0.2$ ns ($\omega = 5 \times 10^9$ rad/s), (green) $\tau_c = 1$ ns ($\omega = 10^9$ rad/s), and (blue) $\tau_c = 5$ ns ($\omega = 5 \times 10^8$ rad/s)

The most effective motions inducing nuclear magnetic relaxation are those with frequency components at the Larmor frequency, and these occur for $\omega_L \tau_c \sim 1$, or $\omega_L \sim 1/\tau_c$. The nearest-neighbor proton dipole-dipole interaction is responsible for proton relaxation. The log-log plot of the spectral densities $J(\omega)$ (Fig. 7.18, bottom) does not resemble a Lorentzian. However, if we make a linear plot of $J(\omega)$, we find that it indeed has the expected shape and that the maximum power available from the lattice to cause relaxation is found at $\tau_c \sim 1/\omega_L$; in Fig. 7.19, the maximum in $J(\omega)$ occurs at $1/\omega = \tau_c$. The frequency bandwidth of the lattice radiation is directly proportional to the rotational correlation time. Figure 7.20 illustrates the increase in the width of the frequency distribution as molecular motion slows from a correlation time of 200 ps to 5 ns.

The direction of the magnetic field in space defines an axis, call it the z-axis, parallel to which the sample magnetic moment is oriented. For a homogeneous magnetic field, rotations about this axis are symmetry transformations that do not alter the energy eigenvalues of the Hamiltonian. As a result, there are only two magnetic relaxation time constants, T_1 for relaxation parallel to the z-axis of the magnetic field and T_2 for relaxation in the transverse x–y plane. From our forgoing exposition of spectral densities, we can now appreciate the difference between these two time constants. T_1 is the time it takes for the magnetically perturbed magnetization to reestablish its parallelism with the z-axis, and since this is stimulated emission, the interaction causing relaxation must have frequency components at the Larmor frequency. On the other hand, T_2 is the time for the magnetization to decay in the x–y plane, and this does not require lattice radiation at the Larmor frequency, but can be understood if we remember that for an NMR signal

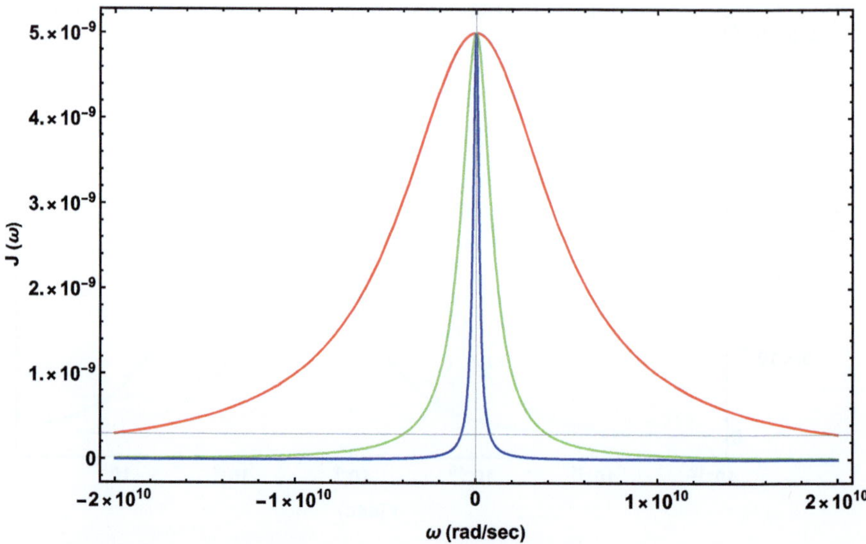

Fig. 7.20 Spectral density functions, $J(\omega)$, for (red) $\tau_c = 0.2$ ns ($\omega = 5 \times 10^9$ rad/s), (green) $\tau_c = 1$ ns ($\omega = 10^9$ rad/s), and (blue) $\tau_c = 5$ ns ($\omega = 5 \times 10^8$ rad/s)

on-resonance in the rotating frame (i.e., at the Larmor frequency), there is no precession in the x–y plane. However, from the NMR equation, $\omega = \gamma B$, we see that if there is a range of magnetic fields, ΔB, throughout the sample, there will also be a range of resonance frequencies, $\Delta\omega = \gamma\Delta B$, and the magnetization vectors will no longer be static in the x–y plane, but will acquire this range of frequencies. The result is that the magnetization vectors from all the nuclei will precess at different frequencies in the x–y plane and will therefore decay as the vector sum approaches zero as these vectors randomize. Since this process just depends on the inhomogeneity in the magnetic field seen by the sample nuclei, rather than on stimulated emission, transverse relaxation is sensitive to the values of the spectral density at zero frequency.

7.15.1   The Longitudinal Relaxation Time T_1

The spin-lattice, or R_1 relaxation rate, $R_1 = 1/T_1$, is proportional to the power of the spectral density function at the resonance frequency coming from the modulation of the lattice by the motion of the molecule, and we can write

$$R_1 = 1/T_1 = \frac{3}{10}K^2\left(J(\omega) + 4J(4\omega)\right)$$

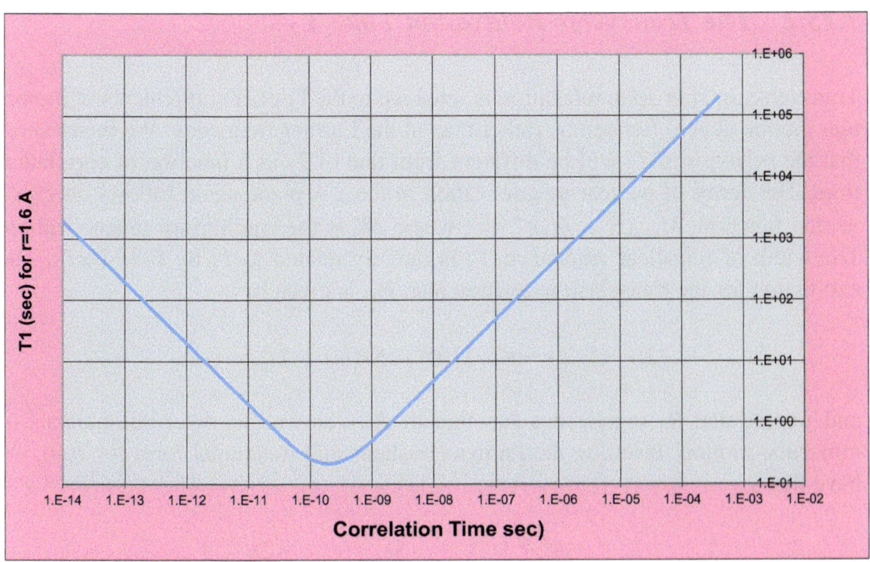

Fig. 7.21 Relationship between the NMR spin-lattice relaxation time, T_1, and the rotational correlation time for a field in which protons resonate at 500 MHz, for two protons separated by 160 pm. The minimum T_1 is at $\sim 1/\omega \sim 3 \times 10^{-10}$ s

where the constant $K = \frac{\mu h \gamma^2}{4\pi r^3}$ depends on the magnetic moment of the nucleus and the internuclear distance. Inserting the form for the spectral densities, $J(\omega)$, we find that

$$1/T_1 = \frac{3\hbar^2 \gamma^4}{10 r^6}\left[\frac{\tau}{1+\omega^2\tau^2} + \frac{4\tau}{1+4\omega^2\tau^2}\right]$$

where τ is the rotational correlation time. Note that the spin-lattice relaxation rate is proportional to $1/r^6$. The nuclear magnetization relaxes back to the z-axis according to the exponential function $M(t) = M_o(1 - 2e^{-t/T_1})$.

A plot of the dependence of T_1 on correlation time (Fig. 7.21) at 500 MHz for a proton relaxed by an adjacent proton 160 pm away shows that the relaxation time is a minimum for $\tau_c \sim 1/\omega = 3 \times 10^{-10}$ s. We can distinguish two regions for the data shown in Fig. 7.21; the first is the region at short correlation times, (i.e., $\tau_c < 10^{-10}$ s) where we observe that $T_1 \sim 1/\tau_c$. This is called the motional narrowing limit because rotational motion in solution leads to narrow NMR signals. The other limit is for long correlation times (i.e., $\tau_c > 10^{-9}$ s) where slow molecular motion leads to long T_1s because molecular motion here does not possess significant Fourier components at the Larmor frequency. This is the region of solid samples, where spin-lattice relaxation times can be as long as weeks.

7.15.2   *The Transverse Relaxation Time* T_2

Transverse, or spin-spin, relaxation is sensitive to the Fourier components of molecular motion at zero frequency, rather than at the Larmor frequency. We then expect that the behavior of T_2 will be different from that of T_1 as a function of correlation time. The decay of nuclear magnetization in the x–y plane again follows an exponential function: $M_{x-y}(t) = M_o e^{-t/T^2}$, where M_o is the equilibrium magnetization. The width of a nuclear resonance, Γ, is directly related to T_2 by $\Gamma = 1/\pi T_2$. The expression for the transverse relaxation rate, R_2, is given by

$$R_2 = 1/T_2 = K^2(3J(0) + 5J(\omega) + 2J(4\omega))$$

and we see that R_2 is indeed a function of $J(0)$, as well as other components of molecular motion. Inserting the known constants and functional form for $J(\omega)$, we have that

$$1/T_2 = \frac{3\hbar^2\gamma^4}{20r^6}\left[3\tau + \frac{5\tau}{1 + \omega^2\tau^2} + \frac{2\tau}{1 + 4\omega^2\tau^2}\right]$$

and again we notice that, just like R_1, R_2 is also proportional to the dipole-dipole interaction term $1/r^6$. T_2 is the time required for the nuclear spin vectors to lose phase coherence in the x–y plane. For a proton relaxed by another proton 0.16 nm away, at 500 MHz, the relationship between T_2 and the rotational correlation time is shown in Fig. 7.22, along with T_1 for the same internuclear distance and Larmor frequency.

The transverse relaxation rate, and hence the NMR linewidth, increases monotonically with correlation time. This feature of T_2 has a direct bearing on the application of NMR to the solution structure of biological macromolecules. The width of a protein resonance increases approximately 1 Hz per 1000 Da of mass, so that a 35,000 Da molecule would have 35 Hz wide lines from approximately 3000 protons. Therefore, resonance overlap becomes a severe issue for macromolecules of this size, and special NMR techniques must be used in order to extract useful structural information. It will also be appreciated that the NMR signals from solids are quite broad, and again other special NMR techniques must be used in order to extract useful structural information from solids. Such techniques have been developed for both large macromolecules (TROSY) and solids (cross polarization and magic angle spinning).

7.15.3   *The Nuclear Overhauser Effect*

A third aspect of nuclear magnetic relaxation concerns the alterations in the relaxation rate of a given nucleus when a neighbor is irradiated with radiofrequency radiation at the neighbor's resonance frequency. The results of such experiments have been known for some time and are shown in Fig. 7.23. Irradiation of the proton

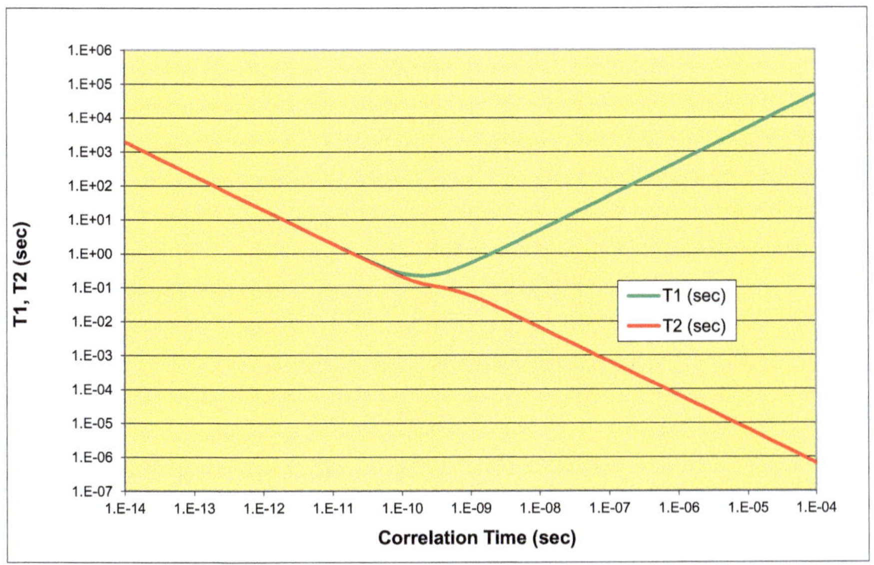

Fig. 7.22 The relationship between the spin-lattice (T_1) and spin-spin (T_2) relaxation times for a pair of protons with $r = 160$ pm at 500 MHz as a function of the rotational correlation time

resonance at 15 Hz causes not only the virtual elimination of this signal (Fig. 7.23, right) but also a marked increase in the intensity of the resonance at 5 Hz from its nearby neighbor. This is a manifestation of the nuclear Overhauser effect (nOe), a relaxation phenomenon that causes an increase in the intensity of the neighbor resonance during the irradiation time. The rate of buildup in the nOe is determined by the T_1. The amount of the nOe is a function of the rotational correlation time. The functional dependence for protons on τ is given by

$$\eta = \frac{5 + \omega^2\tau^2 - 4\omega^4\tau^4}{10 + 23\omega^2\tau^2 + 4\omega^4\tau^4}$$

and it should be noted that the numerator is σ, the cross relaxation rate, and the denominator is ρ, the dipolar relaxation rate; thus, $\eta = \sigma/\rho$. One can observe from the plot of the nOe as a function of the rotational correlation time shown in Fig. 7.24. that the nuclear Overhauser enhancement varies from a maximum of 0.5 for short correlation times, to a minimum of -1.0 for long correlation times.

We have seen that the relaxation rates for a given nucleus depends strongly on the rotational correlation time; T_1, T_2, and the nOe all are functions of motion. It is for this reason that there is great interest in measurements of NMR relaxation times in order to probe the details of molecular motion to give insights into molecular dynamics. We will discover in Chap. 8 how this relaxation information can be used to understand the dynamics and hence important structural details for biological macromolecules like glycogen.

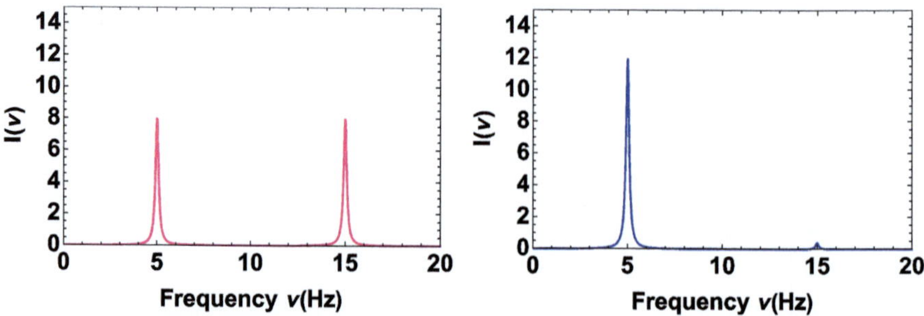

Fig. 7.23 *The nuclear Overhauser effect.* On the left (red) is the NMR spectrum of two nearby, but isolated, protons with resonance frequencies of 5 and 15 Hz. On the right (blue) is the NMR spectrum resulting from irradiation of the proton resonance at 15 Hz. Note the increase in intensity for the signal at 5 Hz and the virtual elimination of the signal at 15 Hz

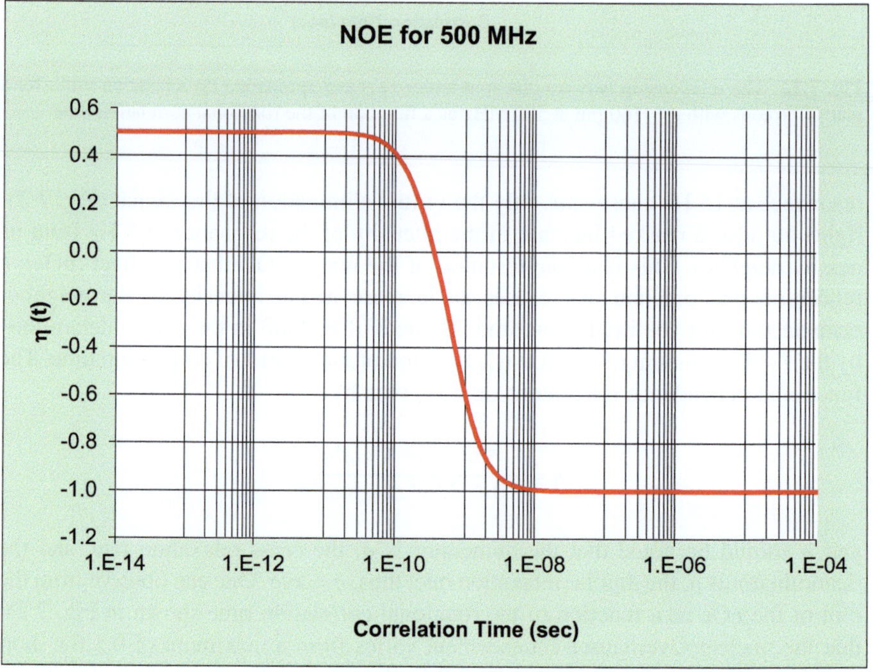

Fig. 7.24 The nuclear Overhauser enhancement for two protons separated by 160 pm at 500 MHz as a function of the rotational correlation time. Note that the nOe is zero for correlation times of ~4 ns

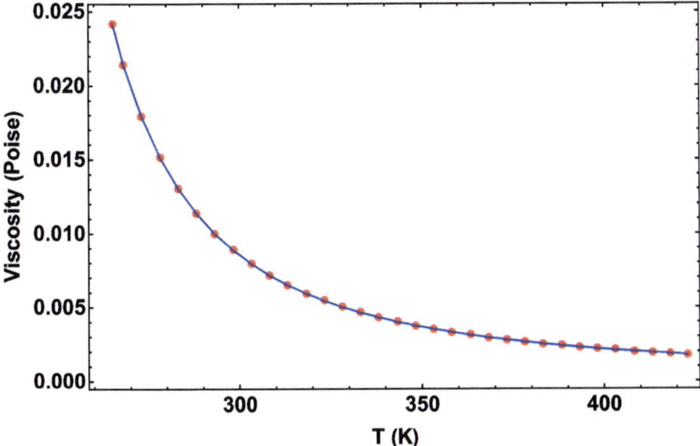

Fig. 7.25 The dependence of the viscosity (poise) of water on the absolute temperature. (Data from Kestin et al. (1978))

7.15.4   *Proton Relaxation in Hemoglobin*

The magnitudes of the relaxation times and nOes for biomolecules can be put into perspective by calculating these parameters for the hemoglobin tetramer. In order to approach such a calculation, we need to first find the rotational correlation time of hemoglobin. Let us assume that the protein is dissolved in water at a temperature of 310 K. For hemoglobin, $M = 68$ kDa (68 kg/mol). The viscosity of water is 0.1002 poise at $T = 293$ K (Fig. 7.25) but it drops to 6.5×10^{-3} poise at 310 K (Kestin et al. 1978). A polynomial fit to the data for the viscosity, $\eta(T)$, of water as a function of absolute temperature is given (for viscosity in poise) by

$$\eta(T) = 40.39 - 67.65 \times 10^{-2}\, T + 47.219 \times 10^{-4}\, T^2 - 17.56 \times 10^{-6}\, T^3$$
$$+ \ 3.670 \times 10^{-8}\, T^4 - 40.84 \times 10^{-12}\, T^5 + 1.889 \times 10^{-14}\, T^6$$

The SI units of viscosity are kg/(m s). The viscosity of water at 310 K is then $\eta = 6.95 \times 10^{-4}$ kg/(m s). The correlation time using Stokes' law is given by

$$\tau = \frac{V\eta}{kT} = \frac{4\pi r^3 \eta}{3kT}$$

where V is the molecular volume, $V = \bar{v}M/N_\mathrm{o}$, and $\bar{v}$ is the partial specific volume for an amino acid, $\bar{v} = 0.73$ cm^3/g, or 7.3×10^{-4} m^3/kg. Then the volume is

$$V = \frac{7.3 \times 10^{-4}\ \frac{\mathrm{m}^3}{\mathrm{kg}}\ 68\ \frac{\mathrm{kg}}{\mathrm{mol}}}{6.024 \times 10^{23}\ \mathrm{mol}^{-1}} = 8.24 \times 10^{-26}\ \mathrm{m}^3$$

and the correlation time becomes

$$\tau = \frac{8.24 \times 10^{-26} \text{ m}^3 \; 6.95 \times 10^{-4} \frac{\text{kg}}{\text{ms}}}{1.38 \times 10^{-23} \frac{\text{kg m}^2}{\text{s}^2 \text{ K}} \; 310 \text{ K}} = 13.4 \times 10^{-9} \text{ s}$$

If we add one layer of water molecules to the surface of hemoglobin, the radius increases by 0.275 nm, from 27.0 to 27.3 nm, the volume increases to $V' = 1.15 \times 10^{-25}$ m^3, and the correlation time increases to $\tau' = 18.7$ ns. This large effect due to the water of hydration occurs because the layer of water is added as far from the center of mass as possible so that it has a maximum contribution to the moment of inertia of the molecule.

Given the correlation time, we can then calculate the spin-lattice relaxation time, T_1, and the spin-spin relaxation time, T_2, for the protons in hemoglobin if the interproton distance is 0.175 nm at a field of 11.75 T. The T_1 and T_2 values are calculated from

$$\frac{1}{T_1} = \frac{3\gamma^4 \hbar^2}{10 r^6} \left[\frac{\tau}{1 + \omega^2 \tau^2} + \frac{4\tau}{1 + 4\omega^2 \tau^2} \right]$$

$$\frac{1}{T_2} = \frac{3\gamma^4 \hbar^2}{20 r^6} \left[3\tau \frac{5\tau}{1 + \omega^2 \tau^2} + \frac{2\tau}{1 + 4\omega^2 \tau^2} \right]$$

At 11.75 T, $\nu = 500$ MHz, or $\omega = 2\pi\nu = 6.28 \times 5 \times 10^8$ rad/s $= 3.14 \times 10^9$ rad/s. Using the two τ–values from above, we have that $\omega\tau = 39.3$–54.9 radians, which puts the hemoglobin molecule to the right of the minimum in the T_1 curve (Fig. 7.22). One would then expect that $T_1 \gg T_2$. The value of $\hbar^2\gamma^4 = 5.688 \times 10^{11} \; -h^2$ Å^6/s^2, so if you measure the distances in Å, then the prefactor in $1/T_1$ is given by $\frac{3\gamma^4 - h^2}{10 r^6} = 5.94 \times 10^9$/sec^2 for $r = 1.75$ Å. These values give $T_1 = 10.4$ s, $T_2 = 8.95 \times 10^{-3}$ s, and a linewidth of $\Gamma = 36$ Hz $(= 1/\pi T_2)$. We also see that $T_1 \gg T_2$; the ratio $T_1/T_2 = 1164$; we are indeed far from the motional narrowing region. The nuclear Overhauser enhancement is found from

$$\eta = \frac{5 + \omega^2 \tau^2 - 4\omega^4 \tau^4}{10 + 23\omega^2 \tau^2 + 4\omega^4 \tau^4}$$

However, we can see from an examination of Fig. 7.24 that for either one or two water molecules hydrating the hemoglobin and for correlation times in the range of 13–18 ns, the nOe is going to be close to -1.0. Insertion of $\omega\tau = 39.3$ and 54.9 radians gives values of the nOe of -0.996 to -0.998, and we indeed observe that the nOe is almost maximally negative.

Let us return to the nuclear Overhauser enhancement. We mentioned above that $\eta = \sigma/\rho$ where the numerator is the cross relaxation rate, and the denominator is the dipolar relaxation rate. The cross relaxation rate, σ, is proportional to the inverse sixth power of the distance between nuclei:

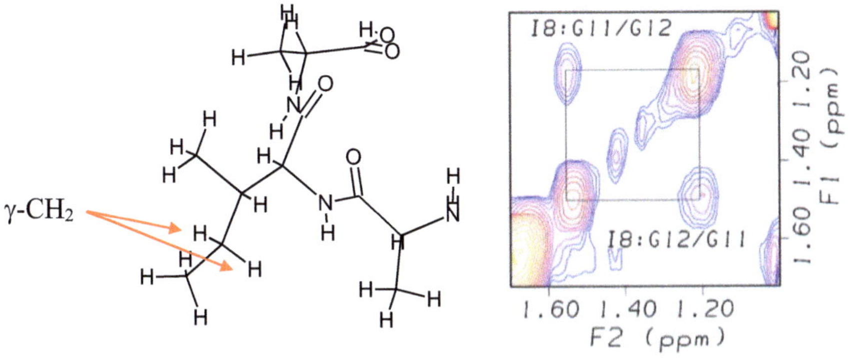

Fig. 7.26 (*Left*) Covalent structure for the tripeptide Ala-Ile-Ala with the pair of γ-CH$_2$ protons indicated on the isoleucine residue that resonate at $\delta = 1.48$, 1.19 ppm (Table A.3) whose distance of 0.175 nm serves as one of the calibrators for the distance dependence of the nOe in peptides and proteins. (*Right*) An expanded region of the 2D NOESY spectrum of a peptide containing Ile at position 8 showing the nOe cross peaks

$$\sigma = \frac{\chi}{r^6}$$

where

$$\chi = \left(\frac{\mu}{4\pi}\right)^2 \frac{\hbar^2 \gamma^2}{10} \left[\frac{6\tau}{1 + 4\omega^2 \tau^2} - \tau\right]$$

This provides a very sensitive way to measure internuclear distances in biological molecules, if this effect can be calibrated by measuring the nOe for a pair of protons with a known separation, for example, if a protein or peptide contains an isoleucine residue with nondegenerate resonances from the γ-CH$_2$ protons in isoleucine (as shown in Fig. 7.26) for the peptide Ala-Ile-Ala. One can measure the nOe between these two protons and determine χ in the above equation based on the fact that these protons are known to be 0.175 nm apart. The rotational correlation time will be determined by the overall rotation of the molecule and will be the same for all protons. All other measured nOes will then use the same χ and their internuclear distances can be determined.

Since the nOe obeys an inverse sixth power relationship with distance, the conversion of measured nOes to distances is rather insensitive to errors in the nOe measurements. However, the nOe results from enhanced relaxation of the observed nuclear spin and any other process which bleeds spin polarization from the observed spin will tend to introduce errors in the value of the nOe. One such mechanism for changing the actual value of the nOe away from its value due to nearby protons is spin diffusion. One can imagine that the nOe arises from "heating up" the nuclear spin population, much like putting a candle under an aluminum plate. The heat from

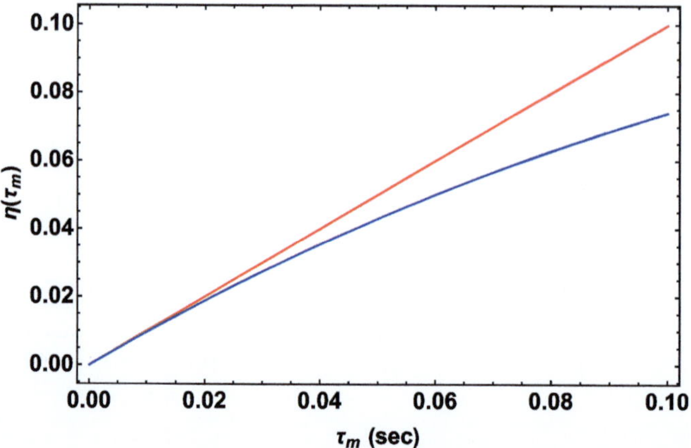

Fig. 7.27 The buildup of the nOe between two protons separated by 0.175 nm as a function of the mixing time between NMR pulses. The red curve is plotted under the assumption that the increase in the nOe is proportional to the mixing time. The blue curve shows the effect of spin diffusion which leaks magnetization and reduces the nOe

the candle flame will not stay at the same spot in the plate just above the candle flame, but will diffuse away from this spot carried by the lattice phonons into the rest of the metal. Such is also the case with the nOe in proteins; the other protons nearby will have resonance frequencies very similar to one which is irradiated by the RF pulse. These other nuclei will serve as sinks for the polarization added by irradiation of a given spin, and the polarization will tend to leak away from an observed site as a function of time. One way around this problem is to acquire the nOe data at various time delays and then to extrapolate the nOe to zero time delay, or to measure the initial slope of the nOe versus time curve. This is known as the initial rate approximation, and its use has been extended to measurements of very small nOes at larger distances (Hu and Krishnamurthy 2006). For short time delays, the nOe is proportional to the delay time, while for longer times, the nOe is reduced by spin diffusion, as shown in Fig. 7.27. For short delay times, all spins behave as if they were isolated; thus, one uses the initial slope of the nOe buildup curve to measure distances among protons.

7.16   Spin-Spin Coupling

For nearby, nonequivalent spins, the energy of one spin is influenced by the orientation of the nearby spins. An example from a two-spin system will serve to illustrate the situation. If we have two spin ½ nuclei close to each other, we have the possibility of four distinct states of total angular momentum, schematically shown by writing the wave function of the two, say protons, as a product of individual spin functions. Let us call spin up the state α and spin down, the state β. The four product states are then, in Dirac notation,

Table 7.1 Rules for NMR multiplets of protons

n	I	N	Intensity ratios
1	1/2	2	1:1
2	1/2	3	1:2:1
3	1/2	4	1:3:3:1
4	1/2	5	1:4:6:4:1
5	1/2	6	1:5:10:10:5:1

$$1. \quad \langle \alpha\alpha |$$
$$2. \quad \langle \alpha\beta |$$
$$3. \quad \langle \beta\alpha |$$
$$4. \quad \langle \beta\beta |$$

The dipole quantum selection rules require that $|\Delta S| = 1$. One can therefore only have transitions between states $3 \rightarrow 1$ and $4 \rightarrow 2$, and these are referred to as spin coupled states. The energy difference that gives rise to the splitting of the states into discrete, observable NMR signals is usually a small number of Hertz for protons, typically <10 Hz, but the splittings can be much larger for heteronuclei, like $^{13}\text{C}-^{1}\text{H}$, where splittings of 100–200 Hz are common.

The number of lines into which an NMR transition is split for a given number of neighboring protons is easily remembered using Pascal's triangle. The total number of distinct energy states is $N = 2nI + 1$, where n is the number of equivalent spins on a given site, and I is the spin of the equivalent member. For example, the resonance from a single proton ($n = 1$; $I = \frac{1}{2}$) will be split into two signals by an adjacent proton, three signals by a pair of adjacent protons, four signals by three adjacent protons, etc. These are given in Table 7.1.

The intensities of the lines in the resulting multiplets will follow the binomial coefficients, also given by Pascal's triangle. Therefore, the intensity ratio for a triplet will be 1:2:1, a quartet will be 1:3:3:1, etc. Besides its use for determining molecular structure for small organic molecules, spin coupling will be seen shortly to have a major role to play in multidimensional NMR spectroscopy.

As an example of spin coupling consider the lactate molecule:

$$\text{CH}_3-\text{CH}-\text{COO}^-$$
$$|$$
$$\text{OH}$$

The three equivalent methyl group protons will be split into a doublet with an intensity ratio of 1:1 by the adjacent methine proton, while the methine proton will be split into a quartet with an intensity ratio of 1:3:3:1 by the adjacent three methyl protons. The spin coupling is referred to in the NMR literature as J-coupling; this coupling is independent of the static magnetic field, and so it is reported in Hz. J-Coupling is a dipole-dipole interaction and is dependent on the bond angle between two atomic nuclei.

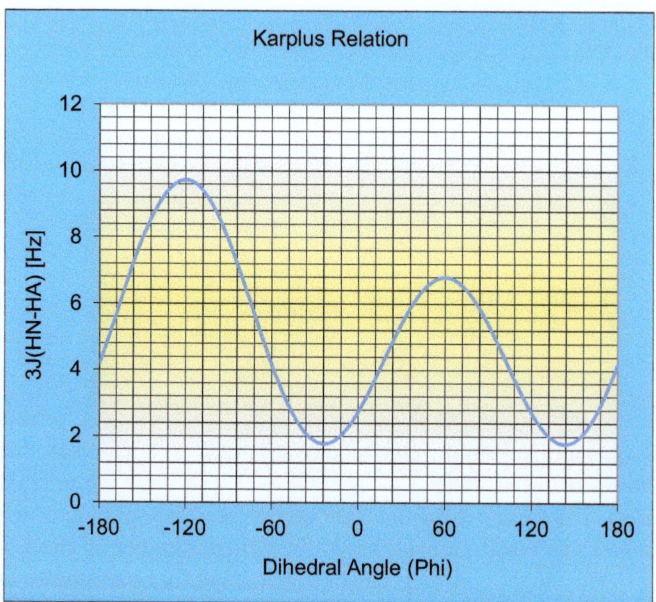

Fig. 7.28 The Karplus relationship between the dihedral angle of the peptide bond and the spin-spin coupling constant

Early work by Martin Karplus led to a quantitative understanding of the relationship between bond angles and J-coupling. For example, in proteins, the peptide bond involves four atoms: the α-proton, C_α, the amide nitrogen, and the amide proton:

H   H
\ Φ /
C_α—N
/   \

The dihedral angle, ϕ, between the pair of protons in the peptide bond determines the J-coupling constant according to the Karplus relationship, as shown in Fig. 7.28. It should be noted that the dihedral bond angle, φ, here is related to the protein backbone bond angle, θ, by $\theta = \varphi - 60°$. The Karplus relation is useful for determining peptide and protein backbone torsion angles, except for the fact that the relationship is multivalued; i.e., a single measured J-coupling constant, such as 6 Hz, can reflect more than one bond angle, and so it is of limited efficacy in protein structure determinations. Many investigators have calibrated the Karplus relation by comparing the measured bond angles from X-ray diffraction data on proteins, and the coupling data have been fitted to the equation:

$$^{3}J_{\text{NH-CH}\alpha} = 6.4 \cos^2\theta - 1.465 \cos\theta + 1.85$$

7.17   Multidimensional NMR

The study of protein structure in solution would not be possible if it were not for the development of multidimensional NMR techniques. The generation of 2-, 3-, 4-, and more-dimensional NMR spectra can be understood by reflecting on how one-dimensional NMR data are actually obtained. Figure 7.29 shows the NMR signal after an RF pulse, where the NMR signal is sampled at 100 discrete time points; the first point is at $t = 0$, the second is at $t = 0.1$ s, the third is at $t = 0.2$ s, ..., and the nth point is sampled at $t = (n - 1)\Delta t$, with the dwell time $\Delta t = 0.1$ s. Notice that the phase of this exponentially damped sine wave is a function of the time at which sampling is initiated, i.e., it is related to the length of the RF pulse. The sampling points are indicated in the figure as the dots. By the Nyquist theorem, in order to determine both the frequency and the phase of a periodically sampled function, the sample rate must be at least twice the highest frequency present in the data stream. For an NMR spectrum, the highest frequency is given by the spectral width dictated by the chemical shift range of the nucleus we are observing. For example, the spectral width would need to be about 6000 Hz in order to capture the range of signals (~12 ppm) for most protons at 500 MHz.

Since the FID is sampled at discrete time points, we can find these time points by using the Nyquist theorem:

$$dp = 2\,SW\,aq$$

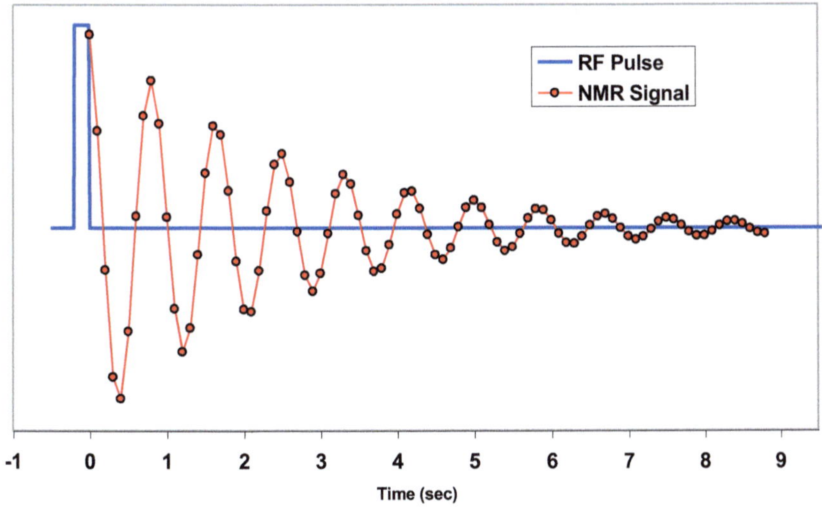

Fig. 7.29 One-dimensional, time domain sampling of an NMR signal (red) after an RF pulse (blue). The initial phase of the signal is determined by the length of the RF pulse

where dp is the number of data points (samples); SW is the spectral bandwidth, in Hz; and aq is the total time needed to sample the FID (the acquisition time). By sampling the FID into 4096 time points we need to sample for aq $= 4096/12{,}000 = 0.34$ s. The dwell time, or time per point, will then be given by aq/dp $= 0.34/4096 = 83.0$ μs per point.

The time domain sampling of the NMR signal suggests additional methods for modulating the measured spectra. The initial phase of the NMR signal is determined by a delay, that of the duration of the RF pulse. We could as easily build up the NMR signal by separating the RF pulse into two pulses with an interspersed incremented interpulse delay (Fig. 7.30). By Fourier transforming the data as a function of both the acquisition time *and* the interpulse delay, one can generate two frequency dimensions. The effect of the variable time delay between pulses is shown in Fig. 7.30. In the first figure, the starting phase of the FID is around $+10°$, while in the second, where the time between RF pulses has been increased, the initial phase of the FID is $-20°$. By acquiring, say, 256 FIDs with a delay between the first and second pulse incremented by 1.328 ms, we can build up an entire second dimension with a spectral bandwidth of 6000 Hz.

The two-dimensional NMR spectrum is generated by considering the entire time domain data set as a two-dimensional matrix; call this matrix $S(n, m)$ where m is the number of data points in the ordinary one-dimensional

FID (e.g., dp $= 4096$), and n is the number of data points in the second dimension (normally much less than m due to time constraints, e.g., 256).

$$S(n, m) = \begin{bmatrix} S_{11} & S_{12} & \cdot & \cdot & S_{1n} \\ S_{21} & S_{22} & \cdot & \cdot & S_{2m} \\ S_{31} & S_{32} & \cdot & \cdot & \cdot \\ \cdot & \cdot & \cdot & \cdot & \cdot \\ S_{n1} & S_{n2} & \cdot & \cdot & S_{nm} \end{bmatrix}$$

The one-dimensional NMR spectra are generated by Fourier transforming the rows of this matrix. Each row in this matrix corresponds to an ordinary one-dimensional NMR spectrum. Its row position is determined by the delay between the pulses and corresponds to a phase shift. When the Fourier transform is performed along the columns, this phase shift will appear as a characteristic frequency in the second dimension and give rise to an off-diagonal correlation peak in the two-dimensional NMR spectrum. One can easily imagine the possibility of extensions to 3, 4, 5, and more dimensions simply by including more RF pulses and more incremented delays. Clearly, if one samples the added dimensions less frequently than the first dimension, the spectral (digital) resolution of the added dimensions will not be as good as that for the first dimension, but in practice, this undersampling of the additional dimensions is not a serious problem because the discrimination of signals improves so dramatically with each added dimension that this compromise is worth pursuing.

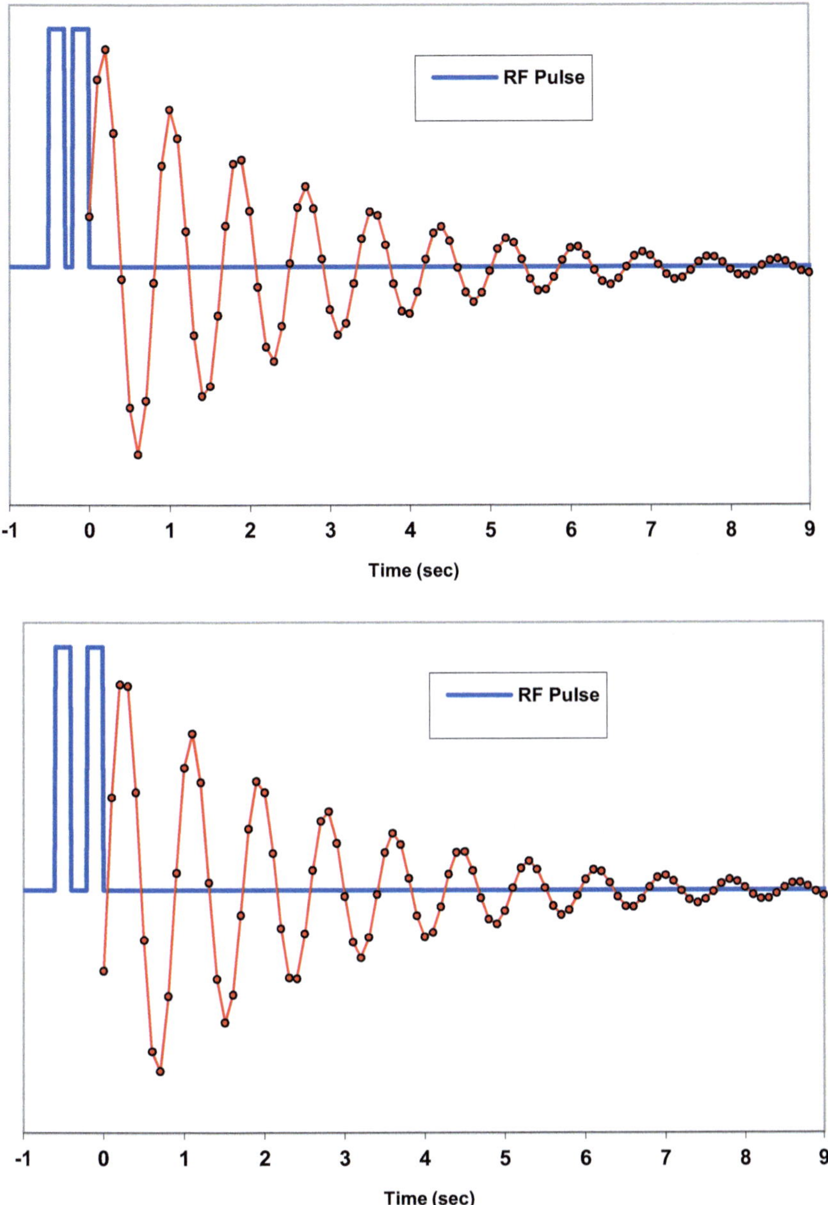

Fig. 7.30 (*Top*) The time domain NMR signal acquired after two RF pulses separated by a short delay. (*Bottom*) The time domain NMR signal acquired after two RF pulses separated by a longer delay. Note that the initial phase of the NMR signal differs between these two plots

Up to now, we have considered only a generic two-dimensional NMR experiment with two RF pulses separated by an incremented delay. What gives rise to the coupling of nuclear spins evident in the Fourier-transformed data? This coupling occurs either via the electrons through covalent bonds (spin-spin coupling) or through space (dipolar coupling). The first 2D NMR experiment was proposed by Jean Jeener at the AMPERE Summer School in Basko Polje, Yugoslavia, in September 1971. This experiment, known as COSY (for correlation spectroscopy), exploited the spin-spin coupling of nearest neighbor spins to produce off-diagonal cross peaks in 2D spectrum. Ernst and collaborators (Aue et al. 1976) then reduced this proposal to practice. Other types of 2D spectra were soon developed in the form of nuclear Overhauser effect spectroscopy (NOESY), J-resolved spectroscopy, exchange spectroscopy (EXSY), and total correlation spectroscopy (TOCSY). Each of these types of 2D spectroscopy involves radiofrequency pulses and delays that are mainly characterized by a common pattern: After initial RF pulse(s) come a pulse-free evolution period, a mixing period (with more RF pulses) and a detection period when the FID is sampled. The COSY pulse sequence is the simplest, consisting of two $\pi/2$ RF pulses separated by an incremented delay. When the time domain data are Fourier transformed, the off-diagonal signals result from pairs of nuclei that share spin-spin coupling. For example, the 1D proton spectrum of 100% ethanol CH_3CH_2OH shows signals at $\delta = 1.175$ ppm ($-CH_3$), $\delta = 2.25$ ppm ($-OH$), and $\delta = 3.645$ ppm ($-CH_2-$). The COSY spectrum has off-diagonal signals at $(\delta_1, \delta_2) = (3.645, 1.175)$, $(2.25, 3.645)$, and $(1.175, 3.465)$ but no cross peak at $(1.175, 2.25)$ because there is no spin-spin coupling between the distant methyl and the hydroxyl group. A TOCSY spectrum of ethanol diluted into water would not show the hydroxyl signal at $\delta = 2.25$ ppm at all due to its rapid chemical exchange with solvent protons and would only show cross peaks for the methyl and methylene groups (Fig. 7.31). TOCSY spectroscopy differs from COSY in that all of the spin correlations in a molecule result in cross peaks, not just the nearest neighbors. In peptides and proteins (Appendix, Table A.3) each amino acid residue contributes a characteristic set of signals in a TOCSY spectrum, but the cross peaks do not connect one residue with any other because the spin coupling is interrupted by the carbonyl of the amide bond. COSY and TOCSY spectroscopy are extremely useful for the assignment of signals to particular nuclei in molecules and are the workhorses for structural studies using NMR.

Two-dimensional NOESY spectroscopy shows cross peaks from nuclei that are dipolar coupled by virtue of their adjacency in space rather than from spin coupling. Measurements of the cross peak volumes as a function of the mixing time then can be used to obtain distance information as long as either the rotational correlation time of the molecule is known, or the nOe from a standard distance can be obtained. When the author was at Yale University, Kurt Wüthrich, who also had been a postdoc in Robert G. Shulman's laboratory, gave a seminar on the use of NMR to probe structural aspects of a small protein, bovine pancreatic trypsin inhibitor. At the end of the seminar I pointed out that since the nOe was a dipole-dipole interaction with a distance dependence of $1/r^3 \times 1/r^3$, one could measure the enhancements and, if one knew the molecule's rotational correlation time, get accurate internuclear

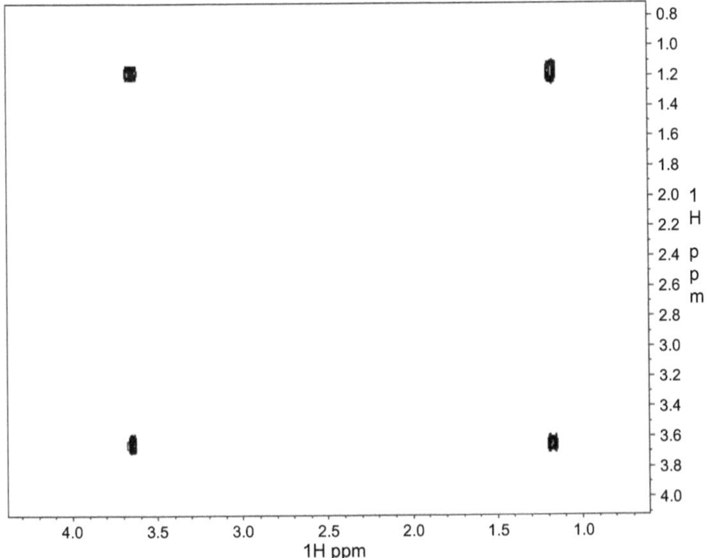

Fig. 7.31 The proton TOCSY spectrum of 100 mM ethanol in D_2O. (Courtesy of the Biomedical Magnetic Resonance Data Bank. https://bmrb.io/ftp/pub/bmrb/metabolomics/entry_directories/bmse000297/nmr/set01/spectra/HH_TOCSY.png, CC0 1.0 Universal (CC0 1.0) Public Domain Dedication)

distances that could constrain models and provide solution structures without the need for crystals. I will never forget his reply, in his distinctive Swiss accent, "Ja, that would be a good idea!" Within a year he had solved the problem by means of calibrated internuclear distances and distance-geometry calculations (Braun et al. 1981) and went on the win the Nobel Prize in chemistry in 2002.

7.18   NMR 3D Solution Structure of the Peptide *Cyclo-CPFVC*

As an example of the use of multidimensional NMR for the determination of the solution 3D structures of biomolecules, let us look at its application to a small cyclic peptide. Peptides containing almost any number of amino acid residues can now be synthesized automatically and purchased for performing many biological studies. One particularly important application of peptides is as inhibitors of protein-protein interaction (Sillerud and Larson 2005). The mutual binding of proteins is the basis of molecular recognition. Inhibitors of this process can, for example, prevent the binding of viral proteins from the capsid to receptors on cells. One such peptide is the cyclic-Cys-Pro-Phe-Val-Cys (cyclo-CPFVC) that inhibits the entry of Sin Nombre virus into mammalian cells (Hall et al. 2007) by binding to the cell surface

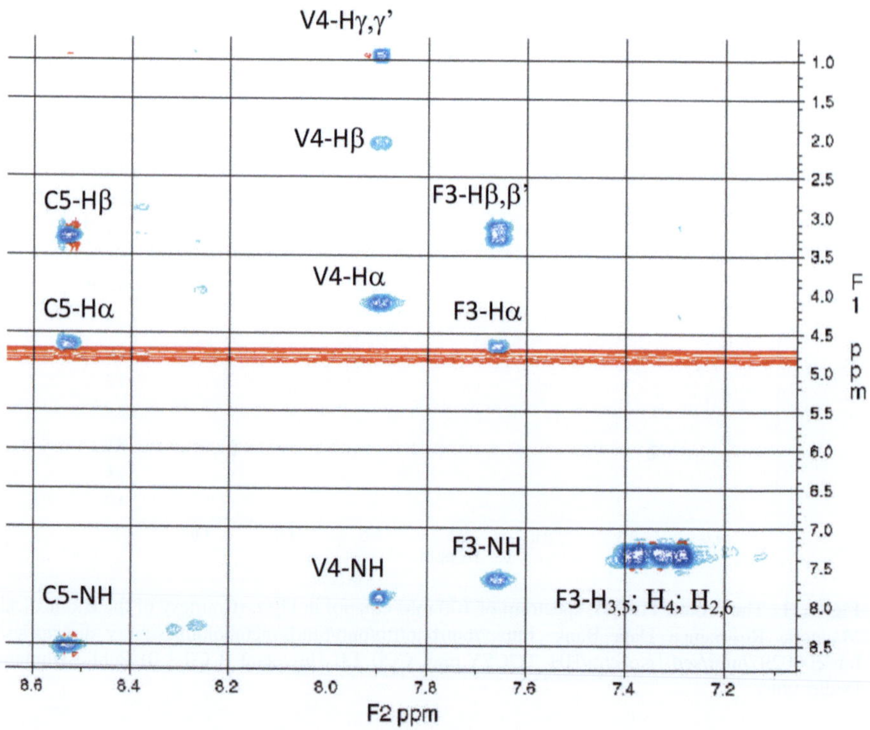

Fig. 7.32 The fingerprint and amide regions of the proton TOCSY spectrum of the peptide cyclo-CPFVC. Note the total spin coupling patterns from amide protons of F3, V4, and C5; C1 and P2 do not contain amide protons

integrin β_3. The 3D structure of (cyclo-CPFVC) in solution was deduced from proton NMR spectroscopy. One can assign most of the proton NMR spectrum by using the data in Table A.3 on the chemical shifts of amino acids in random coil peptides along with data derived from 2D TOCSY, NOESY, and ROESY sequences. We begin with the TOCSY spectrum (Fig. 7.32) where one observes resonance coupling patterns in the fingerprint region from only the three resides that contain amide protons, Phe-3 (7.69 ppm), Val-4 (7.92 ppm), and Cys-5 (8.57 ppm) since Cys-1 and Pro-2 do not have amide protons in this structure. This allowed us to assign the amide proton of Cys-5 (8.57 ppm) without ambiguity. A few of the signals fell significantly outside the expected range (indicated in red in Table 7.2).

One understands that the shifts in the cysteine resonances result from disulfide bond formation during cyclization, transforming the two cysteine residues to a cystine and altering the electron density at the protons next to the disulfide bond. The table of random coil chemical shifts (Table A.3) shows that proline has no resonance at 1.72 ppm, but the TOCSY spectrum of the aliphatic region (Fig. 7.33) indicated that a resonance at this chemical shift was coupled by spin-spin interaction to all of the pro-2 ring protons (as indicated by the red dotted lines) and therefore was

Table 7.2 Assignments for the proton NMR spectrum of cyclo-[CPFVC]

Residue	Proton	δ (ppm) experimental	δ (ppm) random coil	$\Delta\delta$ (ppm)	J (Hz)
CYS-1	Hα	4.48	4.69	**−0.21**	
	Hβ, β'	3.48, 3.19	3.28, 2.96	**+0.20, +0.25**	
PRO-2	Hα	4.31	4.44	−0.13	
	Hβ, β'	2.04, 1.72	2.28, 2.02	**−0.24, −0.30**	
	Hγ, γ'	2.10, 2.10	2.03, 2.03	0.07, 0.07	
	Hδ, δ'	3.63, 3.60	3.68, 3.65	−0.05, −0.05	
PHE-3	HN	7.69	8.23	**−0.54**	8.2
	Hα	4.71	4.66	0.02	
	Hβ, β'	3.29, 3.15	3.22, 2.99	0.07, 0.16	
	H2,6	7.3, 7.3	7.3, 7.3	0	
	H3,5	7.4, 7.4	7.39, 7.39	0.01	
	H4	7.36	7.34	0.02	
VAL-4	HN	7.92	8.44	**−0.52**	6.9
	Hα	4.12	4.18	−0.06	
	Hβ	2.06	2.13	−0.07	
	Hγ, γ'	0.95, 0.95	0.97, 0.94	0.02, 0.01	
CYS-5	HN	8.57	8.31	**0.26**	7.1
	Hα	4.66	4.69	−0.03	
	Hβ, β'	3.26	3.28, 2.96	−0.02, **0.25**	

assigned to Pro-2. The signals from the two β-protons from the single proline residue are shifted upfield by up to 0.3 ppm as a result of secondary structural effects. The nearby aromatic ring on Phe-3 is the likely source of these alterations in the local magnetic field of the Pro-2 β-protons. Aromatic rings contain conjugated electron systems whose circulation produces a current loop and a local magnetic field that augments or subtracts from the local field for distances up to about 35 pm. Observation of these ring currents (Fig. 7.33) places the Phe-3 ring close to the ring of Pro-2, close enough so that an nOe should be measureable between the protons on the two rings. There is free rotation about the two carbon-carbon bonds that attach the Phe ring to the peptide backbone. The existence of a ring current shift indicates that the Phe ring is stacked upon the Pro-2 ring, instead of being oriented at some distance.

We elected to use ROESY spectroscopy (nOe in the rotating frame) to obtain the distance information needed for a structural determination because the signal to noise ratio was better than in the NOESY data. The ROESY spectrum of this peptide (Fig. 7.34) confirmed the proximity of the Phe-3 ring and its orientation with respect to the Pro-2 ring by displaying enhancements between the Phe-3 ring protons H$_{2,6}$ and the stacked Hβ, β' protons of Pro-2. We observed the expected strong enhancements from Phe-3 H$_{2,6}$ protons to the Phe-3 Hβ, β' protons linking the ring to the peptide backbone. Also found were weaker enhancements linking the Phe-3 H$_{3,5}$ protons and the Pro-2 Hβ, β' protons because the Pro-2 nuclei are at a larger distance from the Phe-3 ring H$_{3,5}$ protons than from the Phe-3 ring protons H$_{2,6}$.

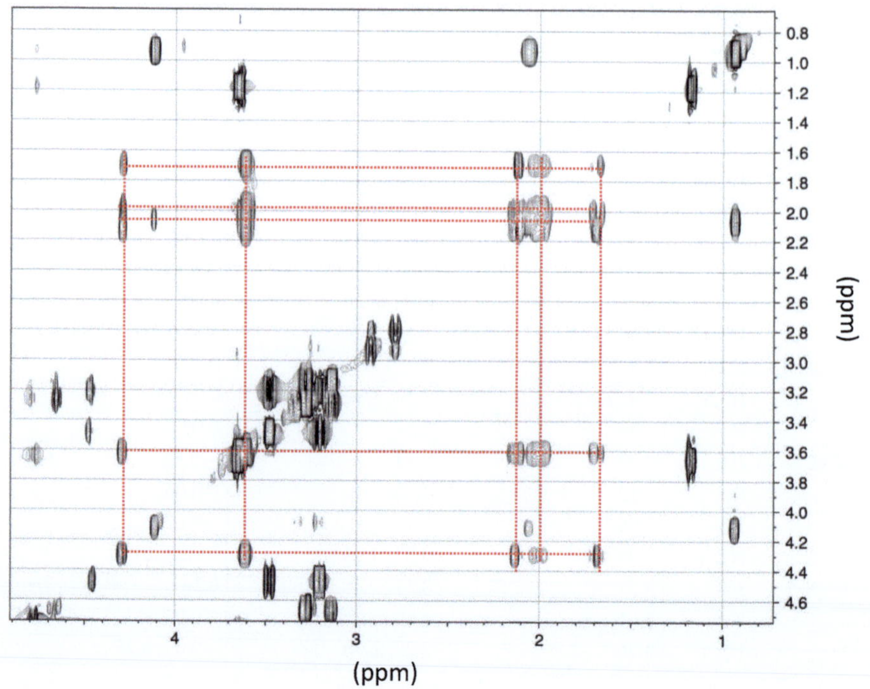

Fig. 7.33 The aliphatic region of the proton TOCSY spectrum of the peptide cyclo-CPFVC. The spin correlations connecting the Pro-2 Hβ resonance at 1.72 ppm to the remainder of the pro-2 spin system are shown in red. See Table 7.2 for resonance assignments and Hall et al. (2007)

The distances between protons were determined from measurements of the time course of the buildup of the Overhauser enhancement in the ROESY spectra taken at mixing times from 50 to 300 ms. This peptide did not contain isoleucine; therefore, the volumes of cross peaks from the β to the β′ protons in phenylalanine ($r = 174$ pm) at 3.29, 3.15 ppm (Table A.3 and Fig. 7.33), were used to calibrate the rOe-derived distances. Data from the ROESY spectra contained 67 pairs of nOes, or ~13 per residue. Additional data for the backbone amide bond angles were derived from measurements of the $^3J_{NH-H\alpha}$ coupling constants for the Phe-3, Val-4, and Cys-5 residues (Table 7.2). The Karplus relation (Fig. 7.28) indicated that the dihedral angles were all around approximately $-120°$. The distance measurements were used as restraints for simulated annealing which produced a consistent set of low-energy structures, the lowest of which (Fig. 7.35) clearly illustrated the relationship of the Phe ring over the Pro-2 Hβ, β′ protons. This structure was supported by the strong Overhauser enhancements between the Phe ring and the Pro ring hydrogens (Fig. 7.34).

The solution structure is the basis for understanding the interaction between this peptide and the integrin β₃, whose structure determined by X-ray crystallography is found in the protein data bank file 1U8C.pdb (Xiong et al. 2004). Docking onto β₃

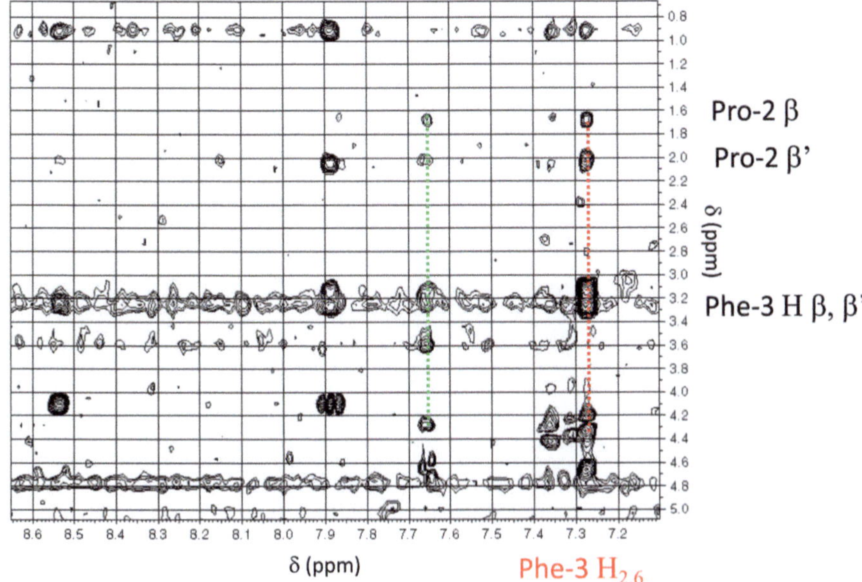

Fig. 7.34 The fingerprint region of the proton ROESY spectrum of the peptide cyclo-CPFVC (mixing time 300 ms). Nuclear Overhauser enhancements between the Phe-3 $H_{2,6}$ ring protons and the β, β′ protons of Pro-2 are indicated in red. Also indicated in green are weaker enhancements between Phe-3 and adjacent Pro-2. See Table 7.2 for resonance assignments and Hall et al. (2007)

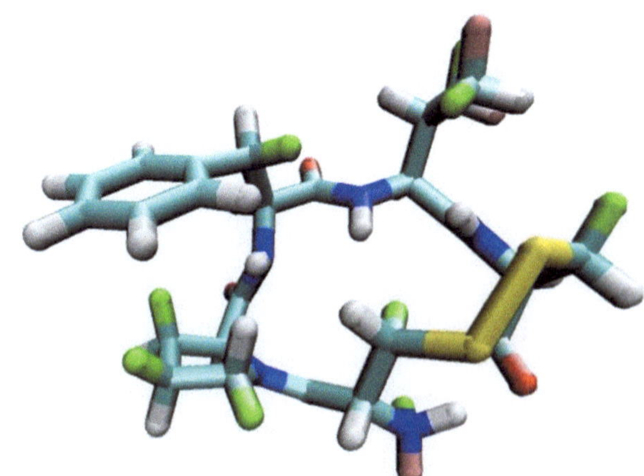

Fig. 7.35 The lowest-energy NMR-derived solution structure for the cyclo-CPFVC peptide showing the stacking of the Phe ring over the Pro residue leading to ring current shifts in the Pro-2 Hβ, β′ protons. The disulfide bond that cyclizes the molecule is shown in yellow

was centered on a niche composed of the residues W129 and C177-C184 (Fig. 7.36). No distortion of the peptide backbone of the lowest-energy structure was required and the side chains of the peptide retained their solution conformation. The interactions adding stabilization energy (increasing the probability) of this state included

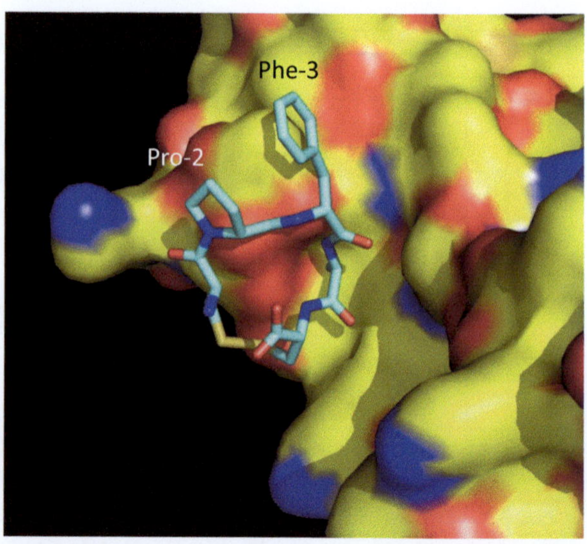

Fig. 7.36 Docking of the cyclo-CPFVC peptide onto the surface of the integrin β_3. The surface colors of the integrin reflect positive (red), negative (blue), and neutral (yellow) portions of the protein surface. Note the positioning of the peptide's Phe-3 ring over the neutral (hydrophobic) region

hydrogen bonding between the main chain carbonyl of both K125 and L128 from β_3 and the peptide nitrogen of Cys-1. In addition, Cys-1 had several van der Waals interactions with W129 as well as these same sites on β_3. Also, Val-4 engaged in van der Waals interactions with K125, along with T183, C184, and Val212 on β_3. The binding energy for the peptide on the surface can be decomposed into hydrophobic and hydrophilic parts. Hydrophobic because the Phe and Val residues have hydrophobic aromatic and aliphatic side chains and hydrophilic from the proline and two cysteine residues. We found that all of the carbonyl oxygens on the peptide's backbone faced opposite (Fig. 7.32) the stacked Phe and Pro rings, giving the molecule a Janus-like property in solution. We have now seen that nuclear magnetic resonance spectroscopy can reveal the structure of biomolecules in detail. It fulfills Faust's search for a method to discover *"Was die Welt im innersten zussamenhält."*

While hydrogens reside at the periphery of molecules and their resonance frequencies reflect intimate details of molecular interaction, another important isotope for biochemical spectroscopy is ^{13}C, which nominally comprises 1.108% of all the carbon in nature (Table A.2). Modern NMR spectrometers can detect this low amount of naturally abundant ^{13}C either by using concentrated samples or by using sophisticated multinuclear methods. ^{13}C is a spin ½ nuclide, as opposed to the biologically predominant ^{12}C, which has spin zero. The six protons and six neutrons in the ^{12}C nucleus each spin pair to produce a total spin of zero, while the addition of the extra neutron to form ^{13}C produces a net spin of ½ (Chap. 6). Just as its radioactive cousin, ^{14}C, has been used as a tracer for mapping biochemical metabolic pathways, ^{13}C NMR is extremely useful because carbon atoms form the backbone of most biomolecules. NMR spectroscopy has the added advantage that one does not usually need to separate the products from the reactants because the large chemical shift range for ^{13}C (200 ppm) means that each is separately observable in a mixture. By incorporating highly enriched ^{13}C atoms into metabolic precursors, the biochemical reactions characterizing the Krebs cycle, glycolysis, gluconeogenesis, pentose phosphate shunt, and many other more exotic bacterial pathways can be observed in real time within the organism (Chap. 8). In the early 1970s, Los Alamos National Laboratory began a project to use cryogenic fractional distillation of gases, like carbon monoxide, ammonia, and hydrogen sulfide, to separate the isotopes of oxygen, carbon, nitrogen, and sulfur, primarily for NMR use in biological studies.

Problems

1. *Two-dimensional peptide NMR spectroscopy.* For the open-chain peptide Arg-Glu-Ser, make a plot of the proton NMR spectrum in H_2O of the expected:

   (a) COSY spectrum
   (b) TOCSY spectrum
   (c) What is the mass of this peptide?
   (d) What would you expect its correlation time in water to be at $T = 280$ K?
   (e) What is its nOe at 11.75 T for an amide-proton/alpha-proton pair?

   Solutions

   (a) The COSY spectrum should have diagonal peaks and cross peaks as shown.

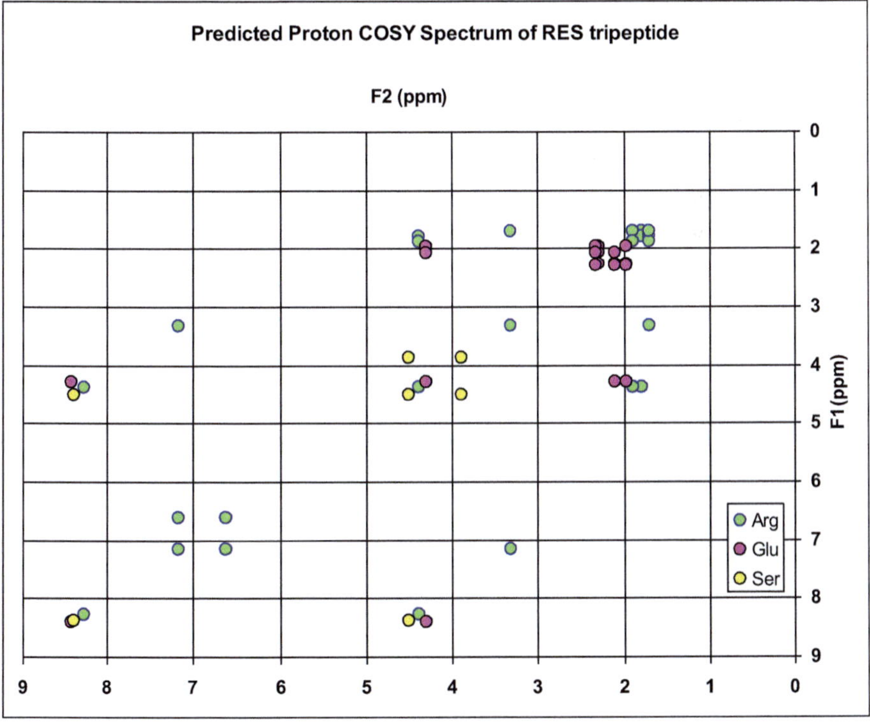

(b) The TOCSY spectrum is shown.

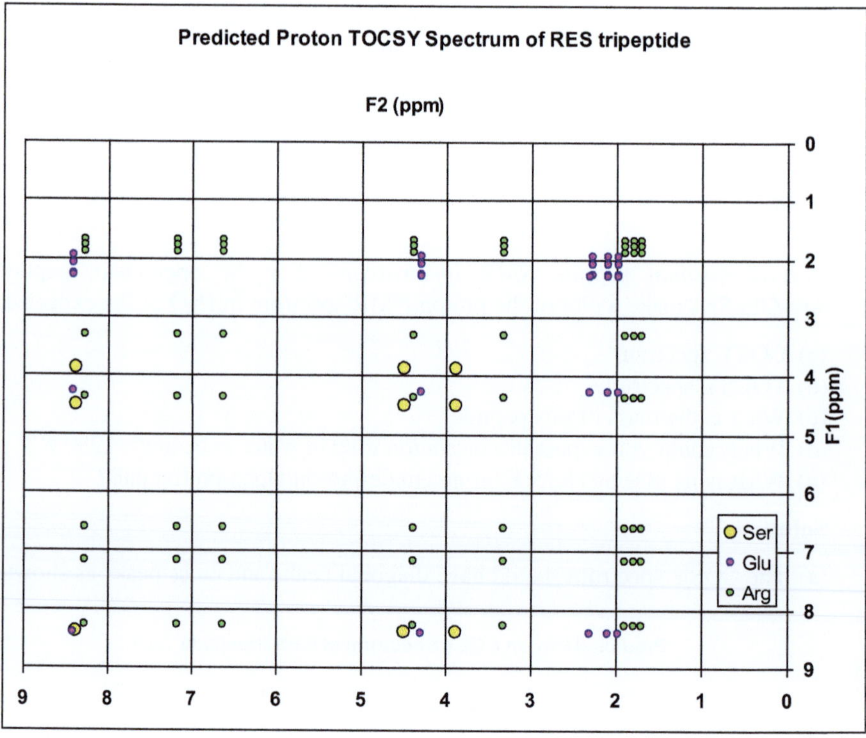

(c) The mass is 390 g/mol.

(d) $\tau_c = V\eta/kT = \nu M\eta/N_o kT =$

$$= \frac{7.4 \times 10^{-4}\,\mathrm{m^3/kg^*}\; 3.91 \times 10^{-1}\,\mathrm{kg/mol^*}\; 1.4 \times 10^{-3}\,\mathrm{m^2\,kg/s}}{6.022 \times 10^{23}\,\mathrm{/mol^*}\; 2.8 \times 10^2\,\mathrm{K^*}\; 1.38 \times 10^{-23}\,\mathrm{J/K}}$$

which gives $\tau_c = 1.74 \times 10^{-10}$ s, or about 0.2 ns.

(e) To find the nOe we first need to calculate $\omega = 2\pi\nu = 6.28.^*$ 500 MHz $= 3.14 \times 10^9$ rad/s. Then, let $x = \omega\tau_c = 0.546$. The nOe is given by

$$\mathrm{nOe} = \frac{5 + x^2 - x^4}{10 + 23x^2 + 4x^4} = 0.30.$$

2. *Two-dimensional peptide NMR spectroscopy.* For the cyclic nonapeptide CLLRMRSIC, where the underlined residues are disulfide linked:

(a) Draw the complete covalent structure of the molecule.

(b) Estimate the COSY and TOCSY spectra (i.e., draw a plot of what you expect).

(c) By assign, I mean associate a peak at a particular chemical shift with a proton in the peptide. This is most conveniently expressed both as a table of proton identities and chemical shifts and as a plot of the spectra with the peaks labeled with their identity.

   Assign the following actual proton NMR spectra:

1. DQ-COSY (double quantum filtered COSY)
2. TOCSY
3. NOESY (Note: the higher magnification figure at the end contains the NOESY data needed for the amide sequential assignments)

(d) What is the mass of this peptide?

(e) What would you expect its correlation time in water to be at $T = 280$ K?

(f) What is its nOe at 11.75 T for an amide-proton/alpha-proton pair?

Hint: The disulfide-linked molecule will have free amino and carboxy termini. Exchange with solvent H_2O renders certain protons invisible. You also need to find the amide/alpha proton distance in a peptide.

Solution

(2b COSY) The COSY spectrum should have diagonal peaks and cross peaks as shown.

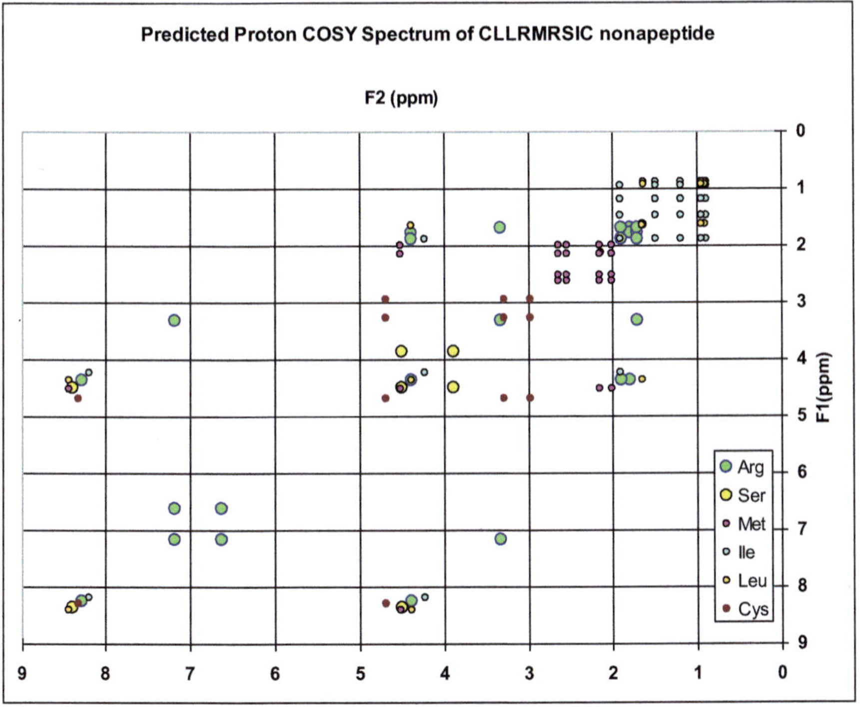

(2b TOCSY) The predicted TOCSY spectrum is shown.

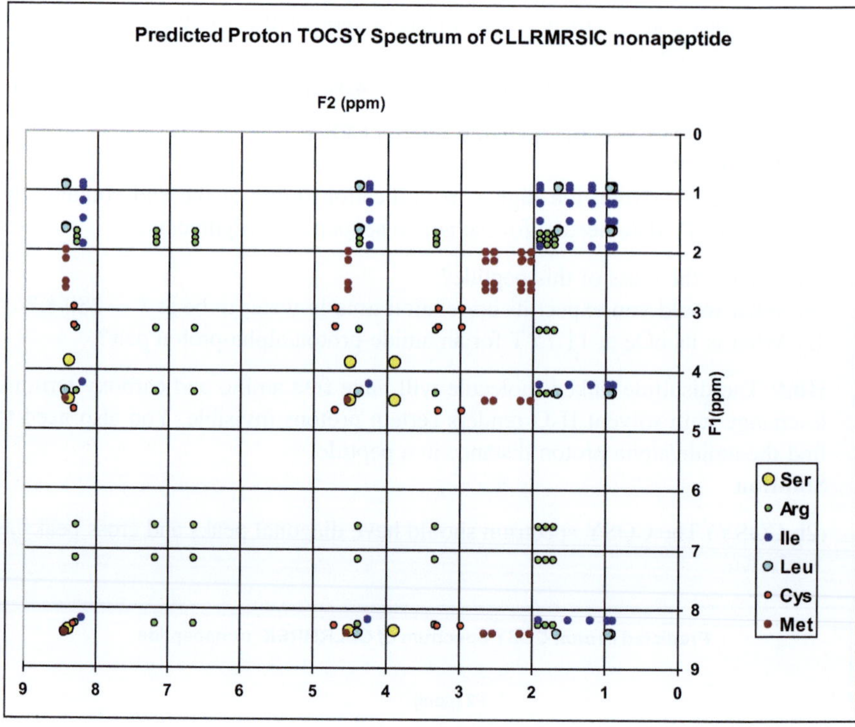

(2c) The assignments for the peptide CLLRMRSIC are given in the paper: Sillerud et al. (2003).

(2d) The mass is 1094 g/mol.

(2e) $\tau_c = V\eta/kT = vM\eta/N_o kT =$

$$= \frac{7.4 \times 10^{-4}\,\mathrm{m^3/kg^*}\ 1.094\,\mathrm{kg/mol^*}\ 1.4 \times 10^{-3}\,\mathrm{m^2\,kg/s}}{6.022 \times 10^{23}/\mathrm{mol^*}\ 2.8 \times 10^2\,\mathrm{K^*}\ 1.38 \times 10^{-23}\,\mathrm{J/K}}$$

which gives $\tau_c = 4.87 \times 10^{-10}$ s, or about 0.5 ns.

(2f) To find the nOe we first need to calculate $\omega = 2\pi\nu = 6.28.^*$ 500 MHz $= 3.14 \times 10^9$ rad/s. Then, let $x = \omega\tau_c = 1.53$. The nOe is given by

$$\mathrm{nOe} = \frac{5 + x^2 - x^4}{10 + 23x^2 + 4x^4} = 0.02.$$

This value of $x = 1.53$ is close to 1.0 so we expect that the nOe will be very small since we are close to the point where the nOe reverses sign at $x = 1.12$.

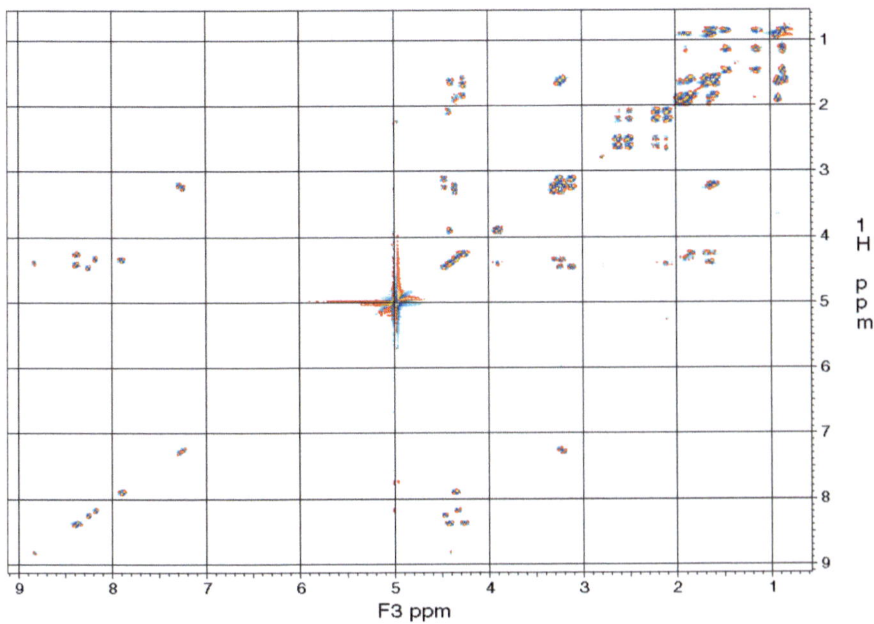

NMR spectrum for problem 2.c.1 Double-quantum filtered COSY of the cyclic nonapeptide CLLRMRSIC

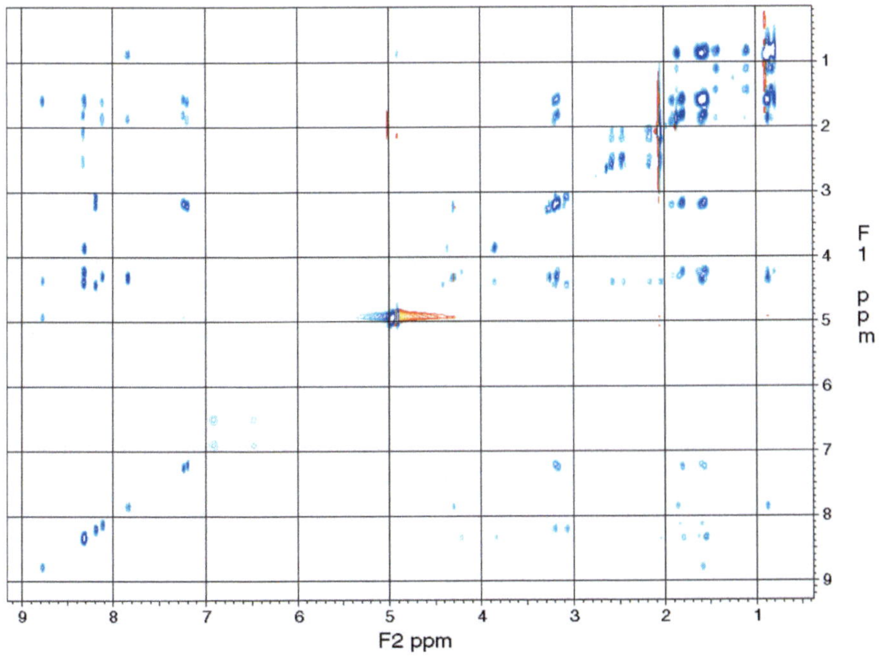

NMR spectrum for problem 2.c.2 Proton TOCSY NMR spectrum of the cyclic nonapeptide CLLRMRSIC

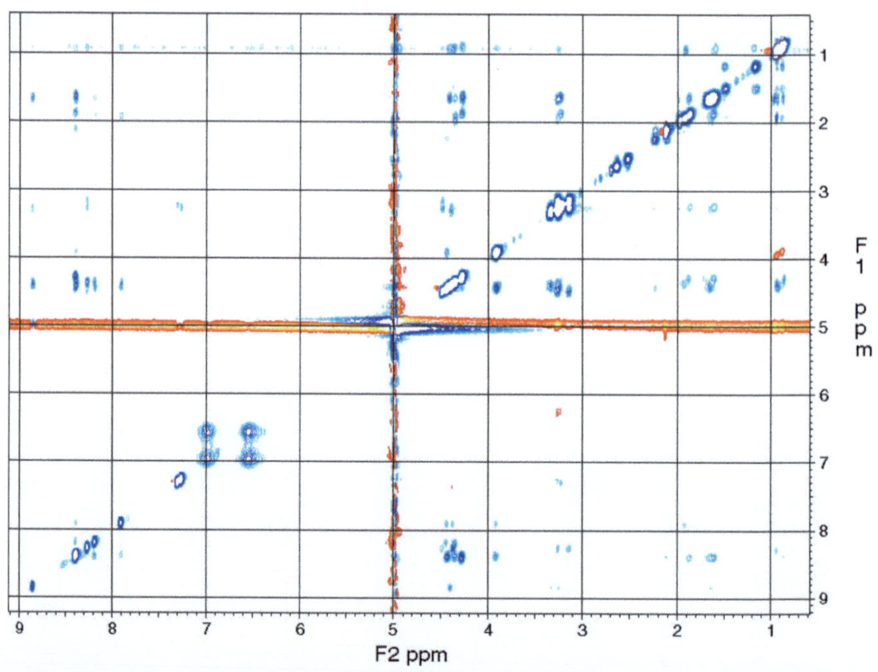

NMR spectrum for problem 2.c.3 Proton NOESY NMR spectrum of cyclic <u>C</u>LLRMRS<u>I</u>C in H_2O

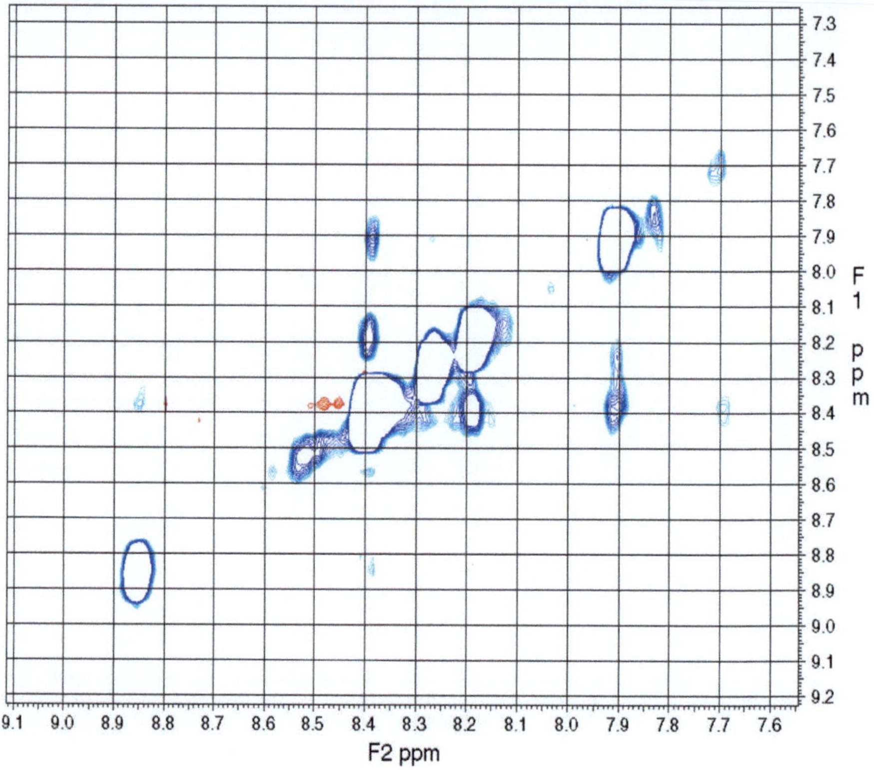

NMR spectrum for problem 2.c.3 Expansion of the amide regions of the proton NOESY NMR spectrum of cyclic <u>C</u>LLRMRS<u>I</u>C in H_2O

3 For 1.0 mL of water in a magnetic field of 10 T at a temperature of 300 K, calculate the magnetic moment of the protons using Curie's law.

3. *Nuclear magnetic moments.* Consider a system of nuclei with spin $I = 1$. Its projection of the spin along the z-axis, m, will then have *three* possible values, since $2I + 1 = 3$, corresponding to $m \in \{+1, 0, -1\}$. The magnetic moment, μ, will then attain one of three values based on

$$\mu = \gamma \hbar m = \{\gamma \hbar, 0, -\gamma \hbar\}$$

which we will call $\{\mu_o, 0, -\mu_o\}$. Suppose this system of nuclei is oriented along the z-axis; there will be a probability, p, that $\mu = \mu_o$ and a probability, p, that $\mu = -\mu_o$; the probability that $\mu = 0$ is $q = 1 - 2p$.

(a) Find the average magnetic moment, $<\mu>$ for a single nucleus.
(b) Find the average value of the square of the magnetic moment, $<\mu^2>$, for a single nucleus.
(c) If we have N nuclei which do not interact with one another, find:

1. The average value of the total magnetic moment, $<M>$, of the sample
2. Its standard deviation, $Sqrt[<(\Delta M)^2>]$
   in terms of N, p, and μ_o, for orientation along the z-axis. **Note:** Only elementary sums are needed to solve this problem.

4. *Nuclear polarization.* In an NMR experiment, the nuclear polarization is defined to be $p = \frac{u}{u+d}$, where u and d are the numbers of u and down spins, respectively. If the energy difference between the up and down states is $\Delta E = 2\mu B_0$, to what temperature must we cool a sample of water so that the polarization $p = 0.5$ in a magnetic field of 10 T?

Hint: For the proton, $\mu_p = 1.41 \times 10^{-26}$ J/T.

$$\mathbf{p} = \frac{u}{u+d} = 0.5 = e^{-\Delta E/kT}; \quad \ln(0.5) = -\Delta E/kT = 2\mu B_0/kT;$$

then solve for T.

$$T = \frac{-\Delta E}{k \ln(0.5)} = \frac{2 \times 1.41 \times 10^{-26} \, \text{Joule/Tesla} \times 10 \, \text{Tesla}}{0.693 \times 1.38 \times 10^{-23} \, \text{Joule/K}} = 29.49$$
$$\times 10^{-3} \, \text{K}, \quad \text{or } 29.49 \, \text{mK}.$$

The point of this problem is to illustrate how low the temperature must be to obtain a significant nuclear polarization given that the tiny energy difference between the spin states has to compete with kT.

5. *Curie's law.* For 1.0 mL of water in a magnetic field of 10 T at a temperature of 300 K, calculate the magnetic moment of the protons using Curie's law as discussed in class.

Curie's law gives the magnetization M_o as a function of B_o and temperature as

$$M_o = \frac{N\gamma^2 \hbar^2 I(I+1) B_o}{3kT}$$

where we are given that $B_o = 10$ T, and $T = 300$ K. For the proton, $\gamma = 2.67522 \times 10^8$ rad/s.T, and $I = 1/2$. The number of protons N is given by $2 \times (1 \text{ g}/18 \text{ g/mol}) \times N_o = 6.69 \times 10^{22}$ protons. [Planck's constant divided by $2\pi] = 1.054 \times 10^{-34}$ J/rad. Then, M_o is computed as follows:

$$M_o = \frac{6.69 \times 10^{22*} \left(2.67522 \times 10^8\right)^{2*} \left(1.054 \times 10^{-34}\right)^{2*} 0.75^* 10 \, \text{rad}^2 \, \text{J}^2 \, \text{T}}{3^* 1.38 \times 10^{-23} \, \text{J/K}^* 300 \, \text{K} \, \text{T}^2 \, \text{s}^2 \, \text{rad}^2}$$

$$M_o = 3.223 \times 10^{-8} \, \text{J/T} * \text{s}^2$$

For the proton, $\gamma = 2.67522 \times 10^8$ rad/s.T, and there are 2 protons per mole of water.

6. Consider a system of nuclei with spin $I = 1$. Its projection of the spin along the z-axis, m, will then have *three* possible values, since $2I + 1 = 3$, corresponding to $m \in \{+1, 0, -1\}$. The magnetic moment, μ, will then attain one of three values based on

$$\mu = \gamma \hbar m = \{\gamma \hbar, 0, -\gamma \hbar\}$$

which we will call $\{\mu_o, 0, -\mu_o\}$. Suppose this system of nuclei is oriented along the z-axis; there will be a probability, p, that $\mu = \mu_o$ and a probability, p, that $\mu = -\mu_o$; the probability that $\mu = 0$ is $q = 1 - 2p$.

(a) Find the average magnetic moment, $<\mu>$ for a single nucleus.

    The average magnetic moment, $<\mu>$, for a single nucleus is defined to be

$$<\mu> = \frac{\sum_{i=1}^{3} \mu_i P(\mu_i)}{\sum_{i=1}^{3} P(\mu_i)}$$

but $\sum_{i=1}^{3} P(\mu_i) = p + p + q = 2p + 1 - 2p = 1$. That is, the probability is properly normalized, so we can ignore the denominator here and just focus on the numerator.

$$<\mu> = \sum_{i=1}^{3} \mu_i P(\mu_i) = \mu_o p + 0q - \mu_o p = 0$$

(b) Find the average value of the square of the magnetic moment, $<\mu^2>$, for a single nucleus.

$$\langle \mu^2 \rangle = \sum_{i=1}^{3} \mu_i^2 P(\mu_i) = \mu_o^2 p + 0q + \mu_o^2 p = 2\mu_o^2 p$$

(c) If we have N nuclei which do not interact with one another, find:

1. The average value of the total magnetic moment, $\langle M \rangle$, of the sample

$$\langle M \rangle = N \langle m \rangle = 0$$

Its standard deviation, $\sigma_M = \mathrm{Sqrt}[\langle (\Delta M)^2 \rangle]$:

$$\langle (\Delta M)^2 \rangle = \sum_{j=1}^{N} \langle (\mu_j - \langle \mu \rangle)^2 \rangle = \sum_{j=1}^{N} \langle (\mu_j)^2 \rangle$$

since $\langle \mu \rangle = 0$. We have from part (b) that $\langle \mu^2 \rangle = 2\mu_o^2 p$, and there are N of these terms, so

$$\langle (\Delta M)^2 \rangle = N 2 \mu_o^2 p$$

or we have our answer that

$$\sigma_M = \mathrm{Sqrt}\left[\langle (\Delta M)^2 \rangle \right] = \mu_o \, \mathrm{Sqrt}[2Np]$$

in terms of N, p, and μ_o, for orientation along the z-axis. **Note:** Only elementary sums were needed to solve this problem.

7. *The time domain NMR signal.* The time domain NMR signal (FID) is given generally for a single resonance as

$$S(t) = Ae^{-(\Gamma + i(\omega_c - \omega_o))t}$$

where A is a constant, Γ is the decay time for the FID, ω_o is the resonance frequency, and ω_c is the radiofrequency carrier.

(a) What does the signal look like if we are on resonance ($\omega_c = \omega_o$)?
   When we are on-resonance, $\omega_c - \omega_o = 0$, and the FID is a single exponential given by $S(t) = Ae^{-\Gamma t}$.
(b) Fourier transform this signal and show that the real part is a Lorentzian, of the form $a/(a + x^2)$, and the imaginary part is a dispersion signal of the form $x/(a + x^2)$.

Hint: You must remember that the NMR signal is complex, so you must rationalize the denominator of your answer.

(b) The Fourier transform of $S(t)$ is defined to be

$$s(\omega) = A \int_0^\infty e^{-[\Gamma + i(\omega_c - \omega_o)]t} e^{-i\omega t} dt$$

Note that this integral must only go from zero to infinity since the signal is not defined for $t < 0$, and in fact the integrand diverges for $t \to -\infty$. We can do this integral with a simple substitution where we let $x = [\Gamma + i(\omega_c - \omega_o - \omega)]t$, and then $dx = [\Gamma + i(\omega_c - \omega_o - \omega)]dt$, or $dt = dx/[\Gamma + i(\omega_c - \omega_o - \omega)]$. Our integral then becomes

$$s(\omega) = \frac{A}{[\Gamma + i(\omega_c - \omega_o - \omega)]} \int_0^\infty e^{-x} dx = \frac{A}{[\Gamma + i(\omega_c - \omega_o - \omega)]}$$

Now, this is not in rational form, so we need to multiply the top and bottom by $\frac{\Gamma - i(\omega_c - \omega_o - \omega)}{\Gamma - i(\omega_c - \omega_o - \omega)}$, which is the complex conjugate of the denominator, to produce

$$s(\omega) = \frac{A(\Gamma + i(\omega_c - \omega_o - \omega))}{\left[\Gamma^2 + (\omega_c - \omega_o - \omega)^2\right]}$$

$$= A \left[\frac{\Gamma}{\left[\Gamma^2 + (\omega_c - \omega_o - \omega)^2\right]} - \frac{i(\omega_c - \omega_o - \omega)}{\left[\Gamma^2 + (\omega_c - \omega_o - \omega)^2\right]} \right]$$

and this is now in the form we sought. The real part is the Lorentzian, pure absorption signal of the form $a/(a^2 + x^2)$, while the imaginary part is the dispersion signal of the form $x/(a^2 + x^2)$.

(c) For $\Gamma = 1$ Hz, $\omega_o = 15$ Hz, $\omega_c = 20$ Hz, and $0 \le t \le 10$ s, plot the real part of the FID as a function of time.

Hint: Remember that $e^{ix} = \cos x + i \sin x$.

The real part of the FID is $\mathrm{Re}[S(t)] = Ae^{-\Gamma t} \cos[(\omega_c - \omega_o)t]$, so if we insert the known values, we get $\mathrm{Re}[S(t)] = Ae^{-t} \cos[(20 - 15)t] = Ae^{-t} \cos(5t)$, and this is shown below:

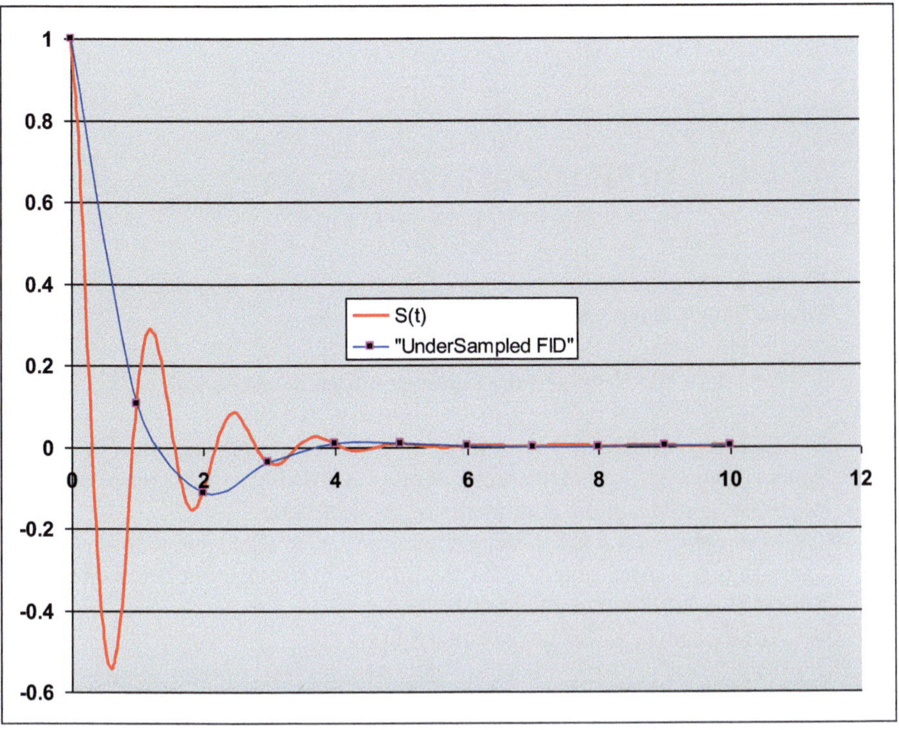

If you only use 10 data points to sample your FID, you will miss most of the higher-frequency wiggles. This data is shown in blue, whereas, if you sample with more data points, you find the FID shown in red. This is an example of the Nyquist theorem.

8. *The rotational correlation time of hemoglobin.* What is the rotational correlation time of hemoglobin in water at a temperature of 310 K?

   For hemoglobin, $M = 68$ kDa. The viscosity of water is 0.1 poise at $T = 293$ K, but it drops to 6.5×10^{-3} poise at 310 K. The cgs units of viscosity are sort of goofy, but in SI units, viscosity has units of kg/(m s). Then the viscosity of water at 310 K is $\eta = 6.5 \times 10^{-4}$ kg/(m s). The correlation time using Stokes' law is given by

$$\tau = \frac{4\pi r^3 \eta}{3kT} = \frac{V\eta}{kT}$$

where V is the molecular volume. $V = vM/N_o$ where v is the partial specific volume for an amino acid, $v = 0.73$ cm^3/g, or 7.3×10^{-4} m^3/kg. Then the volume is

$$V = \frac{7.3 \times 10^{-4}\,\mathrm{m^3/kg^*} 68\,\mathrm{kg/mol}}{6.022 \times 10^{23}\,\mathrm{mol^{-1}}} = 8.24 \times 10^{-26}\,\mathrm{m^3}$$

and the correlation time becomes

$$\tau = \frac{8.24 \times 10^{-26}\,\mathrm{m^3 *} 6.5 \times 10^{-4}\,\mathrm{kg/(ms)}}{1.38 \times 10^{-23}\,\mathrm{kg\,m^2/s^2\,K^*} 310\,\mathrm{K}} = \mathit{12.5\,ns}$$

If we add one layer of water molecules, then the radius increases by 0.36 nm, from 2.7 to 30.6 nm, and the volume increases to

$$V' = 1.15 \times 10^{-25}\,\mathrm{m^3} \text{ and the correlation time increases to } \tau' = \mathit{17.5\,ns}$$

This large effect due to the water of hydration occurs because this layer of water is added as far from the center of mass as possible so that it has a maximum contribution to the moment of inertia of the molecule.

9. *Nuclear relaxation times.* Calculate the spin-lattice relaxation time, T_1, and the spin-spin relaxation time, T_2, for the protons in hemoglobin if the interproton distance is 0.175 nm at a field of 11.75 T.

   The T_1 and T_2 values are calculated from

$$\frac{1}{T_1} = \frac{3\gamma^4\hbar^2}{10r^6}\left[\frac{\tau}{1+\omega^2\tau^2} + \frac{4\tau}{1+4\omega^2\tau^2}\right]$$

$$\frac{1}{T_2} = \frac{3\gamma^4\hbar^2}{20r^6}\left[3\tau + \frac{5\tau}{1+\omega^2\tau^2} + \frac{2\tau}{1+4\omega^2\tau^2}\right]$$

   At 11.75 T, $\nu = 500$ MHz, or $\omega = 2\pi\nu = 6.28 * 5 \times 10^8$ rad/s $= 3.14 \times 10^9$ rad/s. Using the two τ-values from problem 4, we have that $\omega\tau = 39.3$–54.9 radians, which puts the hemoglobin molecule to the right of the dip in the T_1 and T_2 curve. One would then expect that $s_1 \gg T_2$.

   The text gives you the value of $\hbar^2\gamma^4 = 5.688 \times 10^{11}$ Å^6/s^2, so if you measure the distances in Å, then the prefactor is $\frac{3\gamma^4\hbar^2}{10r^6} = 5.94 \times 10^9$/sec^2. Then you just need to plug in τ and ω. You get $\omega = 2\pi\nu = 3.14 \times 10^9$ rad/s.

   These values give $T_1 = 10.4$ s, $T_2 = 8.95 \times 10^{-3}$ s, and the linewidth $= 36$ Hz.

   We also see that $T_1 \gg T_2$; the ratio $T_1/T_2 = 1164$ and we are indeed way out of the motional narrowing region.

10. *Nuclear relaxation of a protein.* Ubiquitin is a protein involved in proteasome function which contains 74 amino acid residues. Its rotational correlation time for overall molecular tumbling in water at 310 K is 4.1 ns. The T_1 measured at 500 MHz for the resonance from the degenerate ring protons in positions 3 of the single tyrosine residue at position 59 is 1.0 s.

(a) Assume that Tyr-59 is buried within the core of ubiquitin and hence rotates with the same correlation time as the whole protein. How close would a single proton have to be to the proton in position 3 to account for the observed T_1?

Here we have to invert the expression for T_1 to find r.

$$\frac{1}{T_1} = \frac{3\gamma^4\hbar^2}{10r^6}\left[\frac{\tau}{1+\omega^2\tau^2} + \frac{4\tau}{1+4\omega^2\tau^2}\right] = \frac{3}{10}\frac{5.688\times10^{11}}{r^6}\left[\frac{\tau}{1+\omega^2\tau^2} + \frac{4\tau}{1+4\omega^2\tau^2}\right]$$

$\omega = 3.14 \times 109$, and $\tau = 4.1$ ns, so $\omega\tau = 12.9$. The term in square brackets is 4.93×10^{-11}, so we have that

$$\frac{1}{T_1} = \frac{3}{10}\frac{5.688\times10^{11}}{r^6}\times 4.93\times 10^{-11}$$

or, $\frac{1}{T_1} = 1 = \frac{3}{10}\frac{5.688}{r^6}\times 4.93 = \frac{8.41}{r^6}$, and we have that r is the sixth root of 8.41, or

$$r = 1.43\ \text{Å}.$$

(b) What would be the linewidth for proton 3 of Tyr-59 in ubiquitin? The linewidth, $\Gamma = 1/(\pi T_2)$, so we just need to find T_2 from the expression:

$$\frac{1}{T_2} = \frac{3\gamma^4\hbar^2}{20r^6}\left[3\tau + \frac{5\tau}{1+\omega^2\tau^2} + \frac{2\tau}{1+4\omega^2\tau^2}\right] = \frac{3}{20}\frac{5.688\times10^{11}}{8.41}\left[3\tau + \frac{5\tau}{1+\omega^2\tau^2} + \frac{2\tau}{1+4\omega^2\tau^2}\right]$$

where we can use the value of r found above. The term in square brackets is 1.24×10^{-8}, so

$$1/T_2 = \left[3\times 5.688\times 1.24\times 10^3\right]/\left[20\times 8.41\right] = 126$$

and we then have that the linewidth is given by $\Gamma = 1/(\pi T_2) = 126/\pi = 40.1$ Hz.

(c) What would be the nOe for proton 3 of Tyr-59 in ubiquitin?

$$\eta = \left[\frac{5+\omega^2\tau^2 - 4\omega^4\tau^4}{10+23\omega^2\tau^2+4\omega^4\tau^4}\right]$$

Now, $\omega\tau = 12.9$, so $\eta = \left[\frac{5+(12.9)^2-4(12.9)^4}{10+23(12.9)^2+4(12.9)^4}\right] = -0.965.$

(d) Can the observed T_1 for the proton in position 3 of Tyr-59 in ubiquitin be accounted for by relaxation induced by the proton in position 2 of the Tyr-59 ring in ubiquitin, or are there other closer protons in the protein that also induce relaxation?

The question here is how far away is proton 2 from proton 3? A C–H bond is 1.09 Å, while a C–C bond in a benzene ring is 1.40 Å. The proton at position 2 on the ring is then 2.49 Å away and is too far away to account for the measured relaxation time. There must be a closer proton(s) than the one at position 2. Note that the relaxation of a given proton in a complex molecule is proportional to the sum of the inverse sixth power of all the N protons in the neighborhood:

$$\frac{1}{T_1} = \frac{3\gamma^4\hbar^2}{10}\left[\frac{\tau}{1+\omega^2\tau^2}+\frac{4\tau}{1+4\omega^2\tau^2}\right]\sum_{i=1}^{N}\frac{1}{r_i^6}$$

and not just due to a single nearest neighbor. This is why the T_1 is so short even though the proton at position 2 is too far away to be solely responsible for the relaxation of the proton at position 3.

Note: The value of $\hbar^2\gamma^4 = 5.688 \times 10^{11}$ Å^6/s^2, so if you measure the distances in Å, then, for example, the prefactor is $\frac{3\gamma^4\hbar^2}{10r^6} = 5.94 \times 10^9/\text{sec}^2$ for $r = 1.75$ Å.

11. *Filtration of NMR signals*. The convolution theorem is the basis for filtration of the NMR signal.

   (a) Use the Mathematica code in the text to plot the NMR spectrum resulting from an unfiltered, *noisy* FID whose decay time is $g = 0.25$ s; i.e., $s(t) = e^{-0.25t}$, which has 10% noise (see Fig. 7.13). Here the linewidth $\Gamma = 1/(\pi\,g)$. Measure the signal to noise ratio and linewidth for this spectrum. Does the measured linewidth agree with $\Gamma = 1/(\pi\,g)$ from above?

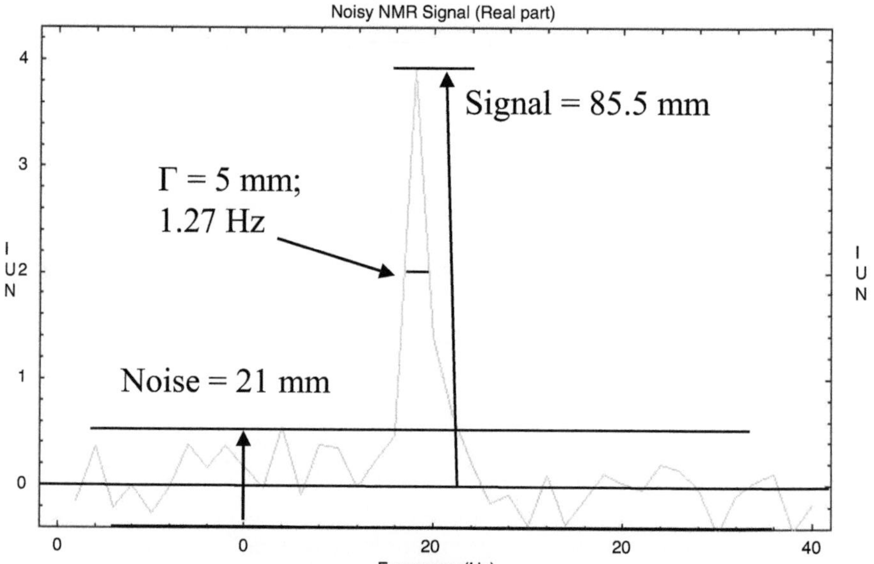

   (b) Then use various filter decay constants, f, ranging from $f = 0.1$ to $f = 1.0$, to determine the optimal exponential filter function decay time by looking for the maximum in the signal to noise ratio as a function of the decay constant. Plot S/N and the width, Γ, of the filtered resonance versus f.

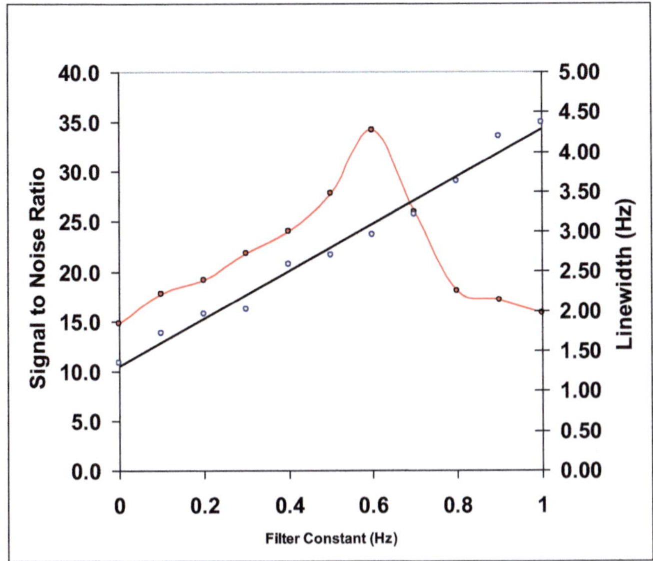

From the plot we see that S/N increases until $f \sim 0.5$ and then falls. The width of the filtered resonance increases linearly with filtration, as it should because filtration adds the filter bandwidth to the width of the unfiltered resonance.

(c) What conclusion can you draw with respect to the relationship between the unfiltered linewidth, Γ, and the filter decay constant, $\Gamma = 1/(\pi f)$, giving the maximum signal to noise ratio?

We can answer this by replotting the signal to noise ratio versus the bandwidth, $1/(\pi f)$, of the filter:

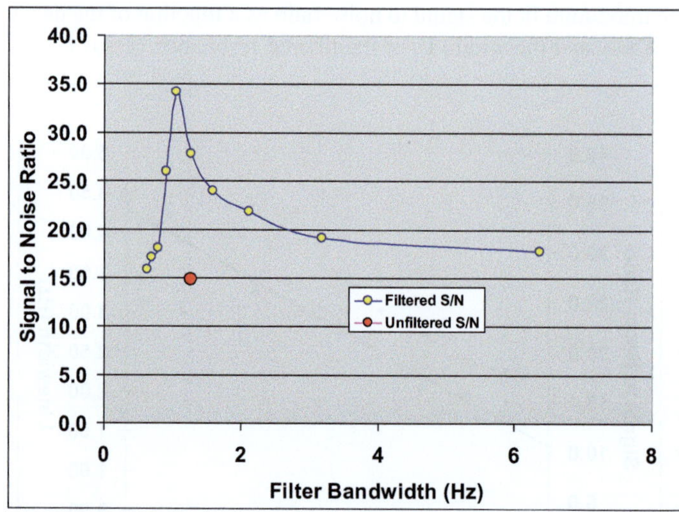

Here we see that filtration increases the signal to noise ratio by more than twofold and that the maximum in S/N occurs when the filter bandwidth equals the width of the unfiltered resonance. This condition is known as a *matched filter*.

(d) Suggest a filter function to increase resolution (i.e., decrease linewidth, Γ). Write a Mathematica notebook to show how this works. What happens to S/N when you use this filter?

One simple way to improve resolution is to multiply the FID by a function which emphasizes the data points later in the FID. If a decaying exponential increases S/N, but broadens the resonances, then an increasing exponential will do the opposite. So one can try changing the sign on the filter coefficient. The effect on the FID is seen to increase the noise later in the acquisition window. The resonance width is now 1.4 instead of 2 Hz, but the S/N has decreased to 7.9.

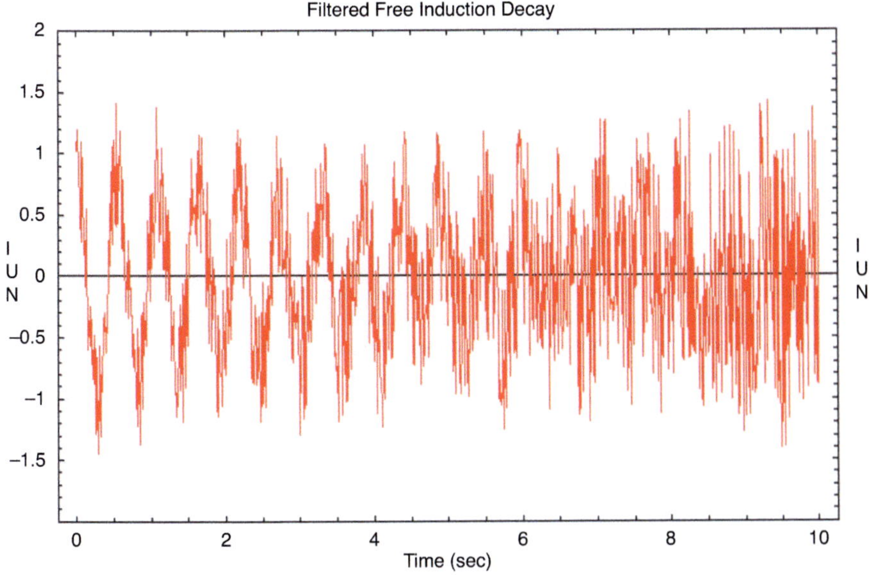

Filtered Free Induction Decay

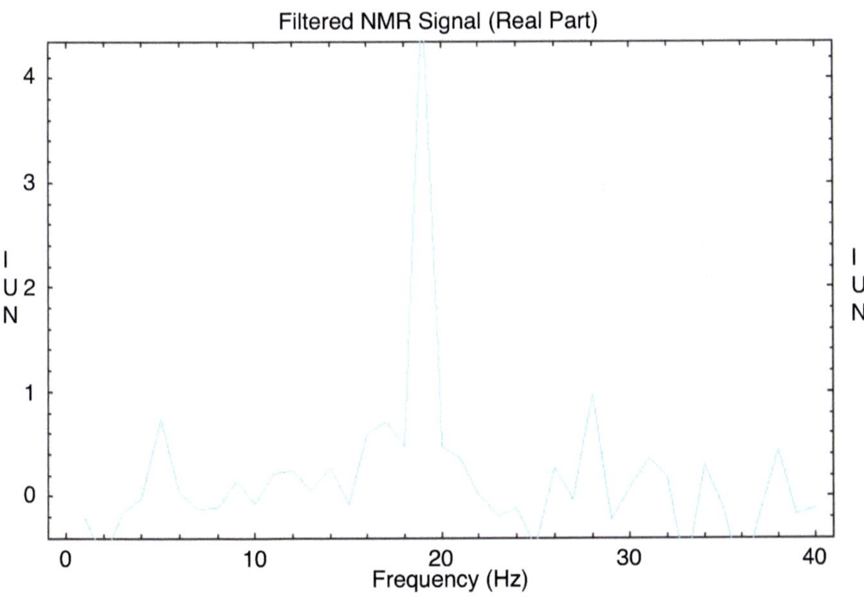

Filtered NMR Signal (Real Part)

References

L.W. Alvarez, F. Bloch, A quantitative determination of the neutron magnetic moment in absolute nuclear magnetons. Phys. Rev. **57**, 111–122 (1940)

W.P. Aue, J. Karhan, R.R. Ernst, Homonuclear broad band decoupling and two-dimensional J-resolved NMR spectroscopy. J. Chem. Phys. **64**, 4226 (1976).

S.D. Bass, The spin structure of the proton. Rev. Mod. Phys. **77**, 1257 (2005)

A. Beyer, L. Maisenbacher, A. Matveev, R. Pohl, K. Khabarova, A. Grinin, T. Lamour, D.C. Yos, The Rydberg constant and proton size from atomic hydrogen. Science **358**, 79–85 (2017)

W. Braun, C. Bösch, L.R. Brown, N. Gō, K. Wüthrich, Combined use of proton–proton Overhauser enhancements and a distance geometry algorithm for determination of polypeptide conformations: Application to micelle-bound glucagon. Biochim. Biophys. Acta **667**, 377–396 (1981)

G. Breit, I.I. Rabi, On the interpretation of present values of nuclear moments. Phys. Rev. **46**, 230–231 (1934)

J. Byrne, *Neutrons, Nuclei and Matter: An Exploration of the Physics of Slow Neutrons* (Dover, Mineola, 2011), pp. 28–31

D.M. Dennison, A note on the specific heat of the hydrogen molecule. Proc. R. Soc. London, Ser. A **115**, 483 (1927)

Y. Dothan, Quark magnetic moments. Physica A **114**, 216–220 (1982)

I. Esterman, O. Stern, Über die magnetische Ablenkung von Wasserstoffmolekülen und das magnetische Moment des Protons. II/Magnetic deviation of hydrogen molecules and the magnetic moment of the proton. I. Z. Phys. **85**, 17–24 (1933)

I. Esterman, O. Stern, Magnetic moment of the deuton. Phys. Rev. **45**, 761 (1934)

M. Gell-Mann, A schematic model of baryons and mesons. Phys. Lett. **8**, 214–215 (1964)

P.R. Hall, L. Malone, L.O. Sillerud, C. Ye, B.L. Hjelle, R.S. Larson, Characterization and NMR solution structure of a novel cyclic pentapeptide inhibitor of pathogenic hantaviruses. Chem. Biol. Drug. Des. **69**, 180–190 (2007)

H. Hu, K. Krishnamurthy, Revisiting the initial rate approximation in kinetic NOE measurements. J. Magn. Reson. **182**, 173–177 (2006)

J. Kestin, M. Sokolov, W.A. Wakeham, Viscosity of liquid water in the range $-8°C$ to $150°$ C. J. Phys. Chem. Ref. Data **7**, 941–948 (1978)

Y. Ne'eman, Derivation of strong interactions from a gauge invariance. Nucl. Phys. **26**, 222–229 (1961)

I.I. Rabi, J.M. Kellogg, J.R. Zacharias, The magnetic moment of the deuton. Phys. Rev. **46**, 163–165 (1934)

R.G. Roberts, *The Structure of the Proton: Deep Inelastic Scattering* (Cambridge University Press, Cambrige, 1990)

L.O. Sillerud, R.S. Larson, Design and structure of peptide and peptidomimetic antagonists of protein-protein interaction. Curr. Protein Pept. Sci. **6**, 151 (2005)

L.O. Sillerud, E.J. Burks, M.J. Wester, D.C. Brown, S. Vijayan, R.S. Larson, NMR-derived model of interconverting conformations of an ICAM-1 inhibitory cyclic nonapeptide. J. Pept. Res. **62**(3), 97–116 (2003)

I.Y. Tamm, S.A. Altshuler, Magnetic moment of the neutron. Dokl. Akad. Nauk SSSR **8**, 455 (1934)

D.D. Traficante, M. Rajabzadeh, Optimum window function for sensitivity enhancement of NMR signals. Concepts Magn. Reson. **12**, 83–101 (2000)

J.P. Xiong, T. Stehle, S.L. Goodman, M.A. Arnaout, A novel adaptation of the integrin PSI domain revealed from its crystal structure. J. Biol. Chem. **279**, 40252–40254 (2004)

W.-X. Zhang, H. Xu, D. Jia, Masses and magnetic moments of hadrons with one and two open heavy quarks: Heavy baryons and tetraquarks. arXiv:2109.07040v3 [hep-ph] (2021)

G. Zweig, An SU(3) model for strong interaction symmetry and its breaking II. CERN Preprint CERN-TH-412 (1964) reproduced In D. Lichtenberg, S. Rosen (eds.), *Developments in the Quark Theory of Hadrons*, vol. 1 (Hadronic Press, Nonatum, 1980), pp. 22–101

Bibliography

T. Axenrod, G. Ceccarelli (eds.), *NMR in Living Systems* (Reidel, Dordrecht, 1986)

A.J. Benesi, *A Primer of NMR Theory with Calculations in Mathematica* (Wiley, Hoboken, NJ, 2015)

J.C. Cavanagh, W.J. Fairbrother, A.G. Palmer III, N.J. Skelton, *Protein NMR Spectroscopy. Principles and Practice* (Academic Press, San Diego, 1996)

R.R. Ernst, G. Bodenhausen, A. Wokaun, *Principles of Nuclear Magnetic Resonance in One and Two Dimensions* (Oxford, New York, 1987)

R.J. Gillies (ed.), *NMR in Physiology and Biomedicine* (Academic Press, San Diego, 1994)

M. Goldman, *Quantum Description of High-Resolution NMR in Liquids* (Oxford University Press, Oxford, 1988)

M.H. Levitt, *Spin Dynamics. Basics of Nuclear Magnetic Resonance*, 2nd edn. (Wiley, West Sussex, 2008)

D. Neuhaus, M.P. Williamson, *The Nuclear Overhauser Effect in Structural and Conformational Analysis* (Wiley-VCH, New York, 2000)

C.P. Schlichter, *Principles of Magnetic Resonance*, 3rd edn. (Springer-Verlag, Berlin, 1990)

K. Wuthrich, *NMR in Biological Research: Peptides and Proteins* (North-Holland, Amsterdam, 1976)

Chapter 8
Monitoring Metabolism with NMR

> *"It was eerie. I saw myself in that machine. I never thought my work would come to this."*
>
> *Isador I. Rabi on NMR (1938)*

Higher animals and plants are complex, highly evolved living systems defined by two universal characteristics: reproduction and metabolism. While reproduction is certainly of fundamental importance to life, its study would take us too far afield from our molecular emphasis on subcellular events, and therefore, we will direct our attention to a brief foray into the study of the latter. Metabolism comprises a network of closely related molecular substrates along with their coupled interconversions catalyzed by enzymes. We showed earlier that enzymes functioned as probability amplifiers that bias the rearrangements of molecular electronic states to favor products over reactants. Cellular metabolism is how organisms extract and utilize energy from molecules in their environment in order to maintain their highly organized state of self-assembly. This process consists of a yin/yang couple— *catabolism* and *anabolism*.

8.1 Catabolism

Catabolism is the stepwise enzymatic disassembly of organic, or inorganic, molecules obtained from the environment as what we commonly refer to as "food," but, for some organisms like bacteria, would not normally be recognized as such. Nutrients would be a more accurate and less ambiguous term. The electronic energy stored in the quantum states of the nutrient molecules is often captured through catabolism and transferred to the pennies in the energy currency of cells, in the form of adenosine triphosphate (ATP) and other energy storage molecules.

© The Author(s), under exclusive license to Springer Nature Switzerland AG 2024

L. O. Sillerud, *Abiogenesis*, https://doi.org/10.1007/978-3-031-56687-5_8

8.2 Anabolism

Anabolism is the construction of new molecules, such as carbohydrates, proteins, lipids, and nucleic acids from simpler precursors, like acetate and pyruvate, that were the products of catabolism. Anabolism is likewise another set of coupled chemical reactions catalyzed by enzymes, but this time, the energy stored in ATP and NADH molecules is used to join smaller molecules into higher-molecular-weight entities. The energy derived from ATP hydrolysis serves to bias the probability of the anabolic reaction in favor of the product's electronic quantum state, much like a magnetic field can bias the probability distribution of spin states.

There are several interesting classes of nutrients. Modern organisms obey the maxim that "You are what you eat" in the sense that most nutrients were components of former or current living systems. Humans, of course, subsist exclusively upon nutrients derived from other organisms, be they plants or animals; however, when life was first developing on the Earth, there were no other living systems to eat. Therefore, other nutrients must have been used by these primordial life forms in order to power cell growth. The first organisms were autotrophs, organisms that could drive their own anabolism using light, water, carbon dioxide, or other nutrients from the environment and produce larger, more complex molecules needed for survival.

8.3 Sunlight Harvesting

One could argue, with good support, that the first source of energy on the Earth that could be used by autotrophic systems came directly from the sun as optical photons, whose wavelengths were in the visible spectrum from 400 to 700 nm. The energy of a green photon of wavelength 555 nm is 2.24 eV (3.58×10^{-19} joules) which is the same order of magnitude as the energy levels of the valence electrons in atoms and molecules (Chaps. 4 and 5). Therefore, optical photons have the correct energy to alter and rearrange molecular electronic states, i.e., to promote primordial and biotic chemical reactions. One of the earliest life forms, autotrophic photon-harvesting cyanobacteria, originated on the Earth ~3.5 Gya, forming stromatolites (layered rocks; literally, *stone pillows*) that are still visible today in shallow waters in Torres del Paine National Park in Patagonia, in Walker Lake, Nevada, in ancient Lake Gosiute in Wyoming, and on the shores of western Australia (Fig. 8.1). The energy from optical photons was captured by pigments in the cyanobacteria (Chap. 10) and used in photosynthesis to convert atmospheric carbon dioxide and water into larger biomolecules, such as sugars. Photosynthesis initially produces sugar phosphates which are converted to polysaccharides such as glycogen for storage when their rates of synthesis exceeds the rate of cellular consumption. Mammals have inherited this pathway, as we will observe later. The mass of glycogen can be as high as 60% of the

The Stromatolites of Hamelin Pool – Hamelin Pool, Australia By Kristina D. C. Hoeppner. CC BY-SA 2.0 https://www.atlasobscura.com/places/the-stromatolites-of-hamelin-pool-australia

https://upload.wikimedia.org/wikipedia/commons/7/70/Stromatolite_28Fort_La clede_Bed2C_Laney_Member2C_Gree n_River_Formation2C_Lower_Eocene3 B_ancient_Lake_Gosiute2C_southwest ern_Wyoming2C_USA 29_1_2815009280980 29.jpg Creative Commons Attribution 2.0 Generic

Fig. 8.1 (**Right**) Stromatolites along the coast of western Australia are formed by photosynthetic cyanobacteria that accumulate a calcium carbonate biofilm that slowly grows over time and produces a characteristic pattern of layers, as seen in this section sliced through the center (The stromatolites of Hamelin Pool—Hamelin Pool, Australia. By Kristina D. C. Hoeppner. CC BY-SA 2.0 https://www.atlasobscura.com/places/the-stromatolites-of-hamelin-pool-australia). (**Left**) Layers of calcium carbonate precipitate around the bacterial colonies as they deplete the carbon dioxide in the water surrounding the colonies. New cyanobacteria grow on top of the crust that forms around the colony. https://cdn-sharkbaywa.pressidium.com/wp-content/uploads/2017/04/ Unknown-stromatolites-1024x768.jpg (https://upload.wikimedia.org/wikipedia/commons/7/70/ Stromatolite_28Fort_Laclede_Bed2C_Laney_Member2C_Green_River_Formation2C_Lower_ Eocene3B_ancient_Lake_Gosiute2C_southwestern_Wyoming2C_USA 29_1_2815009280980 2 9.jpg. Creative Commons Attribution 2.0 Generic)

biomass, an amount that we will compare later with the glycogen stores in mammals. Cyanobacteria also can directly utilize atmospheric nitrogen as a nutrient, completing the set of the abundant elements, C, O, and N, needed in biosynthetic processes.

8.4 Redox Harvesting

Because water significantly attenuates light, particularly lower-energy red photons, stromatolites are limited to shallow bodies of water and are not found at depths greater than ~4 meters. This attenuation of light with depth also implies that even though the surface of the primordial Earth received abundant light and that this light served as a nutrient to foster the evolution of some of the earliest life forms, the rapid decrease in photon flux with depth in water precluded the evolution of photosynthetic organisms in the benthic ocean. Other autotrophic organisms evolved that used an alternative source of nutrients; instead of sunlight, these utilized the energy stored in the electronic states of reduced metals, such as iron, selenium, and manganese, or nonmetals like sulfur, in chemosynthesis in which this energy was used to reduce carbon dioxide, which was combined with water (H_2O) to form carbohydrates and other higher-molecular-weight carbon compounds.

Underwater geysers known as "black smokers" are hydrothermal seafloor vents along oceanic ridges that emit superheated ($\sim$400 °C) water containing high concentrations of dissolved mineral sulfides such as iron, zinc, gold, and copper sulfide, as well as hydrogen sulfide. The water is heated by contact with molten magma released during seafloor spreading due to tectonic activity. The water does not boil because the water pressure is too high (Bischoff and Rosenbauer 1988). At these very high temperatures, the water dissolves large amounts of minerals from the magma. Surrounding these hydrothermal vents are found bacteria that obtained energy by oxidizing the abundant reduced minerals. In chemical terms, reduction is the addition of electrons, while oxidation is the removal of electrons from molecules or atoms. For example, the reduction, or addition of an electron to the trivalent ferric iron ion, Fe^{3+}, produces the divalent ferrous iron ion, Fe^{2+}, while oxidation of the Fe^{2+} to Fe^{3+} releases the energy stored in the electron of Fe^{2+}. The sulfides are a reduced form of sulfur and upon oxidation pure sulfur is released.

Once single-celled bacteria evolved, they formed the basis or niche that supported the evolution of multicellular organisms that would feed upon the bacteria. Almost all modern life feeds upon objects that were once alive, but in the earliest days, prior to the evolution of the first living systems, this was not possible. At this point, we will defer the further study of the origin of primordial life until Chap. 12. For now, let us take for granted that modern living systems are heterotrophic, rather than autotrophic, and derive energy from the catabolism of objects derived from other living systems while bearing in mind that on the early Earth this clearly could not have been the case.

All autotrophs and heterotrophs use the energy derived from the catabolism of nutrients to drive anabolic reactions that produce sugars, primarily glucose (Fig. 8.2), which likely served as the first reduced substrate available to nascent biological organisms because carbohydrate molecules and lipids spontaneously self-assemble from simpler constituents in the environment. Sugars (as well as hydroxy acids) are built up carbon by carbon atom from the autocatalytic polyol condensation of formaldehyde through the formose reaction, discovered in 1861 by Aleksandr Butlerov (Boutlerow 1861) summarized in Fig. 8.3. The reaction is catalyzed by the presence of calcium hydroxide in water. Common terrestrial minerals played a vital role in the abiotic generation of compounds intimately associated with the evolution of living systems on the Earth. Silicates, abundant on the primordial earth, catalyze the formation of 4- and 6-carbon sugars with a specific stereochemistry (Lambert et al. 2010). In addition to glucose, the formose reaction also produces ribose, the sugar portion of both ribonucleic and deoxyribonucleic acids (Ricardo et al. 2004). Glucose, a six-carbon sugar (Fig. 8.2), is naturally stable as the end product of this process, forming as it does, a six-membered heterocyclic ring.

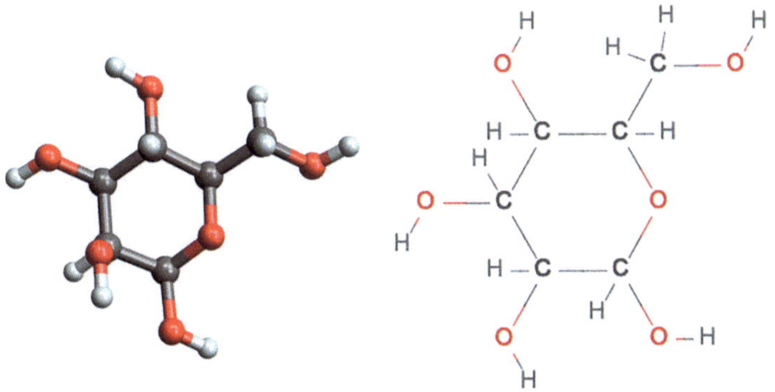

Fig. 8.2 The α-D-glucose molecule in a ball and stick representation on the left and as a chemical graph on the right. Carbon atoms are in gray, and oxygen atoms are red, while hydrogen atoms are light gray

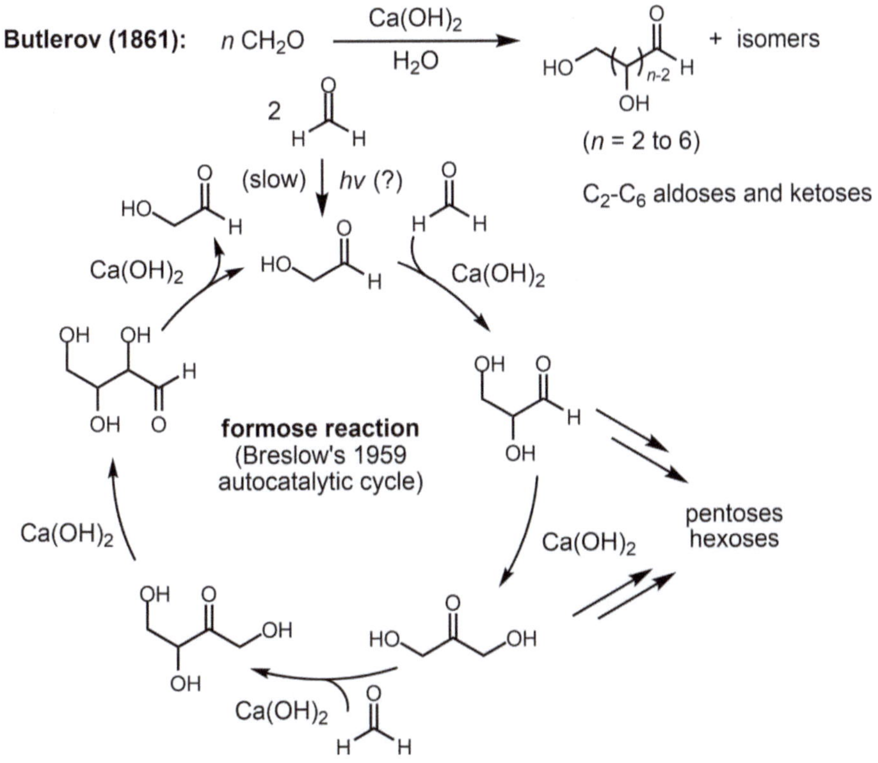

Fig. 8.3 Schematic diagram of the formose reaction of formaldehyde discovered by Butlerov in 1861 showing the autocatalytic cycle proposed by Breslow (Breslow 1959) in the presence of calcium hydroxide. (From: Alsosaid1987 CC BY-SA 4.0, https://commons.wikimedia.org/w/index.php?curid=62998627)

8.5   Glycolysis

The metabolism of glucose is universal among living systems. Let us examine the synthesis, degradation, and storage of this vitally important molecule. The universal enzymatic pathway that evolved for the capture of the energy stored in the electronic states of glucose is called glycolysis (Fig. 8.4) or "sugar splitting," from the Greek, and it is the most primitive, biochemical pathway shared by essentially every living system from *archaea* to *Homo sapiens*. Glycolysis must have been the first metabolic pathway to evolve on the primordial Earth.

The main molecular inputs into the glycolytic pathway are glucose, galactose, and fructose. The intermediate output from glycolysis is a pair of phosphorylated trioses, glyceraldehyde-3-phosphate and dihydroxyacetone phosphate. These are subsequently converted into pyruvate, which then enters the second most primitive biochemical pathway, the tricarboxylic acid pathway, or TCA cycle, which is also known as the citric acid cycle, or the Krebs cycle, after Hans Krebs (Fig. 8.5).

In order to study the metabolism of living systems, scientists in the past were often forced into the extreme reductionist position of *molecule smashing*, grinding up organisms and examining the residues, for many investigations resulted in the destruction of the organisms. Single enzymes were often extracted and purified by separating them from other cellular material and then their rates of conversion of

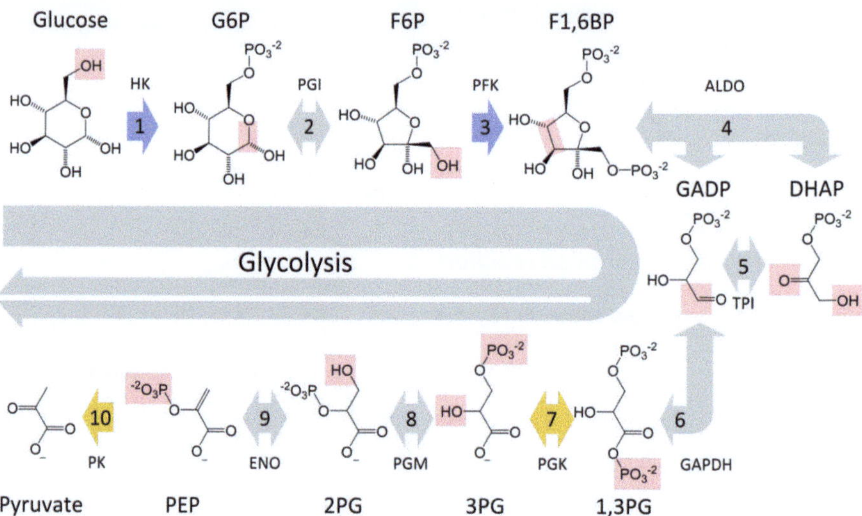

Fig. 8.4 Glycolysis converts glucose into pyruvate through a series of chemical reactions catalyzed by ten different enzymes, indicated by colored arrows: (1) hexokinase, (2) glucose-6-phosphatase, (3) fructose-6-phosphatase, (4) aldolase, (5) triosephosphate isomerase, (6) glyceraldehyde-phosphate dehydrogenase, (7) phosphoglycerate kinase, (8) phosphoglycerate mutase, (9) enolase, and (10) pyruvate kinase. (Author: Thomas Shafee, Licensed under the Creative Commons Attribution 4.0 International license)

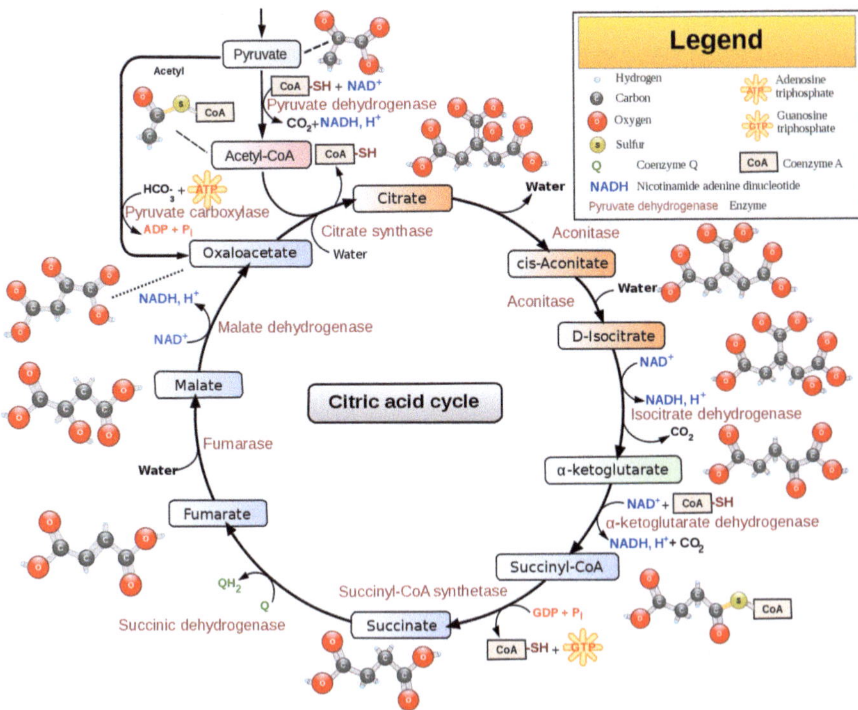

Fig. 8.5 The Krebs cycle of intermediary metabolism. Pyruvate, derived from glycolysis (Fig. 8.4), enters the cycle at the upper left. The Gibbs energy stored in the electronic configuration of pyruvate is used to generate ATP, GTP, and NADH molecules. (Authors: Narayanese, Yassine Mrabet, Toto Baggins, licensed under the Creative Commons Attribution-Share Alike 3.0 Unported license)

precursor substrates into the corresponding products was laboriously determined through clever, but tedious biochemical methods. The development of radioactive tracers, particularly ^{14}C, accelerated these studies, but did not circumvent the need for destructive methods.

8.6   NMR Spectroscopy as a Nondestructive Monitor of Metabolism In Vivo

Nondestructive techniques for the observation of metabolism awaited the development of methods that could measure the time dependence of the concentrations of multiple chemical compounds in cells and tissues in real time. Optical methods would prove to be useless; the light scattering from cells and tissues severely limited the penetration of light to depths of only ~0.1 mm. Radioactive tracers, particularly gamma emitters, are readily detectable from outside the body, but cannot reveal metabolic transformations because nuclear decay is independent of chemistry and the resolution is limited by the acceptable dose of radiation to the organism.

By the end of the Second World War, real-time observation of metabolism in living systems appeared to reside solely in the minds of science fiction writers. However, physicists from the United States, who had worked at MIT's Lincoln Laboratory on the domestic radar project, returned to their home laboratories at Harvard and Stanford and, armed with a sophisticated understanding of both quantum mechanics and electromagnetic theory, performed additional experiments demonstrating that nuclei, when placed into a suitable magnetic field, could absorb radio waves at precisely defined, but modest frequencies in the kHz to MHz range as first found by I. Rabi et al. (1938).

Furthermore, the frequency of absorption was directly related to the electronic structure surrounding the nuclei. An urban legend, *whose veracity remains doubtful,* attends this discovery:

> *The first nuclear magnetic resonance signal was obtained from water because water contains 111 molar, spin ½ protons, as hydrogen nuclei. The NMR signal from water was observed to be a single peak corresponding to all of the equivalent protons in the sample. "Ah, isn't physics wonderful, they all sighed." The investigators next asked what other samples might be of interest and one person in the group, a chemist who was using ethanol to clean an apparatus made from glass, suggested that they replace the water sample with ethanol. To the dismay of the physicists, when the NMR signal from this sample was obtained, it was found to contain, not one, but several complicated and separate signals, some with obvious fine structure! Since the physicists were unschooled in chemistry, they threw up their hands and declared that the new signals were shifted due to chemistry, not physics, and thus the term "chemical shift" was born!*

Organic chemists embraced NMR spectroscopy and rapidly developed large empirical data sets that showed that nuclear resonance frequencies were exquisitely sensitive to the electronic structures of molecules, with particular chemical groups, such as methyl or benzyl groups always resonating in particular regions of the NMR spectrum (Chap. 7). Therefore, one could, simply by taking the NMR spectrum of an unknown compound, obtain excellent structural information without destroying the sample.

Unlike the scattering of light, with its nanometer-scale wavelengths, the radiofrequency waves used in NMR spectroscopy to excite magnetic nuclei have wavelengths on the order of meters and, aside from small dielectric effects arising from the dilute ionic solutions in the cytoplasm of cells, would easily penetrate even rather large living systems, including humans. Now, the stage was set to apply NMR techniques to the study of biological problems.

The first NMR spectrum of a living system was obtained by Jasper A. Jackson in 1967 at Los Alamos (Jackson and Langham 1968) using his laboratory-built, large-bore solenoidal magnet operating at ~10 gauss (with a proton NMR frequency of 40 kHz). After first examining the spectrum of tap water, hen's eggs, a gelatin and water mixture and lard, he put an anesthetized rat into the apparatus and obtained the

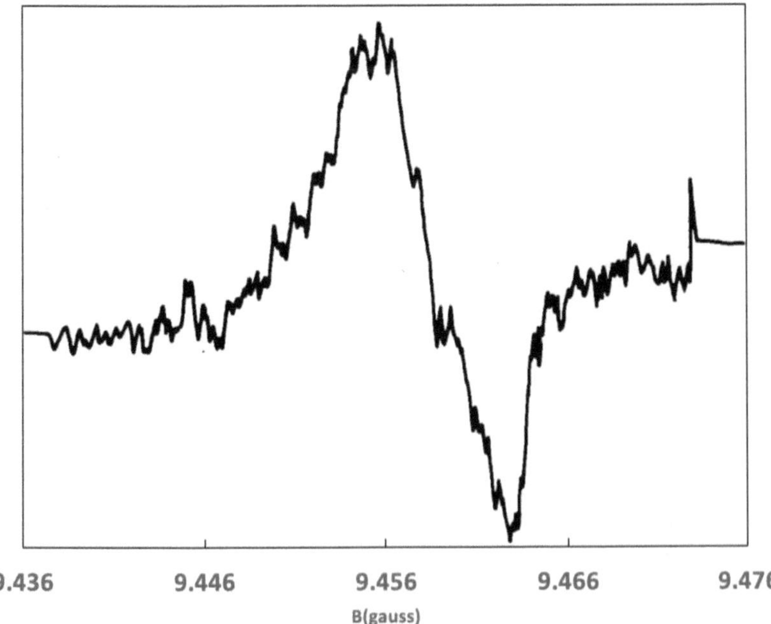

Fig. 8.6 The first NMR spectrum of a living animal obtained by Jasper A. Jackson at Los Alamos National Laboratory in 1967 using a 40-kHz instrument of his own design and construction (Jackson and Langham 1968). This continuous wave spectrum was taken in the derivative mode and shows the resonance signal from water in the blood and cells of an anesthetized 150-g rat. The width of the resonance is ~30 Hz (6.86 mG, or 750 ppm). (Adapted from Rev. Sci. Instr. **39**(1968) 510. © American Physical Society, used with permission)

first NMR spectrum of a living animal (Fig. 8.6). The spectrum is dominated by the signal from water protons in the blood and cells of the animal, although there is a discernible broader component attributed to lipids and proteins.

Both catabolism and anabolism, which we will now simply lump together under the rubric of metabolism, can be directly observed in living systems by means of nuclear magnetic resonance spectroscopy. The resonance frequencies of magnetic nuclei lie in the MHz range, putting them in a frequency window where living systems are essentially transparent to these relatively long (~1 meter) wavelength photons. Since the nuclear resonance frequencies of chemical compounds are determined by the electron density at the nucleus, and this density changes during chemical reactions, NMR spectroscopy can be used to directly measure the concentrations and dynamics of many metabolites in living systems from bacteria (Ugurbil et al. 1978; Ogino et al. 1980) and fungi (den Hollander et al. 1979; den Hollander et al. 1981a, b; Sillerud et al. 1981) to perfused livers (Cohen et al. 1981; Sillerud and Shulman 1983) and animals and man (Alger et al. 1981; Ross et al. 1981). Metabolic transformations involve covalent chemical bond (i.e., electronic) rearrangements of mainly hydrogen, carbon, oxygen, nitrogen, and phosphorus atoms, and each of these nuclei of biological significance can be observed using NMR detection (Chap. 7). Since carbon atoms lie at the backbone of most metabolites, carbon NMR spectroscopy is particularly useful for observing the rates of interconversion of various carbon-containing metabolites.

8.7 ^{13}C NMR Analysis of Carbohydrate Metabolism in the Liver

Multinuclear magnetic resonance spectroscopy can be applied to a wide range of biochemical problems because it offers the ability to monitor each nuclear site in a molecule separately and the low-frequency radiofrequency fields used to excite these nuclei penetrate living systems with ease. The spectral resolution is sufficient to allow the determination of the isotopic composition of individual atoms in a substrate or product. One area of application is to detect and follow the metabolic transformations of substrate molecules in vivo. Metabolism can be directly studied by means of noninvasive, nondestructive methods that allow quantitative measurements of fluxes and pathways in vivo, in real time, without the need for extractions or molecular separations. An example is the application of carbon nuclear magnetic resonance methods to measure the hormonal control of carbohydrate metabolism in the perfused rat liver.

The study of metabolism dates from the very earliest days of biochemistry. Indeed, the elucidation of the pathways by which biological molecules are transformed into muscle, sinew, bone, or energy-rich compounds occupied the attention of biochemists for a large portion of the past century. Much of the tremendous progress made since the Second War in the study of metabolic transformations in vivo had as its basis the exploitation of radioisotope tracers. One could label, and, if one were very clever, extract the radioactively labeled compounds, and follow the biochemical transformation of each substrate atom. At least this is true in principle; in practice the complete analysis was often tedious or difficult due to the lengthy chemical procedures employed.

A better method would be one which employed direct detection of the added, isotopically labeled compound in the organism, without the need for post hoc extractions. Fortunately, the periodic table contains not only radioactive, but also stable isotopes of the biochemically interesting atoms. Some of these isotopes have spin ½ enabling them to be conveniently detected by NMR, which then constitutes this "better method" since it enables one to detect isotopically labeled compounds noninvasively, within the functioning organism. Many metabolites can also be detected, because the chemical shift of a nuclear resonance depends strongly on the chemical identity of the compound; in favorable situations the nuclear resonances from every observed substrate or metabolite atom are distinct.

At present the spin ½ nuclei of ^{1}H (Ogino et al. 1980; Sillerud et al. 1981), ^{13}C (Sillerud and Shulman 1983), ^{15}N (Lapidot and Irving 1977), and ^{31}P (Navon et al.

Table 8.1 Some biochemically interesting, NMR-observable isotopes

Nucleus	Relative sensitivity[*]	Natural abundance (%)	Absolute sensitivity
^{1}H	1.00	99.98	1.00
^{2}H	9.7×10^{-3}	1.5×10^{-2}	1.5×10^{-6}
^{3}H	1.21	0	0
^{13}C	1.6×10^{-2}	1.1	1.8×10^{-4}
^{15}N	1.0×10^{-3}	0.37	3.9×10^{-6}
^{17}O	2.9×10^{-2}	3.7×10^{-2}	1.1×10^{-5}
^{19}F	0.83	100	0.83
^{23}Na	9.3×10^{-2}	100	9.3×10^{-2}
^{25}Mg	$\sim 7 \times 10^{-3}$	10.1	2.7×10^{-4}
^{31}P	6.6×10^{-2}	100	6.6×10^{-2}
^{33}S	2.3×10^{-3}	0.76	1.7×10^{-5}
^{35}Cl	4.7×10^{-3}	75.5	3.6×10^{-3}
^{39}K	5.1×10^{-4}	93.1	4.7×10^{-4}
^{43}Ca	6.4×10^{-3}	0.15	9.3×10^{-6}

[*]At constant field for an equal number of nuclei

1977) have been shown to provide nuclear resonance signals in a sufficiently short accumulation time to be useful for metabolic studies. Signal averaging is used to increase the signal to noise ratio for spectra of ^{13}C, ^{15}N, and ^{31}P in particular. The signal to noise ratio increases as the square root of the observation time, or equivalently, as the square root of the number of FIDs which are co-added. Table 8.1 shows these and some of the other biologically important nuclear isotopes that give rise to NMR signals. ^{1}H and ^{31}P have absolute sensitivities in the range where signals at natural abundance ($\sim 100\%$) can be obtained a few minutes by signal averaging and Fourier transformation from samples with a concentration in the biochemical range, i.e., 1–10 mM. In the cases of ^{13}C and ^{15}N one can isotopically enrich substrates to levels of 99% or greater and bring detection of these nuclei to the same time scale as for ^{1}H and ^{31}P.

The practical limits on the use of NMR in metabolic studies stem from the conflicting demands for both adequate time resolution and signal to noise ratio. The current lower limit for observability given these two constraints is an intracellular concentration of approximately 1 mM for a small molecule. This is not yet a severe limitation since many substrates and metabolites are present in cells at concentrations higher than this. However, it is clear that further progress will be made when detection of more dilute species is possible. Modern pulse sequences which use spin coupling between ^{1}H and either ^{13}C or ^{15}N can be used in order to detect these heteronuclei with proton sensitivity, and this renders natural abundance detection of ^{13}C or ^{15}N relatively easy to do. Indirect observation of ^{13}C resonances is possible with an order of magnitude increase in sensitivity over that possible using direct methods (Sillerud et al. 1981). Both isotopic enrichment methods and spectra taken at natural isotopic abundance have been used to provide complementary metabolic information on liver metabolism.

8.7.1   Liver Perfusion in the NMR

The standard technique for the perfusion of a rat liver inside the 89-mm bore of a commercial, wide-bore superconducting magnet is shown schematically in Fig. 8.7. The recirculating perfusion circuit had a total volume of 80 ml and allowed for continuous monitoring of both pH and oxygen tension in the perfusate. Oxygen and carbon dioxide exchange was handled by a Silastic membrane oxygenator, through which a humidified mixture of 95% O_2 and 5% CO_2 was passed. The CO_2 combined with bicarbonate in the perfusate to act as a pH buffer. The liver from a ~100-g rat was cannulated via the portal vein and placed in a 20-mm NMR tube. During a typical 8-hour experiment, the pH was observed to vary by less than 0.01 units; the O_2 tension stayed constant at 95%.

Because the phosphorus NMR spectrum of the liver is an excellent indicator of its metabolic integrity, ^{31}P spectra were monitored both at the beginning and end of the experiment in order to ensure that the liver was maintained in good metabolic condition throughout. Figure 8.8 illustrates the ^{31}P NMR spectrum of a perfused liver 30 minutes after cannulation and placement into the magnet. Note the large adenosine triphosphate (ATP) resonances, especially the β-phosphate resonance furthest up field near −18 ppm. This signal arises from the middle phosphate in the triphosphate chain and is therefore unique to ATP, while adenosine diphosphate (ADP) contributes to the signals labeled α and γ. The ratio of ATP to ADP can easily be calculated from this spectrum by measuring the integral of the β-phosphate signal, B, and the integrals of the α and γ signals (A and G, respectively). Then the ratio of ATP to that of ATP + ADP is given by

$$\frac{[\text{ATP}]}{[\text{ATP}] + [\text{ADP}]} = \frac{2B}{A + G} = X$$

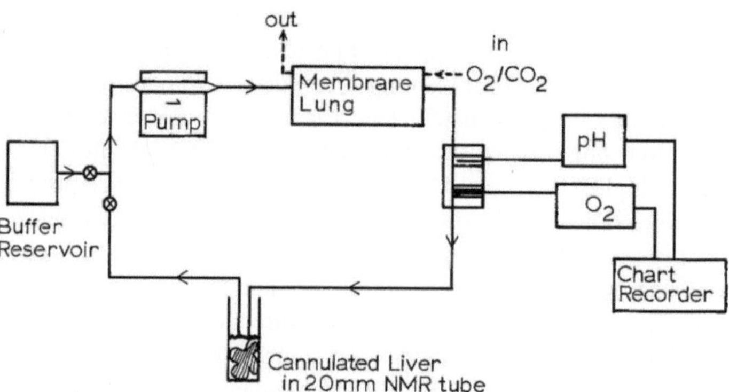

Fig. 8.7 Diagram of the apparatus used for liver perfusion. The perfusion medium consisted of Krebs-Henseleit buffer containing streptomycin, 25 mM MOPS, and heparin. Flow rates of 5 ml/min were used with perfusate recycling. Substrates and sampling took place at the pH sensor block

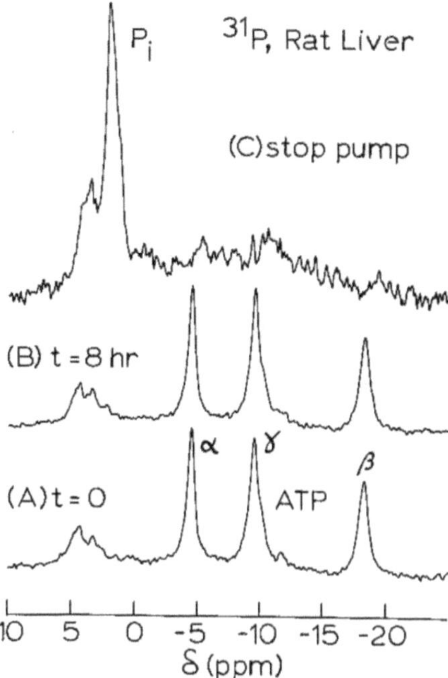

Fig. 8.8 ^{31}P NMR spectra at 145.8 MHz of a perfused rat liver, with 3000 scans, a 90° pulse, 5 minutes total time per spectrum, 15-Hz line broadening. (A) Spectrum of the liver just as it was put into the magnet, about 30 min after cannulation. (B) Spectrum 8 hours after (A). (C) Spectrum taken 30 min after perfusion was curtailed. In addition to the indicated ATP resonances, visible are signals from phosphomonoesters (4.5 ppm), orthophosphate (3 ppm), NAD/NADP (-11 ppm), and UDPG (-12 ppm)

We can solve this for [ATP]/[ADP] $= X/(1 - X)$. Using the measured integrals, we find that the ratio of ATP to ADP is at least 5 in this liver; this is a slight underestimate because the peak attributed to the γ-phosphate of ATP (and the β-phosphate of ADP) around -9 ppm is contaminated by a small contribution from NAD/NADP and UDPG. If we use just integral A, we find that [ATP]/[ADP] $= 9$. In either case, this liver was initially in good bioenergetic status.

It is also interesting to note the small downfield peaks due to sugar phosphates and the very low initial orthophosphate (P$_i$) resonance (visible around ~3 ppm particularly in Fig. 8.8c) in this spectrum taken after 30 min of perfusion (Fig. 8.8a). Prior to ^{31}P NMR studies of live intact tissue, the orthophosphate level was thought to be about 5 times higher than it has subsequently been shown to be.

The second ^{31}P NMR spectrum (Fig. 8.8b) arose from the same liver after 8 hours of perfusion in the bore of the magnet. There is virtually no change in the high-energy phosphorus compounds present within the liver indicating that the life support system maintained the proper biochemical conditions and stability; it was found that [ATP]/[ADP] remained at a value of 9 after 8 hours of perfusion.

Furthermore, the orthophosphate peak is still very low, even after 8 hours. The third ^{31}P spectrum of this liver (Fig. 8.8c) was taken after the perfusion had been curtailed and oxygen delivery ceased. The only resonances left are from some unhydrolyzed sugar phosphates (~4.5 ppm), a small amount of NAD/NADP, and a very large orthophosphate resonance arising from hydrolysis of the high-energy phosphates. The ^{31}P NMR spectrum of this metabolically exhausted liver is characteristic of tissue which is no longer functional.

8.7.2 Liver ^{13}C Resonance Assignments at Natural Abundance

Just as ^{31}P NMR spectra are very informative with respect to the bioenergetic status of perfused livers, the natural abundance ^{13}C NMR spectrum reveals signals from many interesting metabolites. The ^{13}C NMR spectrum of such a liver is shown in Fig. 8.9. Most of the ^{13}C NMR signals arise from carbons contained in the storage

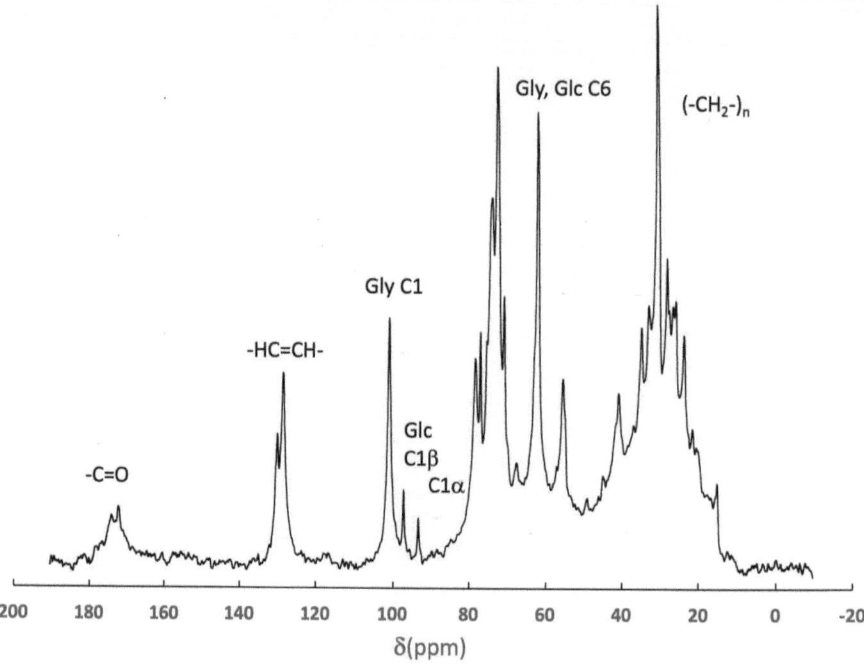

Fig. 8.9 A 90.55 MHz proton-decoupled, natural abundance ^{13}C NMR spectrum of the liver from a rat given ad libitum access to food. The spectral acquisition parameters were 11,718 scans, line broadening 20 Hz, 20 min total time, 60° pulse, T = 310 K. Chemical shifts are shown with respect to DSS at $\delta = 0.00$ ppm

Table 8.2 Fed-rat liver natural abundance ^{13}C NMR assignments

δ(ppm)[a]	Assignment
16.54	-$C_\omega H_3$
19.24	Ala C3
24.60	β-OH-butyrate C1
25.28	-$C_{\omega-1}H_2$-
27.44	-$C_{\alpha+2}H_2$-
28.08	-$C_{\alpha+2}H_2$-
29.70	-HC=CH-$\underline{C}H_2$-HC=CH-
32.40	(-CH_2-)$_n$
34.68	-$C_{\omega-2}H_2$-
36.28	-$C_{\alpha+1}H_2$-
42.10	Phosphatidylethanolamine: -$\underline{C}H_2$-NH_2
49.01	β-OH-butyrate C2
53.43	Ala C2
56.65	Phosphatidylcholine -N-$(\underline{C}H_3)_3$
63.54	Glc, Gly C6
68.29	β-OH-Butyrate C3
72.41	Glc C4α,β
74.24	Glc,Gly C2,5α
75.64	Glc, Gly C3α
77.04	Glc C2β
78.66	Glc C3,5β
80.33	Gly C4
94.94	Glc C1α
98.72	Glc C1β
102.49	Gly C1
130.64	-HC=CH-
132.26	-HC=CH-
174.21	-C=O
176.58	Ala C1

[a]Chemical shifts are referenced to DSS at 0.00 ppm

molecules triacylglycerols and glycogen. For example, Fig. 8.9 shows peaks from the triacylglycerol carbonyl carbons at 174 ppm, the olefinic sites at 129 ppm, and the large methylene carbon peak at 32 ppm (Williams et al. 1973). This liver was taken from a rat given continuous or ad libitum access to food. As a result, there are also resonances in this spectrum from glucose and glycogen between 60 and 103 ppm. Complete assignments are given in Table 8.2.

Perfusion was not performed with this particular liver; it was excised from the rat and placed in D_2O, and the natural-abundance ^{13}C spectrum was recorded every 20 min at 37 °C, in order to establish the changes occurring in the absence of perfusion. The hydrolysis of some glycogen is inferred from the presence of signals from $C_{1\alpha}$ and $C_{1\beta}$ of glucose; a perfused liver, maintained in an energized state,

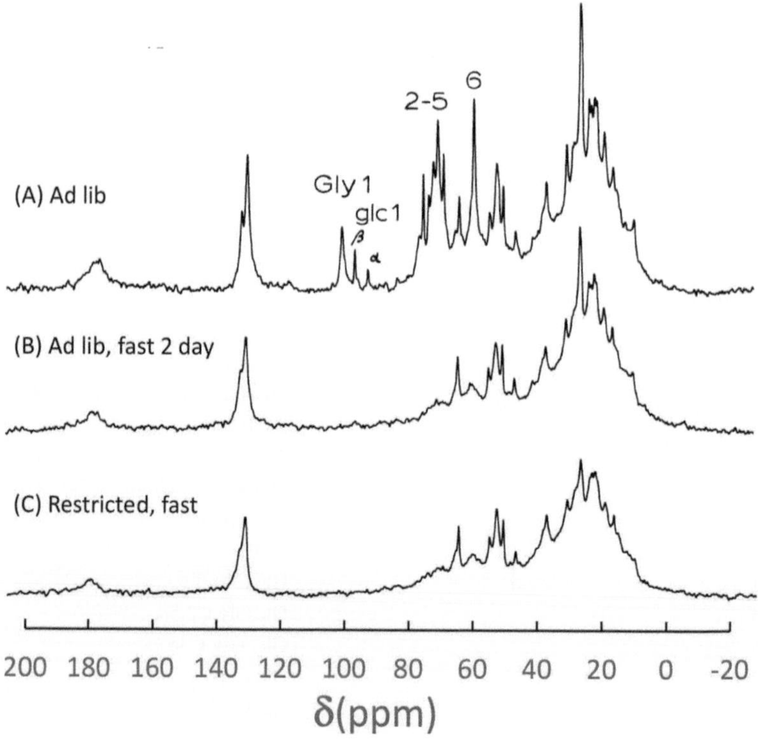

Fig. 8.10 Natural abundance, proton-decoupled ^{13}C NMR spectra at 90.55 MHz from livers of rats subjected to various diets. (**a**) Spectrum from the liver of a rat given ad lib access to food. Note the prominent glycogen, glucose, and triacylglycerol signals. (**b**) Liver from a rat given ad lib access to food that was fasted for 48 hours prior to recording the spectrum. The glycogen and glucose signals are now absent. (**c**) Spectrum from the liver of a rat on a diet restricted to a weight gain of only 25% that of an ad lib fed rat and fasted for 48 hours prior to recording of the spectrum. The prominent triacylglycerol methylene carbon signal at 32 ppm in (**a**) has markedly decreased. The acquisition conditions were the same as in Fig. 8.9

shows no free glucose. Note that the signals from the anomeric carbons of glycogen (C_1 at 102.5 ppm) and glucose ($C_{1\alpha}$ at 94.94 ppm and $C_{1\beta}$ at 98.72 ppm) separately appear in the natural abundance spectrum in a region free from other carbon resonances. The relatively high intensity of these carbohydrate resonances means that we can study their synthesis and degradation at natural abundance, without the need for ^{13}C-enriched substrates. The glycogen C_1 resonance indicates that this liver contained large amounts of this storage carbohydrate.

In a similar way, the other signals in this spectrum (Fig. 8.9) reveal the extent to which natural abundance ^{13}C NMR can be useful in studies of endogenous liver components. For example, Fig. 8.10 shows the effects of dietary restriction on the resonances from the liver triacylglycerols. Spectrum A in Fig. 8.10 was taken from the liver of a rat given continuous (ad libitum) access to food. A rat on this regimen is denoted as a fed rat. Spectrum B (Fig. 8.10) came from the liver of a rat whose ad libitum access to food was stopped entirely for 48 hours before these signals were measured. Note the complete disappearance of the resonances from the anomeric carbon of glycogen and the reduction in the intensity of signals from the methylene, olefinic, and carboxyl carbons of triacylglycerols. This condition is denoted as a "fasted" rat liver. The metabolic demands of the liver during fasting are met by the oxidation of fatty acids.

The third spectrum (Fig. 8.10c) was recorded from the liver of a rat that was fed a limited amount for two weeks and then fasted for 48 hours prior to utilization in this experiment. This condition is denoted as a "restricted" rat. In addition to the disappearance of glycogen already seen in Fig. 8.9, there is a further decline in triacylglycerol signals due to depletion of endogenous lipid stores. It should be mentioned that the signals, in these natural abundance spectra, arise from only those molecules mobile enough to have narrow resonances and that our ability to reduce the background signals by fasting indicates that much of the liver itself does not give rise to high-resolution nuclear resonances.

8.7.3   *Assignments for the ^{13}C NMR Spectrum of Liver Extracts*

There are at least three major sources of molecular information from a perfused liver: (1) the ^{13}C NMR spectrum of the liver itself, (2) the ^{13}C NMR spectrum of the perfusate examined separately from the liver, and (3) the ^{13}C NMR spectrum of a chemical extract of the liver. Note that by varying the extraction procedure and the polarity of the solvents, one can extract molecules, either soluble in water or in nonpolar solvents such as chloroform or hexane, to obtain the lipid fraction. The perfusate can contain both added metabolic substrates and small molecules secreted from the liver during the perfusion. We have observed the ^{13}C NMR spectrum of the liver itself (Figs. 8.9 and 8.10) in various states of nourishment. We primarily observed the major storage molecules, glycogen and triacylglycerols. Now let us turn to examine the ^{13}C NMR spectrum of a perchloric acid extract of the liver. In order to preserve the metabolites in their normal relative abundances, we halted metabolic transformations by rapidly freeze clamping the liver as it was being perfused outside the magnet using two large aluminum plates that had been cooled to liquid nitrogen temperature (77 K). The liver was then immersed in liquid nitrogen in a Dewar flask and allowed to equilibrate. The solidly frozen liver was then broken up under liquid nitrogen in a mortar and reduced to a fine power with the aid of a pestle. The liquid nitrogen was poured off and a dilute (5%) perchloric acid solution in water was added and allowed to freeze. As the mixture melted, the liver paste was ground with the pestle and then put into a centrifuge tube. Additional perchloric acid was used to rinse the mortar and this rinse was added to the centrifuge tube. After

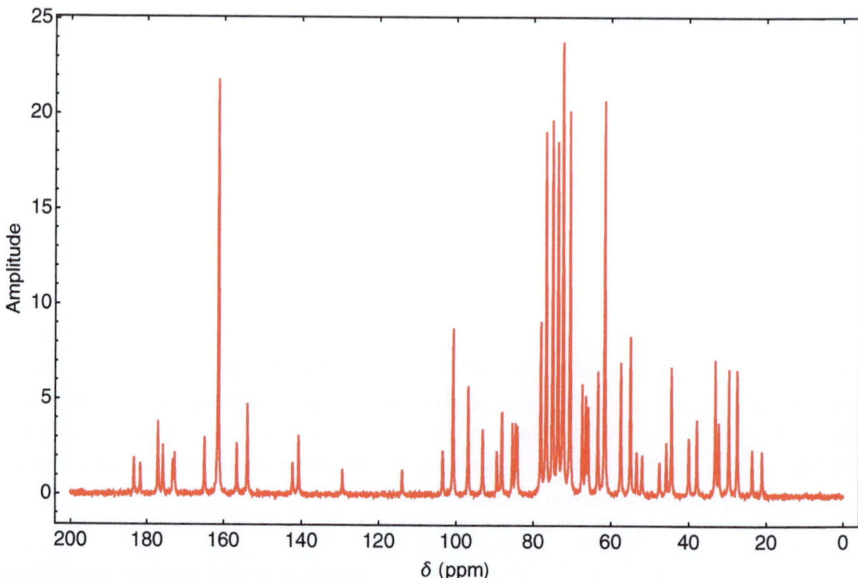

Fig. 8.11 ^{13}C NMR spectrum (24,886 scans) at 90.55 MHz of the perchloric acid extract of a liver from a rat fasted for 8 hours perfused with 15 mM glucose with 1.1% ^{13}C (natural abundance). Signal assignments are given in Table 8.3

centrifugation, the supernatant was pipetted into another tube and neutralized with potassium carbonate. After another centrifugation to pellet the relatively insoluble potassium perchlorate, the supernatant was lyophilized and then dissolved in D_2O for ^{13}C NMR spectroscopy.

The ^{13}C NMR spectrum of the perchloric acid extract of a liver from an 8-hour fasted rat that was perfused with natural abundance glucose (Fig. 8.11) shows large signals from glucose (C1α, 94.8 ppm; C1β, 98.6 ppm; C6, 63.4 ppm) and signals of similar intensity from residual and newly synthesized glycogen (C1, 102.5 ppm). The spectral assignments (Table 8.3) for 49 carbon nuclei greatly exceed the 29 nuclei observed in the ^{13}C NMR spectrum of the fed rat liver (Fig. 8.9 and Table 8.2). The primary differences arise from the 8.5-fold increase in spectral accumulation time which allowed us to observe metabolites present at smaller amounts and from the concentration of the metabolites from the whole liver into a ~7-fold smaller volume (~1.5 mL). Among the metabolites revealed in this process, in addition to the expected glucose and glycogen carbons, are also important cofactors such as ATP, NADH, UDPG, and glutathione, as well as the amino acids glutamate, glutamine, and aspartate and the ketone bodies, β-hydroxybutyrate and acetoacetate (Table 8.3). The spectral resolution is also enhanced because the aqueous extraction did not dissolve the phospholipids, triacylglycerols, cholesterol, and macromolecules removing their broad resonances from the spectrum.

Table 8.3 Assignments for the ^{13}C NMR spectrum of the perchloric acid extract of a liver from a rat fasted for 8 hours perfused with 15 mM glucose

δ (ppm)[a]	Assignment
22.88	Lac C3
25.42	Pyrrolidine (trace solvent)
29.27	Gln C3; Glu C3; glutathione; acetoacetate C4
31.46	Acetone (trace solvent)
34.07	Gln C4; Glu C4
34.93	Glutathione C6
39.68	Aspartate C2
41.76	Malonate
46.18	Cys C3; glutathione
47.50	Citrate C2,4
49.28	β-Hydroxybutyrate C2; acetoacetate C2
53.70	Alanine C2
55.10	Aspartate C3
56.69	Choline; glutathione
59.17	Cys C2; glutathione
63.32	Glucose C6; Gly C6
65.08	Fru C1,6α
67.58	ATP Rib C5'; NADH Rib C5'; UDPG Rib C5
68.20	β-Hydroxybutyrate C3
69.12	NAD,NADH Rib C4"
72.33	Fru C5β; Glc C4α,β; Lac C2
74.14	Glc C2,5α; Gly C2,5
75.43	Glc C3α; Gly C3
76.80	Glc C2β
78.48	Glc C5β; citrate C3
79.85	Glycogen C4
85.86	NADH rib C4"
86.32	NADH rib C4'
87.13	ATP rib C4'
89.88	NADH rib C1'; ATP rib C1'
91.15	UDPG rib C1
94.87	Glucose C1α
98.64	Glucose C1β
102.55	Glycogen C1
105.22	UDPG Uri C5
115.71	Acetonitrile (trace solvent)
131.07	NAD+, NADP+ Nic C1
142.45	NAD+ Ade C4
143.87	ATP Ade C4
144.02	UDPG Uri C6
155.55	NADH Ade C2; ATP Ade C2; NADH Ade C1
158.25	NADH Ade C1; ATP Ade C1;

(continued)

Table 8.3 (continued)

δ (ppm)[a]	Assignment
163.13	HCO3- (from neutralization)
166.74	Uri C2; UDPG Uri C2; acetoacetate C1
174.53	Glutathione
175.02	Acetoacetate C1
177.58	Glutamate C1; glutathione
178.88	Alanine C1; glutathione
183.49	Glutamate C5
185.16	Lactate C1

[a]with respect to DSS at $\delta = 0.00$ ppm

8.8 The Structure of Glycogen

One of the main functions of the mammalian liver is to regulate blood glucose levels. The storage form of glucose is glycogen, a polymer of α-D-glucose whose postprandial anabolism in the liver deposits glucose molecules in covalent, high-molecular-weight ($\sim 10^6$–10^7) particles during times of nutritional sufficiency. Conversely, in times of fasting, the catabolism of glycogen supplies glucose molecules to the blood. Therefore, glycogen exists as a glucose reserve in tissues such as the muscles, brain, and liver. The hydrolysis of liver glycogen stores allows us to get a good night's sleep without the need to get up to eat at 3 AM to overcome the otherwise resulting hypoglycemia. The structure of glycogen must be congruent with this function. Is all glycogen visible with NMR? What are the dynamics of the glycogen polymer in relation to its function as a glucose storage molecule?

The addition of new glucosyl residues to the existing particle is catalyzed by the enzyme glycogen synthetase, while the cleavage of the polymer to release glucosyl units into the blood is catalyzed by glycogen phosphorylase. The storage of glucose in the polymeric form glycogen is a necessary means to avoid the osmotic catastrophe that would ensue if the equivalent number of glucose molecules were to be suddenly present within the cell. The osmotic pressure of a glycogen particle is that of a single object, whereas the osmotic pressure of the glucose molecules, if free in the cytoplasm, would burst the cell.

8.8.1 Assignments for the ^{13}C NMR Spectrum of Liver Glycogen In Situ

^{13}C NMR spectroscopy can be used to monitor the most important aspects of mammalian carbohydrate metabolism, the synthesis and degradation of hepatic glycogen. Although we have not emphasized this point earlier, an NMR spectrum must be assigned, and the origins of the signals must be identified with confidence. One method for assigning NMR signals involves the examination of authentic

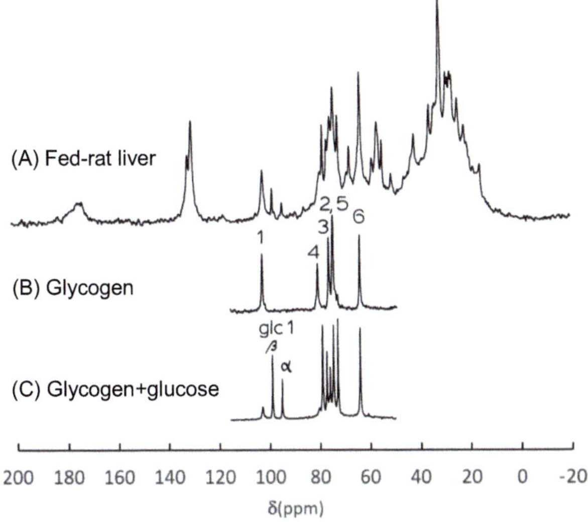

Fig. 8.12 Assignments for the natural abundance, proton-decoupled ^{13}C NMR spectra of (**a**) the liver from a fed rat, (**b**) authentic glycogen dissolved in D_2O, and (**c**) an equimolar mixture of glucose and glycogen in D_2O. (Adapted with permission from Sillerud & Shulman *Biochemistry* **22**(1983)1087. (c) 1983 American Chemical Society)

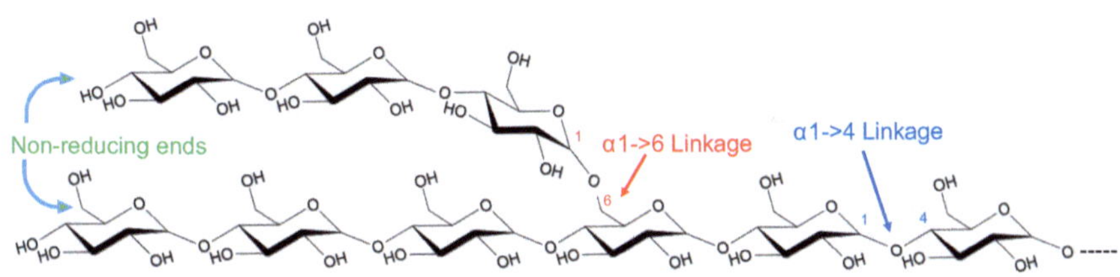

Fig. 8.13 The covalent structure of mammalian liver glycogen showing the terminal branched oligosaccharide as well as the $\alpha(1\rightarrow4)$ and $\alpha(1\rightarrow6)$ linkages of the glucosyl residues

substances as pure compounds and noting their chemical shifts. Figure 8.12 shows a comparison of the natural abundance ^{13}C NMR spectrum of a fed rat liver (Fig. 8.12a), with that from an aqueous solution of pure glycogen (Fig. 8.12b), and an equimolar mixture of glucose and glycogen (Fig. 8.12c). The glycogen spectrum is simpler than that from the glucose because glucose undergoes mutarotation in water, interconverting between its α- and β-anomers, while glycogen solely consists of the α-anomer of glucose. A large fraction (~90%) of the glucosyl residues comprising glycogen are linked through an $\alpha(1 \rightarrow 4)$ glycosidic bond (Fig. 8.13), while the remaining glucosyl residues comprise branches in the polymer

Table 8.4 Assignments for the ^{13}C NMR spectrum of glucose and glycogen

Glucose	α–glc	β–glc	Glycogen α(1 → 4)		Glycogen α(1 → 6)	
C#	δ(ppm)	δ(ppm)	δ(ppm)	Δ (ppm)	δ(ppm)	Δ (ppm)
C1	94.83	98.64	102.51	7.68	101.25	6.42
C2	74.30	76.98	73.99	−0.16	73.99	−0.16
C3	75.58	78.59	75.99	0.41	75.57	−0.01
C4	72.41	72.41	79.76	7.35	72.06	−0.35
C5	74.15	78.63	74.24	0.09	74.24	0.09
C6	63.40	63.55	63.40	0.00	69.91	6.51

chain through α(1 → 6) glycosidic bonds (Fig. 8.13). The chemical shifts of carbons are determined by the electron density at the nucleus and this is sensitive to the type of chemical bonding among atoms. The shielding differences on linkage ($\Delta = \delta_{polymer} - \delta_{monomer}$) therefore are diagnostic for ascertaining the atoms in a molecule that participate in saccharide formation (Colson et al. 1974; Sillerud et al. 1982). The binding of the glucose monomer to the existing glycogen chain shifts the resonance from glucose C1 $\Delta = +7.7$ ppm downfield (Table 8.4), and subsequent linking of another glucose in an α(1 → 4) glycosidic bond shifts the signal from glucose C4 by $\Delta = +7.4$ ppm compared with the chemical shifts found for free α-D-glucose (Table 8.4). The signals from glycogen carbons 2–5 can be assigned by comparison with those from α-D-glucose, since they undergo very small linkage shifts of less than 1 ppm (Table 8.4). Branching in glycogen occurs through the introduction of ~10% α(1 → 6) glycosidic linkages. In these linkages C1 shifts down field by $\Delta = +6.4$ ppm and C6 by $\Delta = 6.5$ ppm (Table 8.4). Note that the signal for C4 in the α(1 → 6) glycosidic linkages appears at the chemical shift of free α-D-glucose (Table 8.4) as it should since it does not participate in the α(1 → 4) linkage.

Two points can be noticed when ^{13}C NMR spectra of glucose and glycogen are examined (Fig. 8.12). First, glucose is in anomeric equilibrium, resonances from both the α- and β-anomers appear, with the β-anomer peak exceeding that from the α-anomer, reflecting its energetic predominance, and second, the glucose signals are much more intense than those from glycogen in the equimolar mixture in Fig. 8.12c. The intensity difference stems partly from the large difference in molecular weights, and consequently rotational correlation times, between the highly polymerized glycogen ($M_r \sim 10^6$ Da; $\tau_c = 4.7 \pm 0.30$ ns) and monomeric glucose ($M_r = 180$ Da; $\tau_c = 41$ ps) resulting in a longer T_1 for the monomer (1.28 sec) compared with that of the polymer (0.22 sec) at 90.55 MHz, as well as a large difference in the nuclear Overhauser enhancements. The nOe for glucose is a full $\eta + 1 = 2.998 \pm 0.2$, while that for glycogen is only $\eta + 1 = 1.22 \pm 0.05$. The correlation time difference, through its influence on T_2, also broadens the spectral lines for glycogen that can be readily seen in the liver spectrum (Figs. 8.9, 8.10, and 8.11). The glycogen C1 resonance has a width of ~45 Hz, while that for glucose is only 12 Hz wide. On this basis the glucose signals should be larger in a proton-decoupled spectrum than those from glycogen; in fact we find that they are over twice as large.

8.8.2   Relaxation in ^{13}C NMR

In Chap. 7 on NMR we covered the relationship between the rotational correlation time and the NMR relaxation parameters, T_1, T_2, and nOe, for protons relaxed by neighboring protons, but here we have a different set of nuclei bound into a molecule. Glucose carbons are covalently attached to hydrogens (Fig. 8.2); the average length for a C-H bond is 109 pm, meaning that the protons are close enough to the carbons that their dipolar, inverse sixth-power magnetic interaction is the dominant feature for carbon nuclear magnetic relaxation. We note that expressions for T_1 and T_2 given in Chap. 7 contain only γ_H^4 and the proton Larmor frequency, ω_H. We must modify these equations to account for the magnetic interaction between ^{13}C nuclei and protons and for their different Larmor frequencies; γ_H^4 needs to be replaced by $\gamma_H^2 \gamma_C^2$ and we must introduce the carbon frequency, ω_C. Here we follow Doddrell (Doddrell et al. 1972) and simply present the results, assuming motion based on a single, isotropic rotational correlation time. The spectral density terms now include both the proton and carbon Larmor frequencies and give, for the T_1, T_2, and nOe, the expressions

$$\frac{1}{T_1} = \frac{K}{10r^6}\chi$$

$$\frac{1}{T_2} = \frac{K}{20r^6}\left(\chi + 8\tau + \frac{6\tau}{1 + \omega_H^2\tau^2}\right)$$

$$K = \hbar^2\gamma_H^2\gamma_C^2$$

$$\eta + 1 = 1 + \frac{\gamma_H}{\chi\gamma_C}\left(\frac{6\tau}{1 + (\omega_H + \omega_C)^2\tau^2} - \frac{\tau}{1 + (\omega_H - \omega_C)^2\tau^2}\right)$$

where

$$\chi = \left(\frac{\tau}{1 + (\omega_H - \omega_C)^2\tau^2} + \frac{3\tau}{1 + \omega_C^2\tau^2} + \frac{6\tau}{1 + (\omega_H + \omega_C)^2\tau^2}\right)$$

Here K is reduced from $\hbar^2\gamma^4 = 5.688 \times 10^{11}$ Å^6/sec^2 (5.688×10^{17} nm^6/sec^2) for proton-proton interactions to $\hbar^2\gamma_H^2\gamma_C^2 = 3.598 \times 10^{11}$ Å^6/sec^2 (3.598×10^{17} nm^6/sec^2) due to the smaller magnetogyric ratio for carbon (γ_C/γ_H) $=$ 67.283/267.52 $=$ 0.2515. Plots of these functions (Fig. 8.14) show that the motion of the glycogen molecule in situ within the liver and as an extracted polymer in vitro is essentially the same and can be described by a single isotropic reorientation and correlation time. This is quite surprising given that the molecular weight of the mature glycogen molecule is quite large, around 10 megadaltons.

From the experimentally measured correlation time, one can estimate the molecular hydrodynamic radius of the motional unit using the Stokes-Einstein relationship (Chap. 9)

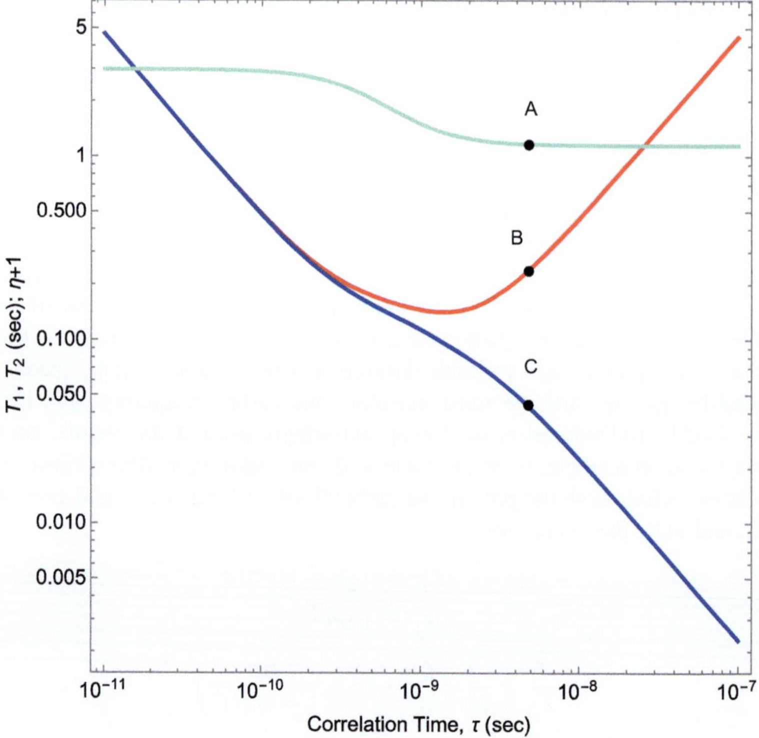

Fig. 8.14 The relationship between the theoretical and the measured relaxation parameters for glycogen at a magnetic field strength of 8.4 tesla and a temperature of 300 K as a function of the isotropic rotational correlation time. The red curve is for T_1, and the blue curve is for T_2, while the green curve reflects the nuclear Overhauser enhancement, $\eta + 1$. The black points labeled with letters are the measured values for glycogen in the perfused rat liver at 310 K: (A) nuclear Overhauser enhancement. (B) T_1, and (C) T_2. Note that all three parameters are consistent with a single isotropic correlation time of 4.7 ns

$$r^3 = \frac{3kT\tau}{4\pi\eta}$$

where k is Boltzmann's constant, T is the temperature, and η is the viscosity. At $T = 300$ K and in water with a viscosity of 1 cPoise, the Stokes radius of 1.66 $\pm$ 0.03 nm can be found. Particles of glycogen have a larger radius of 2.5–5.0 nm when observed in the electron microscope (Drochmans 1962) and this would imply a correlation time of 128 ns and NMR line widths of 1–10 kHz, several orders of magnitude larger than those actually measured. Therefore, it is clear that there is considerable local motion in the glycogen particle and that the NMR parameters are dictated, not by overall tumbling, but by these local reorientations of the glucosyl chains.

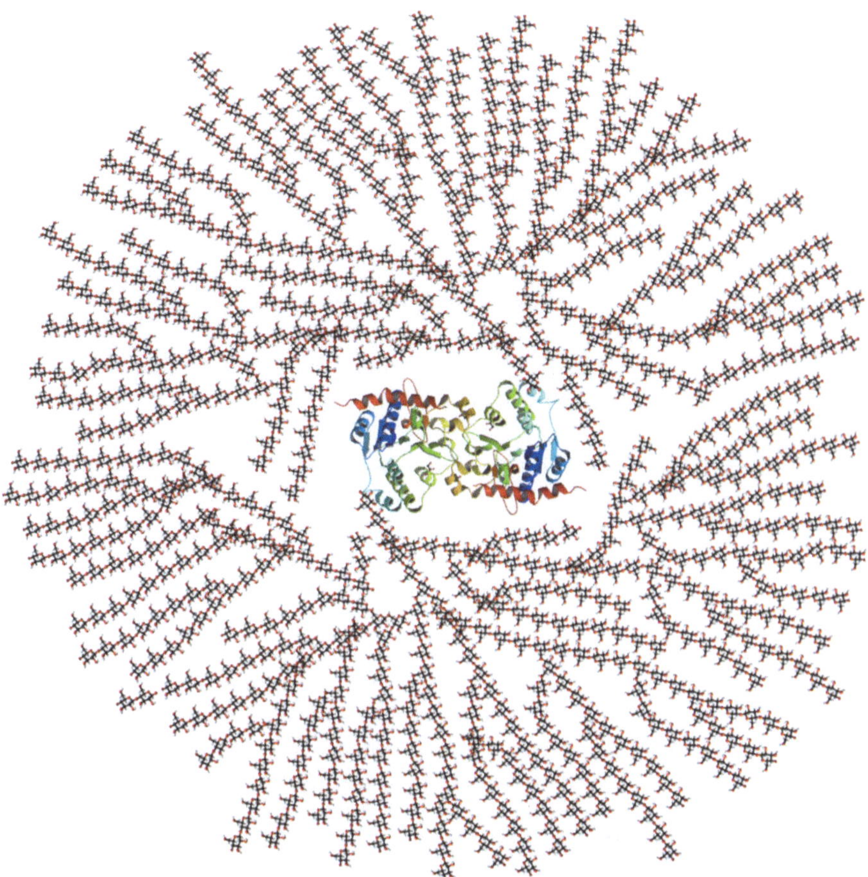

Fig. 8.15 A model for the glycogen particle showing the central glycogenin scaffold protein upon which the initial ($\alpha(1 \rightarrow 4)$-D-glucosyl-)$_4$ primer is attached and then branched using $\alpha(1 \rightarrow 6)$-D-glucosyl-links to extend layers of ($\alpha(1 \rightarrow 4)$-D-linked glucosyl residues. This model omits the peripheral proteins that perform the enzymatic synthesis and cleavage of the glucosyl units in response to hormonal stimuli: glycogen synthetase and phosphorylase, which would move up and down the chains to either add or remove glucosyl units. (Häggström, Mikael (2014). "Medical gallery of Mikael Häggström 2014". *WikiJournal of Medicine* **1** (2). https://doi.org/10.15347/wjm/2014.008. ISSN 2002-4436. Public Domain)

The molecular weight of the local unit can be estimated from the Stokes radius and the volume of a glucose molecule to be 1–10 kdaltons, or approximately 12 residues of oligo-α-D-glucosyl units. A model of liver glycogen consistent with these data is one of a loose structure, open to the movement up and down the chains of glucosyl units of the enzymes, glycogen phosphorylase and glycogen synthetase, needed for the rapid metabolism of this polymer in response to hormonal stimulus (Fig. 8.15) to meet metabolic demands in times of glucose abundance after eating, or in times of fasting to replenish blood glucose levels. This is undoubtedly to be expected for we have neglected the presence of these proteins within the glycogen chains and their presence would certainly distort the close packing of the chains and lead to a much looser structure than in their absence.

8.8.3 Glycogen and Glucose Metabolism

With an understanding of the structure and NMR characteristics of glycogen in hand, including the fact that all of the glucosyl residues of glycogen give rise to high-resolution NMR signals in spite of the large molecular mass of the overall glycogen particle (Sillerud and Shulman 1983; Zang et al. 1990), we now turn to an examination of the metabolic transformations that are at the heart of glycogen metabolism, with the confidence that the ^{13}C NMR spectra of this very large macromolecule accurately reflects its concentration within organs such as the liver. This surprising, but welcome, breakthrough meant that ^{13}C NMR could not only be applied to the study of small molecular metabolites in living systems, but could also be used to examine the transformations of macromolecules, particularly this polymer of glucose that is so important in energy storage and release. Recent studies have used proton MRI to image the metabolism of glycogen in vivo (Zhou et al. 2020).

The metabolism of glycogen can be divided into *glycogenolysis*, the catabolism of existing particles, and *glycogenesis,* the anabolic deposition of new glucosyl units. Fasting, exercise, and sleep lead to hormone secretion that drives the catabolic degradation of stored glycogen in order to maintain the concentration of blood glucose at ~5.5 mM when eating is impractical. During daytime, when blood glucose levels are elevated after feeding, the reverse anabolic process occurs where the polymer is enlarged by the addition of glucosyl units in order to build up this storage form of energy. In addition, other anabolic processes in the liver are also involved in the maintenance of blood glucose levels; in the presence of amino acids generated from the catabolism of dietary, muscle, and other proteins in the gut, the liver can manufacture "new" glucose from these amino acids in a process known as *gluconeogenesis.* We will illustrate the use of ^{13}C NMR spectroscopy to examine each of these processes in turn, beginning with glycogenolysis, the breakdown of existing glycogen particles.

8.8.3.1 Glycogenolysis

The elevation of hormones in the blood, such as glucagon or epinephrine, in response to fasting or exercise stimulates the degradation, or glycogenolysis, of existing liver and muscle glycogen to produce glucose monomers to support the metabolic activity of other organs, particularly the brain. We have seen that the motional characteristics of glycogen are such that all of the glucosyl residues contribute to a high-resolution NMR signal from the polymer. Therefore, we can be assured that NMR spectra of the molecule in vivo reflect the true status of the

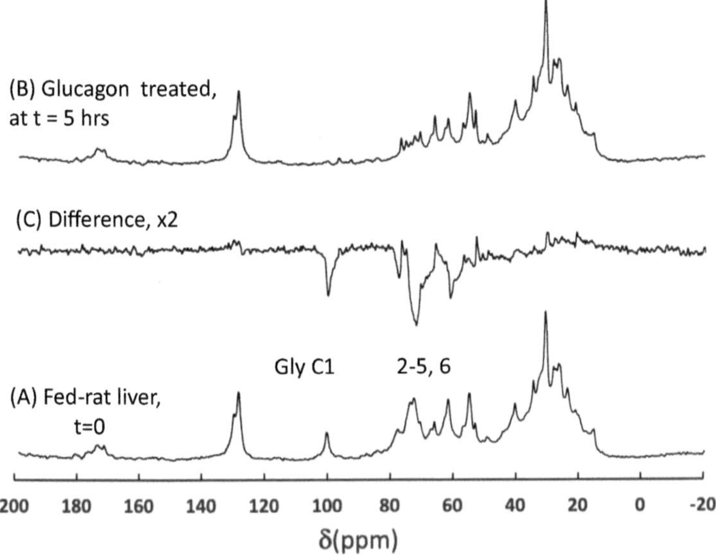

Fig. 8.16 Natural abundance, proton-decoupled ^{13}C NMR spectra of the perfused liver from an ad lib fed rat. (**a**) Control spectrum taken at $t = 0$. Note the prominent signal from glycogen C1 at ~102 ppm. (**b**) The spectrum taken after treatment for 5 hours with 10 nM glucagon. (**c**) The digital difference spectrum (at twice the relative scale) taken between (**a**) and (**b**). Note the disappearance of the glycogen signals in (**b**) compared with (**a**) and the appearance of these negative glycogen signals in the difference spectrum (**c**)

amount of carbon present in the glycogen particle at any given time and that we can utilize the nondestructive nature of NMR to directly interrogate the metabolic transformations of glycogen in living systems. This is particularly important because glycogen is an energy storage molecule not only in the liver, but also in the brain, and in the kidney clear-cell carcinoma and the so-called glycogen shunt is an important source of energy for anaerobic glucose metabolism in other cancers (Rothman and Shulman 2021).

In a liver from a fed rat, nuclear magnetic resonances from glycogen are prominent in the ^{13}C spectra of both the perfused liver (Figs. 8.10, 8.12, and 8.16; see also Tables 8.2 and 8.3) and in a perchloric acid extract of the tissue (Fig. 8.11). Treatment of this liver with a physiological dose of the peptide hormone glucagon (Fig. 8.16) illustrates how useful the natural abundance ^{13}C NMR spectrum can be to measure metabolism in real time. The bottom spectrum (Fig. 8.16a) is a control spectrum taken at $t = 0$, i.e., at the start of the experiment. Signals from glycogen carbons C1, C2-5, and C6 are visible, but we see no free glucose—a fed rat is in an insulin state promoting glycogenesis (Cohen et al. 1978) and any small amount of free glucose was diluted into the perfusion medium outside of the NMR-sensitive volume.

Treatment of this liver with a physiological concentration (10 nM) of the glycogenolytic hormone, glucagon, led to almost complete hydrolysis of the glycogen after 5 hours, as shown in Fig. 8.16b. Figure 8.16c also introduces the digital difference spectrum between the control and subsequent spectra. The long-term stability of modern superconducting spectrometers enables one to subtract the natural abundance liver background from the various later spectra in a given series to produce spectra showing only those changes that took place. In the present example we see that the difference spectrum, Fig. 8.16c, consists primarily of signals from the hydrolyzed glycogen. The other liver resonances have changed very little in 5 hours of perfusion. As noted above, glucose formed from the glycogen is diluted into the perfusate here and exits the NMR coil and is not seen in the liver spectra.

The perchloric acid extract of the liver from a fed rat (Fig. 8.17) shows large signals from the extracted glycogen carbons 1–6. In addition, we observe many signals from low-molecular-weight metabolites (identified in Appendix A.4) including aspartate, glutamate, glutamine, ATP, NAD, β-hydroxybutyrate, UDPG, and glutathione.

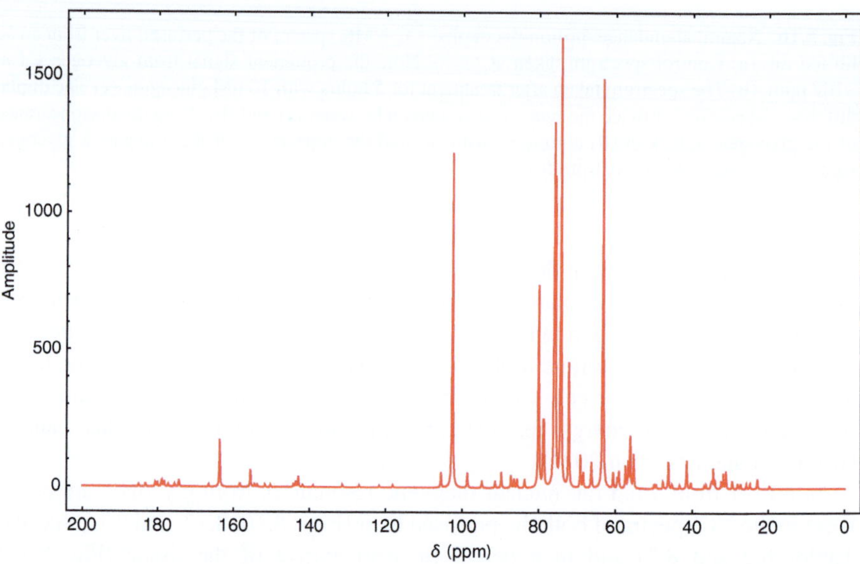

Fig. 8.17 The proton-decoupled, natural abundance ^{13}C NMR spectrum of the perchloric acid extract from a liver from a rat given ad libitum access to food. Note the large signals from glycogen C1–6 between 60 and 103 ppm (see Fig. 8.16 for assignments) as well as many signals from other metabolites (identified in Appendix Four)

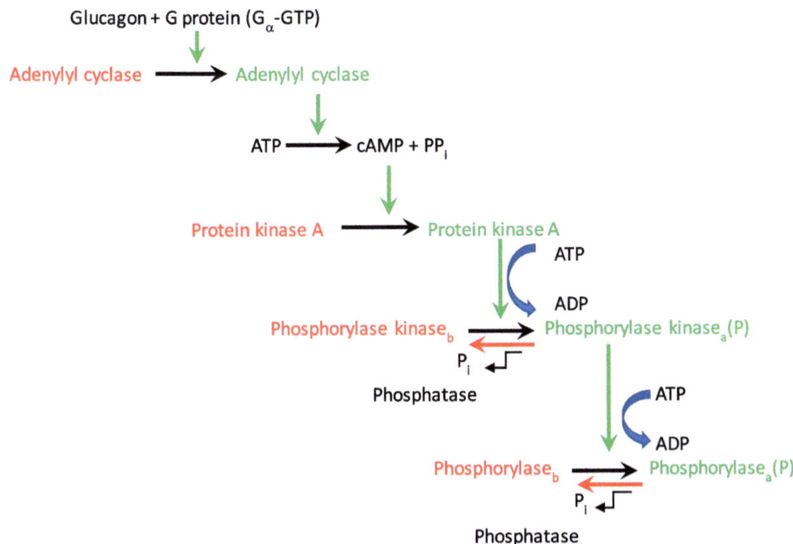

Fig. 8.18 A schematic of the hormone-stimulated regulatory cascade in which an extracellular stimulus, such as glucagon or epinephrine, interacts with a liver cell membrane receptor, in this case a G-protein-coupled protein, to activate adenylyl cyclase, protein kinase A, phosphorylase kinase, and phosphorylase. The inactive enzyme forms are shown in red, while the activated forms appear in green

8.8.3.2   The Stimulation of Glycogenolysis by Glucagon

Human glucagon is a 3485-dalton, 29-amino-acid polypeptide hormone with a primary structure of HSQGTFTSDYSKYLDSRRAQDFVQWLMNT. Glucagon is generated from the cleavage of proglucagon by proprotein convertase 2 in pancreatic islet α-cells. Glucagon activates G-protein-coupled receptors on the plasma membrane of liver cells to trigger a *cyclic*-adenosine monophosphate (cAMP) cell signaling cascade (Fig. 8.18) in response to low blood glucose (Tengholm and Gylfe 2017). The cAMP cascade results in phosphorylation of a serine hydroxyl of the glycogen phosphorylase enzyme, which promotes its transition to the active (relaxed) state. The phosphorylated enzyme is less sensitive to allosteric inhibitors. Thus, even if cellular ATP and glucose-6-phosphate are high, phosphorylase will be active.

The appearance of large glycogen signals in the natural abundance ^{13}C NMR spectrum of the fed rat liver (Fig. 8.9) allows one to observe the response of liver glycogen to physiological regulatory stimuli. Before one embarks upon such studies, however, the unstimulated stability of the glycogen signals must be examined. The time dependence of the glycogen C$_1$ resonance is shown in Fig. 8.19. Control experiments in which the liver from a fed rat was simply perfused with glucose-free buffer showed that without hormonal stimulation the liver glycogen was slowly

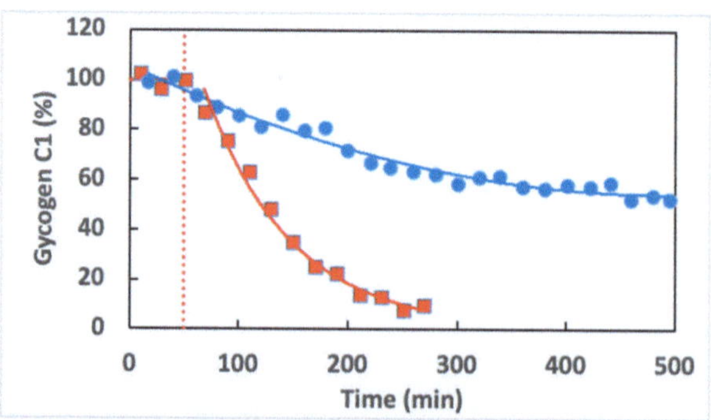

Fig. 8.19 Time course of glycogenolysis in the perfused fed rat liver determined by natural abundance ^{13}C NMR spectroscopy. The blue circles are from a liver that was perfused with only Krebs-Henseleit buffer, with no glucose in the medium. The blue line is an exponential fit with a decay constant of 730 min. The red squares are from a liver perfused with the same buffer containing 10 nM glucagon which was injected into the perfusate at t = 50 min denoted by the dotted red line. The solid red line is an exponential fit with a decay constant of 77 min. The data are from Sillerud (Sillerud and Shulman 1983) and are presented in terms of the percent of the initial glycogen C_1 signal remaining as a function of time after the start of the experiment

hydrolyzed; the time constant for its decay was 730 minutes, or more than 12 hours. This would be certainly enough time to support normal blood glucose levels overnight. However, when a large but still physiologically relevant concentration of 10 nM glucagon, mimicking fasting, is added to the perfusate, the decay constant decreased by a factor of ten to only 77 min. The measured glucose concentration in the perfusate at the end of the experiment was 22 mM, from which we calculated that this liver contained 340 mM glycogen at the start of the experiment. The rate of glucose release was 0.7 μmole per min, a value that compared quite favorably with classical measurements of glucose release.

8.8.3.3 Regulation of Glycogen Metabolism

The physiological regulation of glycogen metabolism involves both the synthesis and the breakdown of glycogen. If glycogen synthesis and glycogenolysis were active simultaneously in a cell, there would be a "futile cycle" with cleavage of one ~P bond per cycle (in the formation of UDP-glucose: Fig. 8.20). It is commonly believed that to prevent such a futile cycle, glycogen synthase and glycogen phosphorylase are reciprocally regulated, both by allosteric effectors and by covalent modification (phosphorylation) so that only unidirectional fluxes of glucose occur depending on whether the metabolic demands of the organism favor storage or breakdown of glycogen.

Glycogen phosphorylase in the liver is subject to weak allosteric regulation by AMP, ATP, and glucose-6-phosphate. AMP, which is increased when ATP is depleted, activates phosphorylase, while ATP and glucose-6-phosphate, which

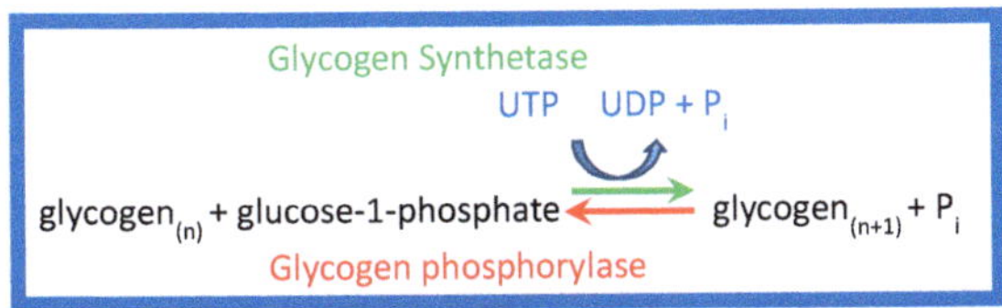

Fig. 8.20 The synthesis and degradation of glycogen. In the expansion of the glycogen molecule, one molecule of intracellular glc-1-P is condensed (indicated by the green arrow) with the existing glycogen glucosyl residues by the enzyme glycogen synthetase using uridine triphosphate (UTP) as a cofactor. Glycogen phosphorylase catalyzes the removal of a single glucose molecule from the existing glycogen polymer

both have binding sites that overlap that of AMP, inhibit phosphorylase, and therefore, glycogen breakdown is inhibited when ATP and glucose-6-phosphate are abundant. On the other hand, glycogen synthase is allosterically activated by glucose-6-phosphate; this is opposite to the effect of glucose-6-phosphate on phosphorylase. Thus, glycogen synthase is active when high blood glucose leads to elevated intracellular glucose-6-phosphate. Glycogen synthase and glycogen phosphorylase are reciprocally regulated also by covalent modification (phosphorylation).

The cAMP cascade has the opposite effect on glycogen synthesis. Glycogen synthase is phosphorylated by protein kinase A as well as by phosphorylase kinase. Phosphorylation of glycogen synthase promotes the less active conformation of this enzyme inhibiting glycogen synthesis. Instead of being converted to glycogen, glucose-1-phosphate in the liver may be converted to glucose-6-phosphate and dephosphorylated for release to the blood. Many aspects of this accepted picture of carbohydrate metabolism in the liver are easily examined by means of ^{13}C NMR spectroscopy of the perfused rat liver.

8.8.3.4 Glycogenesis from Glucose

The process opposite to glycogenolysis is that of glycogenesis, the synthesis of glycogen from glucose (Hers 1976), which can also be followed conveniently by means of ^{13}C NMR. Figure 8.21 shows the results of an experiment in which 30 mM natural abundance glucose, containing 1.8 mM isotopically enriched [90% 1-^{13}C]-glucose, was added to the perfusate of a fed rat liver. A fed rat liver is in a hormonal state where insulin stimulates the uptake and incorporation of glucose into glycogen. The initial spectrum (Fig. 8.21a) shows large signals from the isotopically enriched glucose in the perfusate, as well as signals from 1.1% natural abundance ^{13}C in the ~300 mM glycogen present in this liver at the start of the experiment. After 6 hours, a considerable amount of the initial [90% 1-^{13}C]-glucose had been incorporated into intracellular glycogen (Fig. 8.21b). The difference spectrum, Fig. 8.21c, shows the

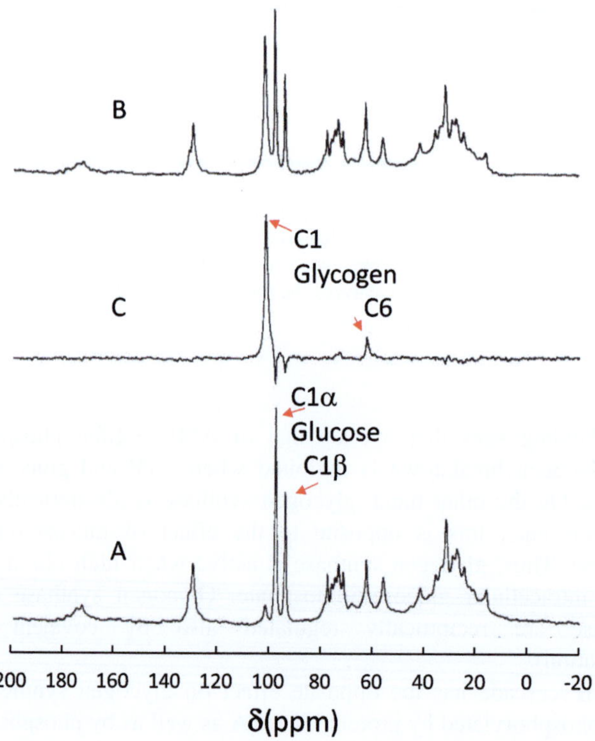

Fig. 8.21 ^{13}C NMR spectra of glycogenesis from [1-^{13}C]-glucose in the perfused, fed rat liver. (**a**) The first 20 min of perfusion with 30 mM glucose containing 1.8 mM [90% 1-^{13}C] glucose showing the separate anomeric signals from glucose C1, as well as the C1 signal from the residual glycogen in the liver. (**b**) Spectrum of the same liver 6 hours later showing the growth of the glycogen C1 signal. (**c**) Difference spectrum (**b–a**) showing the appearance of ^{13}C at glycogen carbons C1 and C6, with a small signal for carbons C2–5 ($\delta \sim 75$ ppm)

substantial incorporation of the ^{13}C from glucose C1 into C1 of glycogen. The signals from the α- and β-anomers of the added [90% 1-^{13}C]-glucose appear negative in the difference spectrum reflecting their uptake by the liver and their incorporation into the newly synthesized glycogen. One also notes a small positive signal near 72 ppm in the difference spectrum from glycogen carbons C2–5 (Table 8.4) reflecting the flow of glucose into glycolysis and the Krebs cycle (vide infra) and its subsequent reformation via gluconeogenesis. There are essentially no changes in the liver background in 6.33 hours of perfusion (Fig. 8.21c).

The production of glycogen from glucose was linear for at least 6 hours in this liver (Fig. 8.22). One also notes that we can monitor the mutarotation of the added glucose here as well. When added as a crystalline solid, the glucose was initially and exclusively in the α-anomeric form, but upon dissolution in the aqueous perfusate, the molecule interconverts between a linear and a 6-membered ring form. The linear form can close into a ring with the anomeric hydroxyl at C1 in either the α- or

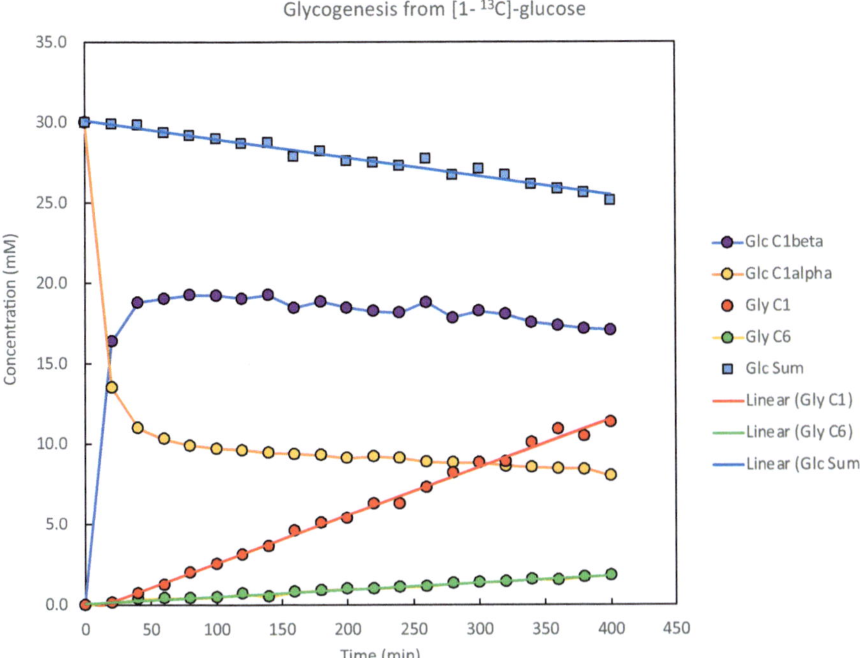

Fig. 8.22 The time course of the ^{13}C NMR signals from glucose and glycogen in a fed rat liver perfused with [90% 1-^{13}C]-glucose. **Blue** squares are total glucose C1. **Purple** and **orange** circles are from the **β**- and **α**-anomers of the added [1-^{13}C]-glucose. The **red** and **green** circles are from glycogen **C1** and **C6**, respectively. Linear least squares fits are also shown

β-form, and each has a unique chemical shift (Table 8.4). This interconversion is called mutarotation and is such a slow process at 310 K that we can easily measure its time course. For example, a fit to the time dependence of the concentration of $c(t) = A(1 - \mathrm{Exp}[-t/\tau])$ gives $A = 2.03$ with a time constant of $\tau = 22.6$ min, yielding the well-known equilibrium ratio of the beta to alpha anomer of 2:1, which is clearly observed for times beyond the first hour of perfusion (Fig. 8.22).

The slope of the total glucose signals from the anomers is negative indicating glucose consumption from the perfusate. From the time course of the glycogen C1 signal shown in Fig. 8.22, one can determine that glycogen was synthesized at a rate of 34.1 μmol glc g^{-1} min^{-1}, which is congruent with results obtained by chemical assay methods (Katz et al. 1979). One also notes that ^{13}C from the added glucose also appears in the glucose/glycogen C6 resonance near 63 ppm (Figs. 8.21 and 8.22), indicating that some ^{13}C initially introduced at glucose Cl has been metabol- ically shifted to C6 by the flow down the glycolytic pathway to the triosephosphate step (Figs. 8.23 and 8.24) where triose phosphate isomerase interconverts the trioses glyceraldehyde-3-phosphate and dihydroxyacetone phosphate and the ^{13}C becomes equilibrated between C1 and C6 of the two molecules. Subsequently, aldolase

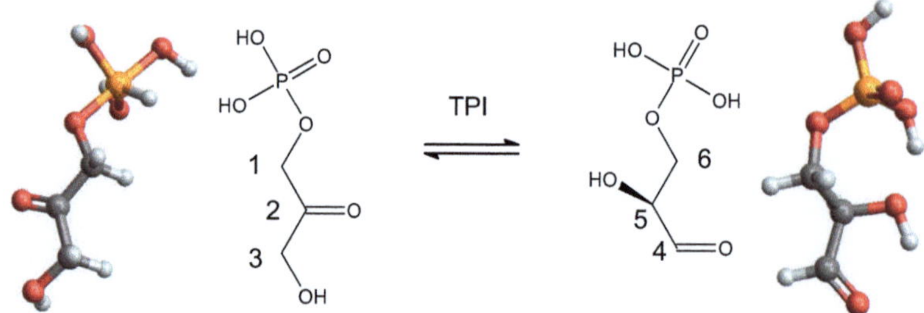

Fig. 8.23 A schematic of the biochemical reaction catalyzed by the glycolytic enzyme triose phosphate isomerase (TPI) that equilibrates the added [1-^{13}C]-label on glucose between C1 of dihydroxyacetone phosphate on the left and C6 of glyceraldehyde-3-phosphate on the right

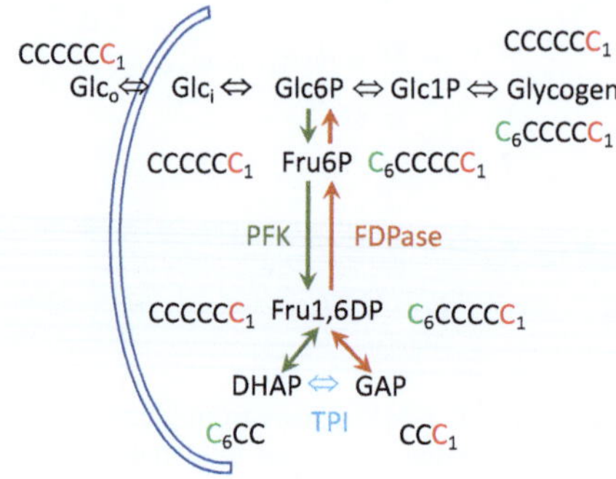

Fig. 8.24 Metabolic pathway for the uptake of extracellular [1-^{13}C]-glucose (Glc$_o$) by a liver cell and its enzymatic conversion directly to glycogen, as well as to the triose phosphates (DHAP and GAP) where ^{13}C initially present at Glc. **C1** is also equilibrated with Glc **C6** by triose phosphate isomerase (TPI), which then sends the ^{13}C label back up to the **C6** position of glycogen. PFK is phosphofructokinase, while FDPase is fructose diphosphatase. The double blue line represents the cell's plasma membrane phospholipid bilayer

rejoins the ^{13}C-labeled glyceraldehyde-3-phosphate and dihydroxyacetone phosphate to form fructose-1,6-diphosphate which is dephosphorylated by fructose diphosphatase to yield fructose-6-phosphate. Phosphoglucomutase then reforms glucose-6-phosphate, now bearing the ^{13}C label at both positions C1 and C6, and this is incorporated into the growing glycogen molecule. The result of this pathway is that the ^{13}C label initially added on C1 of glucose moves to C6 of the newly synthesized glycogen.

We can measure the rate of this cycling to the level of triose phosphate isomerase by means of the ratio of the slopes of the time dependence of the relaxation-corrected integrals of the glycogen C1 and C6 resonances (Fig. 8.22). The measured slope of the C1 curve is 14.8 µmol/min, while that for the C6 line is 2.1 µmol/min indicating that 14% of the glucose incorporated into glycogen cycled into glycolysis through

TPI and aldolase (Fig. 8.23) before reformation as glucose to be incorporated into glycogen in the liver from this fed rat. This is information that would be tedious to obtain using earlier techniques for measuring isotopic incorporation into storage macromolecules, but is obtained with ease using the noninvasive nature of ^{13}C NMR spectroscopy in living systems.

In the fasted rat, the liver is in a hormonal state supported by the elevation of glucagon that stimulates the hydrolysis of glycogen to support blood glucose levels at their normal value of ~5.5 mM. When a liver in this state is perfused with abundant glucose, it will be conditioned to store glucose as glycogen in response to an elevated glucose concentration in the blood, as would occur after a meal. We mimicked this response by withdrawing food from a rat for two days and then perfusing its liver with a large glucose load (30 mM). A comparison of the results obtained from the perfusion of the livers of fed and fasted rats (Fig. 8.25) shows that both livers incorporated glucose into glycogen C1 at the same rate of 14.8 and 15.1 µM/min, respectively, but differed markedly with respect to the rate of accumulation of the ^{13}C label in the C6 position of glycogen. In the fed state this rate was 2.0 µM/min, while in the fasted state this rate increased 3.5-fold to 7.1 µM/min, indicating that in the fasted state the liver glycolytic pathway was significantly elevated over that in the fed state. The ratio of these two rates gives the percentage of glucose that branched into the glycolytic pathway and was subjected to TPI where ^{13}C at position C1 of glucose became C6 as it was reformed from the trioses in the gluconeogenic arm. We found that 14% of the glucosyl units in newly formed glycogen first traversed this pathway prior to incorporation into glycogen in the fed state versus 48% in the fasted state.

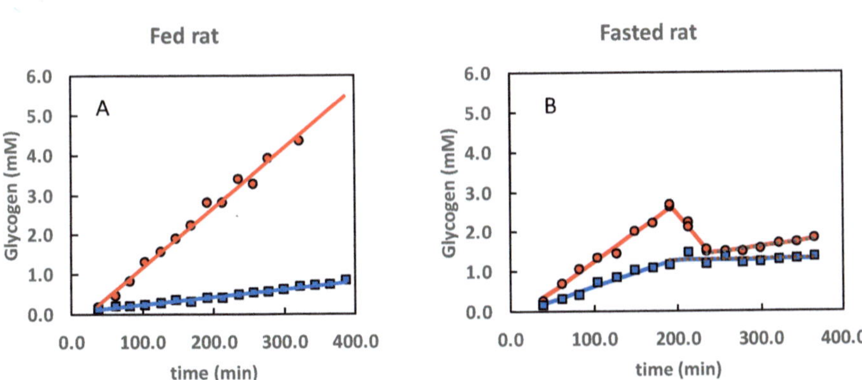

Fig. 8.25 Time course of the incorporation of ^{13}C from [1-^{13}C]-glucose into **C1** (red) and **C6** (blue) of glycogen in the perfused rat liver. (**a**) Liver from a fed rat and (**b**) from a fasted rat. Note that the glycogen is synthesized at the same rate in the fasted rat liver as in the liver from a fed rat and that the rate of incorporation of ^{13}C into **C6** is smaller in the fed rat liver (see text). In the fasted rat liver, the addition of glucagon (10 nM) to the perfusate at 210 min dramatically reduced the rate of ^{13}C incorporation at both the **C1** and **C6** positions

Here is a lesson in metabolism that is readily revealed by ^{13}C NMR spectroscopy; metabolic pathways are not one-way streets; they are networks of coupled reactions that operate simultaneously as both catabolic and anabolic flows with the net result determined by the hormonal status that controls the relevant enzymatic reaction rates.

These features could easily be reversed, even in the fasted rat liver by administering glucagon, as is illustrated (Fig. 8.25b) by its dramatic inhibition of glycogen synthesis and by its attenuation of the flow of glucose into glycolysis, as indicated by the change in slope of both the C1 and C6 time courses at 190 min into the perfusion run. After the introduction of a physiological (10 nM) amount of glucagon, the rates of glucose incorporation into glycogen decreased to 2.8 μM/min for C1 and 0.3 μM/min for C6, a rate ratio of 11%, which is essentially the same as that found for the scrambling rate found above for the fed rat liver.

One of the points to keep in mind is that we observed only a small flow of ^{13}C from the introduced [1-^{13}C]-glucose into any other carbons of glycogen besides C6, which was primarily accounted for through the flow into and out of glycolysis and TPI activity. However, a close inspection of Fig. 8.21c shows a small signal at ~74 ppm, which is the resonance position of the glucose C2,5α carbons (Table 8.4) with an amplitude about 15% of the strength of the signal from glucose C6. This puts limits on the possible flow of ^{13}C from [1-^{13}C]-glucose into the Krebs cycle as we shall see in the next section.

8.8.3.5 Gluconeogenesis from [3-^{13}C] Alanine

The third aspect of hepatic glucose metabolism of interest is the anabolic process of gluconeogenesis. This metabolic pathway for the synthesis of "new" glucose from simpler precursors, such as amino acids, is of vital importance in the maintenance of normal blood glucose levels in between meals and during fasting, especially overnight. Gluconeogenesis comes to the fore when hepatic glycogen supplies decrease and eventually become exhausted, as in a fasted rat after two days without food. The amino acid alanine is the most important precursor of glucose during gluconeogenesis from amino acids (Exton 1972). Other gluconeogenic precursors include lactate, glycerol, and pyruvate. Figure 8.26 shows the results of an experiment where 11 mM [90% 3-^{13}C]-alanine was added to the perfusate of a fasted rat liver. After perfusion for 2.7 hours (Fig. 8.26c) one can observe not only all the six carbons of glucose (see Table 8.4 for peak positions) synthesized from the added alanine, but also fructose carbons C1α,β and C6α,β at ~65 ppm (see Appendix A.4 for assignments) as well as a signal from glycogen C1 deposited in response to the increased cytoplasmic glucose levels in this liver (Fig. 8.26b). The labeling of glutamine carbons C2,3,4 arises from the flow of carbon from alanine into the Krebs cycle.

The difference spectrum (Fig. 8.26b) is again useful here since it enables one to remove the ^{13}C signals from the natural abundance liver background and to therefore observe only those molecules that accumulated ^{13}C from C3 of the added alanine

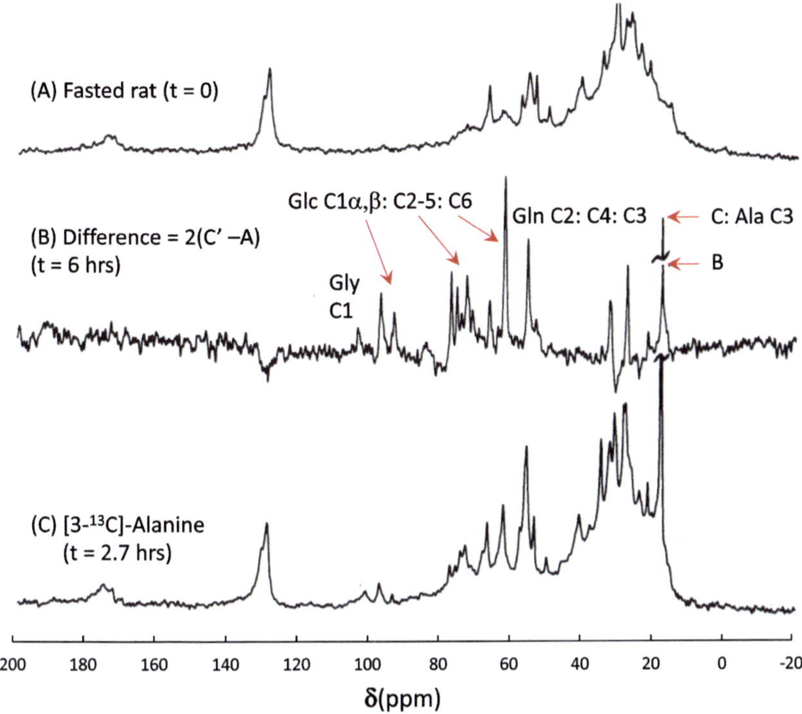

Fig. 8.26 ^{13}C NMR spectra of a perfused liver from a fasted rat supplied with [3-^{13}C]-alanine. (**a**) Proton-decoupled, natural abundance spectrum of a perfused liver from a fasted rat (aq = 20 min). (**b**) The digital difference spectrum 2(C'-A) taken between $t = 0$ and $t = 6$ hr, showing the signals from ^{13}C accumulation into glycogen, glucose, and glutamine, as well as the residual signal from alanine C3 (labeled B with an arrow in (**b**)). Note the appearance of all six carbons of glucose, but not with equal strength (see text). (**c**) The spectrum of this liver 2.7 hours after the addition of 11 mM [3-^{13}C]-alanine; the initial alanine C3 signal is labeled C: Ala C3 in (**c**)

substrate. In this way we can robustly measure the signals from the interesting and important glycolytic/gluconeogenic intermediates, fructose and fructose-1- and fructose-6-phosphates at 67, 86, and 105 ppm, respectively, in Fig. 8.26b.

The flow of ^{13}C from alanine C3 into the gluconeogenic pathway (Fig. 8.27) would introduce the ^{13}C nucleus into both carbons 1 and 6 of *neo*glucose via the action of the TPI/aldolase reactions (Figs. 8.23 and 8.24), but not into carbons 2–5. The ^{13}C NMR spectra (Fig. 8.26b) clearly indicate that the newly formed glucose bears ^{13}C nuclei at carbons 2 and 5, in addition to the appearance of ^{13}C nuclei at carbons 1 and 6, which we have encountered earlier. Comparison of the spectral intensities for the glucose molecule, in its unlabeled, natural abundance state containing a uniform 1.1% ^{13}C at each carbon, with those from the glucose that

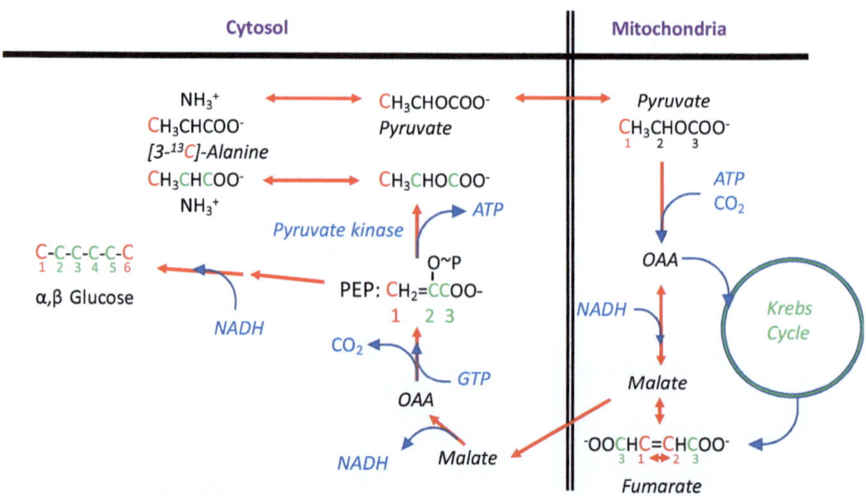

Fig. 8.27 A schematic of the flow of ¹³C from [3-¹³C]-alanine that shows how the carbon label moves through both the cytosol and the mitochondria with the result that the labeling pattern in glucose reflects the relative rates of the different pathways. The carbons in red arise directly from alanine, while the carbons in green arise from Krebs cycle activity. PEP is phosphoenolpyruvate. Cofactors and enzymes are shown in blue

resulted from gluconeogenesis using [3-¹³C]-L-alanine, showed that carbons 1,2,5 and 6 received equal amounts of ¹³C from alanine, but that carbons 3 and 4 of the *neo*glucose received only ~40% of the flux of ¹³C from the alanine substrate.

We observe here another of the inherent strengths of ¹³C spectroscopy for the study of metabolism in situ in that one can follow the fate of each carbon nucleus in the substrate, almost in real time, and that this pattern in metabolic products constitutes a fingerprint for the metabolic pathway(s) through which the substrate passed on its way to the given metabolic product.

Many details of metabolic networks can be probed by examining the time course of ¹³C label flow through the various molecules involved. A temporal map (Fig. 8.28) derived from several 20-min difference spectra of the liver (like Fig. 8.26b) of the flow of ¹³C from alanine C3 into carbons C1(α+β) and C3α of glucose shows two facets of this process. The first is that, during gluconeogenesis, the incorporation of ¹³C from alanine C3 into carbons C1(α+β) of glucose proceeded linearly at a rate of 23.4 μM/min for 350 min as the alanine supply in the perfusate was consumed. We have already commented on the nonuniform labeling pattern of glucose. The amount of ¹³C at glucose carbons 1,6 and 2, 5 is equal, but 2.5-fold greater than that found at carbons 3,4. We observe this pattern in the time course (Fig. 8.28, green squares) but also notice a delay of 60 minutes in the appearance of significant ¹³C in glc C3 due to the need for the reactions in the Krebs cycle to move the ¹³C label to the meso compound malate with subsequent backflow up the gluconeogenic pathway and back to glucose (Fig. 8.27). Then, the incorporation of

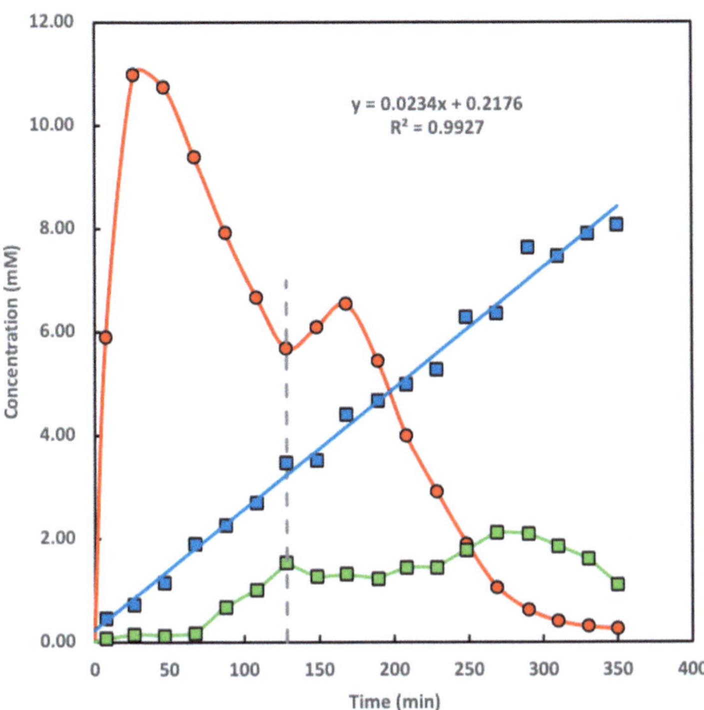

Fig. 8.28 Time course of the ^{13}C NMR signals from a perfused liver from a fasted rat. At time zero a bolus of [3-^{13}C]-L-alanine was added to the perfusate, and sequential spectra were taken every 20 minutes. The red circles are the ala C3 signals; the blue squares are the glucose C1(α+β) peaks, while the green squares are the signals from glucose C3α. Note that both glucose signals have been multiplied by ten to fit on the same scale with the alanine. Glucagon was added to the perfusate at 128 min as indicated by a gray dotted line. Note the constant slope for the glc C1 data, but the abrupt slope change in the glc C3α after the introduction of glucagon

^{13}C into C3 proceeded at a rate of 21.9 μM/min essentially identical to the rate of ^{13}C incorporation into glc C1. The ratio of the signal integrals for glucose C1 to C3 is 2.78 $\pm$ 0.58 which is within the experimental error of the expected 2.5.

As indicated above, the label distribution among the various glucose carbons is not uniform; the amount of ^{13}C at glc C3α is much smaller than at C1 (Fig. 8.28) illustrating one of the unique features of ^{13}C NMR applied to metabolic studies. Not only can we follow the label's disappearance from its initial site, but we can also follow its subsequent appearance in any metabolite concentrated enough to be observed and we can tell from chemical shift information which metabolite carbons pick up labeling.

In the pathway leading from alanine to glucose (Fig. 8.27), only the methyl carbon of alanine initially contains ^{13}C above the natural abundance level. As the pyruvate derived from alanine passes through the TCA cycle (as oxaloacetate, malate, and fumarate), the amount of ^{13}C at C2,3 of fumarate is equalized by scrambling on the enzyme fumarase. Subsequent gluconeogenesis results in equal ^{13}C intensities at carbons 1,2,5 and 6 of glucose and no ^{13}C labeling at carbons 3 and 4. We also see that any activity of pyruvate kinase will convert the phosphoenol-pyruvate formed back to pyruvate and hence alanine.

The initial alanine feeding was consumed at a rate of 17.6 µmole min^{-1} and gave rise to glucose at a rate of 0.34 µmole g^{-1} min^{-1}. This latter value agrees with similar measurements using radioactive alanine by Yount (Yount and Harris 1980). When glucagon was added to the perfusate at 128 min, the flow of ^{13}C into glc C3α abruptly ceased (Fig. 8.28) due to inhibition of label scrambling in the Krebs cycle in the mitochondria (Fig. 8.27). Glucagon treatment also temporarily increased the alanine concentration of the liver, as shown by the signal increase after glucagon infusion at 128 min, because glucagon stimulated amino acid transport from the perfusate into the liver (Mallette et al. 1969), but since this liver was from a fasted rat, in which gluconeogenesis was already maximally stimulated by the higher glucagon levels bound in the fasted rat than in a fed rat, there was no further stimulation by glucagon of the rate of gluconeogenesis. Furthermore, the concentration of alanine used is high enough to saturate the rate of gluconeogenesis.

8.9   Glutamate/Glutamine Synthesis from [3-^{13}C]-Alanine

After perfusion with [3-^{13}C]-L-alanine for 6 hours, one can observe in the difference spectrum (Fig. 8.26b) not only the *neo*glucose formed from gluconeogenesis, but also signals from carbons C2, C3, and C4 of glutamate/glutamine at ~57, 34, and 30 ppm, respectively (see Appendix Four for assignments), arising from the diversion of alanine to amino acids either through the Krebs cycle or via pyruvate dehydrogenase. The dynamics of ^{13}C label flow from [3-^{13}C]-L-alanine into glutamate and glutamine are shown in Fig. 8.29 where one observes that glutamate C2 and C3 rapidly acquire the ^{13}C label essentially equally, but the signal intensity at glutamate C4 is only 6% of the amount found for C2 and C3. The flow of ^{13}C into glutamine C3 is delayed as glutamate must be transported out of the mitochondria for the cytosolic reaction catalyzed by glutamine synthetase to form glutamine from glutamate (Fig. 8.30). The appearance of ^{13}C in glutamate C2 and C3 results from traversal of the fumarase branch of the Krebs cycle, while C4 is alternatively labeled through the condensation of ^{13}C-acetyl-CoA and oxaloacetate to form citrate. The small amount of label in glutamate C4 indicates that more than 90% of the ^{13}C was sequestered in glutamate via the route through fumarase.

Examination of the pathways in Fig. 8.30 shows that after alanine is converted to pyruvate it can take two alternative paths, and as Yogi Berra said, *"When you come to a fork in the road, take it!"* One branch involves oxaloacetate, via pyruvate carboxylase, and on to malate where the ^{13}C is equilibrated between carbons 2 and

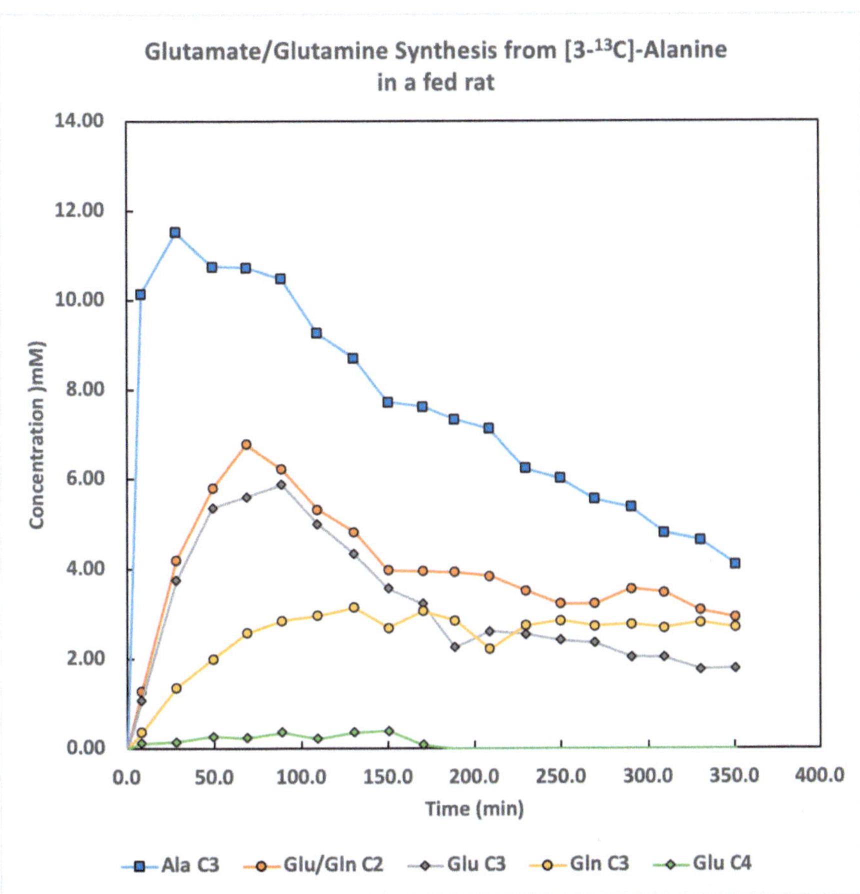

Fig. 8.29 The time course of NMR signals from [3-^{13}C]-alanine injected into the perfusate of the liver from a fed rat. There is substantial labeling of gln/glu carbons C2 and C3, but little ^{13}C is found in C4 of either amino acid

3 by fumarase ending up in glutamate C2 and C3. The other branch involves acetyl CoA, via pyruvate dehydrogenase, with the label ending up in glutamate C4. In the present example most of the alanine was metabolized by the first pathway, indicating that the ratio of pyruvate carboxylase activity to that of pyruvate dehydrogenase activity was at least 7:1.

The hormonal influence on liver metabolism is once again demonstrated by an examination of the fate of carbon nuclei from [3-^{13}C]-alanine in the perfused liver from a fasted rat (Fig. 8.31) where it is observed that ^{13}C is incorporated into carbons C2, C3, and C4 of both glutamate and glutamine equally at an initial rate of 47 μM min^{-1} up to the time (~130 min) when glucagon was added to the perfusate. There was a delay of about 40 min before glutamine C4 began to pick up the ^{13}C (green squares in Fig. 8.31) due to need to transport glutamate from the mitochondria

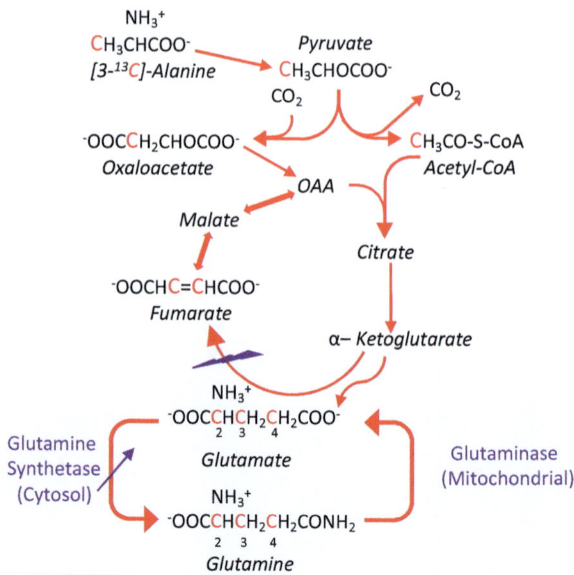

Fig. 8.30 A metabolic scheme showing the flow of ¹³C from alanine **C3** into glutamate and glutamine carbons C2, C3, and C4 in the Krebs cycle. Glu/Gln C2 and C3 are labeled via fumarase, while C4 is labeled through acetyl-CoA and citrate

into the cytosol and the slow activation of cytoplasmic glutamine synthetase which is inhibited in resting conditions. After the addition of a high, but still physiological dose (20 nM) of glucagon, the rate of glutamine/glutamate synthesis declined to almost zero (<0.5 µM min^{-1}) as indicated by the marked decrease in the slope of the C2 and C3 signals around 130 min (Fig. 8.31). The most marked change was found for the rate of incorporation of ¹³C from alanine C3 into glutamine C4, which became negative at -0.017 µM min^{-1} due to glucagon's inhibition of pyruvate dehydrogenase and the subsequent flow through citrate synthetase, while the flow through fumarase was maintained. This is another example of how ¹³C nuclear magnetic resonance spectroscopy can be used to trace the fate of individual carbon nuclei in metabolic pathways and reveal intimate details of hormonal regulation.

8.10   Synthesis of Ketones from [3-¹³C]-Alanine

In addition to gluconeogenesis, ¹³C NMR can reveal the fate of any carbons metabolically generated from [3-¹³C]-alanine whose concentration is large enough to be detected in a reasonable amount of NMR time. In the perfused liver that time frame is around 10–20 min. When the spectrum from the perfusate from a liver perfused with [3-¹³C]-alanine was examined at the end of the 6-hour perfusion run, we found that the liver cells generated additional metabolites of interest and secreted them into the perfusate (Fig. 8.32). We observed not only glucose, but also the ketone bodies, β-hydoxybutyrate and acetoacetate, which were labeled at carbons C2 and C4 (orange circles in Fig. 8.33). These ketone bodies are made in the liver under

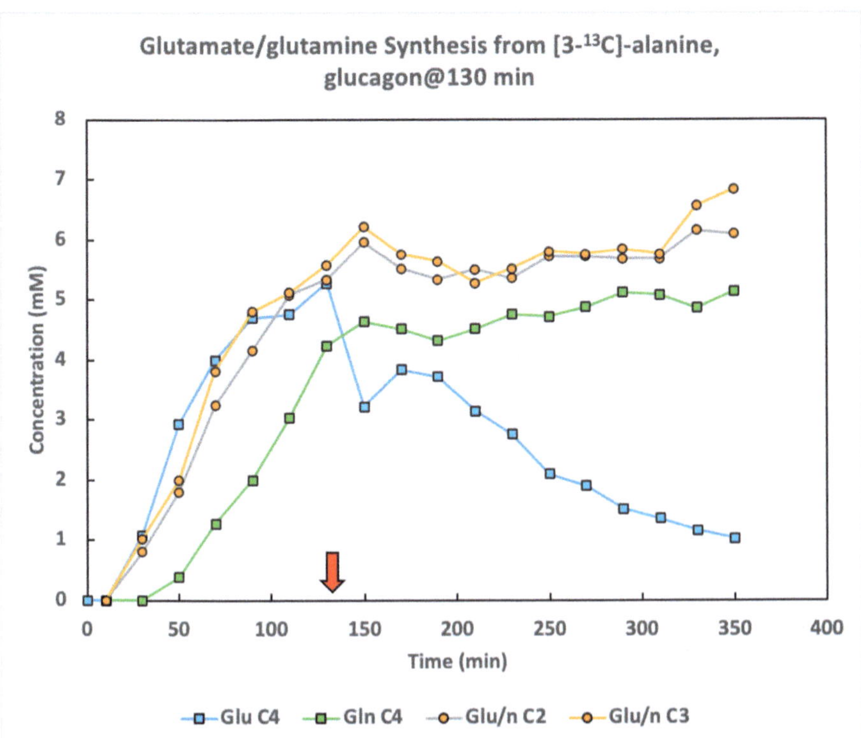

Fig. 8.31 The time course of ^{13}C incorporation into glutamate and glutamine from [3-^{13}C]-alanine (11.3 mM) in the perfused liver from a fasted rat. Glucagon (20 nM) was added to the perfusate (red arrow) at 130 min after the alanine. Note the slope of all the curves is the same (47.5 ± 3.5 μM min^{-1}) up to the addition of glucagon and then drops abruptly afterwards

circumstances of intense gluconeogenesis, such as when glucose is produced from amino acids, such as alanine, and are released into the bloodstream along with the newly synthesized glucose. β-Hydoxybutyrate (CH_3-HCOH-CH_2-COO$^-$) appears in the perfusate of the liver (Fig. 8.33) with signals at $\delta_{C2} = 49.16$ ppm and $\delta_{C4} = 24.51$ ppm (Table 8.5 and Fig. 8.33). These carbon arise from the ^{13}C-carrying methyl group of acetyl-CoA which is elevated under these conditions. The free acetyl groups combine to form a covalent dimer (acetoacetate) which upon reduction gives β-hydoxybutyrate. We also note the presence of β-hydoxybutyrate in the perchloric acid extract from this liver (Figs. 8.32, 8.34, and Table 8.5). The signals from carbons C1 at 183.15 and C3 68.23 ppm of β-hydoxybutyrate (Table 8.5) are below the limits of detection because these carbons are formed from the unlabeled carbonyl carbons of the two acetyl-CoA molecules derived from the added [3-^{13}C]-alanine.

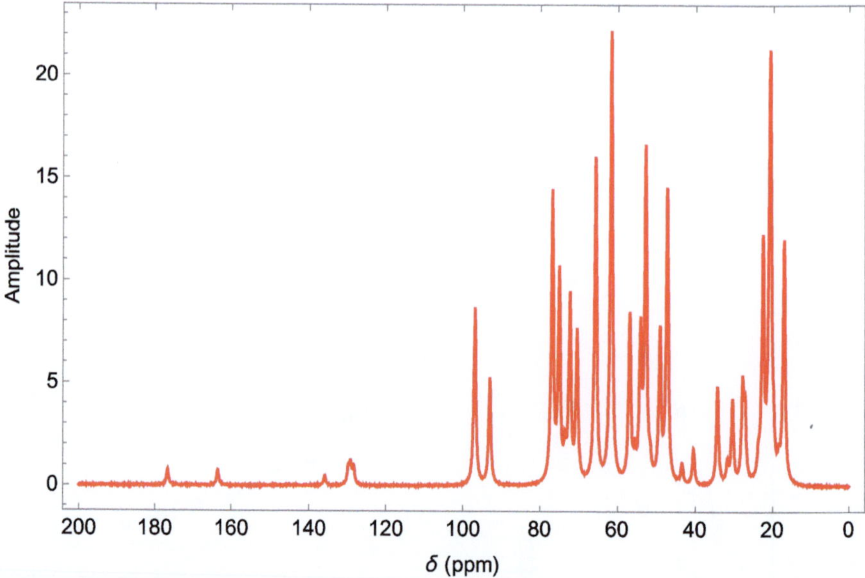

Fig. 8.32 A ^{13}C NMR spectrum of the perfusate from a liver that had been perfused with [3-^{13}C]-alanine. Assignments for the resonances are given in Table 8.5

The labeling dynamics for several of metabolites of interest are shown in Fig. 8.33. As we have previously found, the added [3-^{13}C]-alanine is rapidly taken up by the liver cells and enters the metabolic network when converted to pyruvate (Fig. 8.30). A portion of the added alanine is then deaminated to form lactate. The flow of ^{13}C into lactate C3 parallels the uptake of [3-^{13}C]-alanine into the cytoplasm of the hepatocyte; we measured a linear relationship between the integral of the alanine C3 signal and that from lactate C3, with a slope of 0.158, indicating that lactate is labeled at 16% of the rate of alanine uptake. The conversion of alanine into lactate occurs via alanine transaminase to pyruvate and on to lactate by means of lactate dehydrogenase, preserving the ^{13}C label at lactate C3 and explaining why the signals for lactate carbons C2 and C1 are below the limits of detection (Table 8.5).

The glucose produced by gluconeogenesis from the added [3-^{13}C]-alanine was equally labeled at carbons C1,2 and C5,6 (Fig. 8.33) but contained only 50% as much ^{13}C at positions C3,4 indicating, as we have seen above, that carbons 3 and 4 of the *neo*glucose received only ~50% of the flux of ^{13}C from the alanine substrate. While signals from glycogen produced during gluconeogenesis appear in the perchloric acid extract, none leaves the liver to be found in the perfusate (Fig. 8.34 and Table 8.5).

Most of the metabolites detected in the liver extracts and perfusates can be straightforwardly assigned and understood based on the reaction networks already explored (Figs. 8.23, 8.27, and 8.30) but there are clear signals in the spectrum of the PCA extract (Fig. 8.17) and the perfusate (Fig. 8.32) in the downfield region that

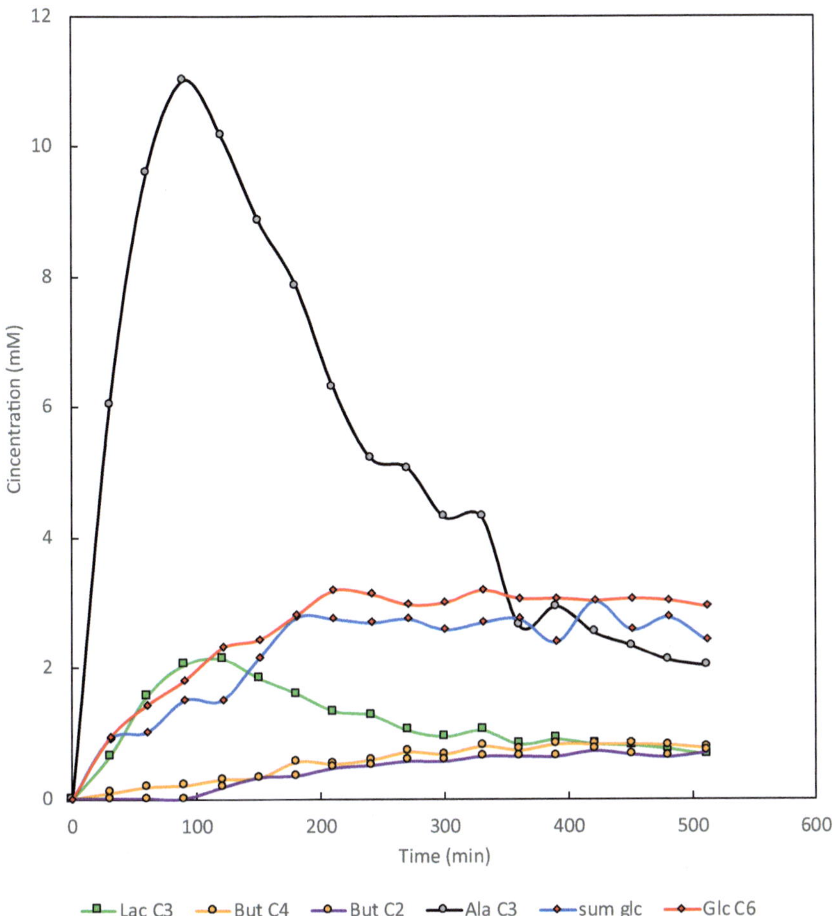

Fig. 8.33 Time course of the flow of ^{13}C from [3-^{13}C]-alanine (grey circles) in the perfused liver from a fed rat. Lactate C3 (green squares) is generated directly from the [3-^{13}C]-alanine by deamination so that its time course parallels that of alanine. In addition to the glucose (red, C1 (α+β); blue, C6) formed, we also notice the production of the ketone body, β-hydoxybutyrate which is labeled at C2 and C4 (orange circles) from acetyl-CoA

require further investigation. In particular, signals at 131, 137, and 165 ppm have chemical shifts characteristic of carbons in the nicotinic acid moiety of nitrogen-containing heterocycles such as NAD$^+$ and NADP$^+$ (Table 8.5). We have seen that ketone synthesis is ongoing (Fig. 8.33) in the fasted rat livers perfused with [3-^{13}C]-alanine. Under these conditions, the oxaloacetate produced can be transaminated to aspartate, which could enter the biosynthetic pathway for the nicotinamide portion of NAD and acquire ^{13}C at several specific ring positions, including C1, C9, and C18 (Fig. 8.35). The NAD in the extract would have been oxidized, because the shifts of the reduced (Table 8.6) form are quite distinct and are not seen in the spectra.

Table 8.5 Comparison of the ^{13}C NMR spectra of a PCA extract from a liver with the perfusate after perfusion with [3-^{13}C]-alanine (see also Fig. 8.35)

Carbon Assignment	Extract δ(ppm)	Perfusate δ(ppm)
Ala C3	18.96	18.95
Lac C3	22.84	22.85
β-OH-But C4	24.51	24.55
Gln C3	29.04	29.13
Glu C3	29.30	29.90
AcAc C4	32.31	32.40
Gln C4	33.63	33.72
Glu C4	36.24	36.28
EtOHNH2	42.23	—
β-OH-But C2	49.16	49.27
Ala C2	53.44	53.58
AcAc C2	56.06	56.10
Gln C2, Glu C2	57.08	57.21
Glc C6α	63.45	63.62
Gly 6	63.49	–
Glc C6β	63.58	63.49
Glc-6-P C6	65.47	—
β-OH-But C3	68.41	68.23
Lac C2	71.27	—
Glc 4α	72.24	72.28
Glc 4β	72.45	72.44
Gly C2,5	74.18	–
Glc 2,5α	74.31	74.08
Gly C3	75.57	–
Glc 3α	75.61	75.86
Glc 2β	77.01	76.81
Glc C3β, 5β	78.63	78.59
Glc C1α	94.86	94.84
Glc C1β	98.69	98.48
Gly 1	102.55	–
NAD$^+$ Nic C1	131.27	131.07
NADP$^+$ Nic C1	131.27	131.66
NAD$^+$ Nic C9	–	137.75
NAD$^+$ Nic C18	–	165.45
Ala C1	178.72	178.61

8.11 NMR Probes of the RedOx Status of Cells

We have previously examined (Chap. 2) the reduction/oxidation reaction in which nicotinamide adenine dinucleotide (Fig. 8.35) participates in order to maintain the RedOx state of the cell. One can measure the ratio of the reduced to the oxidized

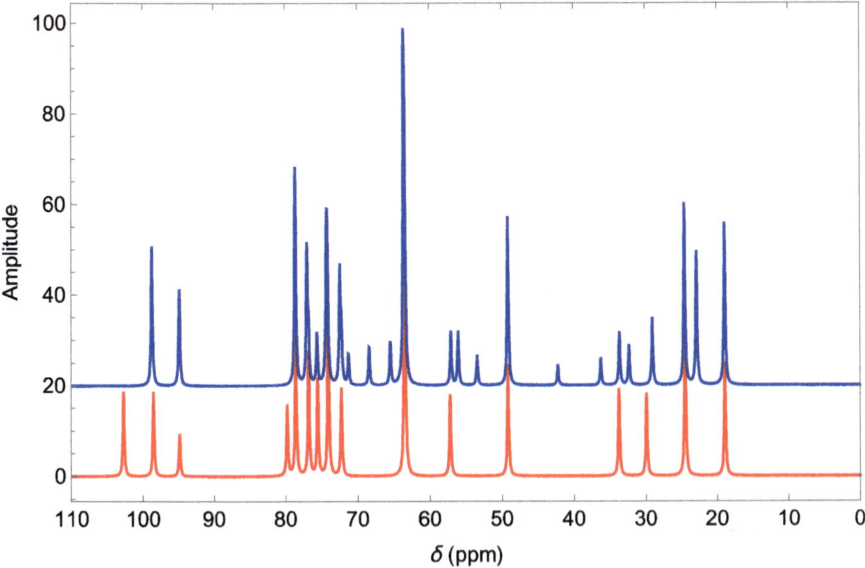

Fig. 8.34 Comparison of the central regions of the ^{13}C NMR spectra of a PCA extract (**red**) from a perfused liver with the perfusate (**blue**) after 6 hours with [3-^{13}C]-alanine. The peak assignments are given in Table 8.5

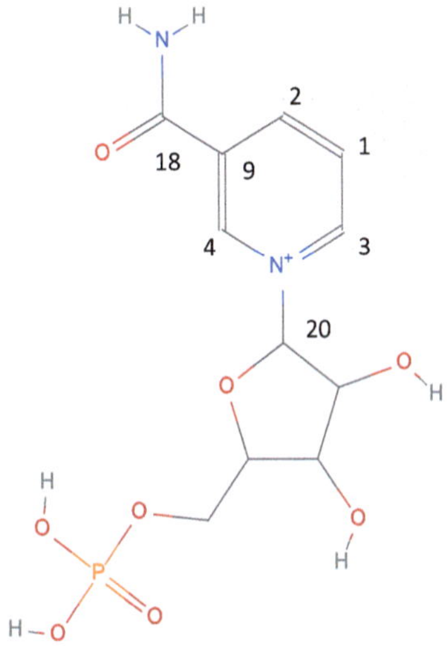

Fig. 8.35 The chemical structure of the nicotinamide riboside portion of NAD+, NADH, NADP+, and NADPH showing the carbon numbering scheme from the BioMagnetic Resonance data bank of the ring and ribose carbons whose chemical shifts reflect the RedOx status of the heterocyclic ring electrons (Table 8.6)

forms of NADH and NADPH using the large changes in the chemical shifts of seven carbons in the nicotinamide and riboside rings (Table 8.6). Proton spectroscopy can also be used in vivo to measure the amounts of NAD+ in the living rat brain (de Graaf and Behar 2014) and in human brains (de Graaf et al. 2017) where the in vivo concentrations are found to be ~300 μM.

Table 8.6 Chemical shifts of the ^{13}C NMR signals of carbons of nicotinamide sensitive to the cellular RedOx state

C#	NADPH	NADP+	$\Delta\delta$(ppm)	NADH	NAD+	$\Delta\delta$(ppm)
Nic C1	108.07	131.35	23.28	108.08	131.33	23.25
Nic C2	24.67	148.38	123.71	24.62	148.43	123.82
Nic C3	126.49	145.02	18.53	126.44	145.05	18.62
Nic C4	141.09	142.89	1.80	141.14	142.65	1.51
Nic C9	102.60	136.30	33.70	102.56	136.30	33.74
Nic C18	175.43	167.85	−7.58	175.29	167.80	−7.49
Rib C1 (C20)	97.74	102.73	4.99	97.83	102.74	4.91

It is useful to summarize at this point the main features of ^{13}C NMR studies of metabolism in perfused liver. The natural abundance liver spectrum is useful for the study of several areas of glucose metabolism, the primary example being hormonal control of glycogenolysis or glycogenesis. The introduction of specific ^{13}C labels enables the tracing of pathways in gluconeogenesis to be accomplished with relative ease. In a broader sense we now see how to probe the biochemical details of how organisms self-assembled the complex interacting network of reactions that enabled living systems to extract energy from their environment in order to compensate for the entropy reduction attendant on the origin of life.

Problems

1. **Isotopic enrichment.** The use of ^{13}C isotopically enriched substrates and NMR spectroscopy as a detection method to study carbohydrate metabolism in the liver provides unique information about the pathways traversed by metabolites in real time.

   A. For glycogenesis in the liver using [1-^{13}C]-glucose as the starting material, explain by tracing the flow of each glucose carbon atom, the appearance of the ^{13}C label in carbons 1 and 6 of glycogen.

   B. For gluconeogenesis in the liver using [3-^{13}C]-alanine as the starting material, explain by tracing the flow of each alanine carbon atom, the appearance of the ^{13}C label in the resulting:

   (a) Carbons 2 and 5 of glucose.
   (b) How many turns of the Krebs cycle are required in order to explain the appearance of ^{13}C in these carbons of glucose?
   (c) Carbons 2 and 3 of glutamate.
   (d) How many turns of the Krebs cycle are required in order to explain the appearance of ^{13}C in these carbons of glutamate?

   C. Can there be ^{13}C label flow from [3-^{13}C]-alanine into carbon 4 of glutamate? Explain how this would happen.

2. **Quantitative NMR spectroscopy.** Compare and contrast the procedures needed to produce quantitative NMR spectroscopic measurements of metabolite concentrations by means of

   A. Proton signals
   B. ^{13}C signals

3. **Glucose metabolism in the human liver.** How would you use ^{13}C NMR spectroscopy to measure the relative rates of glycogen synthesis and glycolysis in the human liver?

4. **Citrate metabolism in the human prostate.** The male human prostate gland has a unique metabolism that shunts substrates from the Krebs cycle into the production of large amounts of citrate which is exported into seminal fluid. The liver, on the other hand, utilizes citrate as a metabolic intermediate in the Krebs cycle that is then consumed by further reactions so that it does not accumulate in this organ. Look up the pathways leading to and away from citrate and find the rate controlling enzymes for the products in order to explain this metabolic shift (a useful suggestion can be found in Sillerud et al. 1988).

   a. If you were to use proton NMR spectroscopy in vivo to measure the amounts of citrate and triacylglycerols in human patients, what chemical shift region would you find most useful?
   b. During tumorigenesis in the prostate, find out which metabolite concentrations are altered and propose a change in enzymatic regulation that explains this shift.
   c. Could you use proton NMR spectroscopy in vivo to diagnose prostate cancer?
   d. The measurement of the serum concentration of human kallikrein III (also known as prostate-specific antigen [PSA]) has been used for several decades as a screening tool for prostate cancer. The concentration of citrate in either the prostate gland or in seminal fluid changes during tumorigenesis. Which correlates better with cancer, [citrate] or [PSA], and why? (Useful suggestions can be found in Averna et al. (2005) and Kline et al. (2006))

Solutions

1A. *Carbons 1 and 6 of glycogen:* Carbon 1 of [1-^{13}C]-glucose will directly label C1 of glycogen through glycogen synthase. To label C6 of glycogen, one needs aldolase to split the [1-^{13}C]-glucose into two trioses, glyceraldehyde-3-phosphate (G3P) and dihydroxyacetone phosphate (DHAP). Triosephospate isomerase activity then equilibrates the ^{13}C label between C3 of each triose. Trace the label flow up the glycolytic pathway back to glucose, which is now labeled, but not equally, at both C1 and C6. The ratio of the ^{13}C label at C1 to C6 tells you the relative rates of direct glycogenesis from glucose and the flow through glycolysis.

1B. (a) ***Carbons 2 and 5 of glucose:*** The first step is to label C1 and C6 of glucose. The direct flow of ^{13}C label from the methyl group of [3-^{13}C]-alanine to C1 and C6 of glucose is understood by considering that ala-C3 directly labels pyruvate C3, which in turn forms [3-^{13}C]-oxaloacetate (OAA) via pyruvate carboxylase. The OAA formed can move out of the mitochondrion via the malate shuttle, be reduced using cytosolic NADH to OAA again, and form [3-^{13}C]-phosphoenol-pyruvate (PEP) through the enzyme phosphoenolpyruvate carboxykinase (PEPCK). Then PEP travels up the glycolytic pathway where two [3-^{13}C]-trioses are joined to form glucose, where both C1 and C6 contain ^{13}C.

The second step, which labels glucose C2 and C5, involves the necessary movement of ^{13}C from C3 to C2 of OAA via the conversion of malate to the symmetric (*meso*) molecule fumarate. Since fumarate has C_{2v} point group symmetry, it can bind to the enzyme fumarase in either of two indistinguishable ways, one simply rotated 180 about the vertical symmetry axis. The product of fumarase activity is then either [2-^{13}C]-malate or [3-^{13}C]-malate, in essentially equal amounts. From these labeled malate molecules is produced either [2-^{13}C]-OAA or [3-^{13}C]-OAA, which can move out of the mitochondrion via the malate shuttle, be reduced using cytosolic NADH back to OAA again, and form [2, 3-^{13}C$_2$]-PEP via PEPCK. This PEP travels up the glycolytic pathway, where two [2, 3-^{13}C$_2$]-trioses are joined to form glucose, with labels at C1, C2, C5, and C6.

1B. (b) How many turns of the Krebs cycle are required in order to explain the appearance of ^{13}C in these carbons of glucose?

Answer $=$ *No full turns are needed at all.*

1B. (c) ***Carbons 2 and 3 of glutamate:*** [3-^{13}C]-Alanine can label glutamate in two separate ways. In the first, path the [3-^{13}C]-alanine is converted into [3-^{13}C]-pyruvate by the enzyme alanine/pyruvic transaminase. Then the [3-^{13}C]-pyruvate enters the Krebs cycle via pyruvate carboxylase, which forms [3-^{13}C]-OAA, [3-^{13}C]-malate and [2, 3-^{13}C$_2$]-fumarate. This fumarate can traverse the Krebs cycle back up to [2, 3-^{13}C$_2$]-OAA, which joins with unlabeled acetyl-Coenzyme A to form [2, 3-^{13}C$_2$]-citrate and eventually, [2, 3-^{13}C$_2$]-α-*keto*-glutarate, the precursor that forms [2, 3-^{13}C$_2$]-glutamate via transamination with alanine. The second pathway is discussed in 1.(C), below.

1B. (d) How many turns of the Krebs cycle are required in order to explain the appearance of ^{13}C in carbons C2 and C3 of glutamate?

Answer $=$ *Not quite one full turn.*

1C. Can there be ^{13}C label flow from ***[3-^{13}C]-alanine into carbon 4 of glutamate***?

Yes, this time [3-^{13}C]-alanine is decarboxylated by pyruvate dehydrogenase to form [2-^{13}C]-acetyl-coenzyme A, which condenses with OAA to form [5-^{13}C]-citrate. Carbon 6 of this citrate is removed as CO_2 by isocitrate dehydrogenase to form [4-^{13}C]-α-*keto*-glutarate, which is once again transaminated to [4-^{13}C]-glutamate. Note that by measuring the label distribution in carbons 2,3 and 4 of glutamate and glutamine, one can determine the relative activities of pyruvate carboxylase and pyruvate dehydrogenase.

2. Compare and contrast the procedures needed to produce quantitative NMR spectroscopic measurements of metabolite concentrations by means of the following:

> 2a ***Proton signals:*** For protons one would need to know both the T_1 of the signal and the number of protons that contribute to the resonance. Upon integrating that resonance, one would compare its area with that of a standard of known concentration contained in the same sample and spectrum. In the absence of such an internal standard, one could compare the unknown signal with the noise in the spectrum, or with internal water, if the sample is dissolved in water, or with another internal solvent signal of known concentration.
>
> 2b *^{13}C signals:* One would need all of the information in 2.A as well as either the nuclear Overhauser enhancement or to acquire spectra where the nOe is suppressed.

3. How would you use ^{13}C NMR spectroscopy to measure the relative rates of glycogen synthesis and glycolysis in the human liver?

> Since ^{13}C is stable, nonradioactive, one can safely ingest nutritional concentrations of ^{13}C molecules, such as [1-^{13}C]-glucose. You would feed your subjects sufficient [1-^{13}C]-glucose. How much? A standard glucose tolerance test uses 100 g of glucose, so it is known that this amount of glucose is safe. This is about the amount of sugar that is contained in one 36-ounce sweetened soft drink. You would then place a coil tuned to the ^{13}C resonance frequency in your magnet over the liver and acquire data as a function of time to determine the disappearance of the C1 signal from glucose and the appearance of the signal from C1 of glycogen, to give you the rate of glycogenesis. The rate of the subsequent appearance of ^{13}C in C6 of glycogen would give you the rate of glycolysis. T_1 measurements would also be needed. Corrections for finite acquisition parameters can be found in Sillerud and Shulman (1983).

References

J.R. Alger, L.O. Sillerud, K.L. Behar, R.J. Gillies, R.G. Shulman, R.E. Gordon, D. Shaw, P.E. Hanley, In vivo ^{13}C NMR studies of mammals. Science **214**, 660 (1981)

T.A. Averna, E.E. Kline, A.Y. Smith, L.O. Sillerud, A decrease in ^1H NMR spectroscopically-determined citrate in human seminal fluid accompanies the development of prostate adenocarcinoma. J. Urol. **173**, 433–438 (2005)

J.L. Bischoff, R.J. Rosenbauer, Liquid-vapor relations in the critical region of the system NaCl-H$_2$0 from 380 to 415 °C: A refined determination of the critical point and two-phase boundary of seawater. Geochemica. et Cosmochimca. Acta **52**, 2121 (1988)

A. Boutlerow, Formation synthétique d'une substance sucrée. Comp. Rendus **53**, 145–147 (1861)

R. Breslow, On the mechanism of the formose reaction. Tetrahedron Lett. **21**, 22 (1959)

P. Cohen, H.G. Nimmo, C.G. Proud, How does insulin stimulate glycogen synthesis? Biochem. Soc. Symp. **43**, 69 (1978)

S.M. Cohen, R. Rongstad, R.G. Shulman, J. Katz, A comparison of ^{13}C NMR and ^{14}C tracer studies of hepatic metabolism. J. Biol. Chem. **256**, 3428 (1981)

P. Colson, H.J. Jennings, I.C.P. Smith, Composition, sequence, and conformation of polymers and oligomers of glucose as revealed by ^{13}C nuclear magnetic resonance. J. Am. Chem. Soc. **96**, 8081 (1974)

R.A. de Graaf, K.L. Behar, Detection of cerebral NAD+ by in vivo ^{1}H NMR spectroscopy. NMR Biomed. **27**, 802–809 (2014)

R.A. de Graaf, H.M. De Feyter, P.B. Brown, T.W. Nixon, D.L. Rothman, K.L. Behar, Detection of cerebral NAD+ in humans at 7 T. Magn. Reson. Med **78**, 828–835 (2017)

J.A. den Hollander, T.R. Brown, K. Ugurbil, R.G. Shulman, ^{13}C NMR studies of anaerobic glycolysis in suspensions of yeast cells. Proc. Natl. Acad. Sci. USA **76**, 6096 (1979)

J.A. den Hollander, K. Ugurbil, T.R. Brown, R.G. Shulman, ^{31}P NMR studies of the effect of oxygen upon glycolysis in yeast. Biochemistry **20**, 5871 (1981a)

J.A. den Hollander, K. Behar, R.G. Shulman, ^{13}C NMR study of transamination during acetate utilization by *Saccharomyces cerevisiae*. Proc. Natl. Acad. Sci. (USA) **78**, 2693 (1981b)

D. Doddrell, V. Glushko, A. Allerhand, Theory of nuclear overhauser enhancement and C13–1H dipolar relaxation in proton-decoupled carbon-13 NMR spectra of macromolecules. J. Chem. Phys. **56**, 3683–3689 (1972)

P. Drochmans, Morphology of glycogen. Electron microscopic study of the negative stains of particulate glycogen. J. Ultrastruct. Res. **6**, 141–163 (1962)

J.H. Exton, Gluconeogenesis. Metabolism **21**, 945 (1972)

H.G. Hers, The control of glycogen metabolism in the liver. Annu. Rev. Biochem. **45**, 167 (1976)

J.A. Jackson, W.H. Langham, Whole-body NMR spectrometer. Rev. Sci. Instr **39**, 510 (1968)

J. Katz, S. Golden, P.A. Wals, Glycogen synthesis by rat hepatocytes. Biochem. J. **180**, 389 (1979)

E.E. Kline, E.G. Treat, T.A. Averna, M.S. Davis, A.Y. Smith, L.O. Sillerud, Citrate concentrations in human seminal fluid and expressed prostatic fluid determined via ^{1}H nuclear magnetic resonance spectroscopy outperform prostate specific antigen in prostate cancer detection. J. Urol. **176**, 2274–2279 (2006)

J.B. Lambert, S.A. Gurusamy-Thangavelu, K. Ma, The silicate-mediated Formose reaction: Bottom-up synthesis of sugar silicates. Science **327**, 984 (2010)

A. Lapidot, C.S. Irving, Dynamic structure of whole cells probed by nuclear overhauser-enhanced ^{15}N NMR spectroscopy. Proc. Natl. Acad. Sci. USA **74**, 1988 (1977)

L.E. Mallette, J.H. Exton, C.R. Park, Control of gluconeogenesis from amino acids in the perfused rat liver. J. Biol. Chem. **244**, 5713 (1969)

G. Navon, S. Ogawa, R.G. Shulman, T. Yamane, ^{31}P nuclear magnetic resonance studies of Ehrlich ascites tumor cells. Proc. Natl. Acad. Sci. USA **74**, 87–91 (1977)

T. Ogino, Y. Arata, S. Fujiwara, Proton correlation NMR study of metabolic regulation and pyruvate transport in Anaerobic *E. coli* cells. Biochemistry **19**, 3684 (1980)

I.I. Rabi, J.R. Zacharias, S. Millman, P. Kusch, A new method of measuring nuclear magnetic moments. Phys. Rev. **53**, 318 (1938)

A. Ricardo, M.A. Carrigan, A.N. Olcott, S.A. Benner, Borate Minerals Stabilize Ribose. Science **303**, 5655 (2004)

B.D. Ross, G.K. Radda, D.G. Gadian, G. Rocker, M. Esiri, J. Falconer Smith, Examination of a case of suspected Mcardle's syndrome by ^{31}P NMR. N. Engl. J. Med. **304**, 1338 (1981)

D.L. Rothman, R.G. Shulman, Two transition states of the glycogen shunt and two steady states of gene expression support metabolic flexibility and the Warburg effect in cancer. Neoplasia **23**, 879–886 (2021)

L.O. Sillerud, R.G. Shulman, Structure and metabolism of mammalian liver glycogen monitored by carbon-13 nuclear magnetic resonance. Biochemistry **22**, 1087–1094 (1983)

L.O. Sillerud, J.R. Alger, R.G. Shulman, High-resolution proton NMR studies of intracellular metabolites in yeast using ^{13}C decoupling. J. Magn. Reson. **45**, 142 (1981)

L.O. Sillerud, R.K. Yu, D.E. Schafer, Assignment of the carbon-13 nuclear magnetic resonance spectra of gangliosides GM4, GM3, GM2, GM1, GD1a, GD1b, and GT1b. Biochemistry **21**, 1260–1271 (1982)

L.O. Sillerud, R.H. Griffey, S. Shepard, K.R. Halliday, In vivo ^{13}C NMR spectroscopy of the human prostate. Magn. Reson. Med. **8**, 224–230 (1988)

A. Tengholm, E. Gylfe, cAMP signaling in insulin and glucagon secretion. Diabetes Obes. Metab. **19**, 42–53 (2017)

K. Ugurbil, T.R. Brown, J.A. den Hollander, P. Glynn, R.G. Shulman, High resolution ^{13}C NMR studies of glucose metabolism in *Escherichia coli*. Proc. Natl. Acad. Sci. **75**, 3742 (1978)

E. Williams, J.A. Hamilton, M.K. Jain, A. Allerhand, E.H. Cordes, S. Ochs, Natural abundance carbon-13 nuclear magnetic resonance spectra of the canine sciatic nerve. Science **181**, 869 (1973)

E.A. Yount, R.A. Harris, Studies on the inhibition of gluconeogenesis by oxalate. Biochem. Biophys. Acta **633**, 122 (1980)

L.H. Zang, D.L. Rothman, R.G. Shulman, ^{1}H NMR visibility of mammalian glycogen in solution. Proc. Natl. Acad. Sci. USA **87**, 1678–1680 (1990)

Y. Zhou, P.C.M. van Zijl, X. Xu, N.N. Yadav, Magnetic resonance imaging of glycogen using its magnetic coupling with water. Proc. Natl. Acad. Sci. USA **117**, 3144–3149 (2020)

Chapter 9
Probabilistic Diffusion Constrains Self-Assembly

> *"The same expression whose abstract properties geometers had considered ... represents as well the motion of light in the atmosphere, as it determines the laws of diffusion of heat in solid matter, and enters into all the chief problems of the theory of probability."*
>
> *Jean-Baptiste-Joseph Fourier*

> *"The absorption of oxygen and the elimination of carbon dioxide in the lungs take place by diffusion alone. There is no trustworthy evidence of any regulation of this process on the part of the organism."*
>
> *August Krogh*

The biochemical events that generated life arose from the random collision of primordial molecules. Therefore, we need to understand the probability that two molecules will collide, their mean free path, for without collision there would be no chemical reaction and no product formation. The molecules need to approach sufficiently close so that their electronic wave functions overlap. While gas-phase collisions were important in generating the chemistry of the atmosphere, it is in the realm of solid- and, particularly, liquid-phase reactions that we must turn to find the laws regarding the limits on the formation of biomolecules since life originated in the presence of liquid water.

Just as Feynman diagrams enable one to calculate the probability of a transition from one quantum state to another and the path integral formulation tells how elementary particles get from point A to point B, the diffusive movement of molecules and atoms in gases and liquids and within cells is a random process that is at the same time an organizing principle of matter. Diffusion is a random walk, and we will use our examination of the laws of probability from in Chap. 1 to provide a quantitative description of its properties. We are once again confronted with the seeming enigma that an organizing principle of life is based on randomness.

Diffusion arises from the random thermal motion of molecules driven by their kinetic energy at finite temperature and tells us how far an atom or molecule moves from one position in a gas or fluid to another in a given time.

9.1 Diffusion Is an Organizing Principle: A Physical Law That Cannot Be Easily Violated

That diffusion is indeed an organizing principle is supported by the fact that the architecture of living systems has evolved in response to the limits imposed by its properties. For example, the rubber raft-like shape of the erythrocyte is optimized for the diffusive loading and unloading of oxygen. The fact that no mammalian cell is more than 100 microns from a capillary is a result of the balance between the diffusive movement of oxygen and the metabolic consumption of oxygen by tissues. Multistep biochemical reactions are arranged on the surfaces of enzymes in order to prevent the products, which are the reactants for the next reaction, from diffusing away which would greatly slow the molecular turnover. Biomolecules originated on the surfaces of minerals in order to facilitate the evolution of multistep reactions and to avoid diffusive losses. A case can be made that diffusion over millisecond timescales sets the optimum size for a living cell. Finally, we will find that since diffusion is a very inefficient means for traversing large distances, living systems have evolved methods for the active transport of biomolecules over the meter-scale lengths of nerve axons.

Closely related to diffusion is the speed with which biophysical processes occur. We will see that diffusion sets the timescale and hence limits the rates for the myriad biochemical reactions that serve to sustain life. Unless there are active means to facilitate the movement of biomolecules, metabolic rates cannot exceed the rate at which the reactants diffuse in solution.

9.2 The Differences Between Solvent Diffusion and Solute Diffusion

There are two forms of diffusion differing only in their focus on the diffusing substance with respect to its background; in the first form, called *self-diffusion*, an object diffuses within a milieu of identical objects, like the diffusion of a water molecule in a glass of water. The second form, denoted *solute diffusion*, involves the movement of objects within a background of dissimilar objects, such as for a glucose molecule dissolved in water. Self-diffusion is an equilibrium process in which there is no bulk movement of matter; the molecules switch position while their concentration does not vary with time. Solute diffusion, on the other hand, begins as a non-equilibrium process driven toward equilibrium by a chemical potential gradient.

For example, given samples of two different gases, say oxygen and nitrogen, in separate containers sharing a common partition, removal of the partition will allow them to mix due to their random thermal motion, and each gas will move down its concentration gradient until, at equilibrium, a uniform mixture of oxygen and nitrogen will obtain. As we have seen in Chap. 2, these mixing and randomization occur because the final state is more probable than the unmixed, initial state. After equilibrium is reached, the gases cannot be easily unmixed. Solute diffusion is an irreversible process. Therefore, we must imbed our study of diffusion within the larger context of irreversible thermodynamics which views solute diffusion as a flux of molecules whose movement is driven by a force; in this case, the force is the gradient of the chemical potential.

9.3 Irreversible Thermodynamics, Forces, and Fluxes

The state of thermodynamic equilibrium is explicitly time-independent, and consequently, equilibrium thermodynamics gives no information with respect to the time-dependent processes by which living systems approach the equilibrium state. Kinetics is the realm of nonequilibrium thermodynamics developed first by Onsager (Onsager 1931a, b) and then more recently refined and expanded by Prigogine. The approach to equilibrium is an irreversible process. Common features of nonequilibrium, irreversible processes involve the time-dependent movement, or transport, of particles or energy under the influence of anisotropic physical forces.

The thermodynamics of transport are determined by fluxes generated in response to generalized forces. In mechanics, the force, $\vec{F}$, is the rate of change of potential energy with distance. In analogy with mechanics, one can define a generalized force vector, $\vec{F}$, as the derivative of a potential function, $\varphi(x, y, z)$, by

$$\vec{F}(x, y, z) = \vec{\nabla}\varphi(x, y, z)$$

where $\vec{\nabla}$ is the spatial gradient operator,

$$\vec{\nabla} = \hat{x}\frac{\partial}{\partial x} + \hat{y}\frac{\partial}{\partial y} + \hat{z}\frac{\partial}{\partial z}.$$

It follows then that if the energy distribution is isotropic, $\varphi(x, y, z) = \text{constant}$, there is no force, $\vec{\nabla}\varphi(x, y, z) = 0$, and no flux. For example, let us take an aluminum plate in air at 25 °C and measure its temperature at every point. The result will be the blue dotted line in Fig. 10.1 (right), which shows that there is no temperature gradient across the plate and subsequently no heat flow. If we now place the center of the plate over a candle flame and wait a short time, a second temperature

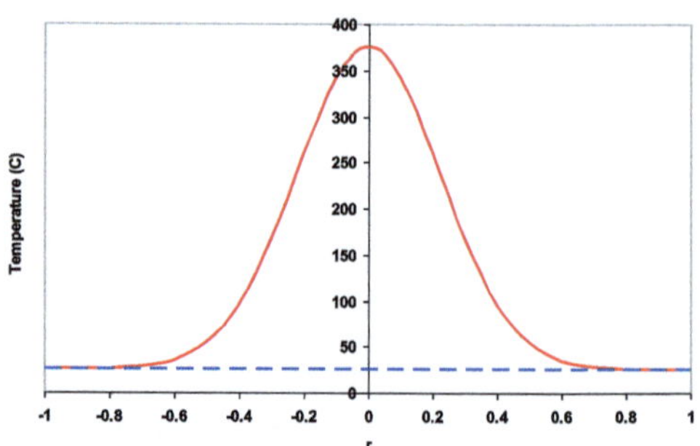

Fig. 9.1 On the left is a candle heating an aluminum plate at its center where $r = 0$. Prior to heating, the temperature would be constant, isotropic, and independent of position as indicated by the blue dashed line at 25 °C on the right. After heating, the resulting temperature profile would resemble the red curve

measurement would show that the center of the plate would be hotter than the edges. The nonequilibrium distribution of heat as a function of the radial distance, r, from the center will resemble the red curve in Fig. 9.1 (right).

If, however, we remove the candle and wait for some time, the temperature will return to its preheating value and produce the dashed blue curve once again. The heat flux will be driven by the anisotropic temperature gradient, and the heat will move away from the center, eventually reaching equilibrium with the surrounding air. This process was described by Fourier in 1822 using the same mathematics as diffusion. The approach to equilibrium is an irreversible process because it is extremely improbable that the reversal of the situation would spontaneously produce a plate with a hot center from an isothermal, room temperature plate.

9.4 Biophysical Fluxes and Their Generalized Potentials

The basic postulate of irreversible thermodynamics is that the flux of a quantity is proportional to the generalized force driving its displacement. The flux is defined as the amount of an object (e.g., heat, electrons, molecules, ions, etc.) crossing a unit area in a unit time. Let $U_i(x)$ be a generalized potential; then the one-dimensional Flux, J_i, associated with this quantity is

$$\vec{J}_i = -L_i \hat{x} \frac{\partial U_i}{\partial x},$$

where L_i is a constant of proportionality known as the generalized conductance. The negative sign enters because the flux is in the opposite direction from the gradient. One of the most important biophysical transport processes is that of diffusion. It, along with several other examples of transport processes, is given in Table 9.1.

Table 9.1 Biophysical transport processes

Process	Potential, U	Flux	Equilibrium condition
Current	Voltage, φ	Electrons	$\nabla\varphi = 0$
Heat	Temperature, T	Phonons	$\nabla T = 0$
Diffusion	Chemical, μ_c	Particles	$\nabla\mu_c = 0$
Electrophoresis	Electrochemical, $\mu_e = \varphi + \mu_c$	Ions	$\nabla\mu_e = 0$
Sedimentation	$\mu_c + g$	Particles	$\nabla T = 0$
Osmosis	Chemical, μ_c	Particles	$\nabla\mu_c = 0$

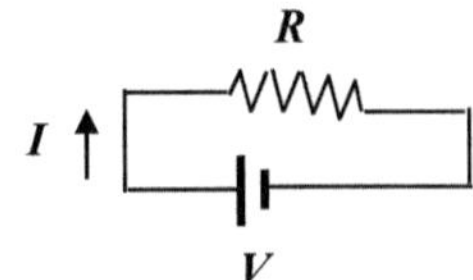

Fig. 9.2 An electrical circuit with a battery of voltage, V, wired to a resistance, R, through which a current, I, flows

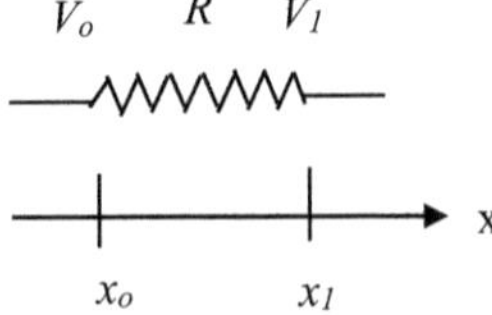

Fig. 9.3 The voltages at each end of a resistor

In electricity, Ohm's law arises from just such an expression. For an electrical circuit formed from a battery of voltage, V, wired to a resistor of value, R (Fig. 9.2), the flux is the electrical current, I, or the number of electrons flowing through the circuit wires per unit time and area given by

$$\vec{I} = -L\,\hat{x}\,\frac{\partial U}{\partial x} = \hat{x}\,\frac{V}{R},$$

and the force is the voltage difference, which is the gradient of the electrical potential difference across the two ends of the resistor

$$\hat{x}\,\frac{\partial U}{\partial x} = \hat{x}\,\frac{V_o - V_1}{x_o - x_1}$$

as shown in Fig. 9.3.

The conductivity is the reciprocal of the resistance, $-L = 1/R = \sigma$, and we have that the current I (flux) is proportional to the voltage difference V (gradient of the potential) or Ohm's law

$$I = \sigma V.$$

9.5   The Frictional Coefficient

The derivation of the generalized flux is illustrated in Fig. 9.4. where we have drawn a rectangular parallelepiped with a length, Δx, and a face area of S, within a fluid. Imagine that the fluid contains a solute which flows through this box along a direction parallel to the x-axis, shown by the arrows. For solute molecules in the fluid with a velocity, v, the distance traveled in a time, Δt, is $\Delta x = v\Delta t$. The number of solute molecules, Δw, passing through the exit face of the volume will be the volume of the box, $S\Delta x$, times the concentration, c, of solute in the fluid given by

$$\Delta w = cS\Delta x,$$

or with Δx given above,

$$\Delta w = cSv\Delta t.$$

Now, the flux, J, is the number of solute molecules passing through S per unit area and unit time,

$$\vec{J} = \frac{\Delta w}{S\Delta t} = c\,\vec{v}$$

but this is also given by our defining equation as

$$\vec{J} = -L\hat{x}\frac{\partial U}{\partial x} = c\,\vec{v}$$

Since the force acting on each solute particle is

$$\vec{F} = -\frac{1}{N_{\mathrm{o}}}\hat{x}\frac{\partial U}{\partial x},$$

where N_{o} is Avogadro's number, we can equate the derivatives to give

$$c\,\vec{v} = N_{\mathrm{o}}L\vec{F}$$

or that the velocity of a solute molecule is

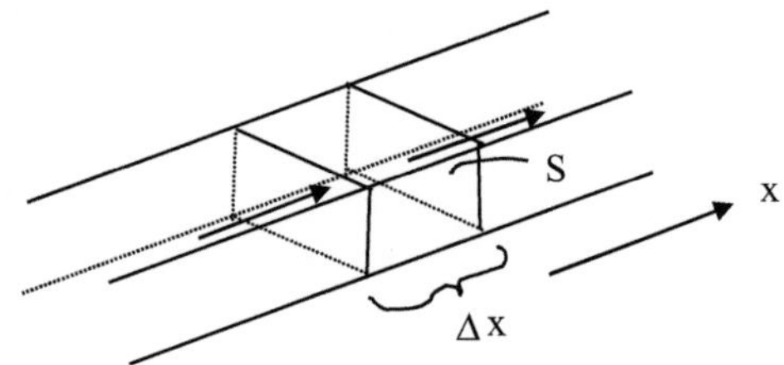

Fig. 9.4 A box of length Δx and a face area of S, within a fluid. The direction of fluid flow is indicated by the arrows parallel to the x-axis

$$\vec{v} = \frac{N_o L}{c}\,\vec{F}.$$

Now, for sedimentation in a gravitational field, the force, $\vec{F}$, is proportional to the velocity,

$$\vec{F} = -f\,\vec{v},$$

where the constant of proportionality, f, is known as the frictional coefficient. The minus sign arises because the frictional force acts to oppose the motion. Any movement of a particle in a liquid is accompanied by resistance due to frictional losses attending collisions with the solvent molecules. The relationship between the conductance, L, of the solution and the frictional coefficient, f, is

$$L = \frac{c}{N_o f}.$$

Here, one can consider L as a conductivity, in the sense of Ohm's law, where it is proportional to the concentration and inversely proportional to f, the "resistance."

9.6 Viscosity

Fluids are bound states of atoms and molecules because the particles attract one another. When a solute attempts to move within a fluid, it encounters friction due to this interparticle attraction. The friction within a fluid that resists the motion of a particle is proportional to the viscosity of the fluid; therefore, the frictional coefficient is also proportional to the viscosity of the fluid. Furthermore, the frictional coefficient depends on the shape of the particle; the minimum frictional coefficient f_o is that of a sphere $f_o = 6\pi\eta r$, where η is the viscosity and r is the radius. The cgs units of viscosity are dyne-sec/cm^2 or Poise (pronounced "pwas") named after Poiseuille, while the SI units are Pascal-sec or Newton-sec/m^2. Water at 20 °C (293 K) has a viscosity of 1.002×10^{-2} poise, or about 1 centipoise (cP) in cgs units, and 1.002×10^{-3} Pascal-sec (=Newton-sec/m^2) in SI units.

Since essentially all biology occurs in water, it is useful to understand the temperature dependence of its viscosity. Values of η for water tabulated in the *CRC Handbook of Chemistry and Physics* for temperatures between 0 and 100 °C show (Fig. 9.5) that water's viscosity decreases in an approximately exponential manner. The data (in centipoise) can be described by a closed-form expression due to Vogel as

$$\eta(T) = e^{A + \frac{B}{C+T}}$$

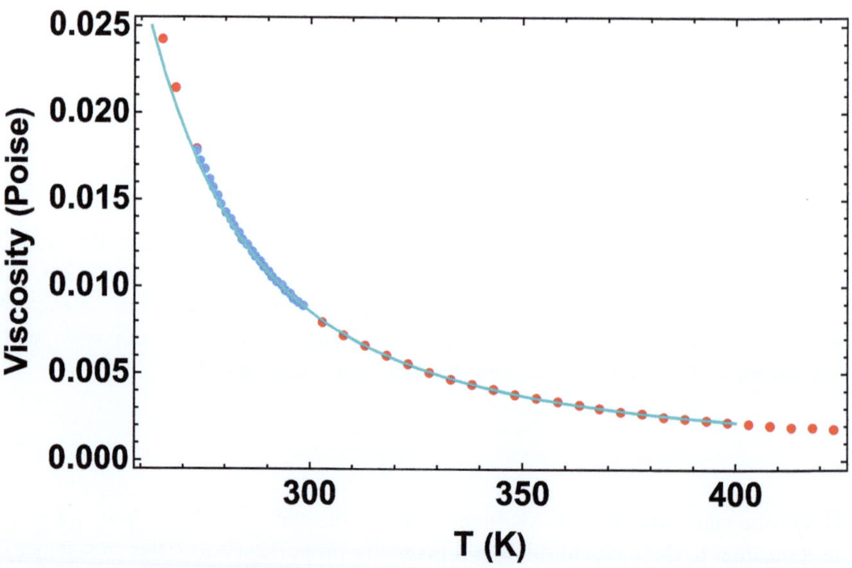

Fig. 9.5 The viscosity of water as a function of the absolute temperature. The green line is the fit using the Vogel equation (see text)

Table 9.2 Frictional coefficients for ellipsoids and rods

Shape	Limits	f/f_o	Equivalent radius, r_e
Prolate ellipsoid (cigar shape)	$c > a = b$	$\dfrac{p^{-1/3}\left(p^2-1\right)^{1/2}}{Ln\left[p+(p^2-1)^{1/2}\right]}$	$\left[\dfrac{3\left(a^2c\right)}{4}\right]^{1/3}$
Oblate ellipsoid (disk shape)	$b = c > a$	$\dfrac{p^{-2/3}\left(p^2-1\right)^{1/2}}{\tan^{-1}\left[(p^2-1)^{1/2}\right]}$	$\left[\dfrac{3\left(c^2a\right)}{4}\right]^{1/3}$
Long rod	$c >> a = b$	$\dfrac{(2/3)^{1/3}p^{2/3}}{Ln[2p]-0.30}$	$\left[\dfrac{3\left(a^2c\right)}{2}\right]^{1/3}$

where $A = -3.7188$, $B = 578.919$, $C = -137.546$, and T is in Kelvins. We also discussed viscosity in Chap. 7 (see Fig. 7.25).

The frictional coefficient for an extended object is always larger than that for a sphere, $f > f_o$. Although the calculation of the frictional coefficient for an arbitrary-shaped object is not usually possible, f for some easy objects has been found analytically. For example, Table 9.2 shows the frictional coefficients for prolate and oblate ellipsoids and for long rods. The volume of an ellipsoid is $V = (4/3)\pi abc$. The equivalent radius, r_e, is the radius of a sphere whose volume is V, so $f_o = 6\pi\eta r_e$. Note that in Table 9.2 $p = c/a$ is the axial ratio, where c is the semimajor axis, a is the semiminor axis, and c for a long rod is half the length.

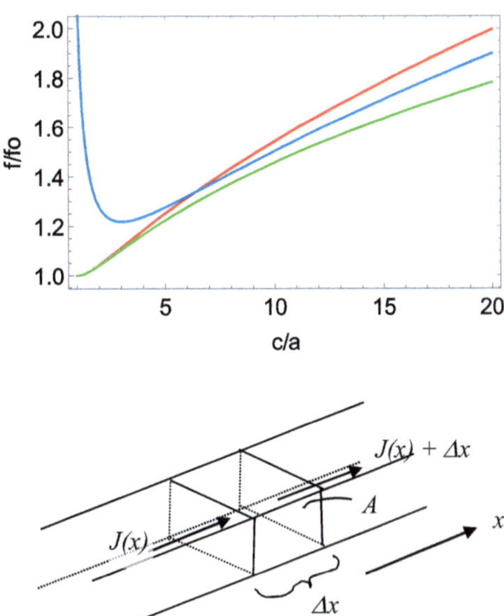

Fig. 9.6 Dependence of the frictional coefficient (f/f_o) on the axial ratio, c/a, for prolate (red) and oblate (green) ellipsoids and for long rods (blue). f/f_o is the ratio of the frictional coefficient of the extended object to that of a sphere

Fig. 9.7 A box of length Δx, and a face of area A, within a fluid. The flux, $J(x)$, of particles is indicated by the arrows parallel to the x-axis entering and leaving this volume

Since the frictional coefficient measures the resistance to flow through a fluid, it makes good physical sense that molecules which are elongated or flattened experience higher drag forces than those experienced by a sphere. This is reflected in the dependence of f on the shape or axial ratio as shown in Fig. 9.6. The frictional coefficients for all extended molecules are rather similar, whether they are rodlike or disclike.

9.7   The Continuity Equation

Although a useful concept for theoretical considerations, the actual fluxes of particles are rarely directly accessible in experiments; a more easily determined quantity is the concentration and its time dependence, $c(t)$. We can relate the flux to the concentration through the conservation of mass. Imagine a rectangular parallelepiped (Fig. 9.7) placed into a flow and that there are no sources or sinks within this volume element, $\Delta V = A\Delta x$. The area of the face orthogonal to the flow is A and the flux through this face is $\vec{J}(x)$. At a distance $x + \Delta x$, the flux will be $\vec{J}(x + \Delta x)$. How does the concentration change in the volume element, ΔV, as a function of time? The change in mass, Δm, in ΔV is given by the difference between the mass entering and exiting the volume in a time, Δt:

$$\Delta m = \vec{J}(x)A\Delta t - \vec{J}(x + \Delta x)A\Delta t.$$

The change in concentration, Δc, is the change in mass per unit volume, or

$$\Delta c = \frac{\Delta m}{\Delta V} = \frac{\Delta m}{A \Delta x}$$

so that

$$\Delta c = \frac{\Delta m}{A \Delta x} = \frac{\left[\vec{J}(x) - \vec{J}(x + \Delta x)\right]\Delta t}{\Delta x},$$

which implies that the change in concentration with respect to time is given by

$$\frac{\Delta c}{\Delta t} = \frac{\left[\vec{J}(x) - \vec{J}(x + \Delta x)\right]}{\Delta x}.$$

Note that the right-hand side resembles the definition of the derivative and can be written as

$$\frac{\Delta c}{\Delta t} = -\frac{\Delta \vec{J}(x)}{\Delta x}.$$

If we take the limit as the differences become zero, $\Delta \to \partial$, we obtain the *continuity equation* in one dimension:

$$\frac{\partial c}{\partial t} = -\frac{\partial \vec{J}(x)}{\partial x}.$$

This states that the concentration changes in our elementary volume if and only if the amount of material flowing into the volume (i.e., the flux) does not equal the amount of material flowing out in a given time interval. In three dimensions, the continuity equation becomes

$$\frac{\partial c}{\partial t} = -\vec{\nabla} \cdot \vec{J}(x).$$

If there are sources of material within the volume, V, then a source term would need to be added to the right-hand side.

9.8　Diffusion

Diffusion is the movement of molecules, or other particles, in a fluid due to random Brownian motion of the solute and/or solvent. Diffusion at equilibrium is known as self-diffusion and is subject to the restraint that the concentration, $c(x, y, z, t)$, of a solute is isotropic and constant independent of time,

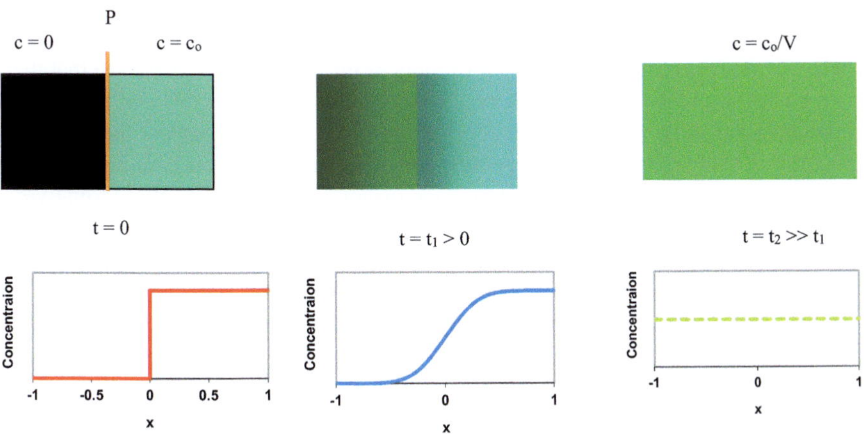

Fig. 9.8 An illustration of the time-dependent approach to an equilibrium distribution of solute molecules (colored green) in a fluid containing two compartments separated by a partition, P at time $t = 0$. In the initial configuration in the left bottom panel, the concentration $c(x, t)$ (red) is given by $c(x, 0) = 0$, $-1 \leq x \leq 0$, and $c(x, 0) = c_o$, $0 \leq x \leq 1$. After a time $t_1 > 0$, the concentration profile would look like the middle panels, (blue) while for long times the concentration would be uniform and constant (right panels, green)

$$\frac{\partial c(x, y, z, t)}{\partial t} = 0 = \vec{\nabla} c(x, y, z, t),$$

since the flux is zero. However, for nonequilibrium anisotropic solutions, the concentration is a function of the spatial coordinates and time, and one or more of the above derivatives will be nonzero. This is the situation where we encounter diffusion in response to a concentration gradient; the driving force arises from the derivative of a potential,

$$\vec{J} = -L\vec{\nabla} c$$

where the concentration serves the role of the potential here.

We can understand the movement of molecules in a qualitative sense by considering a fluid-filled rectangular container of total volume, V, divided into two compartments of equal volume with a removable partition, P, separating the two compartments (Fig. 9.8). If the concentration of a solute is zero in the left compartment and c_o in the right compartment at time $t = 0$ (Fig. 9.8, left panels), what will be the concentrations of the solute in the two compartments at a time $t > 0$ after we remove the partition? At a time shortly after the removal of the partition, we see from the figure (Fig. 9.8, middle panels) that there will be some movement of material from areas close to the previous partition location into the left compartment. At very long times, the solution will be uniform in both compartments (Fig. 9.8, right panels). It is also useful to examine the behavior of the concentration as a function of the distance along an axis orthogonal to the partition; call this dimension x. This behavior is shown in the bottom row of plots in Fig. 9.8. For $t = 0$, we see that there is a sharp boundary at $x = 0$ where the concentration on the left and right of the boundary abruptly varies from $c = 0$ to $c = c_o$. As time passes, the boundary becomes less distinct, until at very long times the concentration is uniform in both volumes at equilibrium and the boundary disappears. At $t = 0$, the derivative of concentration with respect to x is a nonzero, delta function, we are not at equilibrium, and ΔG is a maximum, while ΔS is a minimum. We know from our examination of the behavior of a two-compartment ideal gas problem in Chap. 2 that the entropy tends to increase when the partition is removed and at equilibrium ΔS is a maximum given by the entropy change on mixing. The Gibb's energy is also a minimum at equilibrium at long times, and there is no concentration gradient, $\frac{dc}{dx} = 0$.

Instead of concentration, one should really use the chemical potential, μ_c, so that at equilibrium $\nabla \mu_c = 0$. The approach to equilibrium is driven by the chemical potential of the solute in the boundary region, which provides a force for solute flow according to

$$J = -L\frac{\partial \mu}{\partial x}$$

where the chemical potential of the solute, μ_s, is a function of temperature, pressure, and concentration $\mu_s = \mu_s(T,P,c)$. But biological systems operate at constant temperature and pressure, so

$$\frac{\partial \mu_s}{\partial x} = \left[\frac{\partial \mu_s}{\partial c}\right]_{T,P} \frac{\partial c}{\partial x}$$

where the change in chemical potential with concentration is essentially constant, and we observe that a concentration gradient solely determines the flux for diffusion.

9.9   Fick's First and Second Laws: The Diffusion Equation

If we write the chemical potential for an ideal solution as $\mu = \mu_o + RT \ln(c/c_o)$, then the change in the chemical potential as a function of concentration is given by

$$\frac{\partial \mu_s}{\partial c} = \frac{RT}{c}.$$

And the gradient becomes

$$\frac{\partial \mu_s}{\partial x} = \frac{RT}{c} \frac{\partial c}{\partial x}$$

which when inserted into the above equation for the flux gives the flux with respect to the gradient of the concentration as

$$J = - \left[\frac{LRT}{c}\right] \frac{\partial c}{\partial x},$$

and if we remember that

$$L = \frac{c}{N_\alpha f}, \text{ or}$$

$$\frac{L}{c} = \frac{1}{N_\alpha f},$$

we can rewrite this as

$$J = - \left[\frac{RT}{N_\alpha f}\right] \frac{\partial c}{\partial x}.$$

Here, we see that the flux is proportional to the concentration gradient and that the proportionality constant, L, is

$$L = - \frac{RT}{N_\alpha f}.$$

This states that the diffusional flux is proportional to the concentration gradient and is expressed as *Fick's first law:*

$$J = - D \frac{\partial c}{\partial x}$$

where D (the "conductance" of the solution) is the diffusion coefficient,

$$D = \frac{RT}{N_\alpha f}.$$

For a sphere, the frictional coefficient is $f = 6\pi\eta r$ so that the diffusion coefficient for a spherical molecule becomes

$$D = \frac{RT}{6\pi N_o \eta r}.$$

The flux goes to zero as the concentration gradient vanishes. We can combine this result with the continuity equation

$$\frac{\partial c}{\partial t} = -\frac{\partial J(x)}{\partial x}$$

to give the diffusion equation or *Fick's second law*,

$$\frac{\partial c}{\partial t} = D\frac{\partial^2 c}{\partial x^2},$$

in one dimension. For three dimensions, the spatial second derivative is replaced by the Laplacian

$$\frac{\partial c(x,y,z,t)}{\partial t} = D\left[\frac{\partial^2}{\partial x^2} + \frac{\partial^2}{\partial y^2} + \frac{\partial^2}{\partial z^2}\right] c(x,y,z,t).$$

For ideal solutions, the diffusion coefficient is independent of concentration, and in bulk solvents, such as water, there is a single, isotropic self-diffusion coefficient.

9.10   The Diffusion Tensor

In biology, however, diffusion is often anisotropic due to the presence of multiple compartments within cells, such as the endoplasmic reticulum, the nuclear envelope, the plasma membrane, etc. It is also often restricted by these structures so that the actual distance a molecule can diffuse is limited by barriers through or beyond which molecules are unable to penetrate or are reflected. The central nervous system contains myriad axonal fibers whose diameters (~10μm) are a tiny fraction of their length (up to ~1 m). Under these conditions, diffusion is highly restricted perpendicular to a fiber while significantly less hindered along the axon. As a consequence, the motion of water in cells and fibers cannot be described by a single isotropic diffusion coefficient but must reflect the fact that diffusion varies with direction. We can accommodate this variability by assuming that there exists a diffusion coefficient in each direction by replacing the diffusion coefficient with a diffusion tensor, D_{ij}, and we can generalize Fick's second law to

$$\vec{J} = -D\vec{\nabla}c$$

where D is a 3×3 matrix. In component form, we have that

$$
\begin{bmatrix} J_x \\ J_y \\ J_z \end{bmatrix} = - \begin{bmatrix} D_{xx} & D_{yx} & D_{zx} \\ D_{xy} & D_{yy} & D_{zy} \\ D_{xz} & D_{yz} & D_{zz} \end{bmatrix} \begin{bmatrix} \dfrac{\partial c}{\partial x} \\[2mm] \dfrac{\partial c}{\partial y} \\[2mm] \dfrac{\partial c}{\partial z} \end{bmatrix} .
$$

The three elements along the diagonal are the diffusion coefficients along the x, y, and z axes, while the off-diagonal elements are due to random correlations between pairs of axes. In bulk water, diffusion is isotropic, and all the diagonal components of D_{ij} are the same and equal to the single isotropic diffusion coefficient, $D = D_{xx} = D_{yy} = D_{zz}$, and the off-diagonal elements are all zero. In biological materials, on the other hand, the diagonal components are not all equal, and the off-diagonal elements are nonzero. The diffusion tensor is positive definite and symmetric; therefore, only six components are needed to completely describe its properties. The eigenvalues (λ_l) and eigenvectors ($\vec{\varepsilon}_i$) obtained by diagonalizing D define the principal axes of the diffusion ellipsoid (Fig. 9.9).

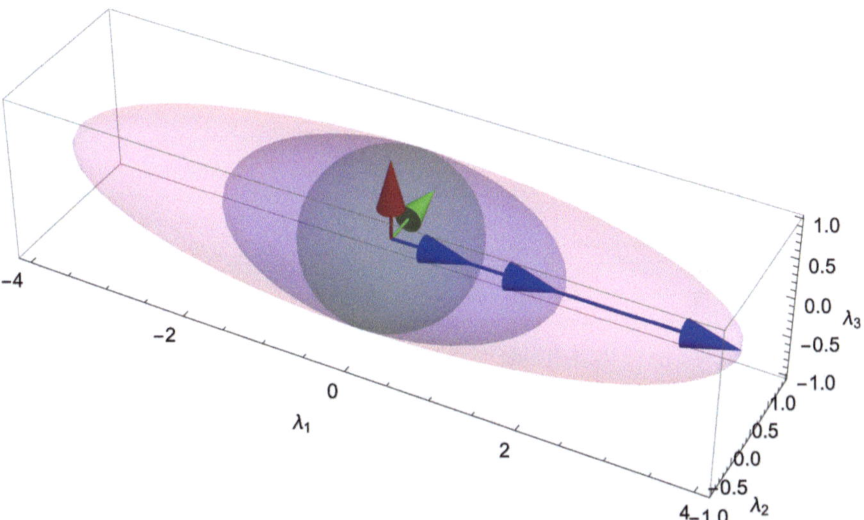

Fig. 9.9   The diffusion ellipsoids obtained by diagonalizing the diffusion tensor in which the largest eigenvalue of the diffusion tensor is equal to (isotropic, green sphere), twice as large as the remaining two eigenvalues ($\lambda_1 = 2\lambda_2 = 2\lambda_3$) scaled by λ_2 (blue ellipsoid), and four times large as the remaining two eigenvalues ($\lambda_1 = 4\lambda_2 = 4\lambda_3$) scaled by λ_2 (pink ellipsoid)

9.11 Solutions of the Diffusion Equation

The diffusion equation is a partial differential relationship between temporal and spatial variations in the concentration of a solute. We can apply the technique of separation of variables in order to find solutions. The diffusion equation in one spatial dimension is

$$\frac{\partial c(x,t)}{\partial t} = D\frac{\partial^2 c(x,t)}{\partial x^2}$$

We will write the concentration, $c(x, t)$, as a product of two separate functions, $c(x, t) = U(x)T(t)$, each, $U(x)$ and $T(t)$, only a function of a single variable. Then we insert this product into the diffusion equation to find that

$$\frac{\partial c}{\partial t} = U\frac{\partial T}{\partial t}, \quad \text{and}$$

$$D\frac{\partial^2 c}{\partial x^2} = DT\frac{\partial^2 U}{\partial x^2}$$

Therefore, we can rewrite the diffusion equation as

$$U(x)\frac{\partial T(t)}{\partial t} = DT(t)\frac{\partial^2 U(x)}{\partial x^2}$$

Now, we divide both sides by $DU(x)T(t)$ to give

$$\frac{1}{DT(t)}\frac{\partial T(t)}{\partial t} = \frac{1}{U(x)}\frac{\partial^2 U(x)}{\partial x^2}$$

Note that here we have the left side of this equation with only a t dependence and the right side with only an x dependence. The only general way that an arbitrary function of t can equal an arbitrary function of x is that both sides of this equation must be equal to a constant. Let us call this constant $-1/\lambda^2$, so that our equations now read

$$\frac{1}{DT}\frac{\partial T}{\partial t} = \frac{1}{U}\frac{\partial^2 U}{\partial x^2} = -\frac{1}{\lambda^2}$$

We can use Mathematica to solve both of these equations, and then we will proceed analytically to show how the solutions are obtained in detail:

```
In[5]:= DSolve[{Y''[x] + Y[x]/L^2 == 0, Y[0] == 0}, Y[x], x]

Out[5]= {{Y[x] → C[2] Sin[x/L]}}

In[6]:= DSolve[{T'[t] + (D/λ^2) T[t] == 0, T[0] == A}, T[t], t]

Out[6]= {{T[t] → A e^(-D t/λ^2)}}
```

First, examine the time equation

$$\frac{1}{DT}\frac{\partial T}{\partial t} = -\frac{1}{\lambda^2}$$

or

$$\int_{A}^{T(t)} \frac{dT}{T} = -\frac{D}{\lambda^2}\int_{0}^{t} dt$$

which shows that the time dependence of the concentration is simply an exponential decay

$$T(t) = Ae^{-\frac{Dt}{\lambda^2}}$$

where λ has units of length and A is the value of T at $t = 0$, $A = T(0)$. We observe that the units of the diffusion coefficient, D, are length^2/time, in order for the units of the exponent to cancel. What is the significance of λ? One can think of it as the average distance a particle will travel between collisions or the *mean free path*.

Now, let us turn our attention to the spatial equation:

$$\frac{1}{U}\frac{\partial^2 U}{\partial x^2} = -\frac{1}{\lambda^2}$$

This can be rearranged as

$$\frac{\partial^2 U}{\partial x^2} + \frac{U}{\lambda^2} = 0,$$

and the solution to this equation is $U(x) = G\cos(x/\lambda) + H\sin(x/\lambda)$. We can then write the full solution to the diffusion equation in Cartesian coordinates as

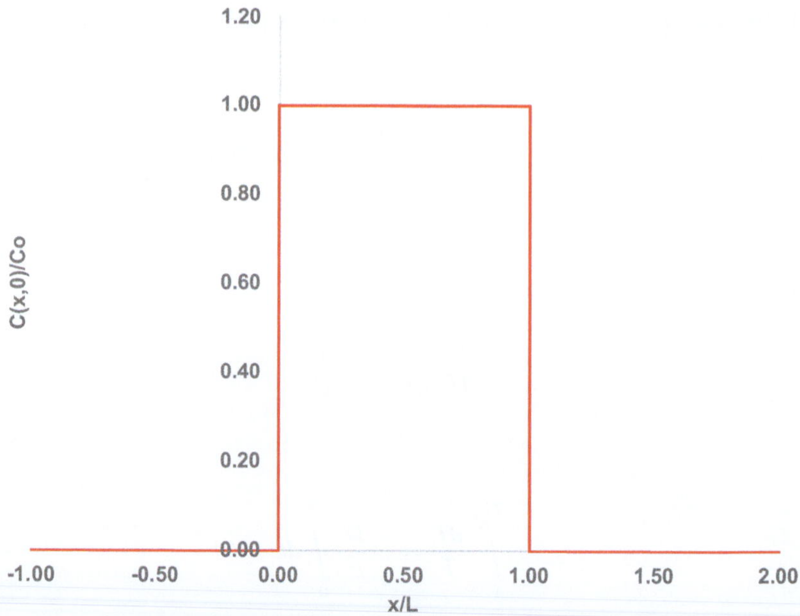

Fig. 9.10 The initial concentration profile for $t = 0$ is a square wave of height C_o and width L

$$c(x, t) = Ae^{-\frac{Dt}{\lambda^2}}[G\cos(x/\lambda) + H\sin(x/\lambda)].$$

The constants G and H are found from imposing boundary conditions on the solutions. For example, for a concentration that is initially constant at C_o within a region L and zero outside this region, as shown in Fig. 9.10, $C(x, t) = C(0 < x < L, 0) = C_o$; $C(0, t) = C(L, t) = 0$.

If we impose the initial condition that $C(0,0) = 0$, we have that

$$G\,\cos\left(\frac{x}{\lambda}\right) + H\,\sin\left(\frac{x}{\lambda}\right) = G\,\cos(0) + H\,\sin(0) = 0$$

which can only be satisfied for $G = 0$. Now, at $x = L$, we have

$$H\,\sin\left(\frac{L}{\lambda}\right) = 0,$$

and since $H \neq 0$, the sine term must be zero. This occurs for

$$\frac{L}{\lambda} = n\,\pi$$

giving us the means to find $\lambda = \frac{L}{n\,\pi}$ and to write $T_n(t) = A_n\,e^{-\frac{n^2\pi^2}{L^2}Dt}$, as well as

$$U_n(x) = H_n \ \sin\left(\frac{n\pi x}{L}\right)$$

and then

$$C(x,t) = B_n \ e^{-\frac{n^2\pi^2}{L^2}Dt} \ \sin\left(\frac{n\pi x}{L}\right).$$

We can find the constant B_n by noting that the initial concentration function $C(0 < x < L, 0) = C_0$, so that the concentration curve at $t = 0$ should give this value

$$C_0 = \sum_{n=1}^{\infty} B_n \sin\left(\frac{n\pi x}{L}\right)$$

which is the Fourier sine series expansion of a constant, where the coefficients, B_n, can be found from the Fourier sine transform of the initial condition

$$B_n = \frac{2}{L} \int_0^L f(\tau) \sin\left(\frac{n\pi\tau}{L}\right) d\tau$$

or in this case with a constant $f(\tau) = C_0$

$$B_n = \frac{2C_0}{L} \int_0^L \sin\left(\frac{n\pi\tau}{L}\right) d\tau.$$

Performing this integration gives

$$B_n = 2C_0 \frac{(1 - \cos(n\pi))}{n\pi}$$

so that our solution now is

$$C(x,t) = \frac{2C_0}{\pi} \sum_{n=1}^{\infty} \frac{1}{n}(1 - \cos(n\pi)) \sin\left(\frac{n\pi x}{L}\right) e^{-\frac{n^2\pi^2}{L^2}Dt}.$$

Now, the term $[1 - \cos(n\pi)]$ is 0 for even n and is 2 for n odd, so we can take this into consideration by replacing n with $k + 1$ and start the sum from $k = 0$ instead of 1. Our solution then reads

$$C(x,t) = \frac{4C_0}{\pi} \sum_{k=0}^{\infty} \frac{1}{(k+1)} \sin\left(\frac{(k+1)\pi x}{L}\right) e^{-\frac{(k+1)^2\pi^2}{L^2}Dt}.$$

We note that the time dependence is one of exponential decay with a rate determined by the diffusion coefficient. The spatial component is an infinite Fourier sine series.

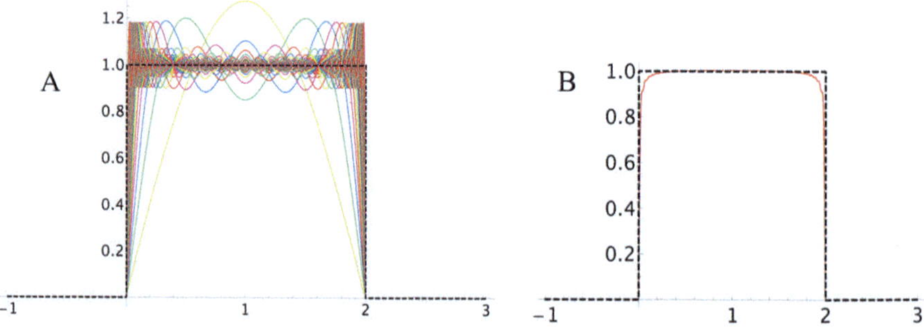

Fig. 9.11 (**a**) The first 100 terms in the Fourier sine expansion of the initial concentration profile. (**b**) The sum of the first 100 terms (red) compared with the initial concentration profile (black dotted line)

We can make this more general by letting the initial condition be a function

$$C(0 < x < L, 0) = f(x),$$

which for the above case is $f(0 < x < L) = C_o$ and zero elsewhere. This is an even function so we can write $f(x)$ as a Fourier sine series representing the initial condition which provides a general solution for the diffusion equation as

$$C(x, t) = \sum_{n=1}^{\infty} \left[\frac{2}{L} \int_0^L f(\tau) \sin\left(\frac{n\pi\tau}{L}\right) d\tau \right] \sin\left(\frac{n\pi x}{L}\right) e^{-\frac{n^2\pi^2}{L^2}Dt}.$$

Our solution for $f(0 < x < L) = C_o$ corresponds to the Fourier sine transform of a square wave of constant height, C_o. It is instructive to plot the individual terms in this series (Fig. 9.11a) and their normalized sum (Fig. 9.11b). The solution in terms of a Fourier series is a reminder that sines and cosines form complete, orthonormal sets, as we saw in our previous studies of the quantum mechanics of the particle in a box (Chap. 4) which bears a close resemblance to this problem.

What then are the properties of the solution as a function of time? We expect that the initial sharp boundaries will spread with time and approach a uniform concentration and that is what we observe (Fig. 9.12). The initial concentration decays exponentially with a time constant on the order of $\tau = \frac{L^2}{\pi^2 D}$.

9.12   Solution for Step Function Initial Conditions

It is instructive to solve the differential equation for Fick's second law under the initial conditions that

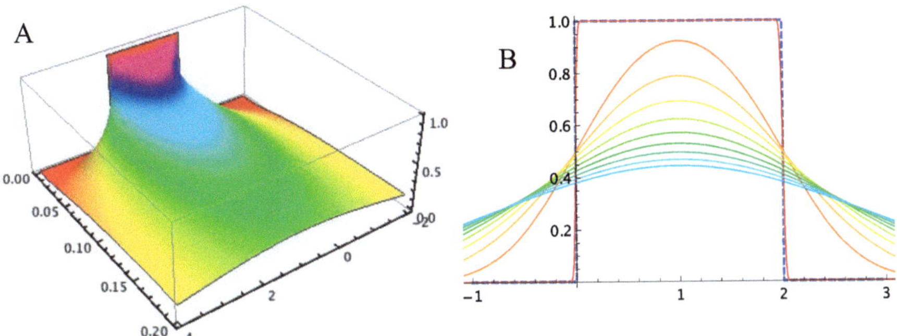

Fig. 9.12 (**a**) A three-dimensional plot of the spatial and temporal dependence of the diffusion of an initial constant concentration for $C(0 \leq x \leq L) = C_o$. (**b**) Profiles of the concentration at times ranging from $t = 0.00001$ to 0.1 in steps of 0.01 showing the Gaussian spread of the initial profile

$$c(x < 0) = 0,$$
$$\text{and} \quad c(x \geq 0) = C_o$$

which are similar to those seen in Fig. 9.8 (upper left). Here, we illustrate the use of Mathematica's ability to solve differential equations, explicitly with the inclusion of the initial conditions:

$$solution = NDSolve\left[\left\{\frac{\partial C[x,t]}{\partial t} == D\frac{\partial^2 C[x,t]}{\partial x^2}, C[x,0] == UnitStep[x-1], C[0,t] == 0, C[2,t]\right.\right.$$
$$\left.\left. == e^{-\frac{t}{5}}\right\}, C, \{x,0,2\}, \{t,0,10\}\right]$$

```
Plot3D[Evaluate[C[x,t]/.solution[[1]]],{x,0,2},{t,0,20},  AxesLabel  →
{x,t,C[x,t]},  PlotLabel  →  "C[x,t]  with  time-dependent  boundary
conditions", ColorFunction → Hue, Mesh → False]
```

The solution, $c(x, t)$, is

$$c(x,t) = \frac{C_o}{2}\left[1 - \frac{2}{\sqrt{\pi}}\int_0^{\frac{x}{\sqrt{2Dt}}} e^{-y^2}\,dy\right]$$

which is the integral of a Gaussian providing the S-shaped curve shown above for intermediate times after removal of the partition. Figure 9.13 shows the spatial and temporal behavior of the concentration, given these initial conditions and a diffusion coefficient of 10^{-5} cm^2/s. The spatial derivative of $c(x, t)$ gives the concentration gradient

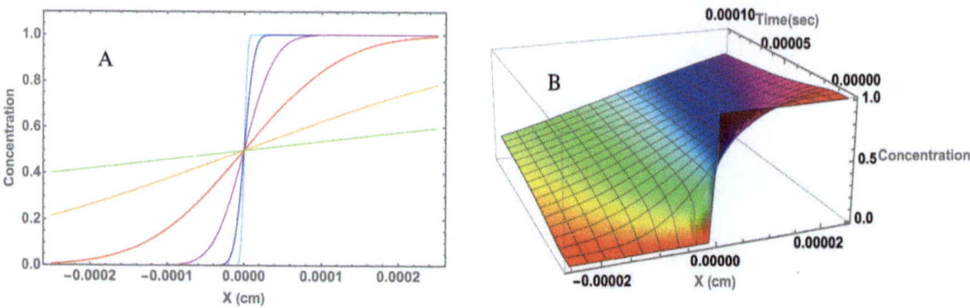

Fig. 9.13 (**a**) The concentration profile for various times after the beginning of diffusion from an initial step function with a diffusion coefficient of 10^{-5} cm^2/s. (**b**) The temporal and spatial dependence of the solution of the diffusion equation with a step function initial condition

$$\frac{\partial c}{\partial x} = \frac{c_o}{2\sqrt{\pi Dt}} e^{\frac{-x^2}{4Dt}}$$

as a Gaussian (Chap. 1) of width $\sqrt{4Dt}$. Note that as t approaches infinity, the concentration gradient vanishes. Figure 9.14 illustrates the concentration gradient as a function of time where we see that as t approaches infinity, the concentration gradient vanishes.

9.13   Solution for Gaussian Initial Conditions

The above treatment of the diffusion equation illustrated a solution for a particular set of step function initial and boundary conditions. There are two other important initial concentration profiles of interest: diffusion with a delta-function initial concentration distribution and that with a Gaussian profile. One can approximate a delta-function initial condition by using a Gaussian with a very narrow width. The use of Mathematica to solve these situations is illustrated by the following with a Gaussian profile first (Fig. 9.15):

```
sol = NDSolve[{∂t v[x, t] == 0.1 Diff ∂x,x v[x, t],
    v[x, 0] == Exp[-((x - 1) / (2 Diff))^2], v[0, t] == 0, v[2, t] == 0},
    v, {x, 0, 2}, {t, 0, 10}]
Plot3D[Evaluate[v[x, t] /. sol[[1]]], {x, 0.5, 1.5}, {t, 0, 20},
    AxesLabel → {X, T, C[x, t]}, PlotRange → All,
    PlotLabel → "C[x, t] with time-dependent boundary conditions",
    ColorFunction → Hue, LabelStyle → Directive[Bold, Medium]]
{{v → InterpolatingFunction[{{0., 2.}, {0., 10.}}, <>]}}
```

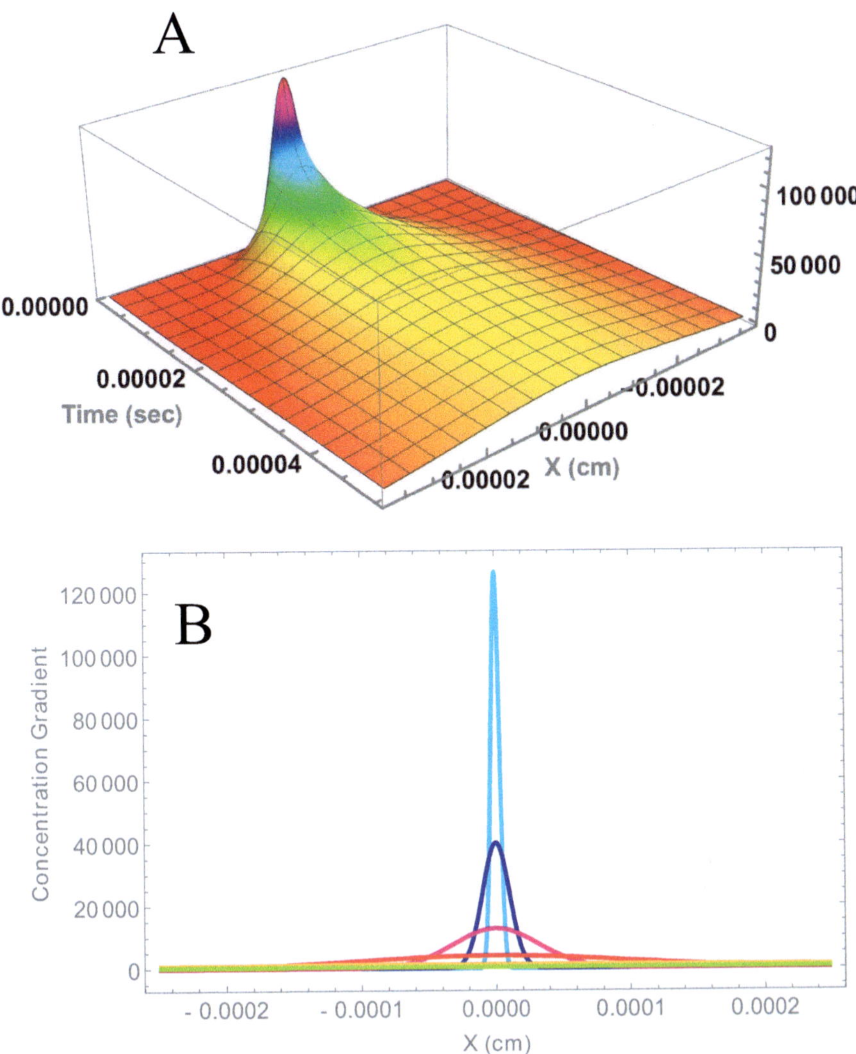

Fig. 9.14 (**a**) The temporal and spatial dependence of the solution of the diffusion equation with a step function initial condition. (**b**) The concentration gradient for various times after the beginning of diffusion from an initial step function with a diffusion coefficient of 10^{-5} cm^2/s

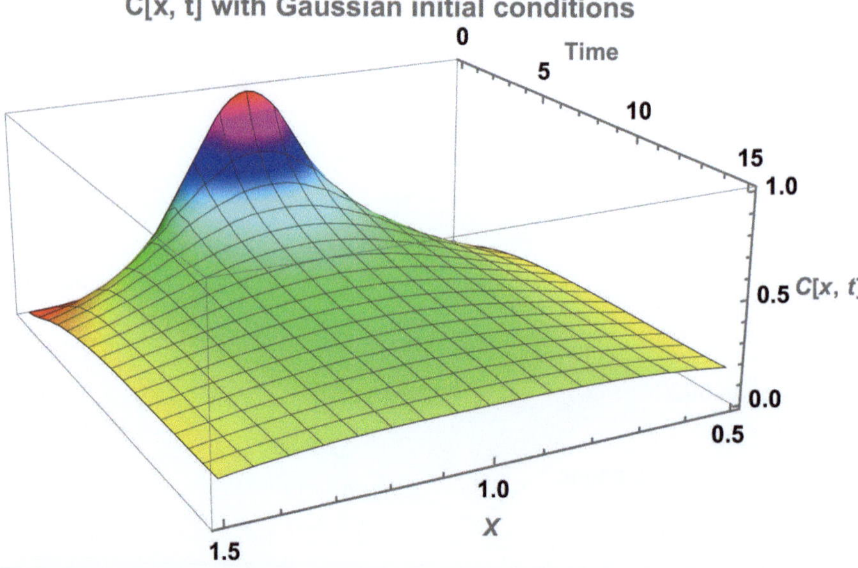

Fig. 9.15 The temporal and spatial dependence of the solution of the diffusion equation with a Gaussian function initial condition. One can approximate a delta-function initial condition by using a Gaussian with a very narrow width

9.14 Solution of the Diffusion Equation in Cylindrical Coordinates

Blood flow occurs in arteries, capillaries, and veins, all of which can be naturally modeled in cylindrical coordinates. We can express the Laplacian in cylindrical coordinates, much as was done for the hydrogen atom where the Laplacian was transformed from Cartesian coordinates to spherical coordinates using the transformation equations (Chap. 4) and their derivatives. The Laplacian in polar coordinates $\{r, \theta\}$ is

$$\nabla^2 = \frac{\partial^2}{\partial r^2} + \frac{1}{r}\frac{\partial}{\partial r} + \frac{1}{r^2}\frac{\partial^2}{\partial \theta^2}$$

so that the diffusion equation becomes

$$\frac{\partial c(r,\theta,t)}{\partial t} = D\nabla^2 c(r,\theta,t) = D\left[\frac{\partial^2 c(r,\theta,t)}{\partial r^2} + \frac{1}{r}\frac{\partial c(r,\theta,t)}{\partial r} + \frac{1}{r^2}\frac{\partial^2 c(r,\theta,t)}{\partial \theta^2}\right],$$

and we seek a solution for $c(r, \theta, t)$ given a set of initial and boundary conditions. We can expect, based on the cylindrical symmetry, that the solution will involve Bessel functions. Let us seek to separate variables by writing the concentration as a product of separate functions of r, θ, t as

$$c(r,\theta,t) = R(r)T(t)\phi(\theta).$$

When we insert this into the diffusion equation, we find that

$$R(r)\phi(\theta)\frac{\partial T(t)}{\partial t} = D\left[T(t)\phi(\theta)\frac{\partial^2 R(r)}{\partial r^2} + \frac{T(t)\phi(\theta)}{r}\frac{\partial R(r)}{\partial r} + \frac{T(t)R(r)}{r^2}\frac{\partial^2\phi(\theta)}{\partial\theta^2}\right],$$

and now, if we divide through by $DR(r)T(t)\phi(\theta)$, we obtain

$$\frac{1}{DT(t)}\frac{\partial T(t)}{\partial t} = \left[\frac{1}{R(r)}\frac{\partial^2 R(r)}{\partial r^2} + \frac{1}{rR(r)}\frac{\partial R(r)}{\partial r} + \frac{1}{r^2\phi(\theta)}\frac{\partial^2\phi(\theta)}{\partial\theta^2}\right],$$

and we observe that the left-hand side is a function only of t, while the right-hand side is independent of t. The only way this can happen is if both sides are equal to a constant, which we will denote, $-\lambda^2$. Then

$$\frac{1}{DT(t)}\frac{\partial T(t)}{\partial t} = -\lambda^2 = \left[\frac{1}{R(r)}\frac{\partial^2 R(r)}{\partial r^2} + \frac{1}{rR(r)}\frac{\partial R(r)}{\partial r} + \frac{1}{r^2\phi(\theta)}\frac{\partial^2\phi(\theta)}{\partial\theta^2}\right]$$

which gives an ordinary differential equation for $T(t)$ as

$$\frac{dT(t)}{dt} + \lambda^2 DT(t) = 0,$$

which has the solution

$$T(t) = T(0)e^{-\lambda^2 Dt}.$$

Turning to the radial and angular parts of the problem, we find that

$$\frac{1}{R(r)}\frac{\partial^2 R(r)}{\partial r^2} + \frac{1}{rR(r)}\frac{\partial R(r)}{\partial r} + \frac{1}{r^2\phi(\theta)}\frac{\partial^2\phi(\theta)}{\partial\theta^2} + \lambda^2 = 0,$$

and on multiplying through by r^2 and bringing the $\phi(\theta)$ term to the right, we have

$$\frac{r^2}{R(r)}\frac{\partial^2 R(r)}{\partial r^2} + \frac{r}{R(r)}\frac{\partial R(r)}{\partial r} + \lambda^2 r^2 = -\frac{1}{\phi(\theta)}\frac{\partial^2\phi(\theta)}{\partial\theta^2} = \mu^2$$

where the left-hand side is only a function of r, while the right is solely a function of θ, and this can only happen if both sides equal another constant which we have called μ. The angular part must then satisfy

$$\frac{\partial^2 \phi(\theta)}{\partial \theta^2} + \mu^2 \phi(\theta) = 0$$

whose solutions are, as we have seen before (Chap. 4) with the periodic boundary condition, that $\phi(\theta) = \phi(\theta + 2\pi)$, $\phi(\theta) = e^{i\mu\theta}$, and $\mu = \{0, 1, 2, 3, \ldots\}$. It remains to address the radial part of the concentration equation

$$\frac{r^2}{R(r)} \frac{\partial^2 R(r)}{\partial r^2} + \frac{r}{R(r)} \frac{\partial R(r)}{\partial r} + \lambda^2 r^2 - \mu^2 = 0.$$

This, when multiplied through by $R(r)$, gives

$$r^2 \frac{\partial^2 R(r)}{\partial r^2} + r \frac{\partial R(r)}{\partial r} + \left(\lambda^2 r^2 - \mu^2\right) R(r) = 0$$

which is Bessel's equation, and its solutions are Bessel functions of the first kind of order μ. Just as we have seen that μ can take on only integral values, the same is true for λ, so that the solutions are indexed by $\{\mu = m, n\} \in \{0, 1, 2, 3, \ldots\}$. The solutions, $R_{m,\,n}(r)$, are eigenfunctions of a Hermitian operator and hence are orthonormal

$$\int_0^1 R_{m,n}(r) R_{m,p}(r) r \, dr = \delta_{n,p}$$

and can be written as

$$R_{m,n}(r) = \kappa_{m,n} \mathrm{BesselJ}(m, \lambda_{m,n} r)$$

where the constants, $\kappa_{m,\,n}$, are gotten from the integrals

$$\kappa_{m,n} = \left[\int \left[\mathrm{BesselJ}(m, \lambda_{m,n} r)\right]^2 r \, dr \right]^{-1}$$

and the eigenvalues, $\lambda_{m,\,n}$, are the zeros of the Bessel function obtained from the boundary condition $R(r_\mathrm{o}) = \text{constant}$, given by

$$\mathrm{BesselJ}(m, \lambda_{m,n} r) = \text{constant}.$$

The complete solution to the spatial and temporal dependence of the concentration is then the product and is written as a Fourier series in the orthonormal Bessel functions,

$$c(r,\theta,t) = R(r)T(t)\phi(\theta) = \frac{2}{\sqrt{\pi}} \sum_{n=1}^{\infty} \left(\sum_{m=1}^{\infty} \kappa_{m,n} C_{m,n} e^{-\lambda_{m,n}^2 Dt} \mathrm{BesselJ}(m,\lambda_{m,n}r) e^{im\theta} \right)$$

where the Fourier coefficients, $C_{m,\,n}$, are found from the initial condition, $f(r,\theta)$, by solving

$$C_{m,n} = \frac{2\kappa_{m,n}}{\sqrt{\pi}} \int_0^{2\pi} \int_0^{r_o} f(r,\theta) \mathrm{BesselJ}(m,\lambda_{m,n}r)\, dr\, d\theta$$

where r_o is some specified radius.

9.15   The Average Displacement in Self-Diffusion

What is the average displacement of a molecule during a self-diffusion process? Diffusion is accurately modeled as a random walk in one, two, or three dimensions. We know from our previous work on random walks that the average displacement from the origin, $<x> = 0$, if the probabilities of left and right steps, $p = q = 1/2$, are equal.

However, the root-mean-square displacement, $(<x^2>)^{1/2}$, is not equal to zero, and we can find it from the definition of the average of x^2 given by

$$<x^2> = \frac{\displaystyle\int_{-\infty}^{\infty} x^2 c(x,t)\,dx}{\displaystyle\int_{-\infty}^{\infty} c(x,t)\,dx}$$

Note that here both integrands are even functions, meaning that we can integrate from zero to infinity instead from minus infinity to plus infinity. Let us provide initial conditions such that the solute is present in a slab of thickness δ, with a constant concentration of c_o (Fig. 9.16). Then the time and spatial dependence of the concentration is given by

$$c(x,t) = \frac{c_o \delta}{2\sqrt{\pi Dt}} e^{\frac{-x^2}{4Dt}}$$

and by inserting this into the integrals above, we have that

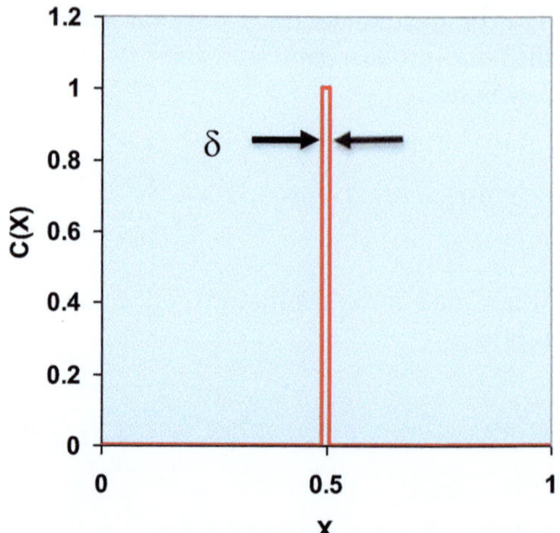

Fig. 9.16 The initial concentration profile used to determine the average displacement of a molecule during self-diffusion. The width of the profile is denoted by δ which is small with respect to x

$$<x^2> = \frac{\int_0^\infty x^2 e^{\frac{-x^2}{4Dt}}\,dx}{\int_0^\infty e^{\frac{-x^2}{4Dt}}\,dx}$$

and the integrals are Gaussian integrals which can be looked up in tables of integrals or can be done in `Mathematica`. The results are that

$$\int_0^\infty x^2 e^{\frac{-x^2}{4Dt}}\,dx = \frac{2\sqrt{\pi}}{\left(\frac{1}{Dt}\right)^{3/2}}, \qquad \text{and} \qquad \int_0^\infty e^{\frac{-x^2}{4Dt}}\,dx = \frac{\sqrt{\pi}}{\left(\frac{1}{Dt}\right)^{1/2}}$$

so that we obtain the important result that

$$<x^2> = \frac{2(Dt)^{3/2}}{(Dt)^{1/2}} = 2Dt.$$

The RMS value of the displacement is then given by the width of the Gaussian, and we can write that

$$x = \sqrt{2Dt}$$

for diffusion in one dimension; for n-dimensions, this needs to be modified to $x = \sqrt{2nDt}$.

Fig. 9.17 A plot of the position at every step for 100 random walks with 2500 steps each using the Mathematica code `Walk2D` (Gaylord and Wellin1995). Each walk was assigned a different random color. The black circle has a radius equal to the square root of the mean squared distance traveled

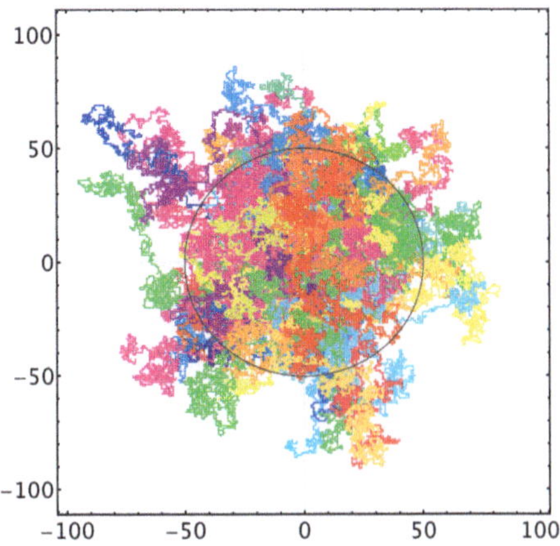

We remember from our study of the random walk that the dispersion, σ^2, is given by $N * L$, where N is the total number of steps of length L, and that the total number of steps is proportional to the time, $N \sim t$, so that $\sigma^2 \sim t$ or $\sigma \sim$ Sqrt[t], which is the same behavior as seen above for diffusion. The random walk is a good model for free diffusion in solution. Let's examine it, therefore, in more detail. A simple Mathematica random walk with a unit step length of n steps is generated by (Gaylord and Wellin 1995)

```
Walk2D[n_]=
FoldList[Plus,{0,0},{{0,1},{1,0},{0,1},{1,0}}[[Table[Random
[Integer,{1,4}],{n}]]]],
```

and the results of generating 100 random walks with 2500 steps each (Fig. 9.17) show that the walks cluster at the origin with a root-mean-square distance traveled that is congruent with our expectations that this distance is given by the square root of the number of steps (Fig. 9.18).

The mean square distance was computed using the Mathematica module (Gaylord and Wellin 1995):

```
MeanSquareDistance[n_Integer, m_Integer] =
Module[{walk2D},
walk2D[s_]= FoldList[Plus, {0,0},
{{0,1}, {1,0}, {0,-1}, {-1,0}}[[
Table[Random[Integer,{1,4}],{s}]]]];
N[Sum[Apply[Plus,Last[walk2D[n]]^2],{m}]/m]]
```

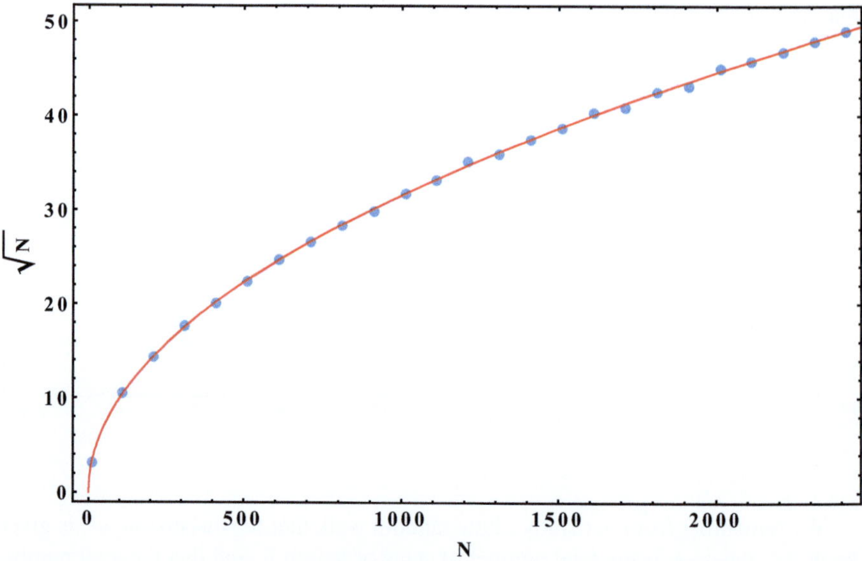

Fig. 9.18 The behavior (blue dots) of the square root of the root-mean-square distance traveled in a random walk as a function of the number of steps for 100 walks computed using the Mathematica code `Table[Sqrt[Mean[Table[MeanSquareDistance[n,25],{500}]]],{n, 10, 2500,100}]` compared with the square root of the number of steps (red line)

9.16 The Time Dependence of the Average Displacement in Self-Diffusion

How far do biomolecules diffuse in one dimension in a given time period? Table 9.3 shows calculated values of Sqrt[$2Dt$] for $t = 3600$ s (1 h) at $T = 298$ K and in water, whose viscosity, η, is (in SI units) 10^{-3} Kg/m-s, for various representative biomolecules. Diffusion coefficients are sensitive to both the temperature and the viscosity of the solution, so these are measured using the adopted standard conditions ($T = 293$ K, water). The size of a typical mammalian cell is 10–15μm giving times for diffusing across a cell ranging from ~40 ms to 6.3 s for the objects in Table 9.3; conservation of energy can be used to estimate the cell traversal time for a water molecule undergoing ballistic motion in a straight line from one edge of a cell to the other as ~20 ns. Diffusion is not a very efficient means for traversing large distances, particularly in fluids.

The data presented in Table 9.3 suggest that the diffusion coefficient is inversely related to the molecular weight. We saw earlier that the diffusion coefficient was inversely proportional to the molecular radius,

Table 9.3 Diffusion distances in 1 h for several biomolecules in water

Particle	Molecular weight (Daltons)	$D_{20,w}$ (cm^2/s)	$<x>^{1/2}$ (mm)
H$_2$O	18	2.00×10^{-5}	3.79
Sucrose	342	4.59×10^{-6}	1.82
Albumin	66,430	5.94×10^{-7}	0.65
Bushy stunt virus	9,100,000	1.15×10^{-7}	0.29

$$D = \frac{RT}{6\pi N_o \eta r}.$$

We can estimate the radius of a protein from its partial specific volume, $\bar{v} = V/M$, where V is the volume and M the molecular weight in grams/mole. The partial specific volume of a protein is approximately 0.73 cm^3/g (Harpaz 1994) and only varies by a few percent as a function of the amino acid composition. The radius (in cm) is

$$r = \sqrt[3]{\frac{3\,\bar{v}\,M}{4\pi N_o}}$$

from which we find that the diffusion coefficient for spherical proteins becomes

$$D = \frac{RT}{6\pi N_o \eta} \sqrt[3]{\frac{4\pi N_o}{3\,\bar{v}\,M}}$$

where we note that the diffusion coefficient is proportional to the inverse cube root of the molecular weight. A fit to the data in Table 9.3 gave the result that $D = e^{-9.868} M^{-0.389}$ where the slope of -0.389 was close to the expected value of -0.333 (Fig. 9.19). From a fit to the diffusion data from 143 proteins (Young et al. 1980), a value for the constant, a, in

$$D = a\frac{T}{\eta M^{1/3}}$$

was found to be 8.34×10^{-8} for the viscosity given in units of centipoise, while a direct calculation using

$$a = \frac{100\,R}{6\pi N_o} \sqrt[3]{\frac{4\pi N_o}{3\,\bar{v}}}$$

with $\bar{v} = 0.73$ results in $a = 11.07 \times 10^{-8}$. This calculated value is about ~33% higher than the above empirical value found by Young et al. (1980) likely because it does not take into account the layer(s) of bound water that surround a protein in aqueous solution and hence the larger hydrodynamic radius of the hydrated protein.

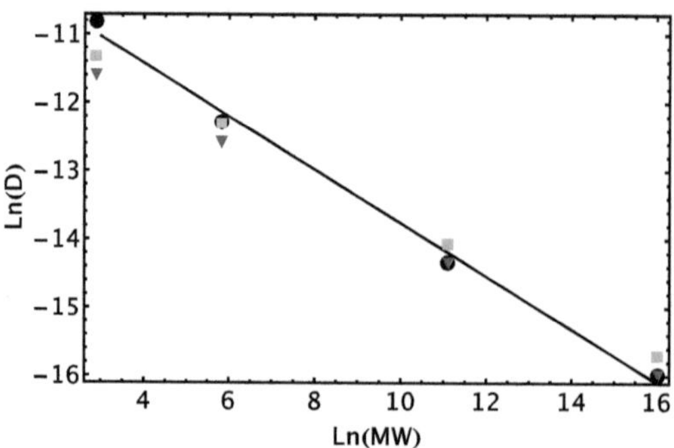

Fig. 9.19 A log-log plot of the diffusion coefficients as a function of the molecular weight for the molecules in Table 9.3 (black points) showing the inverse $\sim M^{1/3}$ relationship expected (black fitted line with a slope of -0.389). The red triangles are estimates based on the work of Young et al. (1980), while the green squares are the calculated values using the equation derived in the text

9.17 Propagators: Diffusion, Schrödinger, and Feynman

When we introduced Feynman diagrams in Chap. 6, we also introduced the concept of the *propagator*, a function that carried an electron from point A to point B in space-time (Fig. 6.9). However, the Schrödinger and Feynman propagators are complex functions, in keeping with their ability to interfere and through this interference generate quantum effects that are not seen for real-valued propagators, such as the diffusion propagator.

What is a propagator? We will find that a propagator is an inverse function that solves the integral equation for a differential operator. But let's begin with the simplest case, illustrated by considering the equation $3x = 1$, which has the obvious solution, $x = 1/3$, or written in another way: $3 * 3^{-1} = 1$, i.e., three times *three-inverse* is one or the identity. For variables, let $xy = 1$, then $y = x^{-1}$, y is the inverse of x, and again $x\,x^{-1} = 1$, i.e., x times *x-inverse* is one or the identity. Now, let's generalize and suppose that x and y are functions such that $y(z) = x^{-1}(z)$ and what do we mean by $y(z)\,x^{-1}(z) = 1$? The right-hand side of this equation cannot be simply a number, but rather it must be some kind of identity function.

This naturally leads us to consider operators as these functions and then the right-hand side of $y(z)\,x^{-1}(z) = 1$ can be interpreted as the identity operator, one that leaves the operand unchanged. For example, a Fourier transform followed by the inverse transform leaves the operand unchanged, except that one needs to use the boundary conditions in order to uniquely recover the input function correctly. The same is true for the differentiation/integration pair of operators. British mathematician George Green used this concept in the 1830s to convert differential equations into integral equations, including the boundary conditions. But just what is the identity operator in these cases? That is, what is the right-hand side of $F\,F^{-1} = \text{Identity}$?

Mathematics proceeds through a process of continuous generalization and abstraction. In this case, let's confront the inverse of a differential operator. If $\mathcal{L}$ is a linear differential operator, such as $\mathcal{L} = \frac{d^2}{dx^2} + x^2$, which operates on $f(x)$ to give the differential equation

$$\mathcal{L} f(x) = \left[\frac{d^2}{dx^2} + x^2 \right] f(x) = 0,$$

then Green's function, $G(x, y)$, is the analog of the inverse of $\mathcal{L}$:

$$G(x, y) \sim \mathcal{L}^{-1} \sim \left[\frac{d^2}{dx^2} + x^2 \right]^{-1}.$$

Now, if we add a source term, $h(x)$, to our differential equation

$$\mathcal{L} f(x) = h(x),$$

we can solve for $f(x)$ using

$$f(x) = \mathcal{L}^{-1} h(x) = G(x, y) \, h(x),$$

and if we operate on both sides of this equation with $\mathcal{L}$, we regain our original operator equation

$$\mathcal{L}\, f(x) = \mathcal{L}\, \mathcal{L}^{-1}\, h(x) = \mathcal{L}\, G(x, y)\, h(x) = h(x).$$

In the solution for $f(x)$ given above in terms of Green's function, we have left out just how to obtain the inverse of a differential operator, but we expect that it must somehow obey $\mathcal{L}\, \mathcal{L}^{-1} \sim \textit{Identity}$, and since the inverse of differentiation is integration, we anticipate that Green's function must involve integrals.

The mathematical way that an operator and its inverse combine to produce the identity function is solved through the use of the Dirac delta function introduced in the chapter on quantum mechanics (Chap. 4). This delta function (really a mathematical distribution) is singular at a particular point, where it is infinite, while for any other value of its argument, it is zero. These properties are embodied in

$$\delta(x) = \left\{ \begin{array}{l} 0, \ \text{for } x \neq 0 \\ \infty, \ \text{for } x = 0 \end{array} \right\}$$

along with

$$\int_a^b \delta(x)\,dx = 1, \text{ if } 0 \in [a, b]$$

and zero otherwise. The delta function has the following useful property

$$\int f(x)\delta(x)dx = f(0)$$

for any given function, $f(x)$. The delta function can also be shifted along the x axis by an amount, x_o, resulting in

$$\int f(x)\delta(x-x_o)dx = f(x_o).$$

This gives us just what we need to define the *Identity* operator in $L\,L^{-1} \sim$ *Identity* as the delta function because the delta function is invertible and is a generalization of the concept of an integer and its inverse.

The inverse of the operator $\mathcal{L}$ is Green's function so that $\mathcal{L}$ operating on Green's function must give the *Identity* operator

$$LG(x-x_o) = L\,L^{-1} = \delta(x-x_o)$$

from which we can formally write the solution of $\mathcal{L}f(x) = h(x)$ as

$$f(x) = \int G(x,x_o)h(x_o)dx_o$$

because the application of $\mathcal{L}$ to $f(x)$ is given by

$$Lf(x) = \int LG(x,x_o)h(x_o)dx_o = \int \delta(x-x_o)h(x_o)dx_o = h(x).$$

Green's functions are most useful in terms of convolutions with the source function. Green's functions are explicitly two-point functions and are also known as propagators in that they carry the effect from the source to a remote site.

9.18　The Diffusion Propagator

We can use the above formalism to determine how a molecule gets from point A to point B (Fig. 9.20) as follows. Once again, the diffusion equation is

$$D\frac{\partial^2 c(x,t)}{\partial x^2} - \frac{\partial c(x,t)}{\partial t} = 0.$$

The differential operator $\mathcal{L}$ in this case is

Fig. 9.20 A space-time diagram illustrating the propagation of a molecule from point A(x, t) to point B (x', t')

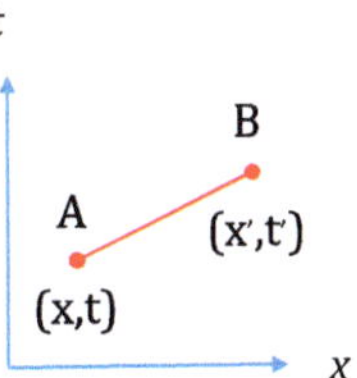

$$\mathcal{L} = D\frac{\partial^2}{\partial x^2} - \frac{\partial}{\partial t}$$

such that $\mathcal{L}c(x,t) = 0$. We have solved this equation before, but let us do it again and find Green's function as an added bonus. Although there are no universal methods for obtaining Green's functions for differential equations, the Green's functions must satisfy at least the following conditions which allow their construction in many useful cases. The function $G(x' - x, t' - t)$ must be discontinuous at $x' = x$ and $t' = t$ in order to give rise to a delta function, while for other values of its arguments, it needs to satisfy

$$\mathcal{L}G = 0. \tag{9.1}$$

It must also satisfy the homogeneous boundary conditions at the ends of the intervals.

One way of solving the diffusion equation is to assume a form of the solution and then to insert this into the diffusion equation and work with the resulting expression. Letting the concentration as a function of space and time $c(x, t)$ be given by its inverse Fourier transform

$$c(x, t) = \int \chi(k, t)e^{ikx}dk,$$

then,
$$\mathcal{L}\, c(x,t) = \left(D\frac{\partial^2}{\partial x^2} - \frac{\partial}{\partial t}\right)\int \chi(k, t)e^{ikx}dk$$

$$= D\int \chi(k, t)\frac{\partial^2}{\partial x^2}e^{ikx}dk - \int \frac{\partial}{\partial t}\chi(k, t)e^{ikx}dk$$

$$= -k^2 D\int \chi(k, t)e^{ikx}dk - \int \frac{\partial}{\partial t}\chi(k, t)e^{ikx}dk,$$

and this must equal to zero, which implies that the sum of the integrands must be zero:

$$-k^2 D \chi(k,t) - \frac{\partial}{\partial t}\chi(k,t) = 0$$

This can be solved with a single integration, giving

$$-k^2 D \int_t^{t'} dt = \int_{\chi(k,t)}^{\chi(k,t')} \frac{d\chi(k,t)}{\chi(k,t)}$$

with the solution for the time dependence of the Fourier transform of the concentration given by

$$\chi(k,t) = \chi(k,t=t')e^{-Dk^2(t'-t)},$$

and then the concentration becomes

$$c(x,t'-t) = \int \chi(k,t=t')e^{-Dk^2(t'-t)}e^{ikx}dk$$

where this is the convolution of the initial condition $\chi(k,t=t')$ with $e^{-Dk^2(t'-t)}$. Assume that we have a point source at $t=t'$ so that the initial condition is that

$$c(x,t'-t) = \delta(x'-x).$$

We can find $\chi(k,t=t')$ from the Fourier transform of the initial condition

$$\chi(k,t=t') = \frac{1}{2\pi}\int_{-\infty}^{\infty}\delta(x'-x)e^{ikx}dk = \frac{e^{-ikx'}}{2\pi},$$

and we can obtain $c(x'-x,t'-t)$ for $t'>t$ from its inverse Fourier transform

$$c(x'-x,t'-t) = \frac{1}{2\pi}\int_{-\infty}^{\infty}e^{-Dk^2(t'-t)}e^{ik(x'-x)}dk.$$

This integral can be directly calculated with Mathematica or can be done by completing the squares; letting $s=t'-t$ and $y=x'-x$, then

$$-Dsk^2 + iky = -Ds\left(k - \frac{iy}{2Ds}\right)^2 - \frac{y^2}{4Ds}$$

so that

$$c(y,s) = \frac{1}{2\pi}\int_{-\infty}^{\infty}e^{-\left[Ds\left(k-\frac{iy}{2Ds}\right)^2 - \frac{y^2}{4Ds}\right]}dk = \frac{1}{\sqrt{4\pi Ds}}e^{-\frac{y^2}{4Ds}}$$

which is a Gaussian in space that decays exponentially with time. The result of this integral is the Green's function (returning to x, t variables):

$$G(x' - x, t' - t) = \frac{1}{\sqrt{4\pi D(t' - t)}} e^{-\frac{(x' - x)^2}{4D(t' - t)}}$$

Does this Green's function satisfy $\mathcal{L}(x' - x) = \delta(x' - x)$? This is left as a useful exercise.

9.19 The Free-Particle Schrödinger Propagator

It is interesting to note the similarity between the diffusion equation

$$\frac{\partial c(x, t)}{\partial t} - D \frac{\partial^2 c(x, t)}{\partial x^2} = 0$$

and the free-particle Schrödinger's equation

$$\frac{\partial}{\partial t} \psi(x, t) - \frac{i\hbar}{2m} \frac{\partial^2}{\partial x^2} \psi(x, t) = 0.$$

These are formally the same except for the fact that the wave function is complex, while the concentration is strictly a real function of space-time. The Schrödinger equivalent to the diffusion coefficient is also complex and can be read from the diffusion equation as

$$D' = \frac{i\hbar}{2m}$$

whose real part for an electron gives a "diffusion coefficient" of 7.25×10^{-4} m^2/s or 7.25 cm^2/s, a large value indeed. We can therefore find the Green's function for the free-particle Schrödinger's equation simply by replacing D with D' to obtain

$$G(x, t) = \sqrt{\frac{2m}{4\pi i\hbar t}} e^{-\frac{mx^2}{4i\hbar t}}.$$

9.20 Green's Function for the Schrödinger Equation with a Potential

The time-independent Schrödinger equation for a particle of mass m in a potential $V(x)$ in a single spatial dimension is given by

$$\left[\frac{d^2}{dx^2} + k^2\right]\psi(x) = \frac{2m}{\hbar^2}\,V(x)\psi(x)$$

where $k^2 = \frac{2mE}{\hbar^2}$. From the work above on the diffusion equation, it should be appreciated that this is once again the diffusion equation, albeit with a source term given by the product of the potential and the (now complex) wave function. The Green's function is a solution of $\mathcal{L}G(x) = \delta(x)$ or

$$\left[\frac{d^2}{dx^2} + k^2\right]G(x) = \delta(x).$$

Again, we take $G(x)$ to be given by the inverse of its Fourier transform, $g(s)$

$$G(x) = \frac{1}{\sqrt{2\pi}}\int g(s)e^{isx}ds,$$

and here we can use one of the several definitions of a Dirac delta function

$$\delta(x) = \frac{1}{2\pi}\int e^{isx}ds$$

which when inserted into the Green's function equation results in

$$\frac{1}{\sqrt{2\pi}}\left[\frac{d^2}{dx^2} + k^2\right]\int g(s)e^{isx}ds = \frac{1}{2\pi}\int e^{isx}ds$$

or

$$\frac{1}{\sqrt{2\pi}}\int\left(-s^2 + k^2\right)g(s)e^{isx}ds = \frac{1}{2\pi}\int e^{isx}ds$$

where this equation only holds if

$$\sqrt{2\pi}\left(-s^2 + k^2\right)g(s) = 1$$

which gives the Fourier transform of the Green's function as

$$g(s) = \frac{1}{\sqrt{2\pi}\left(-s^2 + k^2\right)}.$$

The Green's function is then obtained through inverse Fourier transformation as

Fig. 9.21 Illustration of the semicircular contours of integration for the case where $x > 0$ (left) and $x > 0$ (right)

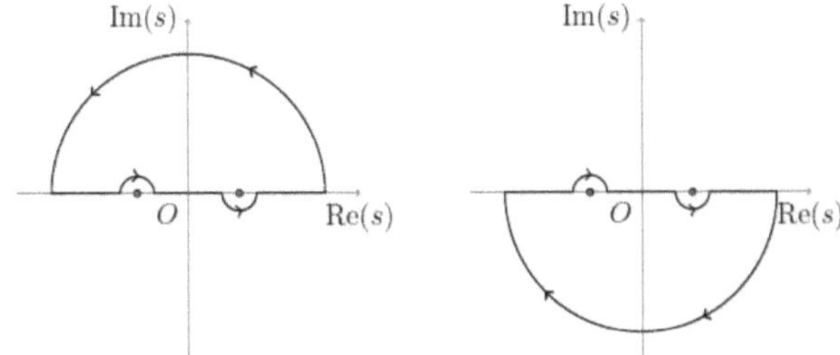

$$G(x) = \frac{1}{\sqrt{2\pi}} \int g(s) e^{isx} ds = \frac{1}{2\pi} \int \frac{1}{(k+s)(k-s)} e^{isx} ds.$$

One can compute this integral using contour integration in the complex s plane by noting that the integrand contains simple poles at $s = \pm k$. For $x > 0$, the closed contour is a semicircle of infinite radius in the upper half plane, while for $x < 0$ the contour is chosen in the lower half plane, as indicated in Fig. 9.21. In either case, the contours are chosen to only include one of the two poles. The choice depends on the sign of x because we wish the term e^{isx} in the integral to vanish at infinity, and this means that the exponent must have an overall negative sign. In the case where $x > 0$, we find the pole at $s = k$ so that application of Cauchy's theorem gives

$$G(x) = -\frac{1}{2\pi} \oint \frac{1}{(k+s)(k-s)} e^{isx} ds = -2\pi i \frac{1}{2\pi} \frac{e^{ikx}}{2k} = -i \frac{e^{ikx}}{2k}.$$

And performing the integration for the other case, $x < 0$ gives

$$G(x) = -i \frac{e^{-ikx}}{2k}$$

which can be combined with the solution for the first case to give the Green's function for the one-dimensional Schrödinger equation:

$$G(x) = -i \frac{e^{ik|x|}}{2k}$$

Let's check that this satisfies

$$\mathcal{L} G(x) = \left[\frac{d^2}{dx^2} + k^2 \right] G(x) = 0$$

by substitution

$$\left[\frac{d^2}{dx^2} + k^2 \right] G(x) = -i \left[\frac{d^2}{dx^2} + k^2 \right] \frac{e^{ik|x|}}{2k} = \frac{-i}{2k} \left[i^2 k^2 + k^2 \right] e^{ik|x|}.$$

We see that the term in brackets on the right vanishes for $x \neq 0$, which is consistent with the definition of the Green's function (9.1).

The solution for Schrödinger's equation in terms of a general potential $V(x)$ is then given by the sum of a solution of the homogeneous free-particle equation ($\psi_o(x)$ with $V(x) = 0$)

$$\left[\frac{d^2}{dx^2} + k^2\right]\psi_o(x) = 0$$

and an integral (convolution) of the potential and the Green's function

$$\psi(x) = \psi_o(x) + \frac{2m}{\hbar^2}\int G(x' - x)V(x')\psi(x')dx'$$

$$\psi(x) = \psi_o(x) + \frac{im}{\hbar^2 k}\int e^{ik|x' - x|}V(x')\psi(x')dx'.$$

Note that the prescription for avoiding the poles in the integral is based on the work of Feynman and hence this Green's function is referred to in the literature as the *Feynman propagator* and corresponds to the probability of a particle observed at one point to be observed at a different point at a later time. The fact that one pole is always included accommodates both forward and time-reversed cases. This is like diffusion except for the complex nature of the wave function which allows interference among different paths. This is seen most clearly in Feynman diagrams for electron-electron scattering in quantum electrodynamics where a photon is the electromagnetic field 4-vector, A_μ, which obeys Maxwell's equation

$$\left(g_{\mu\nu}\partial^2 - \partial_\mu\partial_\nu\right)A^\mu = j_\mu$$

where $g_{\mu\nu}$ is the metric tensor and j_μ is the charge current 4-vector (source term). We can find the photon propagator by the simple trick of preplacing each derivative by ik to find the Fourier transform of the propagator, which is the propagator in momentum space (see Chap. 4) as

$$G_{\mu\nu}(k) = \frac{-i}{k^2}\left[g_{\mu\nu} - \frac{k_\mu k_\nu}{k^2}\right],$$

and each photon line in a Feynman diagram, such as seen in Fig. 9.22, is then replaced by the propagator in the computation of the scattering integrals.

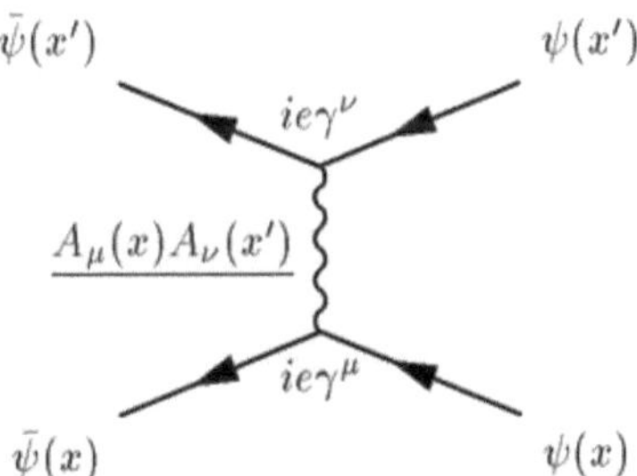

Fig. 9.22 Feynman diagram for electron-electron scattering mediated by the exchange of a photon, A_{μ}, whose propagator appears as the wavy line connecting the two electron lines, $\psi(x)$. The time-reversal invariance of the propagator is indicated by the lack of arrows on the photon propagator. (https://en.wikipedia.org/wiki/Feynman_diagram#/media/File:Feynman-diagram-ee-scattering.png under Creative Commons licensing for reuse and modification)

9.21   The Mean Free Path in Gases and Liquids

Although the diffusion equation gives a quantitative answer to the question of the average distance moved in a given time period, a related and interesting problem is to find how far a molecule moves before it scatters from another. This is the mean free path, λ. Dimensional analysis suggests that the diffusion coefficient, which has units of (length)^2 vs. time, could be written as the mean free path times a velocity:

$$D = \frac{1}{3}\,\lambda\,v$$

We can use the Maxwellian speed distribution for gas molecules to estimate D from the mean thermal speed

$$v = \sqrt{\frac{8kT}{\pi m}}$$

for a molecule of mass m, at a temperature T. The mean free path for molecules governed by the Maxwellian speed distribution is

$$\lambda = \frac{1}{\sqrt{2}\,n\sigma}$$

where n is the number density and σ is the cross section (the effective area presented by a molecule). From the ideal gas law,

$$n = \frac{N}{V} = \frac{P}{kT}$$

and $\sigma = \pi(2r)^2 = \pi d^2$ is the effective cross section (area) for a molecule of radius r. We use the diameter here because it represents the distance of closest approach for the centers of two molecules. Combining these terms leads to the mean free path as

$$\lambda = \frac{kT}{\sqrt{2}\,\pi d^2 P}.$$

And we can estimate the mean free path for a gaseous molecule like H_2 ($d = 240$ pm) at $P = 1$ atmosphere (101.3 kPa) and 300 K as $\lambda \sim 1.6 \times 10^{-7}$ m, which is 674 times the molecular diameter and 48 times the average distance between molecules (3.1×10^{-9} m).

The average density of a gas is much lower than that for the same material in a liquid state. For example, at standard temperature (273 K) and pressure, air has a density of 1.293 kg/m^3, while water under the same conditions has a thousand-fold greater density of 1000 kg/m^3. This results in the mean free path for water molecules in water of about 2.5×10^{-10} m, which is smaller than the average intermolecular spacing (4×10^{-10} m) derived from its specific volume v_{s} as follows. The number density of water is

$$n = \frac{N_0\left(\frac{\text{molecules}}{\text{mole}}\right)}{18\left(\frac{\text{cm}^3}{\text{mole}}\right)} = \frac{6.024 \times 10^{23}}{18.01528} = 3.34 \times 10^{22}\,\frac{\text{molecules}}{\text{cm}^3}$$

and

$$v_{\mathrm{s}} = \frac{1}{n}\,\frac{\text{cm}^3}{\text{molecule}} = \frac{4\pi}{3}r^3$$

which give the average distance between molecules as

$$r = \sqrt[3]{\frac{3v_s}{4\pi}} = 4.13 \times 10^{-10}\ m.$$

These have important consequences for biochemical reactions in the mostly-water cytoplasm of living cells in that the biomolecules basically touch each other all the time; in these same units, the size of a water molecule is 2.75×10^{-10} m which is larger than the mean free path. It also implies that the mean free path for a gas like oxygen dissolved in water is essentially that of the water molecules and not that for a gas like air. We will extensively take up the properties of water in Chap. 11.

9.22   Measurements of Diffusion Coefficients

How do you actually measure diffusion coefficients? The example in Fig. 9.8 using a removable partition suggests one method, but it can be readily appreciated that monitoring a moving boundary is fraught with experimental difficulties. Nevertheless, exactly this method has been used in many ultracentrifuge measurements of macromolecular diffusion coefficients (Fig. 9.23) using the techniques pioneered by Svedberg, for which he got the Nobel Prize in 1926.

Since that time, many other methods have been devised for the measurement of diffusion coefficients. These include the use of radioactive tracers to measure the diffusion of molecules in porous media (Rowe 1969) and several optical techniques, such as dynamic laser light scattering and fluorescence recovery after photobleaching and NMR spectroscopy and imaging, particularly diffusion tensor MRI.

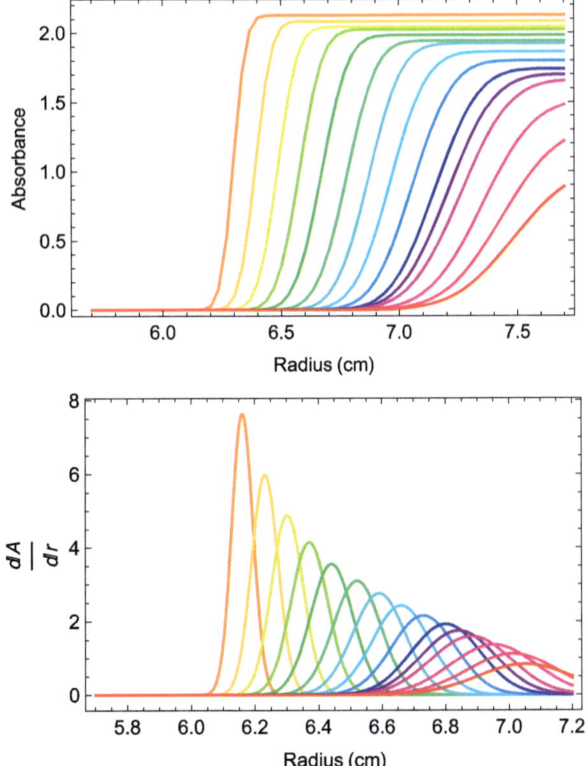

Fig. 9.23 Measurement of the diffusion coefficient of a protein by means of analytical ultracentrifugation. Each successive scan (left to right) is shown in a different color beginning at radius = 6.1 cm. (Top) note that the boundary is the integral of a Gaussian and that it broadens with time as it translates down the cell and at the same time lowers the concentration gradient. (Bottom) the derivative of the absorbance showing the Gaussian and its broadening with time due to diffusion

9.22.1 *Dynamic Laser Light Scattering*

Dynamic laser light scattering is also referred to in the literature as photon correlation spectroscopy or quasielastic light scattering. The translational diffusion of a particle due to random Brownian motion in solution is described by $D = \frac{kT}{6\pi\eta r}$ where r is the hydrodynamic radius, the size of the particle including any bound water, and is defined to be the radius of a sphere that has the same diffusion coefficient as the particle (Table 9.2). The intensity of light scattered from a small object depends on both the radius of the object and the wavelength of the light. For particles that are small ($r \sim \lambda/20$) with respect to optical wavelengths ($\lambda \sim 500$ nm), i.e., for $r \sim 25$ nm, Rayleigh scattering predominates where the intensity of the scattered light is proportional to the inverse fourth power of the wavelength, $I(\lambda) \; \alpha \; \frac{1}{\lambda^4}$, and to the sixth power of the radius, $I(r) \; \alpha \; r^6$. When the particles become larger and their size approximates the wavelength of light, then Mie scattering predominates, and we have a complex relationship between the scattered light intensity and particle size because strong interference effects occur.

The small sizes of most biochemical objects in solution, such as proteins, nucleic acids, carbohydrates, etc., are such that Rayleigh scattering is the dominant contribution to the scattered light. These molecules are, as we have seen, in constant thermal motion, and the laser light scattered from this ensemble is therefore Doppler broadened so that the phases of the scattered photons are modulated by the molecular motion giving rise to decoherence of the scattered wave fronts. The decay characteristics of the autocorrelation function of the scattered light are dependent on the diffusion coefficient of the macromolecules. The normalized, second-order scattered light intensity correlation function $G_2(\tau)$ is given by

$$G_2(\tau) = \frac{\int I(t + \tau)I(t)dt}{\left[\int I(t)dt\right]^2}$$

where $I(t)$ is the intensity at time t. Even though it is not usually possible to know the motion of each particle in solution, the motional correlation between particles is the first order, or electric field correlation function, $G_1(\tau)$ given by

$$G_1(\tau) = \frac{\int E(t + \tau)E(t)dt}{\left[\int E(t)dt\right]^2}$$

where $E(t)$ is the electric field scattered at time t. If the scattering is random and *homodyne*, where the detector only measures the scattered light, Siegert showed that these two correlation functions can be combined (Siegert 1949) into

$$G_2(\tau) = \alpha + \beta |G_1(\tau)|^2.$$

Here, α (~1) is the baseline and β is an instrumental factor. If the sample solution contains a single, monodisperse collection of molecules whose motion is Brownian, the electric field correlation function decreases exponentially with time as $G_1(\tau) = e^{-\Gamma\tau}$ and the Siegert relationship becomes

$$G_2(\tau) = 1 + \beta e^{-2\Gamma\tau}.$$

The translational diffusion coefficient D determines the decay behavior of the scattered light through Γ by $\Gamma = Dq^2$. Here, q is the Bragg wave vector for a given wavelength λ, in a solvent of index of refraction n, at a scattereing angle, θ, and is found from $q = \frac{4\pi n}{\lambda}\sin(\theta/2)$.

The Siegert relationship is finally

$$G_2(\tau) = 1 + \beta e^{-2Dq^2\tau},$$

and this directly relates the fluctuation-dependent decay in the scattered light intensity to the diffusion coefficient.

One can model the effect of the diffusion coefficient on the alteration of the phases of the scattered light that lead to the decay of the autocorrelation function. The Brownian diffusion of the macromolecules generates a random walk of phases that can be estimated using the average displacement in a specified time interval, t. The mean displacement is $x = \sqrt{2Dt}$ in the dimension parallel to the scattering axis. This factor needs to be compared with the wavelength λ of the incident radiation. Motions with physical displacements less than $\lambda/10$ produce phase shifts that are within the random shot noise of the scattered light. Therefore, the phase shift needs to be at least $\lambda/10$, which for 500 nm light is 50 nm. What is the characteristic time to diffuse this distance? It is given by the position fluctuations due to Brownian motion, $t \sim x^2/2D$, from the random walk; letting $x = \lambda/10$, in water with an index of refraction, $n = 1.33$, and a diffusion coefficient of bushy stunt virus, $D = 1.15 \times 10^{-11}$ m^2/s (Table 9.2), we obtain $t = 6.4 \times 10^{-5}$ s, which is about the time where the autocorrelation function falls to one-half its value at Fig. 9.24 (cyan). We can estimate the diffusion coefficient from this time using $D = \frac{-\ln(0.5)}{q^2 t}$.

Further details with respect to the use of laser light scattering to probe the homogeneity, hydrodynamic radii, molecular weights, shapes, and interactions are discussed in the review by Stetefeld et al. (2016).

9.22.2   Fluorescence Recovery After Photobleaching

Direct measurement of boundary movement can be used for the determination of diffusion coefficients in biological samples such as whole cells or cell membranes if the diffusing species can be labeled with a fluorophore. Often, this method may be

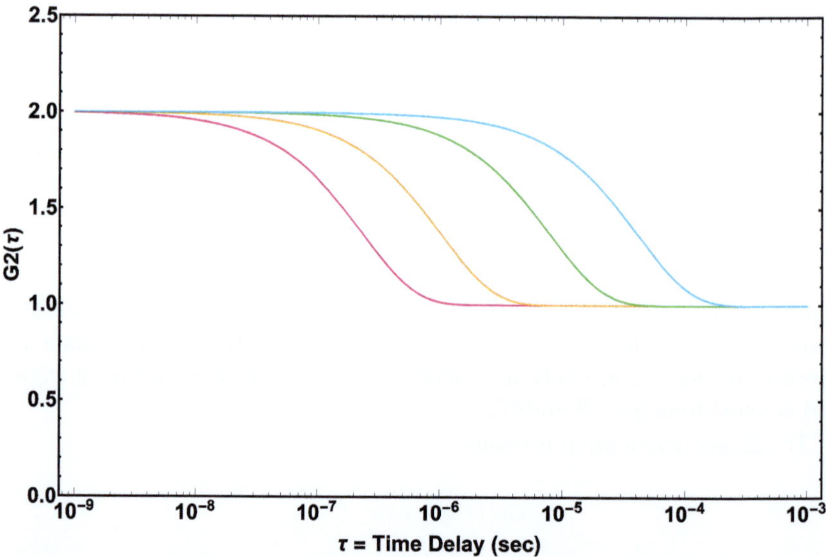

Fig. 9.24 The autocorrelation function for homodyne detected light scattered from a monodisperse solution of biomolecules whose diffusion coefficients are given in Table 9.2, beginning with water (rose) and progressing to sucrose (orange), albumin (green), and bushy stunt virus (cyan). We can read off the diffusion coefficients of the various molecules by noting the time at which the autocorrelation function decays to one-half of its initial value (see text)

the only way to obtain diffusion information from lipid bilayer membranes, concentrated protein solutions, or living cells. The technique illuminates a small patch of an object with an intense pulsed optical source like a laser or filtered broadband mercury or xenon lamp (Fig. 9.25). This permanently bleaches the chromophore in the illuminated area leaving a dark region, after which unbleached molecules will diffuse into this "black hole" and the fluorescence will recover to almost its pre-bleach value.

Early FRAP studies used potentially disruptive methods, such as permeabilization of cells or microinjection for introducing the fluorophore; however, with the discovery of green fluorescent protein (Tsien 1998), this fluorophore could be expressed within the cell, and many proteins could be labeled so that interior molecules within the cell could be monitored. Now, FRAP has become a technique of choice for measuring the movement of molecules in living cells, an area of investigation that is not possible with laser light scattering. The analysis of FRAP data has been made easier recently with the availability of web-based software, *EasyFRAP-web* (Koulouras et al. 2018), which implements the analyses first proposed by Axelrod et al. (1976) and Soumpasis (1983). These authors showed that for bleaching of a circular spot of radius r, the fractional fluorescence recovery curve can be written as

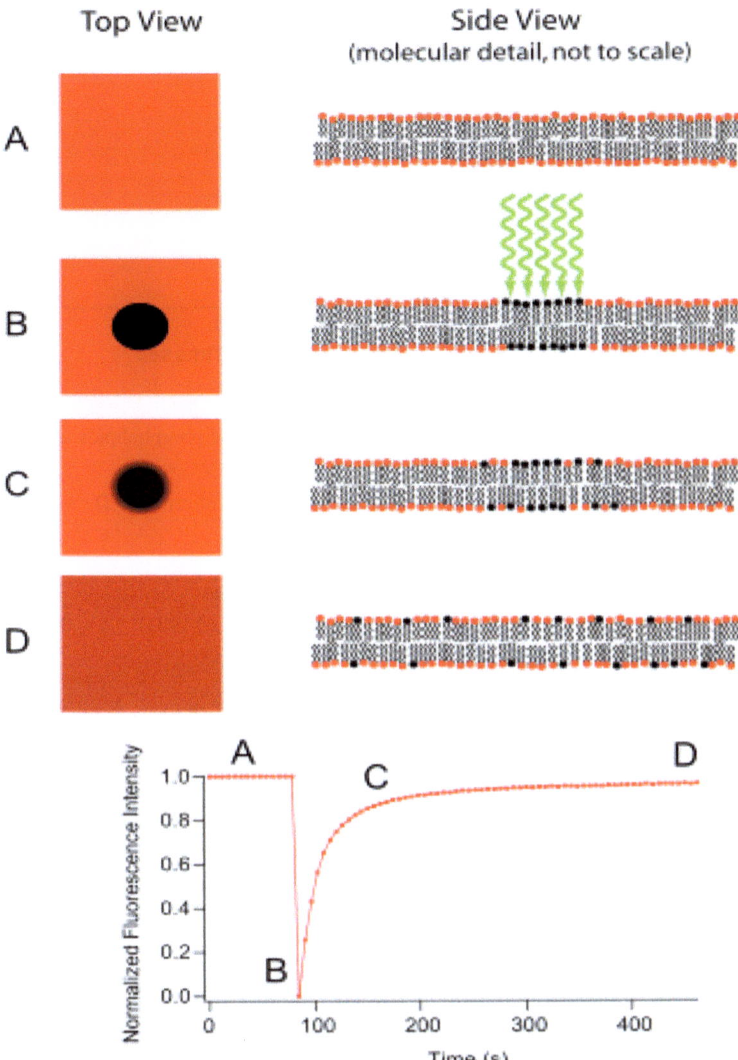

Fig. 9.25 A schematic view of a fluorescence recovery after photobleaching experiment to measure the lateral diffusion of fluorescent phospholipids with red fluorescent head groups in the plane of the bilayer membrane. (**a**) The initial fluorescence observed through a microscope shows a uniform light across the field of view. In (**b**), a red laser beam has bleached a circular spot in the lipids. (**c**) After some time, red fluorescent phospholipids diffuse into the region, restoring some fluorescence. In (**d**), the diffusion of red fluorescent phospholipids has restored the membrane to its initial uniform red fluorescence, albeit with a lowered intensity of the emitted light because some of the fluorescent molecules have been permanently bleached from the laser irradiation in (B). The lower curve shows the time dependence of the normalized fluorescence intensity, from which the phospholipid diffusion coefficient can be extracted. (Author: MDougM https://en.wikipedia.org/ wiki/. Fluorescence_recovery_after_photobleaching (File:Frap_diagram.svg) marked as public domain)

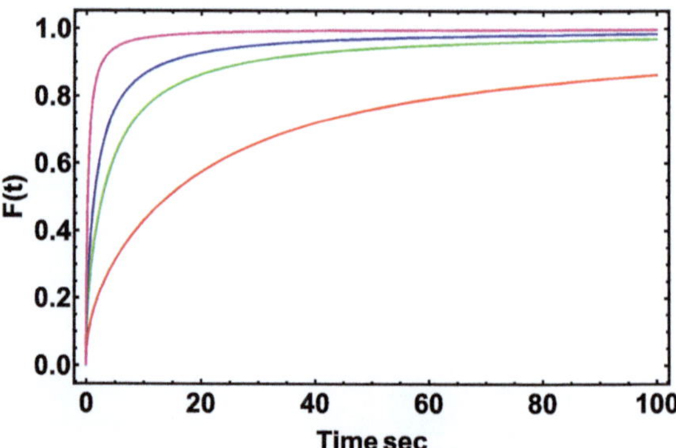

Fig. 9.26 Fractional fluorescence recovery curves for diffusion coefficients of 1×10^{-12} m^2/s (red), 5×10^{-12} m^2/s (green), 1×10^{-11} m^2/s (blue), and 5×10^{-11} m^2/s (violet) with $r = 8 \times 10^{-6}$ m

Table 9.4 Optical measurements of the diffusion coefficients of phospholipids and proteins in bilayers

Molecule	Temperature (°C) (reference)	D ($\times 10^{12}$ m^2/s)
DMPC (1,2-dimyristoyl-*sn*-glycero-3-phosphocholine)	37 (Sterling 2013)	4
DMPC	37 (Soumpasis 1983)	3.9
POPC (1-palmitoyl-2-oleoyl-*sn*-glycero-3-phosphocholine)	30 (Köchy and Bayerl 1993)	4 ± 0.8
DOPC (1,2-dioleoyl-*sn*-glycero-3-phosphocholine)	23 (Benda 2003)	4.2 ± 0.4
DOPC +10% cholesterol	23 (Benda 2003)	2.3 ± 0.2
DOPC +30% cholesterol	23 (Benda 2003)	1.1 ± 0.1
DOPC +60% cholesterol	23 (Benda 2003)	0.5 ± 0.1
Rhodopsin	37 (Wey 1981)	0.3 ± 0.12

$$F(t) = e^{-\frac{2\tau}{t}}\left[I_0\left(\frac{2\tau}{t}\right) + I_1\left(\frac{2\tau}{t}\right)\right]$$

where $\tau = \frac{r^2}{4D}$ is the characteristic diffusion time and I_0 and I_1 are modified Bessel functions.

This function is shown in Fig. 9.26 for diffusion coefficients ranging from 10^{-12} m^2/s to 5×10^{-11} m^2/s. The approximate diffusion coefficients can be obtained from the plots using the numerical solution to $F(t) = 0.5$ to give $D = 0.224 \frac{r^2}{t}$ where t is the time at which the fractional fluorescence recovery curve reaches 0.5. Application of this method to measure the diffusion of a fluorescent lipid analog in multibilayers of dimyristoylphosphatidylcholine gave a value of $D = 3.9 \times 10^{-12}$ m^2/s at 37 °C (Table 9.4). Phospholipids and proteins diffuse in the plane of the lipid bilayer in a manner that is consistent with the fluid mosaic model of the cell membrane (Nicolson 2014). For example, the photopigment rhodopsin has been labeled with iodoacetamide tetramethyl rhodamine, and its diffusion coefficient has been determined using FRAP (Table 9.4) showing that it diffuses within the plane of the bilayer about an order of magnitude slower than that of the phospholipids themselves.

For a cell of diameter 10µm, a lipid can diffuse from one end to the other in around 6 s, while the photosensitive protein rhodopsin (Chap. 10) can laterally diffuse the 2µm diameter of a rod outer segment in about 2.5 s. These times and distances can be used to also estimate the viscosity of the bilayer with the result that it is about 100-fold higher than water; the viscosity inside the bilayer is about that of triolein, which is not surprising given that the fatty acyl chains are of similar length (C_{18}) in both substances.

9.22.3 *Fluorescence Correlation Spectroscopy*

Instead of bleaching an area and monitoring the recovery of the fluorescence, one can also use the techniques of steady-state photon correlation spectroscopy to measure the diffusion coefficient of fluorescent molecules in cells and in solution (Elson 2011). The mean concentrations of fluorescent molecules fluctuate due to Brownian motion causing them to move into and out of the optical detection volume leading to time-dependent variations in the fluorescence intensity. The normalized autocorrelation function for diffusion is

$$g(t) = \frac{1}{1 + \frac{t}{\tau}}$$

where t is the time delay and $\tau = \frac{r^2}{4D}$. This function is shown for several different diffusion coefficients in Fig. 9.27. Measurements of the diffusion coefficients for several phospholipids using fluorescence correlation spectroscopy (Table 9.4) give values that are the same as obtained using FRAP (Macdonald et al. 2013).

9.23 Diffusion in NMR Spectroscopy and Imaging

While optical methods can probe many types of biological samples, they are adversely impacted by light scattering and attenuation within the object, and this limits their applications to optically transparent materials. Longer wavelength electromagnetic radiation, such as radio waves, passes through biological tissues without these interfering constraints. For this reason, nuclear magnetic resonance spectroscopy has been used to measure diffusion coefficients since the early 1950s, and with the development of NMR imaging in the 1970s by Paul Lauterbur (1973) and Peter Mansfield (Garroway et al. 1974) sophisticated methods have been developed to

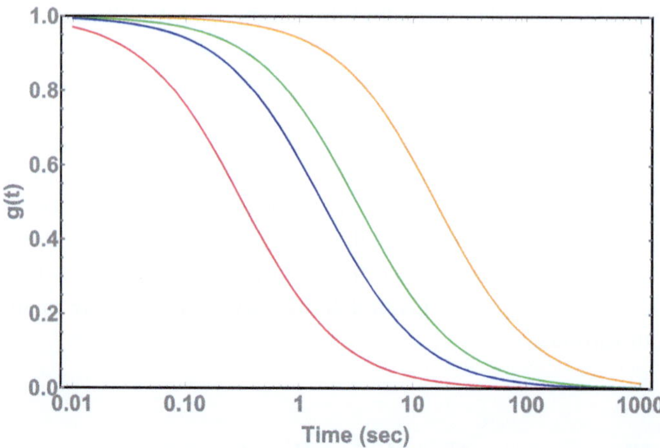

Fig. 9.27 Decay curves for the normalized correlation function for the fluorescence intensity for diffusion coefficients: 1×10^{-12} m^2/s (red), 5×10^{-12} m^2/s (blue), 1×10^{-11} m^2/s (green), and 5×10^{-11} m^2/s (orange) with $r = 8 \times 10^{-6}$ m

measure diffusion coefficients in a wide variety of cells, tissues, and animals and in human subjects, in samples that are not amenable to optical methods. The NMR spectroscopic methods rely on the application of magnetic field gradient pulses, and while this is also true for some of the NMR imaging techniques, one can exploit the spatial nature of imaging in the absence of field gradients to measure boundary diffusion in opaque objects. We will later show a straightforward application of NMR imaging to determine the diffusion coefficient of oxygen by measuring boundary movement.

NMR imaging techniques involve the application of temporal and spatial variations in the local magnetic field experienced by an object. These controlled gradients algebraically add to the static magnetic field, B_o, by the principle of superposition. How do you produce images with NMR? From Chap. 7, we remember the basic NMR equation, $\omega_o = \gamma B_o$, which states that the nuclear magnetic resonance frequency is a linear function of the applied, static magnetic field; the constant γ is the magnetogyric ratio determined by quantum mechanics for each nucleus (Table A.2) and need not further concern us here.

Up to now, we have desired that the static magnetic field B_o be essentially perfectly isotropic, i.e., homogeneous from point to point within the magnet bore. However, if we deliberately introduce a controlled inhomogeneity in the form of a linear gradient G, then the NMR frequency of a nucleus, such as a water proton in tissue, will linearly vary from point to point within the sample volume. For example, let us change the static magnetic field B_o to one that varies linearly along the z-axis of the magnet in the following way,

$$B(z,t) = B_o + zG(t),$$

where $G(t)$ is a time-varying gradient term. Then the NMR equation changes from $\omega_o = \gamma B_o$ to

$$\omega(z,t) = \gamma B(z,t) = \gamma(B_o + zG(t)), \tag{9.2}$$

and we see that the resonance frequency now is a function of the position of the nucleus with respect to the z-axis: $\frac{\partial \omega}{\partial z} = \gamma G(t)$. From the resonance frequency, we can determine the position of the spin along the z-axis

$$z = \frac{\omega - \gamma B_o}{\gamma G}$$

as the variation from the Larmor frequency divided by the frequency shift induced by the gradient. The gradient along the z-direction is defined to be

$$G = \frac{\partial B_z}{\partial z}.$$

We can rewrite the NMR Eq. (9.2) as

$$\omega(z,t) = \gamma B(z,t) = \omega_o + \omega_G(z,t),$$

and we note that the shift in the resonance frequency $\omega_G(z,t) = \gamma z G(t)$ is linear in both position and gradient amplitude and that this encodes the position of a spin into its resonance frequency.

We will recall from Chap. 7 that the time-domain NMR signal resulting from pulsed excitation of nuclear spins is

$$s(t) = \int \sum_{n=1}^{N} A_n(z) e^{i[(\omega_n - \omega_o)t + \varphi_o]} dz$$

for N chemically distinct species. The amplitude $A_n(z)$ is proportional to the sample magnetization (Curie's law) or the spin density for species n at position z. If the radiofrequency pulse, B_1, lasts for a time, τ, then the phase of the signal at a spatial position z is given by

$$\varphi_o(z,\tau) = -\int_0^\tau \omega_o(z,t)dt$$

with the convention that phase increases in a counterclockwise direction. In the case where there are no applied field gradients, $\varphi_o(z,\tau) = -\omega_o(z,t)\tau = -\omega_o\tau$, and the phase due to the RF pulse is constant, position-independent, and can be removed in the frequency domain data by multiplying the spectrum by $e^{-i\varphi_0}$. For MRI, one is primarily concerned solely with the NMR signal from water whose NMR signal is a single peak, so $N = 1$, and we will drop the subscript on A. Then the demodulated ($\omega = \omega_0$) time-domain signal is

$$s(t) = \int_{-\infty}^{\infty} A(z)e^{i(\omega_1 t + \varphi(z,t))} e^{-i\omega_0 t} dz$$

$$s(t) = \int_{-\infty}^{\infty} A(z)e^{i[(\omega_1 - \omega_0)t + \varphi(z,t)]} dz,$$

and if we shift the carrier frequency to be on-resonance, ($\omega_1 = \omega_0$), the signal becomes

$$s(t) = \int_{-\infty}^{\infty} A(z)e^{i\varphi(z,t)} dz.$$

Now, we introduce a spatially constant, time-varying gradient in the magnetic field, $G(t)$, and the time-domain NMR signal acquires an additional, position-dependent phase shift from the action of the gradient up to time t as

$$\varphi(z,t) = -\int_0^t \omega_G(z,t')dt'$$

or inserting the gradient

$$\varphi(z,t) = -\gamma z \int_0^t G(t')dt'.$$

Since $\varphi_0(z,t) = \gamma \int_0^t B_1 d\tau$ is spatially invariant, we have ignored it and concentrated only on the position-dependent phase shift from the action of the gradient, which can be written as

$$\varphi(z,t) = k(t)z$$

where we have introduced the *spatial frequency* $k = k(t)$ which has an implicit time dependence given by the integral of the gradient pulse over time $k(t) = \gamma \int_0^t G(t')dt'$. The NMR signal in spatial frequency space is

$$s(k(t)) = \int A(z)e^{-ik(t)z} dz$$

which shows that the signal is the Fourier transform of the spin density $A(z)$. The spin density (image) of the sample is found by taking the inverse transform

$$A(z) = \int s(k(t))e^{ik(t)z}dk.$$

Note that we have solved this problem in only one spatial dimension z, but the generalization to higher dimensions is obtained by introducing the gradients and positions as vectors.

It is useful and interesting at this point to calculate examples of the k-space data derived from some simple, water-filled objects in a given magnetic field and with specified gradient strengths. These results can easily be scaled, in the main, up or down in either parameter. The first example is that of a 1-cm water-filled cube in a 3 Tesla magnet equipped with gradient coils that deliver a linear field variation of $G = 30$ mT/m. In this magnetic field, water protons resonate at $\omega_o = \gamma B_o = 3$ Tesla $\times$ 2.67664×10^8 rad/(Tesla-sec) $= 8.02991 \times 10^8$ rad/s. The linear frequency is $\nu = \omega_o/2\pi = 128.571$ MHz. For a 1-ms gradient pulse, the gradient in spatial frequencies is $k = \gamma\, G\, t = 8029$ radian/m, and the phase shift across the cube is 80.3 radians. The 2D NMR signal in spatial frequency space is

$$s(k) = A_o \int_{-x_o}^{x_o} \int_{-y_o}^{y_o} e^{-ik_x x} e^{-ik_y y}\, dy\, dx$$

where the base of the cube in the (x, y)-plane is the square $\{\{-x_o, x_o\},\{-y_o, y_o\}\}$. The integral gives

$$s(k) = 4A_o \frac{\sin(x_o k_x) \sin(y_o k_y)}{k_x k_y},$$

and A_o is the constant proton spin density in the cube, which for pure water is 111 moles/l (or 6.7×10^{22} spins/cm^3). The spin density in the sample and its k-space transform (Fig. 9.28) show the expected inverse relationship between conjugate Fourier pairs; the real-space object is mapped into an inverse space whose dimensions are its reciprocal.

Another sample geometry of interest is that of a cylinder. For this, we need to develop the Fourier transform in polar coordinates. Letting us define the transformation from Cartesian (x, y) to real-space polar coordinates (r, θ) by

$$x = r\, \cos\theta$$

$$y = r\, \sin\theta$$

and the similar coordinates in k-space (k_x, k_y) by

$$k_x = \rho\, \cos\varphi$$

$$k_y = \rho\, \sin\varphi,$$

the NMR signal in real space

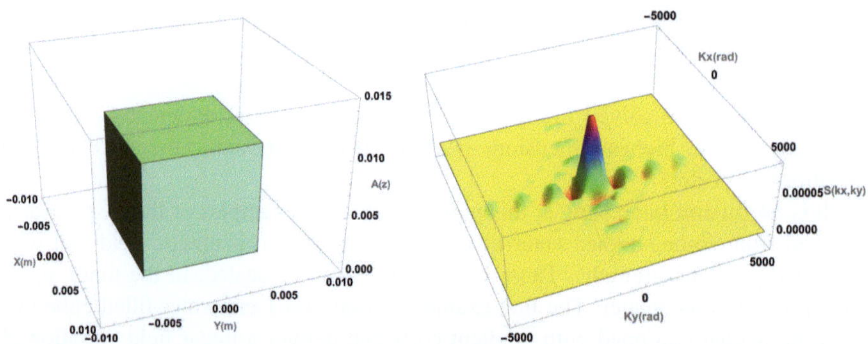

Fig. 9.28 (*Left*) the proton spin density $A(z)$ for a 1-cm cube of water and (*right*) its k-space NMR signal function $s(k_x, k_y)$ in a magnetic field gradient of 30 mT/m

$$s(k_x, k_y) = \int_{-x_0}^{x_0} \int_{-y_0}^{y_0} A(x, y)e^{-ik_x x}e^{-ik_y y}dydx$$

becomes, in polar coordinates but real space (the area element in polar coordinates is $r\, dr\, d\theta$),

$$s(k_x, k_y) = \int_0^{2\pi} \int_0^{\infty} A(r, \theta)e^{-i\rho r\,\cos\varphi\,\cos\theta}e^{-i\rho r\,\sin\varphi\,\sin\theta}r\,dr\,d\theta.$$

The Mathematica function `TrigReduce`$[\cos\varphi\,\cos\theta + \sin\varphi\,\sin\theta]$ comes in handy here, for we can combine the exponents to give

$$e^{-i\rho r\,\cos\varphi\,\cos\theta}e^{-i\rho r\,\sin\varphi\,\sin\theta} = e^{-i\rho r\,(\cos\varphi\,\cos\theta + \sin\varphi\,\sin\theta)} = e^{-i\rho r\cos(\varphi - \theta)},$$

and then the signal is

$$s(k_x, k_y) = s(\rho\,\cos\varphi, \rho\,\sin\varphi) = \int_0^{2\pi} \int_0^{\infty} A(r, \theta)e^{-i\rho r\cos(\varphi - \theta)}r\,dr\,d\theta.$$

For a circularly-symmetric object, $A(r, \theta) = A(r)$ so that the angular integration gives 2π, and we are left with an integral that only depends on the radius

$$s(k_x, k_y) = s(\rho\,\cos\varphi, \rho\,\sin\varphi) = 2\pi\int_0^{\infty} A(r)e^{-i\rho r\cos(\varphi - \theta)}r\,dr.$$

For our pillbox full of water, the spin density is constant, call it A_o, which leaves the integral as

$$s(k_x, k_y) = s(\rho \, \cos\varphi, \rho \, \sin\varphi) = 2\pi A_\mathrm{o} \int_0^\infty e^{-i\rho r \cos(\varphi - \theta)} r \, dr,$$

and this integral is a Bessel function of the first kind of order one

$$\int_0^\infty e^{-i\rho r \cos(\varphi - \theta)} r \, dr = \frac{J_1(2\pi\rho)}{\rho}$$

$$s(\rho \, \cos\varphi, \rho \, \sin\varphi) = s(k_x, k_y) = 2\pi A_\mathrm{o} \frac{J_1(2\pi\rho)}{\rho}.$$

The integral over θ for a function that is not radially symmetric is the Bessel function of order zero

$$J_0(\rho r) = \frac{1}{2\pi} \int_0^{2\pi} e^{-i\rho r \cos(\varphi - \theta)} \, d\theta$$

from which we can construct the Fourier transform of a radially symmetric function in polar coordinates as

$$s(\rho, \varphi) = 2\pi \int_0^\infty A(r) r J_0(\rho r) \, dr.$$

In our second example, the k-space function for the NMR image in the x–y plane in 2D of the cylinder of water is

$$s(k_x, k_y) = 2\pi A_\mathrm{o} \int_0^\infty r J_0(\rho r) \, dr = 2\pi A_\mathrm{o} \frac{J_1(2\pi\rho)}{\rho}.$$

The pillbox and its k-space NMR function are plotted in Fig. 9.29 for a 1-cm-diameter cylinder and using the above gradient strength.

9.24   MR Imaging of Boundary Movement

In the performance of its physiological role as the oxygen carrier in erythrocytes in the blood, hemoglobin cycles from a diamagnetic, oxygenated state to a paramagnetic, deoxygenated state as it releases, in the capillaries, the oxygen gained in passing through the lungs. The altered magnetic susceptibility near deoxygenated erythrocytes in capillaries is sufficient to significantly shorten the transverse

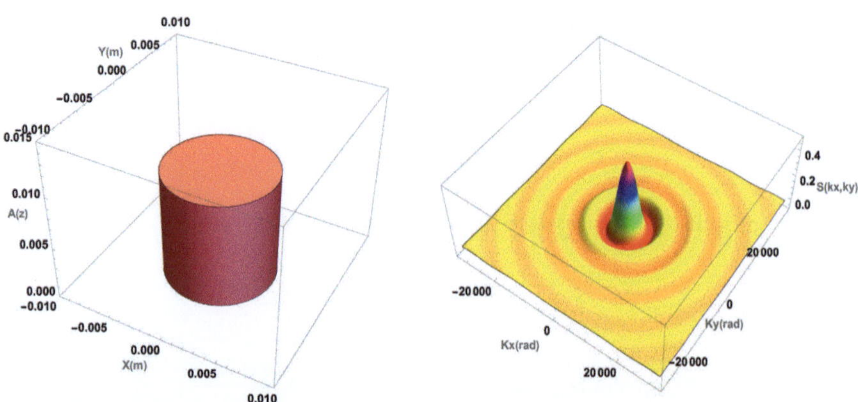

Fig. 9.29 (*Left*) a water-filled cylinder 0.01 m in diameter and (*right*) its *k*-space NMR signal in a 30-mT/m gradient

relaxation time for water protons in the blood (Thulborn et al. 1982), an effect which is widely used as the basis for functional NMR imaging, particularly in the central nervous system (Ogawa et al. 1990; Shulman et al. 1993; Yoshizawa et al. 1996; Kim and Ogawa 2012). The large magnetic susceptibility difference between paramagnetic deoxyhemoglobin and diamagnetic oxyhemoglobin within the erythrocytes in blood directly enhances the transverse NMR relaxation rates of adjacent water protons. This fact can be used to noninvasively image the diffusion, position, and consumption of oxygen in gels composed of *Bos taurus* blood and agarose (1.8% in Krebs-Henseleit buffer).

A cylindrical plastic chamber (whose interior was like Fig. 9.29) was filled with a gel of blood/agarose and sealed with hollow caps that allowed humidified gases to pass over the top and bottom of the gel during the imaging study. When the gas passing over the top of the gel was switched from nitrogen to oxygen, the images showed that oxygen entered the gel and switched the hemoglobin from its paramagnetic deoxygenated to its diamagnetic oxygenated state. This lengthened the transverse relaxation time of the adjacent water protons and produced a bright band (Fig. 9.30a) that was observed to move down the chamber with time (Fig. 9.30b–d). A characteristic of one-dimensional diffusion, seen in this experiment, is the fact that the distance, x, moved by the boundary in a given time interval is proportional to the square-root of the time , $x = \sqrt{2Dt}$, so that a plot of distance vs. $t^{1/2}$ will be a straight line with a slope of $(2D)^{1/2}$. From the measurement of this boundary movement, it was determined that the oxygen diffusion coefficient decreased linearly from its value in water (2.2×10^{-9} m^2/s) to $1.37\ (\pm 0.08) \times 10^{-10}$ m^2/s at a hematocrit of 30% at 17 °C (Fig. 9.31).

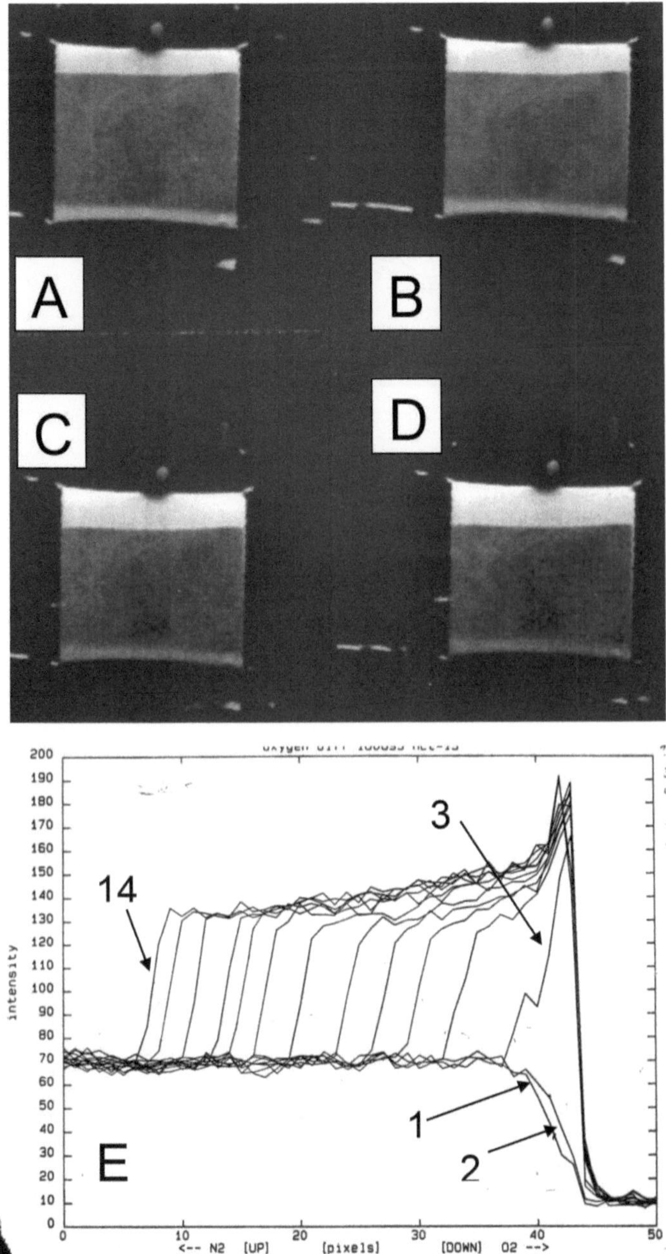

Fig. 9.30 Boundary diffusion of oxygen into a gel formed from blood and agarose. (**a–d**) A set of 4.7 T spin-echo images (TR/TE = 700/50 ms) showing the movement of a boundary of oxygen at a temperature of 17 °C and a hematocrit of 15%, at times of (**a**) 80, (**b**) 105, (**c**) 165, and (**d**) 205 min after switching from nitrogen to oxygen flowing over the top of the gel. (**e**) A stacked plot of the image intensities measured from 14 images like those shown in (**a–d**) with traces #1 and #2 taken prior to the introduction of oxygen into the top compartment; trace #3 was from an image taken just after the introduction of oxygen, while trace 14 was taken 220 min later at the end of the measurement period

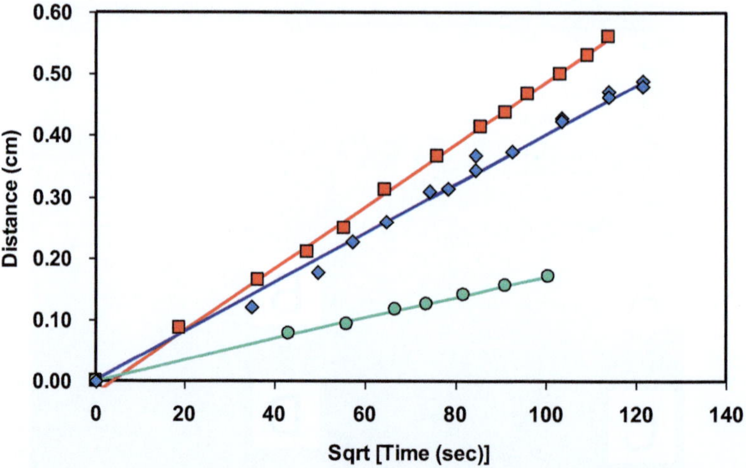

Fig. 9.31 Determination of the diffusion coefficient of oxygen by measurement of the movement of the boundary (as in Fig. 9.29a–d) as a function of the square root of the elapsed time. The least squares fits to the data gave diffusion coefficients of $1.28 \pm 0.03 \times 10^{-9}$ m^2/s (hematocrit 15%, red squares), $0.800 \pm 0.052 \times 10^{-9}$ m^2/s (hematocrit 20%, blue diamonds), and $0.137 \pm 0.0079 \times 10^{-9}$ m^2/s (hematocrit 30%, green circles) where the quoted errors are the standard errors from the fits

9.25   Diffusion Measurements by NMR Spectroscopy

After the postwar demonstration of nuclear magnetic resonance by Bloch (1946), Purcell et al. (1946) and Bloembergen et al. (1947), Carr and Purcell (1954) examined the effects of diffusion on the NMR signal, and Torrey solved the Bloch equations with an added diffusion term (Torrey 1956). It was then shown by Stejskal and Tanner (Stejskal 1965; Stejskal and Tanner 1965) that time-dependent gradient pulses could improve the sensitivity of the NMR experiment to the effects of diffusion and could be used to measure the diffusion coefficients of small molecules in solution. As we have seen in our examination of the physics of MRI, the presence of magnetic field gradients introduces phase and frequency shifts in the NMR signal from molecules in a position-dependent manner. If a molecule moves significantly after a gradient pulse, this is reflected in alterations in its NMR signal when later interrogated using a second gradient pulse and spin echo. The Stejskal and Tanner approach uses a spin echo and a pair of symmetric gradient pulses of strength G and duration δ separated by a time delay τ. The NMR signal decays according to

$$S(\tau) = S_o e^{-bD}$$

where the b-factor

$$b = \gamma^2 G^2 \delta^2 \left(\tau - \frac{\delta}{3} \right)$$

determines the rate of attenuation of the signal as a function of the time delay τ. As long as the first gradient pulse occurs after the $\pi/2$ radiofrequency pulse, but before the π refocusing pulse, and the second gradient pulse occurs after the π refocusing pulse, the exact timing of the gradient pulse pair is not critical. This sequence functions as follows: After the $\pi/2$ *rf* pulse, there is little signal dephasing until the first gradient pulse is applied. This gradient pulse tags the sample nuclei with their position within the field gradient for the duration of the gradient. When the gradient is turned off, again little dephasing occurs. In the absence of diffusion, the application of the π refocusing pulse and the second gradient reverses the direction of the dephasing, while the second gradient pulse produces an opposite but equal rephasing with the net result that there would be no change in the echo amplitude. In the presence of molecular displacement, however, the refocusing will be incomplete, and signal attenuation will occur. If one varies the time delay τ, this technique is particularly well-suited to the study of restricted diffusion or to the study of diffusion in optically opaque samples. For example, early studies of the diffusion of water inside yeast cells showed that the motion was restricted and gave a value for the cell's interior diameter of 4.1 microns, while optical microscopy showed that the cell's outer diameters were 6 microns (Tanner and Stejskal 1968).

9.26 Diffusion Tensor MRI

The ability of radiofrequency waves to penetrate living tissue has given NMR spectroscopy and imaging the very useful feature of being able to noninvasively interrogate the state of water in cells, tissues, organs, and whole organisms (Fig. 8.6). At the micro scale, biological samples are not isotropic but are rather divided into many cellular compartments by membranes; the sizes of these structures are on the order of microns. In a typical MRI acquisition, the duration of the gradient pulses is on the order of milliseconds, and the diffusion times are up to ~1 s. Water molecules will freely diffuse only a short distance during this time and then encounter a relatively impermeable barrier, such as an internal membrane, organelle, the nuclear envelope, the endoplasmic reticulum, or the Golgi apparatus. A mammalian cell is on the order of 10 microns in diameter and is packed with subcellular structures which limit the free diffusion of water in the cytoplasm. The decay of the diffusion-weighted NMR signal is then unlikely to be described by a single diffusion coefficient reflecting isotropic water motion. Diffusion spectroscopy is therefore needed to accurately map the mobility of water in the cell.

One extreme case of restricted motion is found in the movement of water in an axon of a nerve cell (Fig. 9.32). These fibers, which can stretch from the lumbar spinal cord to the foot, have diameters around 10–20 microns and lengths of up to 1 m. A water molecule would require on the order of 25 ms to diffuse across the diameter. For longer diffusion times, it would encounter the plasma membrane and its motion would be restricted. On the other hand, diffusion parallel to the long axis would be significantly less likely to be hindered by collisions with membranes. The

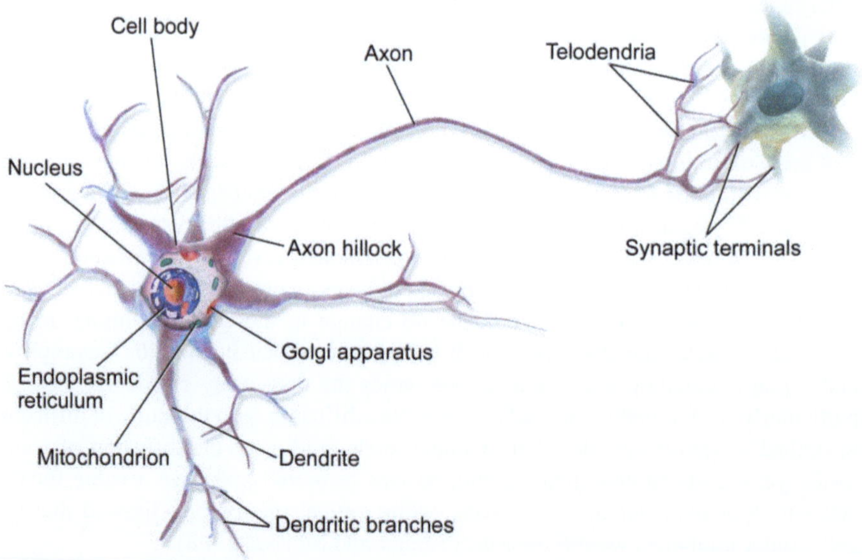

Fig. 9.32 A multipolar neuron showing the long narrow axon that connects one nerve cell to its neighbors. (Wikipedia: Author Thomas Schultz, CC BY-SA 3.0 File:DTI-sagittal-fibers.jpg)

highest density of axonal fibers is found in the white matter of the human brain. The attenuation of the NMR signal in three dimensions will therefore depend strongly on the direction of the gradient field vector $\vec{G} = G_x\hat{x} + G_y\hat{y} + G_z\hat{z}$ with respect to the principal axis of the diffusion tensor. If the environment of a spin is isotropic, gradient pulses along any direction will generate the same attenuation as any other direction, but in an anisotropic system, such as a bundle of white matter tracts in the brain, at least six gradient directions must be used in order to measure the six independent components of the diffusion tensor (Hrabe et al. 2007).

The attenuation will obey the Stejskal-Tanner behavior in each direction with the diffusion coefficient D replaced by the scalar quantity $D = \vec{g} \cdot \boldsymbol{D} \cdot \vec{g}$ where $\boldsymbol{D}$ is the 3×3, symmetric

$$D_{ij} = D_{ji}; i,j \in \{x,y,z\}$$

diffusion tensor,

$$\boldsymbol{D} = \begin{bmatrix} D_{xx} & D_{xy} & D_{xz} \\ D_{yx} & D_{yy} & D_{yz} \\ D_{zx} & D_{zy} & D_{zz} \end{bmatrix}$$

and $\vec{g}$ is a unit vector along the gradient direction

$$\vec{g} = \frac{\vec{G}}{\sqrt{G_x^2 + G_y^2 + G_z^2}} = \frac{\vec{G}}{\left|\vec{G}\right|}$$

and the dot indicates a scalar product. Diagonalization of the diffusion tensor gives its three eigenvalues (λ_1, λ_2, λ_3) (Fig. 9.9) and its eigenvectors $\left(\vec{e_1}, \vec{e_2}, \vec{e_3}\right)$ the orientations and the principal components of D represented by the diffusion ellipsoid (Fig. 9.9). The ellipsoid's major axes are the mean diffusion coefficients along each of the eigenvector directions. Isotropic diffusion is reflected in approximately equal eigenvalues ($\lambda_1 \sim \lambda_2 \sim \lambda_3$), but radial anisotropy is observationally determined by unequal eigenvalues: ($\lambda_1 > \lambda_2 \sim \lambda_3$). It is easy to imagine that pathological processes such as dementia, stroke, ischemia, impact injury, etc. would lead to characteristic changes in the diffusion tensor. For this reason, measurement of the local diffusion tensor in the human brain has assumed clinical significance. If one assumes that the largest eigenvector of the diffusion tensor is parallel to the local direction of the axonal fiber, then one can use computer algorithms that convert DTI MRI data into maps of the white matter pathways.

The apparent (scalar) mean diffusion coefficient is $\mu = $ trace $[D]/3 = (\lambda_1 + \lambda_2 + \lambda_3)/3$, and the *fractional anisotropy* φ is given by

$$\varphi = \frac{\sqrt{3\left[(\lambda_1 - \mu)^2 + (\lambda_2 - \mu)^2 + (\lambda_3 - \mu)^2\right]}}{\sqrt{2(\lambda_1^2 + \lambda_2^2 + \lambda_3^2)}}$$

where $\mu = $ trace $[D]/3$ is the the apparent diffusion coefficient or mean diffusivity. Also of interest is the largest eigenvalue, which is the *axial diffusivity*, $\alpha = \lambda_1$, and the average of the two remaining eigenvalues as the *radial diffusivity*, $\rho = 0.5$ ($\lambda_2 + \lambda_3$). The fractional anisotropy varies from zero for isotropic diffusion to one in the case where the diffusion is purely uniaxial. In the human brain, for example, the measured value of the fractional anisotropy is $\varphi \sim 0.3$ for white matter tracts and only $\varphi \sim 0.1$ in the gray matter; the value is even smaller in the cerebrospinal fluid at $\varphi \sim 0.08$ (Table 9.5). For comparison, free water's fractional anisotropy is zero. The mean fractional anisotropy of the white matter tracts, consisting of myriad axonal

Table 9.5 Diffusion parameters for the human brain

Brain structure (the colors refer to Fig. 9.33)	Fractional anisotropy φ	Mean diffusivity μ (μm^2/ms)	Axial diffusivity α (μm^2/ms)	Radial diffusivity ρ (μm^2/ms)
Cerebrospinal fluid	0.081	1.44	1.60	1.33
Gray matter	0.119	0.84	0.98	0.79
White matter	0.333	0.77	1.10	0.61

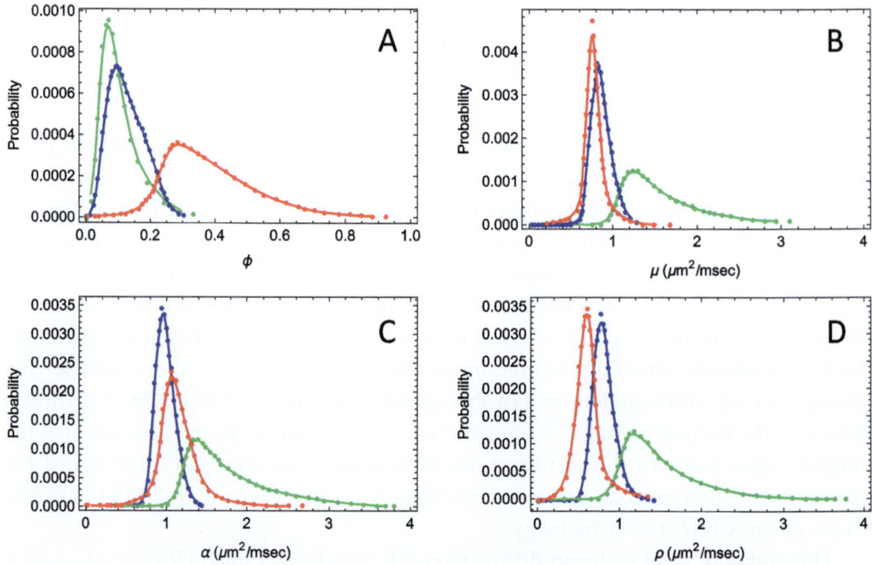

Fig. 9.33 Relative probability histograms of (**a**) the fractional anisotropy φ, (**b**) the mean diffusitivity, μ, (**c**) the axial α, and (**d**) the radial ρ diffusivities for whole human brain gray matter (blue), white matter (red), and cerebrospinal fluid (green). (Based on data from Alexander et al. (2007))

fibers, indicates that water diffusion occurs preferentially along the axonal fiber (Fig. 9.32). These conclusions are further supported by the values of the mean diffusivity, where cerebrospinal fluid has a mean diffusion coefficient of $1.44\,\mu m^2$/ms, a value that is about half of that found for free water of $3.0377\,\mu m^2$/ms at body temperature of 310 K. The water in the gray and white matter in the human brain shows diffusion that is restricted, white more than gray. As a side note, the data from Alexander et al. (2007) used to construct Fig. 9.33 contain relative probability distributions, which are not normalized so that the sum of their "probabilities" is not one. Also, their units are quoted as "mm^2/s," but these are too large by a factor of 1000. The correct units are given in Fig. 9.33 and in Table 9.5.

The vector form of the diffusion propagator (Basser et al. 1994) is given by

$$P\left(\vec{r},t\right) = \frac{1}{\sqrt{(4\pi t)^3 |\boldsymbol{D}|}} e^{-\left(\frac{\vec{r}^{\mathrm{T}} \cdot \boldsymbol{D}^{-1} \cdot \vec{r}}{4t}\right)}.$$

Once the diffusion tensor has been measured, one can diagonalize it (i.e., find its eigenvalues) and use it to follow the axons through the brain, generating images of the white matter fiber tracts. The path of the tract through (x, y, z) space is a

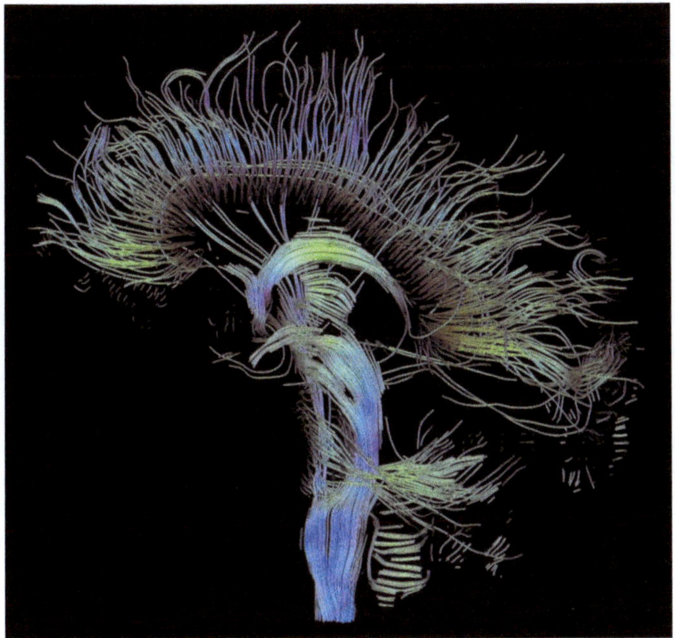

Fig. 9.34 White matter axonal fiber tracts in the human brain constructed using diffusion tensor imaging. The fiber tracts that enter the spinal cord are shown as the downward-trending bundle in the middle, while the fibers at the top connect the left and right cerebral hemispheres by means of the corpus callosum. (Wikipedia: Author Thomas Schultz, CC BY-SA 3.0 File:DTI-sagittal-fibers.jpg)

parametric curve $\vec{r}\,(s)$ where s is the arc length along the curve. The tangent vector $\vec{T}\,(s)$ at point s on this curve obeys the Frenet-Serret equation

$$\vec{T}\,(s) = \frac{d\vec{r}\,(s)}{ds}.$$

And we can use the eigenvector associated with the largest eigenvalue as the tangent vector

$$\frac{d\vec{r}\,(s)}{ds} = \vec{e_1}\left(\vec{r}\,(s)\right);$$

this can be solved numerically to produce space curves that reproduce the paths of axon bundles in the brain (Fig. 9.34), a process that is now referred to as tractography.

9.27 Biological Systems Adaptations to Diffusion Limitations

Because diffusion is a random walk process, it is an inefficient method for moving molecules to substantial linear distances. The distance traveled in a given time is not linear but rather increases only with the square root of the time. The rate of diffusion of molecules in cells sets an upper limit on the rates of chemical reactions involving enzymes and their substrates. The rates of conversion of substrate into product cannot exceed the rate at which these participant molecules can diffuse within the intracellular environment. However, by coupling several biochemical reactions on the surface of an enzyme, like the case discussed above for creatine kinase, living systems can easily beat this diffusion limit on reaction kinetics. It is interesting to speculate that the first reaction networks to evolve during prebiotic evolution resembled glycolysis and that these reactions were catalyzed on the surfaces of rocks, crystals, or inorganic sediments. In this manner, prebiotic reaction networks would have accomplished localized chemistry which would have been entropically impossible if the reactants and products freely diffused away before subsequent reactions in the network could have occurred.

How else is diffusion important in living systems? One of the most important transport processes in nature is the diffusion of oxygen. In mammals, this process is facilitated by the binding of oxygen to hemoglobin in the blood which is contained within specialized cells: erythrocytes. These red blood cells are not rigidly cuboidal or spherical but are flexible and have a disc shape (Fig. 9.35) similar in cross section to the fourth oval of Cassini (Hellmers et al. 2006). How does this shape matter with respect to the diffusion of oxygen? The diameter of an erythrocyte is about 7.8μm, but the thickness is only 2.2μm (Fig. 9.35). An oxygen molecule in the center of the cell needs only to diffuse 1μm, rather than 3.5μm, in order to move from being bound to the intracellular hemoglobin to the surrounding fluid. This difference in

Fig. 9.35 Cross-sectional shape of a human erythrocyte modeled as a fourth oval of Cassini

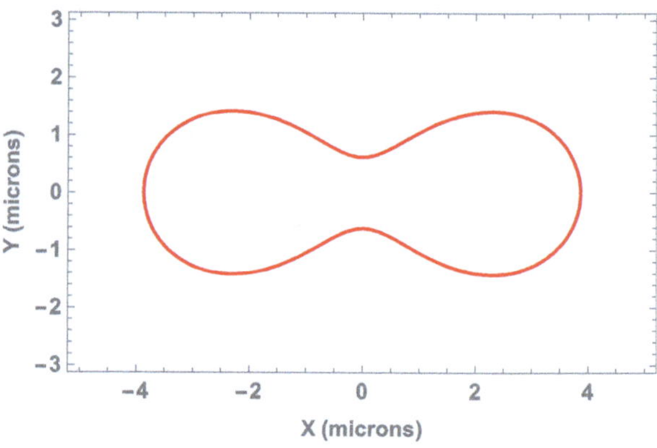

distance translates into a 12-fold rate increase for oxygen unloading and loading into such a disk-shaped cell because oxygen has to diffuse a much smaller distance. Furthermore, since erythrocytes can fold in the ~7-micron-diameter capillaries of the lung and the peripheral circulation, their plasma membranes can touch the endothelium, again shortening the diffusion distance and the loading and unloading times for oxygen.

Problems

1. *Oxygen diffusion into the erythrocyte.* The shape of the erythrocyte as a biconcave disk is puzzling at first sight. One biophysical reason for this interesting shape is that oxygen can reach the interior hemoglobin molecules faster for a disk shape than it could if the erythrocyte were a sphere of equivalent volume. Assume that a three-dimensional model for the erythrocyte is an oblate ellipsoid of revolution with semiaxes:

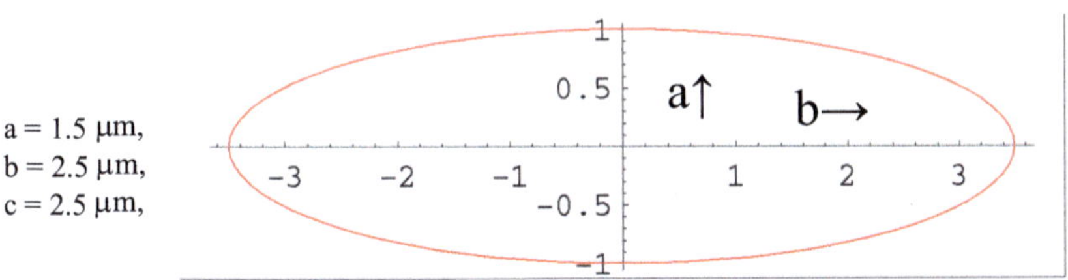

$a = 1.5\ \mu m,$
$b = 2.5\ \mu m,$
$c = 2.5\ \mu m,$

Note that the view from above is that of a circle of radius, $b = c = 2.5\mu m$. The diffusion coefficient for oxygen in concentrated hemoglobin solutions at $T = 310$ K is about 1×10^{-12} m^2/s:

(a) How long does it take oxygen to diffuse via the shortest path into the center of this model erythrocyte?

(b) How long would it take oxygen to diffuse into the center of a spherical erythrocyte of equivalent volume?

(c) How much faster can oxygen reach the center of the disk-shaped erythrocyte compared to a spherical cell of the same volume?

(d) Compare the time found in (a) with the time required for oxygen to diffuse from an erythrocyte in the center of a 10μm capillary to the edge if the diffusion coefficient for oxygen in serum is 3×10^{-5} cm^2/s at $T = 310$ K.

2. *Diffusion coefficients.* Compare the diffusion coefficients of spheres and ellipsoids. The frictional coefficient for a sphere is $1.75 \times 10^{-7}\ \frac{Ns}{m}$, while the frictional coefficient for an ellipsoid is $2.87 \times 10^{-7}\ \frac{Ns}{m}$.

3. *Oxygen fluxes.* If a human consumes 550 L of O_2 per day and the lung has a surface area of 70 m^2,

   (a) What is the flux of oxygen across the alveolus?
   (b) What is the force on an oxygen molecule that is causing this flux if the alveolar wall is 10μm thick and the oxygen tension in venous blood is 5% of the inhaled value?
   (c) What is the conductance of the alveolus?

4. *Ballistic molecular speeds.* What is the ballistic velocity of H_2O at 310 K in cells?
5. *Ballistic molecular times.* Using your answer from 9.5, how much time does it take an H_2O to cross a liver cell?
6. *Diffusion molecular speeds.* How much time does it take an H_2O molecule to diffuse across a liver cell?
7. *Diffusion calculations.* Use the diffusion notebook code in the text to simulate a collection of particles diffusing away from the center of a box as a function of time.

   (a) What happens as you increase the number of particles? Or the amount of noise? Does the apparent rate that the particles diffuse away differ while changing either option?
   (b) Diffusive processes, like diffusion of a protein across a bacterial cell, are characterized as a random walk. Calculate the average displacement, $<x>$, for a molecule moving in water given the diffusion coefficient 2.0×10^{-5} cm^2/s over 10 min.
   (c) Calculate the movement of the molecule in (b). over 10 min but in a sucrose solution instead with diffusion coefficient 4.59×10^{-6} cm^2/s. What pattern do you see as the diffusion constant decreases?
   (d) Because diffusion resembles a random walk, why is two-dimensional diffusion more inefficient than one-dimensional diffusion? Is diffusion an efficient process for the chemical needs of the cell? (Recall how primordial reactions took place in prebiotic evolution).

8. *Diffusion Green's function.* Given Green's function for the diffusion equation

$$G(x' - x, t' - t) = \frac{1}{\sqrt{4\pi D(t' - t)}} e^{-\frac{(x' - x)^2}{4D(t' - t)}},$$

   show that $\mathcal{L}G(x' - x) = \delta(x' - x)$.
9. *Diffusion calculations.* Given a unit step concentration gradient con[x, t] as

$$\text{con}[x\_, t\_] := \frac{10}{2}\left(1 - \frac{2}{\sqrt{\pi}}\int_0^{\frac{x}{\sqrt{2t}}} e^{-y^2}dy\right),$$

(a) Compute and plot the 3D derivative in the (con, x, t) coordinate system for $\{\{x, -1, 1\}, \{t, 0.01, 1\}\}$.

(b) Find the full width at half maximum of the peak in (a). and plot this as a function of time. Fit this to a quadratic.

(c) From the results in (b), what is the diffusion coefficient?

References

A.L. Alexander, J.E. Lee, M. Lazar, A.S. Field, Diffusion tensor imaging of the brain. Neurotherapeutics **4**, 316 (2007)

D. Axelrod, D.E. Koppel, J. Schlessinger, E. Elson, W.W. Webb, Mobility measurement by analysis of fluorescence photobleaching recovery kinetics. Biophys. J. **16**, 1055–1069 (1976)

P.J. Basser, J. Mattiello, D. LeBihan, MR diffusion tensor spectroscopy and imaging. Biophys. J. **66**, 259–267 (1994)

M. Benda et al., How to determine diffusion coefficients in planar phospholipid systems by confocal fluorescence correlation spectroscopy. Langmuir **19**, 4120 (2003)

F. Bloch, Nuclear induction. Phys. Rev. **70**, 460–474 (1946)

N. Bloembergen, E.M. Purcell, R.V. Pound, Nuclear magnetic relaxation. Nature **160**, 475 (1947)

H.Y. Carr, E.M. Purcell, Effects of diffusion on free precession in nuclear magnetic resonance experiments. Phys. Rev. **94**, 630–638 (1954)

E.L. Elson, Fluorescence correlation spectroscopy: Past, present, future. Biophys. J. **101**, 2855 (2011)

A.N. Garroway, P.K. Grannell, P. Mansfield, Image formation in NMR by a selective irradiative process. J. Phys. C Solid State Phys. **7**, L457 (1974)

R.J. Gaylord, P.R. Wellin, *Computer Simulations with Mathematica* (Springer TELOS, Santa Clara, 1995)

Y. Harpaz, M. Gerstein, C. Chothia, Volume changes on protein folding. Structure **2**, 641 (1994)

J. Hellmers, E. Eremina, T. Wriedt, Simulation of light scattering by biconcave Cassini ovals using the nullfield method with discrete sources. J. Opt. A Pure Appl. Opt. **8**, 1 (2006)

J. Hrabe, G. Kaur, D.N. Guilfoyle, Principles and limitations of NMR diffusion measurements. J. Med. Phys. / Assoc. Med. Physicists India **32**, 34–42 (2007)

S.G. Kim, S. Ogawa, Biophysical and physiological origins of blood oxygenation level-dependent fMRI signals. J. Cereb. Blood Flow Metab. **32**, 1188–1206 (2012)

T. Köchy, T.M. Bayerl, Lateral diffusion coefficients of phospholipids in spherical bilayers on a solid support measured by ^{2}H resonance relaxation. Phys. Rev. E **47**, 2109 (1993)

G. Koulouras et al., EasyFRAP-web: A web-based tool for the analysis of fluorescence recovery after photobleaching data. Nucleic Acids Res. **46**, W467 (2018)

P.C. Lauterbur, Image formation by induced local interaction; examples employing nuclear magnetic resonance. Nature **242**, 190–191 (1973)

P.M. Macdonald, Q. Saleem, A. Lai, H.H. Morales, NMR methods for measuring lateral diffusion in membranes. Chem. Phys. Lipids **166**, 31–44 (2013)

G.L. Nicolson, The Fluid-Mosaic Model of Membrane Structure: Still relevant to understanding the structure, function and dynamics of biological membranes after more than 40 years. Biochim. Biophys. Acta **1838**, 1451 (2014)

S. Ogawa, T.M. Lee, A.R. Kay, D.W. Tank, Brain magnetic resonance imaging with contrast dependent on blood oxygenation. Proc. Natl. Acad. Sci. USA **87**, 9868–9872 (1990)

L. Onsager, Reciprocal relations in irreversible processes. Part I. Phys. Rev. **37**, 405 (1931a)

L. Onsager, Reciprocal relations in irreversible processes. Part I. Phys. Rev. **38**, 2265 (1931b)

E.M. Purcell, H.C. Torrey, R.V. Pound, Resonance absorption by nuclear magnetic moments in a solid. Phys. Rev. **69**, 37–38 (1946)

D.S. Rowe, Radioactive single radial diffusion: A method for increasing the sensitivity of immunochemical quantification of proteins in agar gel. Bull. World Health Organ. **40**, 613 (1969)

R.G. Shulman, A.M. Blamire, D.L. Rothman, G. McCarthy, Nuclear magnetic resonance imaging and spectroscopy of human brain function. Proc. Natl. Acad. Sci. USA **90**, 3127–3133 (1993)

A.J.F. Siegert, *On the Fluctuations in Signals Returned By Many Independently Moving Scatterers.* Massachusetts Institute of Technology, Radiation Laboratory Report, 465. (MIT, Cambridge, 1949)

D.M. Soumpasis, Theoretical analysis of fluorescence photobleaching recovery experiments. Biophys. J. **41**, 95 (1983)

E.O. Stejskal, Use of spin echoes in a pulsed magnetic field gradient to study anisotropic, restricted diffusion and flow. J. Chem. Phys. **43**, 597–3603 (1965)

E.O. Stejskal, J.E. Tanner, Spin diffusion measurements: Spin echoes in the presence of a time dependent field gradient. J. Chem. Phys. **42**, 288 (1965)

S.M. Sterling et al., Phospholipid diffusion coefficients of cushioned model membranes determined via Z-scan fluorescence correlation spectroscopy. Langmuir **29**, 7966 (2013)

J. Stetefeld, S.A. McKenna, T.R. Patel, Dynamic light scattering: A practical guide and applications in biomedical sciences. Biophys. Rev. **8**, 409–427 (2016)

J.E. Tanner, E.O. Stejskal, Restricted self-diffusion of protons in colloidal systems by the pulsed-gradient, spin-echo method. J. Chem. Phys. **49**, 1768–1777 (1968)

K.R. Thulborn, J.C. Waterton, P.M. Matthews, G.K. Radda, Oxygenation dependence of the transverse relaxation time of water protons in whole blood at high field. Biochim. Biophys. Acta **714**, 265–270 (1982)

H.C. Torrey, Bloch equations with diffusion terms. Phys. Rev. **104**, 63–565 (1956)

R.Y. Tsien, The green fluorescent protein. Annu. Rev. Biochem. **67**, 509–544 (1998)

C.L. Wey et al., Lateral diffusion of rhodopsin in photoreceptor cells measured by fluorescence photobleaching and recovery. Biophys. J. **33**, 225 (1981)

T. Yoshizawa, T. Nose, G.J. Moore, L.O. Sillerud, Functional magnetic resonance imaging of motor activation in the human cervical spinal cord. NeuroImage **4**, 174–182 (1996)

M.E. Young et al., Estimation of diffusion coefficients of proteins. Biotechnol. Bioeng. **22**, 947 (1980)

Chapter 10
Sunlight as a Driver of Abiogenesis:
The Quantum Mechanics of the Absorption
of Light by Biomolecules

"...that orbed continent, the fire that severs day from night."
Shakespeare, Twelfth Night

The origin of life on Earth took place several billion years ago on a planet bathed in light on a daily basis from our nascent sun. Just a billion years after its formation from the solar nebula, the Earth had already experienced more than 300 billion light/dark cycles. This rich, periodic source of energy consisted of photons with energies of a few eV, just the right amount of energy to drive the self-assembly and chemical transformation of primordial biomolecules. A maxim from ecology is that an energy source is a *niche* into which organisms will evolve. On the early Earth, sunlight served as that energy source. It is also true that the flow of energy through a system organizes it. The oldest organisms found on the Earth were phototrophs, *light harvesters*; archaea such as *Halobacterium salinarum*, the microorganism that lends the purple color of the salt ponds found today adjoining San Francisco bay; and the cyanobacteria, the blue-green algae. These primitive organisms used photosynthetic pigments, including carotenoids, several forms of chlorophyll, and retinal. Bacteriorhodopsin, the light-gated channel protein of *H. salinarum*, relies on retinal to capture photons which are then used to pump hydrogen ions across its membranes, creating an electrochemical gradient that in turn is utilized for ATP synthesis. This important role of light as an energy source for early life places the quantum mechanics of the interaction light with biomolecules at the very center of our investigation of Abiogenesis. It is also quite interesting that the retinal molecule found in *H. salinarum* is the same visual retinal upon which human vision is based.

10.1   Optical Spectroscopy of Biomolecules

How do molecules absorb or emit light? Molecules are constructed from a heavy, positively charged nuclear framework bound into a stable equilibrium structure by the oppositely charged, and much, much lighter, electrons. While nuclear energy levels, as we have seen earlier with our simple particle in a box models, involve

© The Author(s), under exclusive license to Springer Nature Switzerland AG 2024
L. O. Sillerud, *Abiogenesis*, https://doi.org/10.1007/978-3-031-56687-5_10

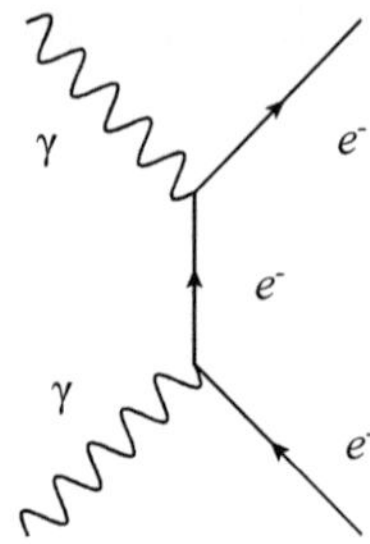

Fig. 10.1 A 2D Feynman diagram showing the interaction of an electron, e^-, with a photon, γ. Time flows from bottom to top, while space flows from right to left

Table 10.1 Molecular energy levels

Type	Frequency (Hz)	Energy (eV)	Regime
Electronic	10^{15}	4.14	Optical, UV
Vibrational	10^{14}	0.41	Infrared
Rotational	10^{10}	41.4×10^{-6}	Microwave
Nuclear	10^{8}	414×10^{-9}	NMR

energies far greater than that found for light visible to the human eye, the *electronic* energy levels we have found for simple molecules, such as H_2, involve transitions with wavelengths in the optical spectrum ($400 < \lambda < 700$ nm). The basic features of the quantum theory of the interaction of a photon with an electron were worked out first by P. A. M. Dirac (~1930) and then given their modern renormalizable form by Feynman, Schwinger, Dyson, and Tomonaga in the early 1950s, for which several shared the Nobel Prize. The complete understanding of this subject is couched in terms of quantum electrodynamics (QED), in which, in addition to the electrons in matter, the electromagnetic (photon) field is quantized through the mechanism of creation and annihilation operators (Chap. 4). QED is the most accurate physical theory ever developed; its perturbation theory predictions agree with the results of extremely high-resolution experiments to within one part in 10^{12}.

The basic interaction between a photon and an electron is embodied in a graph due to Feynman (1948) as shown in Fig. 10.1. This Feynman diagram depicts the interaction between an electron, $\mathbf{e}^-$, and a photon, $\boldsymbol{\gamma}$, by means of solid lines for the electron and wavy lines for the photon. This diagram can be used to generate the integral for the scattering amplitude (whose absolute square is the probability) by using the solid and wavy lines to stand for propagators in the scattering integral.

10.2   Molecular Energy Levels

Virtually all of the electrons we will be concerned with in Physical Biochemistry are involved in bound states in atoms and molecules. The energies of these bound states can be derived from the solution of Schrodinger's equation for the wave functions given the various operators involved (Chaps. 4 and 5). Molecular energy levels can be classified according to their energies (Table 10.1).

The optical transitions we will be concerned with involve mainly the electronic and vibrational energy levels. We will be dealing with the absorption and emission of infrared, visible, and ultraviolet photons. The wavelength region ranges from the ultraviolet at 190 nm to the infrared at 1500 nm. The phenomena of interest include absorption, emission, fluorescence, phosphorescence, and circular dichroism. The energies involved can be derived from Einstein's relationship between energy and frequency, $E = h\nu$, where h is Planck's constant and ν is the frequency in Hertz, or by using the relationship between frequency and wavelength, $\lambda\nu = c$, where c is the speed of light, we can write that $E = hc/\lambda$. The energy of a photon is inversely proportional to its wavelength, and this has given rise to the energy units popular in infrared spectroscopy of wave numbers (cm^{-1}).

What energies are encountered in optical spectroscopy? For a photon of $\lambda = 500$ nm, the frequency is $\nu = c/\lambda = 6 * 10^{14}$ Hz, and the energy is

$$E = h\nu = 6.626 \times 10^{-34}\,\text{Joule sec} \times 6 \times 10^{14}\,\text{Hz} = 4 \times 10^{-19}\,\text{Joule}.$$

If we compare this with the thermal background energy at 300 K, we find that $kT = 4 \times 10^{-21}$ Joule and the energy of a 500-nm photon is 100 times kT. By the Boltzmann distribution, we know that the population difference between the ground state and a state involving the absorption or emission of a 500-nm photon will be

$$\frac{N_i}{N_j} = e^{-\left(\varepsilon_i - \varepsilon_j\right)/kT} = e^{-100} = 2 \times 10^{-42}$$

so that all molecules in any macroscopic sample will be observed to be in the optical ground state at equilibrium. This is in contrast to the situation found for infrared spectroscopy where the vibrational energy levels are more closely spaced. For example, an energy of 200 cm^{-1}, which corresponds to 4×10^{-21} Joule (2.4×10^3 Joule/mole), will have 62% of the molecules in the ground state and 38% in the first excited state. However, a change in the energy splitting by a factor of five to 1000 cm^{-1} will result in 99% of the molecules in the ground state.

Our prior investigations of quantum mechanics (Chaps. 4 and 5) dealt with a single electron atom, the hydrogen atom, and its simple spectrum of energy levels or the hydrogen molecule. The biomolecules found in living systems have more complex electronic, vibrational, rotational, and nuclear energy levels (Table 10.1), and therefore, one can expect transitions among any and all of these states as long as the relevant quantum selection rules are obeyed. Optical transitions involve mainly electronic and vibrational states. An energy-level diagram for a typical molecule (Fig. 10.2) shows that the absorption of an optical photon can occur as long as its energy matches the energy difference between two electronic states with $\Delta m = 1$ in a dipole transition. The ground and excited states will be characterized by many intrinsic rotational and vibrational energy levels that differ in energy by relatively small amounts.

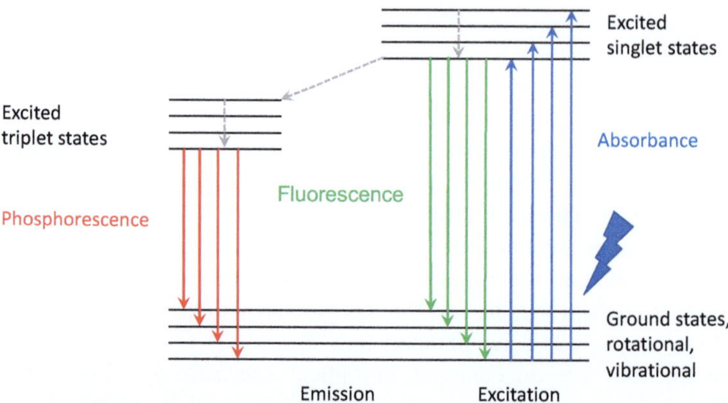

Fig. 10.2 Energy-level diagram for a molecule. Absorption of an optical photon will raise the energy of an electron from any of the manifold of vibrational and rotational ground state energy levels to one of the manifold of excited singlet states. Non-radiative transitions (gray dashed arrows) will drop the energy of the electron to that of the lowest angular momentum ($l = 0$) singlet state. Decay of this state will give rise to the emission of photons at a wavelength much longer than that of the initially absorbed photon in a phenomenon known as *fluorescence*. Conversion of the singlet state to a triplet state will trap the triplet state in a long-lived state whose decay by *phosphorescence* ($\Delta m = 2$) will proceed with times on the order of seconds to minutes

Optical photon absorption will send the electron from the manifold of vibrational and rotational ground states to any one of the manifold of excited singlet states. The lifetime of these states is on the order of 0.1 ns, while the time needed for the molecule to absorb a photon is constrained by the time it takes light to traverse the molecule. We can estimate this time by considering the size of a biomolecule. We know that $d = ct$, where d is the distance and c is the speed of light, while t is the time required. Therefore, for $d = 1$ nm, we find that the time required is on the order of 10^{-17} s, putting a lower limit on the time required for electronic transitions in atoms and molecules. The actual time is longer than this lower limit and is found experimentally by pulsed laser optical studies to have optical lifetimes on the order of $\sim 10^{-15}$ s. The lifetime of the excited state after the absorption of a photon is on the order of 10^{-10} s. The first thing an excited electron can do is simple emission where the wavelength of the emitted photon equals the wavelength of the absorbed photon. Another is to give up small amounts of energy in non-radiative transitions by dropping from the higher angular momentum excited singlet states to the singlet state with zero angular momentum; this too requires about 10^{-10} s. Note that here the energy difference between the ground state and this subsequent electronic state is smaller than that involved with the initial absorption of the photon. Thus, the wavelength of the emitted photon is now longer than that of the absorbed photon. This process is denoted *fluorescence* and is shown schematically in the figure where the color of the various transitions reflects their relative energy, with blue the highest denoting absorbance of the initial photon, while fluorescence is shown in green. The lifetime of the state for fluorescence emission is 10^{-7}–10^{-9} s.

We can use the uncertainty principle to estimate the width of the electronic transitions associated with the absorption and emission of light by a biomolecule. A lifetime of $\sim 10^{-15}$ s corresponds to a transition width of or $\Delta\lambda = 125$ nm, implying that electronic transitions in biomolecules are broad. This is in marked contrast to nuclear spin transitions where lifetimes of 1 s are common, leading to transition line widths of less than 1 Hz.

$$\Gamma \sim h/(2\pi t) = 10^{-19} \ \text{Joule}$$

The electrons in a molecular excited state can also de-excite by radiationless transitions to lower-energy, excited triplet states. These states trap the electron in states whose decay is forbidden by the dipole, $\Delta m = 1$, selection rules (Chap. 4). Their decay involves quadrupole transitions with $\Delta m = 2$, and their lifetime is much longer as a result, often with times in the range of seconds to hours, in a process known as *phosphorescence*.

10.3   Einstein's *A* and *B* Coefficients

The absorption of light by biomolecules can be modeled by means of a two-state transition system with energy levels as shown in Fig. 10.3.

Let N_i be the number of electrons per state i, with energy E_i. What is the probability that a photon will be absorbed or emitted by this molecular system? A is the probability of spontaneous emission, while B is the probability of induced absorption or induced (stimulated) emission. The energy of the absorbed or emitted photon is $E_2 - E_1 = h\nu$. The amount of energy at a given frequency can be approximated by resort to the blackbody spectrum of radiation emitted by a perfect emitter at a temperature T. Let $E = \rho(\nu)$ which is the energy density per unit frequency. The number of molecules absorbing light and making the transition from the ground state (E_1) to the excited state (E_2) per unit time is given by

$$\frac{dN_{12}}{dt} = B_{12}N_1\rho(\nu).$$

Similarly, for stimulated emission, we have that

Fig. 10.3 Electronic
energy-level scheme
defining the amplitudes, A_{21}
for spontaneous emission,
B_{12} for induced absorption,
or B_{21} for induced
(stimulated) emission

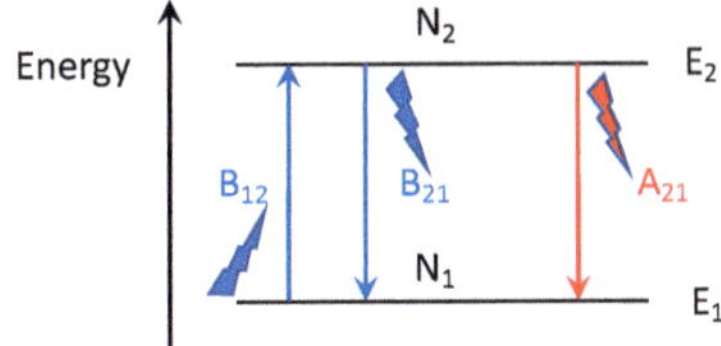

$$\frac{dN_{21}}{dt} = B_{21}N_2\rho(\nu)$$

and there also exists the possibility for spontaneous emission whose probability is given by

$$\frac{dN'_{21}}{dt} = A_{21}N_2,$$

but we note that this term is independent of the incident radiation, $\rho(\nu)$. If we assume that a steady state is present in which the rate of transitions from 1 to 2 is equal to the rate of transitions from 2 to 1 or, expressed mathematically,

$$\frac{dN_{12}}{dt} = \frac{dN_{21}}{dt} + \frac{dN'_{21}}{dt}.$$

Insertion of the definitions of these rates in terms of the A and B coefficients gives

$$N_2A + \rho B_{21}N_2 = N_1B_{12}\rho$$

or

$$N_2A = \rho(N_1B_{12} - N_2B_{21})$$

so that we can solve for ρ to give

$$\rho = \frac{N_2A}{N_1B_{12} - N_2B_{21}}.$$

The energy density can be thought of as arising from a blackbody at a temperature T. The probability of stimulated emission is proportional to the frequency component of the background at an energy of $\Delta E = h\nu$. The energy density from a blackbody at a temperature of T is given by Planck's formula:

$$\rho(v) = \frac{8\pi h\left(\frac{v^3}{c^3}\right)}{e^{\frac{hv}{kT}} - 1}$$

This famous equation is illustrated in Fig. 10.4 for a temperature of 2.7 K, that of the Cosmic Microwave Background radiation left over from the Big Bang (Chap. 6). Note that the x-axis is measured in unit of Hz, so that the radio receivers tuned to the Microwave Background radiation need to have a center frequency of ~170 GHz.

Let us equate this expression for the energy density with that found above from a consideration of probabilities. The result is that

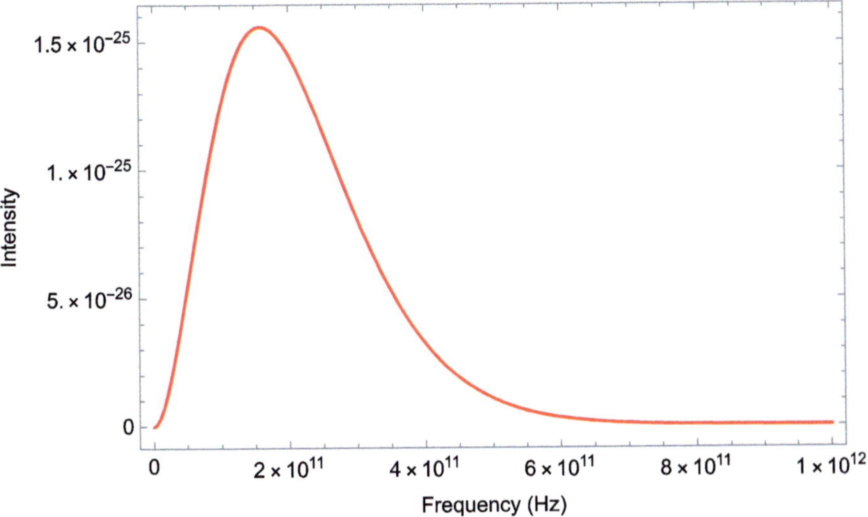

Fig. 10.4 Energy spectrum of radiation emitted by a black body at a temperature of 2.7 K as a function of frequency (Hz)

$$\frac{1}{\rho} = \frac{N_1 B_{12}}{N_2 A} - \frac{B_{21}}{A} = \frac{c^3}{8\pi h \nu^3}\left[e^{\frac{h\nu}{kT}} - 1\right].$$

But we can use the Boltzmann distribution to replace the exponential term

$$\frac{N_1}{N_2} = e^{\frac{h\nu}{kT}}$$

so that

$$\frac{N_1 B_{12}}{N_2 A} - \frac{B_{21}}{A} = \frac{B_{12}}{A} e^{\frac{h\nu}{kT}} - \frac{B_{21}}{A} = \frac{c^3}{8\pi h \nu^3}\left[e^{\frac{h\nu}{kT}} - 1\right],$$

which implies that

$$\frac{B_{12}}{A} = \frac{c^3}{8\pi h \nu^3} = \frac{B_{21}}{A}$$

or that $B_{12} = B_{21} = B(\nu)$. The probability for absorption is the same as the probability for stimulated emission. The probability for spontaneous emission, $A(\nu)$, is related to the probability for absorption by

$$A(\nu) = \frac{8\pi h \nu^3}{c^3} B(\nu).$$

We observe that all three spectral processes can be accounted for. The energies of optical transitions are large with respect to kT so that the probability of occupation of any state except the ground state is essentially zero. Visible and IR transitions are proportional to $B(\nu)$, while the intensity of fluorescence is proportional to $A(\nu)$ because for fluorescence there is only a small excited state population and therefore stimulated emission is much less likely than spontaneous emission.

10.4   Beer's Law

For practical applications of optical spectroscopy to biochemistry, it is necessary to know how the absorption of light by a molecule varies with wavelength and the amount of light absorbed by a given amount of a molecule in solution. Let's assume that we have a beam of light of wavelength, λ, with intensity, I_o, incident on a rectangular cuvette of path length, L. The light leaving the cuvette will have an intensity, $I < I_o$, due to the absorption in the solution. The differential change in intensity within a thin slice of solution of thickness, dL, is dI, and the fractional change in intensity in this slice is—dI/I. This must be proportional to the number of absorbers per unit volume in the light beam times the path length, $N\,dL$,

$$NdL = -\frac{1}{\sigma}\frac{dI}{I}$$

where the constant of proportionality, σ, is the *cross section*, which has units of length squared (area). We can integrate this differential equation

$$N\int_0^L dL = -\frac{1}{\sigma}\int_{I_o}^{I}\frac{dI}{I}$$

to find that

$$\sigma N\,L = \ln[I/I_o].$$

The change in intensity is proportional to the effective area, σ, of the molecule at the frequency used. We can convert this last equation into one more useful to biochemists by using molar concentrations units, c, and common logarithms. The absorbance, a, is defined through the equation (Fig. 10.5).

$$a = \log\left[\frac{I_o}{I}\right] = \sigma NL\log(e) = \varepsilon Lc.$$

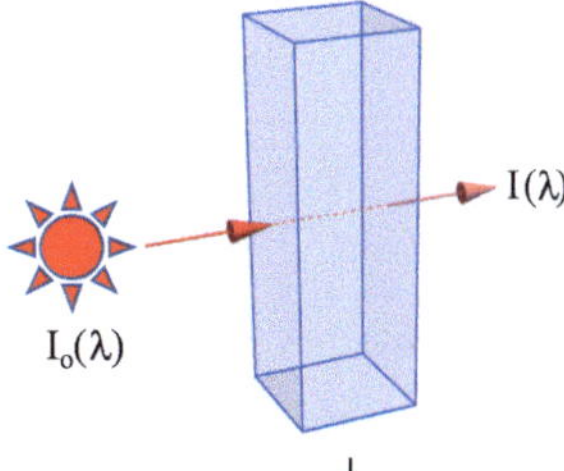

Fig. 10.5 The light path through a cuvette in a spectrophotometer. The incident flux of photons has intensity I_o at wavelength λ and exits with an intensity of I. The light path has a length of L

The absorption varies according to frequency, so we can write *Beer's law* as

$$a(\nu) = \varepsilon(\nu)\, L\, c$$

where $\varepsilon(\nu)$ is called the extinction coefficient, with units of liters/mole.cm:

$$\varepsilon = \sigma N_o \log(e)/1000$$

This states that the absorbance of a solution of biomolecules is proportional to the path length, L; the concentration, c; and a property intrinsic to the molecule (the extinction coefficient). By measuring the extinction coefficient at a fixed frequency, one can determine the optical cross section of a molecule at that frequency through

$$\sigma(\nu) = 1000\ \varepsilon(\nu)/[N_o\log(e)].$$

The cross section is proportional to the probability that a molecule will absorb a photon of frequency, ν, and this probability we just derived as Einstein's B coefficient. Let us find the relationship between $B(\nu)$ and the molecular cross section $\sigma(\nu)$. Consider our derivation of Beer's law again. In traversing the cuvette, the number of photons absorbed out of the incident beam is $-dI(\nu)$, and this is equal to $\sigma N\, dL\, I(\nu)$:

$$dN_{12}/dt = -dI(\nu) = \sigma N\, dL\, I(\nu) = \sigma N_1$$

In our examination of the two-level system, we wrote that the number of molecules in the ground state is $N_1 = N\, dL$ and the number of photons absorbed per unit time is dN_{12}/dt. We also know that $dN_{12}/dt = N_1\, B(\nu)\, \rho(\nu)$. The intensity of the photon beam is number of photons striking the cuvette per unit time given by the energy density, $\rho(\nu)$, divided by their energy, $h\nu$,

$$I(\nu) = \frac{c\rho(\nu)}{h\nu},$$

and from above, we have that

$$-dI(\nu) = \frac{1000\varepsilon(\nu)c\rho(\nu)}{N_o \log(e)h\nu} NdL,$$

which we can rearrange as

$$-dI(\nu) = \frac{1000\varepsilon(\nu)}{\log(e)} \frac{N_1 c}{N_o h} \frac{\rho(\nu)}{\nu} = BN_1\rho(\nu)$$

where the Einstein B coefficient is given by

$$B = \frac{1000\varepsilon(\nu)}{\log(e)} \frac{N_1}{N_o h} \frac{c}{\nu}.$$

Now, note once again that the cross section is $\sigma(\nu) = 1000\,\varepsilon(\nu)/[N_o \log(e)]$. Inserting this into our equation for B gives the desired result that

$$B = \frac{c}{h} \frac{\sigma(\nu)}{\nu}.$$

The Einstein B coefficient is proportional to the molecular cross section at a given wavelength.

We have derived these relationships based on a consideration of the absorption of light of a single frequency by a biomolecule; this situation never obtains in practice. Our discussion of the time scale involved with electronic absorption of light indicated that, by the uncertainty principle, states with lifetimes of $\sim 10^{-14}$ s would give rise to spectral lines of width ~ 12.5 nm. For example, Fig. 10.6 shows a typical optical absorption spectrum from hemoglobin, with its intense heme absorption centered around the $\sim$420-nm *Soret* band, in both oxygenated and deoxygenated states. The width of the Soret band is about 24 nm for either state. The binding of oxygen to the heme iron is shown in Fig. 10.7.

The probability of absorption of a photon by a molecule varies with the wavelength of the photon; in order to find the total probability of absorption, we then must integrate B over the wavelength region encompassing finite absorption:

$$B = \int B(\nu)d\nu = \frac{c}{h} \int \frac{\sigma(\nu)}{\nu} d\nu$$

This gives the probability of both absorption and stimulated emission, because these two probabilities are equal to each other. It remains to find the relationship between the cross section and the probability of spontaneous emission, A. We found that for a single frequency, ν_o,

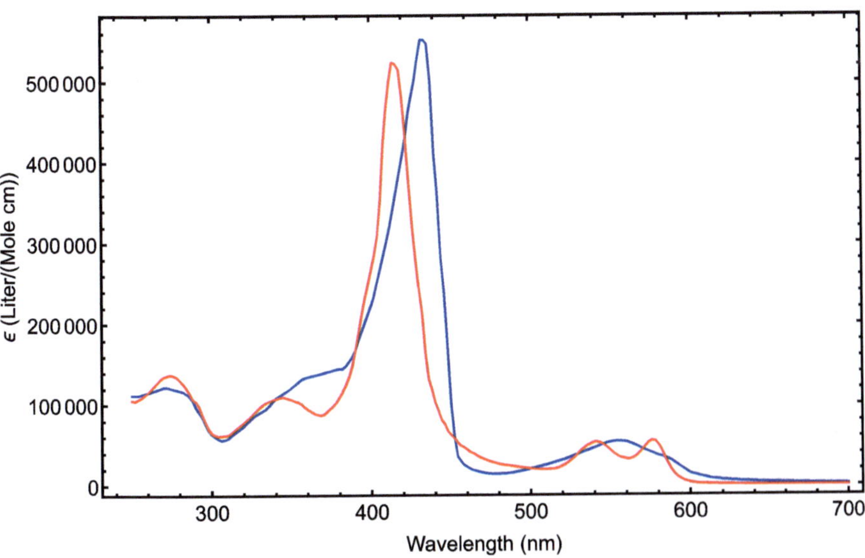

Fig. 10.6 Optical absorption spectrum of hemoglobin. The oxygenated form is shown in red, while that for the deoxygenated form appears in blue. The peak in the absorption cross section is called the Soret band at 414 nm for HbO_2 and 430 nm for deoxyHb. A solution of hemoglobin appears red because it absorbs blue light and reflects red. The spectral shift upon oxygenation arises because O_2 binds to the heme group chromophore; see Fig. 10.7

Fig. 10.7 Schematic of the binding of oxygen by hemoglobin. The O_2 molecule (shown in red) directly binds to the heme iron at the center of the tetrapyrrole ring, which has D_{4h} symmetry. (https://commons.wikimedia.org/wiki/File:Mboxygenation.png by Smokefoot, CC BY-SA 4.0 https://creativecommons.org/licenses/by-sa/4.0)

$$A(\nu) = \frac{8\pi h\nu_o{}^3}{c^3} B(\nu).$$

It is necessary to integrate this expression over the absorption spectrum to get the total probability for spontaneous emission,

$$A = \frac{8\pi h \nu_o{}^3}{c^3} \int B(\nu) d\nu$$

which we can express in terms of the molecular *cross section* as

$$A = \frac{8\pi h \nu_o{}^3}{c^2} \int \frac{\sigma(\nu)}{\nu} d\nu.$$

10.5   Elementary Quantum Mechanical Treatment of the Absorption of Light by Molecules

There are several ways one might proceed with a quantum mechanical calculation of the Einstein A and B coefficients in terms of the molecular wave functions. Consideration of the Feynman diagram (Fig. 10.1) illustrated at the beginning of this chapter suggests that an electron's wave function will not be too radically altered if the energy of the photon is small with respect to the electron's binding energy in a molecule. That is, the absorption of a photon will only slightly perturb the path of the electron. One way of solving this problem then would be to utilize a technique first developed by astronomers who were seeking to account for the slight perturbations of the orbit of Saturn resulting from the gravity due to Neptune. Quantum perturbation theory is an extremely important and useful method for obtaining approximate solutions to Schrodinger's equation if direct techniques prove difficult or impossible. In the case of the interaction of light with atoms, we can illustrate the main features and derive the shape and intensity of a spectral line by using the electric field of the photon as a perturbation on the ground state wave function of an electron.

Let the system in the absence of light be described by the time-dependent Schrodinger equation,

$$i\hbar \frac{\partial \psi}{\partial t} = H_o \psi,$$

with the formal solution for the time dependence given by

$$\psi_n^o = \phi_n e^{\frac{iE_n t}{\hbar}}.$$

We can then describe the interaction with a photon by adding a potential, V, to Schrodinger's equation

$$i\hbar \frac{\partial \psi}{\partial t} = (H_o + V)\psi.$$

At $t = 0$, the atom is in a state with wave function ψ_o; what is the wave function $\psi(q,t)$ for $t > 0$? The solutions for the unperturbed Schrodinger equation are known to form a complete orthonormal set of functions, so we can express the wave function $\psi(q,t)$ for $t > 0$ in terms of the unperturbed wave functions $\psi_n^{\,o}(q,t)$ as a series expansion:

$$\psi(q,t) = \sum_{n=0}^{\infty} a_n(t)\psi_n^o(q,t).$$

Of course, we now have to inquire as to the nature of the coefficients, $a_n(t)$. To find them, we insert our series into

$$i\hbar \frac{\partial \psi}{\partial t} = (H_o + V)\psi$$

to give

$$i\hbar \frac{\partial}{\partial t} \sum_{n=0}^{\infty} a_n(t)\psi_n^o(q,t) = (H_o + V) \sum_{n=0}^{\infty} a_n(t)\psi_n^o(q,t),$$

and by distributing the terms in the Hamiltonian inside the right-hand sum, we have that

$$i\hbar \frac{\partial}{\partial t} \sum_{n=0}^{\infty} a_n(t)\psi_n^o(q,t) = H_o \sum_{n=0}^{\infty} a_n(t)\psi_n^o(q,t) + V \sum_{n=0}^{\infty} a_n(t)\psi_n^o(q,t).$$

The time derivative on the left-hand side of this equation can be evaluated using the product rule from calculus

$$i\hbar \frac{\partial}{\partial t} \sum_{n=0}^{\infty} a_n(t)\,\psi_n^o(q,t) = i\hbar \sum_{n=0}^{\infty} \psi_n^o(q,t) \frac{\partial}{\partial t} a_n(t) + i\hbar \sum_{n=0}^{\infty} a_n(t) \frac{\partial}{\partial t} \psi_n^o(q,t)$$

but here we notice that the last term on the right cancels with the H_o term above

$$i\hbar \sum_{n=0}^{\infty} a_n(t) \frac{\partial}{\partial t} \psi_n^o(q,t) = H_o \sum_{n=0}^{\infty} a_n(t)\,\psi_n^o(q,t)$$

because this is just Schrodinger's equation again. We are therefore left with

$$i\hbar \sum_{n=0}^{\infty} \psi_n^o(q,t) \frac{\partial}{\partial t} a_n(t) = V \sum_{n=0}^{\infty} a_n(t)\,\psi_n^o(q,t).$$

The right side here is just the perturbing potential, V, times our series expansion $\psi(q,t) = \sum_{n=0}^{\infty} a_n(t)\psi_n^o(q,t)$, so we can replace the right-hand side with $V\psi(q,t)$ to give

$$i\hbar \sum_{n=0}^{\infty} \psi_n^o(q,t) \frac{\partial}{\partial t} a_n(t) = V\psi(q,t).$$

In order to proceed, we need to use the orthogonality of the solutions of the unperturbed Schrodinger equation. We multiply this equation, then, by the complex conjugate, $\psi_n^{o*}(q,t)$, and integrate over q

$$i\hbar \sum_{n=0}^{\infty} \psi_n^{o*}(q,t)\psi_n^o(q,t) \frac{\partial}{\partial t} a_n(t) = \sum_{n=0}^{\infty} \psi_n^{o*}(q,t)V\psi_n^o(q,t)$$

$$i\hbar \sum_{n=0}^{\infty} \frac{\partial}{\partial t} a_n(t) \int \psi_n^{o*}(q,t)\psi_n^o(q,t)dq = \sum_{n=0}^{\infty} \int \psi_n^{o*}(q,t)V\psi_n^o(q,t)dq$$

by the orthogonality condition,

$$\int \psi_n^{o*}(q,t)\psi_n^o(q,t)dq = \delta_{nn} = 1,$$

and we have that

$$i\hbar \sum_{n=0}^{\infty} \frac{\partial}{\partial t} a_n(t) = \sum_{n=0}^{\infty} \int \psi_n^{o*}(q,t)V\psi_n^o(q,t)dq;$$

for this to be true, this equation must be satisfied for each term in the sums, so that

$$i\hbar \frac{\partial}{\partial t} a_n(t) = \int \psi_n^{o*}(q,t)V\psi_n^o(q,t)dq,$$

and here we have a differential equation for the expansion coefficients. Now, if the perturbation is small, we can replace $\psi_n^o(q,t)$ by its initial value $\psi_o(q,t)$ (remember that $\psi_n^o = \phi_n e^{\frac{iE_n t}{\hbar}}$) to give

$$i\hbar \frac{\partial}{\partial t} a_n(t) = \int \psi_n^{o*}(q,t)V\psi_o(q,t)dq$$

$$i\hbar \frac{\partial}{\partial t} a_n(t) = \int \phi_n^* e^{-\frac{iE_n t}{\hbar}} V\phi_o e^{\frac{iE_o t}{\hbar}} dq$$

$$i\hbar \frac{\partial}{\partial t} a_n(t) = \int \phi_n^* V\phi_o e^{-\frac{iE_n t}{\hbar}+\frac{iE_o t}{\hbar}} dq.$$

Now, let us use a shorthand for the matrix element in Dirac's bra-ket notation

$$\langle n|V|0\rangle = \int \phi_n^*(q)V\phi_o(q)dq,$$

and we can write our equation for the series coefficients as

$$i\hbar \frac{\partial}{\partial t}a_n(t) = \langle n|V|0\rangle e^{\frac{i}{\hbar}(E_o - E_n)t};$$

we can solve this simple differential equation by direct integration with respect to time to give

$$a_n(t) = \frac{1}{i\hbar}\int_0^t \langle n|V|0\rangle e^{\frac{i}{\hbar}(E_o - E_n)t}dt.$$

Here, we have expressed our expansion coefficients in terms of an integral containing only the potential and the unperturbed solutions of the time-independent Schrödinger equation. Before proceeding, let us first examine the behavior of the expansion coefficients in the case where the potential can be written as $V(q,t) = V(q)\,e^{2\pi i\nu t}$, where the time dependence has been put into an oscillating term at a frequency, ν. Then

$$a_n(t) = \frac{1}{i\hbar}\int_0^t \langle n|V(q)|0\rangle e^{2\pi i\nu t} e^{\frac{i}{\hbar}(E_o - E_n)t}dt$$

$$= \frac{\langle n|V(q)|0\rangle}{i\hbar}\int_0^t e^{2\pi i\nu t} e^{\frac{i}{\hbar}(E_o - E_n)t}dt$$

$$= \frac{\langle n|V(q)|0\rangle}{i\hbar}\delta(2\pi i\nu - \frac{i}{\hbar}[E_o - E_n])$$

where the integral in the middle is a Dirac delta function, and we see that $a_n(t) = 0$ unless

$$2\pi i\nu - \frac{i}{\hbar}[E_o - E_n] = 0$$

or that

$$h\nu = [E_o - E_n];$$

the energy of the photon must match the difference in energies of the two levels. We expect to see this because otherwise the photon could not be absorbed by this quantum system. Once this condition is met, the amplitude of the spectral transition is proportional to the quantity $\langle n|V|0\rangle$ which is called the *transition matrix element*.

The probability that the molecule is in the state characterized by the quantum number n at a time t is given by the square of the wave function

$$P_n(t) = |\Psi_n(q,t)|^2 = \sum_{n=0}^{\infty} a_n{}^*(t)a_n(t)\psi_n^{o*}(q,t)\psi_n^o(q,t),$$

but we know that the unperturbed wave functions on the right side are orthogonal, so

$$\sum_{n=0}^{\infty} a_n{}^*(t)a_n(t)\psi_n^{o*}(q,t)\psi_n^o(q,t) = \sum_{n=0}^{\infty} a_n{}^*(t)a_n(t)\delta_{nn}$$

and only the term for index n survives the summation, and we have the probability that the molecule is in the state characterized by the quantum number n at a time t

$$P_n(t) = |a_n(t)|^2$$

given by the square of the expansion coefficient for the nth term. Let us put in our solution for the expansion coefficients and compute probability that the molecule is in the state characterized by the quantum number n at a time t

$$P_n(t) = \left| \frac{1}{i\hbar} \int_0^t \langle n|V(q)|0\rangle e^{2\pi i\nu t} e^{\frac{i}{\hbar}(E_o - E_n)t} dt \right|^2$$

$$= \left(\frac{1}{i\hbar}\right)^2 \int_0^t |\langle n \mid V(q)|0\rangle|^2 e^{-2\pi i\nu t} e^{2\pi i\nu t} e^{\frac{i}{\hbar}(E_o - E_n)t} e^{-\frac{i}{\hbar}(E_o - E_n)t} dt$$

$$= \left(\frac{1}{i\hbar}\right)^2 \int_0^t |\langle n \mid V(q)|0\rangle|^2 dt$$

$$= \left(\frac{1}{i\hbar}\right)^2 |\langle n \mid V(q)|0\rangle|^2 \int_0^t dt$$

$$P_n(t) = \left(\frac{2\pi}{h}\right)^2 |\langle n \mid V(q)|0\rangle|^2 t.$$

The probability is proportional to the time. The transition rate $R_n(t)$ will be given by the probability per unit time

$$R_n(t) = P_n(t)/t = \left(\frac{2\pi}{h}\right)^2 |\langle n \mid V(q)|0\rangle|^2.$$

We now introduce the specific form of the perturbation potential, $V(q,t)$, as the electromagnetic field from a photon traveling parallel to the x-axis, $\vec{E} = E\hat{x}$, the force on the electron is $\vec{F} = -e\vec{E} = -eE\hat{x}$, and the potential is the integral of the force

$$V(q,t) = \int \vec{F} \cdot dx = -e \int \vec{E} \cdot dx = -eE \int \hat{x} \cdot dx = -eEx.$$

And our matrix element can then be written as

$$\langle n|V(q)|0\rangle = -eE \int \phi_n(x)x\phi_o(x)dx,$$

and this is referred to as the *transition dipole moment*. The transition probability is proportional to the dipole moment of the molecule. Then the probability of absorption or of stimulated emission is the B coefficient

$$B_{no} = \frac{8\pi^3}{3h^2}\left[|\langle n|V_x(q)|0\rangle|^2 + |\langle n|V_y(q)|0\rangle|^2 + |\langle n|V_z(q)|0\rangle|^2\right].$$

Not all pairs of initial and final states will actually absorb a photon. In order to find which pairs of states participate in photon absorption (or emission), we would need to insert actual wave functions, perform the indicated integrals, and, by doing this, develop selection rules for allowed and forbidden transitions. In general, for dipole transitions, we must have $|\Delta m| = 1$. For higher-order (quadrupolar, octupolar, etc.) transitions, we would see that $|\Delta m| = 2$, 3, etc., with correspondingly lower rates. Remember our work on quantum selection rules (Chap. 4) that had to do with the change in symmetry of the wave functions before and after the transition induced by the dipole operator. The operator acting on the initial wave function needed to generate a state of the same symmetry as the final state in order for the transition matrix element to have a relatively large amplitude (see Box 10.1).

Box 10.1 Selection Rules
Absorption of light by a quantum system is only possible if the electric field operator of the photons has the proper symmetry to act on the wave functions to transform the initial quantum state into something closely resembling the final state. Only then will the overlap integral be nonzero. The probability of photon absorption is proportional to the cross section or the effective area that a molecule presents to the incoming beam of light. Let us calculate the probability of absorption of a photon, with the electric field symmetry of a dipole, $E(x) = ex$, by an electron in a box. Remember that the cross section, σ, is given by.

$$\sigma = |<m\,|\,E\,|\,n>|^2$$

(continued)

Box 10.1 (continued)

where $|n>$ and $<m|$ are the initial and final particle in the box wave functions. Use the sine functions and center the box at $x = 1/2$, so that the sides are at $x = \{0, 1\}$.

Dipole Selection Rules
Use $n = 1$ and $m = \{1, 5\}$, and plot the cross section vs m using a BarChart in
  Mathematica.
The Wave Function

```
ψ[x_,n_]=Sin[n π x];
```

The electric field operator is

```
el[x_,d_]=(x-1/2)^d
```

Note the electric field $el[x, d]$ is entered as a polynomial of order d. Let's begin
  with $d = 1$, a dipole, and later progress to a quadrupole, $d = 2$, etc.
Cross Section

```
σ[m_,n_,d_]=Integrate[ψ[x,m]el[x,d] ψ[x,n],{x,0,1}]²;
BarChart[Table[N[σ[m,1,1]],{m,1,5}],ChartLabels->Range
[10],ChartStyle->"Pastel",ChartElementFunction-
>"GlassRectangle",AxesLabel->{Style[m,Medium,Bold,Red],
Style["Cross section(|1> -> |m>)",Medium,Bold,Blue]},
LabelStyle->Directive[Blue,FontFamily->"Helvetica",
FontSize->14]]
```

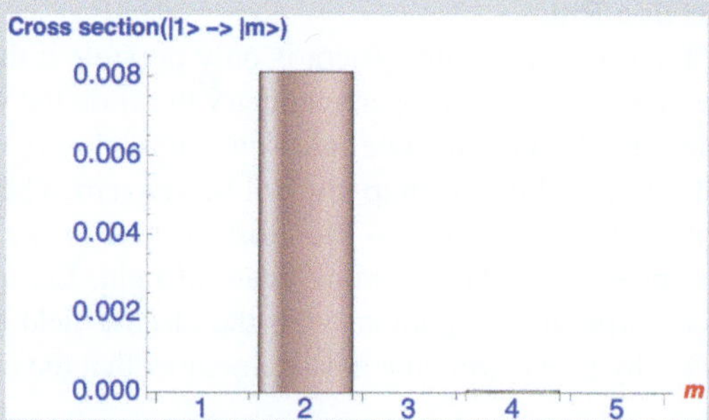

The most probable transition is certainly from $m = 1$ to $m = 2$ as depicted by
  the last graph. The change from $m = 1$ to $m = 4$ ($\Delta m = 3$) is a *forbidden
  transition* in that its matrix element is much smaller than for the $m = 1$ to
  $m = 2$ transition, but it is nevertheless nonzero.

(continued)

Box 10.1 (continued)

What general rule for the relationship between the quantum numbers $\{m, n\}$ can you propose that will summarize the dependence of the probability on the values of $\{m, n\}$?

A hint is provided by the following plot of the integral $<m| (x - 1/2) |1>$ where the zero crossings are important. One can see that the quantum numbers need to change by $\Delta m = 1$ for an allowed transition because only then is the integrand of purely one sign. In the other cases, the integrand has both negative and positive lobes that integrate to ~zero:

```
Table[Plot[ψ[x,m]el[x,1]
ψ[x,1],{x,0,1},PlotRange->All,PlotStyle->Hue[m/5],Frame-
>True,FrameLabel-> {{ "ψ(x)el(x)",None},{"X",Row@{"Δm= ",m-
1}}},LabelStyle-> Directive[Bold, Medium],ImageSize->300],
{m,1,5}]
```

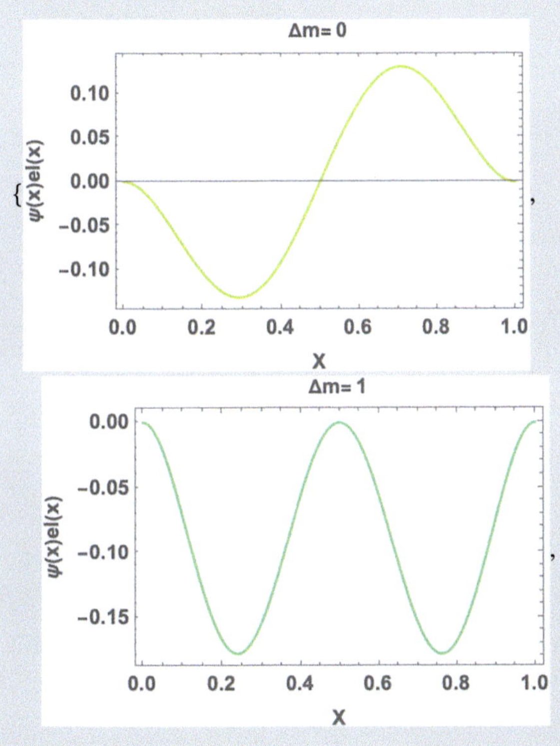

(continued)

Box 10.1 (continued)

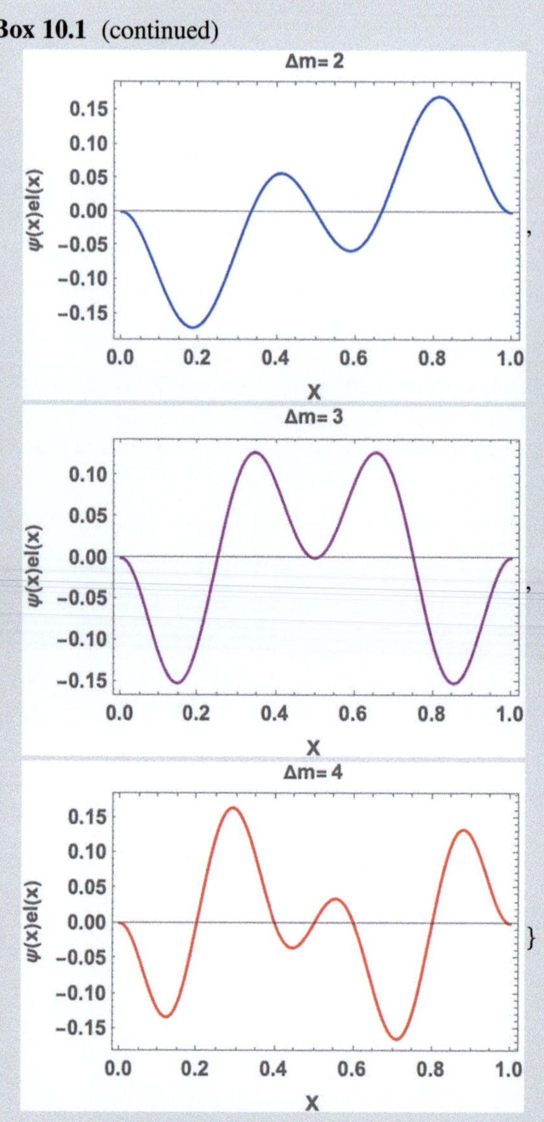

What do you think would happen to your conclusion if the photon field was instead a quadrupole, $E = e\,x^2$?

Quadrupole Selection Rules

Let $d = 2$ in the expression for *el[x,d]*. This will change the parity of the field from odd to even and change the selection rules because the symmetry of

(continued)

Box 10.1 (continued)

the field has changed. For example, now the overlap for $\Delta m = 1$ has odd parity and integrates to zero:

```
N[Integrate[ψ[x,2]el[x,2]ψ[x,1],{x,0,1}]²]0.
```

And the overlap for $\Delta m = 2$ has even parity giving a nonzero integral:

```
N[Integrate[ψ[x,3]el[x,2]ψ[x,1],{x,0,1}]²]
0.000360913
BarChart[Table[N[σ[m,1,2]],{m,1,5}],ChartLabels->Range
[10],ChartStyle->"Pastel",ChartElementFunction-
>"GlassRectangle",AxesLabel->{Style[m,Medium,Bold,Red],
Style["Cross section(|1> -> |m>)",Medium,Bold,Blue]},
LabelStyle->Directive[Blue,FontFamily->"Helvetica",
FontSize->14]]
```

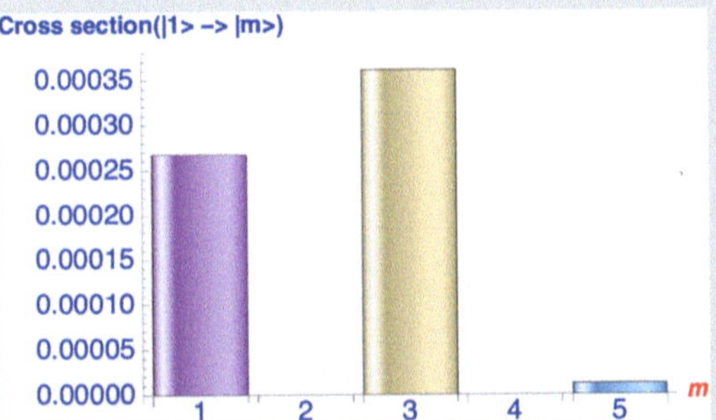

The overlap for $\Delta m = 2$ has even parity, which we can understand by examining the following plots of the integrand:

```
Table[Plot[ψ[x,m]el[x,2]
ψ[x,1],{x,0,1},PlotRange->All,PlotStyle->Hue[m/5],Frame-
>True,FrameLabel->{{"ψ(x)el(x)",None},{"X",Row@{"Δm= ",m-
1}}},LabelStyle->Directive[Bold,Medium],ImageSize->300],
{m,1,5}]
```

(continued)

Box 10.1 (continued)

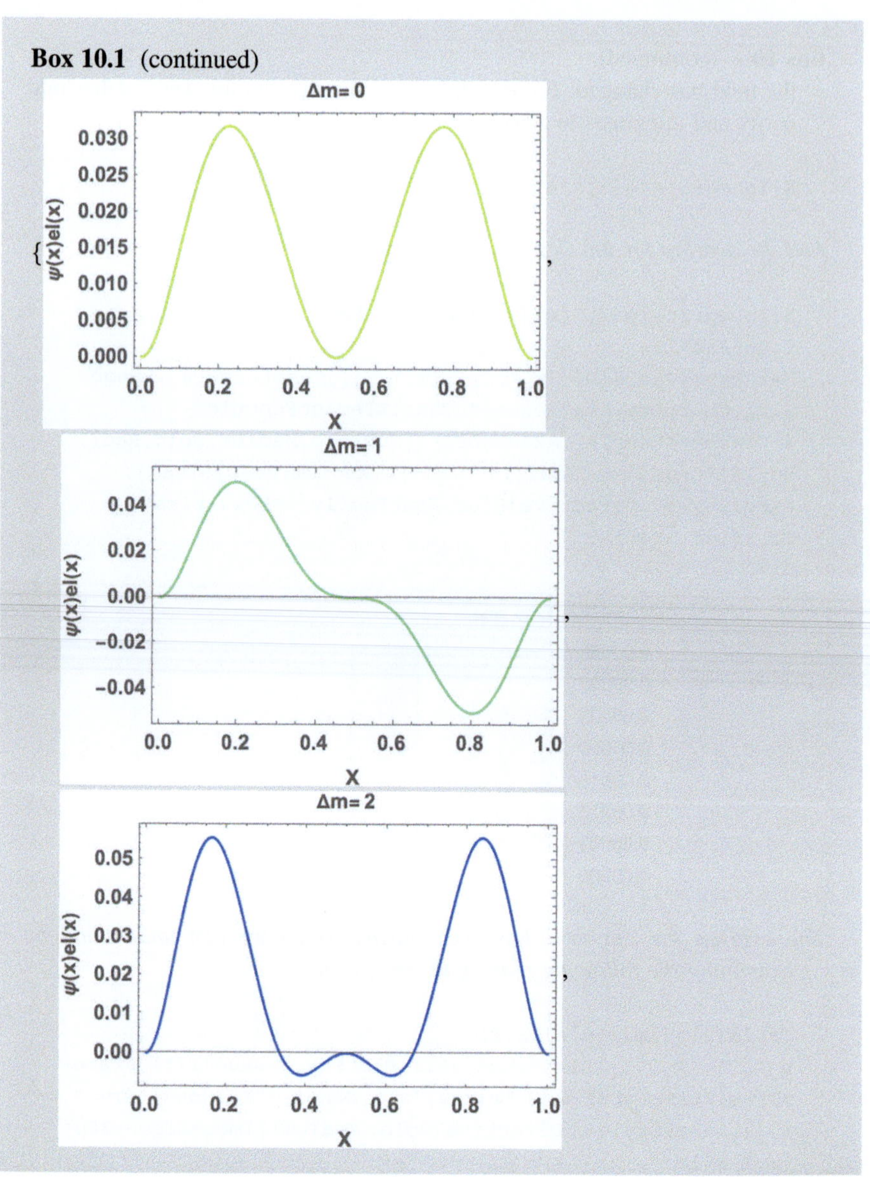

(continued)

Box 10.1 (continued)

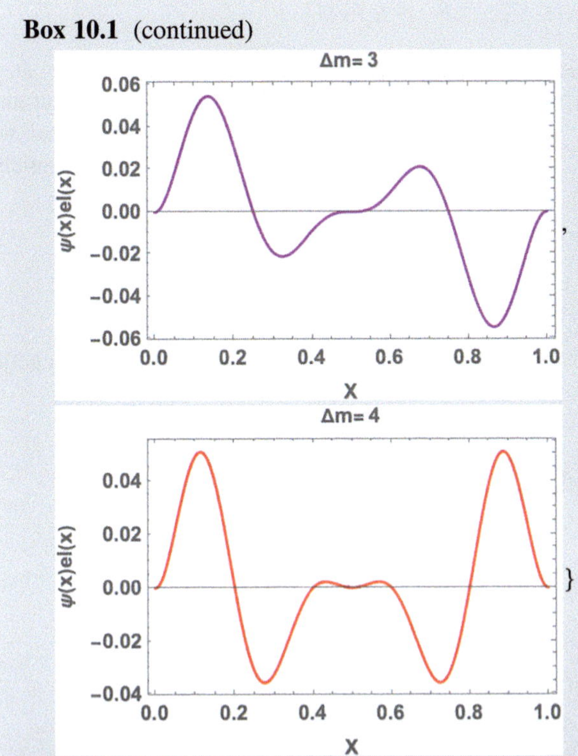

Higher-order selection rules: Feel free to explore higher-order polynomial electric field operators and their influence on the photon absorption selection rules.

For orientation of the electric field vector of the photon along an arbitrary direction with respect to the dipole moment of the molecule, we can write the Hamiltonian as

$$H = H_o - \vec{\mu} \cdot \vec{E},$$

where for n atoms, the dipole moment induced by the light is $\vec{\mu} = \sum_{i=1}^{n} e_i \vec{r}_i$. Let us make several assumptions to simplify the problem. We will place the center of the molecule at the origin of our coordinate system, at $r = 0$, and assume that the wavelength of the incident light beam is much larger than the diameter of the molecule, so that the electric field is constant over the extent of the molecule. The perturbation potential, V, is then

$$V = - \vec{\mu} \cdot \vec{E} \cos(\omega t)$$

in which we have explicitly inserted the sinusoidal time dependence at a frequency, $\omega = 2\pi\nu$. It is useful to remember that $e^{i\omega t} = \cos\omega t + i\sin\omega t$ and that $e^{-i\omega t} = \cos\omega t - i\sin\omega t$, so that by adding these two equations we produce an expression for $\cos\omega t$ which only involves exponentials: $\cos\omega t = (1/2)\,[e^{i\omega t} + e^{-i\omega t}]$. We can rewrite our potential as

$$V = - \frac{\vec{\mu} \cdot \vec{E}}{2}\left[e^{i\omega t} + e^{-i\omega t}\right],$$

and we can insert this into the equation we found above for the expansion coefficients, $a_n(t)$:

$$a_n(t) = \frac{1}{i\hbar}\int_0^t \langle n|\frac{\vec{\mu} \cdot \vec{E}}{2}\left[e^{i\omega t} + e^{-i\omega t}\right]|0\rangle e^{\frac{i}{\hbar}(E_o - E_n)t}dt$$

$$a_n(t) = \frac{\vec{E} \cdot \langle n|\vec{\mu}|0\rangle}{2i\hbar}\int_0^t \left[e^{i\omega t} + e^{-i\omega t}\right]e^{\frac{i}{\hbar}(E_o - E_n)t}dt$$

$$a_n(t) = \frac{\vec{E} \cdot \langle n|\vec{\mu}|0\rangle}{2i\hbar}\left[\int_0^t e^{i\omega t}e^{\frac{i}{\hbar}(E_o - E_n)t}dt + \int_0^t e^{-i\omega t}e^{\frac{i}{\hbar}(E_o - E_n)t}dt\right]$$

The last pair of integrals can be done by substitution. Let $x = \frac{it}{\hbar}[E_n - E_o - \hbar\omega]$ Then $dx = \frac{i}{\hbar}[E_n - E_o - \hbar\omega]dt$, or

$$dt = \frac{dx}{\frac{i}{\hbar}[E_n - E_o - \hbar\omega]}.$$

The integrands become $(e^x\,dx)$, and we can obtain our solution for the time-dependent expansion coefficients

$$a_n(t) = \frac{\vec{E} \cdot \vec{\mu}_{no}}{2}\left[\frac{e^{\frac{i}{\hbar}(E_o - E_n + \hbar\omega)t} - 1}{[E_n - E_o + \hbar\omega]} + \frac{e^{\frac{i}{\hbar}(E_o - E_n - \hbar\omega)t} - 1}{[E_n - E_o - \hbar\omega]}\right]$$

where we have written $\vec{\mu}_{no} = \langle n|\vec{\mu}|0\rangle$. Remember that the transition probability is the square of $a_n(t)$. From the expression found above, we can immediately see that $a_n(t) \ll 1$, unless $E_n - E_o = \hbar\omega$. This expresses the conservation of energy and the Bohr condition for the absorption of light by a quantum system. The above expression for $a_n(t)$ consists of two terms; the first represents stimulated emission, where $E_o > E_n$, while the second term gives the amplitude for absorption, where $E_o < E_n$.

Let us now square our expression for $a_n(t)$ in order to derive the probability that the molecule is in state n after absorption of a photon,

$$|a_n(t)|^2 = a_n^*(t)a_n(t)$$

$$|a_n(t)|^2 = \left|\frac{\vec{E} \cdot \vec{\mu}_{no}}{2}\right|^2 \left[\frac{2 - e^{\frac{i}{\hbar}(E_o - E_n - \hbar\omega)t} - e^{-\frac{i}{\hbar}(E_o - E_n - \hbar\omega)t}}{[E_n - E_o - \hbar\omega]^2}\right]$$

$$|a_n(t)|^2 = E^2 D_{no} \cos^2\theta \left[\frac{\sin^2\left[\frac{t}{2\hbar}(E_n - E_o - \hbar\omega)\right]}{[E_n - E_o - \hbar\omega]^2}\right]$$

where we have replaced the initial dot product of the electric field vector, $\vec{E}$, and the transition dipole moment, $\langle n|\vec{\mu}|0\rangle$, with the square of the electric field vector, E^2; the square of the cosine of the angle, θ, between $\vec{E}$ and $\langle n|\vec{\mu}|0\rangle$; and a quantity, D_{no}, which is called the *dipole strength*

$$D_{no} = \langle n|\vec{\mu}|0\rangle \cdot \langle n|\vec{\mu}|0\rangle = \left|\langle n|\vec{\mu}|0\rangle\right|^2.$$

The probability, $|a_n(t)|^2$, has several interesting properties:

1. Its amplitude depends on the square of the electric field vector, E^2, which is the power emitted by the light source, i.e., the amount of energy emitted per unit time.
2. Since the amplitude of $|a_n(t)|^2$ depends on the square of the cosine of the angle, θ, between $\vec{E}$ and $\langle n|\vec{\mu}|0\rangle$, polarized light can be used to probe the direction of the dipole moment in oriented solid samples. Figure 10.8 shows a plot in polar coordinates of the angular dependence of the probability of absorption of a photon. One can see that if θ is $\pm\pi/2$, then the absorption probability goes to zero.
3. The amplitude oscillates in time between zero and a maximum in phase with the oscillating electric field of the incident photons (Fig. 10.9).

   One can understand this behavior of the transition probability because the electric field of the light is not of constant amplitude but rather varies at the frequency given by $\nu = E/h$.
4. The transition probability is only nonzero for incident photons whose energies, $h\nu$, match the Bohr condition, $h\nu = (E_f - E_o)$. Let us illustrate this point by

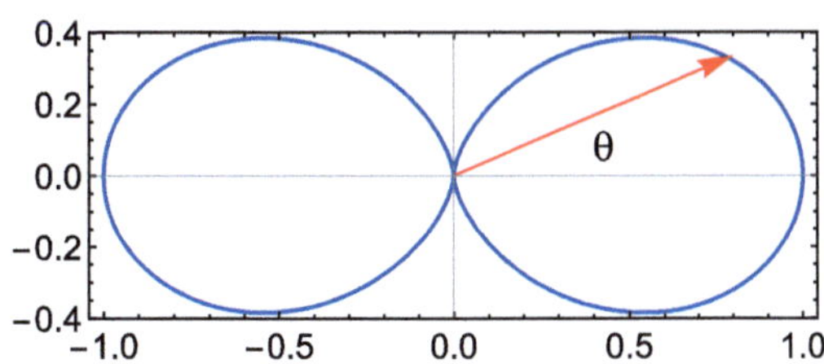

Fig. 10.8 The probability of absorption of a photon as a function of the angle, θ, between $\vec{E}$ and the absorption dipole moment $\langle n|\vec{\mu}|0\rangle$

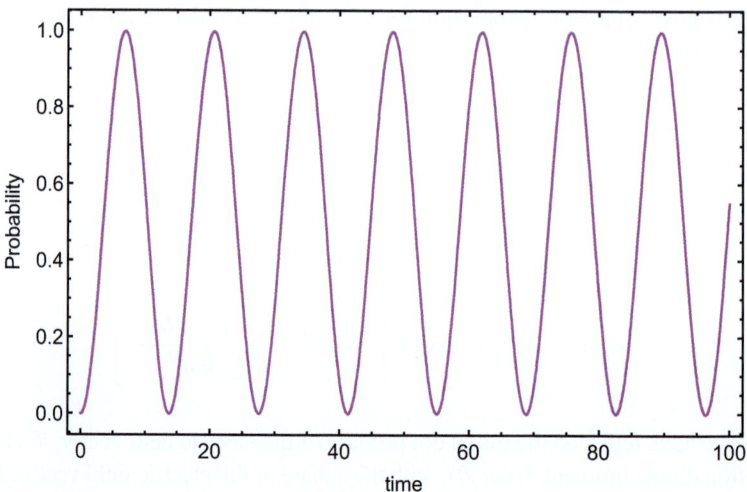

Fig. 10.9 The time dependence of the probability of photon absorption

assuming that $E_f = 4 \times 10^{-19}$ Joules, $E_o = 4 \times 10^{-19}$ Joules, and $h\nu = (E_f - E_o) = 1 \times 10^{-19}$ Joules. Thus, the incident photon must have a wavelength of $\lambda = hc/(E_f - E_o) = 667$ nm in order to induce transitions from E_o to E_f. Figure 10.10 shows that the transition probability peaks at $\lambda = 667$ nm and is essentially zero for other wavelengths.

In reality, unless the light source is a narrow-bandwidth, tunable laser, or some other source of monochromatic visible radiation, the incident photon beam will possess a distribution of frequencies due to its own intrinsic emission spectrum. Let us assume that the light source is a blackbody, whose photons have an electric field distribution given by $E^2 = 8\pi \, \rho(\nu) \, d\nu$, and thus the transition probability will depend on the overlap between the source distribution and that of the quantum energy levels in the absorber (molecule). This will be found by integrating the product of $|a_n(t)|^2$ with $\rho(\nu)$: This integral is the transition probability, $P_n(t)$,

$$P_n(t) = \frac{2D_{no}}{\pi\hbar^2} \cos^2\theta \int \rho(\nu) \frac{\sin^2[\pi(\nu_{no} - \nu)t]}{(\nu_{no} - \nu)^2} d\nu,$$

where we have written $h\nu_{no} = E_f - E_o$. The blackbody spectrum is given by

$$\rho(\nu) = \frac{8\pi h\left(\frac{\nu}{c}\right)^3}{e^{\frac{h\nu}{kT}} - 1},$$

and insertion of this into the integral for $P_n(t)$ would be appropriate at this time. However, note from the plot of $|a_n(t)|^2$ as a function of wavelength above that the

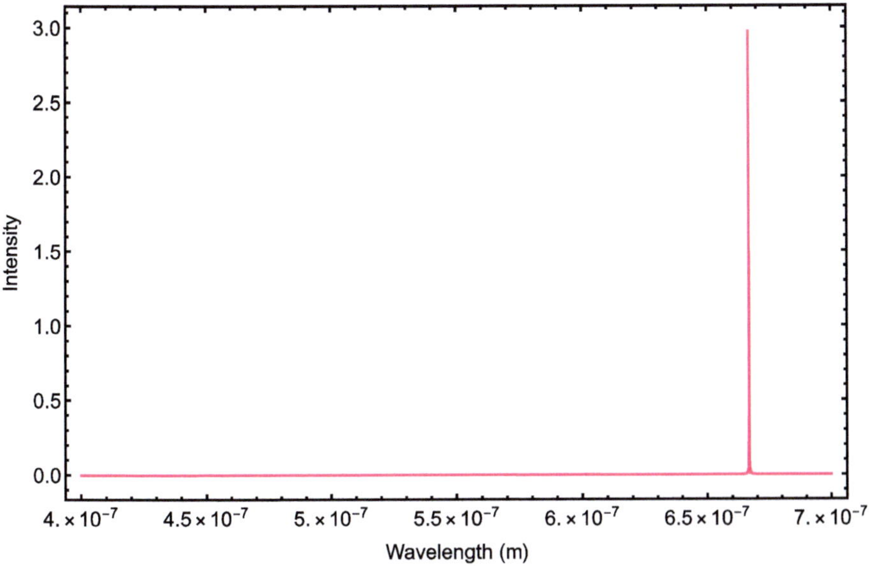

Fig. 10.10 The transition probability as a function of the wavelength of the incident light for $h\nu = (E_f - E_o)$ as given in the text

absorption probability is essentially a delta function at $h\nu = E_f - E_o$, and therefore the overlap integral between the blackbody spectrum $\rho(\nu)$ and the spectral line simply picks out the component of $\rho(\nu)$ at the transition frequency, $\nu = \nu_{no} = (E_f - E_o)/h$. Then, the integral is approximately

$$\int_0^{\infty} \rho(\nu)\,\frac{\sin^2[\pi(\nu_{no} - \nu)t]}{(\nu_{no} - \nu)^2}\,d\nu \approx 2\pi t$$

so that the transition probability per unit time is then

$$\frac{P_f(t)}{t} \approx \left[\frac{2\pi}{\hbar^2} D_{no}\rho(\nu_{no}) \cos^2\theta\right].$$

From our consideration of the two-level system from which we derived the Einstein coefficients (Fig. 10.3), we found that rate of transitions from the ground state (0) to the nth excited state (n) was

$$\frac{dN_{0n}}{dt} = BN_o\rho(\nu),$$

and a comparison of this with the previous equation allows us to identify the B coefficient from our perturbation transition probability equation as

$$B = \frac{2\pi D_{no}}{\hbar^2} \cos^2\theta,$$

or if we assume that we have N_1 molecules with random orientations in a fluid, we can average the $\cos^2\theta$ term over all directions and find that it equals 1/3. Then our equation for B becomes

$$B = \frac{2\pi D_{no}}{3\hbar^2}.$$

Note that here the B coefficient is proportional to the dipole strength, D_{no}, which is the square of the transition matrix element,

$$D_{no} = \langle n|\vec{\mu}|0\rangle \cdot \langle n|\vec{\mu}|0\rangle = \left|\langle n|\vec{\mu}|0\rangle\right|^2$$

so that the B coefficient is $B = \frac{2\pi}{3\hbar^2}\left|\langle n|\vec{\mu}|0\rangle\right|^2$. This shows the direct relationship between the wave functions of the molecule and its probability of absorption of light. The matrix element also contains the selection rules (Box 10.1) which govern the allowed quantum numbers of the transitions.

Let us return to the concept of the molecular cross section, which we found as

$$B = \frac{c}{h}\frac{\sigma(\nu)}{\nu}.$$

Then, from the relationship between B and the matrix element, we can obtain the cross section in terms of quantum wave functions of the molecule

$$B = \frac{c}{h}\frac{\sigma(\nu)}{\nu} = \frac{2\pi}{3\hbar^2}\left|\langle n|\vec{\mu}|0\rangle\right|^2.$$

$$\sigma(\nu) = \frac{8\pi^3}{3h}\left|\langle n|\vec{\mu}|0\rangle\right|^2$$

10.6 Fluorescence Lifetime, Polarization, Anisotropy, and Dipole Moment

One of the irreducible representations of the Poincare group is an object with zero mass, the photon. A special property of massless photons is that they can only have two degrees of freedom; these are called helicity states and involve momentum and spin. A positive helicity state $h = +1$ is one in which the spin and momentum vectors are parallel, while helicity $h = -1$ is associated with spin antiparallel to the

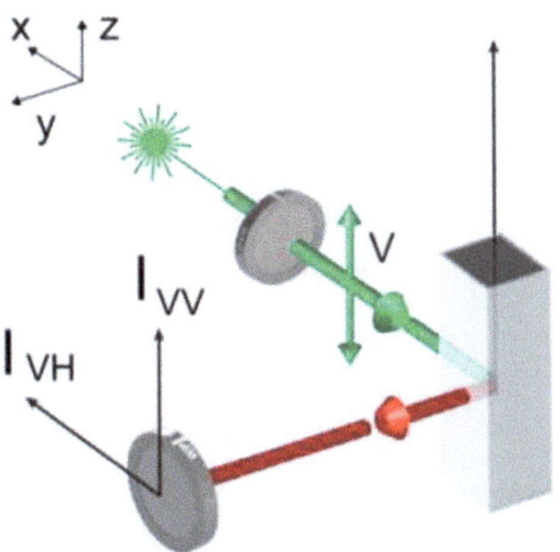

Fig. 10.11 Schematic of a fluorescence anisotropy measurement using excitation and emission polarizers rotated to 0° (vertical, V) and 90° degrees (horizontal, H) orientations. (Figure courtesy of Vinegoni et al. (2019) used by permission)

momentum. If we think of the photon as an electromagnetic wave, then these helicity states correspond to the polarization of the light.

Many molecules of biological interest fluoresce in aqueous solution. Fluorescence is a delayed emission of light after illumination (Fig. 10.2). A molecule may therefore be excited, and when the light source is removed, the molecule will emit longer wavelength photons. This property can be exploited to determine molecular rotational correlation times by utilizing a polarized light source for excitation. Then only those molecules whose absorption dipole moment $\langle n | \vec{\mu} | 0 \rangle$ is parallel to the plane of polarization of the incident light will be excited. When these excited molecules emit fluorescence, the emitted photons will also be polarized. By measuring the time dependence of this polarization, the rate of molecular motion can be obtained.

In practice, the amount of fluorescence polarization is determined as a dimensionless number, the anisotropy. The plane of polarization of the photon emitted during fluorescence is determined by the orientation of the emission dipole moment, and since this is defined by the framework based on the heavy nuclei in the molecule, the plane of polarization reflects the orientation of the molecule at the time of emission. The amount of fluorescence polarization is measured as the dimensionless quantity, the time-resolved polarization anisotropy (Fig. 10.11) defined as

$$r(t) = \frac{I_{VV}(t) - G\, I_{VH}(t)}{I_{VV}(t) + 2G\, I_{VH}(t)}$$

where G is the grating factor given by $G = \frac{I_{VH}}{I_{HH}}$. The polarization anisotropy is largest right after excitation and then decreases with time, as shown in Fig. 10.12 for the laser dye coumarin 6. To make a measurement of the polarization anisotropy, the exciting and emitted photons pass through polarizers. The anisotropy is calculated using the above equation, where $I_{VV}(t)$ is the time-dependent intensity with vertically

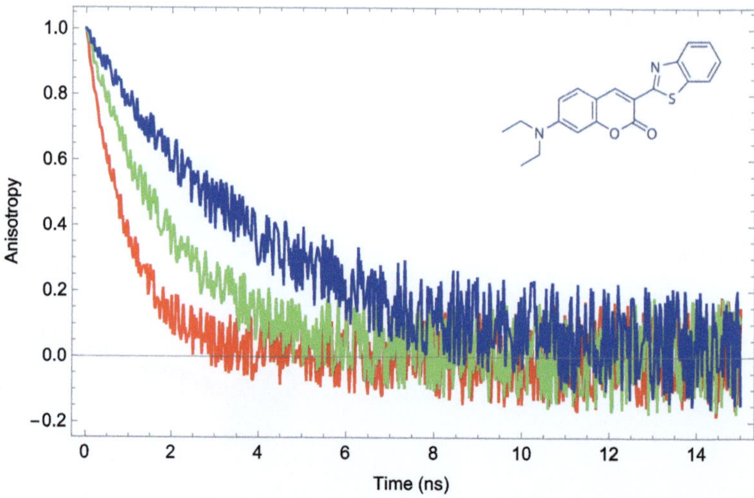

Fig. 10.12 An example of the time-dependent fluorescence anisotropy as a function of temperature for a commonly used laser dye, coumarin 6, whose structure is shown in the inset. The three decay curves were taken at temperatures of 283 K (blue), 298 K (green), and 313 K (red) and illustrate the decrease in rotational correlation time as the thermal energy in the fluid increases

polarized excitation (and detected emission), while $I_{VH}(t)$ is likewise the intensity for vertical excitation and horizontal emission. The grating factor G corrects for the instrument's different transmission of the two orthogonal polarizer orientations.

The absorption dipole moment can be and often is different from that of the emission dipole moment. A moment's reflection aided by an examination of the energy-level diagram generically illustrating the processes of absorption and emission (Fig. 10.2) shows that while absorption requires that the incident photon's energy match a difference in levels between two states, no such requirement ensues for emission, except that the wavelength for emission must always be greater than or equal to that for absorption. There can be non-radiative transitions from the initial excited state to lower-energy states followed by a transition from these lower states to the ground state. Since these lower-energy states can, and often do, have different symmetry properties from the states responsible for the absorption, the optical selection rules come into play and dictate the structure of the emission dipole moment.

For molecules free in solution, collisions due to random molecular motion (Brownian motion) cause the decay of orientation and hence polarization anisotropy. The rotational correlation time, t, is the time it takes for the orientation to decay to $1/e$ of its initial value,

$$P(t) = P_o e^{-t/\tau}$$

where P_o is the initial polarization anisotropy. We can see from Fig. 10.12 that the decay of the fluorescence polarization follows this exponential relationship. The mass of coumarin 6 is 350 g/mol, from which we can estimate its volume

$(5.8 \times 10^{-28}$ m$^3)$ and its rotational correlation time in water at room temperature of around 0.2 ns which is of the same order of magnitude as the decay times seen in Fig. 10.12.

10.7 Light Absorption by Phototrophs

The amount of energy available from sunlight on the primordial Earth was enormous; if we assume that the current solar constant of ~300 Watts/m^2 (300 Joule/(sec m^2)) can be extrapolated back to the epoch of life's origin around 3.5 billion years ago, then we can estimate the number of $\lambda \sim 500$-nm photons that flooded the surface of the Earth during its first billion years. A photon of this wavelength has an energy of 2.48 eV. The solar constant then corresponds to 1.875×10^{21} eV/(sec m^2). The flux of photons is then 7.5×10^{20} per second, per square meter. Over the course of a billion years, this amounts to an integrated flux of 2.38×10^{37} photons with 2.48 eV, impinging on every square meter of the Earth's surface, water and land. This flux was not constant but varied with the day/night cycle, so the daytime flux was ~50% of this value. Nevertheless, this was a large, easily exploitable energy source for abiogenesis. The energy of solar photons is matched to the bond energies of biomolecules; if we think of this energy as a signal and compare it with the random noise from thermal background fluctuations, then at a temperature of 295 K, the signal (energy)-to-noise (kT) ratio was on the order of 100–1000, which is significantly greater than one. For UV photons, this ratio would be even higher. The cyclic nature of this energy source also meant that molecular reaction products formed during the daytime could relax into dark-stable states at night and therefore could be preserved from further photolysis or could diffuse into shadowed areas protected from the next day's solar irradiation.

10.7.1 Retinal: The Primordial Light Harvester

Phototrophs evolved around this time to harvest the energy in light, storing it for later use to synthesize lipids, proteins, sugars, and nucleic acids. An important clue with respect to this scenario comes from the study of the purple archaea, *Halobacterium salinarum*, which uses the transmembrane light-gated channel protein bacteriorhodopsin (Neutze et al. 2002) to capture photons in order to transport protons across its lipid membrane (Braun-Sand et al. 2008). The resulting electrochemical gradient powers ATP synthesis in a direct conversion of light into biomolecular structure. The capture of photons in bacteriorhodopsin is accomplished by the bound retinal molecule, the same retinal (Fig. 10.13) that humans use for vision billions of years later.

Retinal's structure contains five conjugated carbon-carbon double bonds. The π bonds, like in ethene ($H_2C{=}CH_2$), share delocalized electrons spread out over the length of the chain. In ethene, a single double bond absorbs light with a wavelength

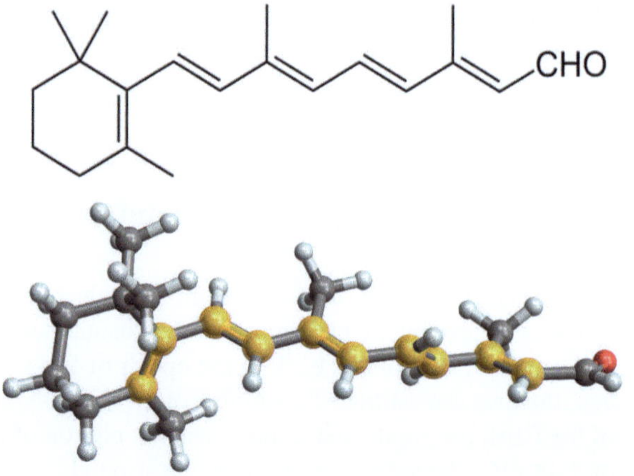

Fig. 10.13 Covalent structure of all-*trans* retinal. (Top) atomic connectivity graph. (Bottom) 3D model of all-*trans* retinal where the yellow color superimposed on the molecule emphasizes the contiguous chain of 5 π carbon-carbon conjugated double bonds over which the electrons can wander, delocalized much as in a metal

around 180 nm, while in 1,3-butadiene, with its two conjugated double bonds ($H_2C=CH-CH=CH_2$), the absorption maximum shifts to the longer wavelength (lower energy) of 217 nm. In 1,3,5-hexatriene, with three conjugated double bonds, the absorption maximum shifts further to 268 nm, etc. One can understand this increase in the absorption maximum by reference to the particle in a box model of quantum states (Chap. 4). The quantization of the energy levels arose from physical confinement of the electron within the box; the amplitude of the wave functions went to zero at the box walls, and this allowed only states with integral numbers of wavelengths to fit into the box. As the box walls were moved apart, the energies decreased, or the wavelengths increased. In the case of the series ethene, butadiene, hexatriene, etc., one finds that the absorption maximum shifts to the red by 48 nm per additional double bond. The particle in a box model suggests for linear conjugated alkenes with m double bonds that the absorption maximum is proportional to m:

$$\lambda_{\mathrm{max}} \sim km.$$

Empirically, k is found in the range 30–50 nm. From this consideration, one can estimate that the five double bonds in retinal would increase the box length and shift the absorption maximum to about 367 nm; indeed, the absorption spectrum of vitamin A aldehyde (retinal) in solution is found to peak at 370 nm, just outside the range of visible light. One would perhaps expect that such a molecule used for capturing photons would have its peak in the absorption cross section more closely aligned with the peak in the visible light spectrum, the spectrum of sunlight.

In order to put this on a quantitative basis, let's look at the distribution of photon energies in sunlight. The sun, as is the case with all stars, radiates energy with a wavelength probability distribution that closely follows that for a blackbody; the small deviations from a blackbody spectrum are primarily of interest to astrophysicists. For our purposes, we are interested in the wavelength dependence of the solar flux. Examination of a plot (Fig. 10.14) of the spectrum of a blackbody whose

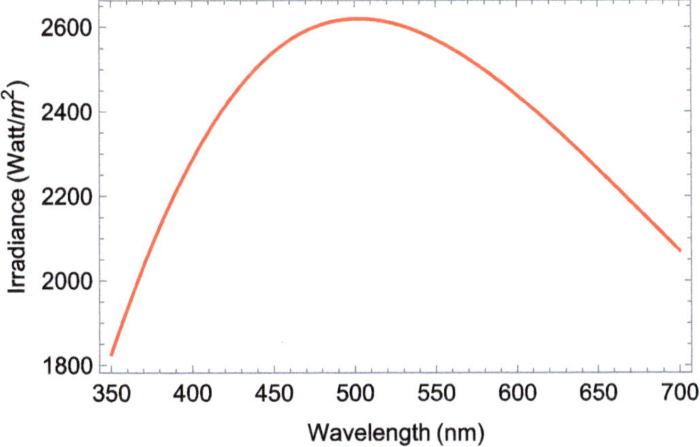

Fig. 10.14 Irradiance of the sun as a function of wavelength assuming that the solar spectrum is that of a blackbody with a temperature of 5775 K. The peak in the flux occurs at 502 nm. The width is about 250 nm. In energy units, the range from 350 to 700 nm corresponds to 3.54–1.77 eV, with a peak at 2.47 eV

temperature matches that of the sun (5775 K) indicates that the peak energy of solar photons is 2.47 eV and that the photon flux is smoothly and broadly spread over the visible region of light with a variation of only ~30% from this peak. This is approximately what primordial organisms had to use as their early energy source, although it is recognized that the sun was somewhat dimmer in the archaic past. But wait, this might not be so important after all because retinal in solution does *not* have an absorption peak in the visible portion of the solar spectrum (vide supra).

10.7.2 *The Pigments Used for Vision Evolved as Energy-Harvesting Molecules*

However, retinal's role in the capture of photons must have been important because it has been found as the chromophore in archaea, microorganisms that date back to the origin of life, some 3.7–4.1 billion years ago (DasSarma and Schwieterman 2018). The first retinal-containing rhodopsins in microorganisms, bacteriorhodopsin and halorhodopsin, were discovered in the archaeon *H. salinarum*.

A major clue that led to the solution of this puzzle came from studies, first of human vision and then of the extracted bacteriorhodopsin from *H. salinarum*. Measurements of the absorption cross section for the frog retina, of what was at that time called "visual purple" (really rhodopsin from rod outer segments), have been made for almost 150 years (Kühne 1879; Chase and Haig 1938) and show an almost identical absorption spectrum to that from the rhodopsin in the human eye (Fig. 10.15). The peak in the spectrum occurs at 502 nm, smack dab in the middle of

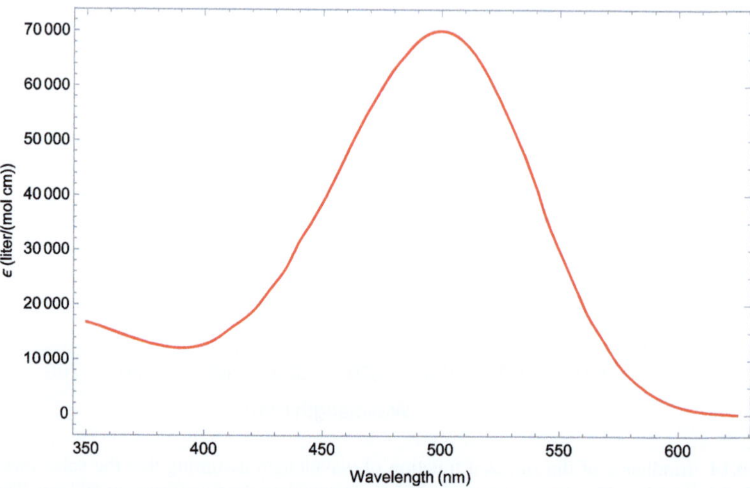

Fig. 10.15 Absorption spectrum of retinal in a human rod cell as a function of wavelength. The peak in the absorption occurs at 502 nm. The bandwidth is about 100 nm. The rod cells in the retina are responsible for night (scotopic) vision; our rods are most sensitive at night to wavelengths around 500 nm

the solar spectrum. What is it then that shifts the absorption peak of retinal from 370 nm (3.35 eV) to 502 nm?

Additional clues appeared when the absorption spectrum was obtained of the bacteriorhodopsin extracted from *H. salinarum.* This hydrophobic protein is a 25-kDa integral transmembrane molecule containing 7 α-helices, which exists in a hexagonal lattice in the lipid membrane of the archaeon. Its retinal chromophore is located in the middle of the protein, linked by a protonated Schiff base (-C=NH+) to the ε-amino group of residue Lys-216 in the seventh helix (Neutze et al. 2002; Hasegawa et al. 2018). These conjugated Schiff bases have large optical cross sections, and the basic imine nitrogen has the properties of a π-acceptor.

In its native form, the retinal in bacteriorhodopsin has an absorption maximum at 553 nm (Sekharan et al. 2006). It was initially thought that the binding of retinal to opsin was responsible for the shift to the red of the peak in the retinal absorption spectrum, and this shift was named an "opsin shift." However, when the absorption spectrum of retinal linked via a protonated Schiff base to lysine was measured in vacuo (Andersen et al. 2005; Rasmussen et al. 2012), it was found that the peak was shifted into the red to 610 nm (2.03 eV). The real "opsin shift" was actually a blue shift of 0.45 eV (108 nm). The amino acid environment of the opsin protein surrounding the retinal clearly plays an important role in moving the maximum of the absorption cross section into the middle of the visible spectrum (Hasegawa et al. 2012).

How then do photons interact with the retinal chromophore? In bacteriorhodopsin, absorption of a photon switches the bound retinal molecule from the all-*trans* to

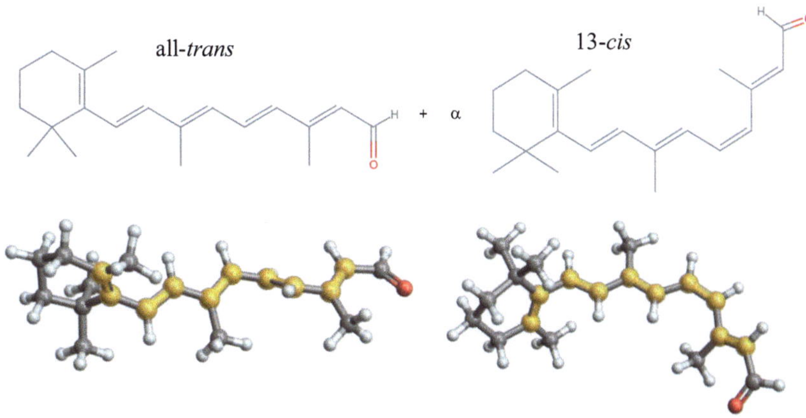

Fig. 10.16 The alterations in molecular structure which occur upon the absorption of a photon with a wavelength of ~500 nm by the retinal chromophore. All-*trans* retinal (Left) in bacteriorhodopsin is converted into 13-*cis* retinal (Right). The conjugated carbon-carbon double bonds along the polyene chain are highlighted in yellow

the 13-*cis* configuration (Fig. 10.16). The quantum states of most interest are S_0, the highest occupied (the so-called HOMO) and S_1 the lowest unoccupied (the so-called LUMO) electronic energy levels; transitions between these states involve energies in the ultraviolet to visible part of the light spectrum. The absorption of a photon promotes the electron in the HOMO S_0 of all-*trans* retinal to the LUMO S_1 in about the time it takes for light to traverse the retinal molecule's conjugated π-system, $\sim 10^{-17}$ s. All-*trans* retinal then isomerizes to the LUMO of 13-*cis* retinal in 100–200 fs, and this relaxes to the HOMO of 13-*cis* retinal (Fig. 10.16) in about 500 fs. The absorption spectrum of retinal is broadened from a single narrow resonance to about 100 nm width (Fig. 10.15) due to the manifold of abundant vibrational and rotational sublevels, so that the energy of the electron in the ground state S_0 is smeared out and the energy of the transition to S_1 varies according to which sublevel the electron was in prior to photon absorption. There is a Boltzmann probability distribution of ground state S_0 energies from which the electron can make the quantum transition to the S_1 state. By the Franck-Condon principle, one treats the molecule as fixed during the photoexcitation because the absorption of the photon occurs so quickly that the molecular framework has little time to reorient. This means that the electron is not directly excited into the ground state of S_1, but to a higher energy state, which then decays in the dark in about 100–200 fs to the ground state of S_1 (Fig. 10.17). Since the ground state of S_1 is still an excited state with higher energy than that of the 13-*cis* ground state of S_0, the S_1 ground state will decay into the 13-*cis* ground state of S_0. This smearing of the energy levels enables retinal to absorb photons with a broader range of wavelengths than if it were, say, a hydrogen atom, with its sharp spectral lines.

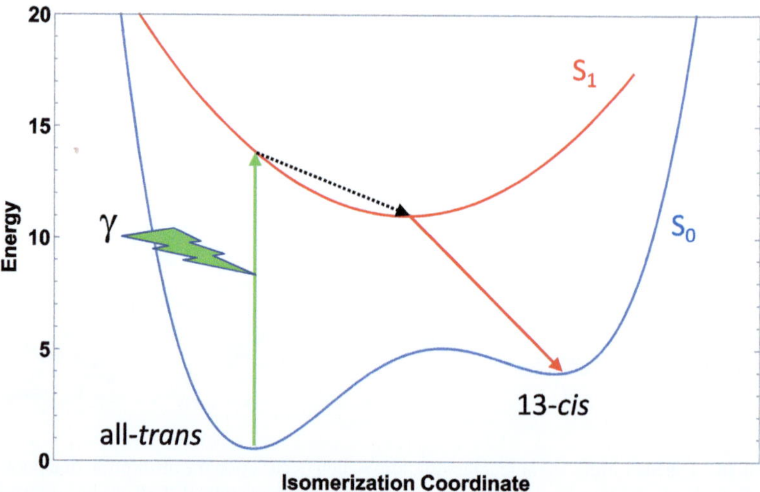

Fig. 10.17 Energy-level scheme for the absorption of ~500-nm (green) photon by the all-*trans* retinal molecule in bacteriorhodopsin. The retinal is excited from its manifold of ground state S_0 vibrational and rotational energy levels to one of the manifolds of excited states in S_1 from which it decays to the ground state of S_1 in about 100–200 fs. Retinal then isomerizes in about 500 fs to the ground state of the 13-*cis* conformation

Fig. 10.18 Molecular wave functions for all-*trans* retinal bound via a protonated Schiff base to lysine. (**a**) Highest occupied molecular state (HOMO). (**b**) Lowest unoccupied molecular state (LUMO). Calculations performed with Spartan

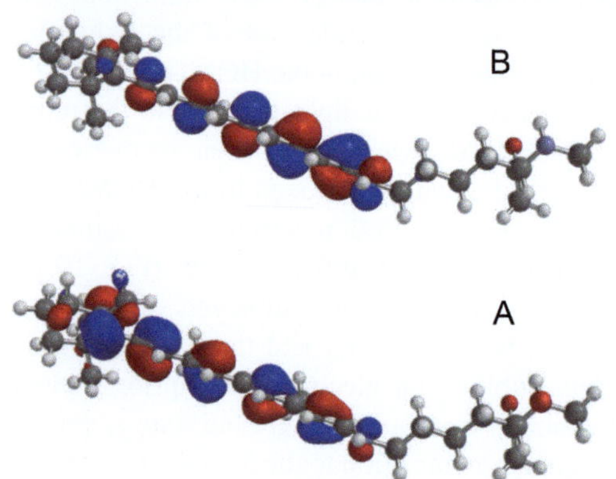

For both the free retinal molecule and the retinal-lysine protonated Schiff base (Fig. 10.18), the optical absorption of bacteriorhodopsin is dominated by the large cross section provided by the retinal polyene chain; there is no absorption of visible light by the lysine (or other protein residues). When a photon is absorbed by all-*trans*-retinal, a p electron in the π bond is promoted to a higher-energy level, temporarily converting the π component of the double bond into a single bond, releasing rotational constraints, so that the ionone ring of retinal swivels by 180°. After several picoseconds, the double bond is restored, and the molecule relaxes into

Fig. 10.19 An illustration of the scheme used by *Halobacterium salinarum* for harvesting photons and using them to generate a proton gradient, which is then used by ATP synthase in a chemiosmotic process to synthesize ATP. (https://commons.wikimedia.org/wiki/File:Bacteriorhodopsin_chemiosmosis.gif. This file is licensed under the Creative Commons Attribution-Share Alike 3.0 Unported license. Author: Darekk2)

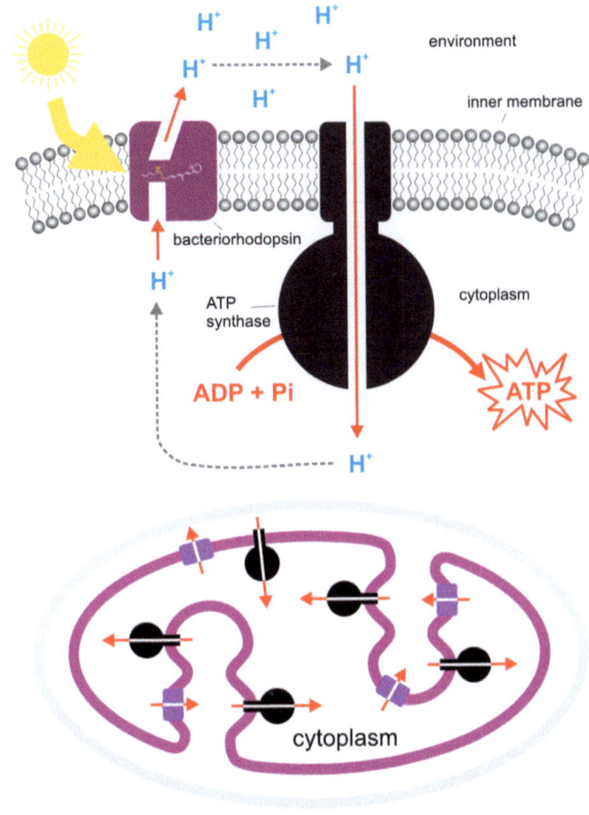

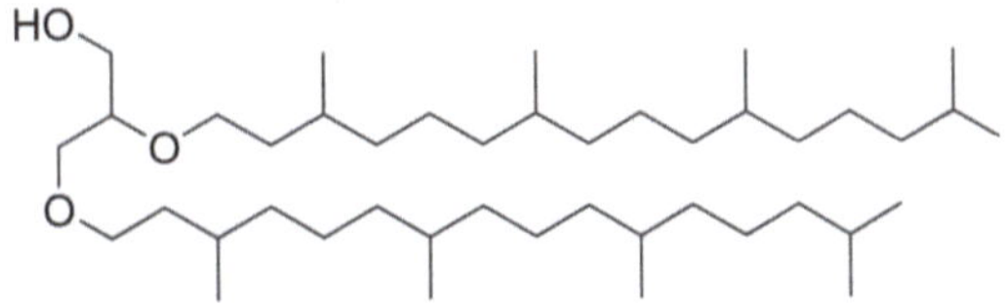

Fig. 10.20 The covalent structure of archaeol, which comprises all of the lipid content of the *H. salinarum* membrane

the stable 13-*cis* conformation (Garavelli et al. 1997). The retinal in bacteriorhodopsin is used by *H. salinarum* for harvesting photons and using their energy to generate a proton gradient (Lanyi 2006; Braun-Sand et al. 2008) which is then used by ATP synthase in a chemiosmotic process to synthesize ATP (Fig. 10.19).

Another relevant aspect of the structure of the lipid membranes of *H. salinarum* is the use of unusual ether-linked phytanyl side chains (Bale et al. 2019) attached to 2,3-sn-glycerol, an example of which is archaeol (Fig. 10.20). Archaea evolved the enzymes needed to synthesize the thermally stable, dialkyl glycerol diether archaeol-based membrane phospholipids containing combinations of long C_{20} and C_{25} isoprenoid alkyl chains (Bale et al. 2019) using a unique variant of the mevalonic acid lipid synthesis pathway. At the same time, they developed the capability to produce retinal (DasSarma and Schwieterman 2018) with its closely related polyene chain; the same enzymes were able to synthesize either of these products. These unique archaeal lipids have been found on the Earth in rocks dated at 3.5 Gya and in sediments from Greenland dated at 3.8 Gya, indicating that the source of chemical energy in the form of light constituted an early niche into which organisms could evolve. In our search for molecular fossils relevant to the origin of life, these archaeal lipids give us a surprising window on the past.

All human vision cells, both rods and cones, use the same retinal chromophore but have differing absorption maxima; the blue cone cells have a λ_{max} of 437 nm, the green have a λ_{max} of 533 nm, while the red cone cells have a λ_{max} of 564 nm. The tuning of the absorption maximum illustrates the large active role played by the electronic interactions between the chromophore and the opsin protein (Lin et al. 1998; Zhou et al. 2014; Fujimoto 2021) and indicates the great sensitivity of retinal to the environment. A shift of a single positive charge moves the absorbance maximum by more than 1 eV (Sekharan et al. 2006).

10.7.3   Chlorophyll-Based Photosynthesis

While there are sound observations supporting the proposal that retinal was a major chromophore and among the first light-harvesting molecules (Zhong et al. 2012; DasSarma and Schwieterman 2018) without a doubt, the other important molecule used for harvesting light is the one we are most familiar with from the color of the green plants that provide all our food: the chlorophyll molecule whose structure is shown in Fig. 10.21. Not only is chlorophyll found in plants, but its evolutionary history predates that of plants since it is also found in more primitive organisms like cyanobacteria, protists (e.g., *Euglena*), and algae. The energy of the photons absorbed by chlorophyll is used in photosynthesis in which light, carbon dioxide, and water are used to produce sugars. This is distinct from phototrophy, the direct conversion of light into a proton gradient and ATP molecules, as in *H. salinarum*.

Chlorophyll contains a tetrapyrrole (chlorin) ring (Fig. 10.21) that is similar to the heme group of hemoglobin (Fig. 10.7), but instead of an iron atom at the center of the ring, chlorophyll uses a magnesium atom. The chlorin ring in chlorophyll *a* is distinguished by the additional α-keto acid methyl ester ($-(O=C)-(C=O)-OCH_3$, but this does not interrupt the π conjugation of the electrons. Note that chlorophyll also contains a long alkane tail that anchors it in the lipid membrane of the plant chloroplast, the plant cell's equivalent to the mitochondria found in animal's cells. The tetrapyrrole ring also contains nine conjugated π bonds, leading to delocalization of the ring's π electrons and giving the chromophore a pair of absorption maxima (Fig. 10.22) B at 429 nm and Q at 640 nm. The B resonance is the same electronic transition that produces the *Soret* band in hemoglobin (Figs. 10.7 and 10.22). The simple, empirical particle in a box model would predict a resonance at

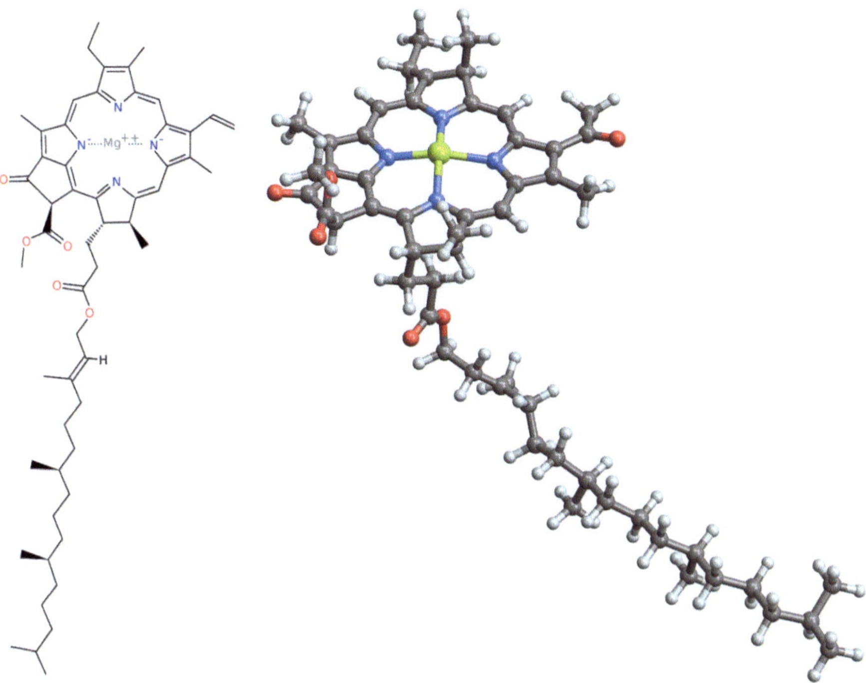

Fig. 10.21 The covalent structure of chlorophyll *a*. Instead of an iron atom in the center of the tetrapyrrole (chlorin) ring, as in the heme group of hemoglobin, chlorophyll has a magnesium atom, colored green at the center of blue quartet of nitrogens. The D_{4h} symmetry of the heme porphyrin is also broken in chlorophyll by the a-keto acid moiety. The long phytane tail of chlorophyll anchors the molecule in the lipid membrane of the plant chloroplast

570 nm, not too far from the observed wavelength of 640 nm. In photosynthesis, the process is initiated by the absorption of a red (~680 nm) photon in photosystem II, which promotes two electrons into the electron transport chain, providing energy for the chemiosmotic synthesis of ATP. These two electrons flow down the chain to photosystem I where the energy from the absorption of a ~700-nm photon is used to synthesize two molecules of NADPH (Fig. 10.23).

Chlorophyll has been the subject of extensive quantum mechanical calculations over the years (Alvarado-González et al. 2013; König and Neugebauer 2012). The wave functions for chlorophyll, as was the case with retinal, indicate (Fig. 10.24) that the primary optical resonance rests squarely with the nine conjugated π bonds on the tetrapyrrole ring. Early work by Gouterman (1961) modeled the absorption spectrum in terms of transitions among the four frontier states, HOMO-1, HOMO, LUMO, and LUMO+1, generating the Q and B bands. These four states transform according to the following irreducible representations of the point group D_{4h}: a pair a_{2u} and a_{1u},

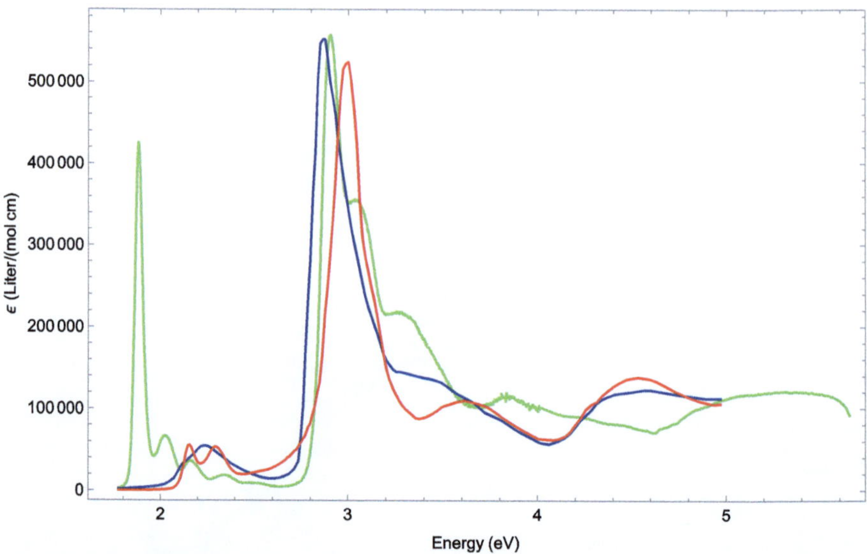

Fig. 10.22 Optical absorption spectrum of chlorophyll *a* in solution (green) showing the two main absorption resonances, *Q* at 1.94 eV (640 nm) and *B* at 2.89 eV (430 nm). Plants appear green because chlorophyll absorbs these red and blue photons and reflects green light. For comparison, this figure also shows the absorption of the heme group of hemoglobin in the deoxy (blue) and oxy (red) forms. The *Soret* band at ~3 eV is characteristic of the tetrapyrrole ring shared by both chromophores. Note that here we have plotted the absorption as a function of photon energy (eV) instead of wavelength. The cross section (extinction coefficient) for chlorophyll has here been multiplied by five to match that for hemoglobin

which are accidentally degenerate, and an unoccupied degenerate pair, e_{gx} and e_{gy}. In chlorophyll *a*, the substituents of the porphyrin ring break the D_{4h} symmetry, lifting the degeneracies so that the transitions separate into unique components commonly denoted *x* and *y*, according to the directions of their transition dipoles with respect to the plane of the tetrapyrrole ring (Fig. 10.25).

One can use the angular dependence of the absorption of polarized light to determine the orientation of the absorption dipole moment for oriented, solid samples. This has been used to find that the emission dipole moment of the tetrapyrrole ring of chlorophyll *b* is raised up at angle of ~49° relative to the plane of the thylakoid membrane surface in plant chloroplasts (Paillotin and Breton 1977; Okamura et al. 1995). Recent quantum mechanical calculations (Sirohiwal et al. 2020) indicate (Fig. 10.25) that the transition moment Q_x is oriented at 45.3°, which agrees with these earlier measurements.

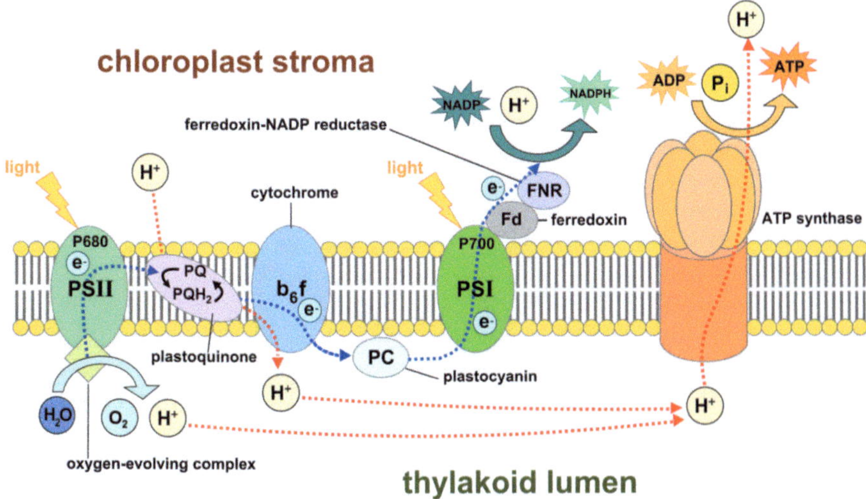

Fig. 10.23 Light-dependent reactions of photosynthesis in the thylakoid membrane of plant cells. PS I and PS II are the two photosystems that absorb red photons. (Wikipedia. This file is licensed under the Creative Commons Attribution-Share Alike 4.0 International license. Author: Somepics

Fig. 10.24 Frontier states of chlorophyll a upon which the Q and B band transitions are based after the Gouterman (1961) model of porphyrins. In parentheses are the symmetry labels of the basic porphyrin (D_{4h}) and chlorin (C_{2v}) point groups. The arrows indicate which pairs of states contribute greatest to the transitions. (From Sirohiwal et al. (2020). Creative Commons Attribution (CC-BY) License)

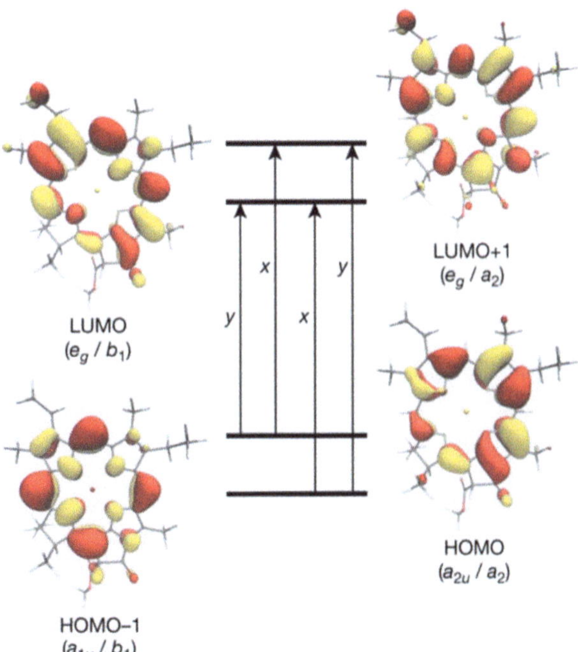

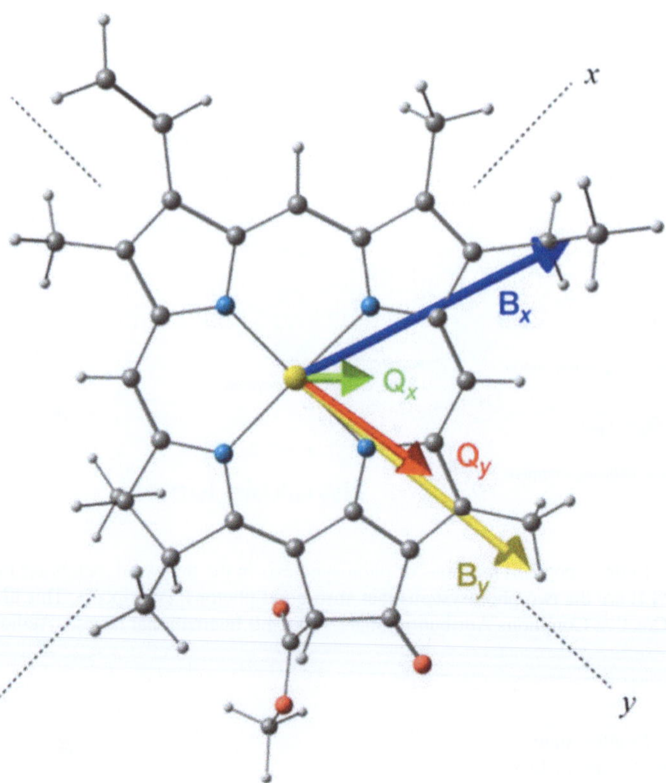

Fig. 10.25 The results of quantum mechanical calculations (Sirohiwal et al. 2020) of the transition dipole moments for the Q and B bands in chlorophyll a. The vector for Q_x forms an angle of 45.3° with respect to the y-axis. (Creative Commons Attribution (CC-BY) License)

Problems

1. *Protein extinction coefficient.* Predict the extinction coefficient at 280 nm of a protein which contains 1 tryptophan, 2 phenylalanines, and 2 tyrosines with a molecular weight of 10,000. Express the results both in units of liters cm^{-1} mol^{-1} and ml cm^{-1} mg^{-1}.
2. *Protein extinction coefficient.* By how much would the extinction coefficient change in Problem 1 if the protein did not contain any tryptophan?
3. *Fluorescence lifetime.* In a fluorescence lifetime measurement, the intensity of the fluorescence is measured following a short excitation pulse of light. A sample data set follows:

Fluorescence intensity	Time (ns)
1.1	0
0.49	5
0.25	15
0.05	20

(a) What is the fluorescence lifetime in this example?

(b) Assume that the fluorescence quantum yield is known to be 0.7. What is the intrinsic rate constant for fluorescence decay?

4. *Ligand binding.* An enzyme binds a ligand which leads to an increase in the extinction coefficient from 0 to 15,000 M^{-1} cm^{-1} at 280 nm. The extinction coefficient of the protein is 12,000 M^{-1} cm^{-1}. A solution of a 1×10^{-4} M protein and 2×10^{-3} M ligand has an absorbance in a 1-cm cell at 280 nm of 1.4.

(a) What is the concentration of the complex in the solution?

(b) What is the binding constant?

(c) When the ligand binds to the protein, the α-helical content of the protein is thought to change. How would you measure this?

5. *Molecular optical cross section.* For a chromophore, Cr, with an absorption maximum at a wavelength of 690 nm, the following data were recorded:

[Cr] Mole/L	A (690 nm)
2.00E-05	0.185
3.93E-05	0.366
5.78E-05	0.542
7.56E-05	0.72
9.28E-05	0.84
1.09E-04	0.998

(a) Make a plot of the absorbance data as a function of the chromophore concentration.

(b) What is the molar absorption coefficient, ε, (the slope of the line)?

(c) What is the effective area of the chromophore at 690 nm? That is, find the cross-section, σ, of the chromophore?

6. *Molecular optical cross section.* Find the Einstein B coefficient for the absorption of light by a molecule whose cross section varies with frequency as $\sigma(\nu) = \nu^3 e^{-\nu^2}$.

7. *Blackbody radiation.* Let us assume that the human body radiates heat as if it were a blackbody at a temperature of $T = 310$ K.

(a) At what frequency would you set a receiver in order to detect the maximum amount of energy from a human?

(b) Plot the blackbody radiation with increasing frequency.

(c) At what wavelength would you set a receiver in order to detect the maximum amount of energy from a human?

(d) Is the frequency found in the radio, infrared, visible, or ultraviolet region?

8. *Optical absorption by chlorophyll.* The human eye can detect light one million times fainter than sunlight. The attenuation of sunlight by seawater is due to the absorption by chlorophyll released by algae. The average concentration of chlorophyll in seawater is 1 nanomolar.

   (a) Given the absorption spectrum of chlorophyll below, at what ocean depth would human sight fail if the attenuation is solely due to the band at $\lambda = 428$ nm?

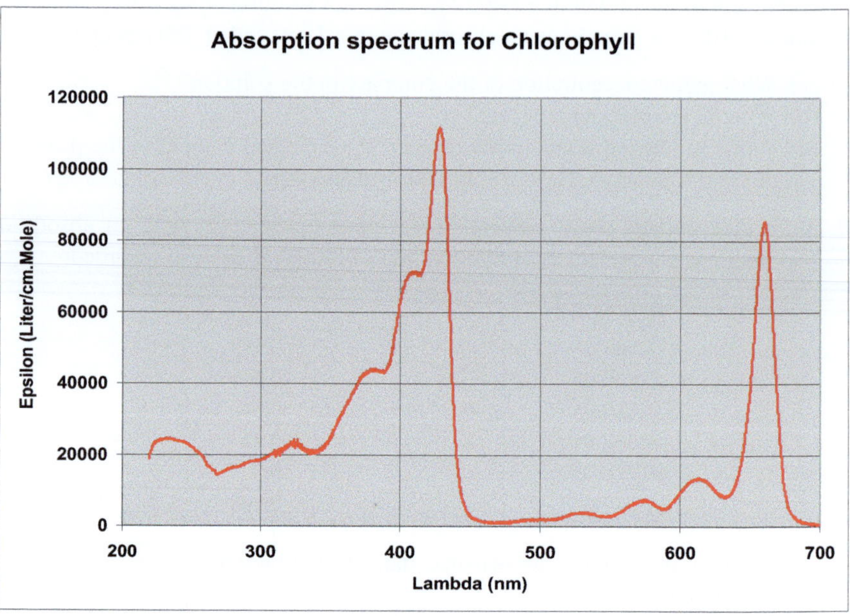

   (b) What does the energy-level diagram of chlorophyll look like if one just considers the strongest spectral lines? Assume that the ground state energy $E_o = 0$ Joules.

   (c) Find the cross section of chlorophyll at $\lambda = 428$ nm. You can estimate ε, from the y-axis on the graph. Estimate the size of the chlorophyll chromophore from the literature. Does your value for the cross section agree with your expectations about the size of the molecule; is the cross section larger or smaller than you expect? By how much?

9. *Species extinction and climate change.* In the benthic (deep) ocean, sunlight does not penetrate due to the absorption of light by chlorophyll from algae in the seawater. The production of algal chlorophyll is directly proportional to the concentration of CO_2 in the air. Due to man's activities, $[CO_2]$ increases each

year, as is shown by the data, on the right, from Mauna Loa, on Hawaii. The concentration of chlorophyll a (written here as [Chla]).

$$[\text{Chla (Moles/Liter)}] = K[\text{CO}_2 \text{ (ppm)}],$$

where $K = 1.36 \times 10^{-12}$ Moles/Liter.ppm.

A quadratic fit to the $[\text{CO}_2(t)]$ curve above gives that.

$$[\text{CO}_2(t)] = 1.2385 \times 10^{-2}\, t^2 - 47.7559\, t + 46{,}341.5.$$

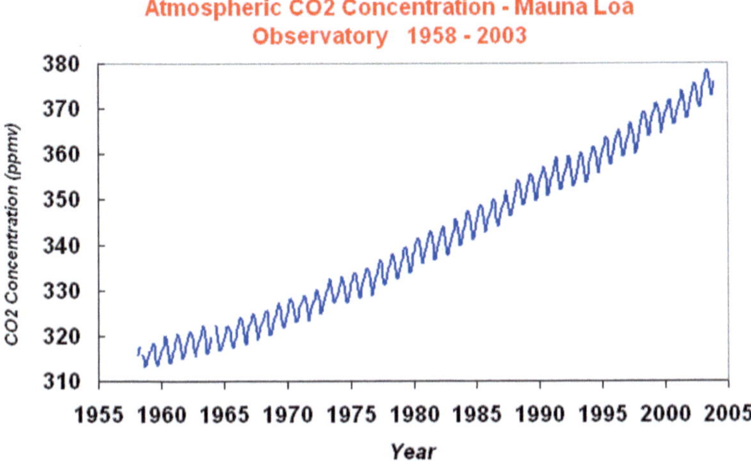

where t is the time in years. Now, benthic fishes live in total darkness, but they need to be able to see each other to mate. Without mating, the species will become extinct. The deep sea fish *Aristostomias* has solved the problem of finding opposite sex mating partners by using bioluminescence. The eyes of *Aristostomias* are not very good, so they can only detect each other's light if it is at least *10%* of its emitted intensity. However, the wavelength of emitted blue light just happens to be *428 nm*, right at the absorbance maximum of chlorophyll a. Currently, the density of *Aristostomias* in the Pacific Ocean is *one fish per million cubic meters*. They can still see each other because the extinction of light by chlorophyll a at a concentration of *0.5 nM* means that, on average, light at 428 nm travels 179 meters before it becomes too dim for another *Aristostomias* to see. If we assume that man's activities will continue to dump CO_2 into the atmosphere at a rate given above, when will *Aristostomias* become extinct? That is, when will the atmospheric CO_2 cause the absorption of light by seawater to rise to the point where the *Aristostomias* can no longer find each other in the dark?

References

M. Alvarado-González, N. Flores-Holguín, D. Glossman-Mitnik, Computational nanochemistry study of the molecular structure and properties of chlorophyll a. Int. J. Photoenergy (2013). https://doi.org/10.1155/2013/424620

L.H. Andersen, I.B. Nielsen, M.B. Kristensen, M.O.A. El Ghazaly, S. Haacke, M.B. Nielsen, M.Å. Petersen, Absorption of Schiff-base retinal chromophores in Vacuo. J. Am. Chem. Soc. **127**, 12347–12350 (2005)

N.J. Bale, D.Y. Sorokin, E.C. Hopmans, M. Koenen, W.I.C. Rijpstra, L. Villanueva, H. Wienk, S. Jaap, Sinninghe Damsté, new insights into the polar lipid composition of extremely halo (alkali)philic Euryarchaea from Hypersaline Lakes. Front. Microbiol. **10**, Article 377 (2019)

S. Braun-Sand, P.K. Sharma, Z.T. Chu, A.V. Pisliakov, et al., The energetics of the primary proton transfer in bacteriorhodopsin revisited: It is a sequential light-induced charge separation after all. Biochim. Biophys. Acta **1777**, 441–452 (2008)

A.M. Chase, C. Haig, The absorption spectrum of visual purple. J. Gen. Physiol. **21**, 411–430 (1938)

S. Das Sarma, E.W. Schwieterman, Early evolution of purple retinal pigments on Earth and implications for exoplanet biosignatures. Int. J. Astrobiol. **20**, 241 (2018)

R.P. Feynman, Space-time approach to non-relativistic quantum mechanics. Rev. Mod. Phys. **20**, 367–387 (1948)

K.J. Fujimoto, Electronic couplings and electrostatic interactions behind the light absorption of retinal proteins. Front. Mol. Biosci. **8**, Article 752700 (2021)

M. Garavelli, P. Celani, F. Bernardi, M.A. Robb, M. Olivucci, The C5H6NH2+ protonated Schiff base: An ab initio minimal model for retinal photoisomerization. J. Am. Chem. Soc. **119**, 6891–6901 (1997)

M. Gouterman, Spectra of porphyrins. J. Mol. Spectrosc. **6**, 138–163 (1961)

J.-y. Hasegawa, K.J. Fujimoto, T. Kawatsu, A configuration interaction picture for a molecular environment using localized molecular orbitals: The excited states of retinal proteins. J. Chem. Theory Comput. **8**, 4452–4461 (2012)

N. Hasegawa, H. Jonotsuka, K. Miki, K. Takeda, X-ray structure analysis of bacteriorhodopsin at 1.3 Å resolution. Nat. Sci. Rep. **8**, 13123 (2018)

C. König, J. Neugebauer, Quantum chemical description of absorption properties and excited-state processes in photosynthetic systems. Chem. Phys. Chem. **13**, 386–425 (2012)

W. Kühne, Chemische Vorgänge in der Netzhaut. in Hermann, L. Handbuch der Physiologie, Leipzig, F.C.W.Vogel **3**, 325 (1879)

J.K. Lanyi, Proton transfers in the bacteriorhodopsin photocycle. Biochim. Biophys. Acta **1757**, 1012–1018 (2006)

S.W. Lin, G.G. Kochendoerfer, K.S. Carroll, D. Wang, R.A. Mathies, T.P. Sakmar, Mechanisms of spectral tuning in blue cone visual pigments: Visible and raman spectroscopy of blue-shifted rhodopsin mutants. J. Biol. Chem. **273**, 24583–24591 (1998)

R. Neutze, E. Pebay-Peyroula, K. Edman, A. Royant, et al., Bacteriorhodopsin: A high-resolution structural view of vectorial proton transport. Biochim. Biophys. Acta **1565**, 144–167 (2002)

E. Okamura, T. Hasegawa, J. Umemura, Quantitative analysis of molecular orientation in chlorophyll a Langmuir monolayer: A polarized visible reflection spectroscopic study. Biophys. J. **69**, 1142–1147 (1995)

G. Paillotin, J. Breton, Orientation of chlorophylls within chloroplasts as shown by optical and electrochromic properties of photosynthetic membrane. Biophys. J. **18**, 63 (1977)

A.P. Rasmussen, E. Gruber, R. Teiwes, M. Sheves, L.H. Andersen, Spectroscopy and photoisomerization of protonated Schiff-base retinal derivatives in vacuo. Phys. Chem. Chem. Phys. **23**, 7227 (2012)

S. Sekharan, O. Weingart, V. Buss, Ground and excited states of retinal Schiff base chromophores by multiconfigurational perturbation theory. Biophys. J. **91**(1), L07–L09 (2006)

A. Sirohiwal, R. Berraud-Pache, F. Neese, R. Izsák, D.A. Pantazis, Accurate computation of the absorption spectrum of chlorophyll a with pair natural orbital coupled cluster methods. J. Phys. Chem. B **124**, 8761–8771 (2020)

C. Vinegoni, P.F. Feruglio, I. Gryczynski, R. Mazitschek, R. Weissleder, Fluorescence anisotropy imaging in drug discovery. Adv. Drug Deliv. Rev. **151**, 262–288 (2019)

M. Zhong, R. Kawaguchi, M. Kassai, H. Sun, Retina, retinol, retinal and the natural history of vitamin A as a light sensor. Nutrients **4**, 2069–2096 (2012)

X. Zhou, D. Sundholm, T.A. Wesołowski, V.R.I. Kaila, Spectral tuning of rhodopsin and visual cone pigments. J. Am. Chem. Soc. **136**, 2723–2726 (2014)

Chapter 11
The Physics of Water

> *"Throughout all the natural sciences, no single other substance approaches water in prominence and indispensibility."*
>
> F. H. Stillinger (1975)

> *"Water is H_2O, hydrogen two parts, oxygen one, but there is also a third thing, that makes it water and nobody knows what it is."*
>
> D. H. Lawrence (1885–1930)

> *"Water, that silvery cloud of atoms that defines us all."*
>
> Laurel O. Sillerud (2024)

An interesting demonstration of the importance of water in life comes from counting the number of papers on PubMed with water and life in their titles: ~86,000 as of 2023—far more than for any other single substance. Even with this intensive research effort, the above quotation from D. H. Lawrence is still very much *a propos*. To this day, no complete theory exists that can account for all of the many unique and important features that make the detailed properties of water yet another organizing principle with respect to the origin of life. At last count, there are 76 properties of water (https://water.lsbu.ac.uk/water/martin_chaplin.html) that make it unique, which differentiate it from other several light molecules comprised of atoms in the lowest nuclear mass portion of the periodic table. Water, with a molecular weight of 18, is one of the lightest of all molecules, but it has several properties not shared by many other light molecules (Shi et al. 2018).

The unique properties of water include its polarity, high heat capacity, density, and ionizability, i.e., the energy required to alter its charge by adding or removing electron(s). Another surprising property of water is that due to strong hydrogen bonding among water molecules, during the phase transition involved in the melting of ice, the density of its solid form (ice) is less than that of the liquid form; ice floats on water. No other common fluid exists that has a density maximum (Fig. 11.1) above its melting point (Cho et al. 1996; Nilsson & Pettersson 2015; Pettersson et al. 2016) and there are in total only a handful of other substances with this property. The volume occupied by 1 mole of liquid water at 0 °C is 18.02 cm^3, while this same number of water molecules expands to 19.66 cm^3 upon freezing, an expansion of

L. O. Sillerud, *Abiogenesis*, https://doi.org/10.1007/978-3-031-56687-5_11

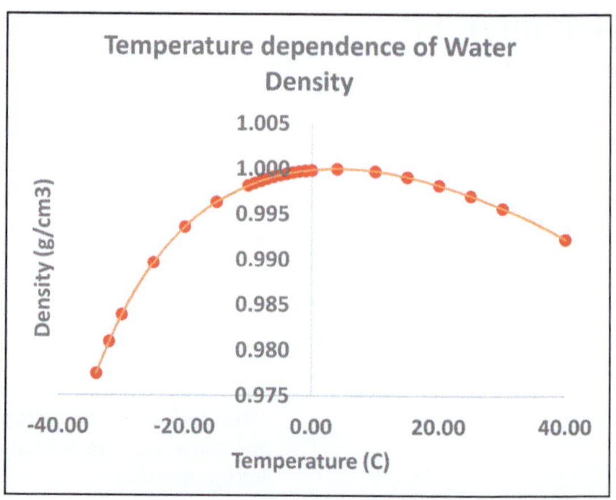

Fig. 11.1 The density of water as a function of temperature near the density maximum of 3.98 °C at 1 atmosphere pressure. (Data drawn from Vedamuthu et al. (1994))

9.1%. This is crucial, as pointed out so eloquently by Kurt Vonnegut in *Cat's Cradle*, because ice denser than ocean water would sink to the ocean floor in Earth's polar regions and the oceans might have frozen from the bottom up preventing the formation of all ocean life. As ice melts at 0 °C, water contracts reaching a density maximum at 3.98 °C at 1 atmosphere pressure (Fig. 11.1).

Given water's importance in the origin of life (Franks 2000; Ball 2017), the search for extraterrestrial life is intimately associated with the search for water in the solar system and on the myriad exoplanets discovered to date. Saturn's moon Enceladus and Jupiter's moon Europa are covered by thick layers of ice that overlie oceans of salty liquid water. The Cassini spacecraft imaged plumes of water erupting as geysers from the surface of Enceladus in 2015. Mass spectroscopic investigations of these plumes (Postberg et al. 2023) also detected phosphorus, one of the lowest abundance elements that is nevertheless essential for life. Thousands of exoplanets have been discovered in the last 25 years. It is estimated that 5–10% of the exoplanets smaller than 1.3 Earth radii that orbit M dwarf stars have sufficient quantities of seawater to enable life to form (Kimura and Ikoma 2022). Recent transit timing and spectroscopic observations of the exoplanet system surrounding the star Kepler-138 (Piaulet et al. 2023) have found that two of the five exoplanets (Kepler-138c&d) have radii 1.5 times larger than Earth, with masses that are only 2.1–2.3 times the mass of the Earth, instead of the ~3.4-fold larger mass expected purely on the basis of size. The inference is that these exoplanets are ~11% water and are covered by a 2000-km-deep ocean. Water has also been detected in the terrestrial planet-forming zone of the protoplanetary disk of the star PDS70 (Perotti et al. 2023).

11.1   What Is Water?

The first attempts to answer this question were made in the 1780s by Henry Cavendish and Antoine Lavoisier who showed that water was made up of hydrogen (ύδρο-γενής; *hydro genes; water producer*) and oxygen (ὀξύς-γενής; *oxys genes;*

acid producer). The formation of water from hydrogen and oxygen is at the foundation of life. The most abundant *molecules* in the universe are H_2 and H_2O. The water molecule self-assembled from hydrogen and oxygen, two primordial constituents of the universe that have very special properties.

The ordinary water molecule ($^1H_2{}^{16}O$) is the stable bound state of (a) two ordinary hydrogen atoms (^1H), each containing a nucleus of a single a spin ½ proton as well as a single spin ½ electron, and (b) an ordinary oxygen atom (^{16}O) containing a nucleus with eight spin ½ protons and eight spin ½ neutrons, as well as eight spin ½ electrons. In the molecule, the ten negatively charged electrons exactly balance the positive charges from the ten protons. The nuclei form the stable scaffold which the electrons decorate giving rise to the nonspherical shape of the molecule.

What we mean by water is ordinarily thought of as $^1H_2{}^{16}O$, but there are more than a hundred isotopomers that were formed from the many known isotopes of hydrogen (^1H, ^2H, ^3H) and oxygen ($^{11}O \ldots {}^{26}O$) derived from the complexities of Big Bang nucleosynthesis and post-Big Bang stellar nucleosynthesis. Since most of these isotopomers involved combinations of unstable (radioactive) nuclei, the resulting mixture of isotopomers present as water today is comprised of their resulting stable decay products, including a small percentage of *heavy* water, $^2H_2{}^{16}O$, or deuterium oxide, and water containing correspondingly small amounts of stable ^{17}O or ^{18}O. The natural abundance of deuterium in the Earth's ocean is 0.015%, while the stable oxygen isotopes are present at ^{16}O (99.8%), ^{17}O (0.0380%), and ^{16}O (0.205%).

11.2   The Production of Hydrogen in the Big Bang

Here, we are going to review several important details of nucleosynthesis, portions of which we have already covered in Chap. 6. Hydrogen is special because it is so abundant in the universe (Table 11.1) and it also readily forms covalent compounds with a myriad of other atoms through the sharing of its single electron. Hydrogen is the simplest atom with the simplest nucleus. It is therefore unsurprising that it was

Table 11.1 Abundance of elements in the Milky Way

Z	Element	Abundance (atom %)
1	**Hydrogen**	73.900
2	Helium	24.000
8	**Oxygen**	1.040
6	**Carbon**	0.460
10	Neon	0.134
26	**Iron**	0.109
7	**Nitrogen**	0.096
14	Silicon	0.065
12	**Magnesium**	0.058
16	**Sulfur**	0.044

generated in the largest amounts by the Big Bang. Its formation from the primordial quark-gluon plasma is well-understood. After the Big Bang, as the temperature of the universe decreased below the Hagedorn temperature, the long-lived bound states of quarks, namely, protons and neutrons, were able to form. The two hydrogen atoms in water were generated during the β-decay of the heavier neutrons into lighter protons, electrons, and (anti)neutrinos,

$$n \rightarrow p + e^- + \overline{\nu_e},$$

with a half-life of $t_{1/2} \sim 880$ s (Nico et al. 2013; Gonzalez et al. 2021).

Nota bene, there is confusion in the literature, even by professional scientists who should be expected to know better, between the half-life ($\sim$880 s) and the mean-life of 610 s, of the neutron. Both concepts are conflated into the "lifetime" of the decaying state. To clear up this confusion, radioactive decay is characterized by the half-life, t_{h}, which is the amount of time required for the number of nuclei in a sample at time zero, $N(0)$, to decay to one-half of its original value according to the equation

$$N(t) = N(0)e^{-\frac{t}{t_{\mathrm{h}}}}$$

so that

$$\frac{N(t)}{N(0)} = e^{-\frac{t}{t_{\mathrm{h}}}} = 1/2$$

or

$$Ln\left(e^{-\frac{t}{t_{\mathrm{h}}}}\right) = Ln\left(\frac{1}{2}\right)$$

which gives the relation between the half-life and the mean life, t_{m}

$$t_{\mathrm{m}} = -t_{\mathrm{h}}Ln\left(\frac{1}{2}\right) = 0.693t_{\mathrm{h}},$$

and we observe that the mean life of $\sim$610 s is indeed 0.693 × 880 s. Free neutrons generated in the Big Bang underwent β-decay to protons in what was essentially the blink of an eye compared with the age of the universe.

As the temperature (energy) of the universe dropped, after $\sim$380,000 years, the electrons and protons generated in the β-decay of neutrons combined, self-assembling into the bound state of the neutral hydrogen atom (Chap. 6, Fig. 6.3). Hydrogen formed $\sim$74% of the universe at this time; the other $\sim$24% was helium (Table 11.1). The single protons in the hydrogen atoms have spin ½, and hence magnetic moments, and can therefore be detected using dipolar nuclear magnetic resonance spectroscopy (Chap 7). The single electrons on each proton are in the 1s quantum state at ~290 K. Of the ten most abundant elements in the Milky Way galaxy, seven (70%) are light elements necessary for the development of life (Table 11.1 in bold).

11.3 Oxygen Production Through Stellar Nucleosynthesis

The formation of oxygen was delayed until the primordial hydrogen clouds underwent turbulent gravitational collapse forming the first stars. Hydrogen was assembled immediately in the Big Bang, while oxygen and all the elements heavier than boron were built up in the thermonuclear reactions occurring in the centers of the early stars. Oxygen nuclei arose, beginning ~100 million years after the Big Bang, from stellar nucleosynthesis in the cores of heavy stars (whose masses were several orders of magnitude larger than that of our present-day Sun), and was subsequently dispersed into the interstellar medium when these massive stars exploded as supernovae. The next generations of stars were then formed from the interstellar medium to which oxygen had been added to the vast amounts of existing Big Bang hydrogen.

Stars in the universe produced the oxygen through the CNO cycle (Chap. 6). The present abundance (~1.04%) of oxygen (^{16}O) in the universe (Table 11.1) makes it the third most common nuclide after only hydrogen and helium, even though oxygen was not directly produced in the Big Bang. We will shortly provide a rationale for why oxygen was so abundantly formed. Note that carbon (^{12}C), also produced in stars in the CNO cycle, is the next most abundant element. As already indicated, carbon, oxygen, and nitrogen, along with hydrogen, are among the essential elements needed for the prebiotic assembly of primordial biomolecules (Table 11.1 in bold).

Why is oxygen so abundant in the universe? The spins of the eight spin ½ protons in the oxygen nucleus pair according to the spin-statistics theorem into four spin-up/spin-down states with total spin zero, as do the eight spin ½ neutrons, resulting in spin zero states for both proton and neutron sectors, leading to an overall spin zero nucleus. According to the shell model of nuclei, two protons fill the s shell, and six protons fill the d shell, and a similar shell occupancy occurs in the neutron sector. With *both* the proton and the neutron shells completely filled, this produces the *doubly magic* ^{16}O nucleus, which has a larger binding energy per nucleon (^{16}O, 7.98 MeV) than its lighter (^{15}O, 7.46 MeV, unstable) or heavier but stable (^{17}O, 7.75 MeV; ^{18}O, 7.767 MeV) nuclear neighbors. The ^{16}O nucleus is particularly stable against nuclear decay, giving it a stability, and hence production, advantage in the CNO cycle in stars. Since the ^{16}O nucleus is stable, without any decay products, its formation in stars is the terminus of a decay chain that begins with heavier, unstable isotopes (Fig. 11.2) that eventually lead to the doubly magic and hence stable ^{16}O. Similar production schemes give rise to small amounts of the stable ^{17}O and ^{18}O isotopes through analogous decay processes (see https://periodictable.com/Isotopes).

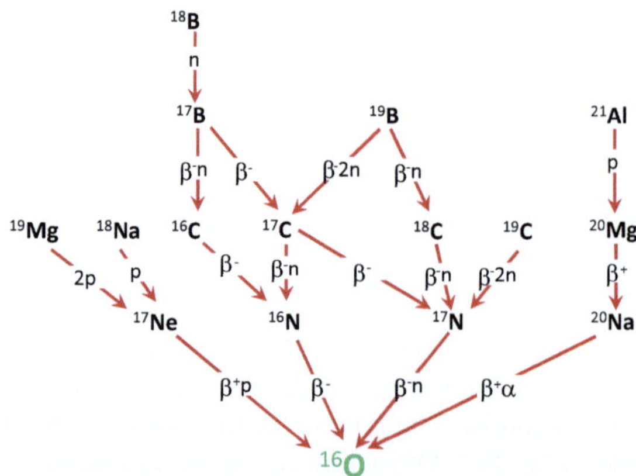

Fig. 11.2 The nuclear decay chain that leads to the formation of the stable ^{16}O nucleus. Here, n is a neutron, p a proton, α an alpha particle, and β an electron (positron). Note that most of these reactions involve the weak force

Incidentally, the other nuclide produced in great abundance in the Big Bang was helium, but the helium nucleus is another *doubly magic* nuclide, with both protons and both neutrons spin paired in closed *s* shells, giving it such great stability that this nucleus is often emitted as the α particle, a discrete unit in radioactive decays of nuclei. The same is true for the electrons in helium which also form a closed $1s^2$ atomic shell with spin paired electrons. Helium, a noble gas, is so chemically unreactive that it cannot serve as the basis for life, although it has recently been reported that helium can actually form exotic stable compounds with sodium under extreme pressure (Liu et al. 2018). These compounds do not result from electron sharing as in all other compounds. Rather, the helium atom appears to simply shield the sodium ions from each other, lowering their Coulomb repulsion.

We now see that water is composed of nuclei in the universe that have special physical properties; hydrogen was generated in massive quantities during the initial phase of the Big Bang, while oxygen survived to the present day due to the quantum mechanics of its closed, doubly magic nuclear shell model states. In addition, water as a substance also has many unique physical properties that imbue it with the special characteristics needed to act as the matrix for life (Fig. 11.3).

11.4 The Unique Properties of Water

One of these special characteristics is that water molecules in the liquid state engage in an extensive intermolecular hydrogen bonding network (Fig. 11.4) whose interaction energy of 0.21–0.34 eV (~10 to 16 kJ/mol) means that relatively stable water dimers and oligomers are formed in liquid water because this energy is 8–13 times kT at 300 K and, by resort once again to the Boltzmann probability distribution, essentially all water molecules participate in this hydrogen bonding network; the number of unbound water molecules is on the order of only 0.01%.

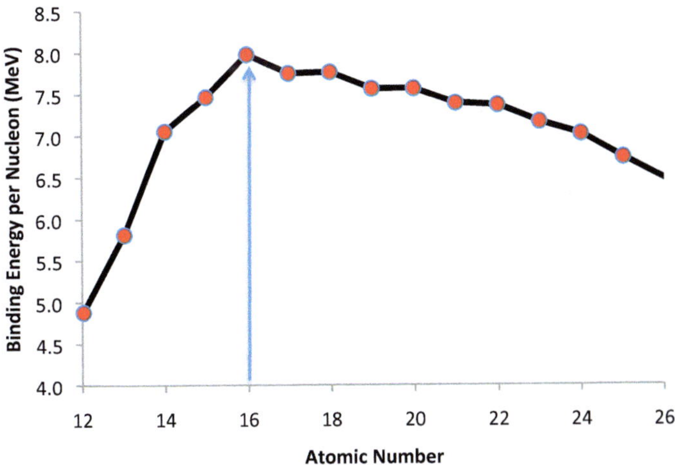

Fig. 11.3 The binding energy per nucleon of the isotopes of oxygen. Note the peak at the doubly-magic nucleus ^{16}O (arrow)

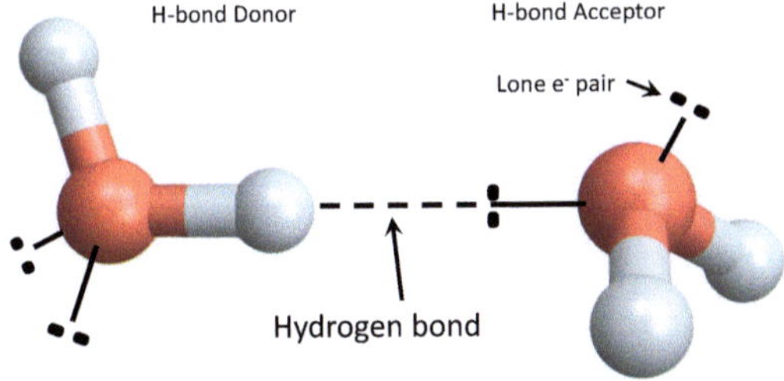

Fig. 11.4 A schematic of the hydrogen bond in a pair of water molecules. One of the four lone electron pairs is indicated. The hydrogen bond is oriented from the hydrogen of the donor molecule to one of the lone pairs of the acceptor water molecule. The nominal hydrogen bond O–H distance is 188 pm, while the distance between oxygens is 282 pm. Note that both of these distances vary as a function of temperature and environment

11.5   The Water Dimer: Hydrogen Bonding Energy

The hydrogen bond is primarily electrostatic in nature. Both the dipole moment and the hydrogen bonds of water are due to the fact that even though the water molecule is neutral (at pH 7) the electronic charge is not uniformly distributed over the molecular surface. This separation of charge renders the hydrogen atoms slightly positive as the $Z = +8$ oxygen nucleus withdraws a portion of electron density from the $Z = +1$ hydrogen nuclei and gives a partial negative charge to the oxygen (Fig. 11.19). Several physical properties of water indicate that hydrogen bonding in water is very important (Popkie et al. 1973). It is also known that, in addition to frozen water (hexagonal ice I_h) at 1 atmosphere pressure, at higher pressures, ice exists in several polymorphic phases (II, III, V, VI, VII, VIII, and IX) which indicates that water's intermolecular interactions are more complex than expected from simple spherical structures.

The large heat capacity of water (75.38 J/(mol K)) is another reflection of the strong hydrogen bonds formed between water molecules. Relatively large amounts of energy with respect to kT are needed in order to disrupt this hydrogen bonding network, with the result that water has the highest heat capacity of all solids and liquids except NH_3 (Table 11.2); its heat of fusion, 334 J/g (the amount of energy needed to melt ice at 0 °C), is also only surpassed by that for ammonia. Water's melting (273 K) and boiling points (373 K) are much higher than other molecules with similar molecular weights (Table 11.3). Its heat of evaporation, 40.65 kJ/mol, is

Table 11.2 Heat capacities for common substances

Substance	Specific heat capacity $C_{p,s}$ (J/g °C)	Molar heat capacity $C_{p,m}$ (J/mol °C)
Air	1.01	29.19
Aluminum	0.89	24.20
Argon	0.52	20.79
Copper	0.39	24.47
Graphite	0.71	8.53
Helium	5.19	20.79
Iron	0.45	25.09
Lead	0.13	26.40
Lithium	3.58	24.80
Mercury	0.14	27.98
Methanol	2.14	68.62
Sodium	1.23	28.23
Ammonia (liquid)	4.70	80.80
Titanium	0.52	26.06
Water (ice, 0°C)	**2.09**	**37.66**
Water (liquid)	**4.18**	**75.38**

Specific (C_s) and molar (C_m) heat capacities at 1 atm pressure and 25 °C

Table 11.3 Thermal properties of light molecules compared with water

Molecule	Molecular weight (g/mol)	Melting point (K)	Boiling point (K)
CH_4	16	91	112
NH_3	17	195	240
H_2O	**18**	**273**	**373**
HF	20	189	293
HCN	27	260	299
N_2	28	63	77
H_2S	35	191	213
CO_2	44	195 (sublimes)	195

the highest of all substances. Water's thermal conductivity, 598 mJ/(s K m) at 20 °C, and its surface tension (0.072 J/m^2 at 20 °C) are the highest of all liquids. Note that the heat capacity of ice is almost exactly half that of liquid water so that there is large a discontinuity in the heat capacity as a function of temperature. Therefore, liquid water is entropically (probabilistically) twice as favorable as the solid state.

11.6   The Water Molecule Is Not a Sphere

The isocontours of the 3D electronic probability distribution define a shape of the water molecule (Fig. 11.5) that is tetrahedral, more like a wedge (actually, a; Fig. 11.6) than a sphere, so that its packing in the fluid is not the hexagonal close-packed structure typically found for spherical substances, such as argon, but rather that of a lattice with the water molecules bound to each other through the mutual embrace of shared hydrogen bonds transiently directed along asymmetric bond vectors as a result of the separation of charge.

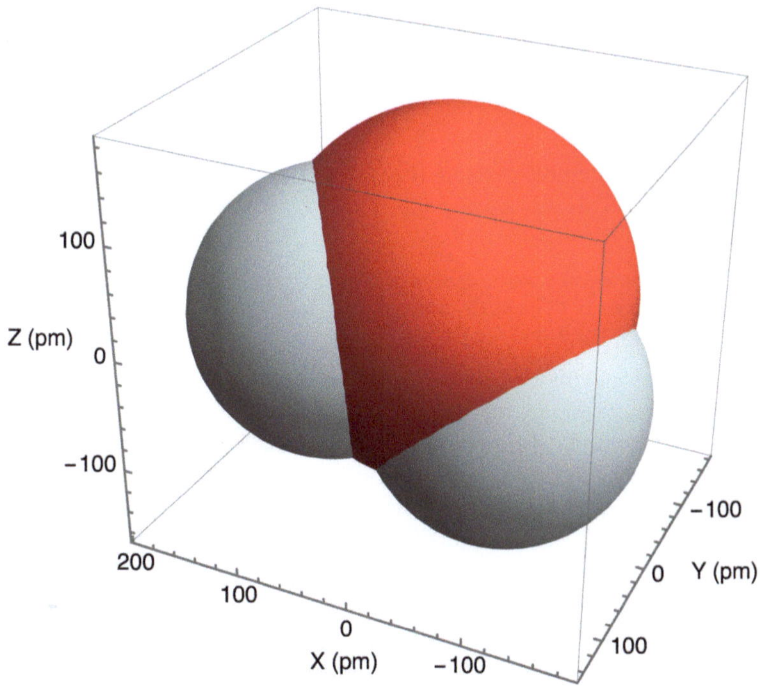

Fig. 11.5 A space-filling model of the water molecule. The two hydrogens are white, while the single oxygen atom is red. The contours of the electron density show that water is not a spherical molecule

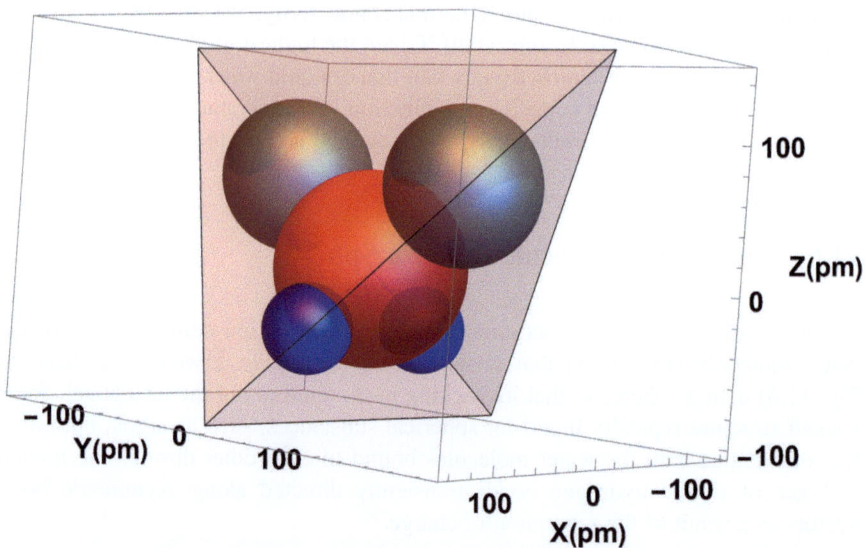

Fig. 11.6 The approximate tetrahedral structure of the electronic distribution surrounding the water molecule, showing the hydrogens as sliver spheres, the oxygen in red, and the lone pair electrons from the oxygen atom in blue

Water has a large dipole moment (Fig. 11.19) and is a very polar (i.e., an easily polarized) solvent. This was already appreciated by Bernal and Fowler in 1933, who wrote that

> ...the unique position of water is due not only to its dipole character but even more to the geometrical structure of its molecule, which is the simplest one that can form extended four-coordinated networks. (Bernal and Fowler 1933)

We can understand these properties of water by examining the electronic structure that results from quantum mechanics. In the water molecule, the electrons in hydrogen are each in the 1s state. Since the ionization potential of oxygen (in the gas phase) is rather high (12.62 eV), it is safe to assume that the eight electrons of oxygen are in a ground state configuration with two in the almost spherical ($1s^2$) state and six in the $2s^2 : 2p^4$ configuration. The two s electrons from the hydrogen atoms pair with two p electrons in oxygen leaving two lone pairs (Fig. 11.5) of electrons in p-states in an approximately tetrahedral structure (Fig. 11.6).

To form the molecular quantum states of water, we need to use the fact that only those wave functions that serve as a basis for the point group formed by the set of transformations of the atoms that leaves the molecular skeleton invariant can combine to make the molecular wave functions. In this manner, we will find how to combine the atomic wave functions of hydrogen and oxygen to form water's molecular wave functions. We will once again use group theory, the mathematical theory of symmetry. However, in this case, the symmetry is not that of spacetime itself but that of point groups, which we will use to pick out the symmetry-adapted linear combinations of atomic wave functions that can contribute to the molecular wave functions.

We begin by revisiting group theory, but this time instead of groups composed of continuous symmetry operators, we need to examine that of point groups, where the transformations are all discrete. Then we will pick bases for the hydrogen and oxygen atoms and show how the transformation properties of the hydrogen and oxygen wave functions restrict the possible linear combinations. These will lead to the *irreducible representations* into which we will insert electrons according to the Pauli principle. While the next sections will delve into the basics of group theory rather extensively, please keep in mind that this is necessary in order to explicate the foundation for the nomenclature used in the spectroscopy literature to label the various quantum states of the water molecule. The effort spent learning this will pay off later as enhanced understanding.

11.7   Group Theory of the Water Molecule Quantum States

While we have earlier dipped into group theory with our exposition of the Lorentz and Poincare groups, these latter are continuous groups, which depend on a parameter that can take on any value. We found that the continuous symmetries of spacetime served as an organizing principle that constrained the types of objects (fermions and bosons) that can exist within it. In a similar fashion, we will see that the discrete symmetries of molecules also limit the ways that atomic wave functions can be combined to produce the wave functions of molecules.

We have previously defined a symmetry as an operation on a set of objects that leaves the members of the set invariant; we could present the set to an observer, ask the observer to leave the room, perform the symmetry operation, and invite the observer back to view the set again. If the operation we performed was a symmetry operation, the observer would say that the two sets were identical, i.e., that no transformation had occurred. In describing molecular wave functions, symmetry operations are performed by geometric operators that leave the spatial relationships among the atoms invariant.

A given molecule is defined by the discrete positions (approximated as points) in space of the heavy atomic nuclei that define the molecular backbone. The sets of transformation operators (excluding translations) that leave this geometric collection of nuclear points invariant form *point groups*. If one expands the set of operators to include translation operators, then the resulting set of symmetry operators form *space groups* which are important in crystallography but will be left untouched here.

While we have already discoursed upon group theory in Chap. 3, it doesn't hurt to remind the reader of the basic axioms that define the subject and how these are separately defined for *point groups*. Remember that group theory is a branch of mathematics devoted to the study of sets of objects that obey the following collection of simple group axioms:

11.7.1   Group

A group G is a *set* of group elements $g_i \in G$ together with a *rule* for the combination of elements, often referred to as "multiplication" and here represented by the symbol ($\cdot$), but this rule can be more general than ordinary multiplication. For point groups, the elements of the group are a set of geometric symmetry operators (matrices) that transform the molecules (matrices of position vectors) into themselves, and the *rule* ($\cdot$) is matrix multiplication.

11.7.2   Closure

The combination of two group elements in G according to the *rule* is also a member of the group G. For example,

$$\forall \{g_i, g_j\} \in G, \text{then } g_i \cdot g_j = g_k \in G.$$

For point groups, this means we can combine two geometric symmetry operators to produce a third operator, which is also a geometric symmetry operator.

11.7.3   Associativity

The rule of combination is associative:

$$\forall \{g_i, g_j, g_k\} \in G, \text{then } g_i \cdot (g_j \cdot g_k) = (g_i \cdot g_j) \cdot g_k.$$

For point groups, this means we can combine three geometric symmetry operators by combining the second and third and then combining the result with the first, and the result is the same as combining the first and second and then combining the result with the third.

11.7.4   Identity

The group G contains the identity element, $e \in G$, such that when combined with any element $g_i \in G$ one recovers that element unchanged:

$$\forall g_i \in G, \qquad e \cdot g_i = g_i \cdot e = g_i.$$

The use of e for the identity comes from the German *einheit* (unity). For point groups, this means that there must exist a geometric symmetry operator that leaves the molecular framework unchanged.

11.7.5   Inverse

The group G contains the inverse element $g_i^{-1} \in G \ni$.

$$\forall g_i \in G, \exists\, g_i^{-1} \ni$$
$$g_i^{-1} \cdot g_i = g_i \cdot g_i^{-1} = e.$$

i.e., the inverse g_i^{-1} of an element $g_i \in G$ combined with the element itself produces the identity element. For point groups, this means that there must exist a geometric symmetry operator that reverses the effect of another geometric symmetry operator, so that when both are applied to a molecule the result is the same as doing nothing to the molecule.

11.7.6   The Commutator

The commutator $[g_i, g_j]$ of two elements $\{g_i, g_j\} \in G$ is defined as

$$[g_i, g_j] = g_i \cdot g_j - g_j \cdot g_i.$$

11.7.7   Groups Are Abelian

If $\forall\, \{g_i, g_j\} \in G$,

$$[g_i, g_j] = g_i \cdot g_j - g_j \cdot g_i = 0,$$

or

$$g_i \cdot g_j = g_j \cdot g_i$$

then the elements are said to commute, and the group is called *Abelian* (after nineteenth-century Norwegian mathematician Niels Henrik Abel). Not all groups are *Abelian*; many groups contain elements whose combination depends on the order in which they are combined. The rotation group SO(3) is an example of a *non-Abelian* group.

11.8   Point Group Symmetry Operators

The sets of operators that transform the molecular framework into itself must obey certain properties. For molecules placed in 3-space such that the center of symmetry of the molecule is centered at the origin of the coordinate system, the symmetry operators must preserve bond angles and lengths, which means that these operators must be unitary and orthogonal, familiar concepts from our previous study of the rotation group (Chap. 3).

Orthogonality of an operator M means that

$$M\,M^{\mathrm{T}} = E,$$

where M^{T} is the transpose of M and E is the identity operator.

The *unitarity* of M means that

$$M\,M^{\dagger} = E,$$

where $M^{\dagger}$ is the complex conjugate transpose (adjoint) of M and E is the identity operator. There are only five kinds of operators that can satisfy these constraints, and they therefore fall into the following five standard classes which we will later illustrate for the water molecule. These operators are as follows:

1. **E**: The identity operator. All point groups must contain the identity operation consisting of no transformation at all. All molecules are invariant under the identity operator.
2. **C$_n$**: Rotations about an axis that leave the skeleton unchanged. A rotation about an axis through an angle $360°/n$ is given the symbol **C$_n$**. A water molecule can be rotated by $180°$ about an axis passing through the oxygen atom and the perpendicular bisector of the line joining the two hydrogen atoms, in which case $n = 2$ (Fig. 11.8). The operator for **C$_1$** is **E**. A molecule may be invariant under rotations around several axes. The axis with the largest value of n is called the principal axis, and the coordinate system is conventionally centered at the intersection of these several axes and is oriented with the z-axis along the axis with the largest value of n.
3. **σ**: Reflection in a mirror plane. This plane may either contain or be perpendicular to the principal axis; when the principal axis contains this plane, it is called a vertical plane, and the operator is written **σ$_v$**. When oriented as in (2), the water molecule is invariant under reflections in two distinct vertical planes. These are reflections in the x–z plane, **σ$_v$** (xz), and reflection in the y–z plane, **σ$_v$′** (yz) (Fig. 11.8).
4. **I**: Inversion through a center of symmetry. This operator is also the parity operator, and it transforms the coordinates into their negatives:

$$\mathbf{I}\{x, y, z\} = \{-x, -y, -z\}.$$

Water does not possess inversion symmetry.

5. **S_n**: An *n*-fold improper rotation about a **C_n** axis. An improper rotation is a rotation followed by a reflection in a plane perpendicular to the rotation axis. Water does not remain invariant under any improper rotations.

We will now find the set of symmetry operators that describe water and show that because this set obeys the group axioms it forms a group. To begin, we need to define a *basis* upon which the symmetry operators act. The basis will be a set of vectors that connect nuclei of the oxygen atom and the two hydrogens to the origin of the coordinate system. Finding the set of symmetry operators that leave this basis invariant is much like going fishing. The basis is the bait which we cast into the finite sea of all possible symmetry operators, and the fish we catch are the operators that leave the basis invariant. We will therefore obtain a basis for the *irreducible representations* of the point group, more about this later.

11.8.1   Group Elements Can Be Matrices

A powerful theorem of group theory states that any element of a group can be represented by a matrix, with the rule of combination that of matrix multiplication. We can therefore construct the group symmetry operators as a set of matrices whose matrix elements are defined in terms of a set of basis vectors, and the group operations transform these basis vectors into themselves. Let's start with a basis representing the water molecule as the set of position vectors: $W = \left\{ \vec{o}, \vec{r}_{\text{H1}}, \vec{r}_{\text{H2}} \right\}$ with hydrogen atoms at the two corners of a triangle in the *y–z* plane and the oxygen atom at the origin as shown in Fig. 11.7. It is convenient to rewrite this three-dimensional basis in terms of a logical matrix of the normalized components of the three atomic position vectors:

$$W = \begin{bmatrix} O_x & r_{\text{H1}x} & r_{\text{H2}x} \\ O_y & r_{\text{H1}y} & r_{\text{H2}y} \\ O_{xz} & -r_{\text{H1}z} & r_{\text{H2}z} \end{bmatrix} = \begin{bmatrix} 0 & 0 & 0 \\ 0 & 1 & 1 \\ 0 & -1 & 1 \end{bmatrix}$$

If we also write the symmetry operators as three-dimensional matrices, we can perform the symmetry operations by matrix multiplication. The identity is given by the 3×3 unit matrix:

$$E = \begin{bmatrix} 1 & 0 & 0 \\ 0 & 1 & 0 \\ 0 & 0 & 1 \end{bmatrix}$$

The set of three reflection operators $\sigma = \{\sigma_x(yz), \sigma_y(xz), \sigma_z(xy)\}$ through each of the coordinate planes in 3D are conveniently labeled by the axis that is reflected

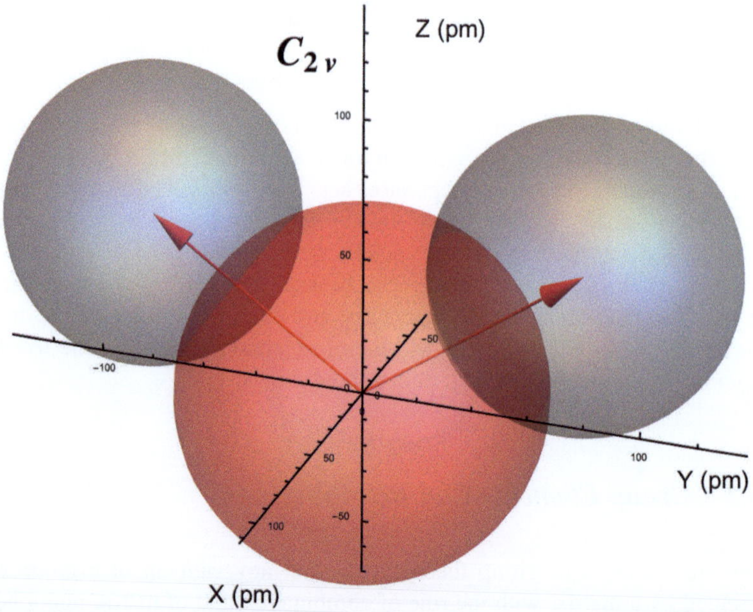

Fig. 11.7 The basis vectors for the water molecule composed of hard spheres drawn to scale. The red sphere at the origin is the oxygen atom $\vec{o} = \{0, 0, 0\}$ with an atomic radius of 66 pm, while the two silver spheres are the hydrogen atoms with a radius of 52.9 pm (the Bohr radius) with their nuclei at position vectors (in pm) of $\vec{r}_{H1} = \{0, 78.4, 55.4\}$, $\vec{r}_{H2} = \{0, -78.4, 55.4\}$, denoted by the two red arrows. Note that the C_{2v} rotation axis is the z-axis here in keeping with standard nomenclature

through the other two, for example, $\sigma_x(yz)$ reflects the x-axis through the y–z plane. These can be written in matrix form as the set of matrices (as a list in Mathematica):

$$\sigma = \left\{ \begin{bmatrix} -1 & 0 & 0 \\ 0 & 1 & 0 \\ 0 & 0 & 1 \end{bmatrix}, \begin{bmatrix} 1 & 0 & 0 \\ 0 & -1 & 0 \\ 0 & 0 & 1 \end{bmatrix}, \begin{bmatrix} 1 & 0 & 0 \\ 0 & 1 & 0 \\ 0 & 0 & -1 \end{bmatrix} \right\}$$

In a similar manner, the rotation operators can be labeled according to the n-fold symmetry axis as, for example, C_{nx} which is an n-fold rotation around the x-axis. The set of rotation operators are obtained from the rotation matrices; an n-fold rotation around the x-axis is represented by

$$C_{nx} = \begin{bmatrix} 1 & 0 & 0 \\ 0 & \cos[2\pi/n] & \sin[2\pi/n] \\ 0 & -\sin[2\pi/n] & \cos[2\pi/n] \end{bmatrix}.$$

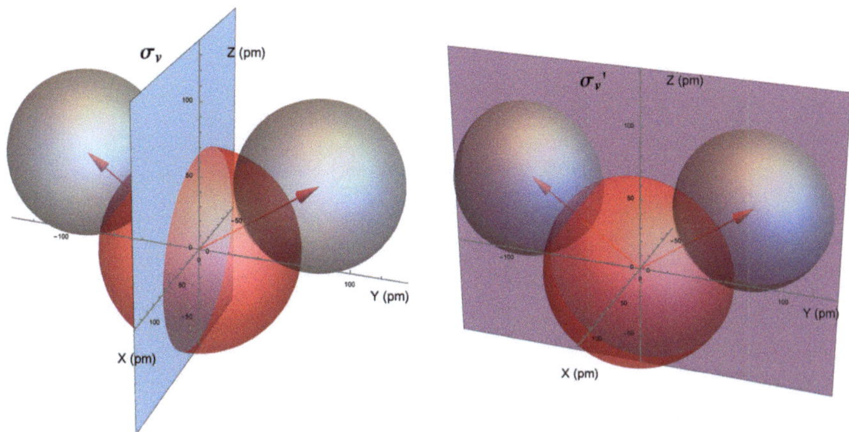

Fig. 11.8 A model of a water molecule showing the mirror reflection planes of symmetry. On the left is the $\boldsymbol{\sigma}_v = \sigma_x(yz)$ plane, and on the right is the $\boldsymbol{\sigma}_v' = \sigma_y(xz)$ plane

Water possesses only a single axis of rotational symmetry; examination of Fig. 11.8 shows that this is the z-axis in our chosen coordinate system with the oxygen at the origin and the z-axis directed along the bisector of the pair of O–H vectors. This is in keeping with the standard symmetry nomenclature which places the z-axis along the molecular axis with the highest symmetry (i.e., the largest value of n above). We can therefore write the set of three rotation operators as $C_2 = \{C_{2x}, C_{2y}, C_{2z}\}$, and in matrix form, they are given by this list:

$$
C_2 = \left\{
\begin{bmatrix} 1 & 0 & 0 \\ 0 & -1 & 0 \\ 0 & 0 & -1 \end{bmatrix},
\begin{bmatrix} -1 & 0 & 0 \\ 0 & 1 & 0 \\ 0 & 0 & -1 \end{bmatrix},
\begin{bmatrix} -1 & 0 & 0 \\ 0 & -1 & 0 \\ 0 & 0 & 1 \end{bmatrix}
\right\}
$$

For completeness, we should mention that there are two other types of symmetry operators that are not used when considering the water molecule: The water molecule is not invariant under either the inversion (parity) operator or the set of improper reflections (composed of the products of the rotation and reflection operators, $\sigma \otimes C_2$) so that we need not examine the effects of these operators. Our set of symmetry operators then only includes rotations and reflections.

We now have several tasks ahead of us. The first is to determine the operation of the reflection and rotation operators on our water molecule in order to find those under which the basis is invariant, i.e., those operators that leave the molecule unchanged. The reason for writing the reflection and rotation operators as vectors above is that we can then use Mathematica to compactly perform the necessary matrix multiplications:

```
sigmaW = σW = Table[sigma[[i]].W, {i,1,3}]
Table[sigmaW[[i]], {i,1,3}]//MatrixForm
```

giving

$$\sigma W = \left\{ \begin{bmatrix} 0 & 0 & 0 \\ 0 & 1 & 1 \\ 0 & -1 & 1 \end{bmatrix}, \begin{bmatrix} 0 & 0 & 0 \\ 0 & -1 & -1 \\ 0 & -1 & 1 \end{bmatrix}, \begin{bmatrix} 0 & 0 & 0 \\ 0 & 1 & 1 \\ 0 & 1 & -1 \end{bmatrix} \right\}$$

from which we see that the water molecule, W, is invariant under both $\sigma_x(yz)$ and $\sigma_y(xz)$, but *not* under the operator σ_z (Fig. 11.8). Note that $\sigma_x W = W$ but that the change in the molecule under the operator σ_y is an interchange of the positions of the two hydrogens, and since they are identical atoms, the molecule does not change upon reflection through the x–z plane. The operations of the π rotations of the water molecule around the three coordinate axes are obtained in Mathematica as

```
c2W = Table[c2[[i]].W, {i,1,3}];
Table[c2W[[i]]//MatrixForm, {i,1,3}]
```

$$C_2 W = \left\{ \begin{bmatrix} 0 & 0 & 0 \\ 0 & -1 & -1 \\ 0 & 1 & -1 \end{bmatrix}, \begin{bmatrix} 0 & 0 & 0 \\ 0 & 1 & 1 \\ 0 & 1 & -1 \end{bmatrix}, \begin{bmatrix} 0 & 0 & 0 \\ 0 & -1 & -1 \\ 0 & -1 & 1 \end{bmatrix} \right\},$$

and we observe that the water molecule is invariant only under C_{2z}, because $C_{2z} W = W$, but $C_{2x} W \neq W$ and $C_{2y} W \neq W$.

The next property to compute is the determinants of the set of operators; if they are symmetry operators (i.e., orthogonal- or length-preserving), their determinants should be unity. Letting ops $= \{E, C_{2z}, \sigma_y, \sigma_z\}$, then

```
Table[Det[ops[[i]]], {i,1,4}]={1,1,-1,-1}.
```

The determinants of this set of operators are $+1$ and -1. The C_{2z} rotation is a proper rotation ($\mathrm{Det}[C_{2z}] = +1$), while the reflections are improper rotations with Det $[\sigma_y] = \mathrm{Det}[\sigma_z] = -1$. The traces, i.e., the sums of the diagonal elements, of the representation matrices are called their *characters*, χ. For a matrix A, with elements a_{ij}, the trace is

$$\chi = \mathrm{Trace}[A] = \sum_{i=1}^{n} a_{ii}.$$

We can find the characters for this group by calculating the traces of the set of operators. We find that

```
Table[Tr[ops[[i]]],{i,1,4}] = {3, −1, 1, 1}.
```

Since this operator set contains only symmetry operators, we observe that all of the characters, except for the identity, are either plus or minus one because symmetry operations can only change the signs of their operands. We now have found that out of the complete set $\{E, C_{2x}, C_{2y}, C_{2z}, \sigma_x, \sigma_y, \sigma_z\}$ of symmetry operators giving reflections and π rotations, the water molecule W is only invariant under the limited set $\boldsymbol{C_{2v}} = \{E, C_{2z}, \sigma_y, \sigma_z\}$. The point group $\boldsymbol{C_{2v}}$ contains four distinct operators, and this implies in turn four symmetry classes consisting of the identity, the C_{2z} rotation, and the two reflections $\{\sigma_y, \sigma_z\}$. This group is Abelian in that the result of applying any two operators in succession does not depend on the order in which they are applied to the basis. Another theorem of group theory then tells us that each element is in its own class by itself. An element B of the group G is said to be *conjugate* to an element A of G if for $X \in G$.

$$B = XAX^{-1}$$

or

$$A = X^{-1}BX$$

11.8.2   A Representation of the Group C$_{2v}$

The set of matrices that reproduce the properties of the group operators are called a *representation* of the group $\boldsymbol{C_{2v}} = \{E, C_{2z}, \sigma_y, \sigma_z\}$, which is explicitly given as

$$C_{2v} = \left\{ \begin{bmatrix} 1 & 0 & 0 \\ 0 & 1 & 0 \\ 0 & 0 & 1 \end{bmatrix}, \begin{bmatrix} -1 & 0 & 0 \\ 0 & -1 & 0 \\ 0 & 0 & 1 \end{bmatrix}, \begin{bmatrix} 1 & 0 & 0 \\ 0 & -1 & 0 \\ 0 & 0 & 1 \end{bmatrix}, \begin{bmatrix} 1 & 0 & 0 \\ 0 & 1 & 0 \\ 0 & 0 & -1 \end{bmatrix} \right\},$$

but this representation might be reducible. The characters (traces) of all group elements in a class must be the same; character equals class. The *dimension* of a representation is the number of rows (and columns) of the matrices so our representation of $\boldsymbol{C_{2v}}$ is three-dimensional.

11.8.3   *The Symmetry Operators in C_{2v} Form a Group*

The water molecule is invariant under the set of transformations generated by the operators in the C_{2v} point group

$$C_{2v} = \left\{ E, C_{2z}, \sigma_{yz}, \sigma_{xz} \right\}.$$

Does this set of operators form a group? The first group axiom is satisfied because the identity E is in this set. The next axiom (closure) suggests that we examine all the pairwise combinations of the elements in order to show that all combinations of operators only produce other operators within the set and no others. This is conveniently done using what is commonly referred to as a group multiplication table. For each element in the group, its inverse must also a member, and for any two group elements, the product must also be an element. We can confirm these properties of the operators by considering the effect of two successive group operations, such as

$$E \cdot C_{2z} = C_{2z}.$$
$$E \cdot E = E.$$
$$C_{2z} \cdot \sigma_{xz} = \sigma_{yz}.$$
$$C_{2z} \cdot C_{2z} = E.$$
$$\sigma_{yz} \cdot \sigma_{xz} = C_{2z}.$$
$$\text{etc.,}$$

and summarize these in the following combination ("multiplication") table (Table 11.4) where we have written the four operators in the first column and first row. The products in each cell are the combinations. From this table, we can verify that these four symmetry operators satisfy the group axioms. The inverses are present because, e.g., by the definition of an inverse

$$C_{2z}^{-1} \cdot C_{2z} = E.$$

and

$$C_{2z} \cdot C_{2z} = E,$$

then multiply through by C_{2z}^{-1} to give

$$C_{2z}^{-1} \cdot (C_{2z} \cdot C_{2z}) = C_{2z}^{-1} \cdot E = C_{2z}^{-1},$$

Table 11.4 Multiplication table for the C_{2v} group

C_{2v}	E	C_{2z}	σ_{yz}	σ_{xz}
E	E	C_{2z}	σ_{yz}	σ_{xz}
C_{2z}	C_{2z}	E	σ_{xz}	σ_{yz}
σ_{yz}	σ_{yz}	σ_{xz}	E	C_{2z}
σ_{xz}	σ_{xz}	σ_{yz}	C_{2z}	E

or regroup the left-hand side to give

$$\left(\mathbf{C}_{2z}^{-1} \cdot \mathbf{C}_{2z}\right) \cdot \mathbf{C}_{2z} = \mathbf{C}_{2z}^{-1} \cdot \mathbf{E} = \mathbf{C}_{2z}^{-1}.$$
$$\mathbf{E} \cdot \mathbf{C}_{2z} = \mathbf{C}_{2z}^{-1} \cdot \mathbf{E} = \mathbf{C}_{2z}^{-1};$$

thus,

$$\mathbf{C}_{2z} = \mathbf{C}_{2z}^{-1}.$$

We see that $\mathbf{C}_{2z}$ is its own inverse. Identical reasoning holds for both $\boldsymbol{\sigma}_{xz}$ and $\boldsymbol{\sigma}_{yz}$; they are their own inverses as well. The reason for this is that this group multiplication table forms a 4×4 matrix and that the entries are symmetric across the diagonal. This means that all the group elements commute with each other, that their commutator is zero, and that the group $\mathbf{C}_{2v}$ is *Abelian* and all the elements are their own inverses. The number of operators in the group is referred to as the *order h* of the group, $h = 4$, in this case. Note that Table 11.4 illustrates another feature in that each operator appears only once in each row and column as a result of the *rearrangement theorem*. Also, the elements $\{\mathbf{E}, \mathbf{C}_{2z}\}$ form a subgroup of $\mathbf{C}_{2v}$, a group by themselves of smaller order, $h' = 2$.

This set of operators is closed under multiplication and contains the identity. As a final check, we can compute the inverse operators using Mathematica

$$\{x, y, z\} = \{1, 2, 3\};$$

$$\text{opsinv} = \{\text{ident}, \text{Inverse}[\text{c2}[[x]]], \text{Inverse}[\text{sigma}[[y]]], \text{Inverse}[\text{sigma}[[x]]]\};$$

$$\text{multableinv} = \text{Table}[\text{opsinv}[[i]].\text{ops}[[j]], \{i, 1, 4\}, \{j, 1, 4\}];$$

```
multableinv//MatrixForm
```

$$
\begin{bmatrix}
\begin{bmatrix} 1 & 0 & 0 \\ 0 & 1 & 0 \\ 0 & 0 & 1 \end{bmatrix} &
\begin{bmatrix} 1 & 0 & 0 \\ 0 & -1 & 0 \\ 0 & 0 & -1 \end{bmatrix} &
\begin{bmatrix} 1 & 0 & 0 \\ 0 & -1 & 0 \\ 0 & 0 & 1 \end{bmatrix} &
\begin{bmatrix} 1 & 0 & 0 \\ 0 & 1 & 0 \\ 0 & 0 & -1 \end{bmatrix} \\[18pt]
\begin{bmatrix} 1 & 0 & 0 \\ 0 & -1 & 0 \\ 0 & 0 & -1 \end{bmatrix} &
\begin{bmatrix} 1 & 0 & 0 \\ 0 & 1 & 0 \\ 0 & 0 & 1 \end{bmatrix} &
\begin{bmatrix} 1 & 0 & 0 \\ 0 & 1 & 0 \\ 0 & 0 & -1 \end{bmatrix} &
\begin{bmatrix} 1 & 0 & 0 \\ 0 & -1 & 0 \\ 0 & 0 & 1 \end{bmatrix} \\[18pt]
\begin{bmatrix} 1 & 0 & 0 \\ 0 & -1 & 0 \\ 0 & 0 & 1 \end{bmatrix} &
\begin{bmatrix} 1 & 0 & 0 \\ 0 & 1 & 0 \\ 0 & 0 & -1 \end{bmatrix} &
\begin{bmatrix} 1 & 0 & 0 \\ 0 & 1 & 0 \\ 0 & 0 & 1 \end{bmatrix} &
\begin{bmatrix} 1 & 0 & 0 \\ 0 & -1 & 0 \\ 0 & 0 & -1 \end{bmatrix} \\[18pt]
\begin{bmatrix} 1 & 0 & 0 \\ 0 & 1 & 0 \\ 0 & 0 & -1 \end{bmatrix} &
\begin{bmatrix} 1 & 0 & 0 \\ 0 & -1 & 0 \\ 0 & 0 & 1 \end{bmatrix} &
\begin{bmatrix} 1 & 0 & 0 \\ 0 & -1 & 0 \\ 0 & 0 & -1 \end{bmatrix} &
\begin{bmatrix} 1 & 0 & 0 \\ 0 & 1 & 0 \\ 0 & 0 & 1 \end{bmatrix}
\end{bmatrix}
$$

to find that $C_{2z}^{-1} = C_{2z}$; $\sigma_y^{-1} = \sigma_y$; and $\sigma_x^{-1} = \sigma_x$ so that the operators in this set are indeed their own inverses. We observe that this set of operators does form a group since they satisfy all of the group axioms. The order of the group (four) is the number of elements (operators) that comprise the group.

One of the uses of Table 11.4, besides showing that the abstract symmetry operators form a group, is to compute a concrete example of the group operators as matrices. Mathematicians refer to such an example as a *representation* of the group elements. One of the theorems of group theory is that the elements of any group can be represented by matrices, where the rule of combination is matrix multiplication. We can find a representation of the group by treating the 4×4 multiplication table (Table 11.4) as a matrix. By placing a one at the position of an element in the matrix and a zero elsewhere, we can build a representation that obeys all the group axioms. This representation is called the *regular representation*; in this case, it is four-dimensional, and we construct it to obtain another representation of $C_{2v} = \{E, C_{2z}, \sigma_{yz}, \sigma_{xz}\}$ as

$$C_{2v} = \left\{ \begin{bmatrix} 1 & 0 & 0 & 0 \\ 0 & 1 & 0 & 0 \\ 0 & 0 & 1 & 0 \\ 0 & 0 & 0 & 1 \end{bmatrix}, \begin{bmatrix} 0 & 1 & 0 & 0 \\ 1 & 0 & 0 & 0 \\ 0 & 0 & 0 & 1 \\ 0 & 0 & 1 & 0 \end{bmatrix}, \begin{bmatrix} 0 & 0 & 1 & 0 \\ 0 & 0 & 0 & 1 \\ 1 & 0 & 0 & 0 \\ 0 & 1 & 0 & 0 \end{bmatrix}, \begin{bmatrix} 0 & 0 & 0 & 1 \\ 0 & 0 & 1 & 0 \\ 0 & 1 & 0 & 0 \\ 1 & 0 & 0 & 0 \end{bmatrix} \right\}.$$

We can use Mathematica to find the group multiplication table for the regular representation as follows: Use a list to collect the group elements

```
c2v = {e, c2z, syz, sxz};
```

and then compute the pairwise products using

```
MatrixForm[Table[c2v[[i]].c2v[[j]],{i,1,4},{j,1,4}].
```

The multiplication table for the matrices in the regular representation (Table 11.5) reproduces the behavior of the group elements in Table 11.4. One can also use Mathematica to show that each operator matrix is its own inverse, and this is left as an exercise.

Table 11.5 The C_{2v} group multiplication table using the regular representation

C_{2v}	E	C_{2z}	σ_{yz}	σ_{xz}
E	$\begin{bmatrix} 1&0&0&0 \\ 0&1&0&0 \\ 0&0&1&0 \\ 0&0&0&1 \end{bmatrix}$	$\begin{bmatrix} 0&1&0&0 \\ 1&0&0&0 \\ 0&0&0&1 \\ 0&0&1&0 \end{bmatrix}$	$\begin{bmatrix} 0&0&1&0 \\ 0&0&0&1 \\ 1&0&0&0 \\ 0&1&0&0 \end{bmatrix}$	$\begin{bmatrix} 0&0&0&1 \\ 0&0&1&0 \\ 0&1&0&0 \\ 1&0&0&0 \end{bmatrix}$
C_{2z}	$\begin{bmatrix} 0&1&0&0 \\ 1&0&0&0 \\ 0&0&0&1 \\ 0&0&1&0 \end{bmatrix}$	$\begin{bmatrix} 1&0&0&0 \\ 0&1&0&0 \\ 0&0&1&0 \\ 0&0&0&1 \end{bmatrix}$	$\begin{bmatrix} 0&0&0&1 \\ 0&0&1&0 \\ 0&1&0&0 \\ 1&0&0&0 \end{bmatrix}$	$\begin{bmatrix} 0&0&1&0 \\ 0&0&0&1 \\ 1&0&0&0 \\ 0&1&0&0 \end{bmatrix}$
σ_{yz}	$\begin{bmatrix} 0&0&1&0 \\ 0&0&0&1 \\ 1&0&0&0 \\ 0&1&0&0 \end{bmatrix}$	$\begin{bmatrix} 0&0&0&1 \\ 0&0&1&0 \\ 0&1&0&0 \\ 1&0&0&0 \end{bmatrix}$	$\begin{bmatrix} 1&0&0&0 \\ 0&1&0&0 \\ 0&0&1&0 \\ 0&0&0&1 \end{bmatrix}$	$\begin{bmatrix} 0&1&0&0 \\ 1&0&0&0 \\ 0&0&0&1 \\ 0&0&1&0 \end{bmatrix}$
σ_{xz}	$\begin{bmatrix} 0&0&0&1 \\ 0&0&1&0 \\ 0&1&0&0 \\ 1&0&0&0 \end{bmatrix}$	$\begin{bmatrix} 0&0&1&0 \\ 0&0&0&1 \\ 1&0&0&0 \\ 0&1&0&0 \end{bmatrix}$	$\begin{bmatrix} 0&1&0&0 \\ 1&0&0&0 \\ 0&0&0&1 \\ 0&0&1&0 \end{bmatrix}$	$\begin{bmatrix} 1&0&0&0 \\ 0&1&0&0 \\ 0&0&1&0 \\ 0&0&0&1 \end{bmatrix}$

11.8.4 Representations of the C_{2v} Symmetry Group

While we have just found two representations of the C_{2v} symmetry group, it should be noted that there are in general an infinite number of representations of the elements in a group, but some, the *irreducible representations*, are much more useful than others. The characters of the representation matrices serve as a summary of the representation and have important properties when it comes to applications. The characters for the four-dimensional regular representation are

$$\{\chi\} = \text{Trace}[\mathbf{C_{2v}}] = \text{Trace}\left[\{\mathbf{E}, \mathbf{C_{2z}}, \boldsymbol{\sigma}_{yz}, \boldsymbol{\sigma}_{xz}\}\right] = \{4, 0, 0, 0\}.$$

The *dimension* of a representation is given by the character of the identity matrix, in this case $\chi = 4$. This should be clear since the n-dimensional identity matrix consists only of 1 on the diagonal and 0 elsewhere; the trace is the sum of n ones. Note that the characters of these operators are now $\chi(E, C_{2z}, \sigma_y, \sigma_y) = \{4,0,0,0\}$, whereas in our earlier 3×3 matrix representation the characters were $\{3,-1,1,1\}$. This set of 4×4 operator matrices still forms a group with the same multiplication table as Table 11.4.

11.8.5 Irreducible Representations

Here, we observe that we can have more than one representation of a group. Are some representations better than others? The fact that we have two representations of different *dimensions* suggests that there may be a relationship between them, and indeed there is; the higher-dimensional representations can be reduced to lower-dimensional representations until this process ends with a set of *irreducible representations*.

We can now write the character table for the group C_{2v}, the point group of the water molecule in the basis W.

In Table 11.6, the name of the point group is given in the upper left-hand corner, followed by the set of group operators, beginning with the identity. The character table is a square matrix in which the columns are the conjugacy classes, symmetry classes (or simply classes) and the rows are the irreducible representations of the group; the number of classes equals the number of *irreducible representations*. The order h of the group can be obtained from the characters by

Table 11.6 Character table for the group C_{2v} ($h = 4$)

C_{2v}	E	C_{2z}	σ_y (xz)	σ_x (yz)	Linear	Quadratic
A_1	1	1	1	1	z	x^2, y^2, z^2
A_2	1	1	-1	-1	R_z	xy
B_1	1	-1	1	-1	x, R_y	xz
B_2	1	-1	-1	1	y, R_x	yz
Γ	4	0	0	0		

$$h = \sum_{i=1}^{N} [\chi_i(E)]^2 = \sum_{R}^{n} [\chi_i(R)]^2$$

where i is an index for the representations and R is an index for the symmetry operations, which in this case is $h = 4 = 1^2 + 1^2 + 1^2 + 1^2$, which is also the trace of the identity operator in the regular representation. The irreducible representations are orthogonal to each other:

$$\sum_{R}^{n} \chi_i(R)\chi_j(R) = \frac{1}{h}\delta_{ij} = 0, \text{unless } i = j.$$

The number of classes equals the number of irreducible representations, so from the first column in Table 11.6 we see that there are four irreducible representations of the group C_{2v}, labeled A_1, A_2, B_1, and B_2.

These are standard labels from the Mulliken (1955) scheme for labeling the irreducible representations, and they are defined as follows: For symbols in the first *column* of the character tables, A is for a singly degenerate (or 1D) symmetry under a rotation about the principal axis, while B is for a singly degenerate (or 1D) antisymmetry under a rotation about the principal axis. The first subscript denotes that the representation is symmetric about the C_n principal axis, but if there is none, then it is symmetric with respect to reflections in a vertical plane σ_v. A second subscript denotes that the representation is antisymmetric about the C_n principal axis, but if there is none, then it is symmetric with respect to reflections in a vertical plane σ_v.

The next four columns of Table 11.6 summarize the behavior of the wave functions under the action of the symmetry operators, with plus one indicating that the wave functions are symmetric (they *do not* change sign under the action of the symmetry operators) while a minus one indicates that they are antisymmetric (they *do* change sign under the action of the symmetry operators). The entries in the first column (E) following the group name describe the degeneracies of the row: A and B have a degeneracy of one, i.e., they are one-dimensional representations, such as x or y or z separately.

We have also listed Γ, the characters for the reducible regular representation whose characters are the sums of the characters for each symmetry class. The A_1 representation is therefore one-dimensional ($E = 1$) and is symmetric with respect to rotations about the principal axis, $\chi(C_{2x}) = +1$, where χ is the character function. A_1 is also symmetric during reflections in the x–z and x–y planes, $\chi(\sigma_y) = \chi(\sigma_z) = +1$, while A_2 is antisymmetric under these same reflections. Representations are subsets of the complete point group and are orthogonal to each other. The character of a representation is a signed integer that summarizes the effect of an operator on the basis set. The reducible representation Γ indicates how the water molecule transforms under the group operators. The irreducible representations are the combinations of representations in the point group whose sum combines to produce the reducible representation.

There is always at least one irreducible representation that is totally symmetric with respect to all the transformations produced by the set of symmetry operators. These totally symmetric irreducible representations are denoted by the letter A, and the group C_{2v} contains one, labeled A_1. According to the Mulliken scheme, the label A is used if the wave function is invariant under the principal rotation (C_{2z}) and its character is plus one, while B labels a representation with character minus one. Other representations, A_2, B_1, and B_2, are one-dimensional representations that are not degenerate, while E is a two-dimensional representation.

The group order, $h = 4$, is the number of operators in the group. In column six, we find the functions (x, y, z) of the coordinates that belong to a given irreducible representation which can be used in combination to produce a molecular wave function that obeys this set of group operations. The meaning of these terms is that the A_1 irreducible representation, for example, transforms a unit vector along the z-axis into itself.

11.8.6   A Molecular Basis for the C_{2v} Symmetry Group of the Water Molecule

If we write unit vectors along the three Cartesian axes as a set of column matrices

$$\{\widehat{x}, \widehat{y}, \widehat{z}\} = \left\{ \begin{bmatrix} 1 \\ 0 \\ 0 \end{bmatrix}, \begin{bmatrix} 0 \\ 1 \\ 0 \end{bmatrix}, \begin{bmatrix} 0 \\ 0 \\ 1 \end{bmatrix} \right\},$$

we can use our symmetry operators to investigate their transformation properties. We can make a table like the character table in which the entries are the results of operations on these vectors by the set of group operators (Table 11.7). Since the set of C_{2v} operators contains only symmetry operators, we know that they can only change the sign of the basis vectors, not their length. Therefore, we can replace the

Table 11.7 Effect of the C_{2v} group operators on the Cartesian basis vectors

Basis vector	E	C_{2x}	σ_y	σ_x	IrrRep
$\widehat{x}$	$\widehat{x}$	$-\widehat{x}$	$\widehat{x}$	$-\widehat{x}$	B_1
$\widehat{y}$	$\widehat{y}$	$-\widehat{y}$	$-\widehat{y}$	$\widehat{y}$	B_2
$\widehat{z}$	$\widehat{z}$	$\widehat{z}$	$\widehat{z}$	$\widehat{z}$	A_1
$\widehat{x}$	1	-1	1	-1	B_1
$\widehat{y}$	1	-1	-1	1	B_2
$\widehat{z}$	1	1	1	1	A_1

entries in Table 11.6 with plus or minus one to produce the bottom three rows in Table 11.7. This table then recapitulates the character table.

We observe that $\hat{z}$ is invariant under all of the group operations

$$C_{2z}(\hat{z}) = \sigma_y(\hat{z}) = \sigma_x(\hat{z}) = \hat{z}.$$

and therefore has the symmetry properties of (transforms according to) the irreducible representation A_1 under the action of the group operators. In a similar manner, we find that the unit vector $\hat{x}$ transforms as B_1 and that $\hat{y}$ transforms as B_2. Just as the three Cartesian unit vectors are orthogonal and form a basis for three-dimensional space, i.e., they span 3-space, and any vector may be expressed as a linear combination of them, the irreducible representations are also orthogonal and span the space of representations. Any reducible representation can be expressed as linear combination of these irreducible representations. Note that there are four symmetry classes, which means that there are four irreducible representations for C_{2v}, but that we have found that the basis $\{\hat{x}, \hat{y}, \hat{z}\}$ transforms according to only three of these irreducible representations. This basis (Fig. 11.9) corresponds to the three-oxygen p-wave functions $\{P_x, P_y, P_z\}$ which have the same transformation properties as $\{\hat{x}, \hat{y}, \hat{z}\}$. However, there must exist an expanded basis that includes objects that transform

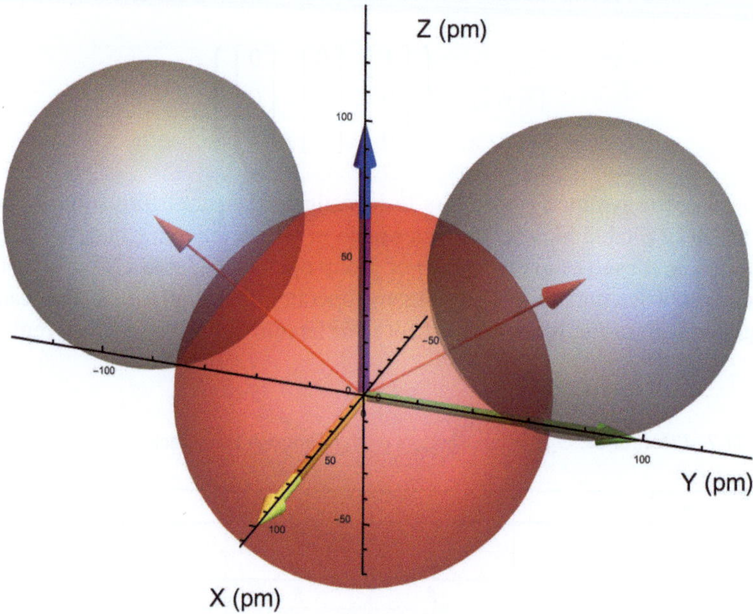

Fig. 11.9 A view of the water molecule with the oxygen atom centered at the origin showing the vector basis $\{\hat{x}, \hat{y}, \hat{z}\}$ (yellow, green, and blue arrows) which transforms as the oxygen atom's P-wave functions $\{P_x, P_y, P_z\}$ that are invariant under all the transformations by the symmetry operators in the group C_{2v}

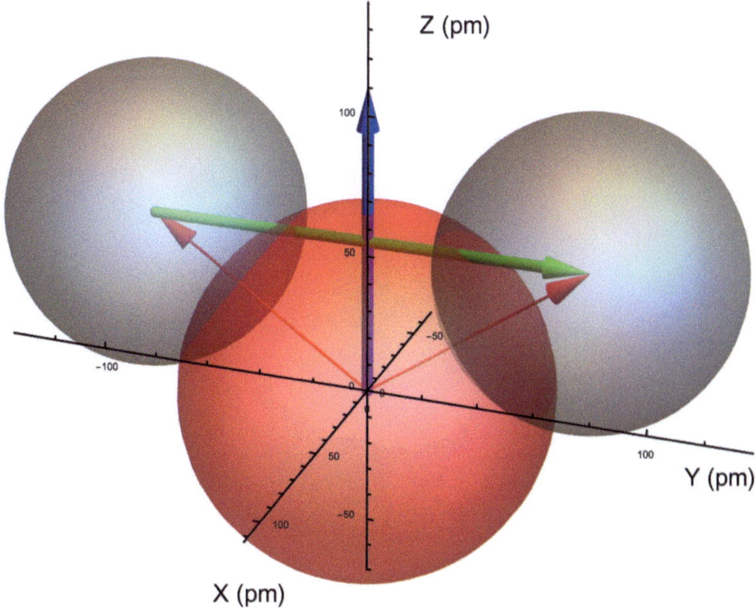

Fig. 11.10 A view of the water molecule with the oxygen atom centered at the origin showing the tensor basis (green, and blue arrows) which now includes the hydrogen atom wave functions and which completes the set of transformations by the symmetry operators in the group C_{2v}

according to A_2 as well. This expanded basis must include not only vectors but also tensors, and the logical tensor to include is the position vectors of the hydrogen atoms (Fig. 11.10) as the matrices

$$\begin{bmatrix} 0 & 0 & 0 \\ 0 & 1 & 0 \\ 0 & 1 & 1 \end{bmatrix} \text{ and } \begin{bmatrix} 0 & 0 & 0 \\ 0 & -1 & 0 \\ 0 & 1 & 1 \end{bmatrix}.$$

We can make a table of the effects of the group operators on this basis just as we did for the unit vector basis (Table 11.5), and this is left as a useful exercise for the interested reader.

If we place the oxygen nucleus at the origin $O = \{0,0,0\}$, we see that it is invariant under any point group operation, including all of the operators in the group C_{2v}. The transformation properties of this set of symmetry operators are summarized in the Character table, computed as follows: We pick as a basis the oxygen atom's P-wave functions $\{P_x, P_y, P_z\}$ which have the same transformation properties as the unit vectors $\{\hat{x}, \hat{y}, \hat{z}\}$ in three-dimensional Cartesian coordinates (Fig. 11.9). To complete the selection of a basis, we choose the vectors locating the two hydrogen atoms $H1s_a = \{0, \hat{y}, \hat{z}\}$ and $H1s_b = \{0, -\hat{y}, \hat{z}\}$ (Fig. 11.10). Next, we compute the result of operating on each of these vectors separately, noting that the

Table 11.8 The results of applying the symmetry operators in C_{2v} on the water molecule basis

C_{2v}	E	C_{2z}	σ_y	σ_z
O2s	O2s	O2s	O2s	O2s
$\widehat{x}\,(O2P_x)$	$O2P_x$	$-O2P_x$	$O2P_x$	$-O2P_x$
$\widehat{y}\,(O2P_y)$	$O2P_y$	$-O2P_y$	$-O2P_y$	$O2P_y$
$\widehat{z}\,(O2P_z)$	$O2P_z$	$O2P_z$	$O2P_z$	$O2P_z$
$H1s_a$	$H1s_a$	$H1s_b$	$H1s_b$	$H1s_a$
$H1s_b$	$H1s_b$	$H1s_a$	$H1s_a$	$H1s_b$

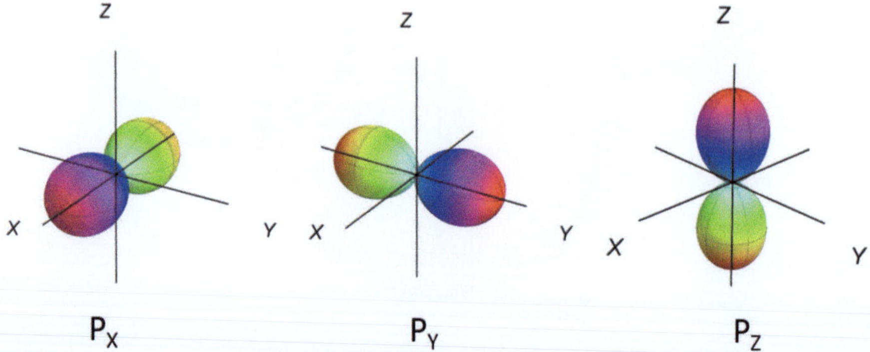

Fig. 11.11 The P_x, P_z, and P_z wave functions for the oxygen atom of water showing their positive (reddish) and negative (greenish) lobes which change sign on rotation by 180° or on reflection by the transformation operators of C_{2v}

result can at most change the sign of the vector or interchange two atoms. The results (Table 11.8) indicate that the signs of the wave functions are determined by their transformation properties under the set of operators comprising C_{2v}.

Chemists are familiar with the basis $\{\widehat{x}, \widehat{y}, \widehat{z}\}$ because its transformation properties are the same as the p_x, p_y, and p_z hydrogen atom wave functions (Fig. 11.11) used for constructing molecular wave functions as linear combinations of atomic wave functions. The full oxygen wave function is the product of this angular function with the radial distribution function, an example of which for the P_z state is given in Fig. 11.12. Unlike the s states for hydrogen and oxygen, the wave functions for the higher angular momentum p and d states in oxygen are zero at the nuclei and would therefore have their electron density displaced from the nucleus, giving, when combined with the hydrogen s states, a molecule with a lobed, nonspherical shape. We will below elucidate these various shapes but, for the time being, need to continue with our group theory in order to understand the nature of the molecular symmetry labels that are used to refer to the various electronic probability distributions and their unique shapes.

Column seven of Table 11.6 gives similar functions for the d series in terms of quadratic rather than linear functions. The symmetry of the wave functions belonging to the irreducible representations A1, A2, B1, and B2 is captured in the

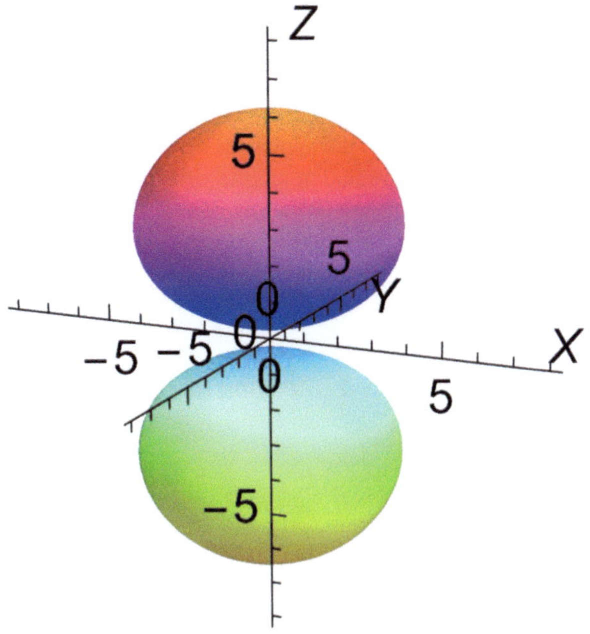

Fig. 11.12 The P_z wave function for the oxygen atom in the water molecule, showing the lack of electron density at the nucleus (origin)

nomenclature for wave functions in polyatomic molecules using lowercase letters a1, a2, b1, and b2; we will see that these labels are widely used when referring to the molecular wave functions of the water molecule. These wave functions span the irreducible representations A1, A2, B1, and B2. The entries in the sixth and seventh columns in Table 11.6 display the symmetry of the p and d atomic wave functions that span an irreducible representation.

We can apply this character table to the wave functions for the water molecule. Oxygen has the atomic wave functions $O1s^2$, $O2p_x^2$, $O2p_y^2$, and $O2p_z^2$, which means that two oxygen electrons are in the innermost 1s wave function and in each of the x, y, and z states. The wave function $O2p_x^2$ from the character table has the symmetry of a vector $\vec{z} = z_0\hat{z}$ parallel to the x-axis.

11.8.7   Irreducible Representations for the Water Molecule

Assuming that the molecular plane is the y–z plane (Fig. 11.8), we can see from Table 11.8 that the orbital $O2p_x$ changes sign under a 180° rotation, C_2, and under the reflection σ_v' but remains the same under the reflection σ_v. Therefore, this orbital belongs to the B_1 irreducible representation. Any molecular orbital built from this atomic orbital will be a b_1 orbital. In a similar way, the $O2p_y$ orbital changes sign under C_2 but remains the same after σ_v'; thus, it belongs to B_2 and can contribute to the b_2 molecular orbital. Similarly, the $O2p_z$ belongs to the A_1 irreducible representation.

The use of the character table for a molecule allows one to determine which atomic orbitals will overlap significantly with each other to produce the bonding in the resulting molecule. Significant overlap only occurs between atomic orbitals with the same symmetry labels. First, we put all the atomic orbitals on oxygen and on the two H-atoms through the symmetry operations indicated in the character table. If the sign of the wave function of the orbitals does not change when the symmetry operation is carried out, then this corresponds to "+1" in the representations in the column below that symmetry operation. If the sign changes, then this corresponds to "−1" in the column of representations.

11.9 The Wave Functions of the Water Molecule

The O–H bond length in water is 96 pm, and the angle between the two O–H bonds is 104.5°. The light hydrogen atoms are bound to the 16-fold heavier oxygen atom (8 protons and 8 neutrons). The water molecule contains ten electrons, one each from the two hydrogens and eight from the oxygen atom. In working out the molecular wave functions for water, the two 1s electrons in oxygen spin pair and form a closed shell leaving six valence electrons. Of these, four can spin pair leaving two that can each pair with a single hydrogen electron resulting in two lone pairs of electrons. The approximate tetrahedral structure of the water molecule (Fig. 11.6) is then understood in terms of repulsion of these lone pairs and results in a decrease in the O–H bond angle. The wave functions for water can be quickly calculated using many molecular quantum mechanics programs, such as Spartan, HyperChem, and others. While the exact energy levels vary somewhat depending on which set of atomic basis functions are used, there is general agreement (Table 11.9) in the ordering of the levels, and these are shown in Fig. 11.13. The energy of the highest occupied molecular orbital ($1b_1$; HOMO) gives water an ionization potential of about 14 eV, while the $1a_1$ pair of 1s electrons are strongly bound in the potential well from the

Table 11.9 The calculated and measured electronic energy levels of the water molecule, the energy difference between the gas and liquid phases, ΔE, and the widths Γ of the spectral lines in the two phases

State	Energy (eV)[a]	Gas (eV)[b]	Liquid (eV)[b]	ΔE (eV)	Γ gas (eV)	Γ liquid (eV)
$1a_1$	−559	540.00	538.23	1.77	0.40	1.67
$2a_1$	−37 (−32.4)	32.28	30.94	1.34	2.82	3.30
$1b_2$	−19 (−18.7)	18.87	17.41	1.46	1.75	2.28
$3a_1$	−15 (−14.8)	14.84	13.50	1.34	1.18	2.42
$1b_1$	−14 (−12.6)	12.85	11.39	1.45	0.30	1.45
$4a_1$	+6	7.61	8.09	0.48	0.81	1.25
$2b_2$	+8	9.36	9.74	0.38	0.70	1.43

[a]The energies were calculated using the restricted Hartree-Fock wave functions using the 6–31** basis set. The measured values for liquid water (in parenthesis) are from Ning et al. (2008)
[b]The measured gas and liquid energies were obtained from Fransson et al. (2016)

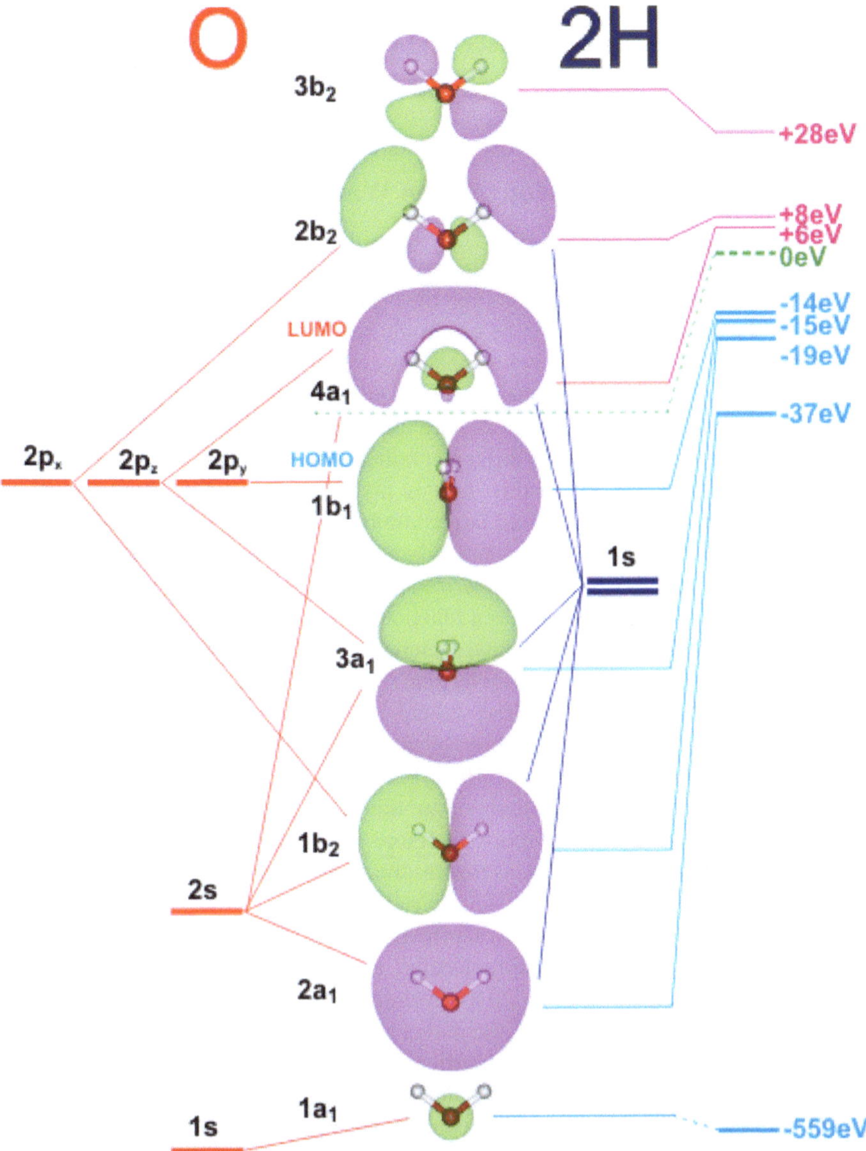

Fig. 11.13 Molecular wave functions for the water molecule calculated using HyperChem. The basis atomic wave functions for oxygen are shown on the left, ordered from bottom (lowest) to top (highest) in terms of relative energy. The hydrogen atom wave functions used are on the right. The wave functions resulting from linear combinations of the oxygen and hydrogen wave functions are labeled according to their irreducible representations as discussed in the text. (Courtesy of M. Chaplin (https://water.lsbu.ac.uk/water/martin_chaplin.html). Used with permission)

$Z = +8$ protons in the oxygen nucleus giving a ground state energy of -559 eV. The lowest unoccupied molecular orbital ($4a_1$; LUMO) sits at 6 eV above the ionization potential.

11.10   X-Ray Spectroscopy of the Water Molecule

With this introduction to the group theoretical foundations for the composition and nomenclature of the wave functions and the energy levels of the water molecule, we are now in a position to explore the electronic spectroscopy of water. The values of the energies of the electronic states makes it clear that we need to turn to the soft X-ray region of the photon regime in order to probe the lowest energy level $1a_1$ at -559 eV centered on the oxygen atom, one that has the greatest likelihood of giving information about the state of hydrogen bonding of the molecule in ice and in liquid water. The binding energy of this state is much more negative than that of the hydrogen atom due to the eightfold larger number of protons in the oxygen nucleus.

A review of the literature by Anders Nilsson and coworkers (Amann-Winkel et al 2019; Fransson et al. 2016) has summarized much of the research up to that date. The absorption of X-rays (Fig. 11.14) as a function of energy and temperature will show the changes in the electronic wave functions as the thermal energy from the environment varies and the intermolecular separation changes. The timescale for electronic processes involving these bound states can be estimated from special relativity as approximately the amount of time that it takes a photon to traverse the atom, and this is on the order of 10^{-18} s. This time is much smaller than that associated with molecular vibrations or the timescale of hydrogen bonding interactions ($\sim 10^{-12}$ s). X-ray spectroscopy therefore allows one to view the water molecule as essentially frozen in time, as long as the measurement time is not the limiting factor (Mariedahl et al. 2018).

Spectroscopic studies of liquid water reveal energy differences from water in the gas phase due to the strong intermolecular interactions that give liquid water its unique properties (Leetma et al. 2010; Weinhardt et al. 2012; Wernet et al. 2004; Winter et al. 2004) . For example, the photoelectron spectra of gaseous and aqueous water (Fig. 11.15) show this energy difference, which arises from the electrostatic interactions of its dipole moments that alter the electronic energies. The widths of the spectral lines also differ in the two phases; the gas phase line is ~ 0.4 eV wide, while the liquid phase peak has a full width at half maximum of 1.67 eV (Winter et al. 2007) indicating that the local hydrogen-bonding interactions in the liquid phase are stronger and more varied than in the gas phase. The valence electrons of the water molecule have energies (-37 to -14 eV; Table 11.9) that lie considerably closer to the ionization threshold than those of the 1s oxygen states. Photoelectron spectroscopy (Fig. 11.16) of water in both liquid and gaseous phases shows that, just as observed for the O1s state, the electronic energies for the liquid are reduced with respect to the gas phase by 1.34–1.46 eV due to the polarization and hydrogen bonding of the water molecules in the liquid. X-ray absorption spectroscopy of

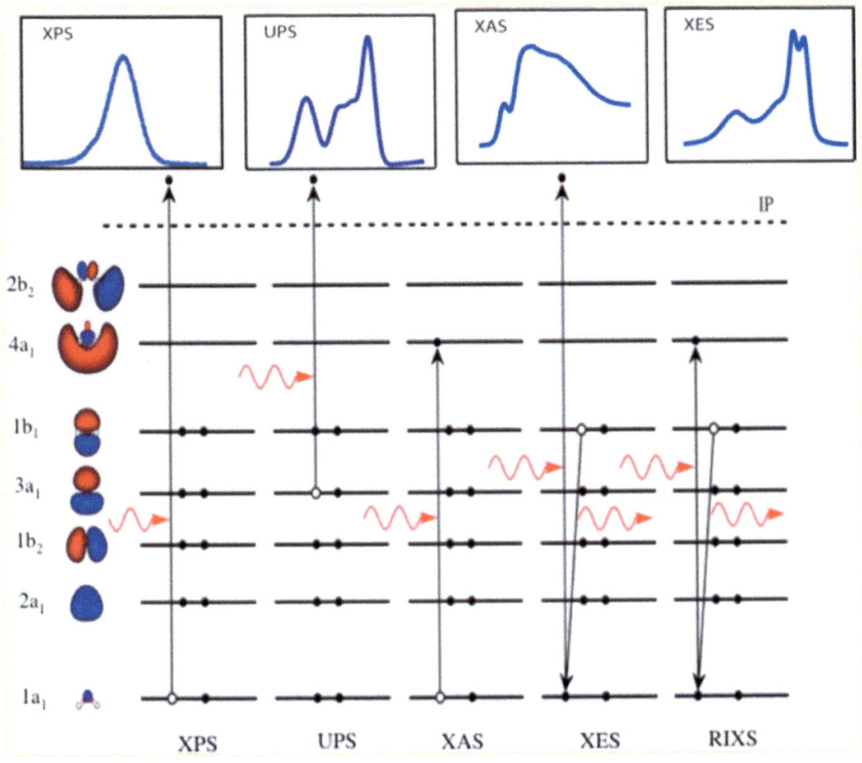

Fig. 11.14 A schematic of the various processes that involve absorption of a photon with subsequent excitation of the electrons in a water molecule. XPS is x-ray photoelectron spectroscopy; UPS is ultraviolet photoelectron spectroscopy; XAS is x-ray absorption spectroscopy; XES is x-ray emission spectroscopy; and RIXS is resonant inelastic x-ray scattering. Here the energy levels are not drawn to scale. The filled and empty circles represent electrons and vacancies. The dashed line, labelled IP, indicates the ionization potential. The photons are represented as red wavy lines. (With permission from Fransson et al. (2016) © 2016 American Chemical Society)

liquid water, water vapor, and ice shows not only the lower-energy bound states but also the first two unbound, excited states, the $4a_1$ and $2b_2$ (Fig. 11.17) at 534.17 and 536.04 eV, respectively, that arise from photon absorption by the O1s electrons of the oxygen atom and promotion to these previously unoccupied levels. These electrons remain bound to the water molecule. The X-ray absorption spectra are usually discussed in terms of a somewhat antiquated nomenclature that refers to "edges" meaning the shoulders of the spectral lines: the "pre-edge" is the feature at 535 eV (Fig. 11.17), the "main edge" is at 537.5 eV, and a "post-edge" is located at 541 eV. As also observed from the photoelectron spectra of the valence states (Fig. 11.16), these states in the X-ray absorption spectra are shifted in energy by +0.77 and +1.43 eV, respectively, in the liquid phase. The transition widths (FWHM) increase from 0.81 and 0.70 eV ($4a_1$ and $2b_2$) to 1.25 and 2.50 eV, but

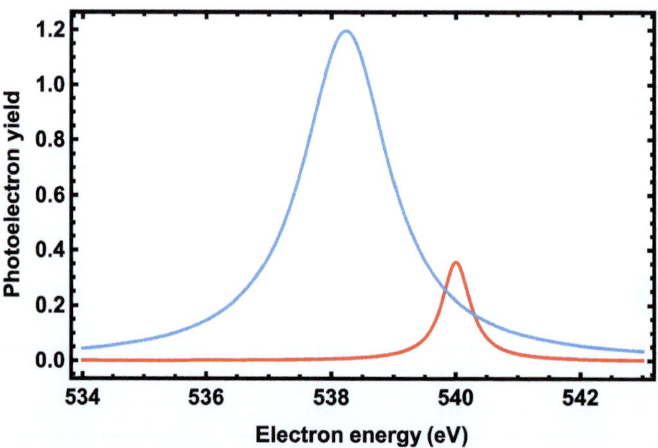

Fig. 11.15 A schematic illustration of the X-ray photoelectron spectra of the $1a_1$ state of the oxygen 1s electrons of water in the liquid (blue) and gas (red) phases shows an energy difference of 1.77 eV due to the dipolar interactions in the liquid that are absent in the vapor

the most interesting feature of the X-ray absorption spectrum is the appearance of a significant "post-edge" signal at 541.0 eV that does not exist in molecules in the gas phase, which is large in liquid water but even larger in the spectrum of ice. The "pre-edge" signals correspond to severed or weakened hydrogen bonds, the "main edge" corresponds to water molecules in the interstices among hydrogen-bonded molecules, and the "post-edge" signal, with a width of 4.11 eV, arises from strong, tetrahedrally bonded molecules, and therefore it is strongest in ice. The extent of hydrogen bonding is a function of temperature. The "post-edge" signal in liquid water decreases as the temperature is raised, and the "pre-edge" and "main edge" signals increase as would be expected given their assignments to weakened hydrogen bonds because the fraction of these weaker bonds would be expected to increase as more thermal energy is present in the fluid.

X-ray emission spectra of the water molecule in the gas phase (Fig. 11.18) show transitions due to the ionization of the O1s electrons and the subsequent filling of this vacancy by the indicated electrons from the valence states with the emission of a photon of the corresponding energy difference. The $1b_2$ state at 521.08 eV does not shift appreciably as one progresses from the gas to the liquid and solid phases. The $3a_1$ state, visible as a shoulder in the liquid spectrum, shifts by -0.95 eV from 525.5 eV in the gas phase to 524.55 eV in the liquid, while the spectrum of ice does not have the resolution needed to unambiguously determine its energy in this phase. The highest-energy occupied $1b_1$ state shifts -0.54 eV from 527.19 eV in the gas phase into two peaks at 526.64 and 525.91 eV in the spectra of both ice and liquid water. The fact that the intensities of these two peaks change from ice to liquid water and that in the liquid their ratio is temperature-sensitive indicates that they arise from water molecules that are hydrogen-bonded, and therefore one can get an idea of the

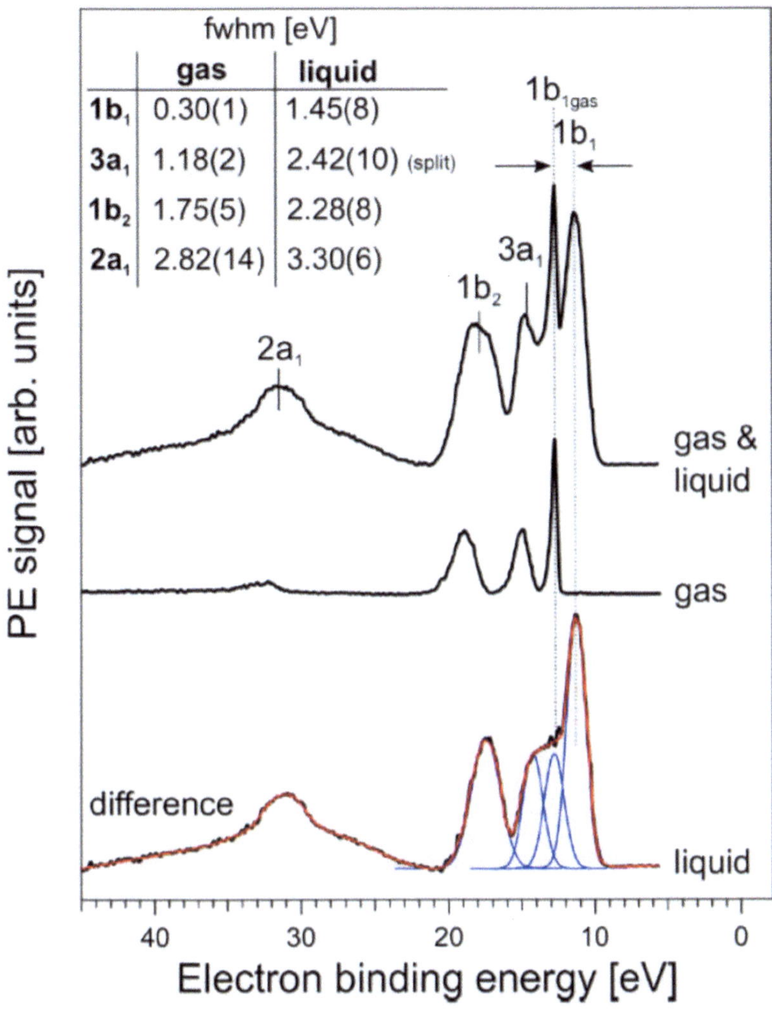

Fig. 11.16 Photoelectron spectra of the valence states of the water molecule in the liquid and gaseous phases. (With permission from Fransson et al. (2016) © 2016 American Chemical Society)

amount of hydrogen bonding in the solid and liquid phases. The hydrogen-bonded state appears at 525.91 eV in both solid and liquid phases and is much more pronounced in the spectrum of crystalline ice I_h where the number of tetrahedrally coordinated water molecules is the largest. The splitting of the signal from the transitions to the lone pair $1b_1$ electronic state is -0.74 eV indicating that the hydrogen-bonded state is lower in energy than the state where interstitial water molecules are randomly oriented. It is also noteworthy that the binding energy of the $1b_1$ electronic state in liquid water is smaller by 0.54 eV than found in ice for the non-hydrogen-bonded state, and the hydrogen-bonded state is again lower in energy

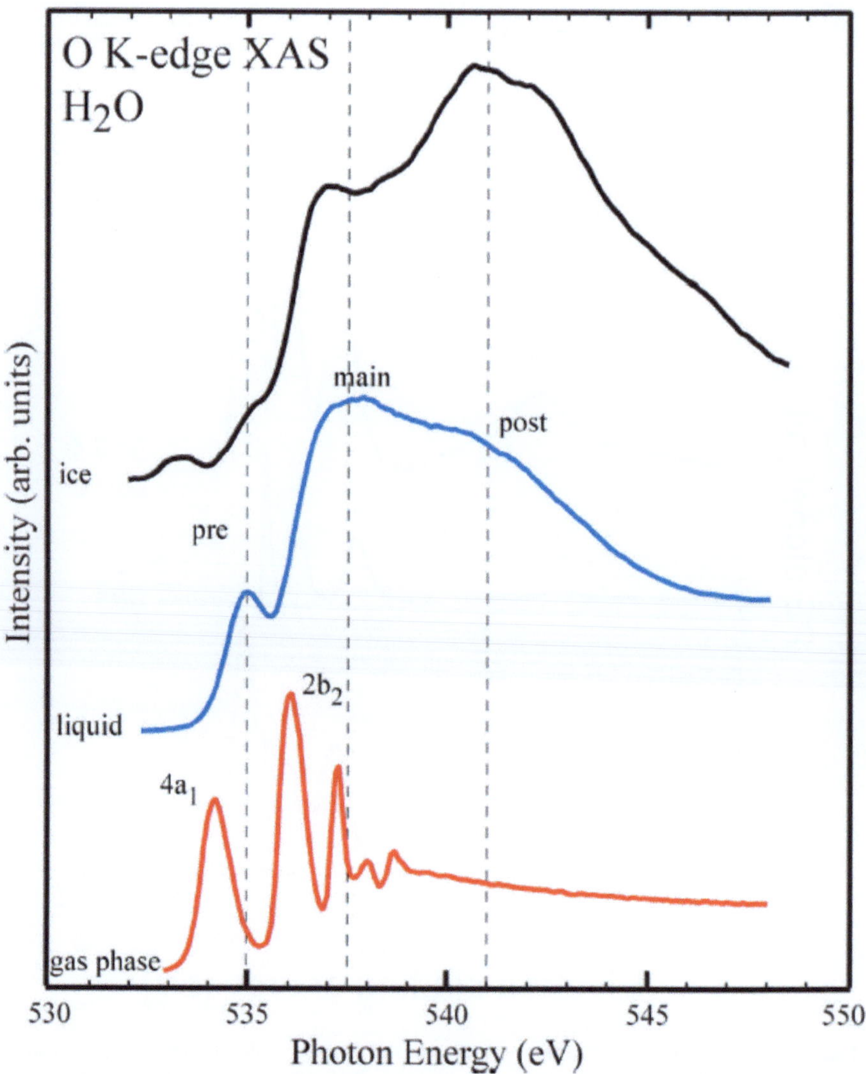

Fig. 11.17 The absorption spectra for water in its three main phases, gas, liquid, and solid (ice) determined using X-rays showing the $4a_1$ and $2b_2$ excited states in the gas phase that broaden and move to higher energy in the liquid and solid phases. (With permission from Fransson et al. (2016) © 2016 American Chemical Society)

by 0.74 eV showing that the tetrahedrally coordinated, hydrogen-bonded state is the most stable form of liquid water.

The splitting of the lone pair $1b_1$ electronic state is a reflection of the existence of the water molecule in two different environments; the lower-energy peak arises from water molecules that are hydrogen-bonded in a lower-density liquid, while the

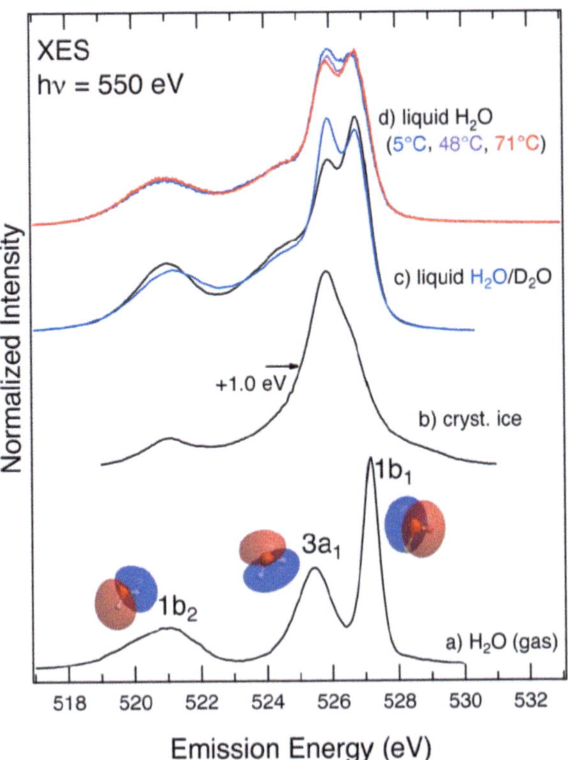

Fig. 11.18 X-ray emission spectra of water in the (**a**) gas, (**b**) solid, (**c**) liquid (H_2O and D_2O compared), and (**d**) pure H_2O (at 5, 48, and 71 °C) liquid phases. (With permission from Fransson et al. (2016) © 2016 American Chemical Society)

higher-energy state comes from the interstitial water molecules that form a higher-density liquid. For more than 100 years, dating from the time of Wilhelm Röntgen, water has been thought of as a mixture of two liquids of varying densities whose relative contributions vary with temperature. The density decrease that occurs during the phase transition from liquid water to solid ice is then a result of the increase in the low-density, tetrahedrally coordinated phase at the expense of the higher-density, but higher-energy, component containing disordered or weakened hydrogen bonds. The key feature of water is that water molecules are not spheres so that they can pack in a structure with low local density but high binding energy. This low-density form is the more ordered.

11.11   The Polarity of the Water Molecule

The water molecule is electrically neutral, but the electrons are not uniformly distributed around the bent skeleton formed by the oxygen atom and the two hydrogens. The center of positive charge is offset by 61 pm (Fig. 11.19) from the center of negative charge, giving water a large electrical dipole moment.

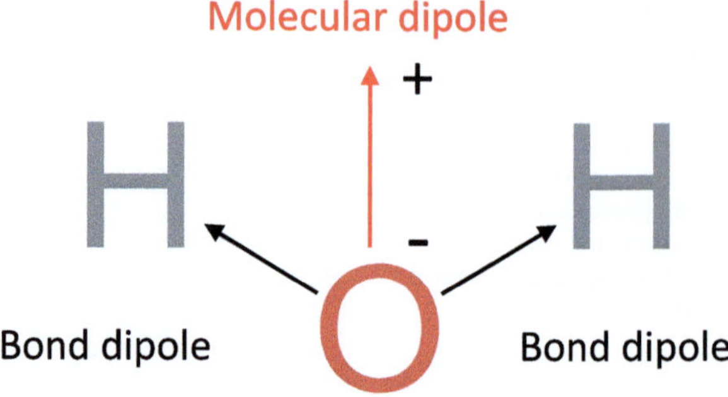

Fig. 11.19 The electric dipole moment of the water molecule as the vector sum of the O–H bond moments

We earlier alluded to the fact that water has a very large dielectric constant of ~80 at 293 K as a result of the mutual polarization among nearest neighbors when water is placed in an electric field and due to its large permanent electric dipole moment of 1.85 Debye. This dipole moment arises because oxygen is more electronegative than hydrogen, shifting the electron density from the hydrogens to the oxygen. The definition of the electric dipole moment $\vec{\mu}$ is

$$\vec{\mu} = \sum_{i=1}^{N} q_i \vec{r}_i$$

where q is the charge at position r. For a proton and an electron separated by $|r| = 100$ pm, the dipole moment is $\mu = qr = 1.6 \times 10^{-19}$ Coulomb $\times 1.00^{-10}$ meter or

$$\mu = 1.6 \times 10^{-29} \text{ C m.}$$

This can be compared with that for water as the vector sum of the moments from each of the O–H bonds and the bond angle of 104.5°. The moments from each O–H bond are obtained from the differences in electronegativity of oxygen and hydrogen for the isolated water molecule in the gas phase:

$$\mu = 2(1.5)\cos(104.5°2) = 1.84 \text{ D}$$

The units of the dipole moment are Debyes, $1 \text{ D} = 3.336 \times 10^{-32}$ Coulomb/m.

"Electronegativity," perhaps more familiar to chemists than to physicists, is a concept first proposed by J. J. Berzelius in 1811 and put on a quantitative foundation by Linus Pauling in 1932, which numerically captures the attraction of an element for a bonding pair of electrons. Electronegativity is not usually calculated but measured.

Fluorine is the most electronegative element, and oxygen is the second most electronegative element in the periodic table. The dimensionless electronegativity of an element is quantitatively given by the difference between the bonding energy of that element bound to another, heteroatom, and a covalently bound pair of homoatoms.

The energy of the atoms of two different elements, e.g., O–H, covalently bound to each other is less than the average of the bond energy of O–O and H–H. This added stabilization is attributed to the additional ionic nature of the bond. For example, the formation of water involves the reaction of hydrogen and oxygen according to

$$2H + O = H_2O + 9.493 \text{ eV}.$$

Indicating that each O–H bond in water has a bond energy of 4.747 eV. The two hydrogens could form dihydrogen, H_2, by

$$H + H = H_2 + 4.52 \text{ eV}.$$

And oxygen combines to form O_2 as

$$O + O = O_2 + 5.15 \text{ eV}.$$

We can use Pauling's expression to find the electronegativity of the O–H bond and thus the dipole moment from

$$|\chi_O - \chi_H| = \sqrt{E_{OH} - 0.5(E_{OO} + E_{HH})} = 1.10.$$

There is no electric dipole moment in a linear molecule because the bond moments point in opposite directions and hence cancel as vectors. This is why the linear carbon dioxide molecule has $\mu = 0$.

The dielectric constant is the ratio of the electrical permittivity of a substance to that of the vacuum, and hence it is a dimensionless number. The dielectric constant of water is strongly temperature-dependent (Fig. 11.20) decreasing with increasing

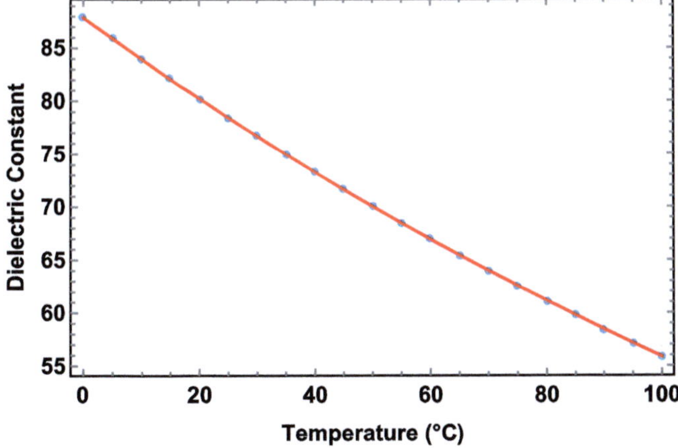

Fig. 11.20 The temperature dependence of the dielectric constant of water. (Data from NIST: Malmberg and Maryott (1956))

temperature because as the temperature increases, the number of hydrogen-bonded (Fig. 11.4) water molecules decreases.

The thermal energy of the water molecules (kT in the Boltzmann probability distribution) increases with temperature, while the fixed hydrogen bond energy of water is a constant. The random thermal motion of the water molecules breaks an increasing fraction of the hydrogen bonds as the temperature increases. The temperature dependence of the dielectric constant is given by

$$\varepsilon(T) = a + bT + cT^2 + dT^3$$

with $a = 87.87$, $b = -0.40008$, $c = 9.4 \times 10^{-4}$, and $d = -1.41 \times 10^{-6}$. Later, we will observe that the shape of this curve closely resembles that of the NMR chemical shift of the water protons as a function of temperature due to the same thermal process and that this and several other measures of the number of hydrogen bonds as a function of temperature enable one to measure the number of hydrogen bonds at each temperature.

11.12   Ice and Liquid Water

The number of hydrogen bonds in liquid water is the largest in any liquid; there are approximately the same numbers of hydrogen bonds (–H—O–H) as covalent (H–O–H) bonds. Water is such a strongly bound liquid because the formation of a hydrogen bond alters the electron distribution. The H-bond donor molecule experiences an increase in electron density in the region of the lone pairs, and the H-bond acceptor has lowered electron density for the hydrogen atoms. This rearrangement gives rise to cooperativity in H-bond networks in water. Each water molecule can form four H-bonds; two H-bond donors and two H-bond acceptors. Cooperativity is so important in creating H-bond networks in water that the binding energy of liquid water is about 2.5 times larger than that solely due to the binding of two water molecules in the dimer (Luck 1998).

11.12.1   Tetrahedral Water

The directed hydrogen bond network in liquid water consists of a tetrahedral arrangement (Russo et al. 2018) of four water molecules surrounding a central molecule (Fig. 11.21) with an H-bond angle of 109.47°. This transient, temperature-dependent structure is supported by the fact that X-ray and neutron scattering measurements have shown that water exhibits a peak in the oxygen-oxygen pair correlation function, $g_{OO}(r)$, at $r = 450$ pm, corresponding to the

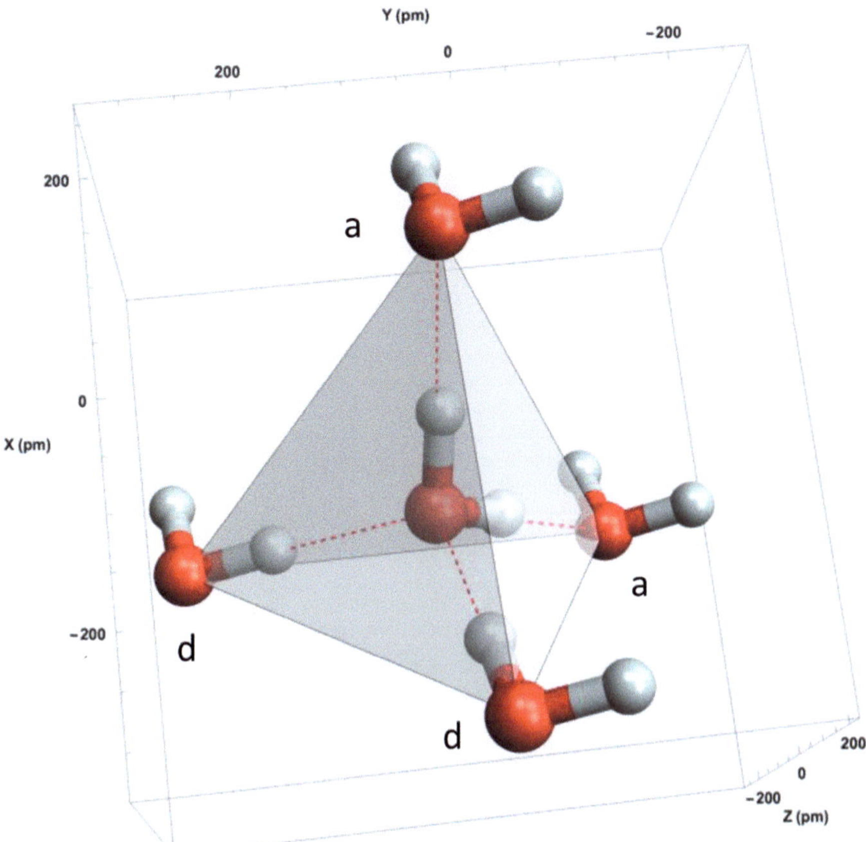

Fig. 11.21 A schematic of the tetrahedral hydrogen bonding in water. The oxygen-oxygen distance in liquid water is about 282 pm (2.82 Å) at ambient temperature. The two molecules labeled a are H-bond acceptors, while those labeled d are H-bond donors. The tetrahedral angle $\psi_{jk} = 109.47°$ is very close to that of the H–O–H bond angle (~105°) in the water molecule

second-nearest neighbor distance in tetrahedral coordination, $r = \sqrt{8/3}a$, where $a \sim 280$ pm is the nearest neighbor oxygen-oxygen distance in liquid water.

Computer simulations of water structure using model intermolecular potential energy functions can be used to examine this tetrahedral arrangement by generating the orientational order parameter, Q, defined (Errington and Debenedetti 2001) as

$$Q = 1 - \frac{3}{8} \sum_{j=1}^{3} \sum_{k=j+1}^{4} \left(\cos \psi_{jk} + \frac{1}{3} \right)^2$$

where the angle ψ_{jk} is formed by the vectors linking the oxygen atom on the central molecule to those of its j nearest neighbor oxygens, with $k \leq 4$. If the central

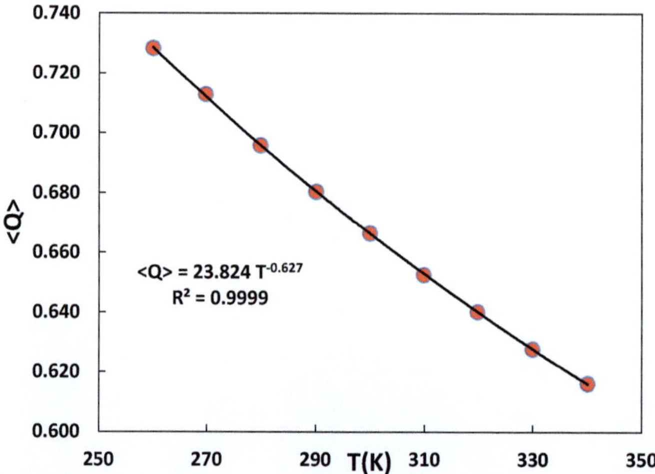

Fig. 11.22 The effect of temperature on Q, the mean tetrahedrality parameter for liquid water. The temperature dependence is $\sim T^{-2/3}$

molecule is located at the center of a tetrahedron and the other four are found at the tetrahedral vertices, then $\psi_{jk} = 109.47°$ for all values of j, k, and $\cos\psi_{jk} = -1/3$, and the squared term in the sum evaluates to zero, giving $Q = 1$. On the other hand, in a random arrangement of molecules, as for an ideal gas, the six angles are uncorrelated, and the average value of Q can be found by integrating over ψ as

$$\langle Q \rangle = 1 - \frac{9}{8} \int_0^\pi \left(\cos\psi + \frac{1}{3} \right)^2 \sin\psi \, d\psi = 0.$$

Thus, Q varies from zero in a random collection of molecules to one for a perfectly tetrahedral structure. It is therefore of significant interest to examine Q for water, particularly with respect to its temperature dependence. Sellberg and coworkers (2014a, b) have found the tetrahedrality of liquid water, as estimated using the TIP4P/2005 (vide infra) potential energy model for water, to be less than one, but still quite large (Fig. 11.22), and decreasing with approximately the inverse 2/3 power of the temperature as the energy due to thermal motion competes with the hydrogen bond energy. How much must the angles of the hydrogen bonds differ from the tetrahedral angle of 109.5° in order to reduce Q from 1.0 to ~0.73? We can use Mathematica to estimate this deviation by writing $Q(X)$ as a function of a random variable X and plotting the resulting Q vs. X (Fig. 11.23). X is a parameter in

```
RandomReal[{1 - x, 1 + x}]
```

that randomly modulates the angles ψ_{jk} between vectors j and k in the above equation for Q as in

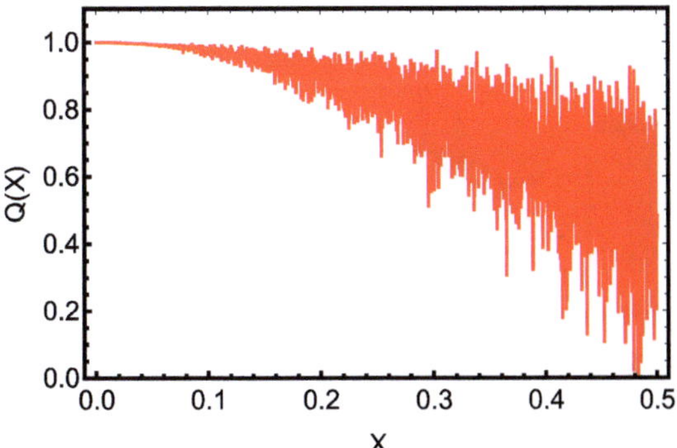

Fig. 11.23 The variation of the mean tetrahedrality $Q(X)$ of water as the O–H hydrogen bond angle randomly deviates from 109.5°

$$Q[X\_] := 1 - \frac{3}{8} \sum_{j=1}^{3} \sum_{k=j+1}^{4} \left(\cos \psi_{jk} + \frac{1}{3} \right)^2$$

where $\psi_{jk} = $ `VectorAngle[vertex[[i]] RandomReal[{1 - x, 1 + x}],vertex[[j]] RandomReal[{1 - x, 1 + x}]]`.

We find (Fig. 11.23) that the mean H-bond angles must deviate from the tetrahedral angle by about 30–35° in order to reduce Q from 1.0 to about 0.7, as observed. Therefore, while most of the molecules in ambient water are involved in tetrahedral coordination of their H-bonds, a substantial fraction is disordered in the liquid.

11.12.2   Ice I_h

It is instructive to compare the values for liquid water with those from ice I_h, whose unit cell is shown in Fig. 11.24. One can also obtain the value of Q and the hydrogen bond angle and length from the X-ray coordinates of the water molecules in ice I_h using Mathematica's `VectorAngle[]` function again, which gives a bond length of 276 pm and an angle of 102.7°.

A view looking down the z-axis of the unit cell (Fig. 11.24, Right) shows the hexagonal arrangement of water molecules in ice I_h. This hexagonal arrangement is shown more clearly in a larger crystal (ice1h.pdb from the Protein Data Bank: https:// www.rcsb.org) in Fig. 11.25 where we have used Mathematica to plot the structure of ice I_h where the coordinates are the real distances (in pm).

Compare this crystal structure with that of a snow flake (Fig. 11.26) which also displays the hexagonal arrangement on a macroscopic basis in photograph number 891 of Wilson Bentley's marvelous series of photomicrographs taken during long, snowy winters in Vermont in the early years of the twentieth century. In the section of his paper on the structure of snow crystals, Bentley's musing that the "...interior details more or less perfectly outline preexisting forms" must have reflected the early

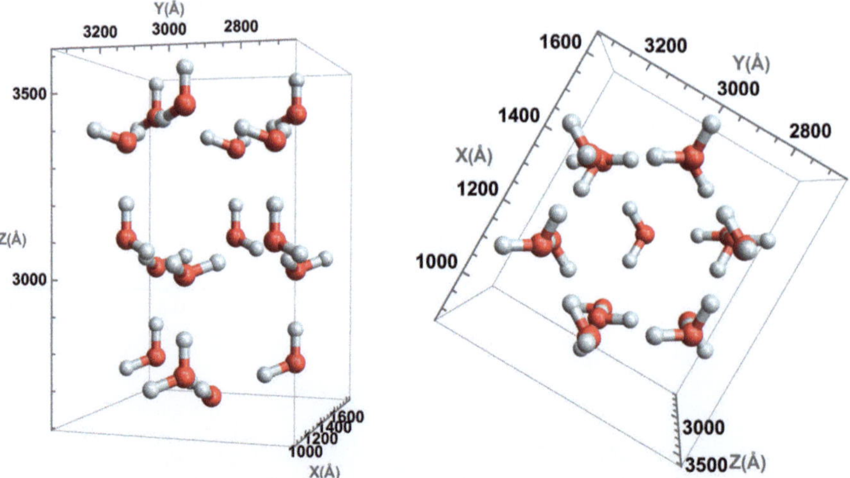

Fig. 11.24 (*Left*) The unit cell for the crystal of ice I_h. (*Right*) A view down the z-axis shows the hexagonal arrangement of the water molecules in ice I_h. Rendered in Mathematica from the protein data bank file: 1h1cel.pdb

recognition that there was an underlying, repeating, basis for these patterns (Bentley 1902). His pioneering work photographing snowflakes occurred prior to the development of X-ray diffraction studies of ice, but nowadays one is tempted to suggest that the nature of these *preexisting forms* was hiding in plain sight in his stunning pictures. The hexagonal unit cell for ice would generate essentially all of the six-sided forms he observed, and could be inferred from the fact that the central portions of many of his snowflakes show this strikingly.

11.13 Models of Water: Intermolecular Potential Energy Function

At Bell Laboratories, in the late 1960s, Ben-Naim and Stillinger were the first investigators to develop an electrical potential model for the water molecule based on the placement of classical point charges at crystallographically suggested positions within the molecular framework (Stillinger and Ben-Naim 1967; Stillinger 1975; Ben-Naim 2009). This BNS model used an intermolecular potential energy model of the form

$$E_{ab} = \sum_{i \in a} \sum_{j \in b} \frac{kq_i q_j}{r_{ij}} + \frac{A}{r_O^{12}} - \frac{B}{r_O^6}$$

which is recognized as the sum of a Coulomb interaction among point charges with the addition of a Lenard-Jones term whose denominator is the separation of the

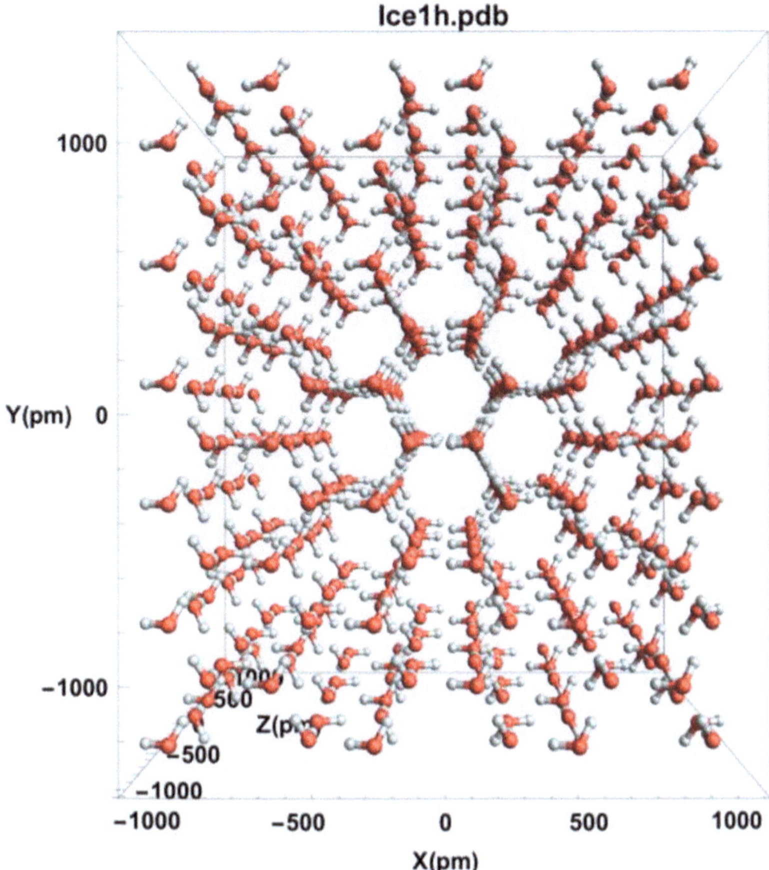

Fig. 11.25 A Mathematica rendering of the protein data bank file Ice1h.pdb showing the hexagonal ordering of the water molecules in ice 1_h along the z-axis of the crystal. The repeating hexagonal motif is quite apparent

oxygen nuclei. The parameters for this model are given in Table 11.10 where one can observe that this model closely observes tetrahedral geometry (as shown in Fig. 11.6) with two negative lone pairs (L) and two positive hydrogen atoms at the vertices of the tetrahedron.

Many have proposed modifications to the positions, strengths, and numbers of these Coulomb centers and have used them as inputs to molecular mechanics simulations in an attempt to reproduce the properties of water, with varying degrees of success (Tan et al. 2003). One of the more successful models is the TIP5P introduced by Mahoney and Jorgensen in 2000 (Mahoney and Jorgensen 2000; Khalak et al. 2018) which uses a five point charge model (hence the 5P in the name) with zero charge on the oxygen atom and equal positive and negative charges on the hydrogen atoms and the lone pairs with a parameter set given in Table 11.11 where the H–O–H bond angle is set to the value in water, not the tetrahedral angle as above,

Fig. 11.26 Photograph number 891 of a snowflake from Plate XIX, page 136 of Wilson Bentley's "Studies among the Snow Crystals ..." Bentley (1902) (https://siarchives.si.edu/sites/default/files/pdfs/WAB_Snow_1902.pdf)

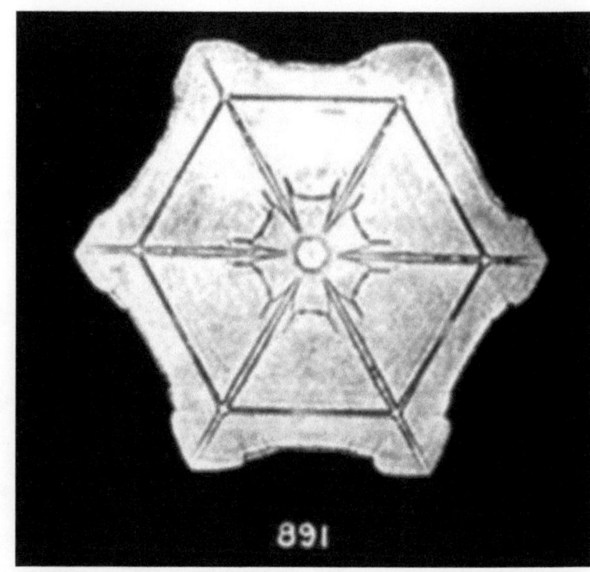

Table 11.10 Parameters for the BNS point charge model of water

Parameter	Value	Units
r_{OH}	100	pm
HOH angle	109.47	degrees (°)
r_{OL}	100	pm
LOL angle	109.47	degrees (°)
A	77.4	10^3 kcal $\mathrm{\AA}^{12}$/mol
B	153.8	kcal $\mathrm{\AA}^{6}$/mol
q_H	+0.196	e
q_L	−0.196	e
k	332.1	$\mathrm{\AA}$ kcal/mol e^2

Table 11.11 Parameters for the TIP5P point charge model of water

Parameter	Value	Units
r_{OH}	95.72	pm
HOH angle	104.52	degrees (°)
$r_{OLone\ pair}$	70	pm
LOL angle	109.47	degrees (°)
q_O	0	e
q_H	0.241	e
q_L	−0.241	e
ε	6.94×10^{-6}	eV
σ_O	312	pm
σ (HH)	100	pm

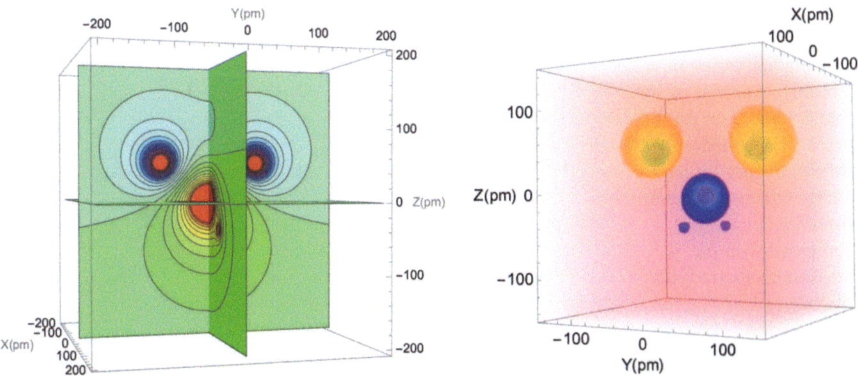

Fig. 11.27 (*Left*) A contour plot of the electrostatic potential from the charge distribution in the TIP5P point charge model of the water molecule. (*Right*) The negative oxygen and lone pairs (purple) and the positive hydrogen (orange) are plotted with the parameters in Table 11.11

while the lone pairs are at the tetrahedral angle. The electrostatic potential of this model (Fig. 11.27) as a function of the inter-oxygen distance r_{OO} is essentially the same as the BNS model:

$$E_{ab} = \sum_i \sum_j \frac{e^2 q_i q_j}{r_{ij}} + 4\varepsilon \left[\left(\frac{\sigma_0}{r_{OO}} \right)^{12} - \left(\frac{\sigma_0}{r_{OO}} \right)^6 \right]$$

The rather small Lennard-Jones energy ε (Table 11.11) indicates that the main source of H-bond energy arises from the dipole moment distribution of charge. This model gives a dipole moment of 2.29 D which is higher than the experimental value of 1.85 D for the measured dipole moment of the gas phase water molecule. It is clear that the quantum mechanical details of the electronic distribution change markedly when water changes from the gas to the liquid phase, with the dipole moment increasing to ~2.5 due to polarization of the cluster environment from nearest neighbors (Gregory et al. 1997) so this may not be an issue for these calculations. The calculated dielectric constant (Fig. 11.28) was also higher than that measured (Fig. 11.21) by Malmberg and Maryott (1956) at every temperature. The density maximum at 3.98 °C was also poorly reproduced (Fig. 11.29). Differentiation of the sixth order fit to the data of Vedamuthu et al. (1994) gave a density maximum of 0.998615 g/mL at a temperature of 4.0711 °C, while doing the same for a second-order fit to the data of (Mahoney and Jorgensen 2000) gave a density maximum of 1.0026 g/mL at a temperature of 10.34 °C.

There are now at least 30 point charge models in the literature (https://water.lsbu. ac.uk/water/water_models.html). It is fair to say that even after many decades of study there is not a uniformly agreed-upon classical electrostatic model that mimics all behaviors of the physical water molecule, but abandonment of the rigid X-ray-

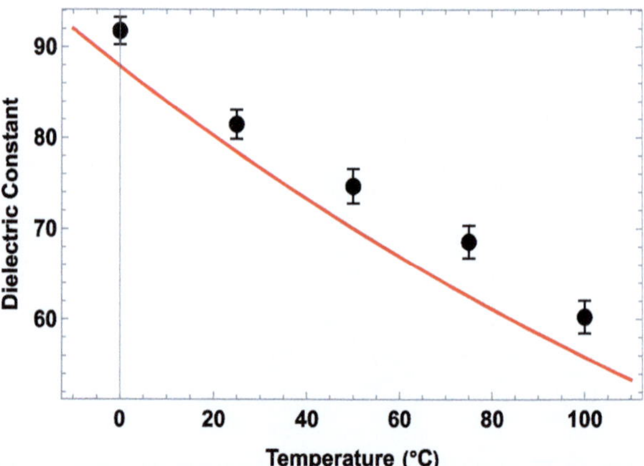

Fig. 11.28 A molecular dynamics calculation of the temperature variation of the dielectric constant of water (black), based on the TIP5P point charge model by (Mahoney and Jorgensen 2000, with estimated errors from the simulation) compared with the measurements (red) of Malmberg and Maryott (1956)

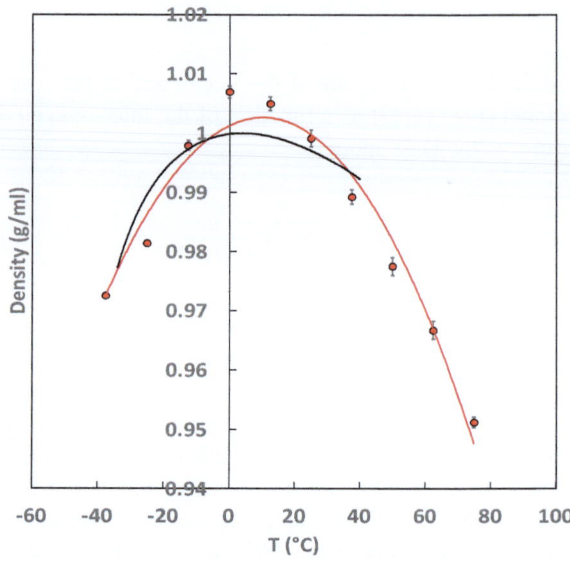

Fig. 11.29 A molecular dynamics calculation of the temperature variation of the density of water based on the TIP5P point charge model by Mahoney and Jorgensen (2000) (red, with estimated errors from the simulation) compared with the measurements (black) of Vedamuthu et al. (1994)

derived structure of point charge models has led to surprisingly better models incorporating the dipole, quadrupole, and octupole moments of the electric charge distribution (Izadi et al. 2014) with good reproduction of the temperature dependence of several of water's physical parameters (Fig. 11.30) including the dielectric constant, the diffusion coefficient, and the density to an accuracy of ~1%, although the calculated temperature of the density maximum (-1 °C) is in error by about 5 °C.

This is hardly surprising given that a realistic model for water needs be based on the quantum mechanical wave functions, but these have not been found to be amenable to rapid computer simulations so that investigators have resorted to classical approximations of water that can be allowed to roam small boxes in silico,

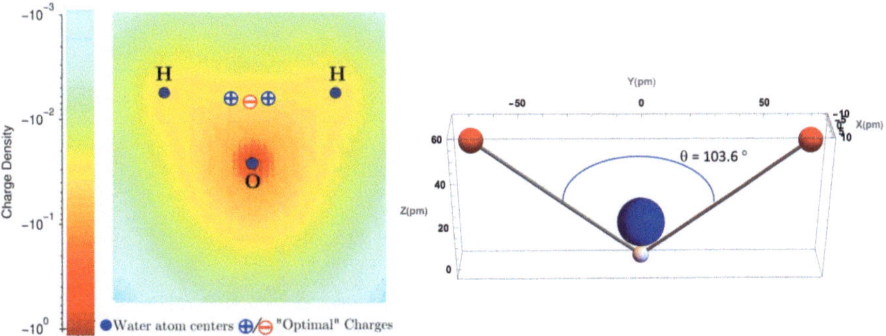

Fig. 11.30 An optimized three-point charge model of water based on the quantum mechanical structure of the gas phase water structure showing the displacement of the charges and the alterations in the rigid structure which differ from the BNS and TIP5P models used in the past. ((*Left*) Reproduced from Izadi et al. (2014), used with permission. © 2014 American Chemical Society. Creative Commons CC BY-NC 4.0.) (*Right*) The parameters include charges and positions (pm) of $\{q\{0, -68.56, 53.95\}, q\{0, 68.56, 53.95\}, -2q\{0, 0, 15.94\}\}$, where $q = 0.6791\ e$

with the hope that refinement of the numbers, strengths, and positions of the classical point charges would serve to provide some reasonable model of water that can serve as a surrogate for the real thing. This is an ongoing process with success measured by the agreement between the computer simulations and the many physical properties that have been measured to great precision. *We aren't there yet.* It would appear that a new way of generating water models is needed in spite of the literally hundreds of publications using these classical models. One would suggest that simulations using the well-known quantum mechanical wave functions (Coolidge 1932; Ceriotti et al. 2016) would do the trick, but this has only recently been attempted (Fig. 11.30).

One can calculate these wave functions using quantum density functional theory to rather good precision, but using them in a simulation has not yet been attempted. Until then, the myriad publications using classical models must remain an interesting first attempt at the real theory. Still, we do know a great deal about this most important molecule whose existence led to life.

11.14   Water Structure

What do we mean when we ask "What is the structure of water?". Many authors, beginning in 1933 with Bernal and Fowler (1933), have examined the crystal structure of ice I_h and have suggested that liquid water retains some of its characteristics, including remnants of its hexagonal pattern of hydrogen bonding (Fig. 11.31). It is difficult to imagine that any liquid could possess sufficient long-range order that qualifies as structure in the sense commonly applied to the time-invariant atomic or molecular structure of solids or crystals for the water molecules

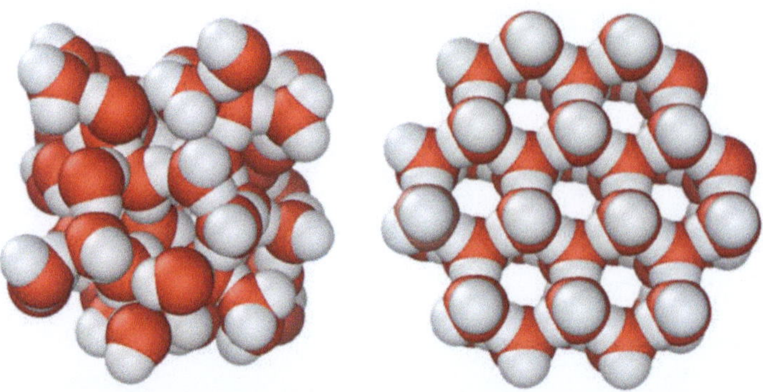

Fig. 11.31 On the *left* is a snapshot of a simulation of liquid water where the molecules are transiently bound by hydrogen bonds. On the *right* is the structure of hexagonal ice I_h which is bound by long-lived hydrogen bonds. These structures have supported the insight from more than 100 years ago by Röntgen that water is a mixture of two components (Tokushima et al. 2010; Gallo et al. 2016), a low-density form of strongly bonded ice-like structures (*right*) and another of the higher-density amorphous weaker bonded liquid (*left*). (From M. Chaplin (https://water.lsbu.ac.uk/water/martin_chaplin.html). Used with permission)

are in constant motion (Rahman & Stillinger 1971; Kalman et al. 1977; Huang et al. 2011; Petkov et al. 2012). Nevertheless, a key feature of water is that water molecules are not spheres so that they can pack in a structure with low local density but high binding energy (ice). This low-density form is the more ordered (Fig. 11.31). It was thought that high resolution X-ray emission spectroscopy of water (Tokushima et al. 2010) directly showed these two these two structural motifs as a doublet in the $1b_1$ x-ray emission spectrum (Fig. 11.16) but recent attosecond measurements in liquid water (Shuai et al. 2024) have shown that this splitting is rather a dynamical phenomenon.

We can determine the timescale for the motions in liquid water using spectroscopic techniques, such as NMR. The NMR spin-lattice relaxation time for water at 25 °C is on the order of ~2.5 s, which translates into a correlation time of about 2 ps, while the diffusion coefficient at this temperature (2.2×10^{-9} m^2/s) indicates that during one correlation time a water molecule diffuses 160 pm or about half its diameter. Therefore, any putative structure must be very short-lived.

11.14.1   The Velocity Autocorrelation Function for Water

These considerations lead one to replace the concept of a static structure with that of temporal and spatial correlations among water molecules that endure long enough to be measured. One of these is the time-dependent velocity autocorrelation function $C(t)$ (Balucaniy et al. 1996)

$$C(t) = \langle v(t') \cdot v(t) \rangle = \frac{1}{N} \sum_{i=1}^{N} v_i(t') \cdot v_i(t)$$

where $v_i(t)$ is the velocity vector for particle i at time t. The vector dot product measures the overlap between the velocity vector at time t with that at time t'. What are the properties of this function? At $t = t' = 0$ or for any other two identical times, the velocity autocorrelation function is the square of the average particle velocity

$$C(0) = \langle v(0) \cdot v(0) \rangle = \frac{1}{N} \sum_{i=1}^{N} v_i(0) \cdot v_i(0) = \frac{1}{N} \sum_{i=1}^{N} v_i(0) \cdot v_i(0) = \langle v^2 \rangle$$

which, using the equipartition theorem, is given by the average thermal kinetic energy in d dimensions as

$$\frac{1}{2} m \langle v^2 \rangle = \frac{d}{2} kT$$

or

$$C(0) = \langle v^2 \rangle = \frac{d\,kT}{m},$$

while, for long times, the correlation between velocity vectors decays to zero:

$$\lim_{t \to \infty} C(t) = 0$$

We have seen similar behavior for the rotational correlation function in our study of nuclear magnetic relaxation (Chap. 7).

The velocity vector is the time derivative of the time-dependent spatial position vector:

$$v(t) = \frac{dx(t)}{dt}$$

This can be integrated to produce the vector displacement of a water molecule in a given time interval as

$$\Delta x(t) = \int_0^t \frac{dx(t)}{dt} dt = x(t) - x(0) = \int_0^t v(t') dt'.$$

As we have seen from an examination of the average displacement in a random walk, the mean distance traveled from the origin of a water molecule undergoing random motions is zero:

$$\langle \Delta x(t) \rangle = \left\langle \int_0^t v(t')\, dt' \right\rangle = 0$$

However, again referring to the random walk, we found that the mean squared distance traveled is nonzero. We can write this as

$$\left\langle [\Delta x(t)]^2 \right\rangle = \left\langle [x(t) - x(0)]^2 \right\rangle = \left\langle \int_0^t v(t')\, dt' \cdot \int_0^t v(t'')\, dt'' \right\rangle$$
$$= \int_0^t \int_0^t \langle v(t') \cdot v(t'') \rangle\, dt'' dt'$$

where, in the last step, we brought the average inside the integral. From the work of Einstein on Brownian motion, the left side of this equation, the rms displacement in d dimensions, is equal to $2dDt$:

$$\left\langle [\Delta x(t)]^2 \right\rangle = 2dDt$$

Therefore, the double integral on the right-hand side must also give this same result

$$2dDt = \int_0^t \int_0^t \langle v(t') \cdot v(t'') \rangle\, dt'' dt'$$

or that the diffusion coefficient is given by

$$D = \frac{1}{2td} \int_0^t \int_0^t \langle v(t') \cdot v(t'') \rangle\, dt'' dt'.$$

The integrand is the velocity autocorrelation function defined above. This is to be understood as taking the limit as t becomes large with respect to the timescale for molecular displacements. For short times, on the order of a few picoseconds, the motion of molecules is ballistic in the sense that they travel linearly within the cage formed by their nearest neighbors, while for longer times the motion switches over to random diffusion.

It is clear from symmetry considerations that the velocity autocorrelation function has three additional properties; the first is that it is symmetric about $t = 0$,

$$C(t) = C(-t),$$

and the second is that it only depends on the time difference $\tau = t'' - t'$ and not on the absolute time. Therefore, we were justified in our initial definition of this function as

$$C(\tau) = \langle v(0) \cdot v(\tau) \rangle.$$

The third property is that $C(\tau)$ is unchanged if the two time arguments are interchanged

$$C(\tau) = C(t''-t') = \langle v(0) \cdot v(\tau) \rangle$$

$$C(-\tau) = C(t'-t'') = \langle v(\tau) \cdot v(0) \rangle = \langle v(0) \cdot v(\tau) \rangle$$

because the dot product of the two velocity vectors is commutative. We can use these properties to find the diffusion coefficient in the long time limit as

$$D = \frac{1}{d} \int_0^\infty C(t)dt.$$

We proceed from

$$\int_0^t \int_0^t \langle v(t') \cdot v(t'') \rangle \, dt''dt' = \int_0^t \int_0^t C(t'-t'') \, dt'dt''$$

$$= \int_0^t \int_{-\infty}^\infty C(t'-t'') \, dt'dt''.$$

Note, here, we extended the limits of integration of the t' integral to infinity since $C(t)$ goes to zero as $|t|$ becomes very large. Then, inserting the time difference $\tau = t' - t''$, with $d\tau = dt'$,

$$= \int_0^t \int_{-\infty}^\infty C(\tau) \, d\tau dt''.$$

The integral over t'' gives a factor of t

$$= t \int_{-\infty}^\infty C(\tau)d\tau$$

$$= 2t \int_0^\infty C(\tau)d\tau.$$

Here, we used the symmetry about zero time to replace the lower limit of infinity with zero with a factor of two:

$$D = \frac{1}{6t} \int_0^t \int_0^t C(t'-t'')dt'dt'' = 2t\frac{1}{6t} \int_0^\infty C(\tau) \, d\tau$$

We have in $d = 3$ dimensions our result that the diffusion coefficient is the integral of the velocity autocorrelation function

$$D = \frac{1}{3}\int_0^\infty C(\tau)d\tau.$$

This has the proper dimensions for a diffusion coefficient of [(Length)^2/time] since $C(t)$ has units of velocity squared (Length/time)^2 and the integral introduces a factor of time. In order to produce the correct magnitude for the diffusion coefficient, this must be multiplied by $C(0)$ from above to give

$$D = \frac{kT}{m}\int_0^\infty C(\tau)d\tau.$$

This is an example of Kubo-Green theory which relates transport coefficients to correlation functions through the fluctuation-dissipation theorem.

It is useful to note that the equation for the velocity autocorrelation function appears to be quite simple,

$$C(\tau) = \langle v(0) \cdot v(\tau)\rangle,$$

but appearances can be deceiving, and in this case they are. In order to calculate this function, one must somehow obtain a table of velocities at two times: the velocity at zero time is a constant vector, but the time-dependent velocity, $v(\tau)$, must be gotten either from a model or from a molecular dynamics simulation.

To what extent can velocity autocorrelation functions inform us about the structure of water? In order to answer this question, let us examine the velocity autocorrelation functions for several substances beginning with noninteracting molecules in a very dilute gas, progressing to denser gases, hard-sphere fluids, Lennard-Jones fluids, water, and crystalline solids. For noninteracting atoms or molecules in a very dilute gas, the motion is purely Newtonian, and since there is no external force acting upon these particles, their velocity is constant, and the velocity autocorrelation function is (in d dimensions)

$$\langle v^2\rangle = \frac{dkT}{m},$$

for all time. As the density of particles increases from this very dilute limit, a gas of hard-sphere particles will start to experience random collisions with other particles, and this will lead to the decay of the velocity autocorrelation function with time. For example, a simple Mathematica code to generate an approximate velocity autocorrelation function begins with a unit vector along the y-direction as $v(0) = \{0,1\}$, then takes the dot product of this with a randomly oriented unit vector as

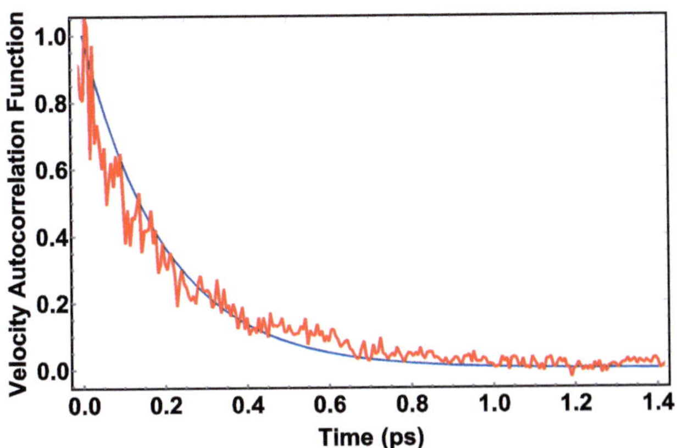

Fig. 11.32 (Red) The normalized velocity autocorrelation function for a unit velocity vector oriented initially along the *y*-axis decays through the addition of a random unit vector in the *x–y* plane. (Blue) An exponential model with a decay constant of 0.2 ps. Note here the zero time limit is {0,1}.{0,1} = 1. The decay is due to random collisions among particles

```
Table[(1/(nv + 1)){0,1}.({0,1}+RandomReal[{-1,1}]),
{nv,0,50,0.35}],
```

and when plotted versus the number (nv) of time steps (Fig. 11.32) shows that the correlation function decays essentially exponentially because the particles undergo only single scattering and over time this randomizes their velocities. If we let the time steps represent collisions, then the velocity decorrelates after about ten collisions.

As the density increases, multiple scattering events and recoil effects come into play which drive the velocity autocorrelation function to zero even faster. Brownian motion is understood as the response of a particle to a randomly oriented force vector. The characteristic decay time τ for the velocity autocorrelation function is proportional to the mass and the diffusion coefficient of the particles. Following Langevin, we can model such a system by assuming that the velocity autocorrelation function is a simple exponential, with a time constant given by the mean time between collisions, such as

$$C(t) = e^{-t/\tau}$$

where

$$\tau = \frac{mD}{kT}$$

so that

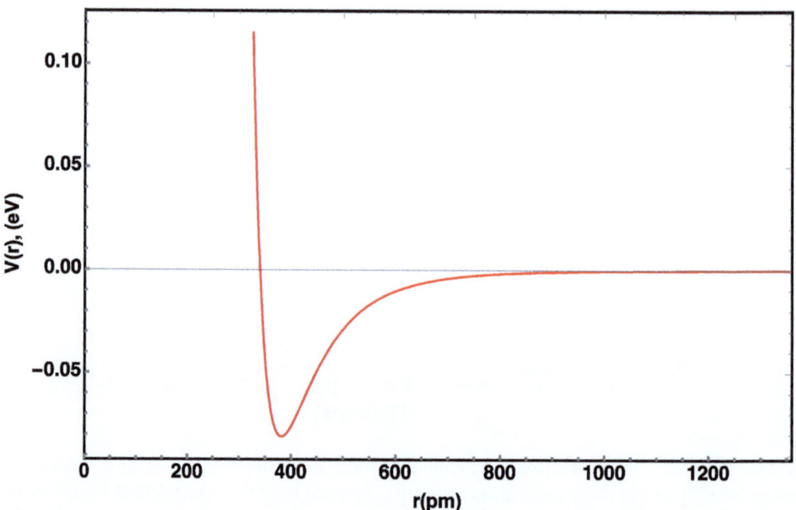

Fig. 11.33 The interatomic potential for liquid argon with $\varepsilon = 1.656 \times 10^{-21}$ J (10.3 meV) and $\sigma = 340$ pm

$$D = \frac{kT}{m} \int_0^{\infty} e^{-t/\tau} dt.$$

And the diffusion coefficient is related to the vibration frequency (the fluctuations in position) of the particle in this Langevin treatment. However, remember that ideal gases are not bound systems and the collisions among gaseous particles are essentially elastic. This ignores the van der Waals and London dispersion forces that arise from the fluctuating multipole moments of atoms and molecules. Particles in fluids, on the other hand, are bound in the liquid state by their mutual van der Waals attraction (and in water by electrostatic attraction). The minimum distance of approach is due to the repulsive Pauli exclusion of the electrons.

For a simple mono-atomic Lennard-Jones fluid, such as liquid argon (Rahman 1964), the interatomic potential function is

$$V(r) = 4\varepsilon \left[\left(\frac{\sigma}{r} \right)^{12} - \left(\frac{\sigma}{r} \right)^{6} \right]$$

with $\varepsilon = 1.656 \times 10^{-21}$ J (10.3 meV) and $\sigma = 340$ pm, as shown in Fig. 11.33. With its negative minimum around 380 pm, liquid argon ($m_p = 84$ K) is more dense ($\rho = 1374$ kg/m^3 at 94 K) than water, whose nominal density is ~1000 kg/m^3 at 277 K. The steep rise at small interparticle separations is the Pauli electronic repulsion, while the attractive part models the net van der Waals force that is responsible for binding the atoms into a liquid.

The velocity autocorrelation function (Fig. 11.34 (left)) for liquid argon is little different from a simple exponential with a slight dip into negative territory around

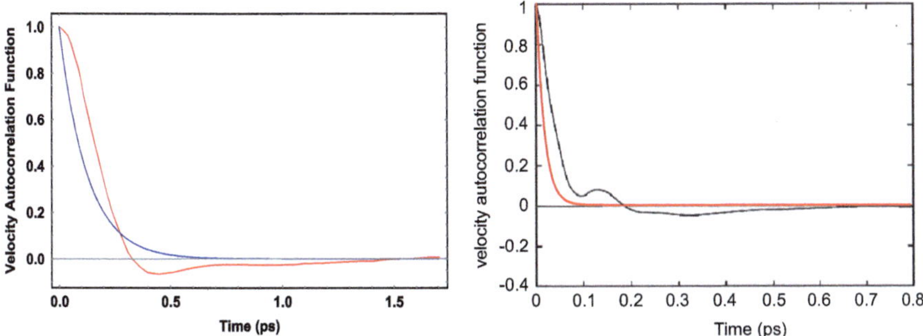

Fig. 11.34 (*Left*) The normalized velocity autocorrelation function for liquid argon (red) computed with molecular dynamics with a Lennard-Jones potential with parameters given in the text (Rahman 1964) compared with an exponential (blue) computed using the Langevin time constant at 94 K of 0.125 ps. (*Right*) The normalized velocity autocorrelation function for liquid water at 293 K (black) computed using molecular dynamics with the TIP4P water pair potential (Zlenko 2012) compared with an exponential (red) with a decay time given by the Langevin time constant at 293 K of 0.167 ps

500 fs arising from the recoil of argon atoms in collisions with other atoms in this dense fluid. The initial decay of the argon velocity autocorrelation function is approximately exponential, with a Langevin time constant at 94 K of

$$\tau = \frac{mD}{kT} = 1.25 \times 10^{-13} \text{ s.}$$

On the other hand, the velocity autocorrelation function for water, as computed using molecular dynamics, shows (Fig. 11.34, right) a pronounced positive peak around 0.120 ps in addition to the main positive peak at zero time. This peak is due to the additional hydrogen bonding in liquid water where a bound water molecule is pulled along with its neighbor in the same direction, but at a later time as a result of its inertia, much like two spheres linked by a short rubber band.

11.14.2   The Spatial Autocorrelation Function for Water

The spatial correlations among water molecules are reflected in the pair correlation function (also known as the radial distribution function) $g(r)$. This contains perhaps the most useful information on the spatial arrangement of water molecules because it is directly measurable via either X-ray or neutron scattering or can be obtained from thermodynamic quantities through inversion of the Kirkwood-Buff theory of solutions. In an isotropic and homogeneous liquid, the mode of packing of the molecules can be understood if we choose a molecule as the center and count the number of

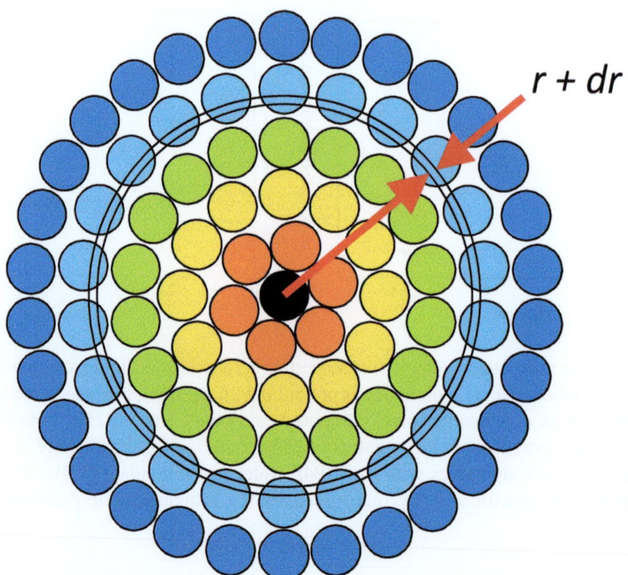

Fig. 11.35 A schematic view of a 2D fluid showing how the radial distribution function is calculated. One begins with a molecule at the origin (black circle) and proceeds outward counting the number of molecular centers encountered as a function of distance in a shell of thickness dr

molecules that are found in a spherical shell at a radius r and thickness, dr (Fig. 11.35), surrounding the central molecule. The volume of each spherical shell is $dV = 4\pi r^2 dr$. Given that the average density of the fluid is $\rho = \frac{N}{V}$, then the number of molecules in dV is $= 4\pi\rho r^2 dr$.

For values of r less than a molecular diameter d, we will find no other molecules due to Pauli electronic repulsion so $g(r \leq d) = 0$. As r becomes large, we expect that the probability of observing two molecules with that separation becomes constant related to the density of the fluid. The pair correlation function is defined to take this into consideration by dividing by the density so that $\lim_{r \to \infty} g(r) = 1$.

We can derive an expression for the molecular correlation functions involving single molecules, pairs of molecules, triplets, etc. by examining a set of N molecules in a total volume of V subject to an intermolecular potential U. The average number density is $\rho = \frac{N}{V}$. Letting the N molecules be located at the vector positions $\{r_i\}$, $i \in \{1, N\}$, then the potential can be written as

$$U = U(N, \{r_i\}) = U(r_1, r_2, \ldots, r_N),$$

and the configuration integral (related to the partition function) Z is given by the N-dimensional integral

$$Z = \int \cdots \int e^{-U/kT} dr_1 dr_2 \ldots dr_N$$

which is essentially an integral of the Boltzmann distribution of probabilities. The probability of finding a particular configuration, with molecule 1 at dr_1 and molecule 2 at $dr_2 \ldots$ and molecule N at dr_N, is

$$P(N, \{r_i\})dr_1 dr_2 \ldots dr_N = \frac{e^{-U/kT}}{Z} dr_1 dr_2 \ldots dr_N$$

if the infinitesimal volumes $\{dr_i\}$ are chosen small enough that essentially only one molecule would be found within it so that the individual probabilities are independent *and* we can multiply them. In any macroscopic volume of water, the number of molecules is enormous, and the probability defined above for $N = 1$ is related to the average density. The probability of a configuration where $n < N$ molecules are specified and the other $N - n$ are free is found by integrating over the remaining coordinates $dr_{n+1} dr_{n+2} \ldots dr_N$

$$P(n, \{r_i\}) = \frac{1}{Z} \int \cdots \int e^{-U/kT} dr_{n+1} dr_{n+2} \ldots dr_N. \tag{11.1}$$

Since all water molecules are identical, we can define the n-molecule density $\rho(n, \{r_i\})$ that gives the probability that any n molecules are found at positions $\{r_i\}, i \in \{1, n\}$ as

$$\rho(n, \{r_i\}) = \frac{N!}{(N-n)!} P(n, \{r_i\})$$

where the factorials correct for the number of identical permutations of N molecules taken n at a time (Chap. 1). The integral of $\rho(n, \{r_i\})$ is the number of combinations of N molecules taken n at a time

$$\int \cdots \int \rho(n, \{r_i\})d\{r_i\} = \frac{N!}{(N-n)!} \int \cdots \int P(n, \{r_i\})d\{r_i\}$$

$$= \frac{N!}{(N-n)!}.$$

The n-molecule correlation function, $g(n, \{r_i\}), i \in \{1, n\}$, is then defined as

$$\rho(n, \{r_i\}) = \rho^n g(n, \{r_i\})$$

which accounts for the correlation among water molecules. If the molecules are all independent, then $g = 1$ ($U = 0; Z = V$), and the n-molecule correlation function reduces to

$$\rho(n, \{r_i\}) = \rho^n = \left(\frac{N}{V}\right)^n.$$

In general, the n-molecule correlation function is then

$$g(n, \{r_i\}) = \frac{\rho(n, \{r_i\})}{\rho^n} = \left(\frac{V}{N}\right)^n \frac{N!}{(N-n)!} P(n, \{r_i\})$$

which, using (Eq. 11.1), is given by

$$g(n, \{r_i\}) = \left(\frac{V}{N}\right)^n \frac{N!}{(N-n)!} \frac{1}{Z} \int \cdots \int e^{-U/kT} dr_{n+1} dr_{n+2} \ldots dr_N.$$

For $n = 1$ for water, a homogeneous, essentially isotropic, liquid, we obtain the singlet distribution or the one-molecule density, which is independent of the position, r_1, of the reference molecule as

$$\rho(1, r_1) = \frac{N!}{(N-1)!} P(1, r_1) = \frac{N(N-1)!}{(N-1)!} P(1, r_1) = NP(1, r_1).$$

Integrating this over the volume of the liquid,

$$\int \rho(1, r_1) dr_1 = \rho(1, r_1) \int dr_1 = \rho(1, r_1) V = N \int P(1, r_1) dr_1 = N.$$

The probability $P(1, r_1)$ is normalized, its integral over the volume of water is one, and we have that

$$\rho(1, r_1) V = N$$

or

$$\rho(1, r_1) = \frac{N}{V} = \rho \tag{11.2}$$

which is the average density of the liquid. The quantity $\rho(1, r_1) dr_1$ is the probability that one of the water molecules is in the volume element dr_1 at r_1 if the arrangement of the liquid is observed. The single-molecule correlation function is then

$$g(1, r_1) = \frac{\rho(1, r_1)}{\rho} = 1.$$

In a similar fashion, we can find the probability of a configuration in which two molecules are at fixed positions r_1 and r_2 and the remaining $N - 2$ are free to assume any position which we write as $\rho(2, r_1, r_2)$

$$\rho(2, r_1, r_2) = \frac{N!}{(N-2)!} \, P(2, r_1, r_2)$$

$$= \frac{N(N-1)(N-2)!}{(N-2)!} \, P(2, r_1, r_2)$$

$$= N(N-1) \, P(2, r_1, r_2).$$

This pair correlation function of the water molecules is obtained by integrating (Eq. 11.1) over the coordinates for the other $N-2$ molecules, $dr_3 dr_4 \ldots dr_N$,

$$g(2, r_1, r_2) = \left(\frac{V}{N}\right)^2 \frac{N(N-1)}{Z} \int \cdots \int e^{-\frac{U}{kT}} dr_3 dr_4 \ldots dr_N,$$

and here we note the factor of $N(N-1)$ which is twice the number of pairs of water molecules. The configuration integral in this case can be simplified because the integrand is independent of r_1 and r_2

$$Z = \int dr_1 \int dr_2 \int \cdots \int e^{-\frac{U}{kT}} dr_3 dr_4 \ldots dr_N$$

$$Z = V^2 \int \cdots \int e^{-U/kT} dr_3 dr_4 \ldots dr_N$$

so that integration over each of these variables just gives factors of the volume of the water, and these cancel V^2 to prevent a divergence in $g(2, r_1, r_2)$ as the volume of water goes to infinity.

All water molecules are identical and form a homogeneous, isotropic fluid at large length scales. Therefore, $\rho(2, r_1, r_2)$ can only depend on the difference in molecular coordinates and not on their absolute positions within the fluid. We can replace r_1, r_2 with their vector difference $r = |r_1 - r_2|$

$$\rho(2, r_1, r_2) = \rho(2, r),$$

and just as the integral of the single-molecule distribution function gives

$$\int \rho(1, r_1) dr_1 = N,$$

the corresponding integral of the pair distribution function gives the number of molecular pairs

$$\int \int \rho(2, r_1, r_2) dr_1 dr_2 = V \int \rho(2, r) dr$$

$$= N(N-1) \int P(2,r)\,dr$$

$$= N(N-1).$$

If the spatial distribution of molecules in the fluid is random, the probability that the first molecule is contained within the volume element dr_1 centered at r_1 is the ratio of dr_1 to the total volume

$$P(1,r_1) = \frac{dr_1}{V}.$$

The same is true for molecules 2 through n; therefore, the joint probability of observing molecule #1 in dr_1 and #2 in dr_2, up to #n in dr_n is the product of the individual probabilities

$$P(n,\{r_i\})\,dr_1 \ldots dr_n = \frac{1}{V^n} \prod_{i=1}^{n} dr_i$$

which can be rewritten as

$$P(n,\{r_i\}) \prod_{i=1}^{n} dr_i = \frac{1}{V^n} \prod_{i=1}^{n} dr_i.$$

Upon factoring out the product of the volume elements, this gives

$$P(n,\{r_i\}) = \frac{1}{V^n},$$

and we have for the general probability density

$$\rho(n,\{r_i\}) = \frac{N!}{(N-n)!}\, P(n,\{r_i\})$$

$$\rho(n,\{r_i\}) = \frac{N!}{(N-n)!}\, \frac{1}{V^n}$$

$$= \frac{N!}{N^n(N-n)!}\rho^n.$$

The special cases of 1- and 2-molecule probability densities (Eq. 11.2) are

$$\rho(1,r_1) = \frac{N!}{V(N-1)!} = \frac{N(N-1)!}{V(N-1)!} = \frac{N}{V} = \rho$$

$$\rho(2, r_1, r_2) = \frac{N!}{V^2(N-2)!} = \frac{N(N-1)(N-2)!}{V^2(N-2)!} = \frac{N(N-1)}{V^2}$$

$$\rho(2, r_1, r_2) = \rho^2\left(1 - \frac{1}{N}\right).$$

The result for $\rho(1, r_1) = \rho$ holds for any fluid, but the result for $\rho(2, r_1, r_2)$ is only true in an imaginary fluid with $U = 0$, i.e., with no binding energy. Water is certainly a bound state of water molecules. We must then modify the above approach to accommodate pairwise and higher correlations in the positions of water molecules in the liquid. This is accomplished by introducing a set of n-molecule correlation functions $g(n, \{r_i\})$ which describe the deviation from independent behavior. We observed above that if the molecules are independent, the joint probability density that the ith molecule is contained within the volume element dr_i centered at r_i, $i \in \{1, n\}$ is the product of the individual probability densities

$$\rho(n, \{r_i\})dr_1 \ldots dr_n = \prod_{i=1}^{n} \rho(1, r_i)dr_i.$$

In the case of water, this joint probability must be modified, and Kirkwood has introduced the n-molecule correlation functions $g(n, \{r_i\})$ through the equation

$$\rho(n, \{r_i\})dr_1 \ldots dr_n = g(n, \{r_i\}) \prod_{i=1}^{n} \rho(1, r_i)dr_i.$$

Now, we know that in general

$$\rho(n, \{r_i\}) = \frac{N!}{N^n(N-n)!}\rho^n$$

so that

$$g(n, \{r_i\}) \prod_{i=1}^{n} \rho(1, r_i)dr_i = \frac{N!}{N^n(N-n)!}\rho^n \prod_{i=1}^{n} dr_i,$$

and since $\rho(1, r_i) = \rho$, the product is ρ^n, and we have that

$$\prod_{i=1}^{n} \rho(1, r_i)dr_i = \rho^n \prod_{i=1}^{n} dr_i$$

so the expression for the n-molecule correlation function for independent molecules is

$$g(n, \{r_i\}) = \frac{N!}{N^n(N-n)!}$$

which can be expanded as

$$\frac{N!}{N^n(N-n)!} = \left[1 - \frac{n(n-1)}{N} + \ldots\right] = 1 + O\left(\frac{1}{N}\right)$$

for small n.

In a real fluid, we must modify (Eq. 11.1) $\rho(n, \{r_i\}) = \rho^n$ to read instead

$$\rho(n, \{r_i\}) = \rho^n \, g(n, \{r_i\})$$

to compensate for the fact that real fluids are bound states due to the attractive intermolecular potential. We then have a general expression for the n-molecule correlation function as

$$g(n, \{r_i\}) = \left(\frac{V}{N}\right)^n \frac{N!}{(N-n)!} \frac{1}{Z} \int \cdots \int e^{-U/kT} dr_{n+1} dr_{n+2} \ldots dr_N,$$

and the correlation function for pairs of water molecules is

$$g(2, r_1, r_2) = \left(\frac{V}{N}\right)^2 \frac{N!}{(N-2)!} \frac{1}{Z} \int \cdots \int e^{-U/kT} dr_{n+1} dr_{n+2} \ldots dr_N. \qquad (11.3)$$

This function is only dependent on the coordinate difference ($r = |r_1 - r_2|$), so that

$$g(2, r_1, r_2) = g(2, r)$$

and

$$\frac{1}{V} \int g(2, r) dr = 1 - \frac{1}{N},$$

or multiplying through by N, we have that

$$\frac{N}{V} \int g(2, r) dr = \rho \int g(2, r) dr = N - 1.$$

The probability of finding a second water molecule in a volume dr at a distance r from the first is

$$\rho g(2, r)dr,$$

and the number of water molecules in a spherical shell of width dr at a distance r is

$$\rho g(2, r)4\pi r^2 dr$$

so that

$$\rho \int g(2, r)4\pi r^2 dr = N - 1 \tag{11.4}$$

which is one less than the number of molecules because we have fixed one molecule at the origin.

From now on, when we refer to the radial distribution function, we will be considering only the pair correlation function $g(2, r)$ which we will denote $g(r)$. It has several interesting properties which we can motivate by examining its limits in terms of fluid density and intermolecular potentials. These will then give way to an interpretation of measurements of its properties in real fluids.

11.14.3   The Radial Distribution Function for a Theoretical Ideal Gas

A system of noninteracting atoms or molecules forms a theoretical ideal gas if the interaction potential vanishes everywhere, so that $U(r) = 0, \forall\, r$. In this case, we can directly compute the radial distribution function by direct integration of (Eq. 11.3) noting that $e^{-\frac{U}{kT}} = e^0 = 1$

$$g(r) = \left(\frac{V}{N}\right)^2 \frac{N!}{(N-2)!} \frac{\int \cdots \int dr_{n+1} \ldots dr_N}{\int \cdots \int dr_1 \ldots dr_N}.$$

The top integral gives $N - 2$ factors of V, while the bottom integral gives N factors of V, so that

$$g(r) = \left(\frac{V}{N}\right)^2 \frac{N!}{(N-2)!} \frac{V^{N-2}}{V^N}$$

which using $N! = N(N-1)(N-2)!$ reduces to

$$g(r) = 1 - \frac{1}{N} \sim 1, N > 1000.$$

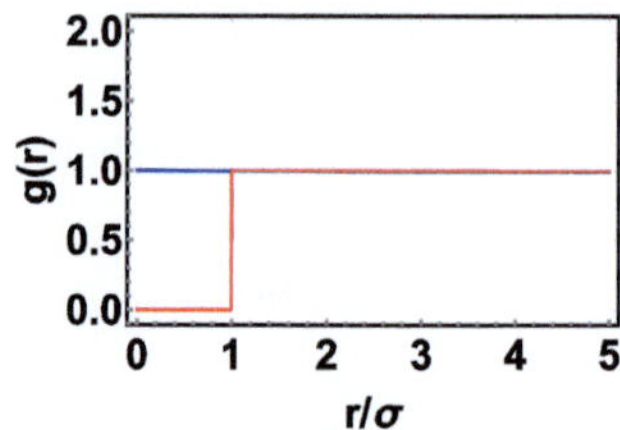

Fig. 11.36 The radial distribution function for a theoretical, noninteracting ideal gas (blue) and a dilute gas of hard spheres (red). Note that the blue and red curves are identical for $r/\sigma > 1$, where σ is the size of the sphere

This is a constant essentially equal to one for any reasonable number N of molecules as seen in Fig. 11.36 for $N = 1000$. There is no correlation among the molecules because there is no interaction among them. This radial distribution function gives the correct number of particles (Eq. 11.4) as

$$\rho \int g(r) 4\pi r^2 dr = 4\pi \rho \int \left(1 - \frac{1}{N}\right) r^2 dr$$

$$= 4\pi \frac{r^3}{3} \frac{N}{V} \left(1 - \frac{1}{N}\right)$$

$$= V \frac{N}{V} \left(1 - \frac{1}{N}\right)$$

$$\rho \int g(r) 4\pi r^2 dr = N - 1. \tag{11.5}$$

For a gas of hard spheres at low density, the interaction potential resembles the particle in a box problem in quantum mechanics (Chap. 4) except here there is a molecule at the origin whose electrons, by the Pauli principle, do not allow the interpenetration of molecules closer than their van der Waals diameters, σ. The potential is given by

$$U(r) = \left\{ \begin{array}{l} \infty, r \leq \sigma \\ 0, r > \sigma \end{array} \right\},$$

and inserting this into the equation for $g(r)$ (Eq. 11.3) splits the integrals in the numerator into a sum of two parts:

$$g(r) = \left(\frac{V}{N}\right)^2 N(N-1)\frac{1}{Z} \left(\int_0^\sigma \cdots \int_0^\sigma e^{-\frac{U}{kT}} dr_{n+1} \cdots dr_N + \int_\sigma^\infty \cdots \int_\sigma^\infty e^{-\frac{U}{kT}} dr_{n+1} \cdots dr_N \right)$$

In the first integral, whose range is $\{0, \sigma\}$, the potential is infinite, and therefore, the exponential is zero, and the integrals in both the numerator and the denominator are also zero. We are left with

$$g(r) = \left(\frac{V}{N}\right)^2 N(N-1)\, \frac{1}{Z} \int\limits_{\sigma}^{\infty} \cdots \int\limits_{\sigma}^{\infty} e^{-\frac{U}{kT}} dr_{n+1} \ldots dr_N$$

where, as above, $U = 0$ and the exponential is one everywhere. The numerator evaluates to V^{N-2} while the denominator is V^N. We once again find that the radial distribution function is

$$g(r) = \left\{ \begin{array}{ll} 0, & r \le \sigma \\ 1 - \dfrac{1}{N}, & r > \sigma \end{array} \right\}$$

and that the integral (Eq. 11.4) gives the correct total number of molecules $(N-1)$. The radial distribution function is shown in Fig. 11.36 (red) showing that it is zero for distances less than the molecular diameter due to this excluded volume.

11.14.4   A Gas of Hard Spheres at Intermediate Density

When the density of a gas of hard spheres is increased to the point where there is one molecule per molecular volume, $\rho = 1/\sigma^3$, the correlation among molecular positions is solely driven by the volume that the pair occupies which excludes a third molecule. Here, we refer the interested reader to the quite lengthy derivation given by Hill, page 210 (Hill 1956), and only quote the results. We can expand the radial distribution function in terms of the density as

$$g(r_1, r_2) = e^{-U(r_1,r_2)/kT}\left(1 + B(r_1, r_2)\rho + C(r_1, r_2)\rho^2 + \ldots\right)$$

where the coefficients, $B(r_1, r_2)$, $C(r_1, r_2)$, etc. are defined as integrals of the function

$$f(r_1, r_2) = e^{-U(r_1,r_2)/kT} - 1$$

such that, for example, the equation for $B(r_1, r_2)$ is

$$B(r_1, r_2) = \int f(r_1, r_3) f(r_2, r_3) dr_3.$$

As the density approaches zero, only the first term in this density expansion of g is important, and we have an important result that

$$g(r) = e^{-U(r)/kT} \qquad (11.6)$$

or that the radial distribution function is given by the Boltzmann probability distribution (Chap. 1). If we then keep the next term in the density expansion, we can accommodate intermediate densities. The radial distribution function in this case must include pairwise interactions not only between molecules 1 and 2 but also between 1 and 3 and 2 and 3 in the approximation that the complicated intermolecular potential in a fluid can be replaced by an interaction energy between pairs of molecules. The result is that the radial distribution function in this case is described by the following integral:

$$g(r_1 - r_2) = \int \left(e^{-\frac{U(r_1 - r_3)}{kT}} - 1 \right) \left(e^{-\frac{U(r_2 - r_3)}{kT}} - 1 \right) dr_3$$

This integral vanishes unless r_{13} and r_{23} are simultaneously less than the molecular diameter σ. The sole volume where the integral is nonzero ($\sigma \leq r \leq 2\sigma$) comes from the volume surrounding molecules one and two that excludes the center of the third molecule. Upon integrating over the intermediate distance, r_3, and letting $r = r_1 - r_2$, the radial distribution function becomes (Hill 1956, p. 210) for the three cases

$$(1) \text{ for } r \leq \sigma : g(r) = 0,$$

$$(2) \text{ for } \sigma < r \leq 2\sigma : g(r) = 1 + \frac{4\pi}{3}\rho\sigma^3 \left[1 - \frac{3}{4}\left(\frac{r}{\sigma}\right) + \frac{1}{16}\left(\frac{r}{\sigma}\right)^3 \right]$$

$$\text{and } (3) \text{ for } r \geq 2\sigma : g(r) = 1,$$

over the density range from $\rho = 0$ to $\rho\sigma^3 = 1$, where there is one molecule for every cell whose volume is the molecular diameter cubed. This forms a cubic lattice in which the molecules are just touching each other along the axes. One can envision slightly higher densities in a hexagonal close-packed lattice, but the basic features of $g(r)$ for densities up to $\rho\sigma^3 = 1$ are shown in Fig. 11.37.

This result is somewhat surprising because it shows an excess of molecules between $\sigma \leq r \leq 2\sigma$ even though at any distance between the molecules $r > \sigma$ the potential is zero and its gradient, the force on a molecule, vanishes. This excess appears to be due to a force acting on molecules one and two as a result of other $N - 2$ molecules. How is this possible? A simple experiment on a piece of paper with a pencil and three equally sized coins, such as quarters, will show that when molecules one and two are in contact, any approach to molecule one is possible only from the direction opposite molecule two. This makes it more likely that collisions of molecule one with the remaining $N - 2$ molecules will take place from this direction. Therefore, $g(r)$ simply reflects this probability difference and does not result from an actual force on molecule one.

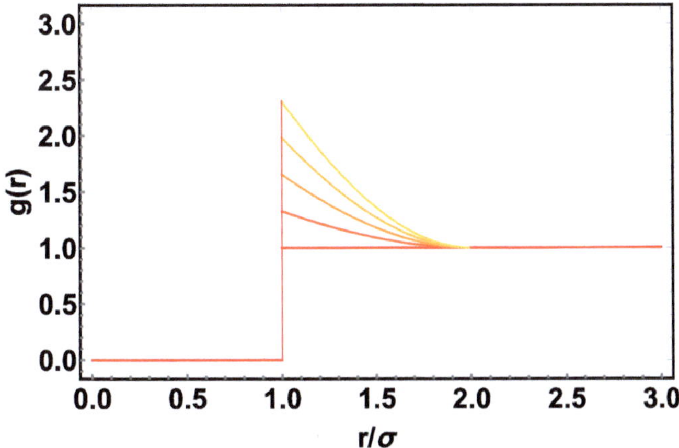

Fig. 11.37 The pairwise radial distribution function for a gas of noninteracting hard spheres of diameter σ for densities $\rho\sigma^3 = 0$ (red) to $\rho\sigma^3 = 1$ (orange) in steps of $\rho\sigma^3 = 0.25$

11.14.5   The Radial Distribution Function for a Lennard-Jones Liquid

We have seen the Lennard-Jones potential several times already (e.g., Fig. 11.33). The probability of finding a molecule next to another molecule is given by the Boltzmann distribution where the molecular energy in the binding region is compared with energy available from the environment, kT. At zero density, (Eq. 11.6) can be used to compute the pair correlation function as the depth of the Lennard-Jones potential varied (Fig. 11.38) for water at 298 K. One notices that as the potential becomes deeper a larger excess of molecules surround the reference molecule at the origin. This is a reflection of the increase in the Boltzmann probability due to the increase in the ratio of the well depth to kT. Note also that the peak in $g(r)$ occurs at the minimum of $U(r)$. At higher densities, both the hard sphere gas and a Lennard-Jones fluid, like liquid argon, will show oscillations in the pair correlation function at distances corresponding to multiples of σ.

11.14.6   Liquid Argon as a Simple Lennard-Jones Fluid

Argon ($_{18}^{40}\text{Ar}$) is a noble gas and therefore contains a filled outer shell of eight electrons, rendering it a spherical atom with a van der Waals radius of 188–192 pm. Its heat of vaporization of 6.53 kJ/mol and its heat capacity of 20.85 kJ/mol.K are about a quarter of the values for water (40.65 kJ/mol and 75.3 kJ/mol.K, respectively) and are indicative of its lack of hydrogen bonding. Argon's Lennard-Jones parameters are $\sigma = 340$ pm and $\varepsilon = 10.3$ meV (Fig. 11.33). At its melting point of

Fig. 11.38 (a) The Lennard-Jones potential $U(r)$ at 298 K for a molecule the size of water ($\sigma = 280$ pm) for values of the energy depth parameter $\varepsilon = \{0.2$ (red), 0.5, 0.8, 1.1, 1.4 (light blue) $kT\}$. The units for $U(r)$ are given in eV. (b) The pair correlation function $g(r)$ at low density for the potentials shown in (a) showing that the buildup of molecules near $r = \sigma$ increases as the potential deepens

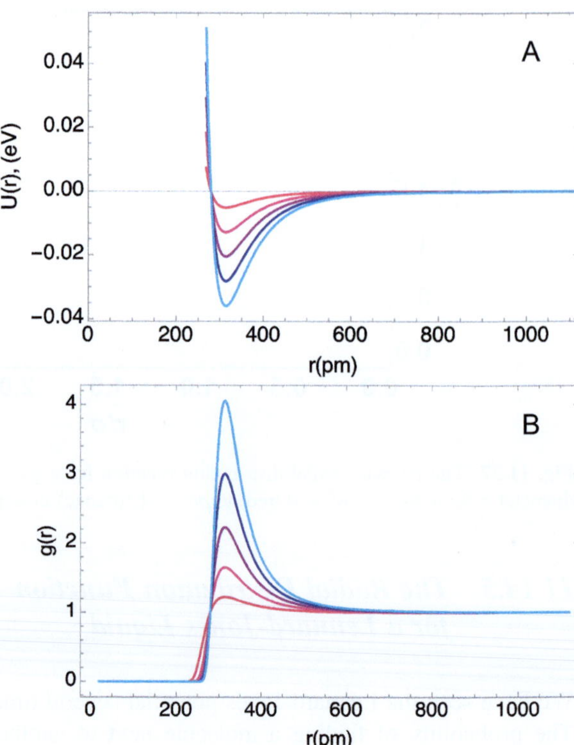

Fig. 11.39 The radial distribution function for liquid argon at 85 K (just above its melting point) as determined through neutron scattering by Yarnell et al. (1973). The successive peaks occur at integral multiples of the van der Waals atomic diameter (368 pm)

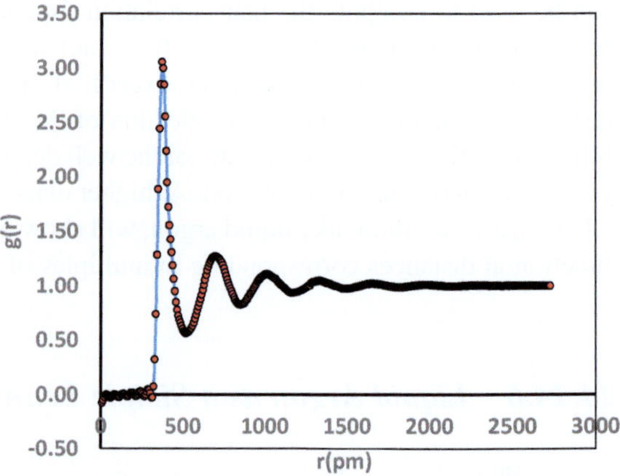

83.8 K, the ratio of the Lennard-Jones potential depth to thermal energy is $\varepsilon/kT = 1.4$, and from plots like Figs. 11.33 and 11.38, one would estimate that its radial distribution function should peak near 380 pm at a value of $g(380) = 2.7$. The experimental data for liquid argon at 85 K determined from neutron scattering is shown in Fig. 11.39 where the first peak occurs at 368 pm with a value of 3.05. One

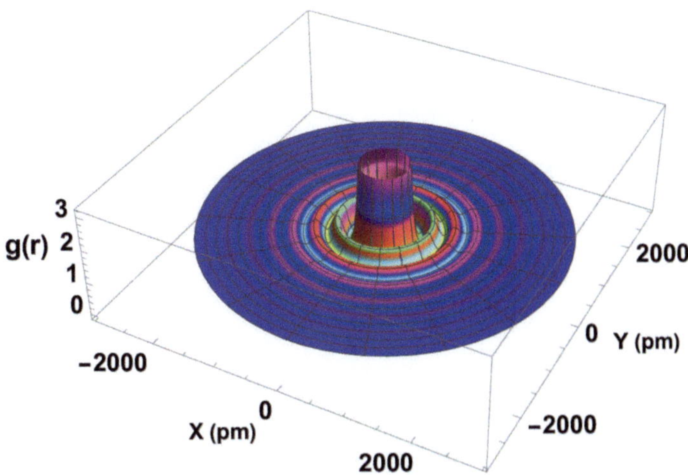

Fig. 11.40 The radial distribution function $g(r)$ for liquid argon, rotated around the $r = 0$ axis showing the numerous circular maxima at radii corresponding to integral values of the van der Waals diameter of this spherical atom and the excluded volume at the center

also notes successive peaks around 715, 1028, 1355, 1662, and 1982 pm; these values are at $r/\sigma = 2.1$, 3.0, 4.0, 4.9, and 5.8, respectively, corresponding to argon atoms in shells surrounding the central atom (such as used in Figs. 11.35 and 11.40 *q.v.*) at distances which are an integral number of atomic diameters and arise from atoms in the third shell bound in the Lennard-Jones potential of the second shell, etc. These successive shells in argon are shown well as rings in the radial distribution function when it is rotated around the $r = 0$ axis (Fig. 11.40).

Since $g(r)$ is a measure of the departure of the local density from the mean density as one moves away from the central reference atom, this function should, and does, approach 1.0 for large radii where the fluid can be viewed as a random continuum.

The radial distribution function can be used to find a number of thermodynamic quantities, such as the energy, the pressure, etc. for a fluid. It can also be integrated using (Eq. 11.5) to find the coordination number, the number of a atoms or molecules present surrounding a given particle b as a function of the distance away from the reference site, as

$$N_{ab}(r_1, r_2) = 4\pi\rho \int_{r_1}^{r_2} g_{ab}(r)r^2 dr, \tag{11.7}$$

because the integral over the entire volume gives the total number of particles surrounding our reference particle at the origin. One can find the number of atoms in the first coordination shell by integrating $g(r)$ from zero up to the first minimum and similarly for the next shells. In order to perform this integration, one must find

Fig. 11.41 A schematic of a plane from a hexagonal close-packed lattice of spheres. The number of nearest neighbor spheres surrounding the red sphere within the plane is six. The black circles represent nearest neighbor spheres from the plane above; there are three of them, and there are also three nearest neighbor spheres from the plane below giving a total of 12 spheres in the first coordination shell in a simple fluid like liquid argon

Table 11.12 The number of argon atoms in each coordination sphere

Shell	r (pm) @ Min	$N_c(r)$
1	519	12
2	854	54
3	1175	142
4	1493	293
5	1815	526
6	2148	873

the density of the fluid in the proper units. If the radius in the pair correlation function is measured in Angstroms, then the integral will have units of $\overset{\circ}{A}{}^3$, and the density would need to have units of particles/$\overset{\circ}{A}{}^3$. One can convert ρ to the usual density, say of liquid argon, in g/mL to particle density using

$$\rho_{\text{part}} = \frac{N_o \rho}{M}$$

where M is the molecule mass in g/mol and N_o is Avogadro's number. For Argon, the density at its boiling point of 87.3 K is $\rho = 1.3954$ g/mL, and its molecular mass is $M = 39.948$ g/mol, giving a particle density of $\rho_{\text{part}} = 0.0210421$ atoms/$\overset{\circ}{A}{}^3$ or 2.1042×10^{-8} atoms/pm^3. The coordination number (Box 11.1) can be found for each successive shell identified as minima in $g(r)$. Figure 11.42 shows the dependence of the coordination number on radius for the first coordination sphere (up to $r = 519$ pm). As we have motivated in Fig. 11.35, the first coordination sphere contains 12 argon atoms, the same as found from stacking spheres in a hexagonal close-packed lattice (Fig. 11.41; Table 11.12):

Box 11.1 The numerical integration of discrete data in Mathematica is a useful technique to know. Here is an example drawn from the present task. Read in a text file in a local directory (use `SetDirectory["dirname"]` to enter the proper directory), then read in the file as, for example

```
argrdf=ReadList["radial.dist.argon.txt",{Number,Number}];
```

This reads in a list of pairs of {position, $g(r)$} values which can be plotted using `ListPlot[argrdf]`. Then, one converts this list into a function using

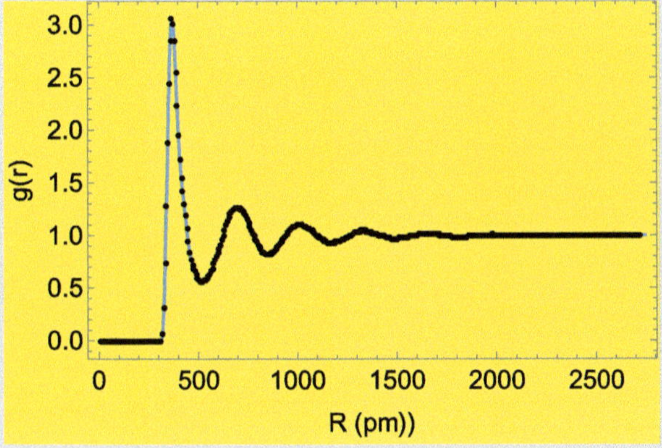

which can also be plotted as a normal function, along with its data points using
`Plot[func[r],{r,0,2500}, Epilog-> Point[argrdf]]`

recapitulating Fig. 11.39. (* Set the density of Argon. *) rhopart = 2.1042×10^{-8}; (*atoms/pm^3*).

Then, the interpolated function can be numerically integrated from zero to any value of r with `NIntegrate` by defining the function

```
coordnum[r_]= NIntegrate[4 π rhopart x² func[x],{x,0.01,r}];
```

This can be plotted from zero to r at a local minimum in $g(r)$ by

```
Plot[coordnum[r],{r,0,519},PlotStyle->Hue[0],Frame->True,
FrameLabel->{"R(A)","Nc(r)"}]
```

which gives the plot shown in Fig. 11.42.

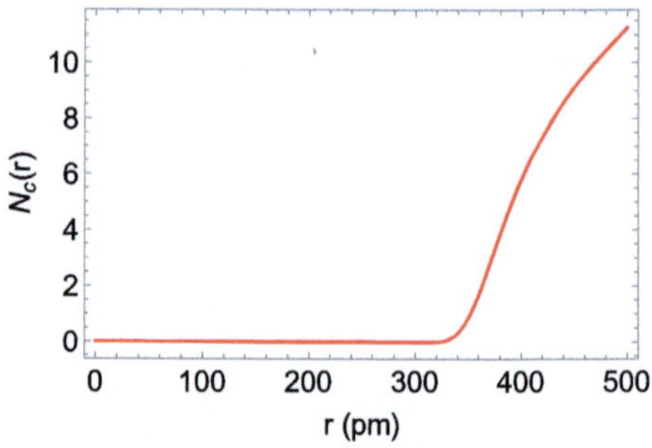

Fig. 11.42 The coordination number for liquid argon at 85 K as a function of distance from a fixed atom in the fluid. The number of atoms in the first coordination sphere is 12 (Table 11.12) as befits hexagonal close packing

11.14.7 The Mean Interatomic Potential in Liquid Argon

Another important feature of fluids is the interatomic potential that binds the particles together in the liquid state. The depth of the Lennard-Jones potential minimum, as we have seen, is related to the height of the first peak in the pair correlation function. The energy at this minimum, through the Boltzmann distribution, gives the probability of finding a particle at that radius. If we assume that the higher-order terms in the density expansion of the pair the correlation function can be ignored, as a first approximation, we have that the intermolecular potential can be obtained from

$$g(r) = e^{-U(r)/kT}$$

by taking the natural logarithm of both sides and rearranging to give the potential as

$$U(r) = -kT \ln[g(r)]$$

which is plotted in Fig. 11.43. The first minimum in $U(r)$ occurs at $r \sim 370$ pm with a value of -8.2 meV or $1.12\ kT$ at $T = 85$ K. The ε value for the Lennard-Jones potential for liquid argon is 10.3 meV. The boiling point (87.15 K) for liquid argon is only a few degrees above its liquefaction temperature of 83.85 K. This is an indication of the weakly bound nature of the liquid.

11.14.8 The Radial Distribution Function for Liquid Water

Unlike argon, the water molecule is not a simple spherical shape. It deviates by about 5% from that of a sphere (Fig. 11.44) allowing for tighter packing of water molecules than found for spherical argon atoms. The particle density for liquid argon at 87 K is 2.1042×10^{-8} atoms/pm^3, while that for water is 2.334×10^{-8} atoms/pm^3. The water molecule also consists of more than one type of atom so that one can define not

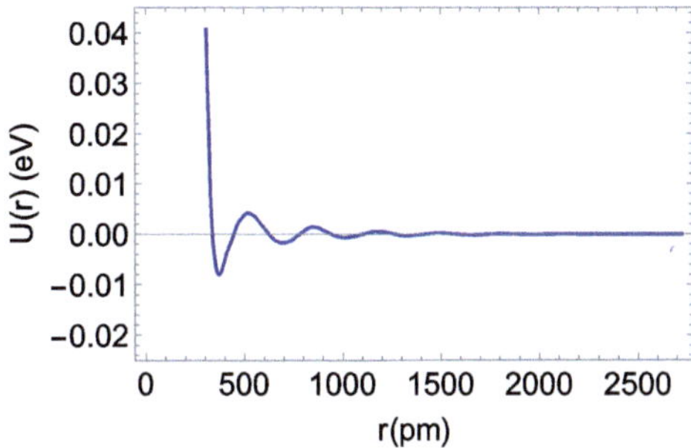

Fig. 11.43 The mean interatomic potential for liquid argon at 85 K. The numerous minima correspond to the radial shells or layers in this hexagonally close-packed liquid. The steep rise for distances smaller than the atomic diameter (380 pm) arises from the repulsion due to the Pauli exchange interaction

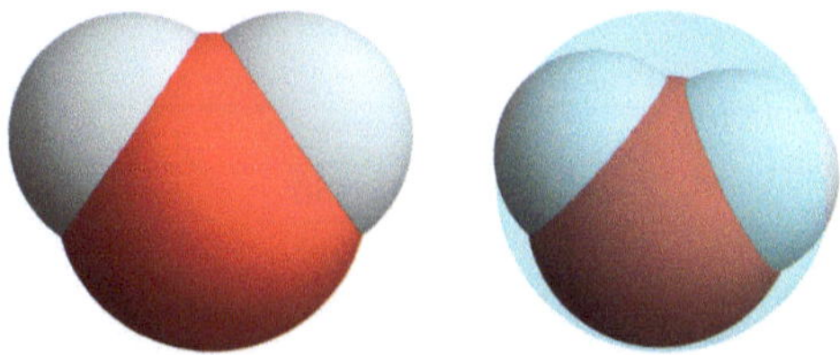

Fig. 11.44 On the left is a space-filling plot of the water molecule, where oxygen is red and the hydrogens are light gray. On the right is a water molecule enclosed in a pale blue sphere whose radius (192.6 pm) is found from the molar volume (18.015×10^{-6} m^3/mol) showing the molecule's deviation from sphericity

just one pair correlation function as was the case with argon but three, denoted $g_{OO}(r)$, $g_{OH}(r)$, and $g_{HH}(r)$, corresponding to the distribution of oxygen-oxygen, oxygen-hydrogen, and hydrogen-hydrogen interatomic distances in the liquid (Soper 2000, 2011; Soper & Ricci 2013; Skinner et al. 2013).

There are several experimental methods for either measuring or estimating the pair correlation functions for water. As we have mentioned earlier, computer simulations using models of the intermolecular potential and molecular dynamics or Monte Carlo methods are popular (Andrés Cisneros et al. 2016) by themselves but need to be compared with experimental data obtained from real samples. For highly ordered samples, such as ice I_h, particle diffraction methods using electrons (Martynowycz and Gonen 2019), X-rays (Bernal and Fowler 1933), or neutrons (Peterson and Levy 1967) can all provide structural details of varying usefulness. Low-energy electrons scatter from the Coulomb potential in the samples (Kalman et al. 1977). X-rays scatter from the electrons, and since the hydrogen in water contains only a single electron, X-ray scattering from water is dominated by scattering from the eight oxygen electrons, and hence X-rays only reveal $g_{OO}(r)$ and

Fig. 11.45 The $g_{OO}(r)$, $g_{OH}(r)$, and $g_{HH}(r)$ radial distribution functions (offset by 3.0 for clarity) for ambient water at 298 K determined from neutron scattering data courtesy of Alan Soper 2000. The peaks A in $g_{HH}(r)$ at 153 pm and B in $g_{OH}(r)$ at 96 pm are from intramolecular covalent bonds, while the other peaks reflect intermolecular distances

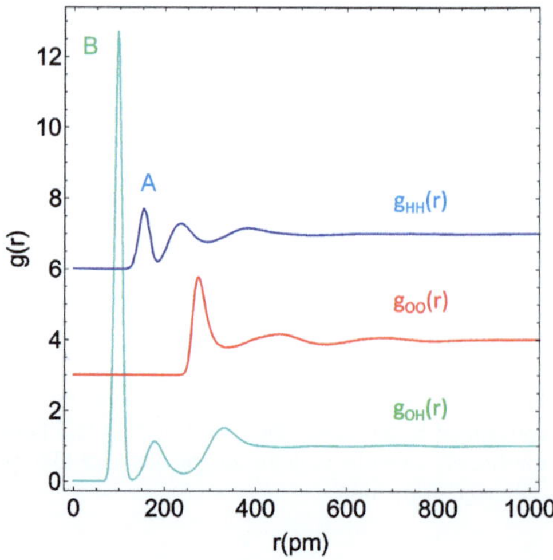

Table 11.13 Peak positions (pm) for the radial distribution functions of ambient water

Peak #	$g_{OO}(r)$	$g_{OH}(r)$	$g_{HH}(r)$
1	282	97	152
2	450	185	242
3	667	328	381
4	861	540	621

mainly miss the positions of the hydrogens (Skinner et al. 2013). Neutrons scatter from nuclei and therefore can be used to measure all three pair correlation functions.

Tabulated data for $g_{OO}(r)$, $g_{OH}(r)$, and $g_{HH}(r)$, generated through neutron scattering (Soper (2000); Fig. 11.45), and $g_{OO}(r)$ from the more limited X-ray measurements (Nilsson et al. 2010) show that, like in liquid argon, water molecules reside at preferred distances surrounding a central molecule (Table 11.13).

The large peaks at 97 pm in $g_{OH}(r)$ and 152 pm in $g_{HH}(r)$ shown in Fig. 11.45 are from the intramolecular, covalently bonded OH and HOH distances, while the remaining peaks arise from intermolecular distances. The distance between the two hydrogens in water is $d_{HH} = 2\,r_{OH}\,\mathrm{Sin}[104.5/2]$, where from Table 11.13 $r_{OH} = 97$ pm, giving a value of $d_{HH} = 153$ pm, which is the position of the first peak in $g_{HH}(r)$. The hydrogen bond length is ~280 pm (peak #1, $g_{OO}(r)$) but this varies with temperature. If we just concentrate our attention on the intermolecular contribution to the radial distribution functions (Figs. 11.46, 11.47, and 11.48), one observes peaks at the distances corresponding to the coordination shells that are more clearly defined and evident when the data are rotated around the $r = 0$ axis (Fig. 11.49) because they are visible as rings against the $g = 1$ background and dramatically reveal the central excluded volume from the molecule at the origin.

It is instructive to directly compare the radial distribution functions for water and liquid argon.

Fig. 11.46 (*Top*) The radial distribution function, $g_{OO}(r)$, for the oxygen-oxygen distances in water at 298 K. (*Bottom*) The effective intermolecular oxygen-oxygen potential for water showing a first minimum at 279 pm with a depth of 23.49 meV

Fig. 11.47 (*Top*) The radial distribution function, $g_{OH}(r)$, for the oxygen-hydrogen distances in water at 298 K. (*Bottom*) The effective intermolecular oxygen-hydrogen potential for water showing a first minimum at 327 pm with a depth of 9.99 meV

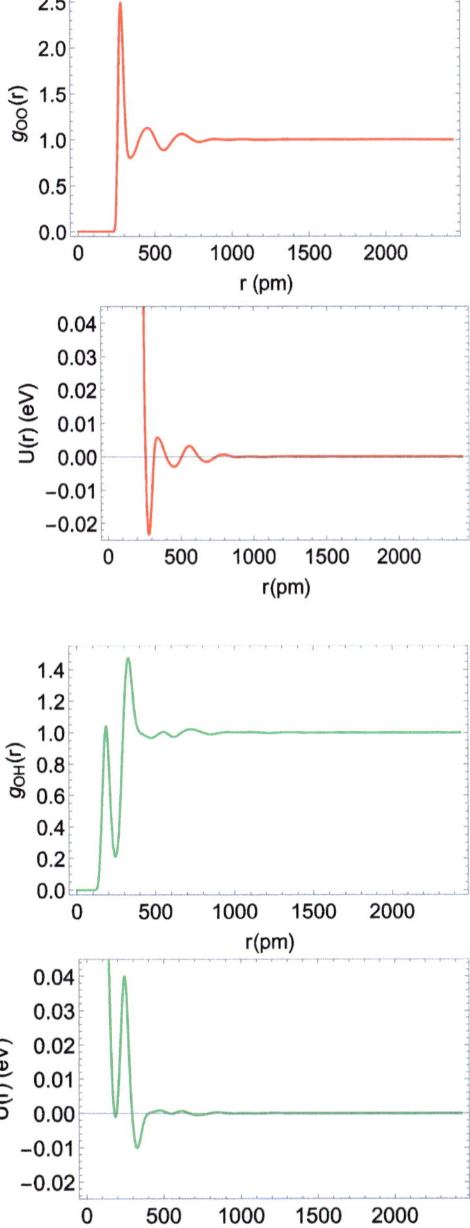

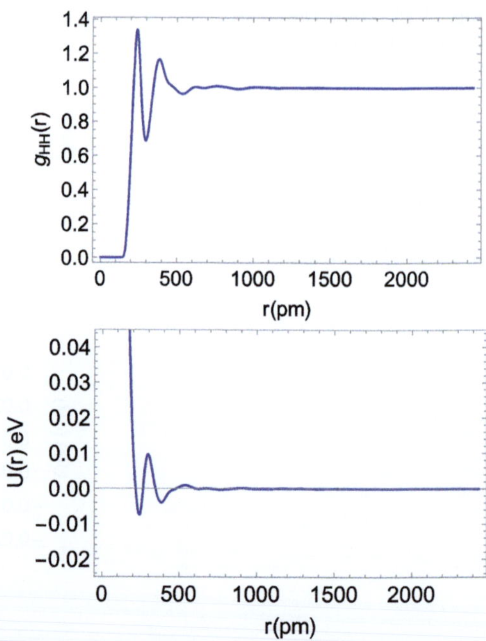

Fig. 11.48 (*Top*) The radial distribution function, $g_{OH}(r)$, for the hydrogen-hydrogen distances in water at 298 K. (*Bottom*) The effective intermolecular hydrogen-hydrogen potential for water showing a first minimum at 243 pm with a depth of 7.49 meV and a second at 384 pm with a depth of 3.93 meV

The $g(r)$ data, plotted as a function of the normalized radius, i.e., versus the radius divided by the Lennard-Jones distance of closest approach (Fig. 11.50), show that the hexagonal close-packed lattice that describes argon with its 12 nearest neighbors (Table 11.12) is less tightly packed than that for water; the peaks in the radial distribution function for argon are found around integral multiples of the atomic diameter (σ, 2σ, 3σ, etc.), while the first few peaks in the radial distribution function for water occur closer to the central molecule at σ, 1.6σ, 2.4σ, 3.2σ, etc., a clear indication that the nonspherical shape of the water molecule allows it to pack into a liquid structure with higher density than that of a hexagonal close-packed lattice of spheres.

Let's compare the effective intermolecular oxygen-oxygen potential of a hard sphere liquid, such as argon, with that of water at 298 K (Fig. 11.51). We note that from water's lower-energy minimum (23.49 meV vs. 8.08 meV) it is more closely organized (tightly bound) than argon; water has its first minimum in the mean intermolecular potential function at $r = \sigma$, but its second minimum occurs at a smaller distance ($r = 1.6\sigma$) than found for argon's hexagonal close-packed limit at ($r = 2\sigma$). How is this possible? A glance at Fig. 11.44 will provide an important clue.

The water molecule is not a sphere; the hcp limit for spheres does not apply. It is, rather, an elongated shape due to the repulsion of the lone-pair electrons on the oxygen. This shape (Fig. 11.44) is sometimes referred to as a "Mickey Mouse" head with its mouse ears protruding at the tetrahedral angle, ~105°. This interesting molecular shape means that water molecules do not pack like spheres because there are intermolecular distances smaller than that found for an hcp liquid, such as argon. Also, unlike in argon, where the number of molecules in the first, second,

Fig. 11.49 The 3D radial distribution functions (r) for water at 298 K showing the excluded central volume corresponding to the molecules at the origin and the successive shells of adjacent molecules: (**a**) $g_{OO}(r)$, (**b**) $g_{OH}(r)$, and (**c**) $g_{HH}(r)$. The peak in the nearest neighbor distances surrounding the central molecule is followed by a rapid decrease in the ordered population. (Data courtesy of Alan Soper (2000))

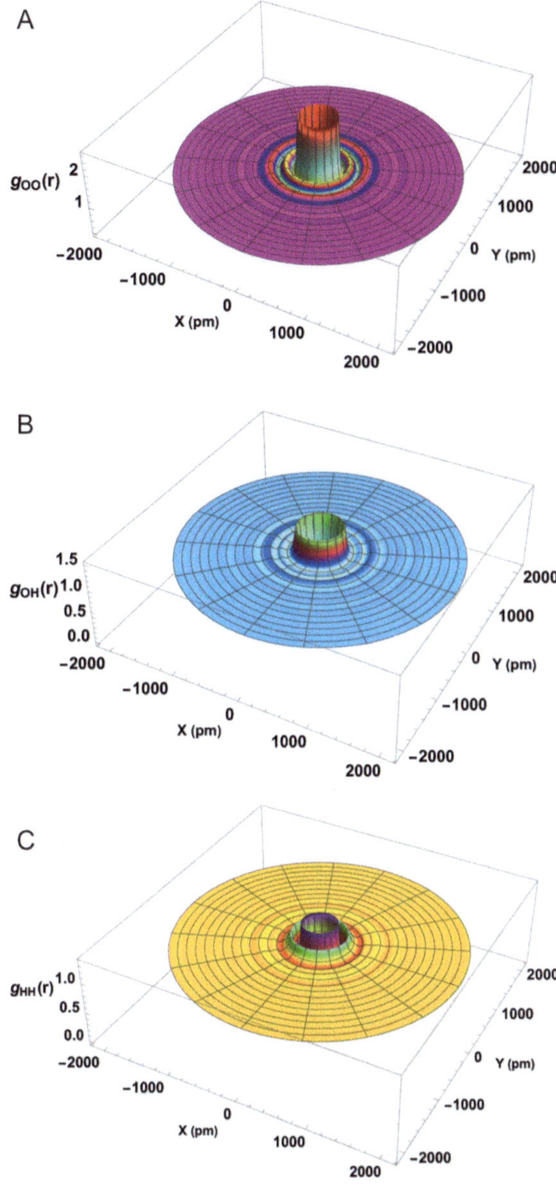

and third coordination spheres is 12, 54, and 142, as found from numerical integration of the coordination number (Fig. 11.52) out to the characteristic radii (Table 11.12) for each sphere, the number of molecules in the first, second, and third coordination spheres for water, whose radii are presented in Table 11.13, gives much smaller values of 4.7, 23.9, and 70.2 molecules. The nonspherical shape of the

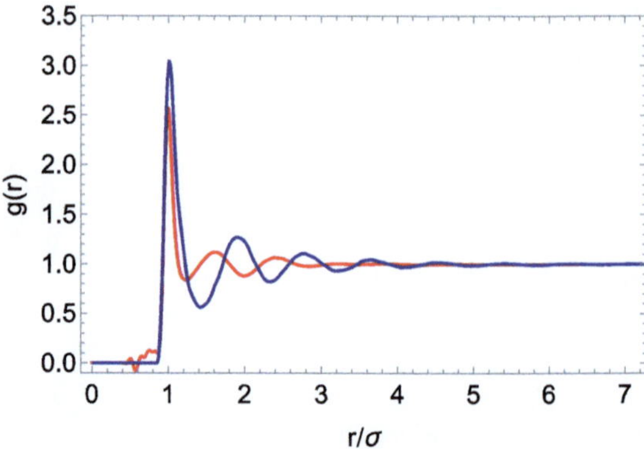

Fig. 11.50 The radial distribution functions for liquid argon (blue) and $g_{OO}(r)$ for water (red) as a function of the reduced radius (the radius divided by the Lennard-Jones σ parameter)

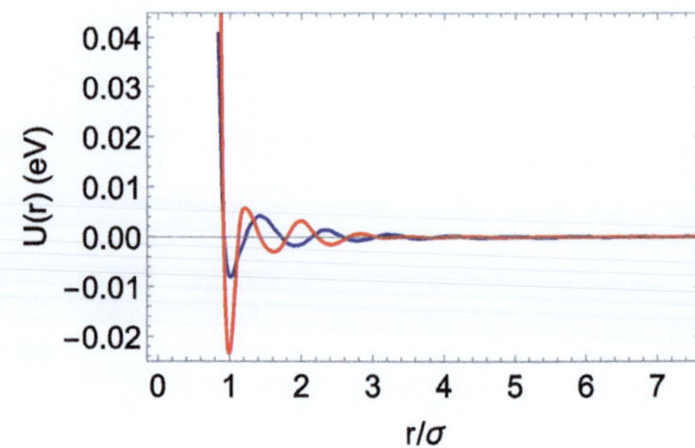

Fig. 11.51 The mean intermolecular potential for liquid argon (blue) compared with that for water for the oxygen-oxygen potential (red)

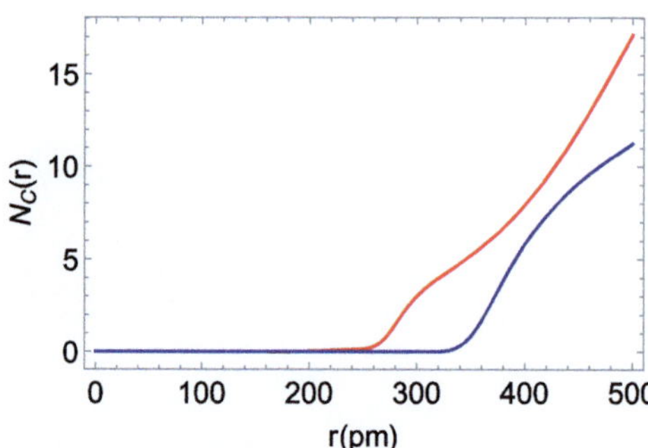

Fig. 11.52 The coordination number as a function of distance from the center of an argon atom in liquid argon (blue) compared with the coordination number for water (red) from the center of a water molecule. The different shapes of the two curves reflect the fact that argon is spherical, while the water molecule is mainly tetrahedral and has the ability to pack more densely than argon. The number of water molecules in the first coordination sphere for water is 4.7 (Table 11.14)

Fig. 11.53 (*Top*) A closer look at all of the radial distribution functions, $g_{OO}(r)$, red; $g_{OH}(r)$, green; and $g_{HH}(r)$, blue, for water at 298 K in the region of the primary maxima. (*Bottom*) The radial distributions for the coordination numbers $N_c(r)$ for O–O, red; for O–H, green; and for H–H, blue in the region around the maxima in $g(r)$ for each distance

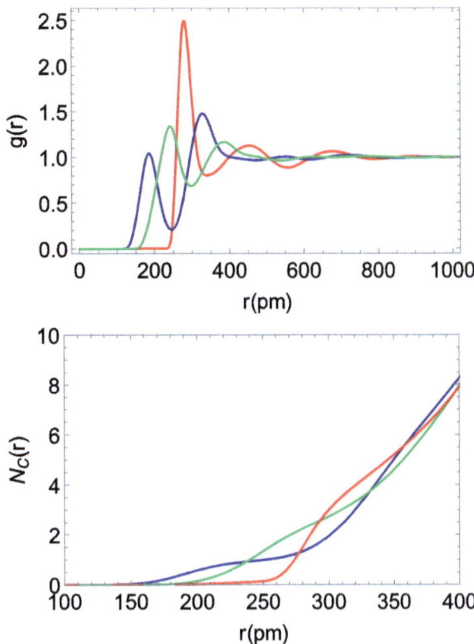

Table 11.14   The number of objects in each coordination sphere of water

Shell	r_{OO} (pm) @ Min	$N_c(r)$	r_{OH} (pm) @ Min	$N_c(r)$	r_{HH} (pm) @ Min	$N_c(r)$
1	338	4.7	244	0.96	295	2.6
2	562	23.9	470	13.7	538	20.9
3	798	70.2	611	30.9	671	41.2
4	968	125.8	843	82.8	905	103
5	1183	230.4	–	–	1140	206

water molecule makes itself apparent once more. A complete presentation of the radial distribution (Fig. 11.53) of the coordination numbers (Table 11.14) in the spheres for $g_{OO}(r)$, $g_{OH}(r)$, and $g_{HH}(r)$ enables one to determine the average numbers of nearest neighbors as one proceeds outward from the atom at the origin. While there are 4.7 water molecules in the first coordination sphere for the oxygen-oxygen separation at the H-bond distance, there is only a single molecule at the minimum hydrogen-oxygen separation, and there are 2.6 molecules at the minimum of the hydrogen-hydrogen distance.

A closer look at the radial dependence of the three pair correlation functions and coordination numbers for water is shown in Fig. 11.53. The maxima in the pair correlation functions for the first coordination shell occur in the order $g_{OO} > g_{OH} > g_{HH}$ at 277, 241, and 183 pm, respectively, corresponding to the distances between oxygen and hydrogen atoms, and these distances also order the relative increases in the radial dependences of the coordination numbers for the nearest neighbors.

11.15   NMR Spectroscopy and the Hydrogen Bond

One of the defining characteristics of water as a liquid is that its structure is largely determined by the intermolecular hydrogen bonding of water molecules (Chaplin 2006, 2007; Rastogi et al. 2011). We reiterate that ordinary water self-assembled from two of the most abundant stable nuclei produced in the early universe: ^{1}H from the Big Bang and ^{16}O from stellar nucleosynthesis. The nucleus of ^{1}H is the single proton, a spin ½ massive particle allowed by Poincare symmetry, which has the largest magnetic dipole moment of all the stable nuclei. Instead of the spin zero ^{16}O nucleus, a small fraction (0.0373%) of oxygen nuclei generated in stellar nucleosynthesis are the magnetic, spin 5/2 ^{17}O isotope. Magnetic resonance spectroscopy of these two nuclei gives us the ability to directly interrogate two physical participants in the hydrogen bonding of water, in either the solid, the liquid, or the gaseous state (Rhim et al. 1979; Cho et al. 2002).

NMR spectroscopy of water is simultaneously straightforward and revealing because the magnetic environment of its nuclei is determined by hydrogen bonding. The response of a hydrogen or oxygen nucleus to these perturbations is encoded in the magnetic shielding tensor, computed from quantum mechanics. The isotropic component of the shielding tensor is the chemical shift (essentially the resonance frequency) of the nucleus which is determined by the local magnetic field at the nucleus, which in turn is determined by the electron density at the nucleus. Since the electron has ~1836-fold larger magnetic moment ($\sim 1/m_e$) than the proton ($\sim 1/m_p$), alterations in the electronic distribution will induce a large shift in the resonance frequency of the water proton. The electronic distribution is altered by the electric field from water's large permanent electric dipole moment as the distance between molecules changes. As a result, of all the spectroscopic methods for the investigation of the hydrogen bond in water, NMR spectroscopy offers a singularly important window because one can directly and precisely monitor the magnetic shielding of the nuclei during a number of experimental interventions and perturbations that alter the local electron density at the nucleus.

Proton magnetic resonance spectroscopy of water is so facile that one often finds oneself in the situation of needing to suppress the solvent water signal during the acquisition of signals from protons in solute(s) of interest. Why is this so? The answer lies in the reply to a question I would often ask my students: "Can you tell me the molarity of protons in water?" The blank stares that greeted my question reinforced my suspicion that one needs to challenge the mind with seemingly complex but, once solved, quite simple puzzles. Water's molecular weight of 18 g/mol and its density of 1 g/mL translate into the quite large proton density of ~111 mol/L, or 6.6×10^{25} protons/L! It takes only a few nanoliters of water to produce a robust μV NMR signal nowadays (Sillerud et al. 2006). Water is the most proton-rich fluid one can put into the NMR spectrometer; it makes great sense then that water was the first substance whose NMR spectrum was obtained after the Second World War by Purcell and Pound (Purcell et al. 1946) and Felix Bloch et al. (1946) in the late 1940s for which they shared the Nobel Prize.

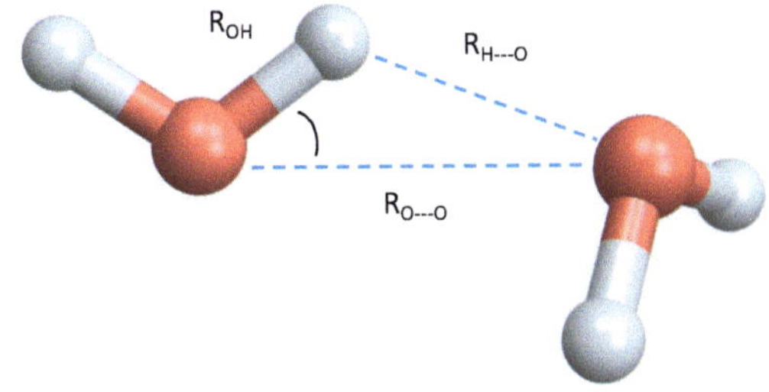

Fig. 11.54 A diagram illustrating the main bonds and angles that define the hydrogen bond between a pair of water molecules

Hydrogen bonding of water involves the 1s electrons (Barbiellini & Shukla 2002) on the ^{1}H atoms (Fig. 11.55). The intermolecular hydrogen bonding of an acceptor A oxygen to a donor D hydrogen

$$R - O - H_D \cdots O_A - R'$$

involves three distances, R_{OH}, the covalent oxygen-hydrogen distance (~96 pm); $R_{H \cdots O}$, the hydrogen bond distance between the donor hydrogen and the acceptor oxygen; and, R_{O-O}, the distance between the two oxygen nuclei (Fig. 11.54). Since the oxygen atom is quite electronegative, variations in these distances alter the electron density on the proton and hence its NMR chemical shift by large amounts, up to about 21 ppm. The nuclear resonance from the H-bonded proton shifts downfield (to larger δ values) as the bond length decreases because, as the hydrogen proton approaches the positive nucleus ($+8e$ charge) of the more electronegative oxygen atom, it is deshielded by the withdrawal of electron density from the exponentially increasing overlap between oxygen and hydrogen wave functions.

11.15.1   *The Proton Chemical Shift of Water Versus Temperature*

At 298 K, the energy of the hydrogen bond in water (~23 meV) is not larger than kT (~25 meV), leading one to expect that as the temperature is increased the fraction of water molecules engaged in hydrogen bonding would decrease according to the Boltzmann probability distribution. Therefore, as the temperature is increased, the average hydrogen bond distance would increase leading to a deshielding of the proton and a shift of its nuclear resonance to higher field. Measurements of this temperature dependence of the proton chemical shift in water as a function of temperature have been pursued since the first studies of Hindman (1966) who confirmed this expectation that the shielding of water protons increased as the temperature rose (Mallamace et al. 2008, 2016).

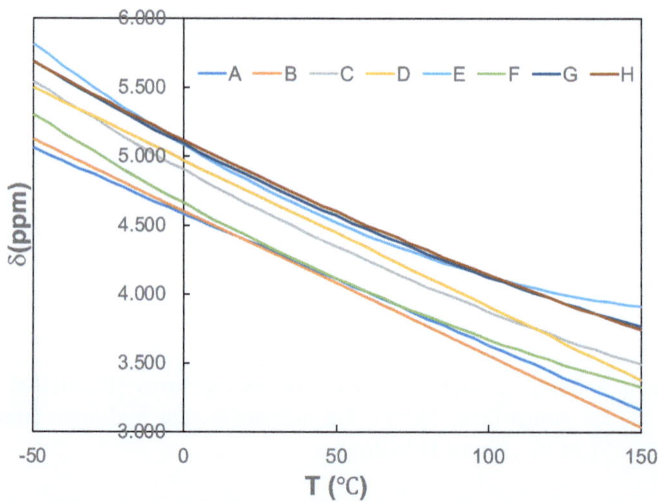

Fig. 11.55 Variation of the NMR chemical shift of liquid water protons as a function of temperature as reported by various investigators. The curves are identified by letter in Table 11.15. The vertical offsets are related to the use of different reference compounds and methods

At first glance, a study of the temperature dependence of the proton chemical shift of water would seem extremely easy. However, the detected nuclear magnetic resonance frequencies of a sample in a practical NMR spectrometer are rarely reported in absolute units (Hz) but rather as frequency differences between a reference material (standard) and that of the resonance frequency of the nucleus of interest. *Aye, therein lies the rub!* One must pick a suitable reference compound under the assumption that its resonance frequency is independent of temperature; of course, different reference compounds will give different chemical shifts. One must also take into consideration the effect of temperature on the magnetic susceptibility of the sample or the reference material, as reflected in Curie's law (Chap. 7). We found eight datasets in the literature that reported the variation of the NMR chemical shift of liquid water protons as a function of temperature; of the eight cited studies, *all* used different references. Some of the plots of the chemical shift of the water protons as a function of temperature were linear, while others exhibited a small quadratic dependence (Fig. 11.55). Nevertheless, all showed that the shielding of water protons increased with the temperature. The lack of a common internal reference, contained within the sample to avoid bulk susceptibility effects, resulted in frequency offsets among the datasets. It is clear from the plots (Fig. 11.55) of these datasets that these problems, which have been routinely encountered in NMR spectroscopy over the past several decades, are manifest in the data. The two issues observed here are that different authors used different reference materials, with slightly different temperature dependencies by themselves, and that the reported chemical shifts are offset with respect to each other due to these differences in reference materials. There are of course technical issues with respect to temperature

Table 11.15 Temperature dependence of the proton chemical shift of water

Curve	References	$a\ (T^2)$	$b\ (T)$	c	$\Delta(\text{ppm})$[a]
A	Matubayasi et al. (1997)	0	-9.50×10^{-3}	4.580	0.503
B	Mallamace et al. (2013)	0	-1.04×10^{-2}	4.597	0.487
C	Modig and Halle (2002)	1.73×10^{-5}	-1.20×10^{-2}	4.897	0.186
D	Angell et al. (1973)	0	-1.06×10^{-2}	4.968	0.115
E	Hoffmann and Conradi (1997)	3.42×10^{-5}	-1.29×10^{-2}	5.079	0.004
F	Hindman (1966)	2.01×10^{-5}	-1.19×10^{-2}	4.656	0.427
G	Chen et al. (2000)	1.60×10^{-5}	-1.12×10^{-2}	5.083	0.000
H	Jaegers et al. (2020)	1.17×10^{-5}	-1.09×10^{-2}	5.109	−0.026

[a]The offset from $\delta_{\text{TSP}} = 0.000$ ppm

measurement and control, as well. The earliest study (Hindman 1966) used an external gaseous reference (CH_4) which was not dissolved in the water itself; therefore, the measured chemical shifts needed to be corrected for the temperature dependence of the density and bulk susceptibility of water. Later investigations used other water-insoluble compounds such as adamantane in solid-state measurements (Jaegers et al. 2020). When the chemical shift data from each report were fitted to polynomials of the form

$$\delta(T) = aT^2 + bT + c,$$

several of the datasets showed linear behavior with temperature, while others were best fit by including a small quadratic term (Table 11.15) which mainly provided upward curvature for $T < 0$ and $T > 100$ °C.

The modern reference material for the NMR study of water-soluble materials is internal TSP (Chap. 7); therefore, I adjusted each dataset to match the chemical shift of water at 0 °C with $\delta_{\text{TSP}} = 5.083$ ppm (Jaegers et al. 2020; Reference H in Table 11.15). The fits in Table 11.15 show the offsets (Δ, ppm) with respect to TSP in the last column.

When these offsets are subtracted from the datasets and the means and standard deviations are computed, the chemical shifts (Table 11.16; Fig. 11.56) can be described by $\delta(T) = 1.2402 \times 10^{-5}\ T^2 - 1.1174 \times 10^{-2}T + 5.083$.

11.15.2 Chemical Shift of Water Protons Versus H-Bond Distances

The main feature of water that results in the decrease in the chemical shift of water protons (Fig. 11.56) as the temperature is raised is the increase in the distance $R_{\text{H}\cdots\text{O}}$, the hydrogen bond distance, between the donor hydrogen and the acceptor oxygen (Fig. 11.54). One also anticipates a smaller but still significant effect from variations

Table 11.16 Mean chemical shift of water vs. temperature

T (°C)	$<\delta(\text{ppm})>$	$\sigma(\text{ppm})$	T (°C)	$<\delta(\text{ppm})>$	$\sigma(\text{ppm})$
−50	5.673	0.083	50	4.555	0.028
−45	5.611	0.072	55	4.506	0.028
−40	5.550	0.062	60	4.457	0.028
−35	5.489	0.052	65	4.409	0.028
−30	5.429	0.043	70	4.362	0.029
−25	5.370	0.034	75	4.315	0.030
−20	5.311	0.026	80	4.268	0.032
−15	5.253	0.019	85	4.223	0.034
−10	5.196	0.012	90	4.178	0.038
−5	5.139	0.006	95	4.133	0.043
0	5.083	0.000	100	4.090	0.048
5	5.027	0.005	105	4.046	0.055
10	4.973	0.010	110	4.004	0.062
15	4.918	0.014	115	3.962	0.070
20	4.864	0.017	120	3.921	0.079
25	4.811	0.020	125	3.880	0.089
30	4.759	0.022	130	3.840	0.099
35	4.707	0.024	135	3.801	0.110
40	4.656	0.026	140	3.762	0.122
45	4.605	0.027	145	3.724	0.028

The data are referenced to $\delta_{\text{TSP}} = 0.000$ ppm

in the bond angle β (Fig. 11.54). For this reason, the chemical shift of hydrogen-bonded protons is a robust measure of the H-bond distance.

By combining data from NMR spectroscopy and crystallography, using either neutrons or X-rays, it has been found that there is an approximately linear dependence (Fig. 11.57) of the hydrogen-oxygen bond distance $R_{\text{H}\cdots\text{O}}$ on the trace of the chemical shift tensor in 21 crystals of small H-bonded molecules, such as ice 1_h ($\delta = 8.3$ ppm), potassium hydrogen malonate ($\delta = 20.5$ ppm), or maleic acid ($\delta = 16.6$ ppm) (Jeffrey and Yeon 1986). These authors measured a 19.3 ppm downfield shift for each 100 pm of bond shortening (Fig. 11.57). A linear fit to the data showed that the chemical shift (with respect to TMS, which resonates within a few hundredths of a ppm of TSP) as a function of the distance is given by $\delta = -0.207\ R_{\text{H}\cdots\text{O}} + 44.7$, which can be inverted to provide the distance as a function of chemical shift: $R_{\text{HO}}(\delta) = -5.19\ \delta + 232$. This gives for water at 298 K ($\delta = 4.811$ ppm with respect to TSP) a hydrogen bond distance of 207 pm at 25 °C, a value which is slightly longer than the 188 pm found in water from neutron scattering at 4 °C. More recent results (Siskos et al. 2017) from a 43-compound sample of phenolic compounds exhibiting intramolecular H-bonds and ionic complexes with both intramolecular and intermolecular H-bonds gave linear fits with a mean value of $\delta = -0.2016\ R_{\text{H}\cdots\text{O}} + 46.99$ ($r^2 \sim 0.98$) which can be inverted to provide the distance as a linear function of chemical shift, $R_{\text{HO}}(\delta) = -$

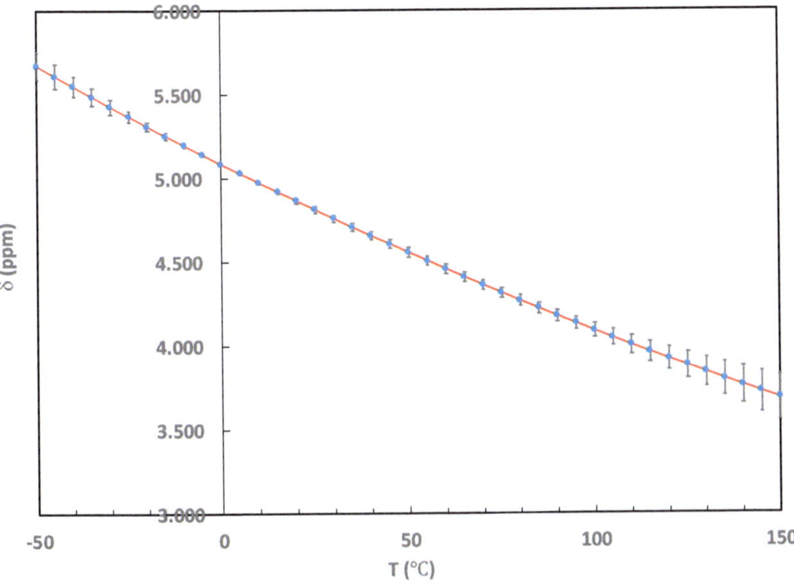

Fig. 11.56 Mean chemical shift of water protons as a function of temperature. The plotted mean values (Table 11.16) are shown here with their standard deviations. Chemical shifts are referenced to internal TSP ($\delta_{\text{TSP}} = 0.000$ ppm)

4.96 δ + 233, which gives essentially the same water hydrogen bond distance of 210 pm at 25 °C (Fig. 11.57). As one can see from Fig. 11.57, there is sufficient spread in the distance measurements that this agreement is within the limits of errors involved. Siskos et al. (2017) also measured the dependence of the chemical shift of protons on the oxygen-oxygen distance R_{OO} in the hydrogen bond and found (Fig. 11.57) that it is described to first order by $R_{\text{OO}}(\delta) = -4.03\,\delta + 323$. Using crystals of enzymes, Harris and Mildvan (1999) performed a similar study of the hydrogen bond lengths and found that the oxygen-oxygen distance R_{OO} varied with chemical shift according to $R_{\text{OO}}(\delta) = 100(-116\ln\delta + 104.47\delta + 5.04)$ giving a R_{OO} distance of 343 pm for water at 298 K. Perhaps this is too extreme an exponential extrapolation to be granted credence; a shorter distance of ~303 pm as suggested by the findings of Siskos et al. (2017) would appear to be more appropriate.

The proton chemical shift also depends on the oxygen-hydrogen covalent-bond distance R_{OH} (Fig. 11.54) in the $-$OH group, which in water is usually given the nominal distance of 96 pm. The O–H binding energy of 10.17 eV (492 kJ/mol) is around 500-fold greater than that of the hydrogen bond (~23 meV) meaning that the hydrogen bond minimally contributes to the O–H binding energy. The data from crystals (Fig. 11.57) indicate that the chemical shift of hydrogen-bonded protons is given by $\delta(R_{\text{OH}}) = 0.802\,R_{\text{OH}} - 67.46$. Applying this to water at 298 K ($\delta = 4.811$ ppm, Table 11.16) gives an R_{OH} of 90 pm, slightly less than the accepted covalent R_{OH} value of ~96 pm. Nevertheless, the data from Siskos et al. (2017)

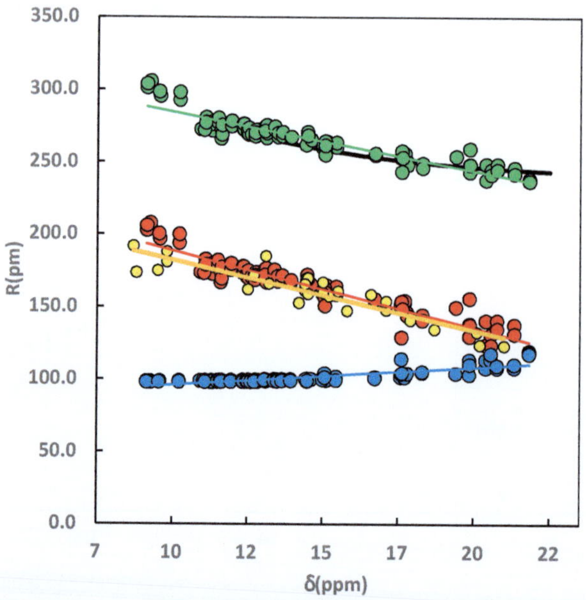

Fig. 11.57 The dependence of the proton NMR chemical shift on hydrogen bond distance in crystals of molecules and in enzymes in aqueous solution. *Blue*, R_{OH}, the oxygen-hydrogen distance as a function of the chemical shift for the *covalent* O–H bond. (Data from Siskos et al. (2017)). The least-squares line is given by $R_{OH}(\delta) = 1.247\ \delta + 84.12$, $r^2 = 0.73$. *Red*, $R_{H\ldots O}$, the distance between the hydrogen donor and the adjacent oxygen atom acceptor in the intermolecular hydrogen bond measured from 43 compounds using X-ray diffraction by Siskos et al. (2017). The least-squares line is given by $R_{H\ldots O}(\delta) = -5.285\ \delta + 238.7$ and $r^2 = 0.90$. The *yellow* data points and the orange least-squares line are $R_{H\ldots O}$ data from crystals of 21 compounds from Jeffrey and Yeon (1986) derived from X-ray and neutron diffraction data. The least-squares line is given by $R_{H\ldots O}(\delta) = -4.826\ \delta + 278.0$, $r^2 = 0.84$. *Green*, R_{OO}, the dependence of the chemical shift of the proton on the distance between the adjacent oxygen atoms in a hydrogen bond from crystals of 43 compounds using X-ray diffraction by Siskos et al. (2017). The least-squares line is given by $R_{OO}(\delta) = -4.028\ \delta + 322.7$, $r^2 = 0.85$. The *black* line is the fit to the data obtained from biological enzymes from Harris and Mildvan (1999) (see text)

(Fig. 11.57) support a small variation in the chemical shift of hydrogen-bonded protons with R_{OH}. We must always keep in mind that molecules are not static, classical physics objects, even though we often use such approximations in our studies. They are an evanescent, time-dependent probability distribution, and our measurements are only ensemble averages over very large numbers of molecules.

11.15.3   *The Chemical Shift of Water Is Phase-Dependent*

Water exists in three phases, solid (ice), liquid, and gas (water vapor), where the strengths and number of hydrogen bonds vary markedly; water molecules in the solid state experience four H-bonds each, and this number decreases in the liquid as a

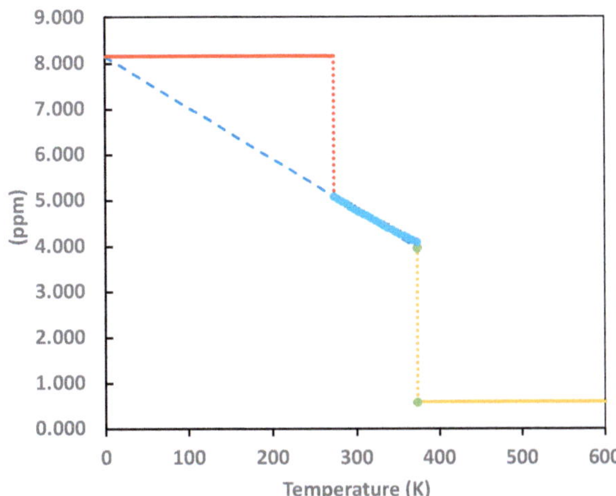

Fig. 11.58 The chemical shift of the protons in water as a function of temperature over the entire range of phases from solid (red) to liquid (blue) to gas (yellow). The blue circles are the measured shifts in liquid water, while the blue dotted line is a linear extrapolation of the liquid measurements to 0 K derived from Table 11.16

function of temperature, while in the vapor state essentially all the H-bonds are broken. As we have just seen, the chemical shifts of water protons are useful not only as precise distance indicators, but we will find that they are also a sensitive indicator of the number of H-bonds and their strengths. The strength of a hydrogen bond, as a dipole-dipole interaction, should vary as an inverse integral power of the distance between the molecules. It is well known that hydrogen bonds decrease approximately linearly with distance with a slope ~20 μeV/pm in the range around $R_{OO} \sim 250$ pm. Then the chemical shift of a hydrogen-bonded proton should be an indicator not only of the bond distance but also of its strength and the number of H-bonds.

As ice is warmed from below 273.15 K, the proton chemical shifts change from that of ice ($\delta = 8.148$ ppm) to that of the liquid ($5.083 < \delta < 4.090$ ppm), and as the heating continues, reaching the shift of the vapor, $\delta = 0.586$ ppm (Fig. 11.58). The chemical shifts reveal a marked change in the magnetic shielding of the water protons whose main cause is the change in H-bond distance with temperature. Thus, NMR of water is an excellent probe of this intermolecular distance as well.

From studies of the temperature dependence of the shielding of the water protons, the relationship between the hydrogen bond distance and the chemical shift of water was found to be described by

$$R_{HO}(\delta) = 238 - 55\,\delta.$$

In water at 25 °C, these distances are around $R_{HO} = 212$ pm and $R_{OO} = 322$ pm; their difference of ~110 pm is about the length of the oxygen-hydrogen bond of 96 pm.

11.15.4   The Average Number of H-Bonds per Water Molecule

The average number of hydrogen bonds in water and its variation with temperature can be derived from the measured NMR chemical shifts. The dependence of the chemical shift of water protons on temperature over the entire range of phases, from ice (Burum & Rhim 1979), to steam (Fig. 11.58), shows that the phase transitions from solid to liquid, and from liquid to gas, are accompanied by large changes in magnetic shielding; the ice-to-liquid phase change shields the protons by $\Delta\delta = -3.065$ ppm, while the liquid-to-gas transition gives an increase in shielding of $\Delta\delta = -3.504$ ppm. Note that in Fig. 11.59 the linear $\delta(T)$ function ($\delta(T) = -9.97 \times 10^{-3}\, T + 5.096$) measured for $-50 < T < 150$ °C extrapolates to 7.82 ppm, approximately the chemical shift of ice at 0 K.

In the temperature range $0 < T < 273.15$ K, each water molecule in the solid is hydrogen-bonded to four other water molecules and the proton chemical shift of ice is constant at 8.148 ppm. Therefore, we can assume that the average number of hydrogen bonds per water molecule remains at four until the ice melts at 273.15 K and the value of the proton chemical shift at the instant of melting ($\delta_0 = 5.083$ ppm) can be used to represent the fully H-bonded chemical shift of the liquid. Then the average number of H-bonded water molecules can be estimated from

$$< n_{HB} > \; = n_0 \frac{\delta(T) - \delta_{\text{vapor}}}{\delta_{\text{o}} - \delta_{\text{vapor}}}$$

with n_0 the number of H-bonded water molecules at $T = 0$ °C. For n_0 in the range $4 \geq n_0 > 3.85$, we find a temperature dependence (Fig. 11.59) with a linear slope of -8.9×10^{-3}/°C, a value which broadly agrees with that found for many other measurements of this quantity, albeit using methods with much larger errors and variations (see Chaplin, M., https://water.lsbu.ac.uk/water/martin_chaplin.html) for many examples. By using the temperature as a parameter on which both δ and n_{HB}

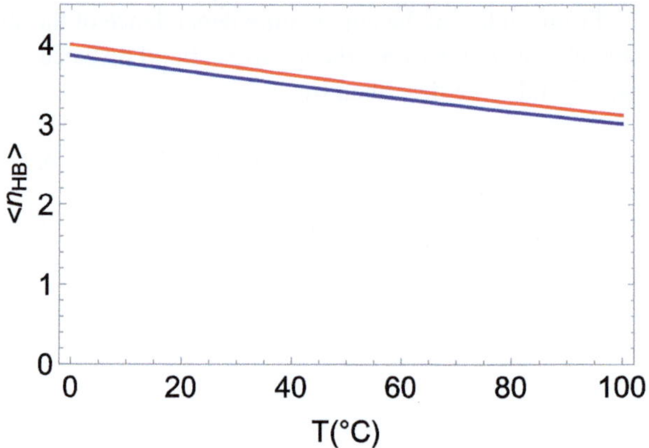

Fig. 11.59 The temperature dependence of the average number of hydrogen bonds per molecule in liquid water determined from the chemical shift of the protons in ice, water, and steam, normalized to values of n_0 the number of H-bonded water molecules at $T = 0$ °C ranging from $n_0 = 4.0$ (red) to $n_0 = 3.86$ (blue)

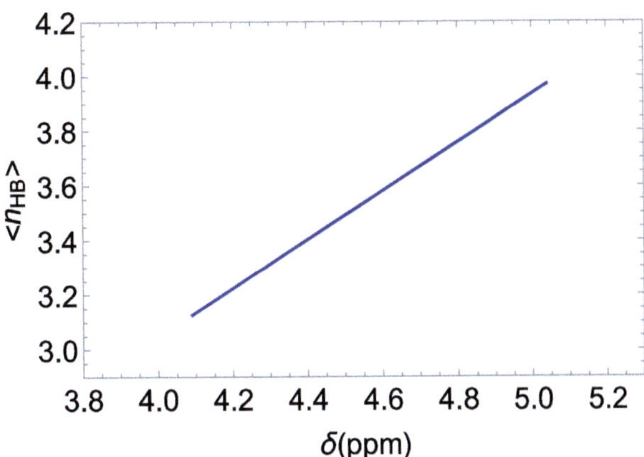

Fig. 11.60 The parametric relationship between the chemical shift of water protons and the average number of H-bonds per molecule liquid water at a given temperature, assuming $n_o = 4$ at 0 °C. A linear fit gives $n_{HB}(\delta) = 0.8895\,\delta - 0.512$

depend, we can also determine the relationship between $<n_{HB}>$ and the chemical shift of water (Fig. 11.59) so that from a simple measurement of the chemical shift of water one can quickly estimate the average number of H-bonds per molecule at any temperature (Fig. 11.60).

11.15.5   *Angular Distribution of H-Bonds*

The water molecule, as we have seen, is a member of the $\mathbf{C_{2v}}$ point group. Water's chemical shift tensor must also obey $\mathbf{C_{2v}}$ symmetry and therefore can be completely determined by only two of its nine elements: the isotropic shielding σ_{iso} and the shielding anisotropy $\Delta\sigma$. Quantum mechanical calculations of the shielding tensor for the water molecule by Pfrommer et al. (2000) generated the isotropic shielding and the shielding anisotropy as a function of both the distances and the angles between the hydrogen and oxygen nuclei in a hydrogen-bonded water molecule, as defined in Fig. 11.54. These indicate that there is a strong dependence of the electron density and hence the magnetic field at the nucleus on the distance and angles between the molecules. For a fixed angle $\beta = 0$ and an oxygen-oxygen distance $R_{O\text{-}\text{-}\text{-}O}$ of ~290 pm, the shielding anisotropy varies approximately as the inverse cube of the distance $R_{H\text{---}O}$ and the isotropic shielding increases (Fig. 11.61 left). The two components of the shielding tensor have a weaker dependence on the geometry of the hydrogen bond, with only the anisotropy varying significantly with the angle β (Fig. 11.61 right). The four-dimensional shielding surface is given by

$$\sigma(r, \alpha, R, \beta) = A + Br + C\alpha + dR^{-3} + E\beta$$

$$\sigma = \{\sigma_{iso} \text{ or } \Delta\sigma\}$$

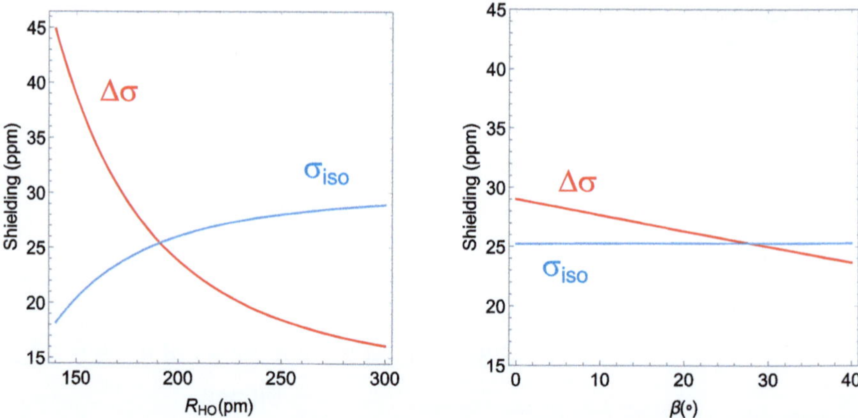

Fig. 11.61 The results of density functional calculations of the NMR shielding tensor for protons in the water molecule using the geometry shown in Fig. 11.54. (*Left*) In red is the shielding anisotropy, $\Delta\sigma$, while in blue is the isotropic shielding σ_{iso} as a function of the distance $R_{\mathrm{H\cdots O}}$. (*Right*) The dependence of the shielding anisotropy, $\Delta\sigma$, and the isotropic shielding σ_{iso} on the angle β. (Curves drawn from data in Modig et al. (2003))

Table 11.17 Magnetic shielding constants for protons in water

Constant (units)	σ_{iso}	$\Delta\sigma$
A (ppm)	64.81	50.00
B (ppm/pm)	-0.3164	-0.3568
C (ppm/°)	-4.24×10^{-2}	-3.03×10^{-2}
D (ppm pm^3)	-32.8×10^{6}	88.0
E (ppm/°)	2.5×10^{-3}	-0.134

where r (R_{OH} in Fig. 11.54) is fixed at the oxygen-hydrogen bond length of 96 pm, α is fixed at the H–O–H bond angle of 104.5°, and R is R_{HO} in Fig. 11.54. The constants calculated by Modig et al. (2003) are summarized in Table 11.17.

Comparison of these calculations (Modig et al. 2003) with experimental measurements of the temperature dependence of the shielding anisotropy (Modig and Halle 2002) showed that the shielding anisotropy decreased linearly with temperature

$$\Delta\sigma(T) = -0.0432 \frac{\mathrm{ppm}}{°\mathrm{C}} T + 28.54 \text{ ppm}$$

with the units of the anisotropy in ppm and the temperature in °C. The isotropic shielding from -15 to 100 °C obeyed (to first order in T)

$$\sigma_{\mathrm{iso}}(T) = 1.188 \times 10^{-2} \frac{\mathrm{ppm}}{°\mathrm{C}} T + 25.406 \text{ ppm}.$$

Note here that the shielding anisotropy is approximately fourfold more sensitive to temperature changes than is the isotropic shielding. The temperature dependence of the hydrogen bond length, R_{HO}, and the average value of the angle β were described by

$$R_{HO}(T) = 0.138 \frac{\text{pm}}{°\text{C}} T + 187.0 \text{ pm}$$

and

$$\beta(T) = 0.1233 \frac{°}{°\text{C}} T + 11.71°$$

which suggest that at room temperature (25 °C) we have that $R_{HO}(25) = 190.45$ pm and $\beta(25) = 14.8°$.

There will clearly be thermal modulation of these average quantities. Of interest therefore is the probability distribution functions $f(\beta)$ that describe the details of the occupation of the various geometrical configurations and temperatures. Ab initio simulations at 27 °C (Sprik et al. 1996) indicate that $f(\beta)$ is described by the product of a $\sin(\beta)$ weighting factor (due to the $\sin \beta$ in the solid angle volume element of the integral) and a Gaussian with a width determined by the constant, c, in

$$f(\beta) \sim \sin \beta \, e^{-c\beta^2}$$

While the Gaussian contribution to $f(\beta)$ will peak at $\beta = 0$, the geometrical $\sin \beta$ weighting factor pushes the peak to finite angles, as shown in Fig. 11.62. The width of these distributions increases with temperature, as would be expected as the competition between the hydrogen bond energy and kT favors the latter.

The width of these distributions (Modig et al. 2003) as a function of temperature is given by

$$\sigma_\beta(T) = 0.064 \, T + 6.1°$$

indicating that there is considerable distortion of the hydrogen bond geometry in water that increases with temperature.

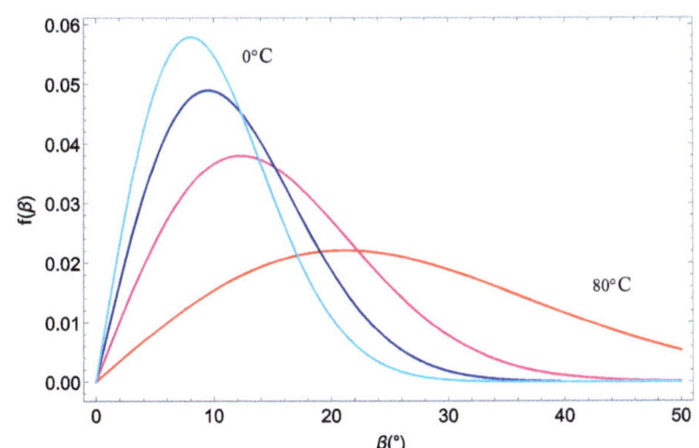

Fig. 11.62 Normalized probability distributions of the angle β of the hydrogen bond in water at various temperatures, ranging from 0 (top, cyan) to 27, 50, and 80 °C (bottom, red). (Redrawn from Modig et al. (2003))

11.15.6 ^{17}O NMR of Water

Water contains two types of nuclei, ^{1}H, which with a nuclear spin of ½ and a natural abundance of almost 100% has been extensively used to probe water's properties, and oxygen, whose only stable magnetic isotope is ^{17}O with spin 5/2 and a natural abundance of only 0.0373% (373 ppm) and a magnetic moment of only 13.56% of the proton. In a magnetic field of 2.35 T where protons resonate at 100 MHz, the Larmor frequency for ^{17}O is 13.556 MHz. The nuclear spin of 5/2 gives ^{17}O an electric quadrupole moment of -0.02578×10^{-24} e cm^2: the negative sign indicating that the nuclear shape is that of an oblate ellipsoid. This nonspherical distribution of charge in the nucleus couples to electric field gradients across the nucleus leading to broad NMR lines for large molecules, such as enzymes, but recent theoretical studies (Zhu and Wu 2011) have provided methods for observing such samples. For very small molecules, such as water, it is possible, using solid-state NMR instrumentation, to easily collect useful narrow lines and to monitor the chemical shift and transverse relaxation rate as a function of temperature to determine how the hydrogen bonding of the oxygen in water varies with temperature (Jaegers et al. 2020). The range of oxygen chemical shifts is much larger than that for the proton, and therefore one could argue that ^{17}O NMR would be the method of choice for measuring the alterations in the chemical shifts as a function of temperature.

As we have seen with respect to the alterations in the electronic shielding of the protons in the water molecule as the temperature of the liquid is raised above 0 °C, a similar increase in shielding (decrease in chemical shift) occurs for the oxygen nuclei in water (Fig. 11.63). The linear change in shielding for ^{17}O as a function of temperature is given by $\delta(T) = -0.0483\,T + 1.4829$ ppm, where the reference for $\delta = 0$ is [^{17}O]-H$_2$O at $T = 25$ °C. The slope of the $\delta(T)$ function for ^{17}O is 4.1-fold larger than that for the temperature dependence of the proton chemical shift, indicating that the magnetic field at the oxygen nucleus changes more with temperature than that at the proton in water.

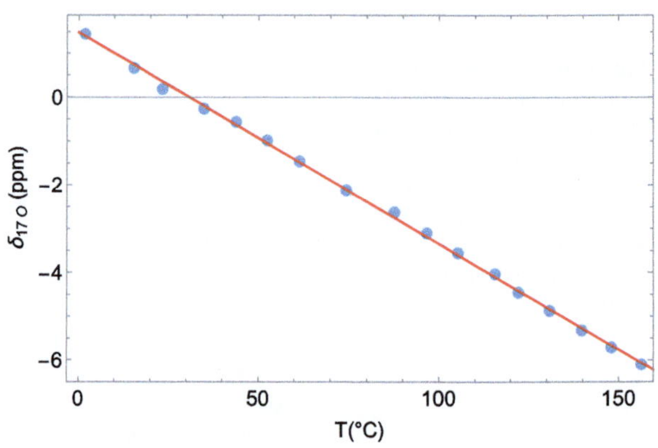

Fig. 11.63 Measurements of the variation of the NMR chemical shift of ^{17}O in water as a function of temperature; the blue points were extracted from Jaegers et al. (2020), and the red least-squares fitted line is given in the text

11.16   The Hydrophobic Effect

The unique properties of water can be summarized in another grand organizing principle with respect to the origin of life: that of the *hydrophobic effect* (Lazaridis 2013) or, in the parlance of Ben-Naim, the hydrophilic effect. We are all familiar with the futility associated with attempting to dissolve oil in water. In 1936, G. S. Hartley eloquently stated (Hartley 1936) that

> The antipathy of the paraffin chain for water is, however, frequently misunderstood. There is no question of actual repulsion between individual water molecules and paraffin chains, nor is there any very strong attraction of paraffin chains for one another. There is, however, a very strong attraction of water molecules for one another in comparison with which the paraffin-paraffin or paraffin-water attractions are slight.

Oil and water differ fundamentally in terms of their polarity, i.e., the deformability of their electronic wave functions in response to the juxtaposition of the electrons from other molecules. Just as it is energetically unfavorable to transfer a charged ion from a high dielectric constant medium, such as water ($\varepsilon \sim 80$) to a medium like oil ($\varepsilon \sim 2$), it is also unfavorable to transfer a water molecule into oil but for entropic, rather than energetic, reasons. The resulting spontaneous phase separation serves to reinforce in our minds the role of amphiphiles in the segregation of primordial biomolecules into structures with topologically distinct interiors and exteriors. Probability arguments then push these molecules in the direction of the self-assembly of closed structures.

The Gibbs energy change upon transfer of a nonpolar molecule from a thermodynamic state where it is surrounded by other nonpolar molecules, such as a neat liquid, to water contains both an energy and an entropy term:

$$\Delta G = \Delta H - T\Delta S$$

At 25 °C, the enthalpy of transfer, ΔH, is essentially zero and can be neglected, but the entropy change, ΔS, is negative. The polarity of the water molecule generates an ordered liquid, and this order is preserved to a large extent when a nonpolar solute is transferred into water. This affinity of the water molecule for itself organizes the water into cage-like structures around the nonpolar solute. Since these cages are a low probability structure, the entropy of transfer is negative, indicating that the transfer has a lower probability than the separated compounds.

From the measured change in entropy associated with bringing a nonpolar solute into an aqueous environment of -26.1 J/mol K, one can use the fact that $\Delta H \sim 0$ to calculate the probability, P, that, for example, hexane will be found in the water using the Boltzmann distribution as

$$P = e^{-\frac{\Delta G}{kT}}$$

where we make the assumption that $\Delta G = \Delta H - T\Delta S$ is dominated by the entropy term since $\Delta H \sim 0$. The probability is then $P = 1.95 \times 10^{-6}$ that a hexane molecule will be found in the water phase.

11.16.1 *Lipids: Inside Versus Outside*

The formation of the lipid bilayer, the basic structure of all cell membranes, relies heavily on the hydrophobic effect. The phosphate head groups of the phospholipids are solvated by water, while the long-chain fatty acyl tails are oriented such that they pack together and make a structure that is symmetric about the midplane (Fig. 11.64). This important structure self-assembles as one can verify by sonicating dipalmitoyl-*sn*-glycero-phosphorylcholine in warm water for a few minutes and then examining the resulting bilayer vesicles with an electron microscope.

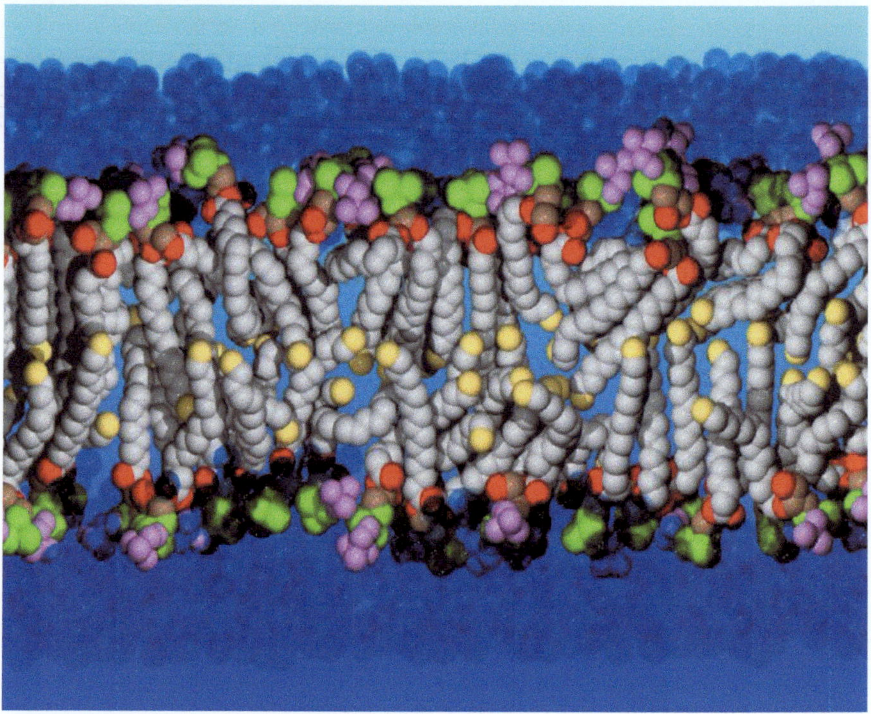

Fig. 11.64 A molecular model of the lipid bilayer structure of cell plasma membranes. The methyl groups at the ends of the fatty acyl chains are shown in yellow, the $(-CH_2-)_n$ methylene chains are shown in gray, and the polar, hydrophilic head groups of the phospholipids are shown in bright colors. The dark blue water molecules surround the bilayer but also interpenetrate within the region of the polar head groups. Not shown are proteins and other molecules that intercalate within the lipid portion of the membrane

Lars Onsager proposed that lipids ordered on an aqueous surface would have their phosphate groups in contact with the water and their nonpolar, methylene chains oriented away from the surface. When a drop of rain would impinge on such a system, the resulting rebound from this drop could easily be imagined to form a spherical arrangement that formed the bilayer structure, and thus we have a primitive method for the construction of the first biological object with an interior that was separated from the exterior by a bilayer membrane that could control the communication, allowing the interior to acquire different properties from the exterior while, at the same time, limiting the transit of molecules to those that could dissolve in the hydrophobic interior of the bilayer. It is clear, therefore, that the hydrophobic effect is crucial to the formation of living systems and must therefore be classified as an organizing principle.

11.16.2   Proteins

It is thought that folding of the nascent polypeptide chain as it emerges from the ribosome is driven by the stabilization of the structure by the burying of hydrophobic amino acid residues into the oil-like interior of the protein, away from water, and that the polar amino acids tend to be found on the surface, exposed to the aqueous environment (Wang & Ben-Naim 1997). Estimates of the hydrophobic effect have relied on studies of the transfer of compounds, such as amino acids and nonpolar compounds from an organic solvent to water, and on studies of the heat-induced denaturation of proteins. In heat denaturation, the unfolded protein exposes its nonpolar residues to the water, and this results in a larger heat capacity for the unfolded protein (Privalov & Makhatadze 1993; Robinson & Cho 1999; Bellissent-Funel et al. 2016). These experiments give a value of the Gibb's energy change for a single methylene group of 4.6 ± 2.1 kJ/mol. Hydrogen bonding among amino acids, and to the peptide backbone, is another factor stabilizing proteins. It has been found that in a folded protein there are about 1.1 hydrogen bonds per amino acid and that each hydrogen bond contributes about the same Gibbs energy (4.6 ± 2.1 kJ/mol) as the hydrophobic effect (Pace et al. 2014).

11.16.3   Hydration of Ions

The polarity of water is an important factor contributing to the dissociation of ionic compounds in aqueous media (Ben-Amotz & Underwood 2008; Seidel et al. 2016). It has been known for some time that the strong electrostatic interactions between a charged ion and water lead to the formation of hydration shells of organized water molecules oriented such that the positive face of water coordinates with the negative ion, and vice versa. This process is referred to as solvation and is illustrated in Fig. 11.65 for the sodium ion in water. Solvation in water involves hydrogen

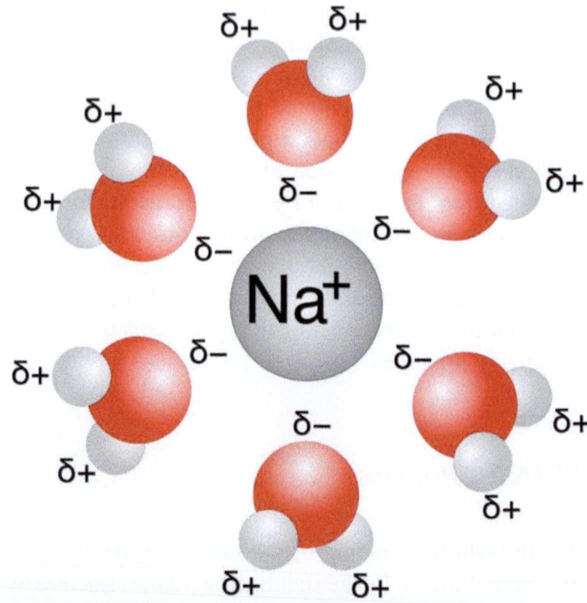

Fig. 11.65 A schematic view of the solvation of a sodium ion (Na+) in water, showing the first hydration shell of water molecules surrounding the ion. (Courtesy of M. Chaplin (https://water.lsbu.ac.uk/water/martin_chaplin.html). Used with permission)

bonding, dipolar interactions, and van der Waals forces. Water is a polar solvent that can both accept and donate hydrogen bonds so it makes an excellent solvent for solutes that either accept or donate (or both) hydrogen bonds. Solvation is a complex process that requires several steps. First, a cavity must be formed in the water in order to accommodate the solute. The subsequent ordering of the water molecules is unfavorable from both an enthalpic and an entropic viewpoint. The strong hydrogen bonding in water implies that this cavity formation is less favorable than in less polar solvents. Then a solute particle must dissociate from the bulk solute, and this is unfavorable from an enthalpic setting since the solute existed in a bound form prior to insertion into the water. Finally, the solute enters the cavity, and its enthalpy decreases, followed by entropy of mixing, which favors solvation. Solvation of macromolecules, such as DNA, RNA, or proteins, is an important factor driving their proper spontaneous folding.

Problems

1. *Tetrahedral order.* In the expression for the orientational order parameter, Q, assume all the angles, ψ_{jk}, are the same so that you may drop the double sum, as well as the factor of 3/8. Then plot $<Q(\psi)>$ versus ψ, and use it to estimate the range of H-bond angles actually found in liquid water at the temperatures shown in Fig. 11.23.

2. *Basis for the water molecule*. Find the transformation properties of the basis

$$\begin{bmatrix} 0 & 0 & 0 \\ 0 & 1 & 0 \\ 0 & 1 & 1 \end{bmatrix} \text{ and } \begin{bmatrix} 0 & 0 & 0 \\ 0 & -1 & 0 \\ 0 & 1 & 1 \end{bmatrix}.$$

using the group operators of $\mathbf{C_{2v}}$ just as we did for the unit vector basis (Table 11.7).

3. *Group theory*. Using Mathematica, show that each operator matrix is its own inverse.

4. *The point group of the triangle*. Find the symmetry group of the equilateral triangle (A) of height 1.0, with the vertices as the set of points in $\{x,y,z\}$ space given by

$$\left\{ \left\{0, \frac{2}{3}, 0\right\}, \left\{\frac{1}{\sqrt{3}}, -\frac{1}{3}, 0\right\}, \left\{-\frac{1}{\sqrt{3}}, -\frac{1}{3}, 0\right\}\right\}.$$

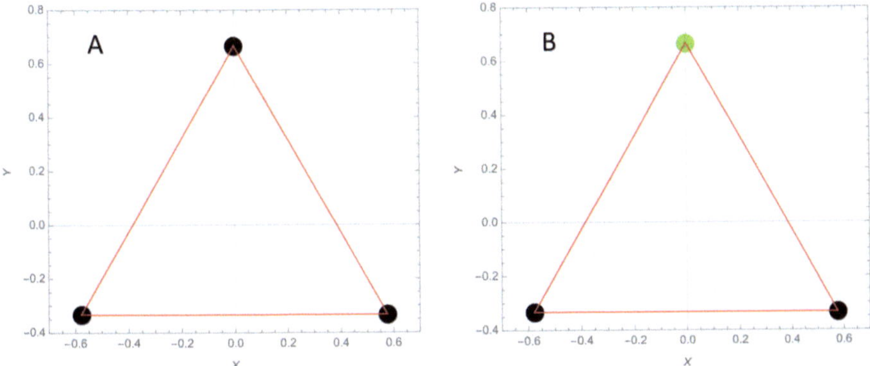

Form the basis vectors linking the vertices to the origin in terms of the unit vectors in 3-space: $\{e_1, e_2, e_3\}$.

(a) What transformation operators leave this basis invariant? That is, find the set of matrices that form a representation of the set of symmetry operators, using the fact that a matrix representation of an operator, T, in the basis $X_k = \{X_1, X_2, \ldots, X_N\}$ is given by

$$T_{ij} = \langle X_i | T | X_i \rangle.$$

(b) Show by explicit matrix calculations that the T_{ij} leave the basis invariant.

(c) Show that the set T_{ij} in (b) form a group, and hence find the group multiplication table. Find the character table.

(d) What is the order of this group? What is the number of operators?

(e) Find the set of commutators $[T_i, T_j]$. Is this group Abelian?

(f) What is the number of classes of operators?

(g) What are the irreducible representations of this group? Find the name of this point group.

(h) If we now change the color of the top vertex from black (A) to green (B), how do the answers to (b)–(h) change? What is the name of this new group? Have we seen this group before?

(i) If we let the vertices each have a different color, how do the answers to (b)–(h) change? Note that we now have something that resembles the colored quarks in Fig. 7.1.

5. *Spectroscopy of water*. Using the energy levels of the water molecule shown in Fig. 11.14, propose spectroscopic measurements that would probe

(a) The core electrons on the oxygen atom.

(b) The highest occupied molecular wave function.

(c) The lowest unoccupied molecular wave function.

(d) Which of these measurements would be most sensitive to hydrogen bonding in the solid and liquid state?

6. *An electrostatic water model*. Using an electrostatic model of the water molecule, calculate its dipole moment. Can you plot the electric field distribution in 3D space using Mathematica?

7. *Tetrahedral order*. The illustration in Fig. 11.22 shows the tetrahedral order of hydrogen bonds in water:

(a) Calculate the dipole moment of liquid water using this as a starting configuration and expanding the space to include enough molecules so that your calculation converges.

(b) Compare your results with the measured dielectric constant at 298 K.

8. *The hydrophobic effect*. If a sodium ion in water is trapped inside a lipid bilayer vesicle, calculate the probability of its escape.

References

M. Amann-Winkel, M.-C. Bellissent-Funel, L.E. Bove, T. Loerting, A. Nilsson, A. Paciaroni, D. Schlesinger, L. Skinner, X-ray and neutron scattering of water. Chem. Rev. **116**, 7570–7589 (2019)

G. Andrés Cisneros, K.T. Wikfeldt, L. Ojamäe, J. Lu, Y. Xu, H. Torabifard, A.P. Bartók, G. Csányi, V. Molinero, F. Paesani, Modeling molecular interactions in water: From pairwise to many-body potential energy functions. Chem. Rev. **116**, 7501–7528 (2016)

C.A. Angell, J. Shuppert, J.C. Tucker, Anomalous properties of supercooled water. Heat capacity, expansivity, and proton magnetic resonance chemical shift from 0 to 38 °C. J. Phys. Chem. **77**, 3092–3099 (1973)

P. Ball, Water is an active matrix of life for cell and molecular biology. PNAS **114**, 13327 (2017)

U. Balucaniy, J.P. Brodholtz, R. Vallaurix, Analysis of the velocity autocorrelation function of water. J. Phys. Condens. Matter **8**, 6139–6144 (1996)

B. Barbiellini, A. Shukla, Ab initio calculations of the hydrogen bond. Phys. Rev. B **66**, 235101 (2002)

M.-C. Bellissent-Funel, A. Hassanali, M. Havenith, R. Henchman, P. Pohl, F. Sterpone, D. van der Spoel, Y. Xu, A.E. Garcia, Water determines the structure and dynamics of proteins. Chem. Rev. **116**, 7673–7697 (2016)

D. Ben-Amotz, R. Underwood, Unraveling water's entropic mysteries: A unified view of nonpolar, polar, and ionic hydration. Acc. Chem. Res. **41**, 957–967 (2008)

A. Ben-Naim, *Molecular Theory of Water and Aqueous Solutions. Part 1: Understanding Water* (World Scientific Press, Singapore, 2009)

W.A. Bentley, Studies among the snow crystals during the winter of 1901-1902 with additional data collected during the previous winters and twenty-two half-tone plates. Annual Summary of the Monthly Weather Review for 1902 (1902), https://siarchives.si.edu/sites/default/files/pdfs/WAB_Snow_1902.pdf

W. Bernal, R.H. Fowler, A theory of water and ionic solution, with particular reference to hydrogen and hydroxyl ions. J. Chem. Phys. **1**, 515 (1933)

J.J. Berzelius, Introduction of electronegativity. As quoted in Jensen, W.B. Electronegativity from Avogadro to Pauling: Part 1: Origins of the electronegativity concept. J. Chem. Educ. **73**(1996), 11–20 (1811)

F. Bloch, W.W. Hansen, M. Packard, Nuclear induction. Phys. Rev. **69**, 127 (1946)

D.P. Burum, W.K. Rhim, An improved NMR technique for homonuclear dipolar decoupling in solids: Application to polycrystalline ice. J. Chem. Phys. **70**, 3553 (1979)

M. Ceriotti, W. Fang, P.G. Kusalik, R.H. McKenzie, A. Michaelides, M.A. Morales, T.E. Markland, Nuclear quantum effects in water and aqueous systems: Experiment, theory, and current challenges. Chem. Rev. **116**, 7529–7550 (2016)

M. Chaplin, Do we underestimate the importance of water in cell biology? Nat. Rev. Mol. Cell Biol. **7**, 861 (2006)

Chaplin, M., Water's hydrogen bond strength. arXiv:0706.1355 (2007).

L. Chen, T. Gross, H.-D. Lüdemann, The T, p-dependence of the chemical shift of the hydroxyl protons in deeply supercooled methanol and water. Z. Naturforsch. **55**, 473 (2000)

C.H. Cho, S. Singh, G.W. Robinson, An explanation of the density maximum in water. Phys. Rev. Lett. **76**, 1651 (1996)

H. Cho, P.B. Shepson, L.A. Barrie, J.P. Cowin, R. Zaveri, NMR investigation of the quasi-brine layer in ice/brine mixtures. J. Phys. Chem. B **106**, 11226–11232 (2002)

A.S. Coolidge, A quantum mechanics treatment of the water molecule. Phys. Rev. **42**, 189 (1932)

J.R. Errington, P.G. Debenedetti, Relationship between structural order and the anomalies of liquid water. Nature **409**, 318 (2001)

F. Franks, *Water: A Matrix of Life* (Royal Society of Chemistry, Cambridge, 2000)

T. Fransson, Y. Harada, N. Kosugi, N.A. Besley, B. Winter, J.J. Rehr, L.G.M. Pettersson, A. Nilsson, X-ray and electron spectroscopy of water. Chem. Rev. **116**, 7551–7569 (2016)

P. Gallo, K. Amann-Winkel, C.A. Angell, M.A. Anisimov, F. Caupin, C. Chakravarty, E. Lascaris, T. Loerting, A.Z. Panagiotopoulos, J. Russo, J.A. Sellberg, H.E. Stanley, H. Tanaka, C. Vega, L. Xu, L.G.M. Pettersson, Water: A tale of two liquids. Chem. Rev. **116**, 7463–7500 (2016)

F.M. Gonzalez, E.M. Fries, C. Cude-Woods, et al., Improved neutron lifetime measurement with UCNτ. Phys. Rev. Lett. **127**, 162501 (2021)

J.K. Gregory, D.C. Clary, K. Liu, M.G. Brown, R.J. Saykally, The water dipole moment in water clusters. Science **275**, 814 (1997)

T.K. Harris, A.S. Mildvan, High-precision measurement of hydrogen bond lengths in proteins by nuclear magnetic resonance methods. Proteins **35**, 275 (1999)

G. Hartley, *Aqueous Solutions of Paraffin-Chain Salts* (Hermann & Cie., Paris, 1936)., as quoted in C. Tanford, *The Hydrophobic Effect: Formation of Micelles and Biological Membranes* (Wiley, New York, 1973), p viii

T.L. Hill, *Statistical Mechanics: Principles and Selected Applications* (Dover, New York, 1956), p. 210

J.C. Hindman, Proton resonance shift of water in the gas and liquid states. J. Chem. Phys. **44**, 4582 (1966)

M.M. Hoffmann, M.S. Conradi, Are there hydrogen bonds in supercritical water? J. Am. Chem. Soc. **119**, 3811–3817 (1997)

C. Huang, K.T. Wikfeldt, D. Nordlund, U. Bergmann, T. McQueen, J. Sellberg, L.G. Pettersson, A. Nilsson, Wide-angle X-ray diffraction and molecular dynamics study of medium-range order in ambient and hot water. Phys. Chem. Chem. Phys. **13**, 19997–20007 (2011)

S. Izadi, R. Anandakrishnan, A.V. Onufriev, Building water models: A different approach. J. Phys. Chem. Lett. **5**, 3863–3871 (2014)

N.R. Jaegers, Y. Wang, J.Z. Hu, Thermal perturbation of NMR properties in small polar and non-polar molecules. Nat. Sci. Rep. **10**, 6097 (2020)

G.A. Jeffrey, Y. Yeon, The correlation between hydrogen-bond lengths and proton chemical shifts in crystals. Acta Crystallogr. **42**, 410 (1986)

E. Kalman, G. Palinkas, P. Kovacs, Liquid water. I. Electron scattering. Mol. Phys. **34**, 505–524 (1977)

Y. Khalak, B. Baumeier, M. Karttunen, Improved general-purpose five-point model for water: TIP5P/2018. J. Chem. Phys. **149**, 224507 (2018)

T. Kimura, M. Ikoma, Predicted diversity in water content of terrestrial exoplanets orbiting M dwarfs. Nat. Astron. **6**, 1296–1307 (2022)

T. Lazaridis, Hydrophobic effect, in *eLS*, (Wiley, Chichester, 2013)

M.L. Leetmaa, M.P. Ljungberg, A.P. Lyubartsev, A. Nilsson, L.G.M. Pettersson, Theoretical approximations to X-ray absorption spectroscopy of liquid water and ice. J. Electron Spectrosc. Relat. Phenom. **177**, 135–157 (2010)

Z. Liu, J. Botana, A. Hermann, S. Valdez, E. Zurek, D. Yan, H.-q. Lin, M.-s. Miao, Reactivity of He with ionic compounds under high pressure. Nat. Commun. **9**, 951 (2018)

W.A.P. Luck, The importance of cooperativity for the properties of liquid water. J. Mol. Struct. **448**, 131–142 (1998)

M.W. Mahoney, W.L. Jorgensen, A five-site model for liquid water and the reproduction of the density anomaly by rigid, nonpolarizable potential functions. J. Chem. Phys. **112**, 8910 (2000)

F. Mallamace, C. Corsaro, M. Broccio, C. Branca, N. Gonzalez-Segredo, J. Spooren, S.-H. Chen, H.E. Stanley, NMR evidence of a sharp change in a measure of local order in deeply supercooled confined water. Proc. Natl. Acad. Sci. USA **105**, 12725–12729 (2008)

F. Mallamace, C. Corsaro, D. Mallamace, C. Vasi, H.E. Stanley, The thermodynamical response functions and the origin of the anomalous behavior of liquid water. Faraday Discuss. **167**, 95 (2013)

F. Mallamace, C. Corsaro, D. Mallamace, S. Vasi, H.E. Stanley, NMR spectroscopy study of local correlations in water. J. Chem. Phys. **145**, 214503 (2016)

C.G. Malmberg, A.A. Maryott, Dielectric constant of water from 0° to 100° C. J. Res. Natl. Bur. Stand. **56**, 1 (1956)

D. Mariedahl, F. Perakis, A. Späh, H. Pathak, K.H. Kim, G. Camisasca, D. Schlesinger, C. Benmore, L.G.M. Pettersson, A. Nilsson, K. Amann-Winkel, X-ray scattering and O−O pair-distribution functions of amorphous ices. J. Phys. Chem. B **122**, 7616–7624 (2018)

M.W. Martynowycz, T. Gonen, MicroED structure of hexagonal ice Ih (2019), https://chemrxiv.org/articles/MicroED_Structure_of_Hexagonal_Ice_Ih/82986, https://doi.org/10.26434/chemrxiv.8298641.v1

N. Matubayasi, C. Wakai, M. Nakahara, NMR study of water structure in super- and subcritical conditions. Phys. Rev. Lett. **78**, 2573 (1997)

K. Modig, B. Halle, Proton magnetic shielding tensor in liquid water. J. Am. Chem. Soc. **124**, 12031 (2002)

K. Modig, B.G. Pfrommer, B. Halle, Temperature-dependent hydrogen-bond geometry in liquid water. Phys. Rev. Lett. **90**, 075502 (2003)

R.S. Mulliken, Report on notation for the spectra of polyatomic molecules. J. Chem. Phys. **23**, 1997 (1955)

J. Nico, A. Yue, M. Dewey, D. Gilliam, G. Greene, A. Laptev, W. Snow, F. Wiefeldt, Improved determination of the neutron lifetime. Phys. Rev. Lett. **111**, 222501 (2013)

A. Nilsson, L.G.M. Pettersson, The structural origin of anomalous properties of liquid water. Nat. Commun. **6**, 8998 (2015)

A. Nilsson, D. Nordlund, I. Waluyo, N. Huang, H. Ogasawara, S. Kaya, U. Bergmann, L.-Å. Näslund, H. Öström, P. Wernet, K.J. Andersson, T. Schiros, L.G.M. Pettersson, X-ray absorption spectroscopy and X-ray Raman scattering of water and ice; an experimental view. J. Electron Spectrosc. Relat. Phenom. **177**, 99–129 (2010)

C.G. Ning, B. Hajgato, Y.R. Huang, S.F. Zhang, K. Liu, Z.H. Luo, S. Knippenberg, J.K. Deng, M.S. Deleuze, High resolution electron momentum spectroscopy of the valence orbitals of water. Chem. Phys. **343**, 19–30 (2008)

C. Pace et al., Evaluating the contribution of hydrogen bonding and hydrophobic bonding to protein folding. FEBS Lett. **588**, 2177 (2014)

L. Pauling, The nature of the chemical bond. IV. The energy of single bonds and the relative electronegativity of atoms. J. Am. Chem. Soc. **54**, 3570–3582 (1932)

G. Perotti et al., Water in the terrestrial planet-forming zone of the PDS 70 disk. Nature **620**, 516 (2023)

S.W. Peterson, H.A. Levy, A single-crystal neutron diffraction study of heavy ice. Acta Crystallogr. **10**, 70 (1967)

V. Petkov, Y. Ren, M. Suchomel, Molecular arrangement in water: Random but not quite. J. Phys. Condens. Matter **24**, 155102 (2012)

L.G.M. Pettersson, R.H. Henchman, A. Nilsson, Water—The most anomalous liquid. Chem. Rev. **116**, 7459–7462 (2016)

B.G. Pfrommer, F. Mauri, S.G. Louie, NMR chemical shifts of ice and liquid water: The effects of condensation. J. Am. Chem. Soc. **122**, 123–129 (2000)

C. Piaulet, B. Benneke, J.M. Almenara, et al., Evidence for the volatile-rich composition of a 1.5-Earth-radius planet. Nat. Astron. **7**, 206–222 (2023)

H. Popkie, H. Kistenmacher, E. Clementi, Study of the structure of molecular complexes. IV. The Hartree-Fock potential for the water dimer and its application to the liquid state. J. Chem. Phys. **59**, 1325 (1973)

F. Postberg, Y. Sekine, F. Klenner, et al., Detection of phosphates originating from Enceladus's ocean. Nature **618**, 489–493 (2023)

P. Privalov, G.I. Makhatadze, Contribution of hydration to protein folding thermodynamics. II. The entropy and Gibbs energy of hydration. J. Mol. Biol. **232**, 660–679 (1993)

E.M. Purcell, H.C. Torrey, R.V. Pound, Resonance absorption by nuclear magnetic moments in a solid. Phys. Rev. **69**, 37–38 (1946)

A. Rahman, Correlations in the motion of atoms in liquid argon. Phys. Rev. **136**, A405 (1964)

A. Rahman, F.H. Stillinger, Molecular dynamics study of liquid water. J. Chem. Phys. **55**, 3336 (1971)

A. Rastogi, A.K. Ghosh, S.J. Suresh, Hydrogen bond interactions between water molecules in bulk liquid, near electrode surfaces and around ions, in *Thermodynamics – Physical Chemistry of Aqueous Systems*, ed. by J.C. Moreno-Piraján, (InTech, Rijeka, 2011) ISBN: 978-953-307-979-0

W.K. Rhim, D.P. Burum, D.D. Elleman, Proton anisotropic chemical shift spectra in a single crystal of hexagonal ice. J. Chem. Phys. **71**, 3139 (1979)

G.W. Robinson, C.H. Cho, Role of hydration water in protein unfolding. Biophys. J. **77**, 3311–3318 (1999)

J. Russo, K. Akahane, H. Tanaka, Water-like anomalies as a function of tetrahedrality. Proc. Natl. Acad. Sci. USA **115**, E3333 (2018)

R. Seidel, B. Winter, S.E. Bradforth, Valence electronic structure of aqueous solutions: Insights from photoelectron spectroscopy. Annu. Rev. Phys. Chem. **67**, 283–305 (2016)

J.A. Sellberg, C. Huang, T.A. McQueen, N.D. Loh, H. Laksmono, D. Schlesinger, R.G. Sierra, D. Nordlund, C.Y. Hampton, D. Starodub, D.P. DePonte, M. Beye, C. Chen, A.V. Martin,

A. Barty, K.T. Wikfeldt, T.M. Weiss, C. Caronna, J. Feldkamp, L.B. Skinner, M.M. Seibert, M. Messerschmidt, G.J. Williams, S. Boutet, L.G.M. Pettersson, M.J. Bogan, A. Nilsson, Ultrafast X-ray probing of water structure below the homogeneous ice nucleation temperature. Nature **510**, 381 (2014a)

J.A. Sellberg, S. Kaya, V.H. Segtnan, C. Chen, T. Tyliszczak, H. Ogasawara, D. Nordlund, L.G.M. Pettersson, A. Nilsson, Comparison of x-ray absorption spectra between water and ice: New ice data with low pre-edge absorption cross-section. J. Chem. Phys. **141**, 034507 (2014b)

R. Shi, J. Russo, H. Tanaka, Common microscopic structural origin for water's thermodynamic and dynamic anomalies. J. Chem. Phys. **149**, 224502 (2018)

L.O. Sillerud, A.F. McDowell, N.L. Adolphi, R.E. Serda, D.P. Adams, M.J. Vasile, T.M. Alam, 1H NMR detection of superparamagnetic nanoparticles at 1T using a microcoil and novel tuning circuit. J. Magn. Reson. **181**, 181–190 (2006)

M.G. Siskos, M.I. Choudhary, I.P. Gerothanassis, Hydrogen atomic positions of O-H···O hydrogen bonds in solution and in the solid state: The synergy of quantum chemical calculations with ^{1}H-NMR chemical shifts and X-ray diffraction methods. Molecules **7**, 415 (2017)

L.B. Skinner, C. Huang, D. Schlesinger, L.G.M. Pettersson, A. Nilsson, C.J. Benmore, Benchmark oxygen-oxygen pair-distribution function of ambient water from x-ray diffraction measurements with a wide Q range. J. Chem. Phys. **138**, 074506 (2013)

A.K. Soper, The radial distribution functions of water and ice from 200 to 673 K and at pressures up to 400 MPa. Chem. Phys. **258**, 121–137 (2000)

A.K. Soper, Water: Two liquids divided by a common hydrogen bond. J. Phys. Chem. B **115**, 14014–14022 (2011)

A.K. Soper, *The Radial Distribution Functions of Water as Derived from Radiation Total Scattering Experiments: Is There Anything We Can Say for Sure?* (ISRN Physical Chemistry, 2013)

A.K. Soper, M.A. Ricci, Structures of high-density and low-density water. Phys. Rev. Lett. **84**, 2881–2884 (2000)

M. Sprik, J. Hutter, M. Parrinello, *Ab initio* molecular dynamics simulation of liquid water: Comparison of three gradient-corrected density functionals. J. Chem. Phys. **105**, 1142–1152 (1996)

F.H. Stillinger, Theory and molecular models for water, in *Advances in Chemical Physics*, ed. by I. Prigogine, S.A. Rice, vol. XXXII, (Wiley, 1975)

F.H. Stillinger, A. Ben-Naim, Liquid—Vapor interface potential for water. J. Chem. Phys. **47**, 4431–4437 (1967)

M.-L. Tan, J.T. Fischer, A. Chandra, B.R. Brooks, T. Ichiye, A temperature of maximum density in soft sticky dipole water. Chem. Phys. Lett. **376**, 646–652 (2003)

T. Tokushima, Y. Harada, Y. Horikawa, O. Takahashi, Y. Senba, H. Ohashi, L.G.M. Pettersson, A. Nilsson, S. Shina, High resolution X-ray emission spectroscopy of water and its assignment based on two structural motifs. J. Electron Spectrosc. Relat. Phenom. **177**, 192–205 (2010)

M. Vedamuthu, S. Singh, G.W. Robinson, Properties of liquid water: Origin of the density anomalies. J. Phys. Chem. **98**, 2222–2230 (1994)

H. Wang, A. Ben-Naim, Solvation and solubility of globular proteins. J. Phys. Chem. B **101**, 1077–1086 (1997)

L. Weinhardt, A. Benkert, F. Meyer, M. Blum, R.G. Wilks, W. Yang, M. Bär, F. Reinert, C. Heske, Nuclear dynamics and spectator effects in resonant inelastic soft X-ray scattering of gas-phase water molecules. J. Chem. Phys. **136**, 144311 (2012)

P. Wernet, D. Nordlund, U. Bergmann, M. Cavalleri, M. Odelius, H. Ogasawara, L.A. Näslund, T.K. Hirsch, L. Ojamäe, P. Glatzel, L.G. Pettersson, A. Nilsson, The structure of the first coordination shell in liquid water. Science **304**, 995–999 (2004)

B. Winter, R. Weber, W. Widdra, M. Dittmar, M. Faubel, I.V. Hertel, Full valence band photoemission from liquid water using EUV synchrotron radiation. J. Phys. Chem. A **108**, 2625 (2004)

B. Winter, E.F. Aziz, U. Hergenhahn, M. Faubel, I.V. Hertel, Hydrogen bonds in liquid water studied by photoelectron spectroscopy. J. Chem. Phys. **126**, 124504 (2007)

J.L. Yarnell, M.J. Katz, R.G. Wenzel, S.H. Koenig, Structure factor and radial distribution function for liquid argon at 85 K. Phys. Rev. A **7**, 2130 (1973)

J. Zhu, G. Wu, Quadrupole central transition ^{17}O NMR spectroscopy of biological macromolecules in aqueous solution. J. Am. Chem. Soc. **133**, 920–932 (2011)

D.V. Zlenko, Computing the self-diffusion coefficient for TIP4P water. Biophysics **57**, 127–132 (2012)

Bibliography

*As **a precaution, I must warn you that the literature on water is vast**. Start with these suggestions as how to proceed without drowning, but as I have stated in the introduction, one cannot possibly read all of the extant publications.*

D. Eisenberg, W. Kauzmann, *The Structure and Properties of Water* (Oxford University Press, 1969)

T. Lazaridis, Hydrophobic effect, in *eLS*, (Wiley, Chichester, 2013)

Martin Chaplin's web site: https://water.lsbu.ac.uk/water/martin_chaplin.html

Shuai, Li Lixin, Lu Swarnendu, Bhattacharyya Carolyn, Pearce Kai, Li Emily T., Nienhuis Gilles, Doumy R. D., Schaller S., Moeller M.-F., Lin G., Dakovski D. J., Hoffman D., Garratt Kirk A., Larsen J. D., Koralek C. Y., Hampton D., Cesar Joseph, Duris Z., Zhang Nicholas, Sudar James P., Cryan A., Marinelli Xiaosong, Li Ludger, Inhester Robin, Santra Linda, Young (2024) Attosecond-pump attosecond-probe x-ray spectroscopy of liquid water Editor's summary Science 383(6687) 1118-1122 10.1126/science.adn6059

Chapter 12
Prebiotic Evolution: The Self-Assembly
of Primordial Biomolecules

"…abiogenesis did clearly occur at least once on Earth…"
Scharf and Cronin (2016)

It is abundantly clear that the primordial constituents of the Earth and the Sun are the same as the components of everything else in the universe and that these arose in a process of energy-directed, probabilistic self-assembly. The molecules of life and, in particular, the molecules used by living systems as both constituents and as metabolites also self-assembled in this manner. In 1828, Friedrich Wöhler (1828) synthesized the *organic* compound urea by heating ammonium cyanate and thereby showed that the molecules of life could be generated without the obligatory participation of living systems, all that was needed was an input of energy. In so doing, he refuted the vitalist theory that organic molecules could only be produced in living systems and that living systems possessed εντελεχηψ or a nebulous vital force that somehow imbued them with life. Of course, here in this work, we explicitly reject such nonscientific and untestable notions and will pursue our investigations into the probabilistic self-assembly of molecules and biomolecules. The main molecular classes of interest are the amino acids, sugars, lipids, and nucleic acids (Fig. 12.1).

12.1 Prebiotic Chemistry as an Interacting Network
of Molecules

Metabolism in living systems cannot profitably be understood in terms of single biomolecules but rather must be described as an interacting network of biomolecules along with their interconversions catalyzed by enzymes. The early search for the chemical pathways whereby biomolecules, such as amino acids, nucleic acids, lipids or sugars and their polymers, etc., were generated on the primordial Earth focused on single molecules or classes of molecules in attempts to discover those substances supporting the origin of life. The conditions discovered that were favorable for the spontaneous generation of one class often were detrimental for other classes. In particular, the RNA world hypothesis resulted in heavy emphasis on the separate

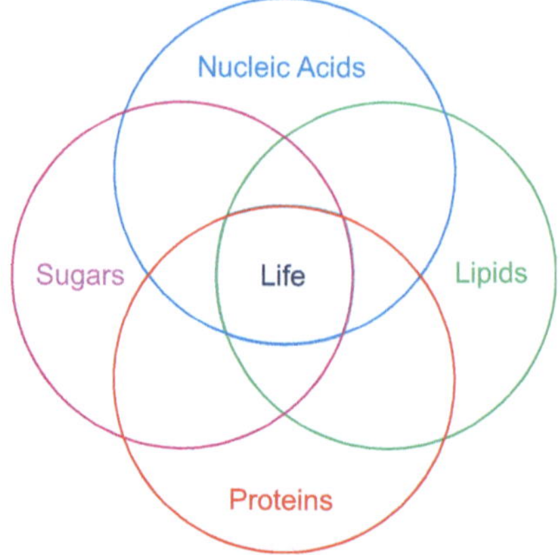

Fig. 12.1 A Venn diagram of the molecular spaces that contributed primordial molecules to the origin of life. Only those interacting molecules at the union of these four sets were selected by evolution as the basis for living systems. The flow of energy through this network organized its constituents into the molecules of life

study of the generation of nucleic acids. Although great progress has been made using this strictly reductionist approach, a moments' reflection on the state of current biochemical knowledge convinces one that it is unlikely that life evolved this way: It did not arise from the convergence of separate classes of fully formed prebiotic molecules. It is much more likely that all manner of primordial molecular classes swam together, were separated and recombined, and were destroyed and reformed with old or new partners. Therefore, a better way of understanding this chemical evolution is to envision it as an energy-driven, open network of interacting prebiotic molecules in a primitive form of metabolism much as we today understand metabolism itself. By reexamining Fig. 12.1, we can now understand the nature of the aqueous milieu in which life developed through the simultaneous presence of all four classes of primordial biomolecules. These interacted in ways that are actively being examined in the laboratory.

I would tell my physical biochemistry students that a single mammalian cell is the most complex molecular system they would ever encounter, but even so, living cells utilize only a limited subset of the potential molecules capable of spontaneously forming through the input of energy (ultimately sunlight) on the primitive Earth. This subset of molecules is, however, universal and conserved among all living systems; metabolism is the same across all phyla, with glycolysis as the *Ur*-metabolism. This molecular universality converged on a single, unique origin for living systems. The molecules of life were selected because they inherently self-assembled into higher structures. From this point of view, life is *built into* the universe. The research in the last ~50 years on prebiotic molecular reactions has shown that the classical reductionist approach, which seeks to understand the origin of amino acids, lipids, sugars, and nucleic acids as separate, unconnected avenues, has now given way to a more profound insight that appreciates their common evolution in a situation where these molecules were all *simultaneously* present and interacting. All of the major classes of biomolecules can self-assemble in an impressively large number of conditions, among them being in giant molecular clouds (Rimola et al. 2010); in meteorites, asteroids, and comets (Menor-Salván et al. 2009; Lee and

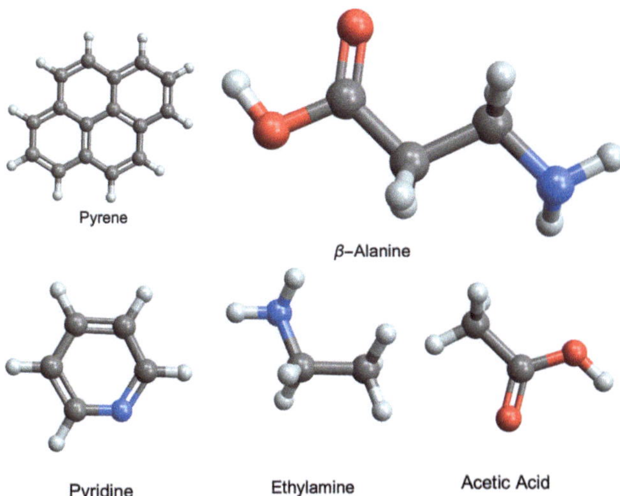

Fig. 12.2 Chemical structures of representative soluble organic molecules detected in surface samples returned from the asteroid Ryugu by the Japanese Hayabusa2 spacecraft. Gray balls are carbon, hydrogen is white, oxygen is red, and nitrogen is blue. Amines are represented by *ethylamine*, nitrogen containing heterocycles by *pyridine*, polycyclic aromatic hydrocarbons (PAHs) by *pyrene*, carboxylic acids by *acetic acid*, and amino acids by *β-alanine*. The background image shows Ryugu in a photograph taken during the approach by Hayabusa2. (Data from Naraoka et al. (2023))

Scandolo 2011; Dworkin et al. 2001; Elsila et al. 2009); in protoplanetary disks; and on nascent planets (Chau et al. 2011; Raulin et al. 2012), including the Earth. For example, mass spectroscopic analysis of the soluble organic molecules in surface samples which returned from the space mission to the carbonaceous asteroid Ryugu (Naraoka et al. 2023) has shown the existence of more than 20,000 organic compounds, including amines, amino acids, nitrogenous heterocycles, polycyclic aromatic hydrocarbons (PAHs), and carboxylic acids (Fig. 12.2).

Furthermore, if we were able to restart the origin of life, it is very likely that the end product would look the same as life today; the same metabolites and molecular structures would form from carbon monoxide, carbon dioxide, hydrogen cyanide, and ammonia because this is simply how they would react, in much the same manner as we have found that the amounts and the nature of nuclei that form the periodic table arose from the nuclear physics of stars and the fusion of hydrogen, the simplest nucleus.

Before we take up a more network-centric approach, however, we must examine the abiotic syntheses of the main classes of biomolecules. Let us also not forget the important roles that energy and entropy must play in this scheme. Physics gives us the playbook—reactions of molecules proceed from states of higher to lower energy, as long as that energy decrease is sufficient to offset the entropy change that accompanies molecular bond rearrangement. It is easy to list the energy sources that drove prebiotic molecular interactions in the direction of lower energy, and lower entropy, into more probable states. These include the ubiquitous solar protons and electrons, the ultraviolet and visible light that bathed the Earth at its origin, the residual heat left over from the great smash of the Mars-like body that formed the Earth-Moon system, some 4 Gya, and a plethora of other natural phenomena with the correct energies to foster molecular rearrangements, such as kT (heat), lightning with its free radical production, volcanoes with their sulfur dioxide, carbonyl sulfide as well as hydrogen sulfide, and the day/night, wet/dry, hot/cold, etc., diurnal cycling of many of these. Entropy reduction accompanied the binding of molecules to catalysts and the sequestration of prebiotic compounds within structures such as pores in rocks and within lipid bilayers.

12.2 Meteorites

Just as there remain what we might refer to as physical *fossils* from the Big Bang as the cosmic microwave background radiation, the signature of baryonic acoustic oscillations, and the origin and abundances of the elements in the periodic table, the abiogenic assembly of molecules has left abundant traces in the oldest material in the solar system, that of meteorites, comets, asteroids, Pluto and the moons of the gas giants, Saturn, Jupiter, Uranus, and Neptune. Meteorites are formed by collisions among bodies in the asteroid belt between the orbits of Mars and Jupiter. The location of the asteroids in a cold region of the solar system has preserved material from the primordial solar/planetary disk as indicated by their measured ages of 4.56 Gyr that match the Sun's age and are slightly older than the Earth or Moon (4.54 Gyr). Although meteorites vary in composition from iron/nickel to stony-iron to carbonaceous chondrites, from the point of view of the origin of biomolecules, the carbonaceous chondrites are the most important meteorites examined so far, consisting of 5–20% water and 1–5% carbon-containing molecules including macromolecules. Perhaps the most studied of these meteorites is the one that fell in Murchison, Australia, in 1969. As the resolution and sensitivity of mass spectrometry have vastly improved since its discovery, the number and variety of organic compounds have correspondingly increased; current studies indicate the presence of (1) amino acids, (2) carboxylic acids, (3) purines, (4) pyrimidines, (5) sugars, and (6) aliphatic and aromatic hydrocarbons such as fatty acids and polycyclic aromatic hydrocarbons (PAHs). It is quite clear from this chemical fossil record that many biomolecules self-assembled during the very early stages of planetary formation in the solar system.

Glycine has been detected in the Murchison and other meteorites along with 85 other amino acids as α, β, γ, and δ amino compounds containing from two to nine carbons including dicarboxyl and diamino groups and nine hydroxylated amino acids (Koga and Naraoka 2017). A major puzzle has been the fact that a chemical synthesis of a chiral amino acid in the laboratory produces a random *racemic* mixture containing 50% *D* amino acids and 50% *L* amino acids. Therefore, it is quite important to note that there is a large enantiomeric excess of *L* versus *D* amino

Table 12.1 *L*-enantiomeric excesses in the Murchison meteorite (Koga and Naraoka 2017)

Amino acid	*L* (ppb)	*D* (ppb)	*% L*
Alanine	732	252	74.4
Serine	942	127	88.1
Aspartic acid	780	193	80.2
Threonine	519	~1	99.8
Valine	419	30	93.3
Glutamic acid	788	142	84.7
Leucine	696	172	80.2

acids in this meteorite (Table 12.1) because living systems on Earth are *homochiral* and use essentially only *L* amino acids to make proteins. However, if one begins a synthesis of a protein with a chiral excess of one enantiomer, the finished protein will be homochiral simply from probabilistic reasons.

12.3 Comets

Comets are the remains of the protoplanetary disk that formed the protosolar nebula; the low temperature of space has preserved these as "dirty snowballs" consisting of rocks and ice. Meteorites are the solid material (rocks) that is released from a comet as the coma forms due to ice sublimation when the comet is heated by solar radiation upon close approach to the Sun and no longer binds the rocks within the icy matrix. The composition of meteorites and cometary comas form a complementary set of materials left over from the formation of the solar system, and their study greatly informs our present knowledge of the origin of many of the biomolecules observed.

Comets arise either from the Kuiper Belt beyond the orbit of Neptune ($\sim6 \times 10^9$ km or ~40 astronomical units (A.U.) from the Sun) or from the Oort Cloud around 10^{12} km (25,000 A.U.) from the Sun. They are the 4.6-Gyr-old icy remains of the protoplanetary disk that once circled the Sun and which condensed to form the Earth and the other planets of our solar system. The central, condensed portion of a comet is called the nucleus and consists of an agglomeration of CO_2, CO, CH_4, NH_3, CH_3OH, HCN, $HCOH$, CH_3CH_2OH, CH_3CH_3, and larger molecules including long-chain hydrocarbons and amino acids, frozen into a water ice clathrate hydrate, along with rock and dust (Greenberg 1998) and a dark coating of carbonaceous material formed by the incessant bombardment of the surface by solar wind particles and cosmic rays. When comets approach the Sun, these nuclear ices sublime and produce the dramatic cometary atmosphere or *coma*, which observers on Earth detect as the comet's tail(s). Optical, infrared, and radiofrequency spectroscopy are widely deployed to monitor the molecular composition of the cometary coma. One of the first molecules spectroscopically identified in a cometary coma was the very reactive molecule cyanogen, CN, from the apparition of Halley's comet (1P/Halley) in 1910 (Bobrovnikoff 1931) that lead to the unfounded supposition that if the Earth passed through the coma our atmosphere would be poisoned. This did

Table 12.2 Abundance (% relative to water) of molecules detected in comet Lovejoy (Biver et al. 2015)

CHO molecules	%	Nitrogenous molecules	%	Sulfurated molecules	%
CO	1.8	HCN	0.09	H_2S	0.5
H_2CO	0.3	HNC	0.004	OCS	0.034
CH_3OH	2.4	HNCO	0.009	H_2CS	0.013
HCOOH	0.028	CH_3CN	0.015	CS	0.043
$(CH2OH)_2$	0.07	HC3N	0.002	SO	0.038
$HCOOCH_3$	0.08	NH_2CHO	0.008	NS	0.006
CH_3CHO	0.047				
CH_2OHCHO	0.016				
C_2H_5OH	0.12				

not obviously happen, even though the Earth did pass through the coma of the comet, but the detection of CN, which arises from the sunlight-induced photodecomposition of HCN, was a major discovery since this molecule, as we will later see, plays a central role in the self-assembly of many important biomolecules, such as the nucleic acid precursor adenine. Glycine, one of the fundamental chemical building blocks of life, is the simplest amino acid and was detected in the material ejected from Comet Wild 2 in 2004, captured by NASA's Stardust probe and returned to Earth in 2006. Glycine was also found in the coma of comets 81P/Wild-2 (Elsila et al. 2009) and 67P/Churyumov-Gerasimenko (Altwegg 2016).

Twenty-one complex organic molecules (Table 12.2) were found in Comet Lovejoy (Biver et al. 2015) including ethanol and the simplest but still important three carbon sugar glycolaldehyde (CH_2OHCHO). The abundances of these molecules with respect to the copious amounts of water produced when the comet approached the Sun (Table 12.2) are similar to the abundances found for other comets, such as Hale-Bopp, and in protoplanetary disks and the Orion nebula starforming region (Fig. 12.3). Since comet Lovejoy originated in the Oort Cloud, it represents pristine, primordial material from the origin of the solar system, and its composition shows that the elements of life self-assembled in space at the dawn of the solar system.

12.4   The Interstellar Medium

We can also find these molecular *fossils* in the interstellar medium, in the gas clouds surrounding dying stars (planetary nebulae), and in the atmospheres of cool stars by means of the infrared and radiofrequency spectroscopy of their rotational and vibrational transitions (Table 12.3). One can readily infer from the large number of small molecules detected in pristine settings, in meteorites, in comets, and in space (Tennyson 2003), that many, more complex, molecules await discovery, limited only by the existing analytical techniques and that as these improve the list

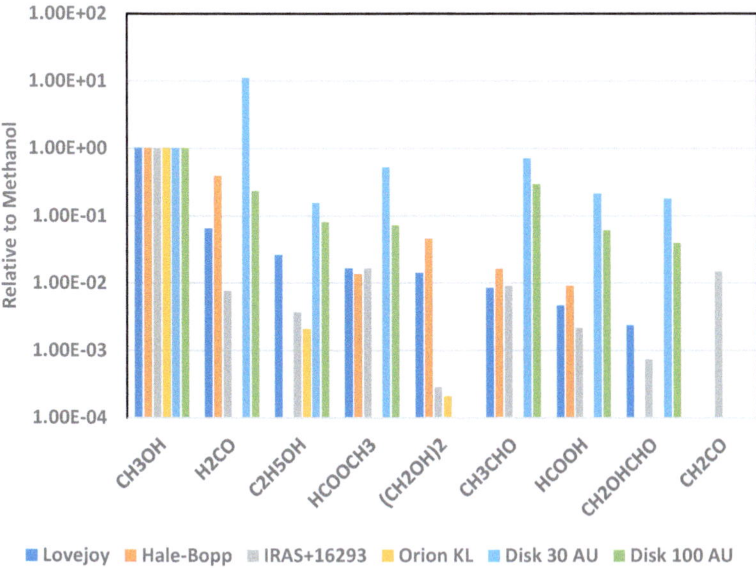

Fig. 12.3 The abundances (relative to methanol) of several complex organic molecules, including ethanol and glycolaldehyde, in the coma from comet Lovejoy (C/2014 Q2) compared with comet Hale-Bopp, protostars and protoplanetary disks. (Data from Biver et al. (2015))

Table 12.3 Assignments for the ^{13}C NMR spectrum of the Murchison meteorite (Fig. 12.6)

Peak #	δ (ppm)	Assignment	Carbon type
1	17	$-CH_3$	Methyl
2	33	$(-CH2-)_n$	Methylene
3	63	$-CH_nO$	Alcohol
4	129	$-CH=CH-$	Aromatic, olefinic
5	163	$-COOH$	Carboxylic
6	195	$-C=O$	Carbonyl

will grow. However, these molecular fossils already include an intriguing array of important biomolecules. For example, the scent of raspberries wafts through the mists of space in the recently discovered ethyl formate molecule. One morning before class, NASA sent me a scratch-and-sniff postcard announcing the discovery of this molecule as the "aroma of space." I brought it to show my students as I was discussing this novel discovery—something that serendipitously happened as I was teaching this section on prebiotic evolution. Everyone enjoyed the scent of raspberries as I passed this card throughout the class of intrigued scholars, for ethyl formate is the ester that evokes the fragrance of this freshly picked fruit.

Appendix Table A.5 contains a wide variety of biologically interesting molecules: among the critical components that we will later encounter as participating in prebiotic molecular evolution are hydrogen, of course, and carbon monoxide, cyanide in its many forms, water, carbon dioxide, sulfur dioxide, hydrogen sulfide, ammonia, methane, formaldehyde, formamide, *iso*-propyl cyanide, ethanol, acetic acid, and glycolaldehyde. What a rich collection of molecules! The upshot is that prebiotic molecular rearrangements are not difficult, but are ubiquitous, even in the deep freeze of outer space, in the regions between stars.

12.5   Polycyclic Aromatic Hydrocarbons

Some of the more interesting molecules recently discovered include large polycyclic aromatic hydrocarbons, including the branched chain alkyl molecule, *iso*-propyl cyanide (Belloche et al. 2014) found in the interstellar medium, and fullerenes C_{60} and C_{70} in a young planetary nebula (Cami et al. 2010). Polycyclic aromatic hydrocarbons (Fig. 12.4) are the most prevalent molecules found in space. These are common in the interstellar medium between galaxies, in comets, meteorites (Clemett et al. 2010), and nebulae, and are a main contributor to the infrared emission spectrum of these objects around 3.28 μm. An infrared image of the Cat's Paw nebula in the constellation Scorpius (Fig. 12.5) taken by the Spitzer satellite shows emission from clouds of polycyclic aromatic hydrocarbons that fluoresce due to excitation of rotational quantum states by ultraviolet light from nearby hot stars. These molecules often also contain nitrogen and undergo chemical transformations through the addition of hydrogen, oxygen, and hydroxyl moieties even within cold (5–100 K) cosmic ice grains (Gudipati and Yang 2012) transforming them into what are commonly expected to be amino acids and nucleotides, precursors of DNA and proteins. These derivatized molecules are planar, amphiphilic (Groen et al. 2012) like phospholipids in cell plasma membranes forming bilayers, and stack like the bases in RNA and DNA. They could have formed a scaffold upon which purines and pyrimidines would attach and perhaps polymerize and then leave the stack to form nucleic acids. It is estimated that widespread polycyclic aromatic hydrocarbons account for at least 20% of the carbon in the universe and may have been important on the pathway of prebiotic evolution (Ehrenfreund 2006) because they appear to have been synthesized within a short

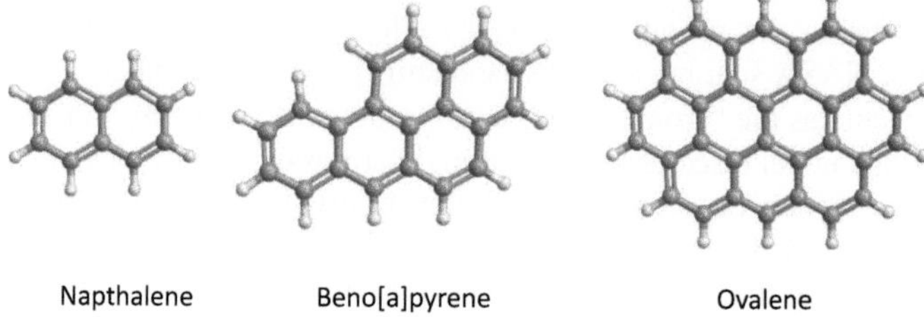

Fig. 12.4 Some examples of polycyclic aromatic hydrocarbons formed from benzene whose emission dominates the infrared spectrum of the Cat's Paw nebula, NGC6334. Note that the PAH pyrene was also recently found as a constituent of the asteroid Ryugu (Fig. 12.2)

Fig. 12.5 An infrared image taken with the Spitzer space telescope of the Cat's Paw nebula NGC6334 in the constellation Scorpius. The green channel is emission at 8 μm from dust grains containing polycyclic aromatic hydrocarbons. (Spitzer space telescope, 2018)

time after the Big Bang. They are found in protoplanetary disks around new stars along with carbon monoxide, cyanide, and other hydrocarbons (Cody et al. 2011). The atmosphere of Titan, Saturn's largest satellite, contains PAHs (Lopez-Puertas et al. 2013; Zhao et al. 2018; Abplanalp et al. 2019). The composition of the macromolecular material isolated from the Murchison meteorite is dominated by polycyclic aromatic hydrocarbons, mainly naphthalene and phenanthrene, as shown by its solid-state ^{13}C NMR spectrum (Fig. 12.6). Also detected are methyl, methylene, hydroxylated, carboxylic, and carbonyl groups.

Yabuta et al. (2005) found that aromatics constitute a large (~55%) fraction of the carbon present. More recent NMR studies have suggested that the PAHs contain at most two rings: naphthalene (Fig. 12.4) has certainly been identified in meteoric material. Amphiphilic PAHs, which can be obtained from carbonaceous chondrites, formed bilayers and vesicles that separated complex reactions within compartments. This was important in the origin of the first protocells (Deamer 2000).

The atmosphere of Saturn's largest moon Titan consists mainly of nitrogen and methane. Ultraviolet solar photons ionize the nitrogen and methane generating reactive species that combine to form larger, more complex organic molecules, such as ethane, propane, butane, polyacetylenes, cyanoacetylenes, and notably benzene, accounting for the haze observed in optical images. Benzene radicals then combine to make large polycyclic aromatic hydrocarbon molecules. Lopez-Puertas et al. (2013) have reported a large variety of these polycycles with 9–96 carbon atoms formed by ~11 benzene rings with approximately 30% as heterocycles containing nitrogen atoms. They determined the concentration of polycyclic aromatic hydrocarbons in Titan's upper atmosphere to be ~2×10^{10} m^{-3}. The fact that one third of these are nitrogen heterocycles provides evidence for the facile prebiotic synthesis of the bases in RNA and DNA.

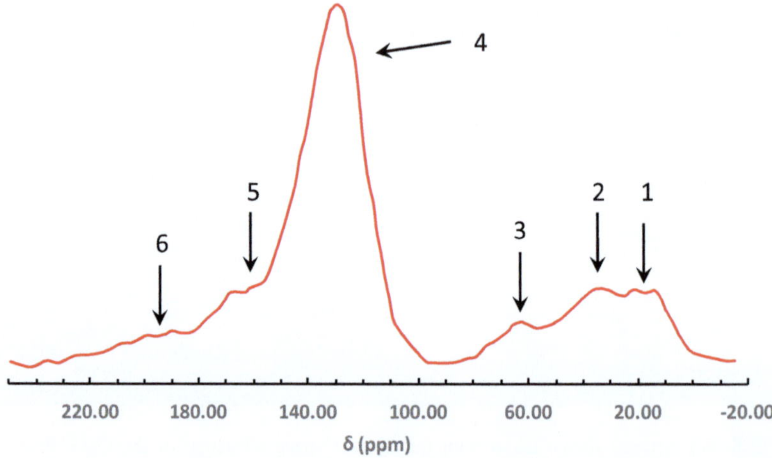

Fig. 12.6 Solid-state ^{13}C NMR spectrum of the macromolecular material from the Murchison meteorite showing that the most abundant carbon-containing compounds are aromatic, such as the polycyclic aromatic hydrocarbons (Fig. 12.4, peak #4, $\delta = 129$ ppm) that are also the most abundant forms of carbon in comets, the interstellar medium, the circumstellar disks, and the atmosphere of Titan. The peak assignments are given in Table 12.3. (Drawn from data in Cody et al. (2002))

Further infrared observations by the Spitzer space telescope of the nascent star Herbig-Haro 46-IR showed that its circumstellar disk contains many molecules of interest to prebiotic molecular formation, with detections of cyanide compounds, hydrocarbons, and carbon monoxide, among others. The first step in the conversion of carbon into more complex aromatic compounds appears to be the synthesis of the very reactive molecules of acetylene (HC≡CH), diacetylene (HC≡C≡CH), and triacetylene (HC≡C≡C≡CH) which then condense into benzene, all of which have been detected in the infrared spectrum of the circumstellar envelope (Cernicharo et al. 2001) of the protoplanetary nebula CRL 618 (Fig. 12.7). Benzene photochemical reactions then lead to the formation of polycyclic aromatic hydro-carbons, fullerenes, and more complex aromatic molecules (Moriconi 2013). Many polycyclic aromatic hydrocarbons, such as isoquinoline or naphthalenecarboxylic derivatives, can form bilayer membranes. The James Webb telescope has recently been used to obtain infrared spectra of the protostellar outflow from another nascent star, HH211, in the Perseus molecular cloud (Ray et al. 2023). Again, it was found that molecules, such as carbon monoxide and SiO, exist in abundance in the jet material being cast into space, even though this star is estimated to be less than 100,000 years old.

Fig. 12.7 The protoplanetary nebula CRLL 618 is formed from the cast-off outer envelope of a red giant star that has ceased thermonuclear fusion and is on its way to becoming a white dwarf. (Source: NASA/ESA https://esahubble.org/images/potw1110a/ (see also Ray et al. 2023))

12.6 Prebiotic Fossils

Pluto lost its designation as our ninth planet, as the result of a faulty size comparison with Ceres the largest of the objects in the asteroid belt between Mars and Jupiter. However, Ceres is a rather dull object indeed compared with the Kuiper Belt object Pluto. The flyby of the NASA probe New Horizons in 2015 found that Pluto's spectacularly colored surface is composed of nitrogen ice, with small amounts of methane and carbon monoxide and with mountains of water ice. The temperature on Pluto averages about 41 K, cold enough to preserve molecular fossils from the origin of the solar system. The color varies from black through red/orange to white. It is likely that the red areas are covered with tholins, complex molecules formed from methane, ethane, ammonia, water, formaldehyde, and hydrogen sulfide (Sagan and Khare 1979). We will have more to say about tholins later in this chapter.

The chemical composition of another Kuiper Belt-like object has also been determined by means of optical and ultraviolet spectroscopy. When a red giant star sheds its outer layers late in its lifespan, the result is the naked hydrogen/helium core of the star, which emerges from the carbonaceous dust cloud as a white dwarf. The spectrum of the light from some of these white dwarves is *polluted* by rocky debris left over from the red giant phase, which orbits the star in a tidally disrupted disk. Since white dwarfs only contain hydrogen and/or helium in their atmospheres, any spectral lines from heavy elements must arise from material present in the disk. Using these ideas, Xu et al. (2017) measured the chemical composition of an extrasolar Kuiper Belt object analog accreted onto the star WD1425+540 and found that it contained the heavy elements C, N, O, Mg, Si, S, Ca, Fe, and Ni in

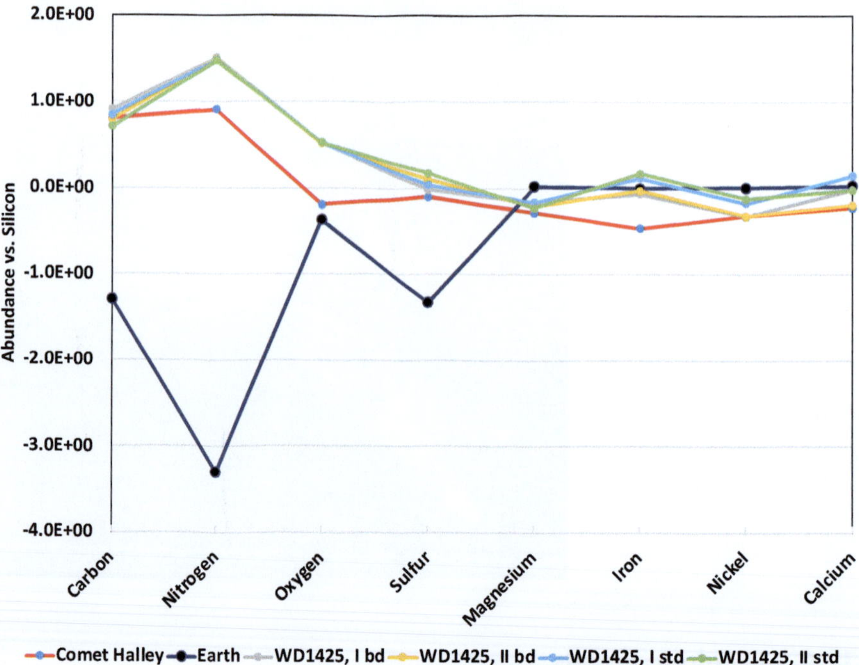

Fig. 12.8 Elemental composition of the extrasolar, Kuiper Belt-like objects accreting onto the white dwarf WD1425+540, compared with Earth and Comet Halley. Note that C, N, O, S, Mg, Fe, and Ca are all elements needed by living systems. The presence of Ni is explained in Chap. 6. (Drawn from data in Xu et al. (2017))

amounts that resemble Comet Halley and other extrasolar Kuiper Belt objects (Fig. 12.8). Of particular interest was the finding that carbon, oxygen, and nitrogen were present in amounts (~10%, ~67%, ~2%, respectively) that were higher than found on Earth or in carbonaceous chondritic meteors.

The composition of comets (and their progeny meteors), asteroids and Kuiper Belt objects, and interstellar medium includes water, carbon monoxide, carbon dioxide, methane, ammonia, hydrogen cyanide, formaldehyde, and a number of other compounds, suggesting an interesting experiment. If a mixture of some or all of these gases were combined into a single container and then a stimulant, such as lightning was added, what would be the result? Harold Urey at the University of Chicago in the 1950s (Urey 1952) postulated that the primordial Earth possessed a reducing atmosphere containing H_2O, H_2, CH_4, and NH_4. His student, Stanley Miller, performed a brilliant set of experiments in prebiotic chemistry (Miller 1953, 1955, 1957) in which a mixture of hydrogen, methane, ammonia (or nitrogen), and water vapor was subjected to a spark discharge and after a week of reaction at 100 °C found that several amino acids, including aspartic acid, alanine, glutamic acid, and glycine, were synthesized (Table 12.4) as well as several hydroxy

Table 12.4 Organic compounds formed from sparking a mixture of CH_4, NH_3, H_2O, and H_2

Compound	Yield (%)[a]
Glycine	2.1
Glycolic acid	1.9
Sarcosine	0.25
Alanine	1.7
Lactic acid	1.6
N-Methylalanine	0.07
α-Amino-*n*-butyric acid	0.34
α-Amino-*iso*-butyric acid	0.007
α-Hydroxybutyric acid	0.34
β-Alanine	0.76
Succinic acid	0.27
Aspartic acid	0.024
Glutamic acid	0.051
Iminodiacetic acid	0.37
Iminoacetic propionic acid	0.13
Formic acid	4.0
Acetic acid	0.51
Propionic acid	0.66
Urea	0.034
N-methyl-urea	0.051

From: Miller (1955)

[a]Based on initial methane carbon input of 59 nmol

acids, such as lactic and acetic acids. The production of these products was rapid (Fig. 12.9) and the added ammonia was depleted after 7 days. A recent reanalysis of Miller's surviving samples (Johnson et al. 2008) using modern techniques and instruments revealed the presence of 23 amino acids in the mixtures instead of the 5 (Table 12.4) originally reported, as well as a greater number of other organic compounds.

Since volcanoes on the primordial Earth would have emitted large amounts of H_2S, and perhaps SO_2, Miller also performed experiments in his apparatus with a mixture of H_2S, CH_4, NH_3, and CO_2, although the vials containing the dried residues (tholins) remained unanalyzed in his lifetime. When these vials were opened and examined by Parker et al. (2011), it was found that sulfur-containing amino acids were also produced in addition to the previously described amino acids (Parker et al. 2011; Bada 2013). The types and amounts of amino acids produced in the H_2S experiments showed a striking similarity with those found in the Murchison meteorite samples (Parker et al. 2011).

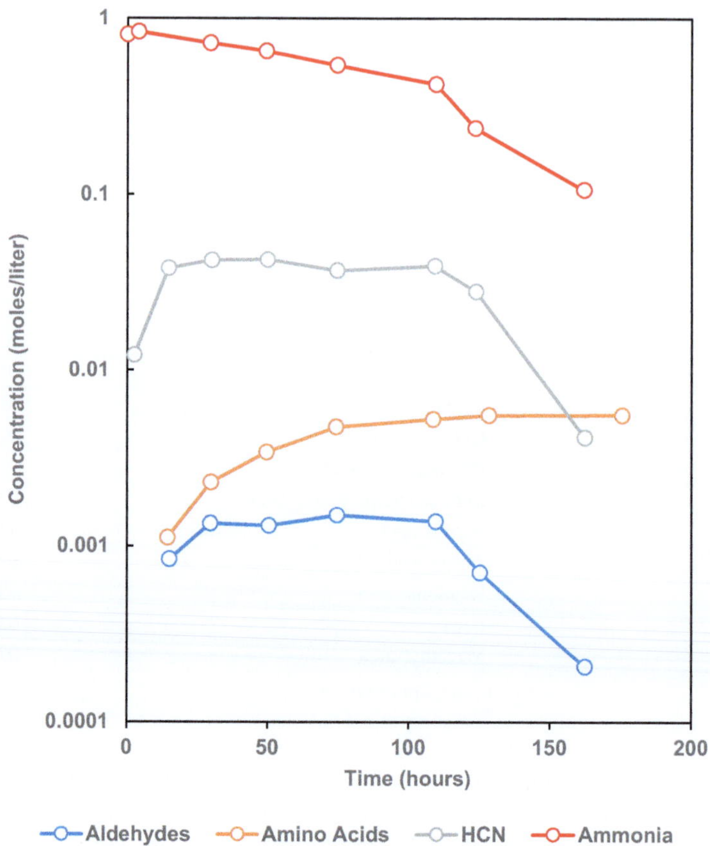

Fig. 12.9 The time course of the compounds present in the aqueous phase of Miller's reaction apparatus after initial charging with CH_4, NH_3, H_2O, and H_2 and sparking and heating at 100 °C for the indicated times. The gaseous phase contained CO, CO_2, and N_2 at the end of the heating period. Based on data from Miller (1957). Similar results were obtained when the ammonia was replaced with N_2, although the yields were smaller

12.7 Formation of Amino Acids

Although we often regard amino acids as having originated through the metabolism of living systems, astronomical observations, the Miller-Urey experiment and many other similar studies have shown that most amino acids can spontaneously self-assemble via naturally occurring chemical reactions on the Earth and in space (vide infra). How does this occur? The components of Miller's mixture produce HCN (hydrogen cyanide), CH_2O (formaldehyde), as well as acetylene, cyanoacetylene, and other organic compounds through the following reactions:

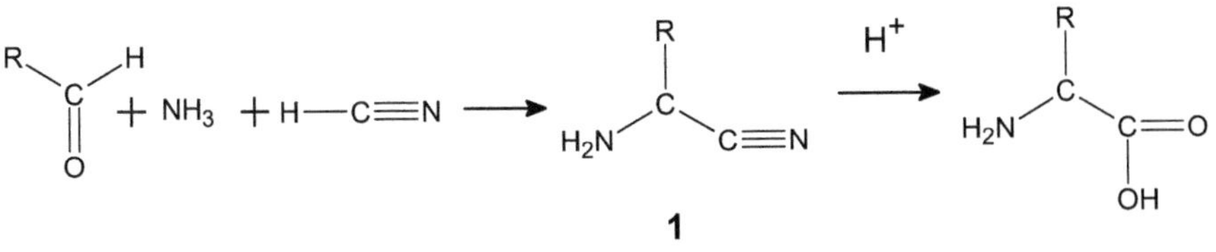

Fig. 12.10 The Strecker reaction in which an aldehyde reacts with ammonia and hydrogen cyanide to form amino acids

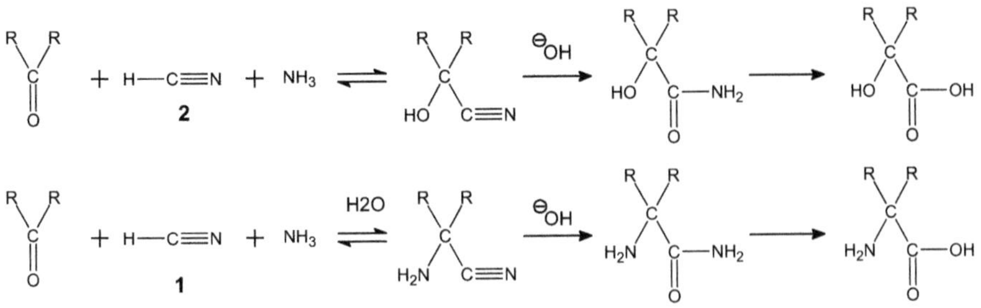

Fig. 12.11 The Strecker reaction can synthesize both amino acids, through pathway **1**, and α-hydroxy acids via pathway **2**. Note here that the radicals denoted R on the starting ketone do not have to be identical

$$CO_2 \rightarrow CO + O$$

$$CH_4 + 2O \rightarrow CH_2O + H_2O$$

$$CO + NH_3 \rightarrow HCN + H_2O$$

$$CH_4 + NH_3 \rightarrow HCN + 3H_2$$

In 1850, Strecker discovered that the reaction between an aldehyde and a ketone with ammonia and hydrogen cyanide can produce amino acids by the scheme in which the nitrile **1** forms in alkaline aqueous solution and is then hydrolyzed in a strong acid. Therefore, in Miller's experiments, the formaldehyde, ammonia, and hydrogen cyanide undergo the Strecker reaction with the formation of amino acids, such as glycine in two steps (Fig. 12.10):

$$CH_2O + HCN + NH_3 \rightarrow NH_2CH_2CN + H_2O$$

$$NH_2CH_2CN + 2H_2O \rightarrow NH_3 + \underset{\textit{Glycine}}{NH_2CH_2COOH}$$

In addition, the formaldehyde can react with water in Butlerov's process resulting in the formation of several sugars, including ribose, the sugar component of RNA. Hydroxy acids were also found in Miller's synthesis, and these can also be formed via the Strecker reaction (Fig. 12.11).

When the Voyager and later Cassini spacecraft probed the hazy upper atmosphere of Saturn's moon, Titan, the data indicated that it contained $\sim$98% N_2 and 2% CH_4 and trace amounts of other molecules including H_2, HCN, CO, and the hydrocarbons ethane, ethylene, acetylene, and cyanoacetylene (Niemann et al. 2005). These are very similar to the gases found on Triton and Pluto (vide supra). Of substantial interest is the fact that when these gases are reacted in the laboratory (Cable et al. 2012; Cleaves II et al. 2014) the resulting tholins (Sagan and Khare 1979) upon hydrolysis share most of the amino acids discovered in Miller's electric discharge products (Neish et al. 2010; Cleaves II et al. 2014). Low-temperature (253–293 K) hydrolysis in 13 wt% NH_3/H_2O for 12 months produced asparagine, aspartic acid, glutamine, and glutamic acid (Neish et al. 2010). The haze in Titan's atmosphere consists of tholin particles that form via the irradiation of the gases with ultraviolet light and cosmic rays (electrons and protons). Compounds such as these are also formed from atmospheric chemistry on Pluto and on Neptune's moon Triton (McDonald et al. 1994) where not only 13 straight-chain amino acids but also the cyclic molecules proline and phenylalanine were detected.

12.8   Formation of Peptides

The fact that the biochemistry of peptides and proteins is central in current biology leads one to conclude that the self-assembly of peptides from amino acids was a key step in the prebiotic evolutionary process. The generation of amino acids by the Miller experiments mainly documents the spontaneous processes that must have taken place in many regions of the primordial Earth and in space, but it says little, if anything, with respect to the next step in prebiotic evolution: the self-assembly of amino acids into larger polymers, peptides and proteins, real biomolecules. Heteropolymers of amino acids (peptides and proteins) are central to life. A first step in the study of the self-assembly of amino acids into peptides was performed by Fox and Harada (1958) where the amino acids glutamate, aspartate, and others were induced to form peptides by melting and cooling. He was able to produce peptides with molecular weights of $\sim$4900, containing about 40 amino acid residues ($\sim$6 Asp, and $\sim$28 Glu) in a nonrandom sequence, but the conditions he used would not likely be applicable to the primordial Earth since they involved heating pure amino acids. These peptides could form spherical, closed structures that could divide asexually via binary fission, form junctions with others, and develop a double membrane like to that of a lipid bilayer, but without any lipid present.

Peptides form from amino acids by dehydration in a process close to equilibrium ($K_{pep} \sim 0.1$ M^{-1}) with a Gibb's energy change of zero, but clearly water competes

$$^N\text{Pept}-\overset{O}{\overset{\|}{C}}-O^- \ + \ ^+\text{H}_3\text{N}-\text{Pept}^C \quad \underset{\pm H_2O}{\overset{K_{pep}}{\rightleftharpoons}} \quad ^N\text{Pept}-\overset{O}{\overset{\|}{C}}-\underset{H}{N}-\text{Pept}^C$$

Fig. 12.12 The participation of Cu(II) in salt-induced peptide bond formation

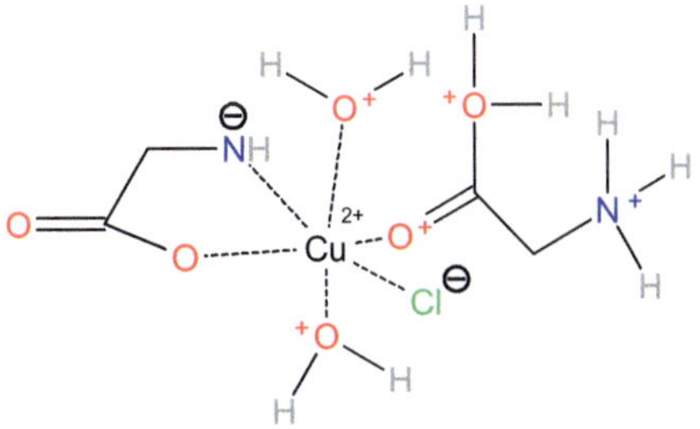

with this dehydration and skews it in the direction of hydrolysis. The peptide bond is quite stable from a kinetic point of view, with a half-life estimated to be hundreds of years at neutral pH and moderate temperatures.

Since water competes with peptide bond formation, investigators have searched for peptide bond formation in situations in which the activity of water is lower than that in bulk water. Sidney Fox's experiments were conducted by melting amino acids in the absence of added water, using only the water of crystallization present in the starting materials. Other ways of lowering the water activity include freezing the mixture, as is relevant to the generation of amino acids in the ice mantle which coats interstellar dust grains (Sandford et al. 2020), or sequestering the amino acids in zeolite pores in which only a few ancillary water molecules are present (Navrotsky et al. 2021). Clays have for many years been studied as peptide condensing agents (Gillams and Jia 2018; Preiner et al. 2018) as have primordial atmospheric molecules such as cyanamide and terrestrial phosphates.

In isolated ponds of ocean water, evaporation can produce both regions of very low water activity inhibiting hydrolysis of the peptide bond, and very high NaCl content, in excess of 3 M. As the salt dissociates into ions and as the water activity drops, there aren't sufficient water molecules to completely fill the six positions in the first hydration shell of the Na^+ ions. Under these conditions, the Na^+ ions act as dehydrating agents to lower the water activity and to promote dehydration of the amino acids through a lowering of the Gibb's energy of the reaction. Divalent cations, which would also be ubiquitous on the primordial Earth, can act as condensing agents under these conditions. Copper (II) has been found to be the most useful metal to participate in peptide formation (Fig. 12.12) through this route (Jakschitz and Rode 2012). It increases the kinetic constants so that the reaction proceeds more quickly. This process of salt-induced peptide formation works with all amino acids and in the presence of clay to make longer peptides.

Since the hydrolysis of the peptide bond in water competes favorably with its formation, it is unlikely that peptide synthesis in water proceeded directly; it required a condensing agent. Several such natural condensing agents have been proposed, including carbonyl sulfide and sulfur dioxide, gases found in volcanic effluents.

Miller considered that volcanic gases could have been important in the prebiotic synthesis of amino acids and used hydrogen sulfide in one of his experimental runs (Parker et al. 2011). Two gases that are also present in volcanic eruptions are sulfur dioxide (SO_2) and carbonyl sulfide (COS). Both of these are condensing agents that promote peptide formation from amino acids. Leman et al. (2004) investigated peptide formation in the presence of carbonyl sulfide and found that mixed peptides up to five residues long could be formed. They indicated the importance of N-carboxy anhydrides

N-carboxy anhydride

in peptide formation, suggesting "nucleophilic attack by a second α-amino acid molecule on the in situ-formed NCA" (Leman et al. 2004). The amount of COS in volcanic gases is small, however, at less than 0.1 mol %, but if amino acids were condensed on surfaces that were subject to wet/dry cycles near a volcano, elongation could have taken place periodically as eruptions and rain occurred. Peptides longer than a certain length bind to mineral surfaces quite well (Lambert 2008). Another potential prebiotic peptide condensing agent is sulfur dioxide, a common component of volcanic gases. Chen and Yang (2007) have proposed that SO_2 could form peptides in aqueous solution.

12.9   Phosphorus

Although the Earth contains large amounts of phosphorus, most of this phosphorus is sequestered in minerals such as apatite [$Ca_5(PO_4)_3OH$] which forms the mineral component of bones and teeth. The solubility product Ksp of calcium and phosphate at 25 °C and pH 7 for the formation of calcium phosphate is 2.07×10^{-33}, which implies that at equilibrium with solid calcium phosphate the amounts of free Ca^{2+} and PO_4^{3-} ions in solution are miniscule and that free phosphorus is only released by the weathering of the solid. Rare on the Earth but common in iron-nickel meteorites, the mineral schreibersite $(Fe,Ni)P_3$ is potentially a source of large amounts of phosphorus due to the liberation of phosphorus upon reaction with water. The late, heavy bombardment of the Earth around 4 Gya is estimated to have delivered many kg/year of phosphorus via this mechanism (Ritson et al. 2020). The phosphorus so liberated would have been important for prebiotic evolution because it would have been in soluble form, whereas the phosphorus in apatite is insoluble at neutral pH. Either ferrocyanide, a readily formed iron salt, or a combination of the abundant magnesium sulfate with sodium cyanide can react with hydroxyapatite to form organophosphates (Burcar et al. 2019).

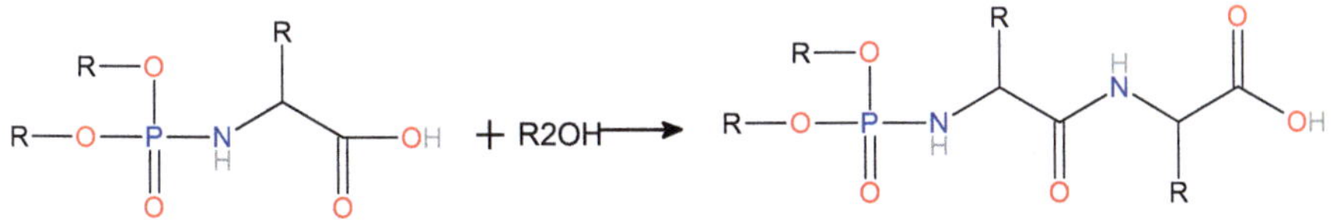

Fig. 12.13 Peptide bond formation via phosphoric-carboxylic mixed anhydrides

Several phosphorus compounds are condensing agents for peptide formation. Phosphoric-carboxylic mixed anhydrides (Fig. 12.13) can form peptides by a reaction that proceeds through diester exchange and was important during the ligation of nucleotides.

N-carboxy anhydride amino acids can polymerize, and their prebiotic synthesis has been suggested through several prebiotic routes (Kricheldorf 2006). They are the simplest form of activated amino acid accessible by the reaction of an amino acid with carbon dioxide followed by dehydration. Polypeptides have been made in this way upon adsorption on the mineral surfaces of montmorillonite (Fig. 12.14), a common layered component of clays formed from the dissolution and subsequent precipitation of aluminosilicates in water. In addition, research by Orgel has shown that peptides of length up to 55 residues can be synthesized in the presence of minerals, such as hydroxyapatite or illite (Ferris et al. 1996). They also found that *polynucleotides* containing more than 50 monomers could be formed in the presence of covalently linked atoms separated by a gap some 5–7 nm wide (Fig. 12.14) that can accommodate many guest molecules and ions, including cations like sodium, potassium, and magnesium, as well as organics important for prebiotic evolution.

12.10   Nucleic Acids

In all living things, amino acids are organized into proteins, and the construction of proteins is mediated by nucleic acids, which are themselves synthesized through biochemical pathways catalyzed by proteins. It is tempting to think that protocellular organization and biochemistry were identical to that of modern cellular processes, but, as indicated above, there is no guarantee that this is correct. There is even evidence that it is not.

12.11   Meteorites and Tholins Again

Just as in the case for compounds of hydrogen, oxygen, and carbon, meteorites also contain nitrogen heterocycles such as pyridine carboxylic acids, piperazinedione, hydantoins, purines, pyrimidines, pyridines, quinolines, carboxy lactams, lactams,

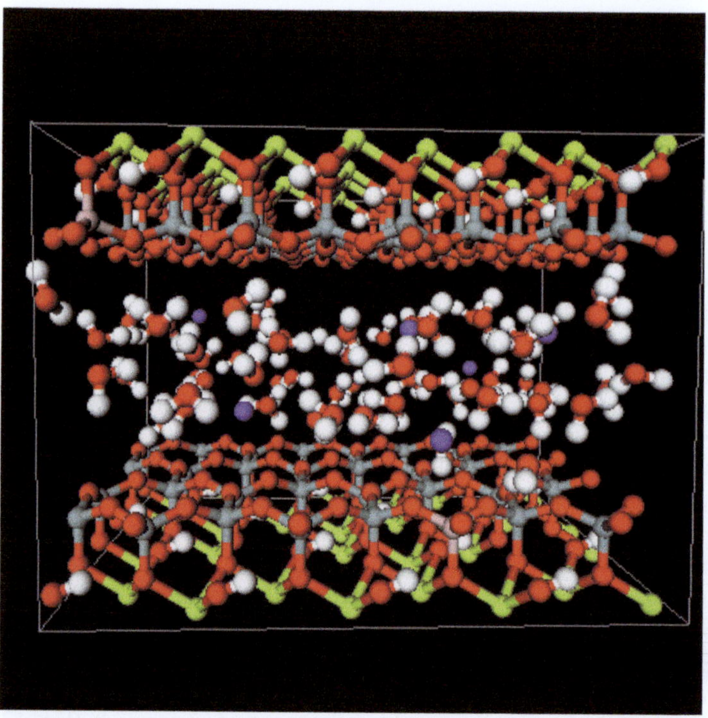

Fig. 12.14 Molecular structure of montmorillonite clay showing the aluminosilicate layers and intercalated water molecules and cations (purple spheres). The layered structure provides interstitial spaces into which a variety of organic molecules and inorganic ions can easily bind. (Visualization courtesy of Dr. G. Sposito, UC Berkeley, and Lawrence Berkeley National Laboratory)

lactims, and the amino acid proline (Martins 2018). Significant is the finding of the purines, adenine, and guanine, components of RNA and DNA, as well as the pyrimidine uracil (Callahan et al. 2011). Irradiation of analogues of astrophysical ices with ultraviolet light has been shown to generate uracil, cytosine, and thymine (Materese et al. 2017 and references therein). Tholins generated in the laboratory under conditions thought to mimic that of Titan have been examined by ^{13}C and ^{15}N solid-state nuclear magnetic resonance spectroscopy (Derenne et al. 2012) and found to contain nitrogenous heterocycles, precursors of nucleic acids. Also, these tholins contain not only amino acids but also the nucleic acid bases adenine, cytosine, thymine, guanine, and uracil (Kawai et al. 2019). Similar studies using gas chromatography-mass spectrometry found both amino acids and all the five nucleotide bases, cytosine, uracil, thymine, guanine, and adenine (Hörst et al. 2012). The Japanese sample-return mission Hayabusa2 to the carbonaceous asteroid Ryugu (Fig. 12.2) also found uracil, along with an abundance of organic molecules, including nitrogen heterocycles, in the pristine material collected (Oba et al. 2023). The recent return of a larger sample of material from the asteroid Bennu by the Osiris-Rex spacecraft should prove to be equally exciting for prebiotic fossil hunters.

12.12   The Structure of Nucleic Acids

One of the striking features of living systems on the Earth is that they use the primary nuclear constituents of the universe to produce molecular polymers to perform the crucial tasks of catalysis and information storage and retrieval; sugars, lipids, proteins, and nucleic acids are all composed of hydrogen, carbon, oxygen, nitrogen, phosphorus, and sulfur nuclei produced by nuclear fusion in stars and distributed throughout the universe by novae and supernovae explosions. Polymers that enable the storage and retrieval of information storage are essential to life if the reproduction of identical copies of an individual is to be generated so that the species can survive. This role is played by the nucleic acids, ribonucleic acid (RNA) and deoxyribonucleic acid (DNA). In order to understand how the prebiotic synthesis of these informational components of living systems is currently viewed, it is necessary to first examine their molecular structure.

The nucleic acids are built in a manner that suggests an assembly of modules; a nucleoside (Fig. 12.15) is assembled from a β-D-ribofuranose sugar and a nitrogenous nucleobase. The nucleobases include the pyrimidines and the purines (Fig. 12.16), heterocycles of nitrogen, carbon, and hydrogen. A typical nucleoside, say adenosine, will have the structure (Fig. 12.15) where the nucleobase is attached to the D-ribose anomeric carbon through a dehydration reaction, i.e., the removal of a water molecule (the OH from the ribose and an H from the base) (Fig. 12.17).

Since a dehydration reaction involves the removal of a water molecule, the chemical potential of water in the reaction mixture determines the Gibbs energy change or the probability of the reaction, and it is no surprise that the attachment of the base to the ribose moiety has a positive ΔG value, and therefore the equilibrium is toward the reactants (ribose + nucleobase) rather than toward the products if the reaction is carried out in dilute aqueous solution, much the same as in the formation of peptides from amino acids, which also involves dehydration of the reactants.

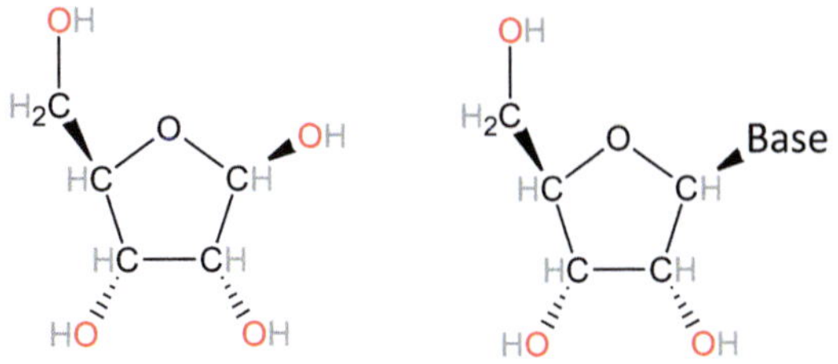

Fig. 12.15 The nucleosides (right) consist of a nitrogenous base covalently bound to the anomeric carbon (C1′) of β-D-ribofuranose (left) through the removal of a molecule of water. The base is a nucleobase, a heterocycle containing nitrogen

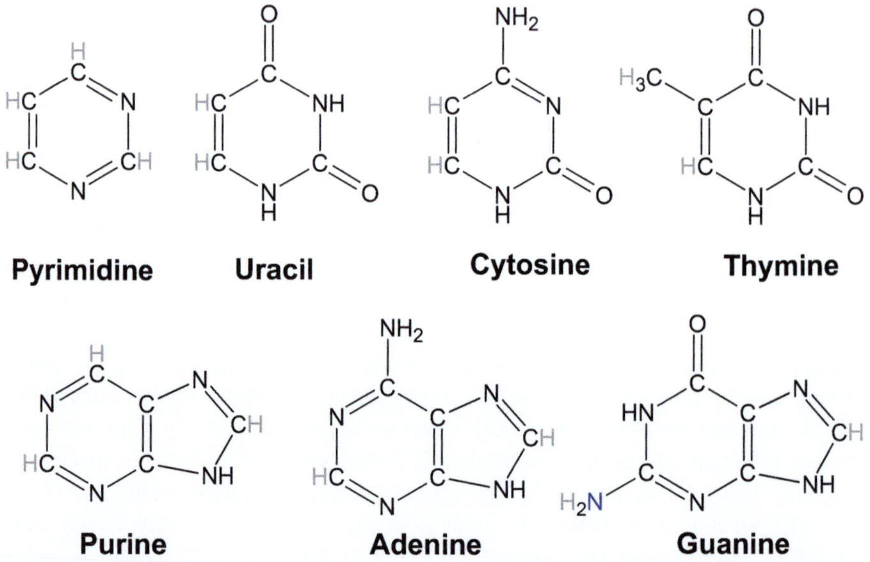

Pyrimidine   **Uracil**   **Cytosine**   **Thymine**

Purine   **Adenine**   **Guanine**

Fig. 12.16 The nitrogenous nucleobases are made up of the *pyrimidines*, uracil (RNA only), cytosine, and thymine (DNA only), as well as the *purines*, adenine and guanine

Fig. 12.17 The nucleosides adenosine (RNA) and deoxyadenosine (DNA) consist of a nitrogenous base (in this case adenine) and β-D-ribose. The nucleosides that form DNA have the hydroxyl on ribose C2 replaced with a hydrogen and are referred to as deoxyribonucleosides

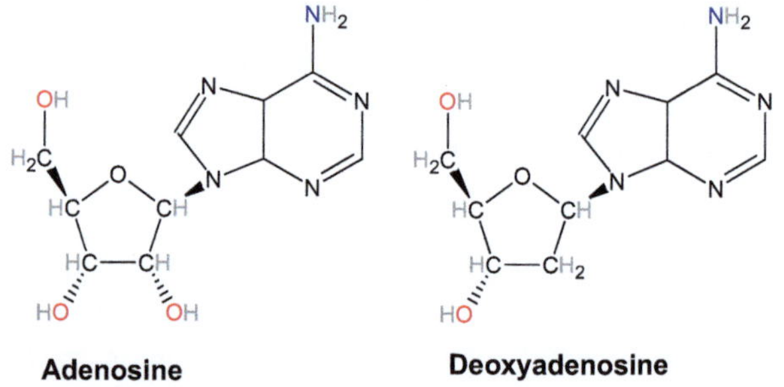

Adenosine   **Deoxyadenosine**

Both RNA and DNA are polymers of nucleotides, phosphorylated nucleosides, where the phosphate module is attached to the hydroxymethyl carbon (C5) of the ribose through another dehydration reaction. The ribonucleotides that form RNA (Fig. 12.18) as well as the deoxyribonucleotides that form DNA are enumerated in Table 12.5. Note that thymine, which is found only in DNA is methyl-uracil and uracil is only found in RNA (Fig. 12.16). Polymerization of the nucleotides in either RNA or DNA proceeds through the covalent attachment of the phosphate on one nucleotide to the 3′ hydroxyl on the next, etc. to form a chain linked through 3′–5′ links, again by a dehydration reaction.

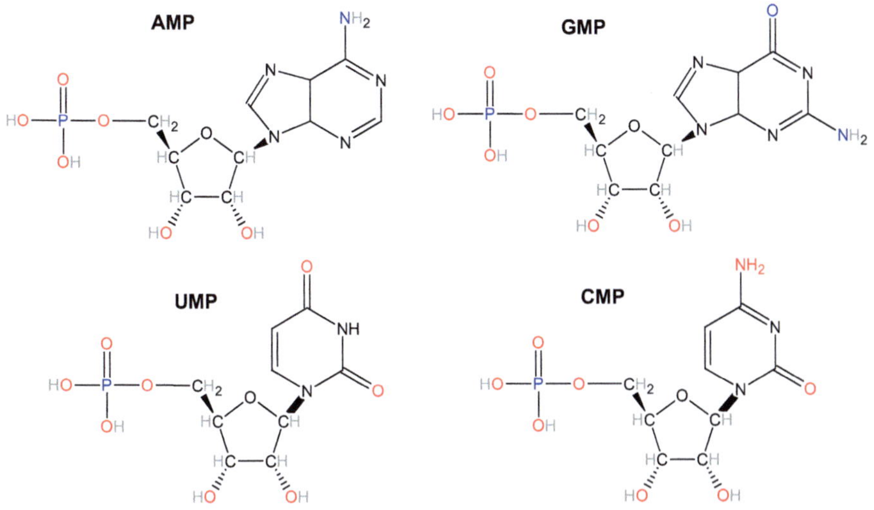

Fig. 12.18 The structures of the ribonucleotides that are found in RNA. Note that all of the bases are covalently bonded to the anomeric carbon (C1′) of ribose through the removal of a water molecule and that phosphorylation occurs at C5′. The names for the abbreviations are found in Table 12.5

Table 12.5 The composition of the informational polymers of living systems

Base	Nucleoside (RNA)	Nucleotide (RNA)	Symbols
Adenine	Adenosine	Adenylate (adenosine 5′-monophosphate)	A, AMP
Guanine	Guanosine	Guanylate (guanosine 5′-monophosphate)	G, GMP
Uracil	Uridine	Uridylate (uridine 5′-monophosphate)	U, UMP
Cytosine	Cytidine	Cytidylate (cytidine 5′-monophosphate)	C, CMP
Base	*Nucleoside (DNA)*	*Nucleotide (DNA)*	*Symbols*
Adenine	Deoxyadenosine	Deoxyadenylate (deoxyadenosine 5′-monophosphate)	A, dAMP
Guanine	Deoxyguanosine	Deoxyguanylate (deoxyguanosine 5′-monophosphate)	G, dGMP
Cytosine	Deoxycytidine	Deoxycytidylate (deoxycytidine 5′-monophosphate)	C, dCMP
Thymine	Deoxythymidine	Deoxythymidylate (deoxythymidine 5′-monophosphate)	T, dTMP

The modular nature of nucleic acids, with their phosphates, riboses (deoxyriboses), and bases suggested to researchers early on that prebiotic reactions proceeded through the straightforward reactions of the bases with the ribose and the phosphates. However, in order to not get too far ahead of ourselves, we need to first examine early studies of the prebiotic synthesis of ribose and the bases.

12.13   The Prebiotic Synthesis of Ribose

It has been known since the work of Buterlow in the nineteenth century (Butlerow 1861) that formaldehyde, a common constituent of interstellar and protoplanetary clouds, of comets and meteorites, can form from the oxidation of methane and can spontaneously condense to form many sugars by repetitively polymerizing to form a large number of polyols, including ribose (Breslow 1959; Pinto et al. 1980). The condensation is called the formose reaction, combining as it does formaldehyde and *−ose* for sugar. This reaction is autocatalytic, that is, it is driven by its products, although the reaction rate is greatly increased by the addition of small, catalytic amounts of glycolaldehyde, another molecule detected in space through radiofrequency spectroscopy. The fact that ribose is produced is illustrated below by the results of the formose reaction (Fig. 12.19). Almost no sugars are produced in

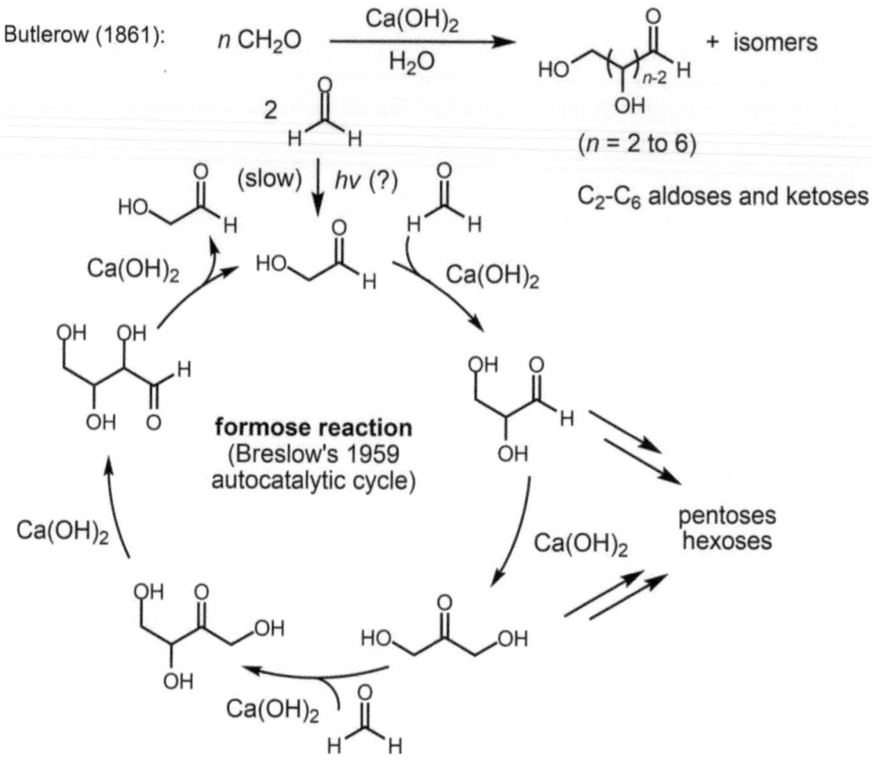

Fig. 12.19 At the top is Butlerow's original scheme for the spontaneous self-assembly of sugars from formaldehyde in the presence of water and a base, calcium hydroxide. Below is Breslow's (1959) proposed autocatalytic cycle in which glycolaldehyde is the catalyst. (From: Alsosaid1987—Own work, CC BY-SA 4.0, https://commons.wikimedia.org/w/index.php?curid= 62998627)

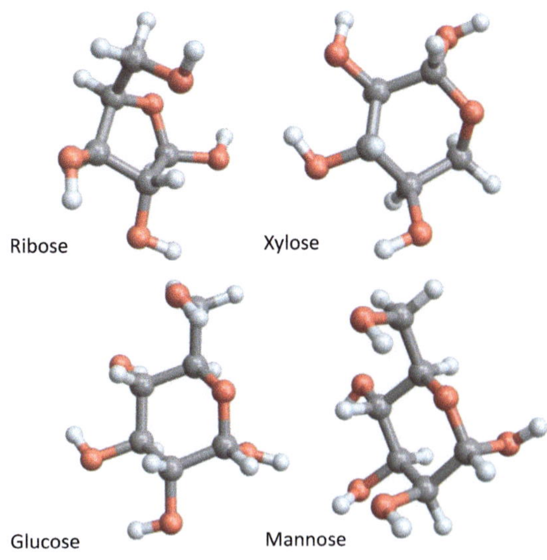

Fig. 12.20 Some of the myriad products of the alkaline formose reaction. In addition to the production of *D*- and *L*-ribose, another five-carbon sugar, xylose is generated. The six-carbon sugars glucose and mannose are found in relatively equal amounts (Decker et al. 1982; Omran 2023)

the absence of a basic (i.e., pH > 7) environment, such as formed when minerals like calcium hydroxide or the borate minerals colemanite ($Ca_2B_6O_{11}5H_2O$) or kernite ($Na_2B_4O_7$) found in evaporated ponds in Death Valley (Ricardo et al. 2004) are present. The ribose so formed is racemic, producing both D- and L-ribose, but the correct stereoisometry is given by the subsequent condensation of glycoaldehyde phosphate with formaldehyde. This leads to the eventual formation of a favorable ribose-2,4-diphosphate, an attractive nucleotide analog.

Other possible routes to ribose are discussed by Biscans (2018) and include formation from cyanide in the presence of ultraviolet light or the stabilization of glycolaldehyde by bisulfate. It is also possible to form large amounts of ribose in laboratory experiments in which interstellar ice analogs containing H_2O, NH_3, and CH_3OH, again, commonly found in space, are irradiated with ultraviolet light (Meinert et al. 2016) in the absence of added calcium or other metallic bases. As an aside that harkens back to metabolism, these authors also found glycerol, glyceraldehyde, lactic acid, and the Krebs cycle intermediates succinate, fumarate, and malate in their irradiated material. It is useful to point out at this stage that the Rosetta cometary mission detected many of these same molecules (Goesmann et al. 2015) and that ribose has been found in several meteorites (Furukawa et al. 2019). Therefore, even though the exact prebiotic pathway for the formation of ribose has not yet been conclusively determined, it is clear that it can be formed from a number of ways that are consistent with spontaneous self-assembly from primordial constituents of the Earth and the universe.

These studies did not answer the question of why ribose, a pentofuranose sugar, was selected to form the backbone of RNA and DNA. The formose process produces a large spectrum of compounds (Fig. 12.20), of which D-ribose is only a minor constituent among the trioses, the tetraoses, the pentoses, and the hexoses (Decker et al. 1982; Omran 2023). Eschenmoser has addressed this problem by examining the base-pairing potential of RNA and DNA made using these alternative sugars as backbone molecules that, together with the phosphate groups, hold the nucleobases in the position for base stacking and base pairing. He has found (Eschenmoser and Dobler 1992) that the energy involved in the hydrogen bonding of base-paired nucleotides is not a consideration driving the selection of ribose over the myriad of other possible three-, four-, and six-carbon sugars. His measurements of the melting temperatures of base-paired polynucleotides indicate that there are alternative structures that are more tightly bonded than RNA or DNA, but these were not selected by chemical evolution for incorporation into life on Earth. The larger number of atoms in pyranosyl, as opposed to furanosyl, nucleic acids introduces steric hindrance that precludes proper nucleic acid structures. Oligonucleotides with $2'$-$5'$ linkages are not present because the melting temperatures of their duplexes were 30–40 °C lower; they were less stable than duplexes formed by $3'$-$5'$ linkages of oligoribonucleotides with the same sequence (Kierzek et al. 1982). Banfalvi (2006) examined the structures formed by pentofuranoses and concluded that only β-D-ribose fits into nucleotides. This conclusion was based on considerations of sugar pucker conformation, steric analysis of the anomeric carbon (C1$'$) where only the β-anomer can support the nucleobase in the proper position for base pairing in a duplex, and inversion of the configuration of the hydroxyl groups on C2$'$ and C3$'$. He further concluded that for helical duplex formation, such as in DNA, the β-D-ribose of RNA is replaced by β-D-deoxyribose because there is insufficient space in DNA helices for the C2$'$ oxygen atom in β-D-ribose (Banfalvi 2006; Dickertson 1982).

12.14   The Prebiotic Synthesis of the Purines

The nucleobases in RNA and DNA are adenine, guanine, uracil, cytosine, and thymine. Recent studies (Hörst et al. 2012; Kawai et al. 2019) of the molecular composition of Titan tholins (vide supra) indicate that all five nucleobases are formed as well as isocytosine and 2,4-diaminopyrimidine. The early studies by Oró, in 1961 (Oró 1961; Oró and Kimball 1961), already found that the polymerization of hydrogen cyanide produced the purine, adenine, in the scheme shown in Fig. 12.21. Subsequent work by Ferris and Orgel (1965, 1966) showed that adenine is formed photochemically in aqueous solutions of ammonia and hydrogen cyanide (Fig. 12.22). We also note that adenine was found in meteorites so that its abiotic synthesis is firmly established.

Since hydrogen cyanide is a constituent of the interstellar medium, of protoplanetary stellar disks, and of the coma of comets, the starting material for this reaction is present in many environments in the universe. A large collection of purine derivatives can be generated by this scheme. In particular, there are facile reactions of the imidazole derivatives to readily form purines such as adenine, guanine, and hypoxanthine. It has been proposed that aqueous hydrogen cyanide could have been

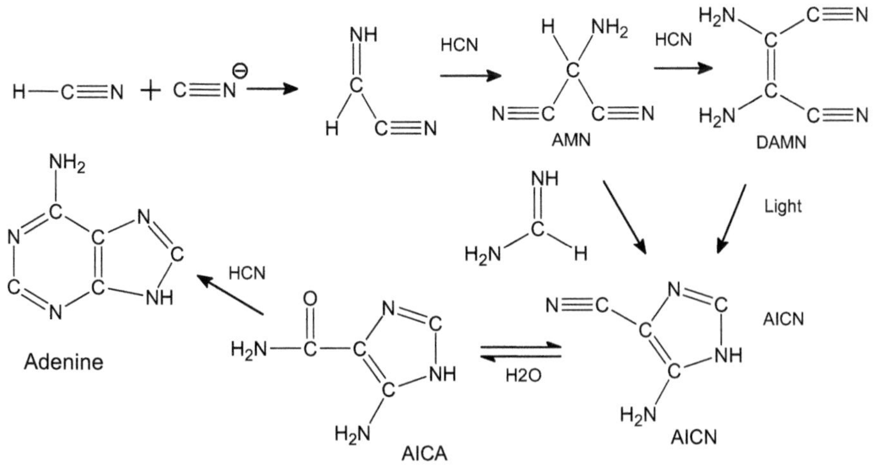

Fig. 12.21 Proposed mechanism of synthesis of adenine from the polymerization of hydrogen cyanide. Based on Fig. 1 of Oró (1961), AMN is aminomalononitrile; DAMN is diaminomaleonitrile; AICN is 4-aminoimidazole-5-carbonitrile; and AICA is 4-aminoimidazole-5-carboxamidine

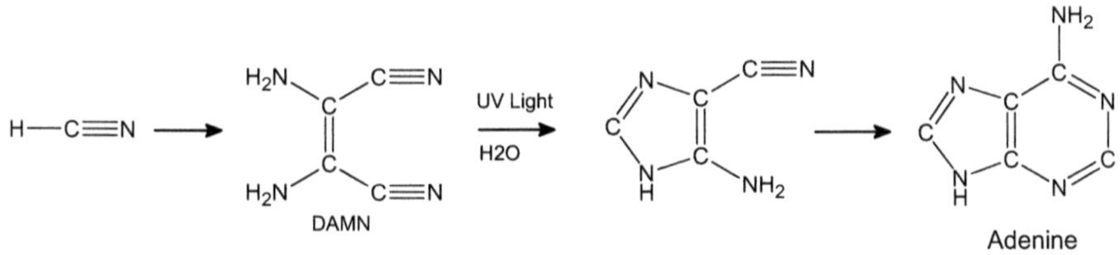

Fig. 12.22 The condensation of hydrogen cyanide to form diaminonitrile and the ultraviolet-induced photochemical rearrangement to imidazoles and purines (adenine) as discovered by Ferris and Orgel (1965, 1966)

concentrated during freezing at interfaces to form a eutectic, thereby raising its local concentration and favoring polymerization rather than hydrolysis to formamide or formic acid. One of the major weaknesses in this proposed scheme is that the yield of purine heterocycles under prebiotic conditions is quite low at ~0.5% (Oró and Kimball 1961) even though the stated yield of adenine in the reaction cited in Fig. 12.22 is ~80% in water at 298 K (Ferris and Orgel 1965, 1966).

12.15   Purine Nucleosides

The next step in the generation of nucleic acid precursors is the synthesis of purine nucleosides (compounds of purines and ribose, Table 12.5) and the synthesis of adenosine has been observed to occur in the molten state when adenine is heated with ribose (Fuller et al. 1971). The resulting complex mixture of products contained about 4% adenosine along with larger amounts (8–9%) of inosine and guanosine and other materials because several of the nitrogens on adenine besides N_6 can react with ribose. This lack of regioselectivity led Becker et al. (2016) to reexamine the prebiotic formation of purine nucleosides with the discovery of a pathway (Fig. 12.23) involving the condensation of formamidopyrimidines with ribose that generates natural purine nucleosides (Robertson and Miller 1995) with good yields (60%). This pathway uses starting materials like HCN, NH_3, glycolaldehyde, and formic acid derivatives that were found by the Rosetta spacecraft mission in comet 67P/Churyumov-Gerasimenko (Goesmann et al. 2015).

12.16   Pyrimidines

The prebiotic synthesis of the pyrimidine bases (cytosine, uracil) is not thought to recapitulate that of the purines because the precursors are likely to be cyanoacetylene (**1** in Fig. 12.24) and cyanoacetaldehyde, along with urea, cyanate, and cyanogen instead of hydrogen cyanide per se. Yields for cytosine of up to 50% have been reported when cyanoacetaldehyde was reacted with urea (Robertson and Miller 1995). Lightning discharges and wet/dry cycles at higher temperatures (~100 °C) are favorable conditions for this synthesis. Uracil is formed through the hydrolysis of cytosine. We reiterate that tholins contain all five nucleobases.

12.17   Pyrimidine Nucleosides

The prebiotic synthesis of ribonucleosides containing pyrimidine bases have previously been thought to involve rare reactions with free ribose moieties. It was recognized (Sanchez and Orgel 1970) that phosphate was needed in order to produce aminooxazoline derivatives upon the reaction of ribose derivatives with cyanamide and that ultraviolet light ($\lambda \sim 250$ nm) was needed to complete the conversion to nucleosides. The low 4% yield and the requirement for free prebiotic ribose rendered this pathway of dubious importance. However, these authors remarked upon the propensity of ribose aminooxazoline to crystallize, which would allow the reaction product to spontaneously separate from other compounds formed so that subsequent reaction steps could occur without interference by the non-interesting side reactions. Research performed in Sutherland's laboratory (Xu et al. 2017) has recently shown

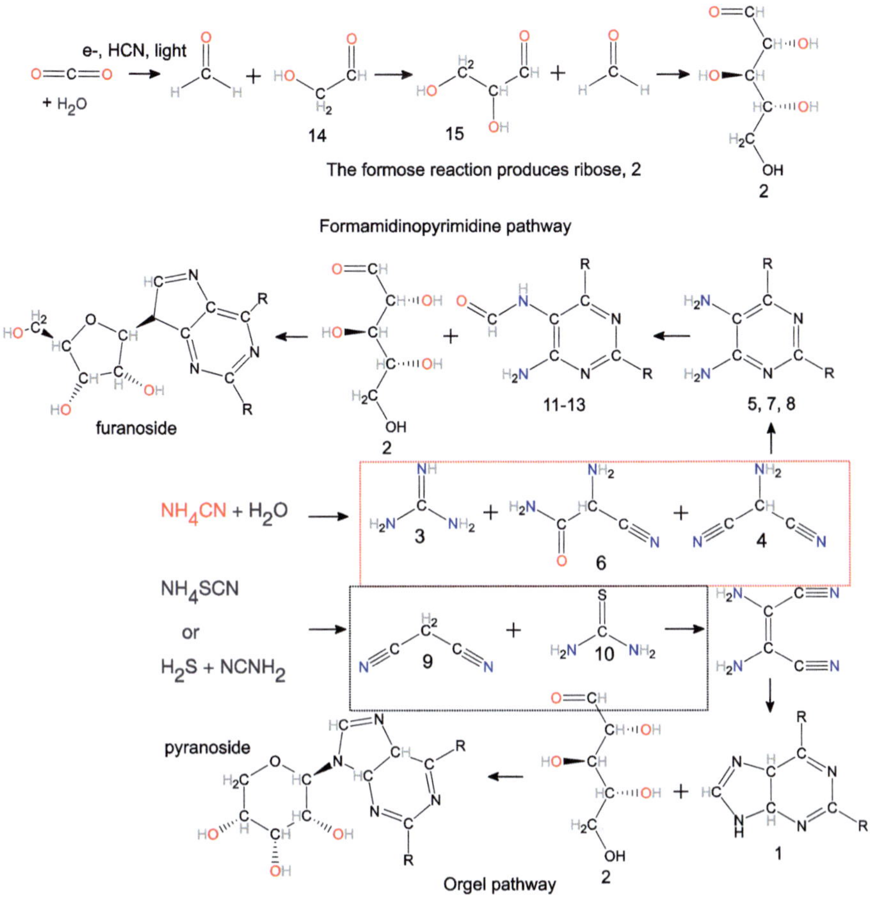

Fig. 12.23 A prebiotic molecular network devised by Becker et al. (2016), for the prebiotic synthesis of purine nucleosides involving simple prebiotically available precursors (in the red and rectangles) such as malononitrile **9** and thiourea **10**, along with guanidine **3**, aminomalononitrile **4**, and amino cyanoacetamide **6**, which (in red) are derived directly from NH_4CN. These react to form the aminopyrimidines (black) **5, 7, 8** which react with either formamide or formic acid to generate formamidopyrimidines **11–13**. Condensation with ribose **2** then produces the purine nucleosides. The prebiotic synthesis of ribose **2** arises from the condensation of glycolaldehyde **14** and formaldehyde, formed by sunlight, electric discharge (lightning, e⁻) from an early humid atmosphere containing hydrocyanic acid (HCN), and CO_2. Orgel's earlier pathway required the coupling of the preformed nucleobases, such as adenine **1**. Note that for **5** and **11**, R = NH_2; for **7** and **12**, R = OH, NH_2; and for **8** and **13**, R = NH_2, H

that the requirement for free ribose is not necessary; they discovered an indirect synthetic route (Fig. 12.25) that formed ribose aminooxazoline through the reaction of glyceraldehyde with 2-aminooxazole in 44% yield. The route to pyrimidine β-ribonucleosides progressed by means of a high-efficiency photoanomerization of α-2-thioribocytidine. Again, it was found that ribose aminooxazoline selectively

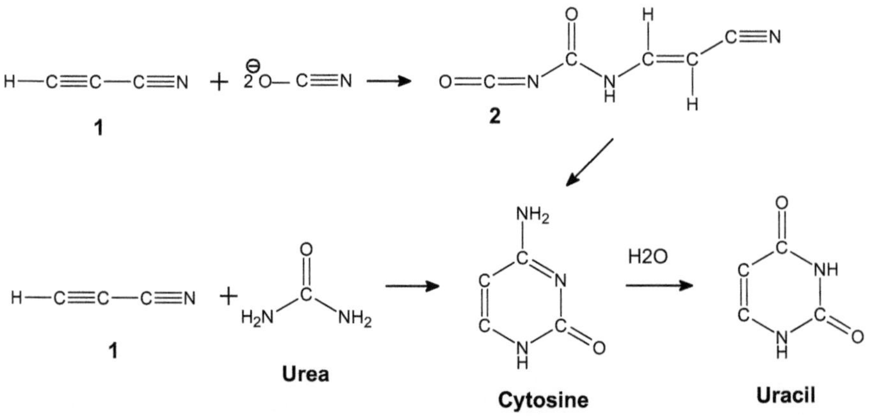

Fig. 12.24 Schemes for the prebiotic synthesis of the pyrimidines cytosine and uracil upon heating starting from cyanoacetylene (**1**) cyanate ($^-$O-CN) and urea. The interesting intermediate (**2**) spontaneously cyclizes where the rightmost nitrile nitrogen becomes N7 of cytosine, which is converted by hydrolysis in water to O7 of uracil

crystallized from the mixture of products. It was not lost on Sutherland that the reactions of prebiotic synthesis were often improbable, with positive ΔG values indicating that they were thermodynamically unfavorable. One way to spur reactions to increase their probability is found in extant biology, that of the coupling of one reaction, such as the hydrolysis of ATP, to the other in order to change the energy landscape. These couplings would have been unlikely on the primitive Earth, but there are other sources of energy available in heat, lightning, shock waves from impacts, and above all ultraviolet light from the Sun. UV light has sufficient energy (a 250-nm photon has ~5 eV) to easily alter molecular bonds. The primordial Earth was bathed in UV light in a night-day cycle over many millions of years, and this light reached the surface relatively unaltered because the atmosphere of the early Earth did not contain oxygen, and hence there was no ozone layer to absorb it.

12.18   Purine and Pyrimidine Nucleotides

The final step on the pathway to the prebiotic synthesis of nucleic acid monomers involves the addition of phosphate moieties to the nucleosides to form the nucleotides, such as adenosine monophosphate (Fig. 12.26). Nucleotides are the monomeric units of the nucleic acid polymers—deoxyribonucleic acid (DNA) and ribonucleic acid (RNA). While phosphorylations are again dehydration reactions and consequently thermodynamically unfavored in aqueous media, high-yielding routes to cytidine and uridine ribonucleotides can be built from small two- and three-carbon fragments such as glycolaldehyde, glyceraldehyde or glyceraldehyde-3-

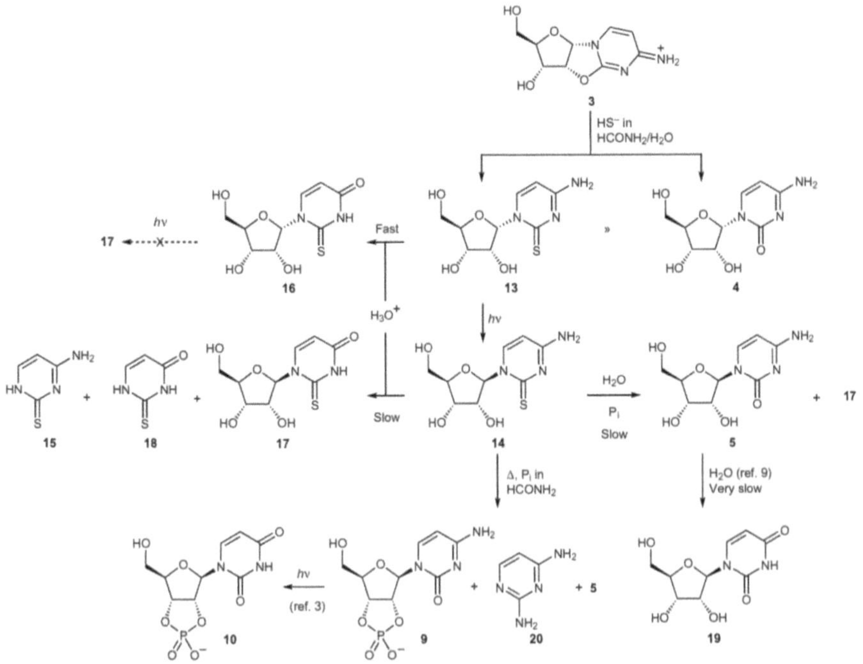

Fig. 12.25 Synthesis of pyrimidine β-ribonucleosides and β-ribonucleotides involving photoanomerization of α-2-thioribocytidine **13**. The reaction scheme incorporates the new synthesis and the known hydrolysis of β-ribocytidine **5** to β-ribouridine **19**. Thiolysis of the ribose anhydronucleoside **3** gives α-2-thioribocytidine **13** in high yield along with a small amount of α-ribocytidine **4**. In marked contrast to **4**, **13** undergoes a remarkably efficient photoanomerization to give β-2-thioribocytidine **14** in excellent yield. The canonical pyrimidine ribonucleosides **5** and **19** can be produced from **14** by hydrolysis. Phosphorylation of **14** is accompanied by conversion of the 2-thiocytosine nucleobase into cytosine, which provides a direct route to the canonical pyrimidine ribonucleoside-2′,3′-cyclic phosphates **9** and thence **10** and an alternative route to the ribonucleoside **5**. 2-Thioribouridines **16** and **17** can also be produced by hydrolysis of **13** and **14**, respectively, under acidic conditions. (Reproduced from Xu et al. (2017) by permission)

phosphate, cyanamide, and cyanoacetylene (Lohrmann and Orgel 1968). One of the steps in this sequence allows the isolation of enantiopure ribose aminooxazoline if the enantiomeric excess of glyceraldehyde is 60% or greater. This can be viewed as a prebiotic purification step, where the said compound spontaneously crystallized out from a mixture of other pentose aminooxazolines. Ribose aminooxazoline can then react with cyanoacetylene in a mild and highly efficient manner to give the alpha cytidine ribonucleotide. Photoanomerization with UV light allows for inversion about the 1′ anomeric center to give the correct beta stereochemistry. In 2009, they showed that the same simple building blocks allow access, via phosphate-controlled nucleobase elaboration, to 2′,3′-cyclic pyrimidine nucleotides directly, which are known to be able to polymerize into RNA.

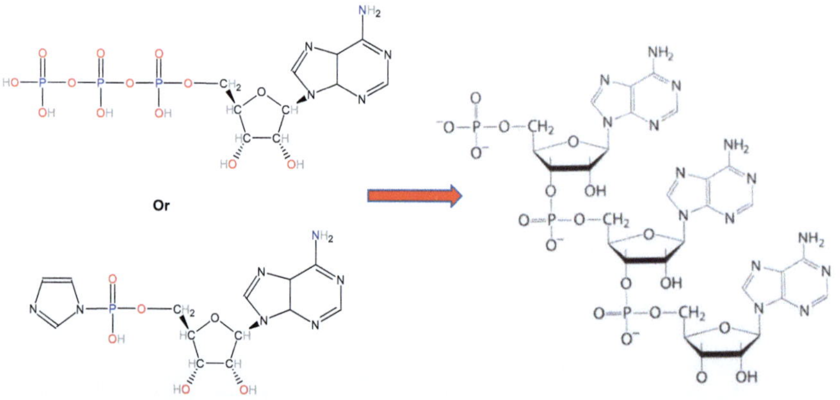

Fig. 12.26 A scheme proposed by Orgel (2002) for the condensation of a nucleotide, such as adenosine triphosphate, or an imidazole-activated nucleotide into a trinucleotide

12.19   The Generation of Nucleic Acid Polymers

The storage and retrieval of biological information is possible only through the medium of polymers whose sequence specifies the characteristics of the information, much as the sequence of bits in a computer word specifies the nature of the information contained within it. In living systems on Earth, biological information is stored in DNA and RNA. James Ferris' studies have shown that clay minerals of montmorillonite (Fig. 12.14) will catalyze the formation of RNA in aqueous solution by joining activated RNA mononucleotides to form longer chains, up to 50-mers (Joshi et al. 2009). Although these chains have random sequences, the possibility that one sequence began to nonrandomly increase its frequency by increasing the speed of its catalysis means that it is possible to "kick start" biochemical evolution. This is only one example of the important role played by minerals as catalysts that alter the probability of prebiotic reactions in favor of otherwise improbable products (Francisco et al. 2019). Dimensional reduction of diffusion (Chap. 9) by the binding of molecules to mineral surfaces also alters the reaction probability (Hashizume 2015). Compartmentalization serves a similar purpose in that it allows a means to circumvent the limitations of three-dimensional diffusion.

12.20   Prebiotic Syntheses of Lipids

The structure of lipids can also be viewed as the assembly of modules; almost all lipids contain one or more long chains of *methylene groups* as found in the higher alkanes like dodecane, etc. These long $(-CH_2-)_n-CH_3$ hydrocarbon chains have 12–24 methylene units ($n = 12$–24) in phospholipids. The methylene chains

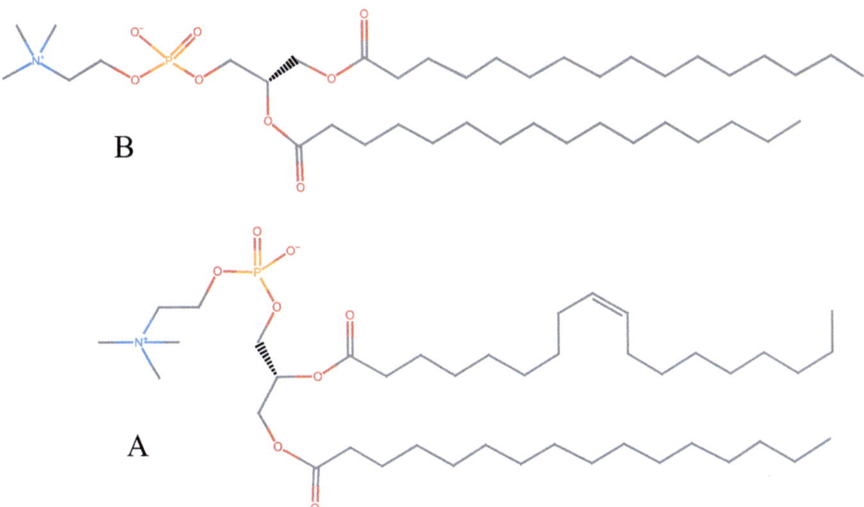

Fig. 12.27 (a) The structure of 1-palmitoyl-2-oleoyl-*sn*-glycero-3-phosphatidyl-choline showing its different $C_{16:0}$ palmitate, and $C_{18:1}$ oleate fatty acyl side chains, and its negatively charged phosphate and positively charged choline head groups. (b) The structure of 1,2-palmitoyl-*sn*-glycero-3-phosphatidyl-choline showing its identical $C_{16:0}$ palmitate fatty acyl side chains

terminate in a polar group such as a carboxylic acid, amide, or alcohol, which is then linked via an ester, amide, or ether bond to a short (~3 carbon) moiety, such as glycerol, which is in turn bound to a phosphate group that is derivatized with another small molecule like choline, ethanolamine, serine, glycerol, etc. An example is the phospholipid 1,2-diacyl-3-*sn*-glycerophosphatidyl-choline (Fig. 12.27) where the fatty acyl side chains are identical in terms of lengths or the state of saturation of the carbon-carbon bonds. The amphiphilic nature of long chains of hydrophobic methylene carbons terminated by hydrophilic carboxy or phosphate groups gives lipids the unique characteristic of being able to spontaneously self-assemble into structures in which there is a spatial separation of properties. The methylene carbon chains are buried within micelles and vesicles (Fig. 12.28), while the polar head groups are in contact with the aqueous medium. Lipid vesicles form spontaneously and separate the world into "inside" and "outside" to provide a means for molecules inside to collide frequently without diffusing away. The lipid bilayer (Fig. 12.28) forms the basic structure for all cellular membranes.

These unique molecules then are central to the formation of cells in living systems that encapsulate and partition the biological milieu and allow the cell to control its internal constitution. It is clear that lipids are an indispensable part of life on Earth. How then did lipids form? Fossils from the earliest living systems on Earth date back to the Eoarchean era, some 3.6–4 billion years ago (Gya), and therefore, the lipid bilayer membrane must have formed prior to that time.

Although fatty acids are amphiphiles, they are also detergents and would therefore serve to disrupt vesicle formation. However, single-chain amphiphiles have now

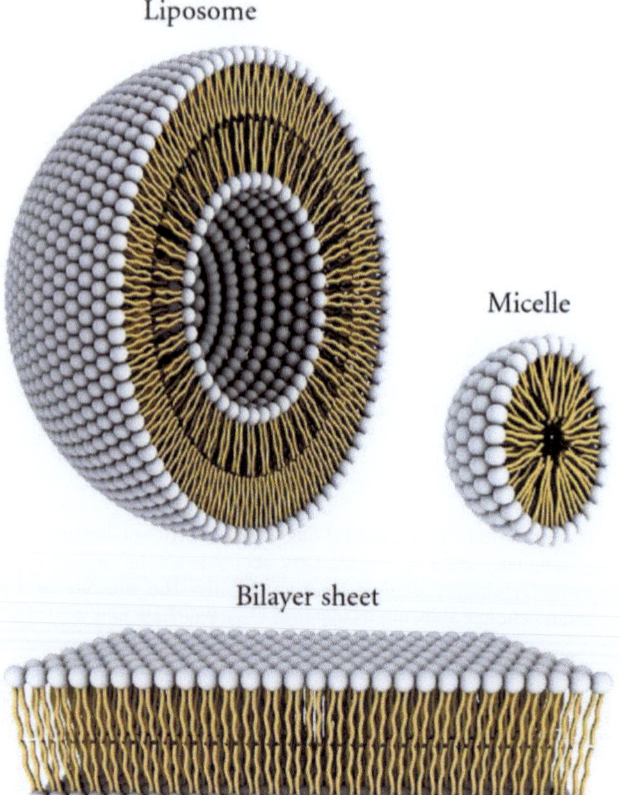

Fig. 12.28 Amphipathic phospholipids, such as lecithin (1,2-diacyl-3-*sn*-glycerophosphatidyl-choline), can form at least three different structures in aqueous solution: micelles, liposomes, and bilayers. (From Bitounis et al. (2012). Creative Commons Attribution License. https:// creativecommons.org/licenses/by/3.0/)

been found to stabilize vesicle assembly. Fatty acids will form bilayers at concentrations in the mM range, while phospholipids will form bilayers at micromolar concentrations, a thousand-fold lower. The phospholipid bilayer structure is the basis for all cellular membranes, so it behooves us to use biological logic to suggest that the prebiotic synthesis of lipids containing fatty acids esterified to glycerol is the place to begin. It is true that Archeal phospholipids are ether-linked, such as the purple membrane lipids of *Halobacterium salinarum* (Chap. 10), but the formation of ethers from two alcohols requires reaction conditions that are not easily understood from a prebiotic perspective and more than likely were formed later in evolution through the action of enzymes. We therefore propose that the development of bilayer-forming glycerophospholipids would have been a necessary step toward cellular compartmentation and the sequestration of molecular reaction networks.

As with nucleic acids and amino acids, the first attempts to generate phospholipids involved putting the modules (glycerol, phosphate, and fatty acids) together in a pot and heating them. In 1977, Deamer and colleagues (Hargreaves et al. 1977) were able to synthesize diacylphosphatidylglycerol from just three precursors, glycerol, a fatty acid or aldehyde, and phosphate in the presence of clays or silicates, and to show that these formed vesicles in aqueous solution. The fatty acids could be generated in a Fischer-Tropsch-type synthesis from aqueous solutions of formic acid or oxalic acid (McCollom et al. 1999). Additional studies by Oró (Rao et al. 1987) produced phosphatidylethanolamine under conditions thought to be possible on the primitive Earth. Note that ethanolamine has recently been discovered in the interstellar medium (Rivilla et al. 2021) in the molecular cloud G+0.693–0.027 which is part of the Sagittarius B2 complex in the Galactic Center of the Milky Way. Fatty acids, alcohols, and phosphonic acids have been found in meteorites (Pizzarello and Shock 2010; Sephton 2002; Cooper et al. 1992). In experiments utilizing irradiation of interstellar ice analogs (Layssac et al. 2020; Zhu et al. 2020), the glycerol phosphate group of phospholipids has been generated. The Almahata Sitta meteorite contained ethanolamine (Glavin et al. 2010). Recent studies by Gibard et al. (2018) have shown that diamidophosphate

> ...efficiently (amido)phosphorylates a wide variety of (pre)biological building blocks (nucleosides/tides, amino acids and lipid precursors) under aqueous (solution/paste) conditions, without the need for a condensing agent.

Diamidophosphate is produced from trimetaphosphate and is a potential prebiotic phosphorylation agent. It phosphorylates sugars, glycerol, and fatty acids and promotes polymer (oligonucleotides, peptides, and liposomes) formation (Fig. 12.29). The prebiotic synthesis of phosphoenolpyruvate has also been reported using this agent (Coggins and Powner 2017). Therefore, the first membrane-forming amphiphiles must have been generated prior to this time. Other components of living systems, the amino acids, nucleobases, and sugars, would have also been produced at the same time in our interacting network of primordial biomolecules. It was against this background that Sutherland outlined his cyanosulfidic scheme for the reactions of this molecular network that produced precursors of RNA, amino acids and lipids from available inputs such as HCN, H_2S, and ultraviolet light (Patel et al. 2015) all using the same common precursors (Fig. 12.30). The energy source was ultraviolet light, and the reactions were kinetically enhanced by Cu(I)–Cu(II) photoredox cycling.

The work of Miller in the 1950s was rapidly followed by a large number of studies of prebiotic syntheses whose main findings in the subsequent years are that all the predominant low molecular weight compounds that are found in living systems can be generated through spontaneous self-assembly. We have seen that α-amino acids, sugars, lipids, and nucleobases (Fig. 12.1) are not rare molecules but are found to be made in a variety of environments *in many different ways* through the application of available forms of energy (light, heat, electrical discharges, etc.). Since these processes are observed to be common throughout the universe, we can conclude that the preconditions for the evolution of life occur essentially everywhere

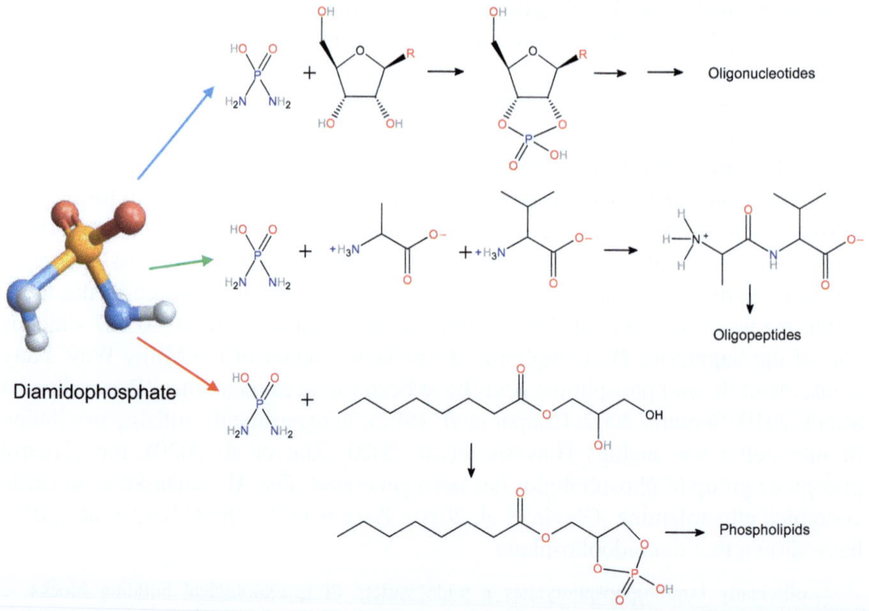

Fig. 12.29 The central role played by diamidophosphate in the phosphorylation and polymerization of nucleotides, peptides, and phospholipids

and that life is not a unique feature of our home planet but is likely ubiquitous among the many billions of exoplanets that populate our Milky Way galaxy and, by logical extension, to the myriad galaxies that surround us in space. Recent support for these ideas comes from the detection by the Webb telescope (Madhusudhan et al. 2023) of CO_2, CH_4, and perhaps the biologically derived dimethyl sulfide, in the hydrogen-rich atmosphere of the candidate Hycean world Kepler 2-18b, an exoplanet about eightfold more massive than the Earth, with a density only half that of our planet, which orbits a 2.4 Gya M dwarf star 124 light years from us. The discoveries of prebiotic synthesis indicate that since the chemistry is the same everywhere, it is therefore likely that life in other parts of the universe would resemble life on Earth in its many forms.

I close this book on the physical basis of life with the words of John Banville (1981) who wrote in *Kepler* that:

The real mystery & miracle is not that numbers have an effect upon things (which they do not!), but that they can express the nature of things; that the world, vast & various & seemingly ruled by chance, is amenable in its basic laws to the rigorous precision & order of mathematics.

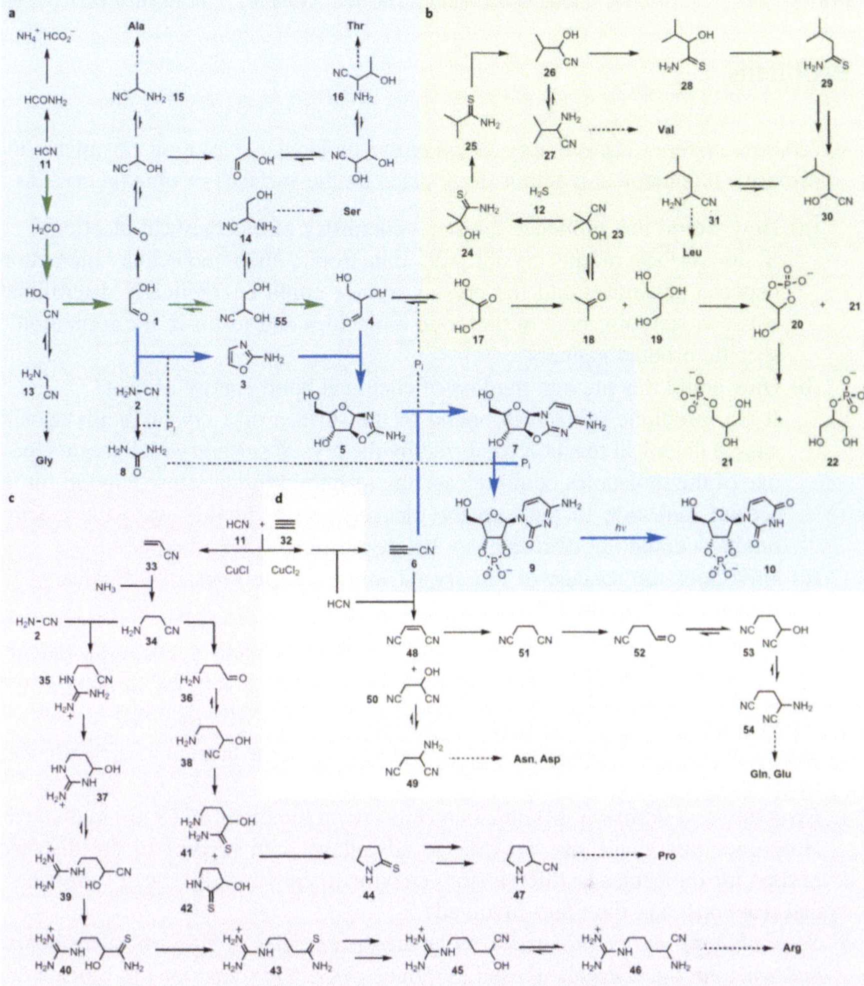

Fig. 12.30 A network of molecular reactions that leads to lipid precursors, amino acids, and RNA. This overview illustrates the extent of interconnections among the syntheses of these three important classes of biomolecules. Note that this scheme produces only biomolecules found in living systems and not a wide variety of irrelevant compounds, implying that the photochemical reactivity of hydrogen cyanide (11) and hydrogen sulfide (12), and not biology, was the main selection process at work. The molecules produced served as the basis for living systems. The four colored sections refer to the following: (**a**) Reductive homologation of hydrogen cyanide (11) (bold green arrows) provides the C_2 and C_3 sugars—glycolaldehyde (1) and glyceraldehyde (4)—needed for subsequent ribonucleotide assembly (bold blue arrows) but also leads to precursors of Gly, Ala, Ser, and Thr. (**b**) Reduction of dihydroxyacetone (17) (the more stable isomer of glyceraldehyde (4)) gives two major products, acetone (18) and glycerol (19). Reductive homologation of acetone (18) leads to precursors of Val and Leu, whereas phosphorylation of glycerol (19) leads to the lipid precursor glycerol-1-phosphate (21). (**c**) *Copper*(I)-catalyzed cross-coupling of hydrogen cyanide (11) and acetylene (32) gives acrylonitrile (33), reductive homologation of which gives precursors of Pro and Arg. (**d**) *Copper*(II)-driven oxidative cross-coupling of hydrogen cyanide (11) and acetylene (32) gives cyanoacetylene (6), which serves as a precursor to Asn, Asp, Gln, and Glu. (From Patel et al. (2015). Used by permission)

Problems

1. *Mineral surfaces.* One theory of prebiotic molecular evolution postulates that primitive biochemical reactions took place on the surfaces of mineral crystals.

    (a) How would this influence the stereochemistry of biochemical reactions?
    If the surface of the crystal was anisotropic, then molecular interactions between substrates and the crystal surface could energetically discriminate between stereoisomers of prebiotic molecules and result in the formation of specific product isomers.
    (b) How could this prevent the loss of chemical bond energy as heat?
    If two prebiotic molecules bound to the surface of a crystal at juxtaposing sites, a chemical reaction catalyzed by the crystal surface which might cleave one of the molecules could release the reaction product close enough for the second molecule to pick up the cleaved group. In this way, the reaction products could not diffuse away before forming the product bond.
    (c) How does the surface of the crystal influence the rates of diffusion of the reactants and products?

2. *Dielectric trapping.* What is the probability of a calcium ion leaking out of a liposome as shown in Fig. 12.27 at 298 K? Is this the same as the probability of a calcium ion entering the low dielectric constant center of a micelle?

3. *Dielectric traversal.* Does the lipid bilayer of a liposome present a barrier to water penetration?

4. *Evolution and diffusion.* Mammalian cells are on the order of 15 μm in diameter. How does this small size provide an advantage with respect to the distances needed for molecules to diffuse from enzyme to enzyme, the speed of biochemical reactions, and the times involved?

5. *Solar photons and biomolecules.* What are the energies of 500-nm solar photons, and what are the energies needed to induce electronic transitions in biochemical reactions? What conclusions can you draw from these concepts? Can you estimate the annual flux of photons on the early Earth? Remember to account for the diurnal variation and the seasons.

References

M.J. Abplanalp, R. Frigge, R.I. Kaiser, Low-temperature synthesis of polycyclic aromatic hydrocarbons in Titan's surface ices and on airless bodies. Sci. Adv. **5**, eaaw5841 (2019)

K. Altwegg, Prebiotic chemicals—Amino acid and phosphorus—In the coma of comet 67P/Churyumov-Gerasimenko. Sci. Adv. **2**, e1600285 (2016). https://doi.org/10.1126/sciadv.1600285

J.L. Bada, New insights into prebiotic chemistry from Stanley Miller's spark discharge experiments. Chem. Soc. Rev. **42**, 2186 (2013)

G. Banfalvi, Why ribose was selected as the sugar component of nucleic acids. DNA Cell Biol. **25**, 189 (2006)

J. Banville, *Kepler* (David R. Gordine, Boston, 1981), p. 145. Used by permission

S. Becker, I. Thoma, A. Deutsch, T. Gehrke, P. Mayer, H. Zipse, T. Carell, A high-yielding, strictly regioselective prebiotic purine nucleoside formation pathway. Science **352**, 833 (2016)

A. Belloche, R.T. Garrodholger, S.P. Muller, K.M. Menten, Detection of a branched alkyl molecule in the interstellar medium: Iso-propyl cyanide. Science **345**, 1584–1587 (2014)

A. Biscans, Exploring the emergence of RNA nucleosides and nucleotides on the early Earth. Life **8**, 57 (2018)

D. Bitounis, R. Fanciullino, A. Iliadis, J. Ciccolini, Optimizing druggability through liposomal formulations: New approaches to an old concept. ISRN Pharm. **2012**, 738432 (2012). https://doi.org/10.5402/2012/738432

N. Biver, D. Bockelée-Morvan, R. Moreno, J. Crovisier, P. Colom, D.C. Lis, A. Sandqvist, J. Boissier, D. Despois, S.N. Milam, Ethyl alcohol and sugar in comet C/2014 Q2 (Lovejoy). Sci. Adv. **1**, e1500863 (2015). https://doi.org/10.1126/sciadv.1500863

N.T. Bobrovnikoff, Halley's comet in its appearance of 1909-1911. Publ. Lick Obs. **XVII**, 309–482 (1931)

R. Breslow, On the mechanism of the formose reaction. Tetrahedron Lett. **1**, 22–26 (1959)

B. Burcar, A. Castañeda, J. Lago, M. Daniel, M.A. Pasek, N.V. Hud, T.M. Orlando, C. Menor-Salván, A stark contrast to modern earth: Phosphate mineral transformation and nucleoside phosphorylation in an iron-and cyanide-rich early earth scenario. Angew. Chem. Int. Ed. **58**, 16981–16987 (2019)

A. Butlerow, Bildung einer zuckerartigen substanz durch synthese. Liebigs Ann. Chem. **120**, 295–298 (1861)

M.L. Cable, S.M. Hörst, R. Hodyss, P.M. Beauchamp, M.A. Smith, P.A. Willis, Titan tholins: Simulating titan organic chemistry in the Cassini-Huygens era. Chem. Rev. **112**, 1882–1909 (2012)

M.P. Callahan, K.E. Smith, H.J. Cleaves, J. Ruzicka, J.C. Stern, D.P. Glavin, C.H. House, J.P. Dworkin, Carbonaceous meteorites contain a wide range of extraterrestrial nucleobases. Proc. Natl. Acad. Sci. USA **108**, 13995–13998 (2011)

J. Cami, J. Bernard-Salas, E.L.S. Peeters, S.E. Malek, Detection of C60 and C70 in a young planetary nebula. Science **329**, 1180 (2010)

J. Cernicharo, A.M. Heras, A.G.G.M. Tielens, J.R. Pardo, F. Herpin, M. Guélin, L.B.F.M. Waters, Infrared space observatory's discovery of C_4H_2, C_6H_2, and benzene in CRL 618. Astrophys. J. **546**, L123 (2001)

R. Chau, S. Hamel, W.J. Nellis, Chemical processes in the deep interior of Uranus. Nat. Commun. **2**, 203 (2011)

F. Chen, D. Yang, Condensation of amino acids to form peptides in aqueous solution induced by the oxidation of sulfur(iv): An oxidative model for prebiotic peptide formation. Orig. Life Evol. Biosph. **37**, 47–54 (2007)

H.J. Cleaves II, C. Neish, M.P. Callahan, E. Parker, F.M. Fernández, J.P. Dworkin, Amino acids generated from hydrated Titan tholins: Comparison with Miller–Urey electric discharge products. Icarus **237**, 182 (2014)

S.J. Clemett, S.A. Sandford, K. Nakmura-Menger, F. Hoan, D.S. McKay, Complex aromatic hydrocarbons in Stardust samples collected from comet 81P/Wild 2. Meteorit. Planet. Sci. **45**, 701 (2010)

G.D. Cody, C.M.O.'.D. Alexander, F. Tera, Solid-state (^{1}H and ^{13}C) nuclear magnetic resonance spectroscopy of insoluble organic residue in the Murchison meteorite: A self-consistent quantitative analysis. Geochim. Cosmochim. Acta **66**, 1851–1865 (2002)

G.D. Cody, E. Heying, C.M.O. Alexander, L.R. Nittler, A.L. David Kilcoyne, S.A. Sandford, R.M. Stroud, Establishing a molecular relationship between chondritic and cometary organic solids. Proc. Natl. Acad. Sci. USA **108**, 19171–19176 (2011)

A.J. Coggins, M.W. Powner, Prebiotic synthesis of phosphoenolpyruvate by α-phosphorylation-controlled triose glycolysis. Nat. Chem. **9**, 310–317 (2017)

G.W. Cooper, W.M. Onwo, J.R. Cronin, Alkyl phosphonic acids and sulfonic acids in the Murchison meteorite. Geochim. Cosmochim. Acta **56**, 4109–4115 (1992)

D.W. Deamer, Membrane compartments in prebiotic evolution, in *The Molecular Origins of Life: Assembling Pieces of the Puzzle*, ed. by A. Brack, (Cambridge University Press, Cambridge, 2000), pp. 189–205

P. Decker, H. Schweer, R. Pohlamnn, Bioids: X. Identification of formose sugars, presumable prebiotic metabolites, using capillary gas chromatography/gas chromatography—Mass spectrometry of n-butoxime trifluoroacetates on OV-225. J. Chromatogr. A **244**, 281 (1982)

S. Derenne, C. Coelho, C. Anquetil, C. Szopa, A.S. Rahman, P.F. McMillan, F. Corà, C.J. Pickard, E. Quirico, C. Bonhomme, New insights into the structure and chemistry of Titan's tholins via ^{13}C and ^{15}N solid state nuclear magnetic resonance spectroscopy. Icarus **221**, 844–853 (2012)

R.E. Dickertson, Base sequence and helix structure variation in B- and A-DNA. J. Mol. Biol. **166**, 419–441 (1982)

J. Dworkin, D. Deamer, S. Sandford, L. Allamandola, Self-assembling amphiphilic molecules: Synthesis in simulated interstellar/precometary ices. Proc. Natl. Acad. Sci. USA **98**, 815–819 (2001)

P. Ehrenfreund, S. Rasmussen, J. Cleaves, L. Che, Experimentally tracing the key steps in the origin of life: The aromatic world. Astrobiology **6**, 49 (2006)

J.E. Elsila, D.P. Glavin, J.P. Dworkin, Cometary glycine detected in samples returned by stardust. Meteorit. Planet. Sci. **44**, 1323–1330 (2009)

A. Eschenmoser, M. Dobler, Warum Pentose- und nicht Hexose-Nucleinsauren Teil I. Einleitung und Problemstellung, Korformationanalyse für Oligonucleotid Ketten aus 2′-3′--Dideoxyglucopyranosyl-Barsteinen (Homo-DNS) sowie Betrachtungen zur Kon-formation von A- und B-DNS. Helv. Chim. Acta **75**, 218–259 (1992)

J.P. Ferris, L.E. Orgel, Aminomalononitrile and 4-amino-5-cyanoimidazole in hydrogen cyanide polymerization and adenine synthesis. J. Am. Chem. Soc. **87**, 4976–4977 (1965)

J.P. Ferris, L.E. Orgel, An unusual photochemical rearrangement in the synthesis of adenine from hydrogen cyanide. J. Am. Chem. Soc. **88**, 1074 (1966)

J.P. Ferris, A.R. Hill, R. Liu, L.E. Orgel, Synthesis of long prebiotic oligomers on mineral surfaces. Nature **381**, 59 (1996)

S.W. Fox, K. Harada, Thermal copolymerization of amino acids to a product resembling protein. Science **128**, 1214 (1958)

J. Francisco, C. Mayén, J. Rydzewski, N. Szostak, J. Blazewicz, W. Nowak, Prebiotic soup components trapped in montmorillonite nanoclay form new molecules: Car-Parrinello *ab initio* simulations. Life **9**, 46 (2019)

W.D. Fuller, R.A. Sanchez, L.E. Orgel, Studies in prebiotic synthesis: VI. Synthesis of purine nucleosides. J. Mol. Biol. **67**, 25–33 (1971)

Y. Furukawa, Y. Chikaraishi, N. Ohkouchi, N.O. Ogawa, D.P. Glavin, J.P. Dworkin, C. Abe, T. Nakamura, Extraterrestrial ribose and other sugars in primitive meteorites. Proc. Natl. Acad. Sci. USA **116**, 24440–24445 (2019)

C. Gibard, S. Bhowmik, M. Karki, E.-K. Kim, R. Krishnamurthy, Phosphorylation, oligomerization and self-assembly in water under potential prebiotic conditions. Nat. Chem. **10**, 212–217 (2018)

R.J. Gillams, T.Z. Jia, Mineral surface-templated self-assembling systems: Case studies from nanoscience and surface science towards origins of life research. Life **8**, 10 (2018). https://doi.org/10.3390/life8020010

D.P. Glavin, A.D. Aubrey, M.P. Callahan, J.P. Dworkin, J.E. Elsila, E.T. Parker, J.L. Bada, P. JEenniskens, M.H. Shaddad, Extraterrestrial amino acids in the Almahata Sitta meteorite. Meteorit. Planet. Sci. **45**, 1695–1709 (2010)

F. Goesmann et al., Organic compounds on comet 67P/Churyumov-Gerasimenko revealed by COSAC mass spectrometry. Science **349**, aab0689 (2015)

J.M. Greenberg, Making a comet nucleus. Astron. Astrophys. **330**, 375 (1998)

J. Groen, D.W. Deamer, A. Kros, P. Ehrenfreund, Polycyclic aromatic hydrocarbons as plausible prebiotic membrane components. Orig. Life Evol. Biosph. **42**, 295–306 (2012)

M.S. Gudipati, R. Yang, In-situ probing of radiation-induced processing of organics in astrophysical ice analogs—Novel laser desorption laser ionization time of flight mass spectroscopic studies. Astrophys. J. Lett. **756**, L24 (2012)

W.R. Hargreaves, S.J. Mulvihill, D.W. Deamer, Synthesis of phospholipids and membranes in prebiotic conditions. Nature **266**, 78–80 (1977)

H. Hashizume, Adsorption of nucleic acid bases, ribose, and phosphate by some clay minerals. Life **5**, 637–650 (2015)

S.M. Hörst, R.V. Yelle, A. Buch, N. Carrasco, G. Cernogora, O. Dutuit, E. Quirico, E. Sciamma-O'Brien, M.A. Smith, A. Somogyi, C. Szopa, R. Thissen, V. Vuitton, Formation of amino acids and nucleotide bases in a Titan atmosphere simulation experiment. Astrobiology **12**, 809 (2012)

T.A.E. Jakschitz, B.M. Rode, Chemical evolution from simple inorganic compounds to chiral peptides. Chem. Soc. Rev. **41**, 5484–5489 (2012)

A.P. Johnson, H.J. Cleaves, J.P. Dworkin, D.P. Glavin, A. Lazcano, J.L. Bada, The Miller volcanic spark discharge experiment. Science **322**, 404 (2008)

P.C. Joshi, M.F. Aldersley, J.W. Delano, J.P. Ferris, Mechanism of montmorillonite catalysis in the formation of RNA oligomers. J. Am. Chem. Soc. **131**, 13369–13374 (2009)

R. Kierzek, L. He, D.H. Turner, Association of $2'$-$5'$ oligoribonucleotides. Nucleic Acids Res. **20**, 1685 (1982)

T. Koga, H. Naraoka, A new family of extraterrestrial amino acids in the Murchison meteorite. Nat. Sci. Rep. **7**, 636 (2017)

H.R. Kricheldorf, Polypeptides and 100 years of chemistry of α-amino acid N-carboxyanhydrides. Angew. Chem. Int. Ed. **45**, 5752–5784 (2006)

J.F. Lambert, Adsorption and polymerization of amino acids on mineral surfaces: A review. Orig. Life Evol. Biosph. **38**, 211–242 (2008)

Y. Layssac, A. Gutiérrez-Quintanilla, T. Chiavassa, F. Duvernay, Detection of glyceraldehyde and glycerol in VUV processed interstellar ice analogs containing formaldehyde: A general formation route for sugars and polyols. Mon. Not. R. Astron. Soc. **496**, 5292–5307 (2020)

M.S. Lee, S. Scandolo, Mixtures of planetary ices at extreme conditions. Nat. Commun. **2**, 185 (2011)

L. Leman, L. Orgel, M.R. Ghadiri, Carbonyl sulfide–mediated prebiotic formation of peptides. Science **306**, 283 (2004)

R. Lohrmann, L.E. Orgel, Prebiotic synthesis: Phosphorylation in aqueous solution. Science **161**, 64–66 (1968)

M. Lopez-Puertas, B.M. Dinelli, A. Adriani, B. Funke, M. Garcıa-Comas, M.L. Moriconi, E. D'Aversa, C. Boersma, L.J. Allamandola, PAH's in Titan's upper atmosphere. Astrophys. J. **770**, 132 (2013)

N. Madhusudhan, S. Sarkar, S. Constantinou, M. Holmberg, A.A.A. Piette, J.I. Moses, Carbon-bearing molecules in a possible Hycean atmosphere. Astrophys. J. Lett. **956**(1), L13 (2023)

Z. Martins, The nitrogen heterocycle content of meteorites and their significance for the origin of life. Life **8**, 28 (2018). https://doi.org/10.3390/life8030028

C.K. Materese, M. Nuevo, S.A. Sandford, The formation of nucleobases from the ultraviolet photoirradiation of purine in simple astrophysical ice analogues. Astrobiology **17**, 761–777 (2017)

T.M. McCollom, G. Ritter, B.R. Simoneit, Lipid synthesis under hydrothermal conditions by Fischer-Tropsch-type reactions. Orig. Life Evol. Biosph. **29**, 153–166 (1999)

D. McDonald, D. Gene, W. R. Thompson, M. Heinrich, N. B. Khare, C. Sagan et al. Chemical Investigation of Titan and Triton Tholins Icarus. **108** (1), 137–145 (1994); https://doi.org/10.1006/icar.1994.1046

C. Meinert, I. Myrgorodska, P. de Marcellus, T. Buhse, L. Nahon, S.V. Hoffmann, L.S. d'Hendecourt, U.J. Meierhenrich, Ribose and related sugars from ultraviolet irradiation of interstellar ice analogs. Science **352**, 208–212 (2016)

C. Menor-Salván, D.M. Ruiz-Bermejo, M.I. Guzmán, S. Osuna-Esteban, S. Veintemillas-Verdaguer, Synthesis of pyrimidines and triazines in ice: Implications for the prebiotic chemistry of nucleobases. Chemistry **15**, 4411–4418 (2009)

S.L. Miller, A production of amino acids under possible primitive earth conditions. Science **117**, 528 (1953)

S.L. Miller, Production of some organic compounds under possible primitive earth conditions. J. Am. Chem. Soc. **77**, 2351 (1955)

S.L. Miller, The formation of organic compounds on the primitive Earth. Ann. N. Y. Acad. Sci. **69**, 260 (1957)

M. Moriconi, E. D'Aversa, C. Boersma, L.J. Allamandola, Large abundances of polycyclic aromatic hydrocarbons in Titan's upper atmosphere. Astrophys. J. **770**, 132 (2013). https://doi.org/10.1088/0004-637X/770/2/132

H. Naraoka et al., Soluble organic molecules in samples of the carbonaceous asteroid (162173) Ryugu. Science **379**, 789 (2023)

A. Navrotsky, R. Hervig, J. Lyons, D.-K. Seo, E. Shock, A. Voskanyan, Cooperative formation of porous silica and peptides on the prebiotic Earth. Proc. Natl. Acad. Sci. USA **118**, e2021117118 (2021). https://doi.org/10.1073/pnas.2021117118

C.D. Neish, A. Somogyi, M.A. Smith, Titan's primordial soup: Formation of amino acids via low-temperature hydrolysis of tholins. Astrobiology **10**, 337–347 (2010). https://doi.org/10.1089/ast.2009.0402

H. Niemann, S. Atreya, et al., The abundances of constituents of Titan's atmosphere from the GCMS instrument on the Huygens probe. Nature **438**, 779–784 (2005)

Y. Oba et al., Uracil in the carbonaceous asteroid (162173) Ryugu. Nat. Commun. **14**, 1292 (2023)

A. Omran, Plausibility of the formose reaction in alkaline hydrothermal vent environments. Orig. Life Evol. Biosph. **53**, 113–125 (2023)

J. Oró, Mechanism of synthesis of adenine from hydrogen cyanide under possible primitive Earth conditions. Nature **191**, 1193–1194 (1961)

J. Oró, A.P. Kimball, Synthesis of purines under possible primitive Earth conditions I. Adenine from hydrogen cyanide. Arch. Biochem. Biophys. **94**, 217–227 (1961)

E.T. Parker, H.J. Cleaves, J.P. Dworkin, D.P. Glavin, M. Callahan, A. Aubrey, A. Lazcano, J.L. Bada, Primordial synthesis of amines and amino acids in a 1958 Miller H_2S-rich spark discharge experiment. Proc. Natl. Acad. Sci. USA **108**, 5526 (2011)

B.H. Patel, C. Percivalle, D.J. Ritson, C.D. Duffy, J.D. Sutherland, Common origins of RNA, protein and lipid precursors in a cyanosulfidic protometabolism. Nat. Chem. **7**, 301–307 (2015)

J.P. Pinto, G.R. Gladstone, Y.L. Yung, Photochemical production of formaldehyde in Earth's primitive atmosphere. Science **210**, 183–185 (1980)

S. Pizzarello, E. Shock, The organic composition of carbonaceous meteorites: The evolutionary story ahead of biochemistry. Cold Spring Harb. Perspect. Biol. **2**, a002105 (2010)

M. Preiner, J.C. Xavier, F.L. Sousa, V. Zimorski, A. Neubeck, S.Q. Lang, H.C. Greenwell, K. Kleinermanns, Serpentinization: Connecting geochemistry, ancient metabolism and industrial hydrogenation. Life **8**, 41 (2018). https://doi.org/10.3390/life8040041

M. Rao, J. Eichberg, J. Oró, Synthesis of phosphatidylethanolamine under possible primitive earth conditions. J. Mol. Evol. **25**, 1–6 (1987)

F. Raulin, C. Brassé, O. Poch, P. Coll, Prebiotic-like chemistry on Titan. Chem. Soc. Rev. **41**, 5380–5393 (2012)

T. P. Ray, M. J. McCaughrean, A. Caratti o Garatti et al. Outflows from the youngest stars are mostly molecular. Nat. **622**, 48–52 (2023). https://doi.org/10.1038/s41586-023-06551-1

A. Ricardo, M.A. Carrigan, A.N. Olcott, S.A. Benner, Borate minerals stabilize ribose. Science **303**, 196 (2004)

A. Rimola, M. Sodupe, P. Ugliengo, Deep-space glycine formation via Strecker-type reactions activated by ice water dust mantles. A computational approach. Phys. Chem. Chem. Phys. **12**, 5285–5294 (2010)

D.J. Ritson, S.J. Mojzsis, J.D. Sutherland, Supply of phosphate to early Earth by photogeochemistry after meteoritic weathering. Nat. Geosci. **13**, 344–348 (2020)

V.M. Rivilla, I. Jiménez-Serra, J. Martín-Pintado, C. Briones, L.F. Rodríguez-Almeida, F. Rico-Villas, B. Tercero, S. Zeng, L. Colzi, P. de Vicente, S. Martín, M.A. Requena-Torres, Discovery in space of ethanolamine, the simplest phospholipid head group. Proc. Natl. Acad. Sci. USA **118**, e2101314118 (2021)

M.P. Robertson, S.L. Miller, An efficient prebiotic synthesis of cytosine and uracil. Nature **375**, 772–774 (1995)

C. Sagan, B. Khare, Tholins: Organic chemistry of interstellar grains and gas. Nature **277**, 102–107 (1979)

R.A. Sanchez, L.E. Orgel, Studies in prebiotic synthesis. V. Synthesis and photoanomerization of pyrimidine nucleosides. J. Mol. Biol. **47**, 531–543 (1970)

S.A. Sandford, M. Nuevo, P.P. Bera, T.J. Lee, Prebiotic astrochemistry and the formation of molecules of astrobiological interest in interstellar clouds and protostellar disks. Chem. Rev. **120**, 4616–4659 (2020)

C. Scharf, L. Cronin, Quantifying the origins of life on a planetary scale. Proc. Natl. Acad. Sci. USA **113**, 8127–8132 (2016)

M.A. Sephton, Organic compounds in carbonaceous meteorites. Nat. Prod. Rep. **19**, 292–311 (2002)

J. Tennyson, Molecules in space, in *Handbook of Molecular Physics and Quantum Chemistry*, ed. by S. Wilson, vol. 3, (Wiley, Chichester, 2003), pp. 356–369

H.C. Urey, On the early chemical history of the earth and the origin of life. Proc. Natl. Acad. Sci. USA **38**, 351 (1952)

F. Wöhler, Ueber könstliche bildung des harnstoffs: Annalen der Physik und Chemie. **88**, 2 253–256 (1828)

S. Xu, B. Zuckerman, P. Dufour, E.D. Young, B. Klein, M. Jura, The chemical composition of an extrasolar Kuiper-Belt-Object. Astrophys. J. Lett. **836**, L7 (2017)

H. Yabuta, H. Naraoka, K. Sakanishi, H. Kawashima, Solid-state ^{13}C NMR characterization of insoluble organic matter from Antarctic CM2 chondrites: Evaluation of the meteoritic alteration level. Meteorit. Planet. Sci. **40**, 779–787 (2005)

L. Zhao, R.I. Kaiser, B. Xu, U. Ablikim, M. Ahmed, M.M. Evseev, E.K. Bashkirov, V.N. Azyazov, A.M. Mebel, Low-temperature formation of polycyclic aromatic hydrocarbons in Titan's atmosphere. Nat. Astron. **2**, 973–979 (2018)

C. Zhu, A.M. Turner, M.J. Abplanalp, R.I. Kaiser, B. Webb, G. Siuzdak, R.C. Fortenberry, An interstellar synthesis of glycerol phosphates. Astrophys. J. Lett. **899**, L3 (2020)

Bibliography

C.S. Cockell, *Astrobiology: Understanding Life in the Universe*, 2nd edn. (Wiley, 2020)

D.W. Deamer, J.W. Szostak, *The Origins of Life*, Cold Spring Harbor Perspectives in Biology (Cold Spring Harbor Laboratory Press, 2010)

I. Fry, *The Emergence of Life on Earth* (Rutgers University Press, New Brunswick, 2000)

D.E. Greenwalt, *Remnants of Ancient Life: The New Science of Old Fossils* (Princeton University Press, Princeton, 2022)

R. Parthasarathy, *So Simple a Beginning: How Four Physical Principles Shape Our Living World* (Princeton University Press, Princeton, 2022)

D.A. Rothery, I. Gilmour, M.A. Sephton (eds.), *An Introduction to Astrobiology*, 3rd edn. (Cambridge University Press, Cambridge, 2018)

J.W. Schopf (ed.), *Life's Origin: The Beginnings of Biological Evolution* (University of California Press, Berkeley, 2002)

E. Schrodinger, *What Is Life?* (Cambridge University Press, Cambridge, 2018)

E. Smith, H.J. Morowitz, *The Origin and Nature of Life on Earth: The Emergence of the Fourth Geosphere* (Cambridge University Press, Cambridge, 2016)

Appendices

Appendix 1: A Brief Introduction to the Syntax
of `Mathematica`

A complete description of the **Mathematica** computer mathematics system can be found in *The Mathematica Book*, 5th Edition, by Stephen Wolfram published by Wolfram Media (ISBN: 1-57955-022-3). For the purposes of working the problems, the following will likely get you started. An inexpensive student edition of **Mathematica** is available via download from http://www.wolfram.com. This web site also contains a large amount of help and many other useful bits, such as detailed descriptions of the syntax of all of the functions. One can access the help system by highlighting the function of interest and copying it into the search bar.

Once the software is invoked on your computer, you are presented with a blank screen, which can be regarded as an intelligent piece of paper, or workspace, onto which you can write equations that **Mathematica** will solve for you. You can also produce plots of your equations and data. The input into **Mathematica** is labeled with line numbers

```
In [1] =2 + 2
```

and is divided into cells, whose boundaries are shown on the right as light blue brackets. The resulting output

```
Out [1] =4
```

is also similarly labeled. One causes **Mathematica** to evaluate an expression by typing Shift-Enter with the cursor within the cell. There are a number of important special characters which perform arithmetic operations. A partial list includes the following:

^ = **exponentiantion**, **X^2 is X**^{2}.
/ = **division**, a/b is$\frac{a}{b}$.
* = **multiplication**, **a*b is a times b**.
blank = **multiplication**, **a b is a times b**.
( ) = grouping of terms to enforce the order of computation.
[ ] = argument of a function; sin(x) is written as **Sin[x]**.

Note that **sin(x)** will not work and that all **Mathematica** functions begin with an upper case letter:

{ } = comma separated lists, **Plot[Sin[x],{x,0,Pi}]**, or **{1,2,3,4}**.

% = results of the previous calculation, **Show[%]**, also Out[3] works.

```
In [2]=2 + 2
Out [2]=4
In [3]= %^2
Out [3]=16
In [4]= Out [3]/2
Out [4]= 8
```

! = factorial, 6! = 6 5 4 3 2 1 = 120.
=means assign, **a** = 5 is assign the value 5 to **a**.
=means assign in a function, **f [x_]= Sin[x]/x,** this is how to make a user-defined function.
==means test if the left hand side equals the right hand side, and produce a logical result,

```
In [5]:= a = 5;
In [6]:= a == 5
Out [6]:= True
```

!= **means test if not equal to,**

```
In [7]:= a != 6
Out [7]:= True
```

; **means do not display the result,**

```
In [8]:= b = 7;
In [9]:= c = 8
Out [9]:= 8
```

Now, **Mathematica** can perform both numerical and symbolic computations. For example,

```
In [10]:= 27 + 8
Out [10]:= 35
In [11]:= 27 x + 8 x
Out [11]:= 35x
```

Or

```
In [12]:= (3 + 7)^5
Out[12]:= 100000
In [13]:= (a + b)^5
Out[13]:= (a + b)^5
In [14]:= %//Expand
Out[14]:= a^5 + 5a^4 b + 10a^3 b^2 + 10a^2 b^3 + 5ab^4 + b^5.
```

Here, we see an example of the use of **%** to recall a prior result and the use of the postfix operator, **//**, to perform additional calculations on a result. We could also have written this as **Expand[%]**, with the same result.

We will often make use of **Mathematica's** abilities to perform analytic integration and differentiation. For example,

```
In [15]:= Integrate[1/(1 + x^2),x]
Out[15] = ArcTan[x]
In [16]:= Integrate[1/(1 + x^2),{x, 0, Infinity}]
Out[16]:= Pi/2.
```

Mathematica will often produce a symbolic result, when we actually wish to have a numerical answer. We can use the **N[ ]** function to convert one into the other:

```
In [17] = Sqrt[12]
Out[17] = 2√3
In [18] = N[%,12]
Out[18] = 3.46410161514
```

Here, we requested 12 significant figures for the result of **N[ ]**.

Plotting can be done to produce 1, 2, and 3D postscript plots.

```
In [19], theplt=Plot[Sin[5 x]Exp[-x/10],{x,0,25},PlotStyle
         -> Red]
```

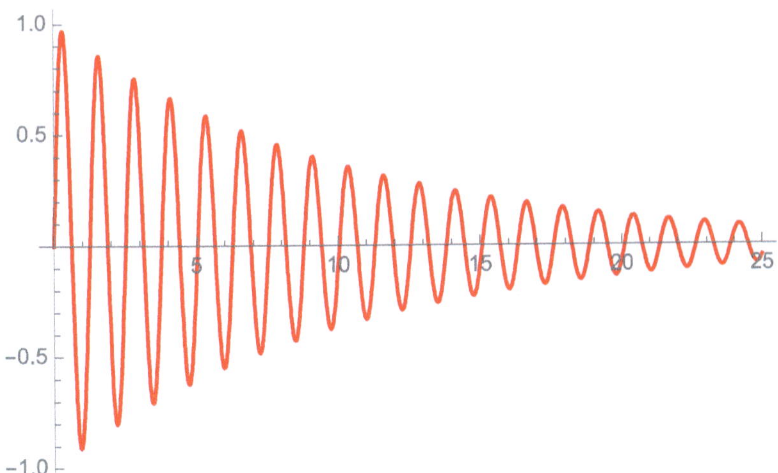

```
Out [19]= -Graphics- (which can be exported in a wide variety of formats)
In [20]= Export ["/space/user/docs/theplt.tiff",theplt]
Out [20]= /space/user/docs/theplt.tiff.
```

Plots can be extensively customized in terms of labels, colors, overlays, fonts, axes, etc.

Help can be gotten in several ways. If one just wishes to know the syntax of a function, like **Plot**, then one can use

```
In [21]= ?Plot
```

```
Plot [f, {x, xmin, xmax}] generates a plot of f as a function of x from
xmin to xmax. Plot [{f1, f2, ... }, {x, xmin, xmax}] plots several
functions fi. More...
```

and more information is available by clicking on the More... or by using **??Plot**. One can also use the online help system and the help from the web at http://www.wolfram.com.

Finally, the construction, **?*Plot***, will treat the * as a wild card and produce a list of all **Mathematica** functions with **Plot** contained within them: **ParamericPlot, ContourPlot, Plot3D, etc.**

User-defined functions are often quite useful in computation (note the use of square brackets [ ], and not parentheses ( ) here, as well as the "set delayed" =, rather than just =, and the important underscore **x_** that tells **Mathematica** to treat x on the right-hand side as a dependent variable):

```
In [21]=f [x_] = Cos [x] + I Sin [x]
In [22]:= f [Pi]
Out [22]:= -1
```

Units are understood as well. The volume of a cylinder is $V = \pi r^2 h$:

```
In [23]= vol [r_, h_]= Pi r^2 h
In [24]= vol [3,5]
Out [24]= 4 Pi
```

```
Let r and h be measured in meters, then
```

```
In [25]= vol [3 meters,5 meters]
Out [25]= 4 Pi meters^3
```

Equations can be solved using the **Solve []** function:

```
In [26]= Solve [x^2 - 2 x == 0,x]
Out [26]= {{x -> 2},{x -> 0}}
```

And we can perform integration and differentiation. Find $\int x^2 dx$

```
In[27]= Integrate[x^2,x]
```
$$\text{Out}[27]=\tfrac{x^3}{3}.$$

And what is $\frac{d}{dx}x^2$?

```
In[28]= D[x^2,x]
Out[28]= 2x
```

We can also take limits:

```
In[29]= Limit[Sin[x]/x,x -> 0]
Out[29]= 1
```

Solving differential equations can also include the boundary or initial conditions. In this case, the initial condition is $v[0] = 0$, and the differential equation and its initial condition are surrounded by curly brackets, as a list:

```
In[30]= DSolve[{m v'[t] ==F - f v[t], v[0]==0},v[t],t]
```
$$\text{Out}[30]=\left\{\left\{v[t] ->\frac{e^{-\frac{ft}{m}}\left(-1+e^{\frac{ft}{m}}\right)F}{f}\right\}\right\}$$

```
In[31]= %//Simplify
```
$$\text{Out}[31]=\left\{\left\{v[t] ->\frac{F-e^{-\frac{ft}{m}}F}{f}\right\}\right\}$$

One can also read in empirical data, plot it, and fit it to a trial function. For example, the viscosity of water is quite temperature-dependent in the range of temperature between the freezing point of water and slightly above body temperature, 310 K. Let us read in a data file containing pairs of points consisting of {temperature, viscosity} stored on a local disk file:

```
In[32]= viscosity = ReadList["/user/data/viscosity.h2O.
   txt", {Number, Number}]
```

Here, the construct {Number, Number} is used to explicitly tell **Mathematica** to parse the data as a list of pairs of real numbers. The output is the list of lists:

```
Out[32]={{273.16, 0.01787}, {274.16, 0.01728}, {275.16,
   0.01671}, {276.16,
   0.01618}, {277.16, 0.01567}, {278.16, 0.01519}, {279.16,
   0.01472}, {280.16, 0.01428}, {281.16, 0.01386}, {282.16,
   0.01346}, {283.16, 0.01307}, {284.16, 0.01271}, {285.16,
   0.01235}, {286.16, 0.01202}, {287.16, 0.01169}, {288.16,
   0.01139}, {289.16, 0.01109}, {290.16, 0.01081}, {291.16,
   0.01053}, {292.16, 0.01027}, {293.16, 0.01002}, {294.16,
   0.009779}, {295.16, 0.009548}, {296.16, 0.009325}, {297.16,
   0.009111}, {298.16, 0.008904}}
```

Now, let's plot this data as an x–y plot:

```
In[33]= plotviscosity = ListPlot[viscosity, PlotStyle ->
    Red, AxesLabel -> {"Temperature K", "Viscosity cP"}]]
```

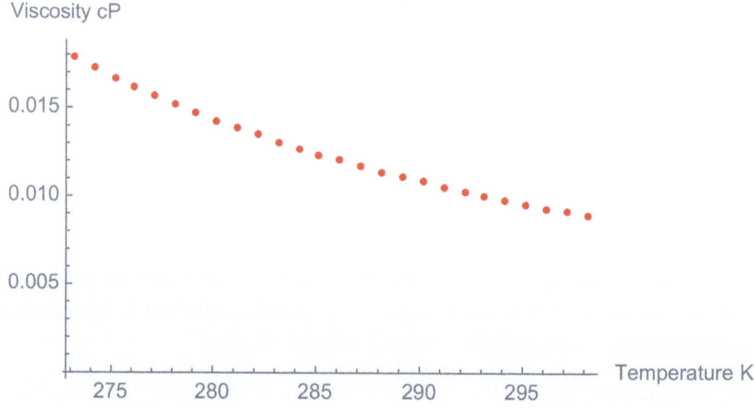

```
Out[33]= -Graphics-
```

Then, we wish to fit this data to a third-order polynomial and plot the fit and the data together on one plot. First, we compute the fit to the polynomial:

```
In[34]= cubicfit[x_]= Fit[viscosity, {1, x, x^2, x^3}, x]
Out[34]=4.042195270698858   -   0.03946020454023558   x   +
    0.0001292919179796497 x^2- 1.4192198772089717*10^-7 x^3
In[35]= tempplot = Plot[cubicfit[T], {T, 273, 299}]
```

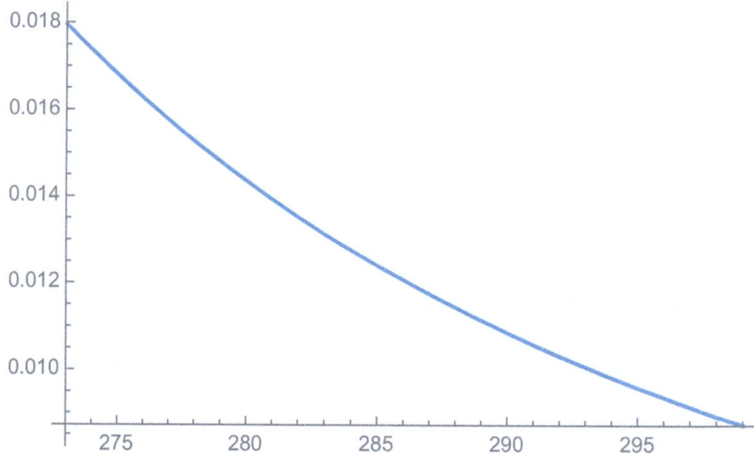

```
Out[35]= -Graphics-
```

Now, we can overlay the data onto the fit as

In[36]= **Show[tempplot, plotviscosity]**

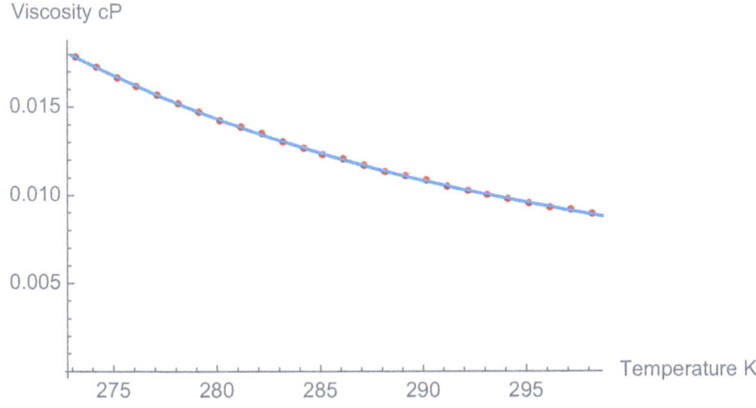

Out[36]= **-Graphics-**

and we can use the parameters of the fit to compute the viscosity for any given temperature within the defined range:

In[37]= **cubicfit[295.16]**
Out[37]= **0.00955286**

We see that **Mathematica** is capable of handling the wide variety of mathematical problems encountered in the study of abiogenesis. The student is urged to use this software extensively; many of the chapter problems cannot be done without it, and I have not yet found a problem that cannot be solved with this software. There are also many books with ample examples. A partial list of these includes (and of course there will be updated editions that supersede these) the following:

T.W. Gray, J. Glynn, *The Beginner's Guide to Mathematica*, version 2 (Addison-Wesley, Menlo Park, 1992). ISBN 0-201-58221-X

B.F. Torrence, E.A. Torrence, *The Student's Introduction to Mathematica*, 2nd edn. (Cambridge, New York, 2009). ISBN 978-0-521-71789-2

G. Baumann, *Mathematica for Theoretical Physics*, 2nd edn. (Springer, New York, 2005). ISBN 0-387-21933-1

R.E. Crandall, *Mathematica for the Sciences* (Addison-Wesley, Redwood City, 1991). ISBN 0-201-51001-4

S. Mangano, *Mathematica Cookbook* (O'Reilly, Sebastapol, 2010)

P.T. Tam, *A Physicists Guide to Mathematica* (Academic Press, San Diego, 1997). ISBN-13: 978-0-12-683190-0

R.L. Zimmerman, F.I. Olness, *Mathematica for Physics*, 2nd edn. (Addison-Wesley, Redwood City, 2002). ISBN 0-8053-8700-5

R. Maeder, *Programming in Mathematica*, 3rd edn. (Addison-Wesley, Redwood City, 1997). ISBN 0-201-85449-X

Appendix 2: NMR Frequency Table

One can find these on the web at, for example, http://web.mit.edu/speclab/www/ nmrfreq.html, and their frequencies, nuclear spins, and natural isotopic abundances are given in Table A.2.

Table A.2 NMR frequency table

Isotope	Spin	Abundance (%)	NMR frequency (MHz) at field (T)		
			5.8717	7.0460	11.7434
^{1}H	1/2	99.98	250.000	300.000	500.000
^{2}H	1	1.5×10^{-2}	38.376	46.051	76.753
^{3}H	1/2	0	266.658	319.990	533.317
^{3}He	1/2	1.3×10^{-4}	190.444	228.533	380.888
^{6}Li	1	7.42	36.789	44.146	73.578
^{7}Li	3/2	92.58	97.158	116.590	194.317
^{9}Be	3/2	100	35.133	42.160	70.267
^{10}B	3	19.58	26.866	32.239	53.732
^{11}B	3/2	80.42	80.209	96.251	160.419
^{13}C	1/2	1.108	62.860	75.432	125.721
^{14}N	1	99.63	18.059	21.671	36.118
^{15}N	1/2	0.37	25.332	30.398	50.664
^{17}O	5/2	3.7×10^{-2}	33.892	40.670	67.784
^{19}F	1/2	100	235.192	282.231	470.385
^{21}Ne	3/2	0.257	19.736	23.683	39.472
^{23}Na	3/2	100	66.128	79.353	132.256
^{25}Mg	5/2	10.13	15.298	18.358	30.597
^{27}Al	5/2	100	65.143	78.172	130.287
^{29}Si	1/2	4.7	49.662	59.595	99.325
^{31}P	1/2	100	101.202	121.442	202.404
^{33}S	3/2	0.76	19.174	23.009	38.348
^{35}Cl	3/2	75.53	24.495	29.395	48.991
^{37}Cl	3/2	24.47	20.389	24.467	40.779
^{39}K	3/2	93.1	11.666	13.999	23.333
^{41}K	3/2	6.88	6.403	7.684	12.806
^{43}Ca	7/2	0.145	16.820	20.184	33.641
^{45}Sc	7/2	100	60.735	72.882	121.470
^{47}Ti	5/2	7.28	14.092	16.910	28.164
^{49}Ti	7/2	5.51	14.095	16.914	28.191
^{50}V	6	0.24	24.926	29.911	49.852
^{51}V	7/2	99.76	65.720	78.864	131.440
^{53}Cr	3/2	9.55	14.130	16.956	28.260
^{55}Mn	5/2	100	61.661	73.993	123.322
^{57}Fe	1/2	2.19	8.078	9.693	16.156

(continued)

Table A.2 (continued)

Isotope	Spin	Abundance (%)	NMR frequency (MHz) at field (T)		
			5.8717	7.0460	11.7434
^{59}Co	7/2	100	59.035	70.842	118.071
^{61}Ni	3/2	1.19	22.340	26.808	44.681
^{63}Cu	3/2	69.09	66.262	79.515	132.525
^{65}Cu	3/2	30.91	70.958	85.183	141.972
^{67}Zn	5/2	4.11	15.635	18.762	31.271
^{69}Ga	3/2	60.4	60.008	72.009	120.016
^{71}Ga	3/2	39.6	76.238	91.485	152.476
^{73}Ge	9/2	7.76	8.721	10.465	17.442
^{75}As	3/2	100	42.817	51.380	85.634
^{77}Se	1/2	7.58	47.669	57.203	95.338
^{79}Br	3/2	50.54	62.633	75.160	125.267
^{81}Br	3/2	49.46	67.515	81.018	135.031
^{83}Kr	9/2	11.55	9.619	11.543	19.238
^{85}Rb	5/2	72.15	24.138	28.965	48.276
^{87}Rb	3/2	27.85	81.803	98.163	163.606
^{87}Sr	9/2	7.02	10.834	13.001	21.669
^{89}Y	1/2	100	12.248	14.697	24.496
^{91}Zr	5/2	11.23	23.325	27.991	46.651
^{93}Nb	9/2	100	61.107	73.328	1222.214
^{95}Mo	5/2	15.72	16.287	19.544	32.574
^{97}Mo	5/2	9.46	16.630	19.957	33.261
^{99}Ru	3/2	12.72	8.474	10.169	16.949
^{101}Ru	5/2	17.07	12.353	14.824	24.707
^{103}Rh	1/2	100	7.868	9.442	15.737
^{105}Pd	5/2	22.23	11.440	13.728	22.881
^{107}Ag	1/2	51.82	10.116	12.139	20.233
^{109}Ag	1/2	48.18	11.630	13.956	23.260
^{111}Cd	1/2	12.75	53.013	63.616	106.027
^{113}Cd	1/2	12.26	55.457	66.548	110.914
^{113}In	9/2	4.28	54.666	65.600	109.333
^{115}In	9/2	95.72	54.785	65.742	109.570
^{115}Sn	1/2	0.35	81.749	98.0999	163.498
^{117}Sn	1/2	7.61	89.063	106.875	178.126
^{119}Sn	1/2	8.58	93.181	111.817	186.362
^{121}Sb	5/2	57.25	59.826	71.791	119.652
^{123}Sb	7/2	42.75	32.398	38.878	64.796
^{123}Te	1/2	0.87	65.519	78.623	131.039
^{125}Te	1/2	6.99	78.992	94.790	157.984
^{127}I	5/2	100	50.018	60.021	100.036
^{129}Xe	1/2	26.44	69.151	82.981	138.302
^{131}Xe	3/2	21.18	20.499	24.598	40.998
^{133}Cs	7/2	100	32.792	39.351	65.585

(continued)

Table A.2 (continued)

Isotope	Spin	Abundance (%)	NMR frequency (MHz) at field (T)		
			5.8717	7.0460	11.7434
^{135}Ba	3/2	6.59	24.835	29.802	49.670
^{137}Ba	3/2	11.32	27.783	33.339	55.566
^{138}La	5	0.089	32.982	39.579	65.965
^{139}La	7/2	99.91	35.315	42.378	70.631
^{141}Pr	5/2	100	73.227	87.872	146.454
^{143}Nd	7/2	12.17	13.594	16.313	27.188
^{145}Nd	7/2	8.3	8.364	10.036	16.727
^{147}Sm	7/2	14.97	10.320	12.384	20.640
^{149}Sm	7/2	13.83	8.224	9.868	16.446
^{151}Eu	5/2	47.82	62.001	74.401	124.002
^{153}Eu	5/2	52.18	27.378	32.854	54.757
^{155}Gd	3/2	14.74	9.549	11.458	19.097
^{157}Gd	3/2	15.68	11.935	14.323	23.871
^{159}Tb	3/2	100	56.695	68.035	113.391
^{161}Dy	5/2	18.88	8.236	9.883	16.471
^{163}Dy	5/2	24.97	11.458	13.750	22.917
^{165}Ho	7/2	100	51.282	61.538	102.564
^{167}Er	7/2	22.94	7.266	8.671	14.451
^{169}Tm	1/2	100	20.679	24.814	41.358
^{171}Yb	1/2	14.31	44.032	52.839	88.069
^{173}Yb	5/2	16.13	12.130	14.556	24.261
^{175}Lu	7/2	97.41	28.518	34.222	57.036
^{176}Lu	7	2.59	19.822	23.786	39.644
^{177}Hf	7/2	18.5	7.801	9.361	15.602
^{179}Hf	9/2	13.75	4.674	5.609	9.349
^{181}Ta	7/2	99.98	29.925	35.910	59.850
^{183}W	1/2	14.4	10.402	12.483	20.805
^{185}Re	5/2	37.07	56.284	67.541	112.569
^{187}Re	5/2	62.93	56.861	68.233	113.722
^{187}Os	1/2	1.64	5.758	6.909	11.515
^{189}Os	3/2	16.1	19.397	23.276	38.794
^{191}Ir	3/2	37.3	4.296	5.156	8.593
^{193}Ir	3/2	62.7	4.678	5.614	9.357
^{195}Pt	1/2	33.8	53.747	64.497	107.495
^{197}Au	3/2	100	4.281	5.138	8.563
^{199}Hg	1/2	16.84	44.568	53.481	89.136
^{201}Hg	3/2	13.22	16.499	19.799	32.998
^{203}Tl	1/2	29.5	142.873	171.448	285.747
^{205}Tl	1/2	70.5	144.270	173.124	288.540
^{207}Pb	1/2	22.6	52.304	62.765	104.609
^{209}Bi	9/2	100	40.174	48.208	80.348
^{235}U	7/2	0.72	4.475	5.371	8.951

Appendix 3: Amino Acid Random Coil Proton Chemical Shifts (Table A.3)

Table A.3 Amino acid random coil proton chemical shifts for assignment of residue spin systems ($\delta_{TSP} = 0.00$ ppm)

Residue	NH	αH	βH	Others
Gly	8.39	3.97		
Ala	8.25	4.35	1.39	
Val	8.44	4.18	2.13	γCH$_3$ 0.97, 0.94
Ile	8.19	4.23	1.9	γCH$_2$ 1.48, 1.19
				γCH$_3$ 0.95
				δCH$_3$ 0.89
Leu	8.42	4.38	1.65, 1.65	γH 1.64
				γCH$_3$ 0.94, 0.90
Pro$^{b(trans)}$		4.44	2.28, 2.02	γCH$_2$ 2.03, 2.03
				δCH$_2$ 3.68, 3.65
Ser	8.38	4.5	3.88, 3.88	
Thr	8.24	4.35	4.22	γCH$_3$ 1.23
Asp	8.41	4.76	2.84, 2.75	
Glu	8.37	4.29	2.09, 1.97	γCH$_2$ 2.31, 2.28
Lys	8.41	4.36	1.85, 1.76	γCH$_2$ 1.45, 1.45
				δCH$_2$ 1.70, 1.70
				εCH$_2$ 3.02, 3.02
				εNH$_3$ 7.52
Arg	8.27	4.38	1.89, 1.79	γCH$_2$ 1.70, 1.70
				δCH$_2$ 3.32, 3.32
				NH 7.17, 6.62
Asn	8.75	4.75	2.83, 2.75	γNH$_2$ 7.59, 6.91
Gln	8.41	4.37	2.13, 2.01	γCH$_2$ 2.38, 2.38
				δNH$_2$ 6.87, 7.59
Met	8.42	4.52	2.15, 2.01	γCH$_2$ 2.64, 2.64
				εCH$_3$ 2.13
Cys	8.31	4.69	3.28, 2.96	Oxidized
Trp	8.09	4.7	3.32, 3.19	2H 7.24
				4H 7.65
				5H 7.17
				6H 7.24
				7H 7.50
				NH 10.22
Phe	8.23	4.66	3.22, 2.99	2,6H 7.30
				3,5H 7.39
				4H 7.34
Tyr	8.18	4.6	3.13, 2.92	2,6H 7.15
				3,5H 6.86
His	8.41	4.63	3.26, 3.20	2H 8.12
				4H 7.14

Appendix 4: Assignments for the ^{13}C NMR Spectrum of a Perchloric Acid Extract of a Fed Rat Liver (Table A.4)

Table A.4 Assignments for the ^{13}C NMR spectrum of a perchloric acid extract of a fed rat liver, referenced to DSS

δ (ppm)	Assignment	δ (ppm)	Assignment
19.57	Ala C3	75.86	IsoCit C2
22.72	Lac C3	75.86	Glycogen C3
22.72	Leu C6	78.51	ATP Rib C2′
22.72	Thr C4	78.51	Fru C3α
24.66	β-OH-Butyrate C4	78.51	Glc C3β
24.66	Leu C5	78.51	Citrate C2
24.66	NADH Nic C3	78.51	Fru C4α
27.09	Cys C3	78.51	Glc C2β
27.09	Leu C4	78.69	Glc C5β
27.09	Proline C5	78.69	NADP+ Rib C2″
27.09	Arginine C4	79.91	Glycogen C4 (α1→4)
27.87	Glutathione C10	79.91	NADPH Rib C2″
29.30	Gln C3	83.55	Fru C5α
29.30	Glu C3	85.54	NADH Rib C4″
29.30	Glutathione C7	86.32	NADH Rib C4′
31.00	Acetaldehyde 2	86.32	UDPG Rib C4′
31.00	Arginine C3	87.13	ATP Rib C3′
31.70	Proline C4	87.13	Uridine
32.27	AcAc C4	89.64	ATP Rib C1′
33.94	Gln C4	89.64	NAD+ Rib C4″
34.40	Glutathione C6	89.64	NADH Rib C1′
36.28	Glu C4	91.23	UDPG Rib C1
36.77	Succinate C2,3	94.66	Glc C1α
40.25	IsoCit C4	98.48	Glc C1β
40.25	Aspartate C2	98.48	UDPG Glc C1
43.15	Arginine C5	102.48	Glycogen C1
43.15	Leu C3	102.48	NADH/NADPH Nic C9
43.15	PhosphoEtNH	102.48	Fru C2β
45.13	Mal C3	105.41	Fru C2α
46.10	Glutathione C2	105.41	UDPG Uri C5
47.50	Citrate C1	118.01	Tyr C5,9 ?
47.50	Citrate C3,4	121.54	ATP Ade C5
49.31	β-OH-Butyrate C2	121.54	NADH Ade C5
49.31	Proline C3	126.82	NADH Nic C3
49.82	Malonate C2	131.27	NAD+/NADP+ Nic C1
49.82	IsoCit C3	141.73	NADH Nic C1
55.16	Aspartate C3	142.86	ATP Ade C4
55.16	Ala C2	143.57	NAD+/NADP+ Nic C4

(continued)

Table A.4 (continued)

δ (ppm)	Assignment	δ (ppm)	Assignment
56.05	AcAc C2	143.57	NAD+/NADP+ Nic C4
56.05	Leu C2	144.19	NAD+/NADP+ Nic C3
56.59	Phosphatidylcholine: $-N(CH_3)_3$	144.19	UDPG Uri C6
57.29	Arginine C2	151.76	ATP Ade C3
57.29	Gln C2	151.76	NADH Ade C3
57.29	Glu C2	154.48	UDPG Uri C2
57.29	Glutathione C8	154.48	Uridine
59.01	Cys C2	155.54	ATP Ade C2
59.01	Glutathione C4?	155.54	NADH Ade C2
59.42	Diethanolamine	158.23	Arginine C6
60.52	Ethanolamine	158.23	ATP Ade C1
60.52	UDPG	158.23	NADH Ade C1
63.19	Proline C2	163.49	HCO_3^-
63.19	Glc C6β	166.40	UDPG Uri C2
63.19	Glc C6α	166.40	Urea C1,3
63.19	Glycogen C6	174.33	Glutathione C3
63.19	Thr C2	175.46	Cys C1
63.19	UDPG Glc C6	175.46	NADH Nic C6; C=O
66.13	Fru C6β	175.46	Thr C1
66.13	Fru C6α	177.13	AcAc C3
66.13	Fru C1β	177.13	Arginine C1
66.13	Fru C1α	177.13	Aspartate C4
68.23	ATP Rib C5′	177.13	Gln C1
68.23	β-OH-Butyrate C3	177.13	Glu C1
68.23	NAD+ Rib C4″	177.13	IsoCit C5
68.99	NADH Rib C4″	177.13	Proline C1
68.99	Thr C3	178.18	Glutathione C5
72.03	Fru C3β	178.18	Leu C1
72.03	Fru C4β	178.72	Ala C1
72.03	Glc C4β	178.91	Glutathione C1
72.03	Glc C4α	178.91	IsoCit C6
72.03	ATP Rib C4′	180.02	Aspartate C1
72.03	Fru C5β	180.02	IsoCit C1
72.03	Lac C2	180.56	Citrate C5
72.03	UDPG Glc C4	180.56	Citrate C6
72.03	Uridine	180.56	Gln C5
74.14	Glc C2α	183.15	β-OH-Butyrate C1
74.14	Glc C5α	183.15	Glu C5
74.14	Glycogen C2	183.15	Glu C5
74.14	Mal C2	183.15	Mal C1
74.14	Glycogen C5	183.15	Mal C4
75.57	Glc C3α	184.92	Lac C1
75.57	UDPG Glc C3	184.92	Succinate C1,4
75.57	UDPG Glc C5	–	

Appendix 5: Molecules Found in Space Through Radio and Infrared Spectroscopy (Table A.5)

Table A.5 Molecules found in space through radio and infrared spectroscopy

Molecules with 2 atoms			
H_2 Hydrogen	CO Carbon monoxide	CSi Carbon monosilicide	CP Carbon monophosphide
CS Carbon monosulfide	NO nitric oxide	NS nitric sulfide	SO Sulfur monoxide
HCl Hydrogen chloride	NaCl Sodium chloride	KCl Potassium chloride	AlCl Aluminum monochloride
AlF Aluminum monofluoride	PN Phosphorus mononitride	SiN Silicon mononitride	SiO Silicon monoxide
SiS Silicon monosulfide	NH Imidyl radical	OH Hydroxyl radical	C_2 Diatomic carbon
CN Cyanide radical	HF Hydrogen fluoride	FeO Iron monoxide	LiH Lithium hydride
SH Mercapto radical PO Phosphorus monoxide	CH^+ Methylidyne radical CH Methylidyne	CO^+ Carbon monoxide radical	SO^+ Carbon monofluoride cation
AlO Aluminum monoxide	O_2 Oxygen	N_2 Nitrogen	CF^+ Carbon fluoride radical
Molecules with 3 atoms			
H_2O Water	CO_2 Carbon dioxide	NaCN Sodium cyanide	HNC Hydrogen isocyanide
SO_2 Sulfur dioxide	H_2S Hydrogen sulfide	HCN Hydrogen cyanide	MgNC Magnesium isocyanide
CH_2 Methylene	N_2O Nitrous oxide	MgCN Magnesium cyanide	OCS Carbonyl sulfide
C_2O Ketenylidene	N_2H^+	SiNC Silicon isocyanide	c-SiC_2 Silicon dicarbide
NH_2 Amidogen radical	HCO Formyl radical	C_3 Triatomic carbon	C_2H Ethynyl radical
SiCN Silicon monocyanide	C_2S Thioxoethenylidene	AlNC Aluminum isocyanide	HNO Nitrosyl hydride
HCO^+ Formyl cation			
Molecules with 4 atoms			
HOC^+	HCS^+	H_3^+	HNCO Isocyanic acid
NH_3 Ammonia	HCP Phosphaethyne	CCP	C_3N^-
H_2O_2 Hydrogen peroxide	HCNO		
C_3S Tricarbon monosulfide	H_2CO Formaldehyde	H_2CS Thioformaldehyde	C_2H_2 Acetylene

(continued)

Table A.5 (continued)

HCCN	HNCS Thio-isocyanic acid	H_3O^+ Hydronium ion	SiC_3 Silicon tricarbide
$HCNH^+$ HSCN	H_2CN Methylene amidogen	c-C_3H Cyclopropenylidene	l-C_3H Propylidene
CH_3 Methyl radical	$HOCO^+$ Protonated CO_2	C_2CN Cyanoethynyl radical	C_3O Fulminic acid
	CH_4 Methane	HCCN	

Molecules with 5 atoms

CH_2CO Ketene	SiH_4 Silane	CH_2NH Methyleneimine	NH_2CN Cyanamide
c-C_3H_2 Cyclopropenylidene	HCOOH Formic acid	HCC-CN Cyanoacetylene	HCC-NC Isocyanoacetylene
l-C_3H_2	HC(O)CN Cyanoformaldehyde	CH_2CN Cyanomethyl radical	H_2COH^+ Protonated formaldehyde
C_5		C_4H^-	C_4Si
C_4H^-		HNCCC	C_4H

Molecules with 6 atoms

CH_3OH Methanol	CH_3SH Methanethiol	C_2H_4 Ethylene	$H(CC)_2H$ Diacetylene
HC_3NH^+ Protonated cyanoacetylene	CH_3NC Methyl isocyanide	$HC(O)NH_2$ Formamide	HCC-C(O)H Propynal
H_2CCCC Butatrienylidene	HC_4N Cyanopropenylidene	c-H_2C_3O Cyclopropenone	
CH_3CN Methylcyanide	C_5N and C_5N^-		
C_5H	H_2CCNH Keteneimine		

Molecules with 7 atoms

CH_3NH_2 Methylamine	c-C_2H_4O Ethylene oxide	$HC(O)CH_3$ Acetaldehyde	H_3C-CC-H Methylacetylene
C_6H^- Hexatriyne anion	CH_2CH (CN) Acrylonitrile	HCC-CC-CN Cyanobutadiyne	C_6H Hexatriynyl radical
$CH_2CH(OH)$ Vinyl alcohol			

Molecules with 8 atoms

CH_3COOH Acetic acid	H_3C-CC-CN Cyanomethylacetylene	CH_2CCHCN Cyanoallene	H_2C_6 Hexapentaenylidene
H_2NCH_2CN Aminoacetonitrile	$HC(O)OCH_3$ Methyl formate	$HOCH_2C(O)H$ Glycolaldehyde	C_7H
$H_2C=CH-C(O)H$ Propenal	$H(CC)_3H$ Triacetylene		

Molecules with 9 atoms

$(CH_3)_2O$ Dimethyl ether	CH_3C_4H Methylbutadiyne	CH_3CH_2CN Ethylcyanide	CH_3CH_2OH Ethanol
	HCC-CC-CC-CN Cyanohexatriyne	$CH_3C(O)NH_2$ Acetamide	C_8H^-

(continued)

Table A.5 (continued)

Molecules with 10 atoms			
$(CH_3)_2CO$ Acetone	CH_3CHCH_2 Propylene	CH_3OCH_2OH Methoxymethanol	H_3C-CH_2-$C(O)H$ Propanal
$HOCH_2CH_2OH$ Ethylene glycol			
Molecules with 11 atoms			
HCC-CC-CC-CC-CN Cyanooctatetrayne	$CH_3(CC)_2CN$ Methylcyanodiacetylene		
Molecules with 12 atoms			
C_6H_6 Benzene	CH_3C_6H Methyltriacetylene	$HC(O)OCH_2CH_3$ Ethyl formate	
Molecules with 13 atoms			
HCC-CC-CC-CC-CC-CN Cyanodecapentayne	$CH_3CH_2CH_2CN$ n-Propyl cyanide	$(CH_3)_2CHCN$ iso-Propyl cyanide	
Molecules with 60–70 atoms			
C_{60} Fullerene	C_{70} Fullerene		

Data from J. Tennyson, Molecules in space, in *Handbook of Molecular Physics and Quantum Chemistry*, vol. 3, ed. by S. Wilson (John Wiley & Sons, Ltd, Chichester, 2003), pp. 356–369. Note that a current and continuously updated list of molecular detections is maintained at the Wikipedia site: https://en.wikipedia.org/wiki/List_of_interstellar_and_circumstellar_molecules, which currently (2023) lists 225 protonated and deuterated molecules

Index

A

Abiogenesis, 1–60, 418, 627, 657, 831
Amino acids, 29, 46, 53, 55, 57, 60, 99–102,
    113, 124, 125, 127, 460, 469,
    478–480, 495, 496, 522, 530, 540,
    544, 547, 589, 660, 771, 781–786,
    788, 792–801, 815, 835
Archaeol, 663
Argon
    coordination number, 746, 748, 754
    interatomic potential, 730, 748, 749
    as a Lennard-Jones fluid, 730, 743–748
    radial distribution function, 744, 745,
      752, 754
    velocity autocorrelation function, 730, 731

B

Bacteriorhodopsin, 627, 657, 659, 660,
    662, 663
Beer's law, 357, 634–638
Big Bang
    Baryon acoustic oscillations, 334,
      358–359
    expansion of the universe, 333, 337, 339,
      341, 342
    nucleosynthesis, 2, 347, 379–381, 386, 388,
      401, 402, 679, 756
Binomial coefficients
    Pascal's triangle, 20, 473
Black body radiation, 633
Boltzmann distribution, 37, 46, 82, 85, 90, 91,
    114–121, 125, 126, 279, 293, 407,
    434, 436, 629, 633, 733, 743,
    748, 769

Born-Oppenheimer approximation, 284,
    286–288, 320, 325, 326
Bosons, 44, 45, 131, 207, 227, 343, 346, 363,
    365, 372, 373, 375, 381, 412, 418,
    425, 685

C

Calorimetry, 96–97, 127
Carbon
    ^{12}C, 333, 388, 395–397, 399, 404, 405,
      484, 679
    ^{13}C, 395, 396, 407, 408, 442, 484, 514–524,
      530, 531, 535–549, 552–555,
      800, 832
    ^{14}C, 374, 375, 484, 511
    ^{13}C NMR, 514–531, 533–537, 539–541,
      543, 546, 548, 550–553, 836
Casimir effect, 370
Chemical potential, 2, 45, 87, 94, 381, 382, 560,
    561, 570, 801
Chemical shift, 439–443, 446, 475, 480, 484,
    487, 512, 514, 519, 525, 526, 537,
    543, 549, 551–553, 714, 756–765,
    768, 835
Chlorophylls, 627, 664–666, 670, 671
Comets
    Churyumov-Gerasimenko, 786, 808
    Hale-Bopp, 786, 787
    Halley's comet, 785
    Lovejoy, 786, 787
    Wild-2, 786
Commutators, 142, 161, 168–170, 172, 173,
    200–202, 205, 250–252, 257–259,
    362, 687, 695, 774